ÉGALEMENT PARUS AUX ÉDITIONS TECHNIP

- Les grès du Paléozoïque inférieur au Sahara. Sédimentation et discontinuités. Évolution structurale d'un craton. **S. BEUF, B. BIJU-DUVAL, O. de CHARPAL, P. ROGNON, O. GARIEL et A. BENNACEF.**
- Symposium International. Histoire structurale du golfe de Gascogne. Rueil-Malmaison, 14-16 décembre 1970.
- Advances in Organic Geochemistry 1973. Actes du 6e Congrès International de Géochimie Organique.
- Formation des gisements de pétrole. Étude des phénomènes géologiques fondamentaux. **Cl. SALLE et J. DEBYSER.**
- Gravimétrie appliquée aux recherches structurales et à la prospection pétrolière et minière. **J. SCHOEFFLER.**
- Éléments de pétrologie dynamique des systèmes calcaires. 1. Description macro et microscopique. Diagenèse. Applications. 2. Atlas photographique. **L. HUMBERT.**
- Microfaciès et ordinateur. **J. CHAROLLAIS et E. DAVAUD.**
- Symposium International. Geodynamics in South-West Pacific — Géodynamique du Sud-Ouest Pacifique. Nouméa (Nouvelle-Calédonie), 27 août-2 septembre 1976.

XXVe CONGRÈS-ASSEMBLÉE PLÉNIÈRE
DE LA
COMMISSION INTERNATIONALE POUR
L'EXPLORATION SCIENTIFIQUE DE LA MÉDITERRANÉE

Symposium International

STRUCTURAL HISTORY OF THE MEDITERRANEAN BASINS

HISTOIRE STRUCTURALE DES BASSINS MÉDITERRANÉENS

SPLIT – YOUGOSLAVIE

25-29 OCTOBRE 1976

Publié sous la direction de

B. BIJU-DUVAL et L. MONTADERT
Institut Français du Pétrole

Sous le patronage de

Comité de Géologie et Géophysique Marines-CIESM
Committee of Marine Geology-IUGS
Inter - Union Commission on Geodynamics
International Association
for the Physical Sciences of the Ocean-IUGG
European Geophysical Society

ÉDITIONS TECHNIP • 27, RUE GINOUX 75737 • PARIS CEDEX 15

ISBN 2-7108-0321-6

avant-propos

Tous les deux ans dans le cadre des réunions du Comité de Géologie et Géophysique Marines de la Commission Internationale pour l'Exploration Scientifique de la Mer Méditerranée (CIESM), les géologues et les géophysiciens de nombreux pays peuvent confronter leurs travaux et suivre ainsi l'avancement des connaissances géologiques et géophysiques en Méditerranée. Dans le cadre de ces réunions un symposium spécial est en général organisé et les communications sont présentées sous forme de condensés dans les "Rapports et Procès-Verbaux des Réunions" publiés sous les auspices de la CIESM sous la direction de J.Y. Cousteau.

Du 24 au 29 octobre 1976, la Commission s'est réunie à Split, Yougoslavie, sous l'aimable invitation du gouvernement yougoslave. Un symposium spécial sur l'Histoire Structurale des Bassins Méditerranéens s'est tenu du 25 au 28 octobre sous le patronage de :

- le Comité de Géologie Marine de l'Union Internationale des Sciences Géologiques qui a fortement appuyé l'idée d'une telle réunion suivant une recommandation d'un groupe de travail du "Scientific Committee on Oceanic Research" (SCOR) ;
- la Commission Inter-Union de Géodynamique spécialement représentée par son groupe de travail n° 3 ;
- l'Association Internationale des Sciences Physiques de l'Océan de l'Union Internationale de Géodésie et Géophysique ;
- la Société Européenne de Géophysique.

Au cours de ce symposium on a pu constater que les interprétations proposées faisaient de plus en plus appel à l'ensemble des disciplines des Sciences de la Terre et s'appuyaient aussi bien sur les données marines que terrestres. D'autre part il est bien apparu que la Méditerranée était une zone privilégiée où, au-delà des aspects régionaux, des problèmes fondamentaux des Sciences de la Terre mis en évidence depuis la révolution apportée par l'exploration des Océans et le concept de tectonique globale qui en est découlé, pouvaient être étudiés.

Compte tenu de cette évolution, outre la publication normale des "Rapports et Procès-Verbaux des Réunions", il a paru utile de regrouper un certain nombre de communications sous forme d'un volume spécial. Il n'a pas été possible de publier de façon exhaustive l'ensemble des communications présentées durant la réunion du Comité mais l'intention des responsables de la publication est néanmoins de fournir au lecteur un moyen de situer l'état d'avancement des travaux en Méditerranée et de connaître les principales idées ou approches nouvelles des scientifiques de toutes disciplines travaillant sur cette région. Leurs commentaires devraient permettre, nous l'espérons, de compléter dans une certaine mesure la lecture des différents chapitres qui constituent ce volume en précisant le cadre dans lequel ils s'inscrivent et en signalant les principales références bibliographiques.

B. BIJU-DUVAL et L. MONTADERT

foreword

Every two years, during the meeting of the Marine Geology and Geophysics Committee of the International Commission for the Scientific Exploration of the Mediterranean Sea (CIESM), geologists and geophysicists from many countries meet together to compare their research and thus to keep abreast of progress in the geological and geophysical understanding of the Mediterranean. As part of these meetings, a special symposium is usually organized, and papers are presented in condensed form in the "Rapports et Procès-Verbaux des Réunions" published under the auspices of CIESM under the direction of J.Y. Cousteau.

From 24 to 29 October 1976, the Commission met in Split, Yugoslavia, upon the kind invitation of the Yugoslav Government. A special symposium on the Structural History of Mediterranean Basins was held from 25 to 28 October, under the sponsorhip of :

- the Committee of Marine Geology of the International Union of Geological Sciences, which strongly supported the idea of such a meeting upon the recommendation of a Scientific Committee on Oceanic Research (SCOR) working group ;
- the Inter-Union Commission for Geodynamics represented by its Working Group No 3 ;
- the International Association for the Physical Sciences of the Ocean of the International Union of Geodesy and Geophysics ;
- the European Geophysical Society.

During this symposium, one could verify that the interpretations proposed now combined more and more all disciplines of Earth Sciences and based themselves on both offshore and onshore data. Morever, it became patent that the Mediterranean was a privileged zone in which, over and above regional aspects, fundamental problems of Earth Sciences revealed since the revolution brought on by ocean exploration and the resulting Global Tectonics concept could be examined.

In the light of this evolution and in addition to the normal publication of the "Rapports et Procès-Verbaux des Réunions", it appeared of use to regroup various papers in the form of a special volume. It was not possible to make exhaustive publication of all the papers presented during the meeting of the Committee, but the intention of the editors is nevertheless to provide readers with a means of situating the state of progress of research on the Mediterranean so as to be aware of the leading ideas and/or new approaches of scientists from all disciplines working in this area. We hope that the editors'comments will, to a certain extent, add to the contents of the different chapters making up this volume by specifying the setting within which they are situated and by indicating the leading bibliographic references.

B. BIJU-DUVAL and L. MONTADERT

table des matières
contents

first part

geological evolution from the tethys to the mediterranean from the mesozoïc to the present

second part

recent tectonics in the mediterranean area

third part

small ocean basins of the mediterranean

TABLE DES MATIERES. CONTENTS

fourth part

the messinian event and recent sedimentation

INTERNATIONAL SYMPOSIUM ON THE STRUCTURAL HISTORY OF THE MEDITERRANEAN BASINS. SPLIT (YUGOSLAVIA) 25-29 OCTOBER 1976. B. BIJU-DUVAL AND L. MONTADERT, Eds. EDITIONS TECHNIP, PARIS 1977, pp. 1-12

INTRODUCTION TO THE STRUCTURAL HISTORY OF THE MEDITERRANEAN BASINS

B. BIJU-DUVAL and L. MONTADERT (1)

INTRODUCTION

Since the last century, the Mediterranean area, including the Alpine belt, has been studied by a great many geologists from different countries. In the beginning studies were restricted to onland areas, especially to the folded belts from which many fundamental tectonic concepts were derived such as subduction or geosynclinal development. But it has only been since about twenty years ago that the submerged parts of the Mediterranean area have been studied. Due to technological advances in geophysics, and especially in seismics, the deep structure of the margins and of the deep basins is becoming better and better understood. Information is now available not only on the shelf where oil exploration has provided numerous data but also in the deep areas. Here our knowledge of the geometry, stratigraphy and sedimentology of the infilling series in the different basins has been considerably enhanced by the holes drilled during legs 13 and 42 A of the Deep Sea Drilling Project (DSDP). The development of knowledge of this complicated Mediterranean-Alpine area appears clearly in the successive publications over the last ten years (Ryan et al., 1966; Ryan et al., 1971; *Rap. Proc. Reun. CIESM*, 1973; Biju-Duval et al., 1974; *Rap. Proc. Reun. CIESM*, 1975). The tectonic pattern of the present Mediterranean, schematically drawn in fig. 1, can be interpreted only as the result of a long geological evolution from the beginning of the Mesozoïc up to the present. The present Mediterranean seas cover nearly 2 000 000 sq. km located between the European and African continental blocks; they are subdivided into several small oceanic basins with more or less broad continental margins and numerous small continental fragments now dislocated and inserted between these basins. Several attempts to synthesize marine geophysical data have recently been proposed (Finetti and Morelli, 1973; Biju-Duval et al., 1974; Morelli, 1975; Finetti, 1976; Mulder, 1973; Mulder et al., 1975; etc.), but relatively few take into account both geophysical and geological data. We will briefly present here a summary of the data concerning the present knowledge of the Mediterranean basins. The reader will find more detailed information and a large bibliography in a recent paper by Biju-Duval et al. (1977).

THE MEDITERRANEAN BASINS ARE OF DIFFERENT AGE, STRUCTURE AND ORIGIN

It must be pointed out that our knowledge of the present Mediterranean basins is mainly based on seismic reflection surveys and on drilling results which give data mainly on the recent geological history, especially the Cenezoïc in the western Mediterranean and part of the Mesozoïc in the eastern Mediterranean. Considering that this recent history is related to the geological events that have occurred over the last 200 MY between Africa and

(1) Institut Français du Pétrole, 1 et 4 avenue de Bois-Préau, 92506 Rueil-Malmaison, France

Europe, it is necessary to take them in account. This is done in other papers in this volume (Biju-Duval et al.; Bernoulli; Channell and Horwath) so here we will only bring up some major problems after looking at some general geological cross-sections in the Mediterranean.

Figs 2 and 3 show several interpretative cross-sections accross the different basins. These cross-sections are based on very numerous land sections proposed by different authors (Durand-Delga, Paquet, Caire, Debelmas, Mattauer, Aubouin, Blumenthal, Elter, Sungurlu, Poisson) and on reflection profiles made by french Institutions (see Biju-Duval et al., 1974) or on profiles published by OGST (Finetti and Morelli, 1973). The geometry of the superficial sedimentary layers is clearly established, but the drawing of the major deep structures has been interpreted in the light of plate tectonics. Hence it is possible to suggest interpretation in terms of subduction zones, marginal basins and old stable continental margins.

As discussed in other papers (Biju-Duval et al., 1976, 1977), because of the small size of the deep basins the nature of the crust is somewhat difficult to charaterize clearly. Nevertheless the oceanic nature of the abyssal plains in the western Basin and in the Tyrrhenian Sea is now well established. The refraction data show a great thickness of sediments in the eastern Mediterranean as well as in the Black Sea basins, with a thin crust. In the Aegean Sea and the Pannonian Basin, the crust seems to be attenuated. In the Adriatic Sea the thick pile of sediments rests on a typical continental crust. As has been illustrated in Figs 2 and 3, the age of formation of these basins is certainly different (see Biju-Duval et al., in this volume).

Sections 1, 2 and 3 of Fig. 2 shows :

1) The North Balearic-Provençal Basin with steep slopes corresponding to margins of the rifted type. As pointed out initially by different authors (Le Pichon et al., 1971; Alvarez, 1972; etc.) the age of formation is Oligocene; more recent data and especially the results of DSDP borehole 372 show a late Oligocene rifting followed by an early Miocene opening and subsidence.

2) The large extension of the continental basement under the abyssal plain, especially west of the Corsica-Sardinia block. The exact extent of the oceanic crust in the central part of the basin is unknown. Taking into account the extent of the continental basement and the magnetic anomalies, the initial fit before opening is still under discussion (see in Biju-Duval et al., 1977).

3) The thick infilling of Miocene to recent deposits (> 7 km), with the thick Messinian salt layer with halokinesis phenomena restricted to a basinal position (Mauffret et al., 1972) and the Upper Evaporites (see Montadert et al., 1977) extend on the margin where they onlap an erosion surface.

This basin could be considered as part of an extensional and intruded marginal basin behind an active Oligo-Miocene subduction zone which was running from the Apennines to the Maghreb (fig. 4). As opposed to the Tyrrhenian and South-Balearic Basin, it has been inactive since the Lower Miocene, marked only by subsidence due to cooling and loading of sediments.

Sections 4, 5, 6 and 7 of Fig. 2 crossing the so-called South Balearic-Algerian Basin show that :

1) The basin is located in a divergent orogene. Northward is the Betic range which extends from Gibraltar to Mallorca; southward the Maghrebin ranges run from Gibraltar to Sicily-Calabria. Eocene and Miocene thrusting, nappes and superficial gravity slidings affect both sides. The southward range which includes numerous early Cretaceous to Eocene flysches, with very scarce pieces of ophiolites and then miocene flysches, can be interpreted as the result of accretion along an old subduction zone in a previously very narrow oceanic domain (subduction B of Bally, 1975). They also include some continental slices detached from Iberia (Rif, Kabylia, Peloritain massives). Overthrusts in Spain and the Balearic Islands should be purely continental (subduction A of Bally). Thus the basin could be called an episutural basin. Interpretation of the Gibraltar Arc is still under discussion; it was probably formed during Eocene times (Paquet, 1974); more recently, since the Miocene, it could correspond to the subduction of the Atlantic below an Alboran sub-plate (Andrieux et al., 1971). Large nappes including gravity slidings extend widely into the Atlantic abyssal plain (Fig. 1).

2) The basin can be considered as a marginal basin with a suboceanic crust and numerous volcanic intrusions. Its age should be Lower Miocene as proved by stratigraphic data. The subsidence was active during the Miocene, while compression and gravity slidings affected both the north and south ranges. The Messinian

salt layer and upper Evaporites are widely represented. Such evaporites are known in uplifted onshore basins (Cheliff Basin, Murcia Basin).

3) The Gulf of Valencia looks like an aborted rifted arm of the North Balearic-Provençal Basin.

Sections 1 and 2 of Fig. 3 illustrate the Calabrian Arc and the relatively recent formation of the Tyrrhenian Basin behind this arc :

1) The tectonic pile of thrusts constituting the Apennines, including pieces of ophiolites from a previous oceanic domain located between Iberia and Apulia, is now resting directly, on continental Apulia (section 1). In the Adriatic Sea a thick sedimentary sequence includes thin Upper Miocene evaporites and Triassic salt below the thick mainly carbonate Mesozoïc series. Eastward the Dinaric front is marked by several overthrusts including ophiolites, relicts of the Tethys, in the internal eastern zones.

2) In Calabria the tectonic pile of overthrusts with metamorphic rocks as old as Upper Cretaceous extends to the Ionian abyssal plain under the "Messina Cone" which is also marked by very recent superficial gravity sliding affecting the thick Messinian salt layer in the Ionian Sea.

3) To the NW, between the external arc and the volcanic arc, an inter-arc basin can be seen, *i.e.* the Calabrian Basin. The Tyrrhenian Basin, itself located farther westward, is marked by young volcanism (Messinian to present) related to the subduction zone. This young subsided marginal basin is also characterized by a relatively thin sedimentary cover. The southern part has an oceanic crust as confirmed by DSDP hole 373 (Barberi et al., 1977). But its recent development is probably superimposed on a more ancient marginal basin previously connected with the South Balearic Basin during the Oligo-Miocene as a part of the Western Basin (Fig. 4) (Biju-Duval et al., 1977, and this volume).

Sections 3 to 9 of Fig. 3 summarize the structure of the different basins in the eastern Mediterranean :

1) In the whole area the thickness of sediments is very great. The entire North African margin is characterized by the northward extent of the Mesozoïc and probably Paleozoïc series which initially extended to Apulia and Anatolia as a part of Gondwana.

2) The southern part of the eastern Mediterranean from the Malta escarpment to the Levantine margin corresponds to a continental subsided passive margin which may have been initiated since the Triassic or Liassic times. In this interpretation, Adriatic, Greece and Turkey (Apulian-Aegean-Anatolian plates) have been separated from Africa since the Mesozoïc; this is in opposition to the concept of the Adriatic promontory of Argand (see Channell and Horwath, 1976 and this volume; Bernoulli, this volume) and the concept of the recent foundering of the basin (Burollet, 1975). This point of view is supported by geophysical and geological data, *i.e.* the scarcity of recent volcanism, the existence of an attenuated crust and the development of old margins since the Triassic or Liassic (Gvirztman and Bein, this volume; Biju-Duval et al., this volume).

3) In front of the active Aegean subduction zone (Hellenic trenches) which is very well marked by strong seismicity (McKenzie, 1972), the Mediterranean ridge could correspond to a tectonized pile of sediments corresponding to the front of an accretionary prism, in which Miocene (pre-evaporitic) slidings could also have played a role (combination of both the Ryan and Mulder hypotheses : see discussion in Finetti, 1976, and in Biju-Duval et al., 1977).

4) The Aegean Sea behind the external arc (Crete) is characterized by a continental crust, with some limited attenuation (northern Crete Basin), and by a recent volcanic arc (Santorin). Its development is recent, from Miocene to present, with Messinian evaporitic deposits. Allochthonous ophiolites in Greece and southwestern Turkey should come from a previous northward oceanic domain, *i.e.* the Tethys which was consummed between the Jurassic and Late Cretaceous.

5) The easternmost part (sections 6 to 9) is marked by the continental collision between Anatolia and Africa-Arabia. Allochthonous ophiolites in Cyprus and southeastern Turkey (Troodos-Kizil Dag-Kevan nappe, Biju-Duval et al., 1976 b) are interpreted as obducted oceanic material from the Mesozoïc oceanic area which was created between Africa-Arabia and Aegean-Anatolia. Neogene infilling is very thick (Baroz et al., 1977), especially the salt layer. It has been affected by recent compressional tectonics until recent times (Misis-Kyrenia ranges, South Cyprus-Florence Rise overthrust).

6) Sediment influxes from the Nile delta and North-Levantine deltas (Adana, Iskenderun, Antalya) are very important.

OLD MARGINS, MARGINAL BASINS, CONTINENTAL COLLISION

As shown by the general map and the sections, the Mediterranean basins can be divided in two main types :

1) A basin with a relict of an old oceanic crust created by spreading during the Mesozoic, *i.e.* the Eastern Mediterranean Basin with its original stable continental margin to the south.

2) Small ocean basins of Cenozoic age which are interpreted as marginal basins created behind subduction zones, some of which are still active today. The sections clearly show active overthrusting in different periods until the present. The creation of these small basins implies the consumption of older oceanic crusts which have now completely disappeared. This means we have to find the relicts not only of the oceanic crust but also of the initial associated stable margins now included in the Alpine folded belt. These old margins have been effectively reconstructed now in different parts of the Alps (Bernoulli et al., 1974; Bourbon et al., 1973, 1975). The interpretation of ophiolites is still under discussion because some of them could be relict of previous marginal basins (Blanchet, in this volume; Miashiro, 1973).

True continental collision exists in wide areas, *i.e.* between Arabia and Turkey (sections 8-9, Fig. 3), north of the Adriatic and Europe, Spain and Morocco.

Thus in recent time we observe the combination of two different tectonic styles :

- the tectonic style classicaly associated with subduction zones with linear narrow active belts;
- the tectonic style associated with continental collision as suggested by McKenzie and Molnar and Taponnier with extensive tectonic effects in the opposite continental plates.

INFILLING OF THE BASINS

As shown by the different sections, the sedimentary infilling of these basins is extensive except in the Tyrrhenian Sea where sediments are trapped on the margins by ridges and reliefs.

The age of this infilling is probably as old as the Mesozoic in the eastern Mediterranean and the Cenozoic in other basins.

Good data on the age and facies are scarce below the Messinian evaporites because of drilling problems. Information on the pre-evaporite Miocene in the western and eastern basins indicates relatively deep conditions. On all the sections we can follow the extent of the Messinian evaporites whose geographic distribution is shown on the map (Fig. 5). We can clearly see the existence of a thick salt layer in the deepest part of the basins. It pinches out on the margins or on highs. An erosional event occurred on the margins while the salt was deposited in the basin (see sections 1, 2, 3, Fig. 2). It has been demonstrated that the "Upper Evaporites" are shallow water or subaerial deposits (Hsü, Montadert et al., 1977) and that they onlap the erosional surface. These facts are generally interpreted as due to a large regression corresponding almost to the closing off of the Mediterranean (Hsü, Montadert et al., 1977). Clearly these evaporites are different from the evaporites often deposited after the phase of rifting in the oceanic basins of the Atlantic type. The Messinian evaporites are related to a sedimentological event in the entire Mediterranean superimposed on a geodynamic evolution which is very different according to the areas. It is stable in the North-Balearic Provençal Basin, and active in the eastern Mediterranean where overthrusting had been active before, during and after evaporite deposition (see sections 3 to 9 of Fig. 3). Subsidence was active due to the cooling of the oceanic crust and to loading by active sedimentation.

CONCLUSION

The different basins making up the Mediterranean area are now better understood from the seismic surveys done on the deep areas, the DSDP boreholes, the exploration of the shelves and a new look at geological data from the margins and the folded belts on land. Because of all this new information on land and at sea, a modern view of the geological development of the Mediterranean area is now emerging, using the concepts of plate and intraplate tectonics resulting from continental collision. This makes it necessary to take a new look at the data concerning the facies on the different margins, including these of folded belts, superimposed tectonics and associated metamorphism, age and overthrusting of ophiolites, significance of magmatism, seismicity, etc.

ACKNOLEDGMENTS

The authors are indebted to their friends with whom the Research Project on the Mediterranean Sea was done at *IFP* : V. Apostolescu, J. Letouzey, P. Courrier and R. Gonnard. This paper takes into account data or hypotheses discussed with various colleagues in other papers published with them, and we thank them for their fruitful collaboration, especially J.F. Mugniot, J. Sancho, E. Winnock, R. Byramjee, J. Dercourt and X. Le Pichon.

BIBLIOGRAPHY

ALVAREZ (W.).- (1972). Rotation of the Corsica-Sardinia Microplate. Nature (Phys. Sci.), 235, pp. 103-105.

ANDRIEUX (J.), FONTBOTE (J.M.), MATTAUER (M.).- (1971). Sur un modèle explicatif de l'arc de Gibraltar. Earth Planetary Sci. Lett., 12, pp. 191-198.

BALLY (A.W.).- (1975). A Geodynamic Scenario for Hydrocarbon occurrences. 9th World Petrol. Congr., Tokyo, vol. 2, pp. 33-44.

BARBERI (F.), BIZOUARD (H.), CAPALDI (G.), FERRARA (G.), GASPARINI (P.), INNOCENTI (F.), JORON (J.L.), LAMBRET (B.), TREUIL (M.), ALLEGRE (C.).- (1977). Age and Nature of Basalts from the Tyrrhenian abyssal Plain. In : Initial Reports of the Deep Sea Drilling Project, vol. XLII, Washington (U.S. Government Printing Office), in press.

BAROZ (J.F.), BERNOULLI (D.), BIJU-DUVAL (B.), BIZON (G.), BIZON (J.J.), LETOUZEY (J.).- (1977). Correlations of the Neogene Formations of the Florence Rise and of Northern Cyprus : Paleogeographical and Structural Implications. In : Initial Reports of the Deep Sea Drilling Project, vol. XLII, Washington (U.S. Government Printing Office), in press.

BERNOULLI (D.), JENKYNS (H.C.).- (1974). Alpine, Mediterranean and central Atlantic Mesozoic facies in relation to the early evolution of the Tethys. In : Modern and Ancient Geosynclinal sedimentation, Dott (R.H.) and Shaver (R.H.), edit. Tulsa, Soc. Econ. Pal. Mineral., 129-160 (Spec. Bull. n° 19).

BIJU-DUVAL (B.), LETOUZEY (J.), MONTADERT (L.), COURRIER (P.), MUGNIOT (J.F.), SANCHO (J.).- (1974). Geology of the Mediterranean Sea Basins. In : The Geology of Continental Margins. Burck (C.A.), Drake (C.L.) Ed., New-York, Heidelberg, Berlin, Springer Verlag, pp. 695-721.

BIJU-DUVAL (B.), DERCOURT (J.), LE PICHON (X.).- (1976 a). La genèse de la Méditerranée. La Recherche, 7, n° 71, pp. 811-822.

BIJU-DUVAL (B.), LAPIERRE (H.), LETOUZEY (J.).- (1976 b). Is the Troodos Massif (Cyprus) allochtonous ? Bull Soc. Geol., France, 7, XVIII, pp. 1347-1356.

BIJU-DUVAL (B.), LETOUZEY (J.), MONTADERT (L.).- (1977). Structure and Evolution of the Mediterranean Sea Basins. In : Initial Report of the Deep Sea Drilling Project, vol. XLII, Washington (U.S. Government Printing Office), in press.

BOURBON (M.), de GRACIANSKY (P.C.), MEGARD-GALLI (J.), LEMOINE (M.).- (1973). L'évolution paléogéographique du domaine briançonnais du Mésozoïque : carbonates de plate-forme subsidente, révolution du Lias supérieur, instauration progressive d'un régime pélagique profond. C.R.Ac.Sc., Paris, 277, pp. 769-772.

BOURBON (M.), de GRACIANSKY (P.C.), LEMOINE (M.), MEGARD-GALLI (J.) et MERCIER (D.).- (1975). Carbonates de plate-forme et séries pélagiques condensées dans le Mésozoïque de la zone briançonnaise (Alpes françaises). Excursion 5. IXème Congrès

International de Sédimentologie, Nice, 1975.

BUROLLET (P.F.) and BYRAMJEE (R.).- (1974). Réflexions sur la tectonique globale, exemples africains et méditerranéens. Notes et Mém., Compagnie Française des Pétroles, n° 11, pp. 71-120.

CHANNEL (J.E.T.), HORVATH (F.).- (1976). The African-Adriatic Promontory as a palaeogeographical Premise for alpine Orogeny and Plate Movements in the Carpatho-Balkan Region. Tectonophysics, 35, pp. 71-101.

Commission Internationale pour l'Exploration Scientifique de la Mer Méditerranée.- (1973). Rapports et Procès-Verbaux des Réunions. Symposium sur la Géodynamique de la région méditerranéenne (Athènes, 3-11 nov. 1972), 210 p.

Commission Internationale pour l'Exploration scientifique de la Mer Méditerranée.- (1975). Rapports et Procès-Verbaux des Réunions. Symposium Géodynamique (Monaco, 6-14 déc. 1974), 313 p.

FINETTI (I.), MORELLI (C.).- (1973). Geophysical Exploration of the Mediterranean Sea. Boll. Geofis. Teorica. Appl., 15, n° 60, pp. 263-341.

FINETTI (I.).- (1976). Mediterranean Ridge : a young submerged Chain associated with the Hellenic Arc. Boll. Geofis. Teorica. Appl., 18, n° 69, pp. 31-65.

HSU (K.J.), MONTADERT (L.) et al..- (1977). Initial Reports of the Deep Sea Drilling Project, vol. XLII, Washington (U.S. Government Printing Office), in press.

LE PICHON (X.), PAUTOT (G.), AUZENDE (J.M.), OLIVET (J.L.).- (1971). La Méditerranée Occidentale depuis l'Oligocène. Schéma d'évolution. Earth Planetary Sci. Lett., 13, n° 1, pp. 145-152.

McKENZIE (D.).- (1972). Active Tectonics of the Mediterranean Region. Geophys. J., 30, n° 2, pp. 109-185.

MAUFFRET (A.), FAIL (J.P.), MONTADERT (L.), SANCHO (J.) and WINNOCK (E.).- (1973). Northwestern Mediterranean sedimentary Basin from Seismic Reflection Profile. Amer. Assoc. Petrol. Geol. Bull., 57, n° 11, pp. 2245-2262.

MIYASHIRO (A.).- (1973). The Troodos ophiolitic complex was probably formed in an island arc. Earth Planet. Sci. Letters, vol. 19, n° 2, pp. 218-224.

MONTADERT (L.), MAUFFRET (A.), LETOUZEY (J.).- (1977). Messinian Event : Seismic evidence. In : Initial Reports of the Deep Sea Drilling Project, vol. XLII, Washington, op. cit.

MORELLI (C.).- (1975). Geophysics of the Mediterranean. Rapp. Comm. Int. Mer. Medit., Spec. Issue 7, pp. 29-111.

MULDER (C.J.).- (1973). Tectonic Framework and Distribution of Miocene chemical Sediments with Emphasis on the Eastern Mediterranean. In : Messinian Events in the Mediterranean, Drooger (C.W.) Ed., Amsterdam, London, North Holland Publ. Co., pp. 44-59.

MULDER (C.J.), LEHNER (P.) and ALLEN (D.C.K.).- (1975). Structural Evolution of the Neogene Salt Basins in the Eastern Mediterranean and the Red Sea. Geol. Mijnbouw, 54, pp. 208-221.

PAQUET (J.).- (1974). Tectonique éocène dans

Fig. 1 STRUCTURAL MAP OF THE MEDITERRANEAN AREA

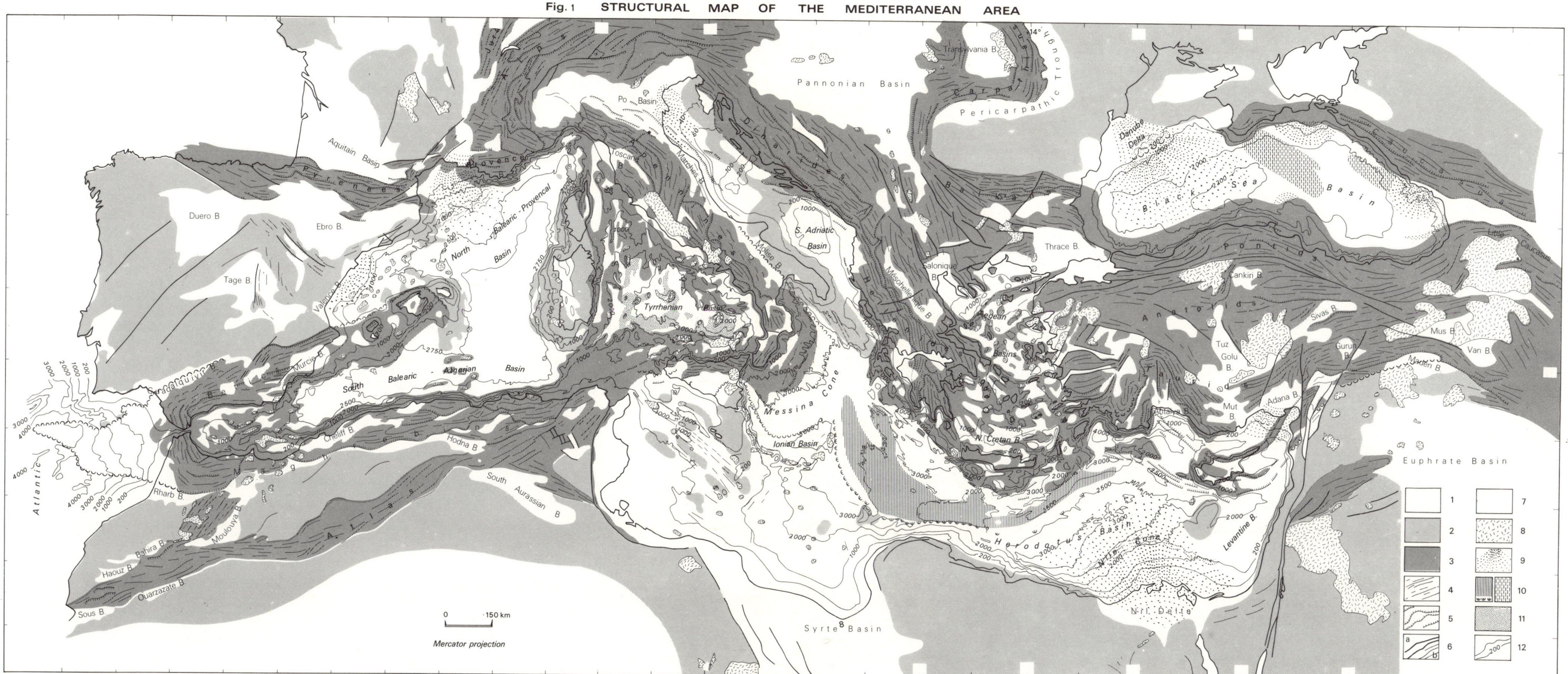

1 Sedimentary Basins 2 Stable Areas 3 Orogenic Belts 4 Main directions of folding 5 Major overthrusts, fronts of nappes 6 Major faults a) onshore b) offshore 7 Front of gravity slidings 8 Volcanism 9 Major recent deltas 10 Mediterranean Ridge – Central Black Sea High 11 Tyrrhenian abyssal plain 12 Bathymetry

les Cordillères Bétiques : vers une nouvelle conception de la paléogéographie en Méditerranée occidentale. Bull. Soc. Geol. France, (7), XVI, pp. 58-73.

RYAN (W.B.F.), OLAUSON (E.), FAIRBRIDGE (R.W.).- (1966). Mediterranean Sea. In : The Encyclopedia of Oceanography. Fairbridge (E.W.) Ed., New York, Reinhold Publ. Corp., pp. 490-498.

RYAN (W.B.F.), STANLEY (D.J.), HERSEY (J.B.), FAHLQUIST (D.A.), ALLAN (J.D.).- (1971). The Tectonics and Geology of the Mediterranean Sea. In : The Sea by Maxwell (A.E.) p[t] Ed., Wiley-Interscience, vol. IV, (II), pp. 387-492.

Fig 2 : INTERPRETATIVE GEOLOGICAL SECTIONS ACROSS THE NORTH BALEARIC–PROVENÇAL BASIN

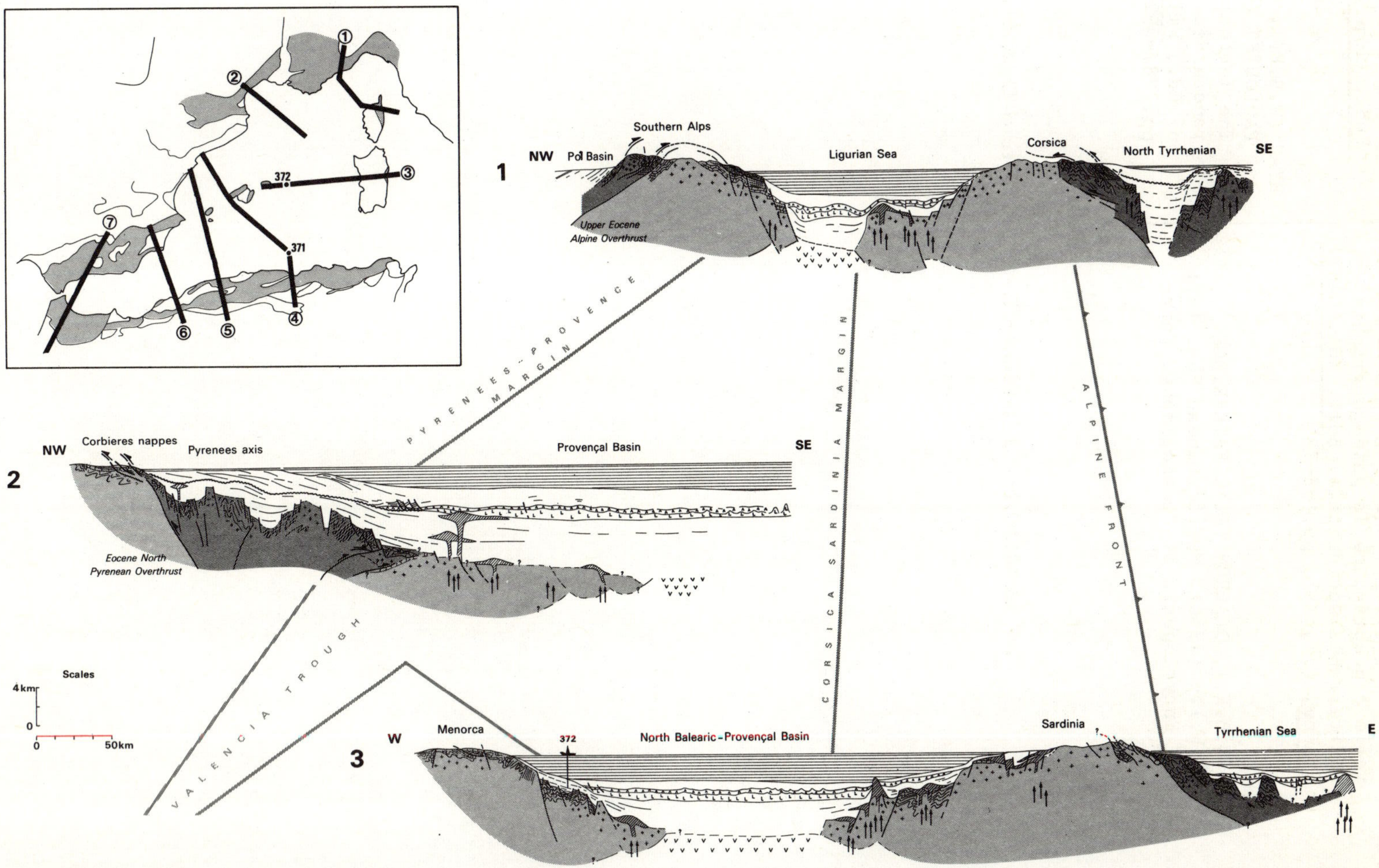

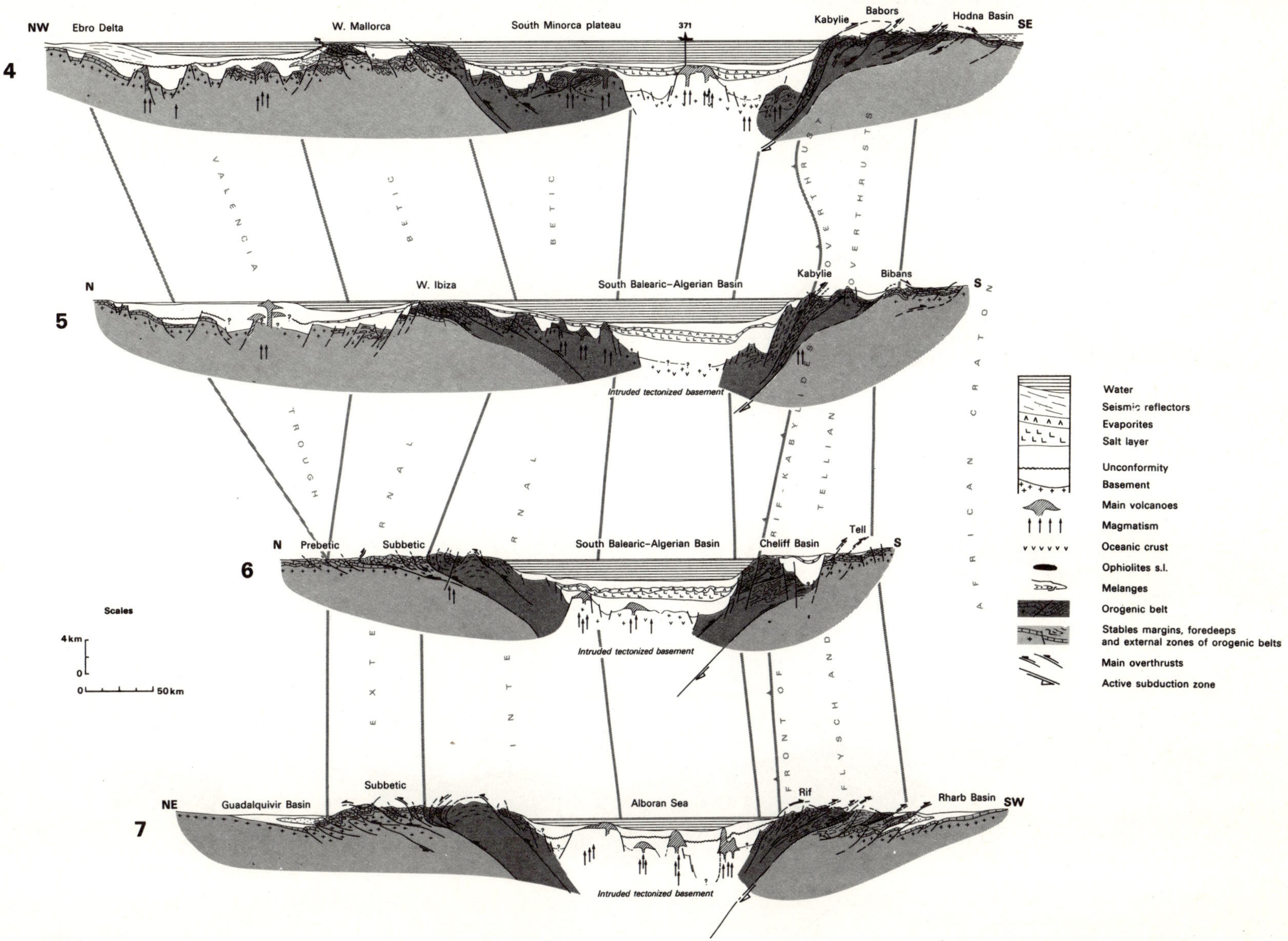

4
NW
Ebro Delta
W. Mallorca
South Minorca plateau
371
Kabylie
Babors
Hodna Basin
SE
VALENCIA
BETIC
BETIC
OVERTHRUSTS
OVERTHRUSTS
5
N
W. Ibiza
South Balearic–Algerian Basin
Kabylie
Bibans
S
Intruded tectonized basement
TROUGH
EXTERNAL
INTERNAL
RIF-KABYLIDES
TELLIAN
AFRICAN CRATON
6
N
Prebetic
Subbetic
South Balearic–Algerian Basin
Cheliff Basin
Tell
S
Intruded tectonized basement
EXTE
INTE
FRONT OF
FLYSCH AND
7
NE
Guadalquivir Basin
Subbetic
Alboran Sea
Rif
Rharb Basin
SW
Intruded tectonized basement
Scales
4km
0
0
50km
Water
Seismic reflectors
Evaporites
Salt layer
Unconformity
Basement
Main volcanoes
Magmatism
Oceanic crust
Ophiolites s.l.
Melanges
Orogenic belt
Stables margins, foredeeps and external zones of orogenic belts
Main overthrusts
Active subduction zone

Fig 3 : INTERPRETATIVE GEOLOGICAL SECTIONS ACROSS THE EASTERN MEDITERRANEAN

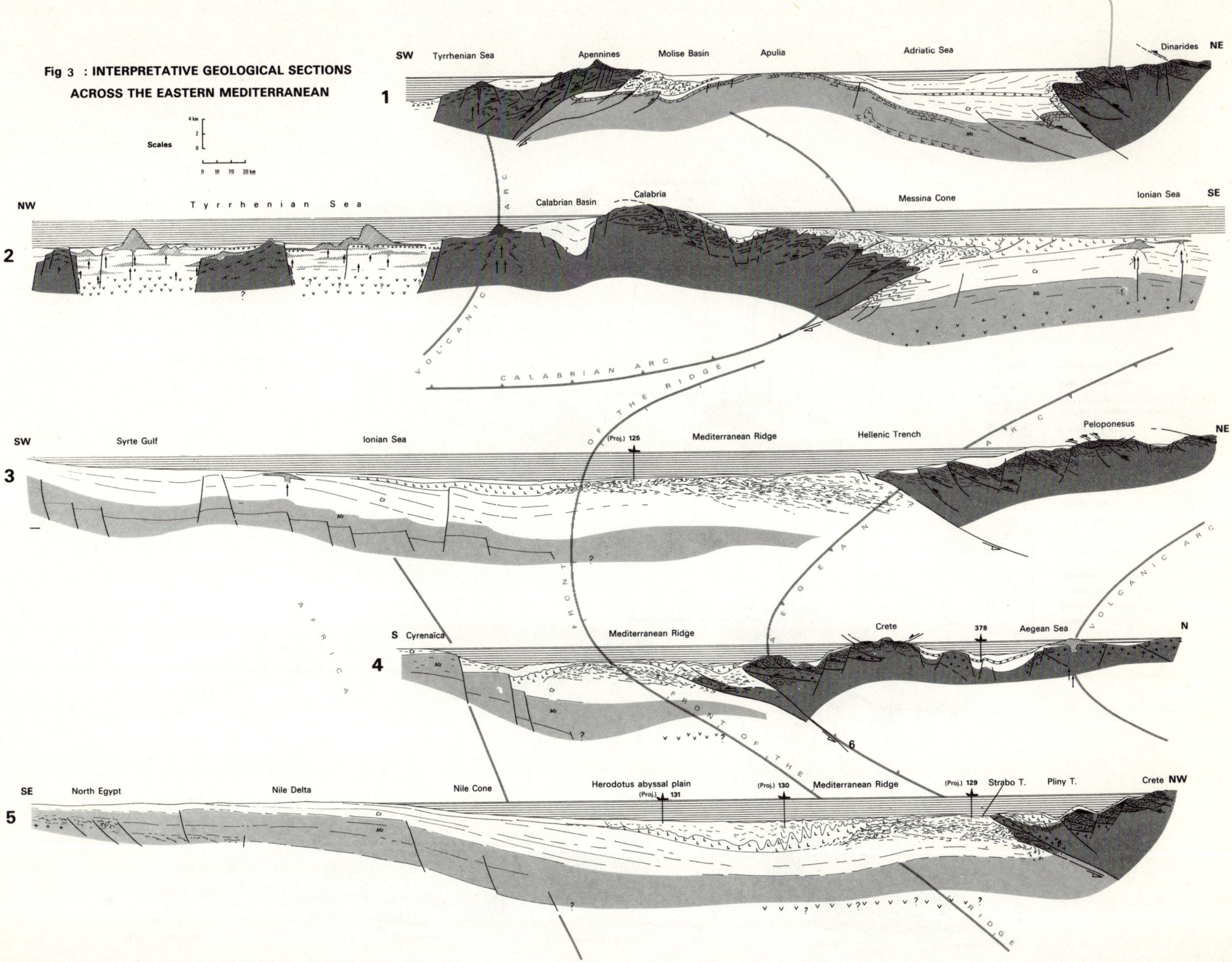

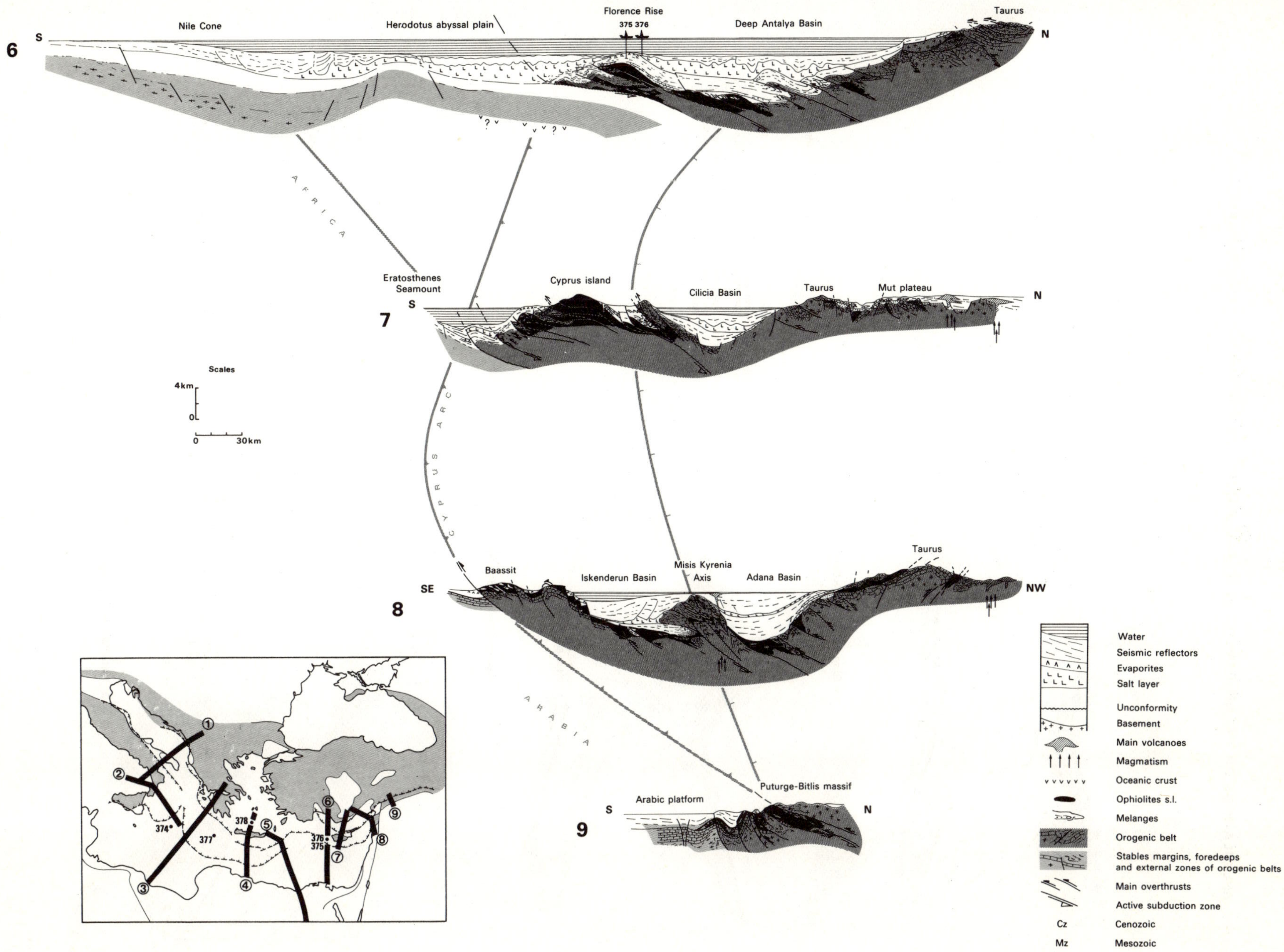
6
S
Nile Cone
Herodotus abyssal plain
Florence Rise
375 376
Deep Antalya Basin
Taurus
N
AFRICA
Eratosthenes Seamount
7
S
Cyprus island
Cilicia Basin
Taurus
Mut plateau
N
Scales
4km
0
0
30km
CYPRUS ARC
8
SE
Baassit
Iskenderun Basin
Misis Kyrenia Axis
Adana Basin
Taurus
NW
ARABIA
9
S
Arabic platform
Puturge-Bitlis massif
N
374
377
378
376
375
Water
Seismic reflectors
Evaporites
Salt layer
Unconformity
Basement
Main volcanoes
Magmatism
Oceanic crust
Ophiolites s.l.
Melanges
Orogenic belt
Stables margins, foredeeps and external zones of orogenic belts
Main overthrusts
Active subduction zone
Cz Cenozoic
Mz Mesozoic

Fig. 4 **SPECULATIONS ON THE NEOGENE STRUCTURAL HISTORY OF THE WESTERN MEDITERRANEAN**

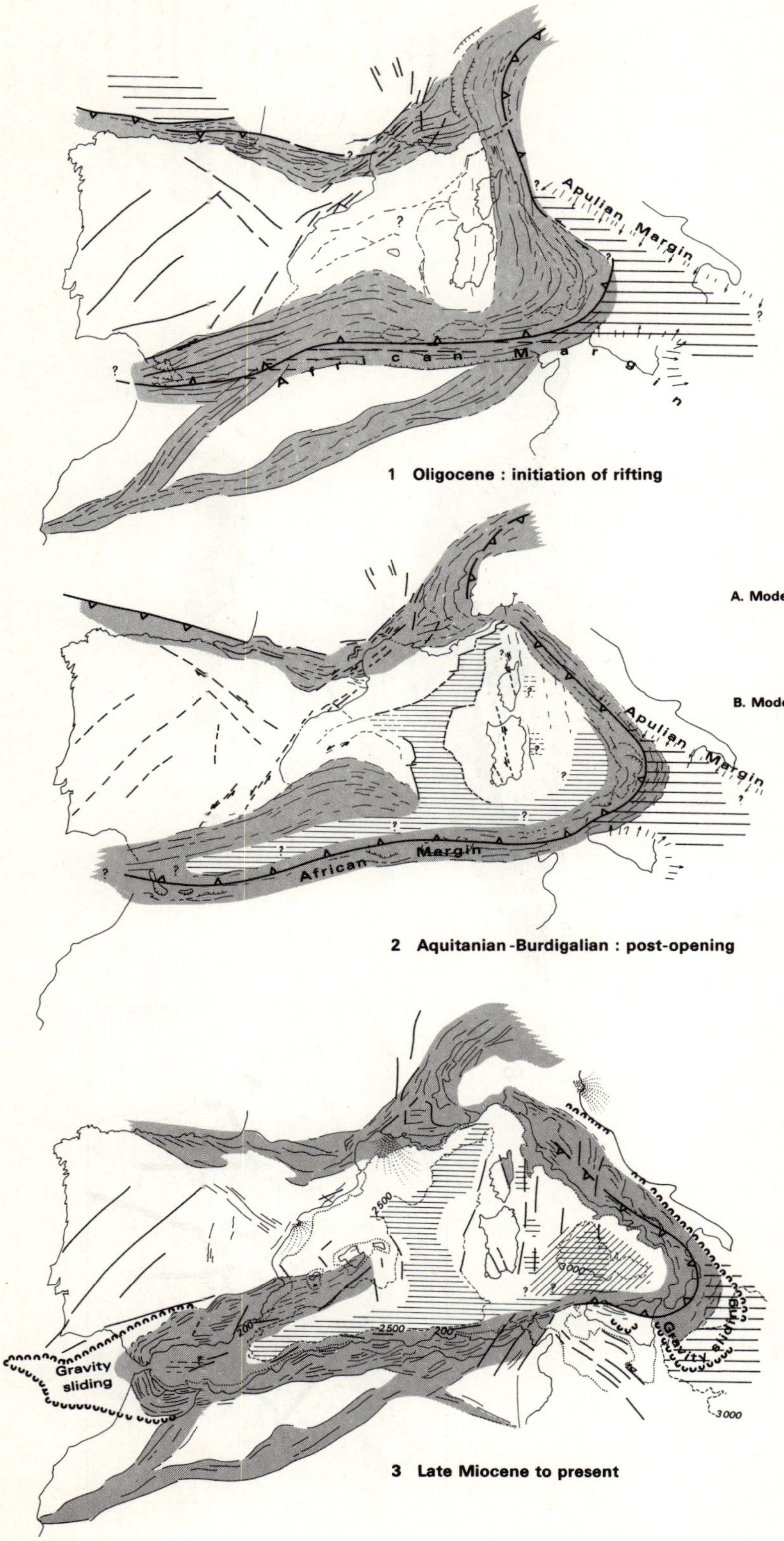

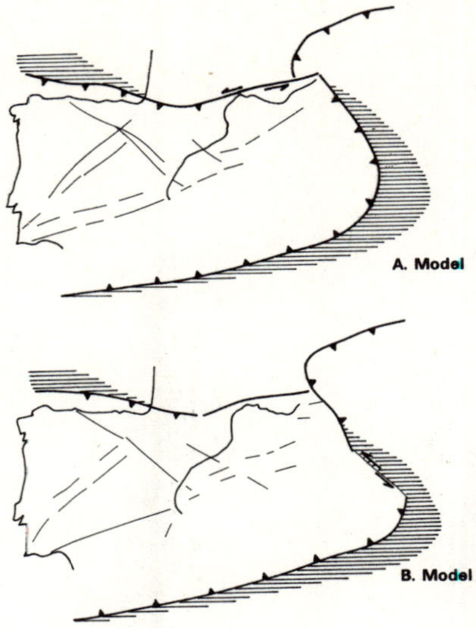

A. Model Priority to Pyrenees-Alps Relations
for : continuity of Apennine subduction during time and role of liguria lineament
against : Alpine corsica must be interpreted as back over thrust; eocene subduction in Northern Apennine unknown

B. Model Priority to Alpine Corsica-Alps Relations

Legend

- Orogenic belt and associated subduction zone
- Mesozoic oceanic crust
- Early miocene opening
- Upper miocene-pliocene opening
- Limit of gravity sliding
- Major fractures
- Deltas
- 200 — Extent of the shelf
- -2500 — Boundary of present abyssal plains (depths in meters)

Fig. 5 MESSINIAN EVAPORITES IN THE MEDITERRANEAN AREA

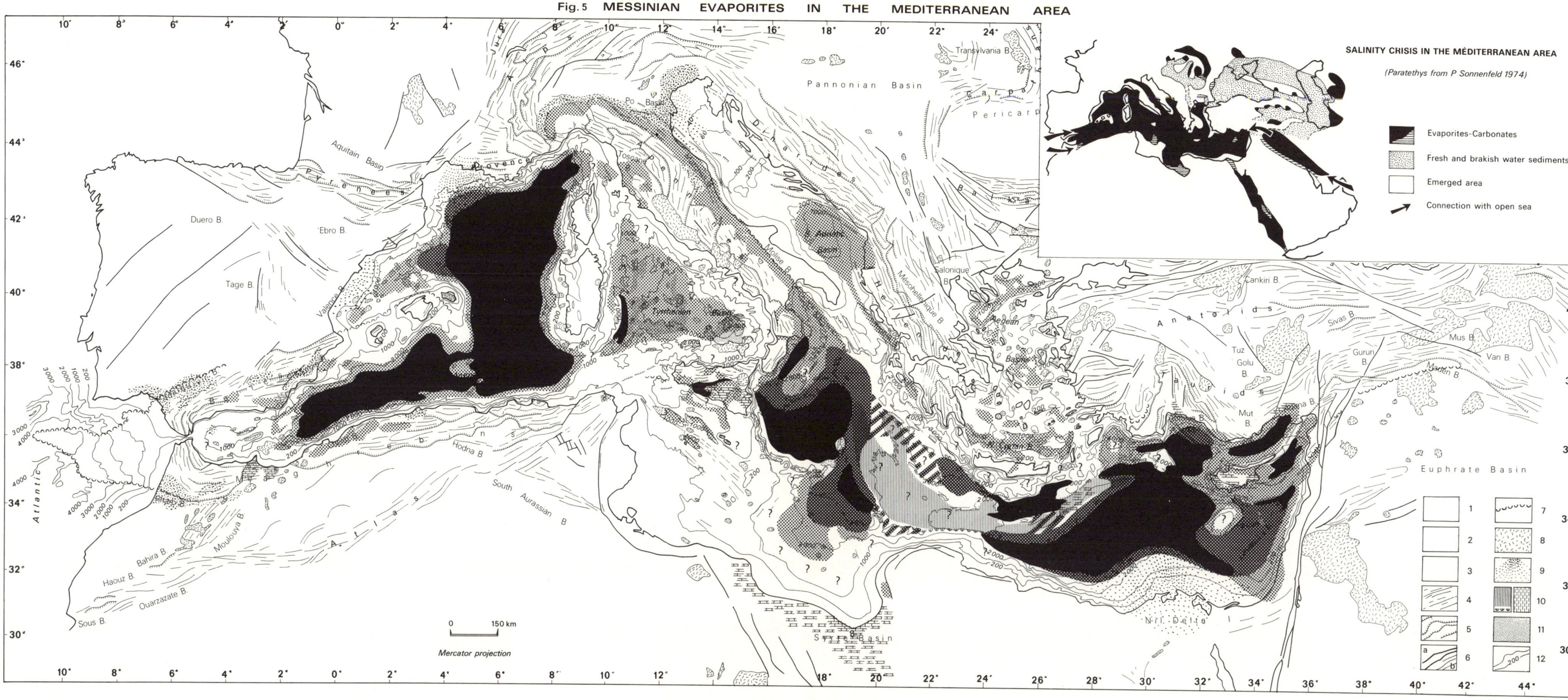

1 Sedimentary Basins 2 Stable Areas 3 Orogenic Belts 4 Main directions of folding 5 Major overthrusts, fronts of nappes 6 Major faults a) onshore b) offshore 7 Front of gravity slidings 8 Volcanism 9 Major recent deltas 10 Mediterranean Ridge – Central Black Sea High 11 Tyrrhenian abyssal plain 12 Bathymetry

Thick evaporites including thick salt layers and Upper evaporites

Evaporites including thin salt layers

Thin evaporites (> 100 m at sea)

Carbonates

Terrigenous

INTERNATIONAL SYMPOSIUM ON THE STRUCTURAL HISTORY OF THE MEDITERRANEAN BASINS. SPLIT (YUGOSLAVIA) 25-29 OCTOBER 1976. B. BIJU-DUVAL AND L. MONTADERT, Eds. EDITIONS TECHNIP, PARIS 1977, pp. 13-18

first part

geological evolution from the tethys to the mediterranean from the mesozoïc to the present

EDITORS' COMMENTS

An initial fundamental advance in our understanding of the Mediterranean basins was the recognition of previous continental margins of the Atlantic type presently tectonized in the Alpine folded belts. This was made possible by comparing results from DSDP holes and land data (Bernoulli, 1972). This led many geologists to adopt a new approach to the field in tectonized areas as well as in unfolded ones, as has been illustrated by several publications during recent years and also during the Split Symposium. Bourbon et al. (in this volume) have presented data and discussed the development of an oceanic Tethys south of the European margin in the Alps; Chorowitz*, Charvet* and Cadet* ([1]) have clarified the structural history of the Dinaric Mesozoic margin along different transects, while Blanchet (in this volume) has discussed the mechanism of the formation of such a margin. Delaune-Mayere et al., and Bein and Gvirtzman (in this volume) have also provided data to interprete old margins in the eastern part of the Mediterranean.

Along the same line of thought, the ophiolite belts in the Mediterranean area, which have long been compared with an oceanic crust (Brunn, 1959), have been thoroughly studied along with their associated sediments in the light of increasing knowledge about the oceanic crusts (see special issue of *Bull. Soc. Geol. Fr.*, 1976). However, there are still many uncertainties

---

([1]) Names with the symbol* correspond to papers given during the Symposium, for which an abstract will be published in the "Rapports et Procès-Verbaux des Réunions" de la Commission Internationale pour l'Exploration Scientifique de la mer Méditerranée (CIESM) (in press).

concerning : 1) the attribution of ophiolites to mid-oceanic ridge or marginal basin crust (see Blanchet, this volume) ;
2) their age (see Delaloye et al., this volume) and their initial geographic setting due to their allochthony (see Celet et al., this volume); nevertheless they are widely used to reconstruct former oceanic areas.

With these new concepts and using the kinematics of European and African plates deduced from the magnetic anomalies of the Atlantic, several partial or global reconstructions of the geological history of the Mediterranean have recently been proposed (Dercourt, 1970; Smith, 1971; Laubscher, 1971; Hsü, 1971; Boccaletti et al., 1972; Alvarez, 1972, 1974; Dewey et al., 1973; Channell and Horvath, 1976; Laubscher and Bernoulli, 1976; Biju-Duval et al., 1976, 1977). In spite of the obvious need for more field data and precise paleogeographic reconstructions for the different periods (see Burollet and Winnock, in this volume), it is a nice surprise to see that the different schemes presented during the Split Symposium (Bernoulli, Biju-Duval et al., Channell and Horvath, in this volume) are generally in good agreement. With the exception of the relationship (continental or not) between the Adriatic subplate and Africa (Adriatic subplate considered as a promontory of Africa or not), basically the authors agreed about :

- the existence of an eastern Tethys at the beginning of the Mesozoic;
- a Liassic-Jurassic distension between Europe and Africa with the creation of new continental margins and narrow oceanic areas connected with the Atlantic opening;
- the existence of compression with probable subduction zones in the NE area, at least since the Upper Jurassic;
- the necessity to introduce some intermediate small plates between Europe and Africa;
- the closure of the eastern Tethys between the Late Cretaceous to Early Cenozoic and a more and more accentuated state of continental collision;
- the creation of small marginal basins in relation to subduction zones;
- the evidence of a volcanism associated with the different active or passive margins, as illustred by Giunta et al.*

In spite of much more complicated features than previsouly expected from early studies of island arcs and marginal basins, some attempts at global interpretation are nevertheless now possible (see review by Bellon and Letouzey, Wezel, in this volume).

In this field of plate tectonics, two topics would have been very interesting to discuss because of their very important implication in active margins and in absolute plate motion :

1) The very difficult problem of Alpine metamorphism known from eastern Turkey (Bitlis massif) to Spain, for which few synthetic reviews have been made (see Dal Piaz, 1976, for example). The exact position in the tectonic pile, the age and the relationship with ophiolite overthrusts are still not clarified. Many "Paleozoic or Precambrian" massives have now been demonstrated to be "Mesozoic metamorphosed" series such as, for example, the Menderes or Cyclades massives, and new chronological data are now available (Aubouin et al., 1976; Seidel and Okrusch, 1976; Durr, 1976, etc.).

2) The second area where research is in progress is in paleomagnetism. Until recently little good data could be used (see review by Vander Voo, 1972). But in the last few years data about Spain, the Corsica-Sardinia block and especially Italy now provide good guidelines for reconstructions (Coulon et al., 1974; Vandenberg and Wonders, 1976; Vandenberg et al., 1977; Klootwiz, 1976). These data have not yet been fully used in the reconstruction schemes proposed.

To complete this approach to the absolute motion of plates, geological data (such as the distribution of facies) may also give some interesting data, as pointed out by Sonnenfeld*.

BIBLIOGRAPHY

ALVAREZ (W.).- (1972). Rotation of the Corsica Sardinia Microplate. Nature (Phys. Sci.), 235, 103-105.

ALVAREZ (W.), COCOZZA (T.) and WEZEL (F.C.).- (1974). Fragmentation of the Alpine orogenic Belt by Microplate Dispersal. Nature (Phys. Sci.), 248, 309-314.

AUBOUIN (J.), BONNEAU (M.), DAVIDSON (J.), LEBOULENGER (P.), MATESCO (S.) et ZAMBETAKIS (A.).- (1976). Esquisse structurale de l'Arc égéen externe : des Dinarides aux Taurides. Bull. Soc. Geol. France, 7ème ser., XVIII, n° 2, 327-336.

BERNOULLI (D.).- (1972). North Atlantic and Mediterranean Mesozoic facies : a Comparison-Initial Reports of the Deep Sea Drilling Project, vol. XI, 801-871, Washington D.C., US Government Printing Office.

BIJU-DUVAL (B.), DERCOURT (J.), LE PICHON (X.).- (1976). La genèse de la Méditerranée. La Recherche, 7, n° 71, 811-822.

BIJU-DUVAL (B.), LETOUZEY (J.), MONTADERT (L.).- (1977). Structure and Evolution of the Mediterranean Sea Basins. In : Initial Report of the Deep Sea Drilling Project, vol. XLII, Washington, D.C., US Government Printing Office, in press.

BOCCALETTI (M.), GUAZZONE (G.).- (1972). Remnant Arcs and marginal Basins in the Cainozoic Development of the Mediterranean. Nature, (Phys. Sci.), 242, 18-21.

BRUNN (J.H.).- (1959). La dorsale médio-atlantique et les épanchements ophiolitiques. Soc. Geol. France Comp. Rend., 234-236.

CHANNEL (J.E.T.), HORVATH (F.).- (1976). The African-Adriatic Promontory as a palaeogeographical Premise for Alpine Orogeny and Plate Movements in the Carpatho-Balkan Region. Tectonophysics, 35, 71-101.

COULON (C.), DEMANT (A.), BOBIER (C.).- (1974). Contribution du paléomagnétisme à l'étude des séries volcaniques cénozoïques et quaternaires de Sardaigne nord-occidentale. Tectonophysics, 22, 59-82.

DAL PIAZ (V.).- (1974). Le métamorphisme de haute pression et basse température dans l'évolution structurale du bassin ophiolitique alpino-apenninique (1ère partie) : considérations paléogéographiques. Boll. Soc. Geol. Ital., 93, 437-468.

DERCOURT (J.).- (1970). L'expansion océanique actuelle et fossile; ses implications géotectoniques. Soc. Géol. France, Bull., 7ème ser., XII, 261-317.

DEWEY (J.F.), PITMAN (W.C.) III, RYAN (W.B.F.) and BONNIN (J.).- (1973). Plate Tectonics and the Evolution of the Alpine System. Geol. Soc. Amer. Bull., 84, n° 10, 3137-3180.

DÜRR (St.).- (1976). Über das Menderes-Kristallin und seine Aequivalente in Griechenland. Soc. Geol. France Bull., 7ème ser., XVIII, 429.

HSU (K.J.).- (1971). Origin of the Alps and Western Mediterranean. Nature, 233, n° 5314, 44-48.

KLOOTWIJK (C.T.), VAN DEN BERG (J.).- (1975). The rotation of Italy : preliminary palaeomagnetic Data from the Umbrian Sequence, Northern Apennines. Earth Planetary Sci. Lett., 25, 263-273.

LAUBSCHER (H.P.).- (1971). The large scale Kinematics of the Western Alps and the Northern Apennines and its Palinspatic Implications. Amer. J. Sci., 271, 193-226.

SEIDEL (E.) and OKRUSCH (M.).- (1976). Regional distribution of critical metamorphic minerals in the Southern Aegean. Soc. Geol. France Bull., 7ème ser., XVIII, 347-350.

SOCIETE GEOLOGIQUE DE FRANCE.- (1976). Vème Colloque sur la Géologie des régions égéennes (Orsay, 1-3 février 1975). Soc. Géol. France Bull., 7ème ser., XVIII, 197-567.

SOCIETE GEOLOGIQUE DE FRANCE.- (1976 b). Création de la lithosphère océanique (Brest, 27-28 nov. 1975). Soc. Geol. France Bull., 7ème ser., XVIII, 791-947.

SMITH (A.G.).- (1971). Alpine Deformation and the oceanic Areas of the Tethys, Mediterranean, and Atlantic. Geol. Soc. Amer. Bull., 82, n° 8, 2039-2070.

VAN DEN BERG (J.), WONDERS (A.A.H.).- (1976). Paleomagnetic evidence of large Fault Displacement around Po Basin. Tectonophysics, 33, 301-320.

ZIJDERVELD (I.D.A.) and VAN DER VOO (R.).- (1973). Paleomagnetism in the Mediterranean Area. In : Implications of Continental Drift to the Earth Sciences. Tarling (D.H.) and Runcorn (S.K.) Ed., London, New York, Academic Press, 133-161.

INTERNATIONAL SYMPOSIUM ON THE STRUCTURAL HISTORY OF THE MEDITERRANEAN BASINS. SPLIT (YUGOSLAVIA) 25-29 OCTOBER 1976. B. BIJU-DUVAL AND L. MONTADERT, Eds. EDITIONS TECHNIP, PARIS 1977, pp. 19-34

MESOZOIC EVOLUTION OF THE WESTERN ALPS : BIRTH AND DEVELOPMENT OF PART OF THE SPREADING OCEANIC TETHYS AND OF ITS EUROPEAN CONTINENTAL MARGIN

M. BOURBON [1], J.-M. CARON [2], P.-Ch. de GRACIANSKY [1], M. LEMOINE [1], J. MEGARD-GALLI [3] and D. MERCIER [1]

1. INTRODUCTION

The major part of the present-day Mediterranean sea is surrounded by folded belts which underwent folding, thrusting and metamorphism during Cretaceous and Tertiary times. Since the beginning of this century, numerous authors such as E. Suess, E. Argand and others have assumed that these Perimediterranean Alpine chains result from the contraction of a former Mesozoic Tethyan Ocean, the shape and width of which remain till now a matter of conjecture and discussion. Of course, the present-day Mediterranean basins cannot be considered as remnants of this Mesozoic ocean : they result from late-Tertiary and Quaternary (i.e. late or post-Alpine) spreadings and/or collapses, as stated by Argand as early as 1924. The birth of recent Global Tectonics models and above all the great development of Oceanography and Sedimentology provided new tools of research to Alpine Geology, so that both our present knowledge of the Tethyan ocean and margins and our prospects for future results are now completely renewed.

Of course, when trying to reconstruct the evolution of the Tethyan ocean and margins, we are confronted with the problem of uniformitarianism. Briefly speaking, two main "uniformitarian" or "actualistic" models may be used, the Atlantic one and the Pacific one -- that is to say, a spreading ocean with "passive" ("stable") margins, and a contracting, more complex, one, with "active" margins and with or without marginal basins. But such a simple choice may be debatable and has to be discussed : first, there are indeed some other possible actualistic models (e.g. the Bay of Biscay) ; second, as will be emphasized later in this paper, the Tethyan ocean very likely had its own features, so that it may need other models, whether its spreading mechanism or the evolution of its margins are considered ; such models may be possibly non-actualistic.

Nevertheless, the classical "actualistic" models appear at the present state of our knowledge to be very useful at least as a first approach ; they allow four main successive stages of the evolution of a given Alpine orogen to be considered : (1) "pre-oceanic" stage ; (2) beginning of rifting ; (3) spreading (possibly an "Atlantic" stage or model, but more likely some other kind of mechanism) ; (4) contraction, i.e. a kind of "Pacific stage" but with or without subductions and with or without marginal seas, finally ending in continent collision. Of course, from one part to the other of the different Perimediterranean chains, the evolutionary models may differ in some details, and above all when the ages of the different stages are considered. We shall now only deal with a single chain, the Alps, and more especially the Western Alps, the history of which is linked with the evolution of only part of the whole complex Tethys.

(1) Centre de Géologie Générale, Ecole Nationale Supérieure des Mines, 60 boulevard Saint Michel, 75272 Paris Cédex 06, France

(2) Laboratoire de Géologie Structurale, Université Louis Pasteur, 1 rue Blessig, 67084 Strasbourg Cédex, France

(3) Laboratoire de Géologie Structurale, Université des Sciences et Techniques du Languedoc, Montpellier, place Eugène Bataillon, 34060 Montpellier Cédex, France

2. THE WESTERN ALPS : FROM A KNOWN STRUCTURAL FRAMEWORK TO A CONJECTURAL PALINSPASTIC SKETCH

The structural framework of the Alps cannot be described here in detail (see Debelmas and Lemoine, 1970). Briefly, three major structural units can be recognized, (1) an Helvetic one (= external, called Dauphiné in the Western Alps), (2) a Penninic one, and (3) an Austro-Alpine and South-Alpine one. Rocks belonging to these overthrust units originate from specific paleogeographic domains (Table I, figs. 1 and 2), and some of these domains are also linked with structural units in the Northern Apennines.

The sedimentary material involved in the Western Alps' folds and thrusts is mainly of Mesozoic and Tertiary age, and a Sialic basement (Variscan and/or older gneisses and granites), i.e. a continental crust, is known everywhere in the Western Alps, either in outcrops or inferred from geophysical data. But, as is well-known, certain overthrust Mesozoic-Tertiary sedimentary series cannot be stratigraphically linked with any sialic basement : one may therefore wonder whether a certain amount of oceanic crust might have existed among the Alpine paleogeographic domains.

It is here assumed as a working hypothesis that most ophiolites may be more or less considered as remnants of a certain kind of oceanic crust and of the underlying upper mantle. On the other hand, it is well-known that true ophiolites which at some places bear their own sedimentary cover occur in certain structural units of both the Western Alps (Piémont-Ligurian : Table I) and the Northern Apennines (Ligurian). Thus, we are led to envision an ancient area with some kind of oceanic crust, which may be called the Piémont-Ligurian "ocean". This "ocean" (or "oceanic crust-bearing area", Lemoine, 1975), which existed during part of the Jurassic and of the Cretaceous, is to be regarded as part of the Tethys, and belonged to what will become the Alps-Corsica orogen and the Apennines orogen (Table I, figs. 1 and 2).

Furthermore, assuming the existence of an oceanic crust-bearing area during Mesozoic times, this area had therefore two continental margins, an European one and an African (Apulian) one. The purpose of this paper is to present some information about the European margin of this Piémont-Ligurian "ocean", using some structural and sedimentological studies which were carried out in the Western Alps.

Of course, the rocks which initially belonged to this margin are now folded, thrust, and at some places metamorphosed, so that it is sometimes rather difficult to undertake sedimentological studies (hence the selection of the Briançonnais in § 4, instead of the Piémont domain), and also to build up a complete palinspastic reconstruction of the whole margin and of its connexions with the oceanic crust-bearing area previous to Alpine folding. Briefly speaking, the main difficulties are : (1) the existence of "blanks", i.e. unknown parts of the palinspastic reconstruction, due to the fact that certain units are now either eroded or concealed below other ones, or even subducted ; (2) the existence of intense polyphase folding and metamorphism in certain zones (e.g. Piémont) ; (3) the difficulty (mainly in the Piémont zone) to decipher the original mutual positions of more or less overthrust structural units which now have very complex geometrical relationships as a result of polyphase folding and thrusting.

When dealing with the Briançon traverse (§ 3), we shall be obliged to make an assumption when facing the choice between several solutions for the reconstruction of the Piémont domain. But, on a larger scale, when considering only the largest structural paleogeographic units (that is to say, Helvetic, Valais, Briançonnais, and Piémont), we are now, in the present state of our knowledge, almost certain of their primitive gross relationships (but for some uncertainty when dealing with the original positions and connections of the ophiolitic belts). Thus the palinspastic image shown in fig. 2 is not devoid of likelihood.

Anyway, whether this palinspastic image or another one are chosen, a fact remains : there are important longitudinal changes in the palinspastic network of the margin, when seen on a large scale (of the order of several hundred kilometres : compare traverses 1-4 of fig. 2). Similar longitudinal irregularities of the margin's framework appear also on smaller scales.

3. THE BRIANCON TRAVERSE : GENERAL

At the latitude of the city of Briançon, from the outer (W) towards the inner (E) part of the chain, we cross several structural zones which are folded, pressed together and thrust upon each other : the Dauphiné zone (chaines subalpines), the Subbriançonnais-Briançonnais units, and finally the many, strongly folded and metamorphosed, Piémont units. On this somewhat large scale, palinspastic reconstruction is still rather easy ; for example, at Upper Jurassic times (see below, § 4 and 5, fig. 4), the Dauphiné domain may be more or less compared with the continental shelf, the Subbriançonnais-Briançonnais domains were very probably located on the continental slope, whereas part of the Piémont domain (namely, the Piémont-Ligurian), had a true oceanic crust (ophiolites). But, when dealing with a more detailed paleogeographic reconstruction of the Piémont structural units, the most probable solution has to be chosen among others (Caron, 1977) : the innermost domain is very likely the Piémont-Ligurian one, with its ophiolitic basement (oceanic crust) overlain by Jurassic-Cretaceous deep-sea sediments ; between this domain and the Briançonnais one, several Piémont units have to be fitted in a certain order (fig. 3) ; they all seem to have possessed a sialic (continental) crust covered by Triassic platform carbonates (see below) ; thus the corresponding paleogeographic domains (Piémont s.str. in Table I) appear to have belonged to the European margin --at least in the case of the palinspastic model here chosen.

This model (figs. 2, 3, 4) will now serve as a basis for the reconstitution of the continental margins' paleotectonic and sedimentological evolution, which is of course closely related to the evolution of the oceanic crust-bearing area. As far as is known, the sedimentary cover overlying the Piémont-Ligurian ophiolites is believed to be of Upper Jurassic and Lower Cretaceous age, as is the identical post-ophiolitic sedimentary series of the Apennines' Ligurian units, which is both fossiliferous and less metamorphosed. Therefore, the Piémont-Ligurian oceanic crust domain appears to have come into existence little before the Upper Jurassic, i.e. probably during Middle, or even Lower, Jurassic times : when trying to reconstruct the paleotectonic evolution of the margin, this time-limit must be kept in mind.

This reconstruction has to be based upon a sedimentological analysis of the Mesozoic deposits. Such an analysis cannot easily be carried out in the nearly always metamorphosed Piémont sediments. Thus, what happened in the Briançonnais domain will be first examined, as this domain, in the Briançon traverse, underwent less tectonical disturbances and a weaker metamorphism.

4. SEDIMENTOLOGICAL AND PALEOTECTONIC EVOLUTION OF THE BRIANCONNAIS DOMAIN IN THE BRIANCON TRAVERSE : A KEY TO THE MARGIN'S EVOLUTION

We shall deal here only with the Mesozoic sediments of the Briançonnais, which overlie either Permo-Carboniferous Variscan molasses or a Variscan sialic basement, and which comprise a thick series of Triassic-Lower Liassic platform sediments, followed by a relatively thin, mainly pelagic, Jurassic-Cretaceous series. Stratigraphical and sedimentological analyses of these series lead to recognize four main stages in the paleotectonic evolution of the area, which reflect more wide-spread events (see § 5) : (1) pre-oceanic (pre-rifting) phase (Triassic-Lower Lias) ; (2) early rifting phase (Middle or Upper Lias) ; (3) further evolution of the "passive" margin (and spreading of the neighbouring oceanic area) (Upper Jurassic-Lower Cretaceous) ; (4) Upper Cretaceous paleogeographic change, linked with the beginning of contraction of the oceanic area.

4.1. THE PRE-RIFTING (PRE-OCEANIC) PERIOD : SUBSIDING TRIASSIC AND LOWER LIASSIC PLATFORMS ; EARLY BUT ABORTED INCIPIENT RIFTING.

Lower Triassic times correspond to a slightly subsiding detrital platform extending on the greater part of Western Europe and especially on what will become the Alpine realm. At that time, the levelling out of the late Variscan reliefs had resulted in a very flat surface upon which, immediatly following, were emplaced several superimposed carbonate platforms which evolved during Middle Triassic, Upper Triassic and Lower Liassic times.

As for the Middle and Upper Triassic beds (Baud and Mégard-Galli, 1975), six main sedimentary cycles are recorded over the whole Briançonnais domain, from the Swiss Prealps to the Ligurian Alps, i.e. along a length of 500 km. These cycles are controlled by wide-spread paleotectonic events, and each cycle corresponds to the infilling of the whole sedimentary area, so that it ends with an overall emergence.

The first three cycles (Anisian-Ladinian) correspond to rythmical shallow subtidal to supratidal carbonate deposits with their typical sedimentological features. Sequential analysis strongly suggests a discontinuous rate of subsidence during cycle I and a more continuous one during cycle II, both giving birth to infilling sedimentary sequences each one several hundred metres thick.

Cycle IV (coming after a well-characterized but shorter cycle III) is partly Carnian in age and shows somewhat different features : (1) sudden appearance of detrital material (Equisetum-bearing shales and sandstones), (2) very restricted carbonate deposits grading into evaporites (two main levels, each one several metres thick), (3) thick algal-mat deposits affected by numerous fault-scarp breccias and slumpings. All these features indicate a temporarily differentiated morphology linked with distensional tectonics.

When this period of tectonic instability ends, the beginning of a new cycle (V, Norian) may be defined ; then a tectonically stable, but rapidly subsiding, basin gave birth to the Hauptdolomit formation which corresponds to a new carbonate platform, extending over several hundred kilometres, beyond the proper Briançonnais domain up to a large part of the South-Alpine and Austro-Alpine domains. The Hauptdolomit deposition shows a renewed pattern of subsidence, trending to the South-East, whereas the Middle Triassic one was North-East trending.

Cycle VI (Rhaetian) is marked by a great transgression which is known over much of Western Europe. This transgression accompanies a new synsedimentary distensional stage which initiates or heralds the great paleogeographic modifications of the Liassic, and especially the Upper Liassic period.

Briefly speaking, during Triassic times, the specific feature of the forthcoming Alpine realms, and especially of the Briançonnais one, is the superimposition of several sedimentary cycles, all characterized by relatively high rates of sedimentation and subsidence (about one thousand metres for the Middle and Upper Triassic beds). Together with the peculiar evolution of cycle IV, these facts provide some evidence for a succession of several but rather faint distensional phases, some of them being coeval with some volcanic activity in the Southern Alps.

Nevertheless, the following must be emphasized : (1) in those parts of the Tethyan domain and during the Triassic period, there is nowhere evidence of any occurence of oceanic crust ; (2) a stable continental crust would very probably have underlain all the Triassic sedimentary basins, the Briançonnais one as well as the Piémont s.str. (as defined in Table I), or the Austro-Alpine ones (Trümpy, 1971) ; (3) the environment remained steadily very shallow. We may therefore conclude that, even if several episodes of distensional tectonics occurred during Triassic and Lower Liassic times, such episodes did not result in true rifting, at least in this part of the Tethyan realm. The true rifting initiating oceanic opening occurred here only from and after the Upper Liassic, along a wholly different paleogeographic pattern.

4.2. UPLIFT AND SUBSEQUENT SINKING OF THE BRIANCONNAIS : PROBABLE CONSEQUENCE OF RIFTING AND OF INCIPIENT OCEAN SPREADING (UPPER LIASSIC TO OXFORDIAN).

In this part of the Tethyan area which corresponds to the Western Mediterranean chains, Liassic and especially Upper Liassic times mark the beginning of individualization of the several paleogeographic domains, as a result of the birth of a highly differentiated topography, either submarine or emerged ("Upper Liassic paleogeographic revolution" ; Bourbon et al., 1973) ; this is the time of birth of synsedimentary fault-scarps, leading to individualization of submarine highs, knolls and troughs (Bernoulli and Jenkyns, 1975).

In the Helvetic-Dauphiné domain of the Western and Central Alps, both subsidence and sedimentary supply resulted in the deposition of thick Ammonites-bearing marls and marly limestones, whereas the Piémont domain was the site of thinner marly and calcareous sedimentation with many interbedded breccias. But between these two domains, a complex system of emerged lands appears, which corresponds to our Briançonnais domain ; this broad "Briançonnais island", located between the Dauphiné and Piémont basins (fig. 4), was then the site of more or less deep erosion (locally reaching down to the pre-Triassic basement), accompanied by local deposition of conglomerates with well-rounded pebbles, of coal seams, and of bauxitic layers. But this island did not remain long emerged (Upper Liassic to Lower Bathonian) and has been thereafter invaded by shallow-sea water : the platform carbonate sedimentary processes resumed, but for a short time only, leading to deposition of less than 50 metres of sediments ranging from restricted to outer shelf environments (Upper Bathonian to Lowermost Upper Jurassic ; de Graciansky and Mercier, 1976).

Immediatly after, i.e. during early Upper Jurassic (but before Middle or even Lower Oxfordian) times, the temporarily emerged Briançonnais area rapidly sank down to deep-sea environment, leading to deposition of pelagic sediments, below the Calcite Compensation Depth (CCD) in the deeper sites, or in the vicinity of the Aragonite Compensation Depth (ACD) for the shallower ones. At the same time, breccias, olistoliths and slumped beds are deposited ; they reflect the presence of a rather intricate submarine topography with fault-scarps and slopes.

This event does imply thus an important phase of active synsedimentary faulting which we regard as linked with a prominent step in the opening of the Piémont-Ligurian "ocean". The Briançonnais domain henceforth belongs to the European margin of this "ocean", and its rapid sinking may be regarded as a classical stage of the evolution of a passive or stable continental margin. But its preliminary uplift and emergence is a special feature of this part of the margin ; one may remark here that the evolution of the opposite (African-Apulian) margin of the same oceanic area was somewhat different : for example, when dealing with the Southern Alps, the rapid sinking, from platform carbonate environment down to pelagic environment, occurred as early as the Middle or Upper Lias (Ammonitico Rosso lying upon Lower Liassic carbonates).

4.3. FURTHER EVOLUTION OF THE PASSIVE MARGIN (UPPER JURASSIC AND LOWER CRETACEOUS).

During Upper Jurassic-Lower Cretaceous times, in the traverse here studied, the Briançonnais domain was the site of deposition of condensed pelagic sequences (e.g., the thickness of the Upper Jurassic beds may range from a few decimetres or less, to a maximum of 50 metres) ; another feature is the presence of hard-ground surfaces with mineralized crusts or nodules (P, Si, Fe, Mn). All these features may be explained by noting that during this period the sea bottom was relatively deep (often near the CCD but for some sea-mounts), and that it exhibited differentiated reliefs directly inherited from the previous phase of collapse. Thus, hard-grounds, crusts and nodules may be linked with non-depositional environments, either on the top of sea-mounts or on steep slopes ; the very low rate of sedimentation may be related with the vicinity of both the ACD and the CCD, with the remoteness of any source area and with the presence of slopes ; the great lateral variability of this sedimentation rate would then result from the much differentiated bottom morphology, with presence of areas of erosion or of accumulation, etc.

Finally, it is recalled that these events were coeval with the deposition of oceanic sediments upon the Piémont-Ligurian oceanic (ophiolitic) crust.

4.4. UPPER CRETACEOUS : END OF SPREADING AND BEGINNING OF CONTRACTION.

In the Briançonnais domain, pelagic sedimentary conditions continued throughout the whole Cretaceous period and even in Paleocene and Lower Eocene times, with deposition of a few hundred metres of foram-nanno-ooze (implying both carbonatic and argillaceous supplies). But this does not necessarily mean that paleotectonic conditions remained unchanged.

As a matter of fact, a complete modification and reorganization of the bottom morphology began during Cenomanian times. It resulted in submarine rejuvenation of the reliefs and in birth of new ones, in deep erosions (down to the pre-Triassic basement in some areas) linked with fault-scarps and accompanied by deposition of siltstones, sandstones, breccias, and olistoliths, especially during Cenomanian and during Lower Senonian times.

One might inquire whether this resumption of synsedimentary tectonical activity reflects a continuation of the rifting or anorther kind of event. As a matter of fact, it is now known that at the same time the more inner parts of the Alpine domain (probably the inner Piémont-Ligurian "ocean" and its African-Apulian margin) were subjected to compression : this is the time of the first clearly recorded Alpine compressional movements, which gave birth to the first overthrusts immediately followed by High Pressure-Low Temperature (Blue-schist) metamorphism, as evidenced by 78 to 100 m.y. radiometric ages (Bocquet et al., 1974) ; this is also the time of early flysch deposition (Upper Cretaceous Simme and Helminthoid flysches, etc.) in the inner zones. Thus the Upper Cretaceous paleotectonic events which are recorded in the still-pelagic sediments of the Briançonnais domain may have been a remote consequence of the beginning of Alpine compressional tectonics, i.e. of the shrinking of the Piémont-Ligurian "ocean". As the active margin was probably not located at that time on the European side, one may wonder whether the tectonical conditions of the European margin were still distensional or already compressional : there is up to now no final answer to this question ; anyway, the margin still existed but could not be compared at that time with present-day, either "passive" or true "active" margins.

5. EXTENSION TO THE WHOLE MARGIN AND SUMMARY

As already sketched in the preceding paragraph, sedimentological analysis in the Briançonnais domain provided paleotectonic data which are not of local importance only, but also reflect more extensive events. Thus, using both the Briançonnais data and what is now known about the other neighbouring paleogeographic domains, we may then attempt a rough sketch of the margin's Mesozoic evolution along the here dealt with Briançon traverse (3 in figs. 1 and 2). A brief general view of what happened in the whole Western and Central Alps will then be given.

5.1. BIRTH AND EVOLUTION OF THE EUROPEAN MARGIN ALONG THE BRIANCON TRAVERSE.

As already seen (§ 1), the major difficulty encountered here is the paucity of informations, both stratigraphical and sedimentological. Enough data are available about the Dauphiné, Subbriançonnais and Briançonnais domains, but not for the metamorphic Piémont units (where only scarce Triassic and Liassic fossils are known, in the outermost units only).

Nevertheless, a provisional sketch can still be drawn, using both established facts and some reasonable hypotheses : (1) More or less thick Triassic platform carbonates and evaporites exist everywhere in all paleogeographic domains (with the obvious exception of the Piémont-Ligurian ophiolitic one, which did not exist at that time). (2) The calcareous and marly Jurassic and Cretaceous deposits of the Dauphiné zone are thick (2000 to 4000 m.) and are often rich either in Ammonites (aragonitic shells) or in benthonic shallow-water fossils ; therefore they may have been deposited in a rather rapidly subsiding area, certainly above the ACD, and at some places in very shallow environments. (3) Very often, fine-grained to coarse-grained breccias are interbedded in the calcareous, marly and argillaceous sediments of the Piémont s.str. domain, which are assumed to be of Liassic, Middle Jurassic and even early Upper Jurassic age (beds which might be of late Upper Jurassic and Cretaceous age are not yet known in this domain). (4) The oceanic sediments which overlie the ophiolitic basement of the Piémont-Ligurian domain are dated (Upper Jurassic and Lower Cretaceous) by comparison with the post-ophiolitic fossiliferous sediments of the Northern Apennines which exhibit identical facies and lithostratigraphic successions ; these deposits are rather thin (a few score metres) and comprise radiolarites, limestones, then argillites with intercalated beds of turbiditic limestones ; thus the whole seems to have been deposited either below or in the vicinity of the CCD.

Using such data, the margin's birth and evolution can be outlined as follows (fig. 4) :

a. The Triassic period (upper section of fig. 4) may be termed "pre-oceanic" in this part of the Alpine Mediterranean area : despite some faint distensional trends (which are recorded for example in the Briançonnais Carnian beds), platform carbonate, and sometimes evaporitic, sedimentation occurred everywhere, with more or less rapid subsidence from one place to the other.

b. <u>During Liassic and Middle Jurassic times</u>, the tectonic evolution which will result in the Tethyan opening is marked in the Piémont domain by active faulting leading to deposition of breccias, whereas in the Dauphiné domain it resulted in the progressive birth and separate evolution of both highly and poorly subsiding areas. The topography of the Briançonnais and Piémont domains became rougher in Upper Liassic and early Middle Jurassic times, i.e. during the emergence of the Briançonnais island (middle section of fig. 4). One may suggest that this reflects the beginning of rifting between the Piémont and Austro-Alpine domains.

c. <u>Upper Jurassic</u> is the time both of the spreading of the Piémont-Ligurian oceanic crust area and of the rapid sinking of the Briançonnais continental margin. Thus, at the end of Jurassic times (lower section of fig. 4), three main areas may be recognized : to the Northwest, the Dauphiné domain was a not so deep area (above ACD), still subsiding and receiving a rather important sedimentary (argillaceous and calcareous) supply ; on the other side, to the Southeast, the Piémont-Ligurian area was probably relatively deep (in the vicinity of the CCD) and was the site either of submarine volcanism (the products of which were also dispersed over part of the Piémont s.str. domain) or of incipient radiolarites deposition. These two very different areas may be easily compared with a continental shelf and an "ocean" ; they were linked by a continental slope which had a very complex topography (Subbriançonnais, Briançonnais and Piémont s.str. domains). A similar situation prevailed during <u>Lower Cretaceous</u> times.

Thus, from the Upper Liassic or Middle Jurassic up to the Lower Cretaceous, all data point towards the birth and evolution of a spreading "ocean" and of its passive continental margin.

d. The beginning of the <u>Upper Cretaceous</u> marks the end of this type of paleotectonic evolution ; the contraction of the "ocean" then began, as evidenced by thrusting and High-Pressure metamorphism in the inner zones, and by reorganization and rejuvenation of the paleogeographic pattern in the more outer ones.

5.2. LONGITUDINAL MODIFICATIONS OF THE MARGIN : FROM THE WESTERN ALPS TO THE CENTRAL ALPS.

Our attempt to reconstruct the Mesozoic evolution both of the Piémont-Ligurian "ocean" and of its European margin applies only to the Briançon traverse (3 in figs. 1 and 2) ; already the data are different when dealing with the Maritime Alps and with the Corsica traverses (4 in figs. 1-2) ; but the most significant change in the paleomorphology and the paleotectonic evolution of this European margin appears when the Central Alps, between Western (traverse 2) and Eastern Switzerland (traverse 1) are considered.

Of course, no important changes appear when the previously emphasized main features are considered, i.e. the wide-spread pre-oceanic Triassic platform sedimentation, the Jurassic oceanic opening and spreading, and the coeval collapse of the thus created continental margin. Nevertheless, some differences, which are not minor ones, may be encountered : for example, the Upper Jurassic limestones of the Briançonnais domain in Switzerland (Median Prealps, Sulzfluh nappe, etc.) exhibit a somewhat greater thickness and were probably deposited at much shallower depths, so that the rapid Upper Jurassic sinking which has been recorded in the Briançon traverse did not occur, or occurred later. Such a difference may be linked with an important modification of the paleotectonic singificance of the Briançonnais domain.

As a matter of fact, the most important change results from the appearance of a new paleogeographic unit (fig. 2), the Valais domain, which is inserted between the Helvetic (= Dauphiné) and the Subbriançonnais-Briançonnais domains. This new domain has been the site of early (Upper Cretaceous) flysch deposition, and its Jurassic-Lower Cretaceous sediments are associated with some scarce ophiolitic bodies.

Thus the Valais domain may have been an area either with thinned-out continental crust or, at least at some places, with a certain kind of oceanic crust. If it is assumed, at least in Eastern Switzerland (Trümpy, 1975), that the Valais domain as well as the Piémont-Ligurian one were both oceanic crust-bearing areas, then the Briançonnais domain will here appear as an elongated "ribbon micro-continent" ; actualistic points of comparison may perhaps be found in the Afar-Red Sea complex (with the intermediate Danakil continental block), or in some parts of the Atlantic margins such as Northeastern Newfoundland (with the Orphan knoll) or the Florida-Blake traverse (Sheridan 1974, figs. 3 and 10).

However it may be, the margin's framework here becomes more complicated, and a continental slope might have been located between the Helvetic and the Valais domains, perhaps in the Ultrahelvetic domain, the extension of which is not known in the Western Alps proper.

6. CONCLUSIONS

6.1. PRELIMINARY REMARK : PROBLEMS OF SEA-FLOOR DEPTHS.

In § 6.2., rates of sedimentation and of subsidence will be both discussed. Of course, when dealing with shallow water (infra-, inter- and supra-tidal) environments, as subsidence was steadily balanced by sedimentation, both rates are equal and can be easily calculated by taking into account both the present-day thickness of the sedimentary series and a certain compaction factor. But in the case of pelagic sediments, it is much more difficult to ascertain the depth of the sea-floor, the successive values of which are to be used when trying to determine a rate of subsidence.

As for the Jurassic and Cretaceous deposits of the Briançonnais domain, they have been compared here with some modern equivalents which are being deposited above or below certain characteristic depths such as ACD or CCD (Berger and Winterer, 1974 ; Bosellini and Winterer, 1975). For example, pure deep-sea siliceous or argillaceous deposits such as Radiolarian cherts are assumed to have been laid down deeper than the CCD, whereas coeval, rather thick, limestones which are known in the vicinity were probably deposited in less deep environments ; in a similar manner, limestones which contain Aptychus shells (calcitic) but are devoid of Ammonites (aragonitic) may have been deposited between ACD and CCD.

But it is also known that the use of such criteria must be carefully discussed case by case ; for example, Ammonite shells may be encountered within sediments deposited below the CCD if these shells were quickly enough brought down and buried (as might be the case with turbiditic flows) ; contrarily, lack of such shells never indicates with certainty the depth of deposition of the corresponding sediment, because numerous processes other than leaching may explain their absence.

Thus it is only after careful weighting of the evidence that it has been assumed (Bourbon et al., 1975 ; Bourbon, 1977) that in the Briançonnais domain, (1) radiolarian cherts and shales have been deposited below the CCD in Oxfordian and in early Upper Cretaceous (Cenomanian ?) times, (2) Upper Tithonian - Lower Cretaceous cherty limestones have been laid down below the foraminifera-CCD, and (3) most Upper Jurassic limestones below the ACD (but for some shallower areas). Similarly, we assume that the Jurassic-Cretaceous Ammonites deposits of the Dauphiné domain were deposited above the ACD, whereas the Upper Jurassic - Lower Cretaceous radiolarites, argillites and probably turbiditic limestones of the Piémont-Ligurian "oceanic" domain were laid down either below or in the vicinity of the CCD.

In fact, it is well-known that the ACD and CCD values may greatly vary according to the bottom topography, to the geological epoch under consideration, and to the latitude corresponding to this epoch. Thus the absolute depths of deposition cannot be fixed without some preliminary hypotheses, as will be seen further on.

6.2. PROBLEMS OF RATES OF SUBSIDENCE AND OF SEDIMENTATION.

In the traverse studied, and especially in the Briançonnais domain, three main phases of birth and evolution of the "passive" European margin of the Tethys have been recorded : (1) Subsidence of a Triassic carbonate platform ; (2) sinking of the sea-floor down to below the CCD for its deeper parts during the early rifting phase (early Upper Jurassic) ; (3) deposition of pelagic oozes on the deep-sea floor (Upper Jurassic-Lower Cretaceous). Such a succession of events has been recorded in several continental margins both in Tethyan and

Atlantic realms, but at different epochs from place to place.

As for the first phase (Triassic), in the Briançonnais domain of the Western Alps, both rates of subsidence and of sedimentation are rather easily calculated, at least if only an order of magnitude is desired. Chosing a reasonable factor of sediments compaction (3/2), these rates, in the Briançonnais domain near Briançon, are of the order of 45 mm/1000 y. for the Middle Triassic, and of 80 to 120 mm/1000 y. for the Norian ; but these figures may vary from place to place, as a result of differential subsidence ; for example, the Norian carbonates of the Lasseron Mountain East of Briançon, in the Piémont domain, are thicker than in the Briançonnais proper, leading to a rate of 150 to 250 mm/1000 y.

Contrarily, during the second phase (sinking of the margin linked with the oceanic opening), the subsidence rate was very different from the sedimentation rate, as a more or less deep pelagic environment has been very rapidly reached. In the Briançonnais domain of the Western Alps, the sedimentation rate was particularly low : 0.1 (or even less) to 1 mm/1000 y. But evaluation of the rate of subsidence during the sinking of the sea-floor in the early rifting phase (between Upper Bathonian and Lower or Middle Oxfordian for the Briançonnais) is much more difficult, especially because of the uncertainty as to the CCD in the Tethyan area during Jurassic times. If the hypothesis of Bosellini and Winterer (1975), assuming 2500 metres for the Jurassic CCD, is accepted, then a subsidence rate of about 250 mm/1000 y. is obtained ; a figure of the same order of magnitude can be calculated for the sinking of the Southern (Apulian-African) margin of the Tethys, as recorded in the sediments of the southern Alps (as already pointed out, this sinking occurred earlier, i.e. during Middle and Upper Liassic times) ; but in both cases this is a minimal figure, because the maximum time interval is used in the calculation : there is in fact nothing against a much more rapid sinking.

Again according to Bosellini and Winterer (1975), the Tethyan CCD might have changed from 2500 metres at Upper Jurassic times down to 4500 metres at Aptian-Albian times ; thus we may envisage an average rate of subsidence of 30 mm/1000 y. for the same time interval, when considering the deepest deposits in the Briançonnais, which were probably laid down in the vicinity of the CCD.

When comparing these values, and especially the subsidence rates, they seem to be of the same order of magnitude : as a matter of fact, the subsidence rate appears to have remained nearly the same for the shallow water pre-rifting deposits and for the syn- or post-rifting deep-sea deposits ; nevertheless, we must remark that in the case of the Briançonnais (not in the case of the Southern Alps) these two episodes with high rates of subsidence are separated by a short but important phase of uplift (emergence and more or less deep erosion of the area). We must also remark that these are maximum figures when dealing with the Triassic platform carbonates, whereas they are minimum figures in the case of the Jurassic sinking of the margin. Thus an acceleration of the subsidence rate linked with oceanic opening appears as a possibility. If this is accepted, it may explain, at least partly, the rapid grading from platform carbonates to pelagic sediments, because a sudden increase of sinking may stop lime production : but other reasons may also, and simultaneously, be invoked, as for example (1) the birth of barriers (linked with the block-faulted tectonics at the beginning of the rifting) which prevent the shelf-originated detrital or biogenic sedimentary supply to reach to bottom of the Briançonnais domain, and also (2) a climatic change, or even (3) a rapid decrease of the temperature of sea-water linked with the oceanic opening.

6.3. COMPLEX BOTTOM TOPOGRAPHY, ABUNDANCE OF SUBMARINE FAULT-SCARPS AND OF BRECCIAS.

When trying to reconstruct this European "stable" margin of the Tethys, some special features must be taken into consideration before attempting any comparison with present-day continental margins.

First, the topography of the sea-floor was certainly both complicated and steadily changing. For example, in the Briançonnais zone near the city of Briançon, sedimentological evidence (Bourbon, 1977) leads to reconstitute a highly intricate Jurassic-Lower Cretaceous submarine paleomorphology, featuring sea-mounts, more or less steep slopes, cliffs, and fault-scarps. When the abundance and the areal distribution of breccias are both considered, similar complex topographic features are likely to have existed also in the Piémont domain during Lower and Middle Jurassic times (Caron, 1977).

Present-day "stable" or "passive" continental margins do not all show the same topographic and structural features, even when dealing only with those of the Atlantic ; some margins have rather smooth, regular continental slopes, descending slowly and continuously from the shelf edge down to the abyssal plain, whereas other margins feature a much more intricate submarine topography, and also a complicated internal structure, with uplifted blocks, rapid variations of the thickness of the sediments, etc. As pointed out by Renard and Mascle (1974 ; see especially fig. 2 and discussion p. 289), both types of margins are not all located at random : as a matter of fact, margins with a smooth topography often occur in areas where the spreading direction acts perpendicular to the first break (rifting) and thus more or less to the present-day continent's limit ; on the contrary, both the topography and the internal structure appear to be much more complicated in "stable" margins located in areas where the tectonic movement is more or less parallel to the coast-line, i.e. if it is a strike-slip movement more or less linked with the activity of transform faults. This of course may not always be the rule. But it must be noted here that, according to certain hypotheses (e.g. in Dewey et al, 1973), the relative motion between the European and the African blocks might have long been (especially during Jurassic and Lower Cretaceous times, i.e. between -180 and -100 m.y.) a left-lateral strike-slip displacement. Thus we may suggest, as a working hypothesis, that the synsedimentary tectonics which acted in the margin here described might have rather resulted from a strike-slip movement than from a pure distensional one, thus leading to the possibility of displacement and of deformation of "ribbon-blocks", and even to some kind of "compressional" folding (e.g. as shown by Royant, 1976, in the Briançonnais domain of the Maritime Alps) coeval with "distensional" events ; such a tectonic activity would have resulted in a complicated and changing submarine topography (fig. 5).

One may also emphasize the great abundance of olistoliths and breccias, with all possible sizes of boulders, pebbles or grains, and different degrees of sorting, which are known to be interbedded in some Jurassic and Cretaceous deposits of all domains belonging to our margin (Dauphiné, Subbriançonnais-Briançonnais, Piémont s. str.). As far as is known, olistoliths and breccias do not appear to be of such importance in the coeval or younger sediments belonging to the margins of the Atlantic ; of course, one must remain cautious in such a comparison, because of important differences in the nature of the knowledge of both types of margins, Atlantic and Tethyan ; the first ones are known through seismic profiles and some scarce boreholes and dredgings, whereas in the second ones the sediments may be observed in outcrops in the mountain ranges, where the chances of finding and identifying breccias or olistoliths are a thousand times greater. Nevertheless, if the difference is significant to some extent, this would perhaps mean that the European margin of the Tethys was the site of a much more important, and especially much more longer, tectonic activity ; once more we are tempted to link this with a postulated long, active, strike-slip motion.

6.4. FINAL REMARKS AND CONCLUSION.

The previous discussion suggests that besides some striking similarities, there exist also some obvious and important differences between Tethys and Atlantic, whether the oceanic-crust areas or the continental margins are considered.

Besides the important and steady tectonical mobility of the Tethyan margin, another important difference is obvious : the Tethyan contraction began 100 m.y. ago and was completed for the past one or two score m.y., whereas the Atlantic ocean is still now spreading. Otherwise, the Tethyan spreading mechanism and thus the corresponding spreading rate are only matters of conjecture, and the successive positions of Europe relative to Africa are not accurately known ; nevertheless, in this Western sector of the Tethyan realm the two continental blocks might well have never been very far from each other.

It may therefore be suggested that this part of the Tethys never has been very wide, perhaps only a few hundred kilometres. This is the reason why we always put the word "ocean" between quotation marks when dealing with the Piémont-Ligurian domain.

When regarding the European continental margin of this domain, the most striking feature is that this margin, even during the "distensional" phase (which may have been linked with a general strike-slip deformation) underwent a tectonic activity which has been both rather important and steady for a relatively long time : so that, stricktly speaking, this margin was neither "stable" nor "inactive", at least just a bit "passive".

REFERENCES

BAUD, A., and MEGARD-GALLI, J., 1975. Modèle d'évolution d'un bassin carbonaté du domaine marin alpin durant la phase préocéanique : cycles et rythmes dans le Trias de la zone Briançonnaise des Alpes Occidentales et des Préalpes. IX Int.Congr. Sedim., Nice, 5, 45-50.

BERGER, W.H., and WINTERER, E.L., 1974. Plate stratigraphy and the fluctuating carbonate line. Spec.Publ.Int.Ass.Sedim, 1, 11-48.

BERNOULLI, D. and JENKYNS, H.C., 1974. Alpine, Mediterranean and Central Atlantic Mesozoic facies in relation to the early evolution of the Tethys. Soc.Econ.Pal.Min. Spec.Publ., 19, 129-160.

BOCQUET, J., DELALOYE, M., HUNZIKER, J., and KRUMMENACHER, D., 1974. K-Ar and Rb-Sr dating of blue Amphiboles, Micas and associated minerals from the Western Alps. Contrib.Mineral.Petrol., 47, 7-26.

BOSELLINI, A., and WINTERER, E.L., 1975. Pelagic limestone and radiolarite of the Tethyan Mesozoic : A genetic model. Geology, May 1975, 279-282.

BOURBON, M., 1977. Reconstitution paléomorphologique de fonds marins sur la marge nord-téthysienne : le Jurassique supérieur et le Crétacé briançonnais dans les unités de Roche Charnière et de Champcella et dans le massif Galibier-Cerces. Colloque final A.T.P. Géodyn.Médit., Montpellier (in press).

BOURBON, M., 1977. Mesozoic evolution of Western North-Atlantic and North-Tethyan margins : a comparison. Init.Rept. D.S.D.P., 44, in press.

BOURBON, M., GRACIANSKY, P.C. de, LEMOINE, M. and MEGARD-GALLI, J., 1973. L'évolution paléogéographique du domaine briançonnais au Mésozoïque : carbonates de plate-forme subsidente, révolution du Lias supérieur, instauration progressive d'un régime pélagique profond. C.R.Ac.Sci. Paris, 277 D, 769-772.

BOURBON, M., GRACIANSKY, P.C. de, LEMOINE, M., MEGARD-GALLI, J., and MERCIER, J., 1975. Carbonates de plate-forme et séries pélagiques condensées dans le Mésozoïque de la zone briançonnaise (Alpes françaises). Guide-book Exc. 5, IX Int.Congr. Sedim., Nice.

CARON, J.M., 1977. Evolution paléogéographique et tectonique de la zone piémontaise dans les Alpes occidentales. Colloque final A.T.P. Géodyn.Médit., Montpellier (in press).

DEBELMAS, J., and LEMOINE, M., 1970. The Western Alps : palaeogeography and structure. Earth Sci.Rev., 6, 221-256.

DEWEY, J.F., PITTMAN, W.C., RYAN, W.B.F., and BONNIN, J., 1973. Plate tectonics and the Evolution of the Alpine System. Geol. Soc.Am.Bull., 84, 3137-3180.

GRACIANSKY, P.C. de, and MERCIER, D., 1976. Séquences sédimentaires dans le "Dogger" briançonnais et ouverture téthysienne. C.R.Ac.Sci. Paris, 283 D, 475-478.

LEMOINE, M., 1975. Mesozoic sedimentation and Tectonic evolution of the Briançonnais zone in the Western Alps. Possible evidence for an Atlantic-type Margin between the European craton and the Tethys. IX Int.Congr.Sedim., Nice, 4, 211-216.

RENARD, V., and MASCLE, J., 1974. Eastern Atlantic Continental Margins : Various structural and Morphological types. The Geology of the Continental Margins, C.A. Burk and C.L. Drake, ed., Springer, 285-292.

ROYANT, G., 1976. Paléotectoniques jurassiques, aux confins des domaines briançonnais et piémontais, dans les Alpes ligures. C.R.Ac.Sci. Paris, 282 D, 1845-1848.

SHERIDAN, R.E., 1974. Atlantic Continental margin of North-America. The Geology of Continental Margins, C.A. Burk and C.L. Drake, ed., Springer, 391-408.

TRÜMPY, R., 1971. Stratigraphy in mountain belt Quart.Journ.Geol.Soc.London, 126, 293-318.

TRÜMPY, R., 1975. Penninic-Austro-Alpine boundary in the Swiss Alps : A presumed former continental Margin and its problems. Am.Journ.Sci., 275 A, 209-238.

TABLE I

Structural units of the Alps	Paleogeographic domains	Structural units of the Apennines
Helvetic-Dauphiné (external)	Helvetic (1)	
	Ultrahelvetic (1)	
Penninic :		
- North Penninic	Valais (1) (2)	
- Middle Penninic	Briançonnais s.lat. (1) (3)	
- South-Penninic :		
-Piémont s.str.	Piémont (1)	
-Piémont-Ligurian	Piémont-Ligurian oceanic area (4)	Ligurian
Austro-Alpine and South-Alpine	Austro-South-Alpine, Italian-Apulian (5)	Tuscan, Marche-Umbria, etc. (6)

(1) Belonging to the European craton (sialic crust)

(2) Scarce ophiolitic bodies indicates the presence of possible oceanic crust, or at least of thinned-out continental crust

(3) Subdivided into Subbriançonnais, and Briançonnais s.str.

(4) An oceanic crust-bearing area (ophiolites) which is believed to be part of the Mesozoic Tethys.

(5) Belonging to the African-Apulian craton (sialic crust)

(6) Due to opposite vergences (fig. 1), the "African" units are the highest overthrust units in the Alps, whereas in the Apennines they are the lowest, more or less autochtonous or parautochtonous below the Ligurian nappes.

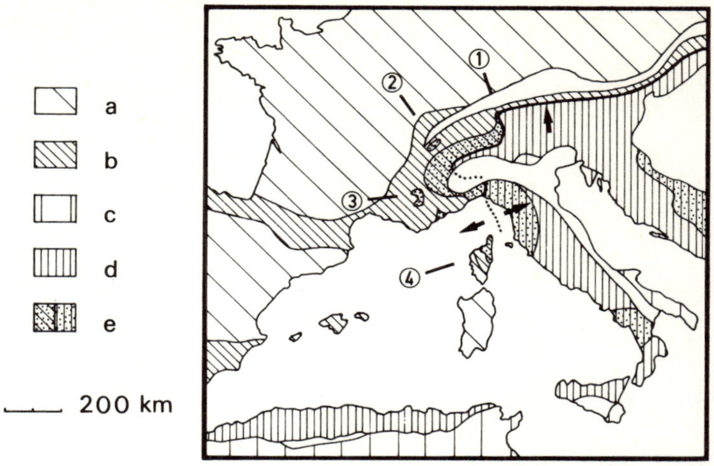

Fig. 1 : Simplified sketch-map of the Alps, of the Apennines and of the surrounding areas.

a, b : European continental block, and related minor ones (Iberia, Corsica-Sardinia).-
c, d : African continental block, including the Apulian (Italian) one.
a and c are those parts of these blocks which remained tabular, i.e. not effected by Alpine folding.
b and d are those parts which were affected by Alpine folding, thrusting and metamorphism ; they more or less correspond to the former continental margins, deformed by subsequent continent collision.
e : the parts of the latter over which are scattered more or less extensive tectonic outliers of ophiolites (Tethyan oceanic crust) and/or of the oceanic sediments which have been deposited upon these ophiolites.
The main molassic basins are left blank.
Arrows : Vergence of folds and thrusts.
1-4, same traverse as in Fig. 2.

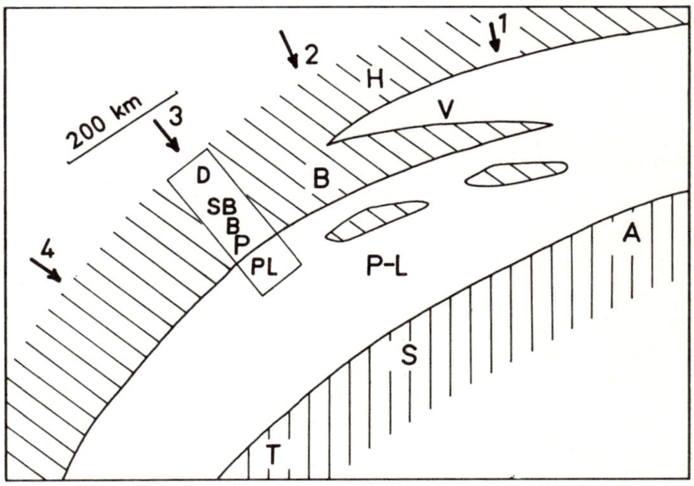

Fig. 2 - Tentative palinspastic sketch of the Alpine and Apenninic paleogeographic domains during Upper Jurassic - Lower Cretaceous times.

Blank : domains where a certain kind of oceanic crust (ophiolites) is assumed to have existed at that time ; P-L, Piémont-Ligurian ; V, Valais.

Hatching : areas with continental crust (i.e. continental margins of the former). Oblique hatching, European block (and the postulated Margna and Sesia microcontinents) ; vertical hatching, African-Apulian block.

D, Dauphiné ; H, Helvetic ; SB, Subbriançonnais ; B, Briançonnais ; P, Piémont s.str.

T, Tuscan (Apennines) ; S, South Alpine ; A, Austro-Alpine.

1, Grisons traverse (Eastern Switzerland), at the limit between the Eastern and the Central Alps ; 2, Lake of Geneva traverse (Western Switzerland), at the limit between the Central and the Western Alps ; 3, Briançon traverse ; 4, Corsica traverse.

The scale is approximate, and concerns the longitudinal distances (e.g. between 1, 2, 3, 4 traverses), not the broadness of the oceanic-crust-bearing areas.

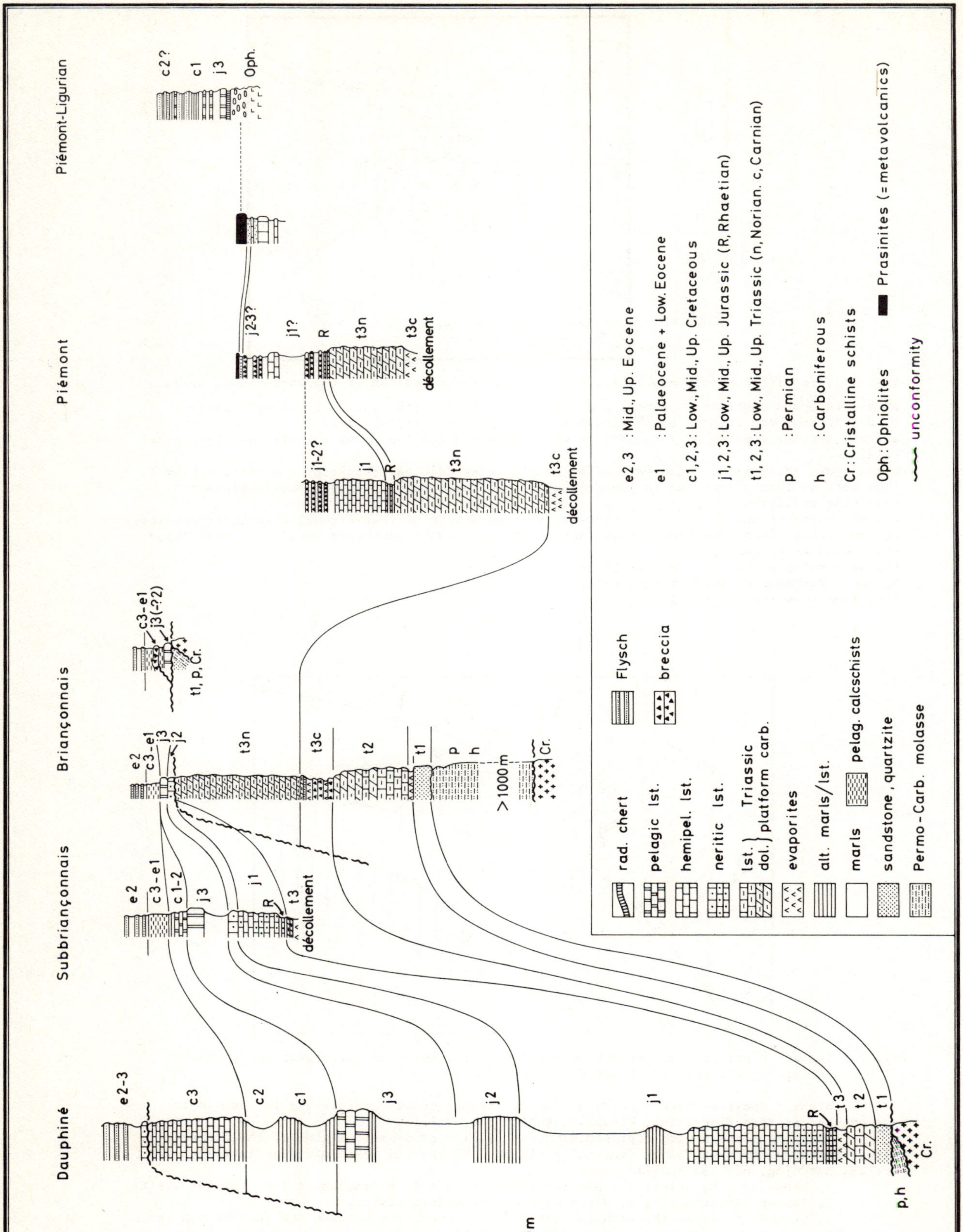

Fig. 3- Stratigraphic columns of the main paleogeographic domains of the Western Alps in the Briançon traverse. These columns are highly schematized, and they do not take into account lateral variations inside of each domain.

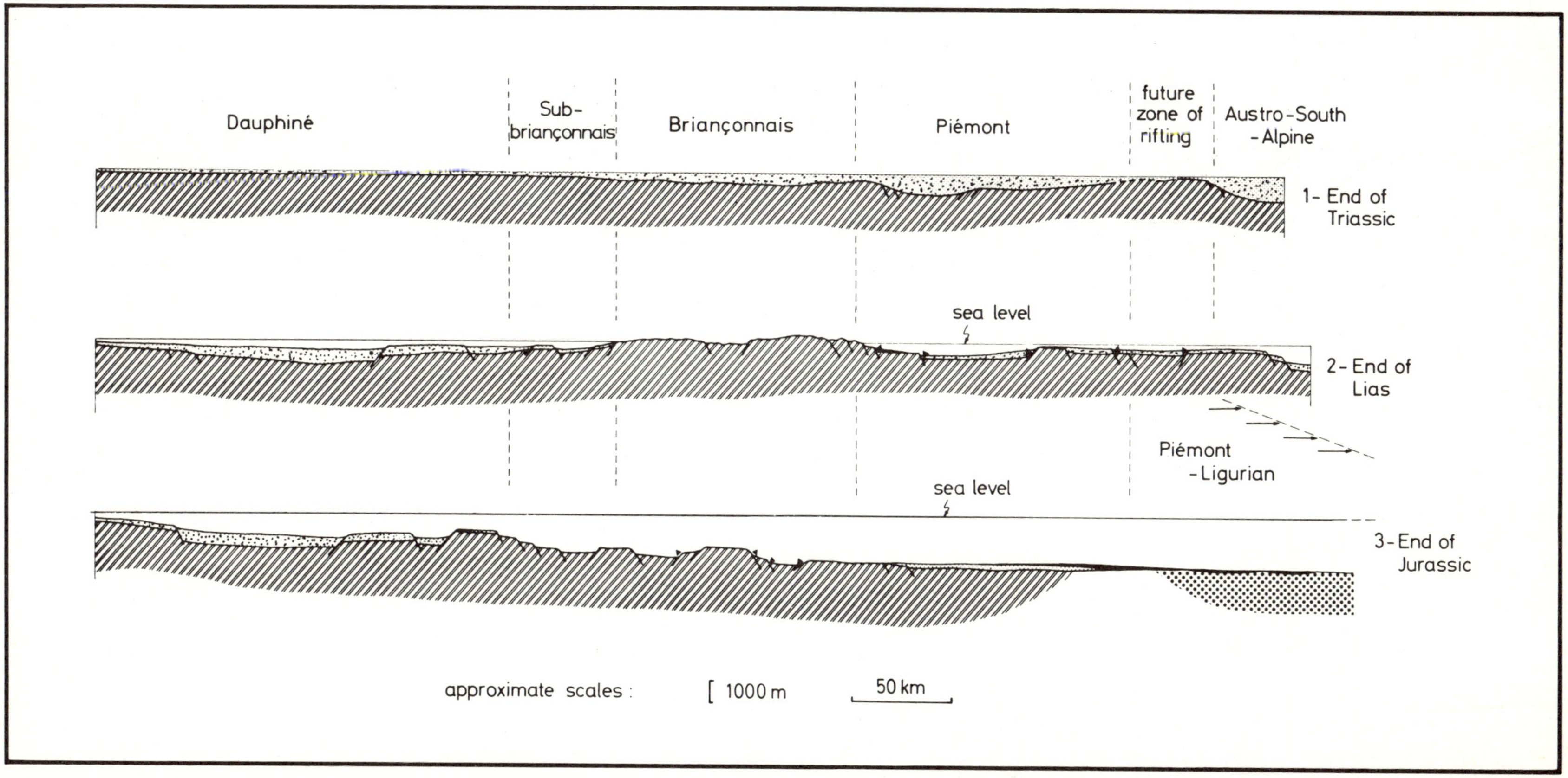

Fig. 4 - Tentative palinspastic sections of the European passive continental margin along the Briançon traverse.

These sections are of course only rough sketches. Their sole purpose is to illustrate the main steps of the birth and evolution (sinking) of the margin ; they are not intended to represent all known details, but only to express the main features which are discussed in the text : (1) pre-oceanic Triassic platform carbonate sedimentation with different rates of subsidence from place to place ; (2) incipient rifting with emergence of the Briançonnais domain ; (3) oceanic spreading, and sinking of the continental margin.

Hatching : Basement (continental crust and sediments) : pre-Norian in section 1 ; pre-Jurassic in section 2 ; pre-Middle Jurassic in section 3.

Heavy dots : Oceanic basement

Dotted : sediments (Norian in 1 ; Lias in 2 ; Middle and Upper Jurassic in 3) ; the absence of a dotted layer in parts of the sections mainly corresponds to the fact that the deposits are too thin to be represented at such a scale.

Triangles : submarine breccias

Black : Volcanites (lavas or tuffs).

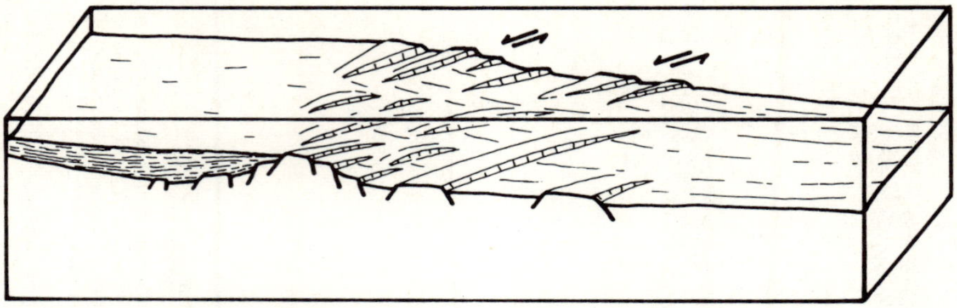

Fig. 5 - Theoretical block-diagram illustrating the idea of an intricate margin linked with strike-slip faulting.

This sketch is not intended to represent any specific area, only to illustrate an hypothetical idea suggested in the text. To the left of the diagram would be the shelf, i.e. the subsiding Dauphiné domain, whereas the part shown to the right would be located near the oceanic-crust bottom ; in between, an attempt has been made at representing the Subbriançonnais-Briançonnais-Piémont (s.str.) zones with intricate submarine topography.

INTERNATIONAL SYMPOSIUM ON THE STRUCTURAL HISTORY OF THE MEDITERRANEAN BASINS. SPLIT (YUGOSLAVIA) 25-29 OCTOBER 1976. B. BIJU-DUVAL AND L. MONTADERT, Eds. EDITIONS TECHNIP, PARIS 1977, pp. 35-46

VOLCANO-SEDIMENTARY AND VOLCANO-DETRITIC PHENOMENA OF MESOZOIC AGE IN DINARID AND HELLENIC RANGES : A COMPARISON

P. CELET [1], J.-P. CADET [2], J. CHARVET [1] and J. FERRIERE [1]

I. MAIN FEATURES

Mesozoic dinaric and hellenic series are frequently characterized during Trias and Jurassic by associations of magmatic elements with marine sedimentary formations. These phenomena are important because they reflect fundamental crustal or sub-crustal transformations linked with the evolution of the whole range.

However, despite apparent similitudes, these episodes are very different in their lithological associations and B. CIRIC clearly established in 1954 the distinctions between these formations.

The association of middle triassic magmatic and sedimentary rocks is widespread in Yugoslavia where it is known under the name of "Porphyrit-Radiolarite Formation" (CIRIC 1954) ; it's a volcano-sedimentary formation with equivalents in Albania and Greece.

The association of upper Jurassic, known in the Dinarids as the "Diabase-radiolarite formation " (CIRIC 1954) is of a volcano-detritic nature and mainly made of volcanoclastic and terrigenous elements associated with radiolarites. More recently the terms of "Olistostrome Melange" (DIMITRIJEVIC and DIMITRIJEVIC 1973), "Mélange de type volcano-sedimentaire" (CELET 1976, CELET et al 1976) have been proposed. To account for the chaotic aspect of the polygenetic material included in this type of deposit, we'll take the general term of "mélange" for the volcano-detritic formation of the upper Jurassic.

Although there exist similar and subcontemporary formations in more inner zones (i. e. Vardar zone, Mercier et Vergely 1972), we won't talk about them at present, our purpose being essentially to collect the characters which draw together or differenciate these volcanogenic formations after having recalled their display. At last we we'll attempt to situate these manifestations in their geotectonic situation.

II. DISTRIBUTION AND LOCALISATION OF TRIASSIC AND JURASSIC VOLCANOGENIC FORMATIONS

The witnesses of middle triassic and upper Jurassic magmatism are widespread in the dinaro-hellenic ranges and mainly in the areas we study here (Fig. 1).

1. <u>The volcano-sedimentary triassic formation</u> is found as well in inner as in outer areas (BEBIEN et al 1975) :

- Budva trough, High-Karst platform, Bosnian trough, Serbian set, and Golija rise in the

(1) Laboratoire de Géologie Dynamique, Université des Sciences et Techniques de Lille, B.P. 36, 59650 Villeneuve d'Ascq, France

(2) Laboratoire de Géologie, Université d'Orléans, 45045 Orléans Cédex, France

Yugoslavian Dinarids.

- Mirdita zone (N.E. of Tirana) in Albania.
- Pindus zone (unit of Megdovas), Vardoussia subzone (west of Parnassus) Maliac zone (Othrys), Pelagonian zone (Attica, Euboea, Argolis (?), Northern Pindus in Greece.

This large scattering underlines the ubiquity of this middle Triassic phenomenon.

2. The upper jurassic volcano-detritic "melange" is essentially linked with the ophiolitic field. Its outcropping most of the time can be found together with those of the ophiolitic nappe in the Serbian zone (northern Montenegro and Bosnia, southern Serbia) in Golija zone, Mirdita zone or Maliac, Pelagonian zone (Geranées, Iti, Kallidromon, Western Othrys and Pindus). However this relation is not constant ; in some areas, the "mélange" is separated from the ophiolitic suite (Ceotina, Durmitor in Yugoslavia, central Othrys in Greece).

So in the dinaric branch of the alpine system, there is a geographical differenciation in the distribution of the volcanic phenomena during Trias and Jurassic. This opposition is still more apparent in the composition of the formations themselves.

III. STRATIGRAPHIC POSITION, PETROGRAPHICAL COMPOSITION AND SEDIMENTARY ENVIRONMENT.

The age and deposit conditions of the volcanic material can be analysed thanks to the stratigraphic context. Here are the main components.

A. TRIASSIC VOLCANO-SEDIMENTARY SERIES

1. In the Dinarides, in all paleogeographic zones, a volcano-sedimentary formation is known during middle Trias. It is constituted by the association of sedimentary (Radiolarites, pelites, sandstones) and volcanic elements (flows and projections). The substratum of this series is made of anisian neritic limestones crowned by nodulous red limestones Han Bulog facies) of upper anisian age (Fig. 2, II et III), except in the Budva zone where there is a thick flysch during the Anisian (Fig. 2 - I). It is capped by Ladinian and upper Triassic limestones in the zones with having a meaning of ridge and by chert limestones in the zones that will evolve into troughs.

A preliminary analysis of the volcanic elements and their repartition (BEBIEN, BLANCHET et al 1975, 1976) allows to distinguish four sets : rhyolitic tuffs and lavas, andesites, diabases and basalts ; the rhyolitics tuffs can be found in all zones, andesite, of calco-alkaline nature, are mainly known in the outer zones and in the serbian zone south of the Sarajevo transverse sector whereas basalts are limited to the serbian zone north of the same transverse sector and to the bosnian zone.

In short, the middle triassic volcano-sedimentary formation of the Dinarides, excorting the differenciation of the paleogeographical zones, is characterized by its wide repartition and varied volcanic phenomena mainly of calco-alkaline nature.

2. In the Hellenides, the triassic volcano-sedimentary outcrops are scarcer and more scattered ; it seems that this is due to tectonic phenomena (shearings and slides often occuring at the basis of the nappes).

However there seems to be some differenciations corresponding to two types of volcanic floods.

a. An alkaline series (HYNES, 1974) predominently lavic with various interstratifications : fine tuffs, breccias and hyaloclastites, very often pillow-lavas and slaggy spilitic lavas in alternance with radiolarites, siliceous pelites and platy limestomes with cherts Conodonts or Halobia (Pindus) dating the top from Carnian-Norian. Such outcroppings can be observed, up to now in Othrys in the area of Neochorion (series of Garmeni and Loggistion, FERRIERE 1972-1974) and in the northern Pindus (area of Panagia, TERRY, 1971).

We must equally mention a series of breccias and effusive rocks with pillow lavas flows of spilitic to basaltic nature, petrographically fairly similar to the preceding one but situated in Arkanania at the front of the Pindus nappe and thrust over the ionian flysch (unit of Megdovas, FLEURY, 1976). At the top, this volcano-sedimentary series turn to an Ammonitico-Rosso of upper Scythian to Carnian age and Halobia limestones.

b. Another volcanic formations with pyroclastites, rhyolitic and dacitic light green tuffs ("Pietra verde" type) can be found in Attica, Parnis,(CLEMENT, 1968) central Euboea (GUERNET, 1971, KATSIKATSOS, 1971) also associated with sedimentary rocks (pelites, sandstones, radiolarites). In Parnis this volcano-sedimentary formation lies over ladinian cherts limestomes and microbreccias and bear carnian chert limestones.

3. Remark.

In Albania (Mirdita zone, N.E of Tirana) a Triassic volcanism has been noticed (A. PAPA, 1970). It seems similar to the Ladinian-Carnian one of the Dinarides and Hellenides.

4. Conclusion

In Dinarides and Hellenides, the triassic volcanism is bonded with the paleogeographical differenciation of the isopic zones this evolution can be followed up to the end of the Jurassic (cf. AUBOUIN et al, 1970) and its first stages-with notably the middle Trias volcano-sedimentary formation - recall the rifting known in the continental margins but it raises the problem of the paleoceanographic organisation of the Dinarides and Hellenides (cf conclusion).

B. JURASSIC VOLCANO-DETRITIC MELANGE

1. Nature

Despite great similarities of facies and petrology, the hellenic and Dinaric volcano-detritic formations are located in various stratigraphical and tectonic surroundings. They display constant characteristics similar to the classical "melanges" (cf BAILEY et al 1950, HSU 1971 et 1974 etc) : absence of stratification, disordered and chaotic aspect of the formation, pelitico-sandy matrix sometimes with radiolarites wrapping various exogenous elements such as blocks of limestones of different nature and age (Permian with Neoschwagerina, upper Trias with Conodonts,etc) pieces of quartzitic sandstone, Radiolarites. The eruptive matter, very abundant is mainly composed of fragments belonging to the ophiolitic cortege : basic and ultrabasic magmatic rocks from gabbros to peridotites and serpentines ; veins stones (dolerites, diabases) effusive rocks (basaltes or spilites with pillow lavas structure ; metamorphic rocks with diverse meaning (amphibolite, gneiss mica schists) are associated there as well as granitic blocks (Iti).

The recents works, in Yugoslavia (BLANCHET, 1970, 1975, CADET, 1970-1976, CHARVET, 1970, RAMPNOUX, 1970-1973) as in Greece give precise details about the composition and different aspects of this formation and we send the reader back to these publications (CELET, 1976).

The volcano-detritic formation constantly lies upon radiolarites, often with pelitic intercalations ; flyschoïd series separate these formations in some cases. The radiolarites and the matrix of the mélange are seldom dated, however Protopeneroplis of Dogger - Malm have been reported (FERRIERE, 1974 ; TERRY, 1975 ; WIGNIOLLE, 1975) together with neritic microfauna of Kimmeridgian-Portlandian (CADET,1967 ; RAMPNOUX, 1970). So, if the age of these formations is still imprecise, its normal stratigraphic support's is better known :

a. In the Dinarides, the melange and its sole lies upon, - either Dogger-Malm condensed or pelagic carbonated series, their chert-limestones progressively turn to pelites and radiolarites (Rogatica, Bosnia, fig 3, III) or are capped by alternances of microbreccias silstones and radiolarites,

- or neritic limestones topped by liassic ammonitico-rosso and, locally, by oxfordian limestones (Romanija, Bosnia, fig 2, IV).

b. In the Hellenides, the mélange lies conformably on various substrata, in Argolis (fig. 2, V) the transition between the norian Megolodonte limestones and a reduced terrigenous series occurs through a Lias-Dogger ammonitico-rosso as in the Yugoslavia Golija zone.

In Othrys the mélange lies either upon neritic series in the pelagonian zone or upon pelagic series in the maliac zone (fig. 3, VI) (FERRIERE, 1974 et 1976).

In Iti (NW of Parnassus) (fig. 3, VII)and in the Kallidromon, the pelagonian limestones series including two horizons of bauxites at different stages of middle-upper Jurassic, dated at the top by Cladocoropsis ends with limestones getting richer in silica and progressively passing to radiolarites and the flyschoïde detrital layers of the mélange basis.

Theses examples shows that the volcano-detritic phenomena normally take place in the sedimentary evolution of the time and is part of it. The subsequent tectonisation doubtlessly increased the chaotic aspect of these formation but is not their fundamental cause. The layers overthrusting the melange can be dated back to Tithonian or Beriasian in the place where they have been spared by the ophiolite overthrusts or others. Keeping in mind the upper or late Jurassic age of the substratum and the thickness of the volcano-detritic deposits, one realizes the brevity and the extent of the event in the history of a range, showing at this time a very important tectonic activity and a major phase in the geosynclinal evolution of the Dinarides and Hellenides.

2. Conclusions

Thus the mélange appears as a stratigraphic formation closely bounded to its substratum in relation with the ophiolitic complex whose material is inherited for a great part and where the resedimentation play a great role. Nevertheless a few differences between Dinarides and Hellenides can be noted, particularly :

- in the lithological composition of the deposits, the Dinaric volcano-detritic seems richer in volcanic products whereas the granitic elements are more abundant in the Hellenides ;
- in the nature of the infra-radiolaritic substratum better diversified in the Dinarides where a transition between the serbian basin in the S W and the Golija ridge in the N E, can be observed in central Bosnia. The same tendency can be noted in Greece in the Iti, on the pelagonian platform, where the carbonate deposits rise to upper jurassic while an oceanization begins in the maliac zone;
- in the early age of the limestone cover (fig. 3, IV) or the flysch like cover (BLANCHET et al, 1970), lateral passage to Bosnian flysch, (fig. 3, III) in the Dinarides;
- finally in the greater thickness of the mélange in Yugoslavia than in Greece.

IV. GENERAL CONSIDERATIONS ABOUT THE ORIGIN AND GENESIS OF THE DEPOSITS

The nature of the material and its stratigraphic context show a contrast between the triassic volcano-sedimentary manifestations and the upper jurassic volcano-detritic phenomena which leads to contemplate an origin and a genetic meaning specific to each of these episodes.

1. Triassic volcanism

The vast platform with a detritic sedimentation during lower Trias and a neritic one during Anisian was first differenciated during upper Anisian in the Dinarides. Marks of this differenciation can be observed in the inner zones as well as in the outer ones (cf BEBIEN et al 1975 for the Dinarides). We can suppose that these strains have fostered a system of fractures that caused the rise of the magma and started the volcanism with its diverse manifestations (explosive volcanism or lava flows above or under the sea). A rifting-like segmentation of a thinned (continental ?) crust can be considered in the case of marginal areas in the adriatic or apulian subplate (cf DERCOURT 1970, DEWEY et al 1973 etc). The first model (fig. 4, A) shows up such a fragmentation with neritic platform and the beginning of subsiding basins some of whick will turn short and become shallow troughs while others will evolve toward genuine oceanic basins. The development of such opening processes consistent with the alkaline characteristic of triassic lavas has been suggested to situate the origin of the oceanic crust begetting the ophiolitic nappes in eastern Mediterranea (HYNES, 1974, LAPIERRE et ROCCI 1976). However, in the Dinarides, the chemistry of the volcanic products seems rather to indicate a calco-alkaline tendency inconsistent with the generalisation of the former model. The facts as they have been observed in these areas would rather induce us to imagine the existence of a subduction zone on the edge of the plate, admitting the possibility of an interplate opening system according to a second model (fig. 4 B). But the geochemical knowledge of the volcano-sedimentary products is too rudimentary to allow us to propose now a simple explanation of these phenomena or to choose between either of these hypothesis.

Anyhow, the triassic volcanism is the beginning of an important stage in the geodynamic evolution of the dinaro-hellenic system.

2. Volcano-detritic mélange

The nature and heterogeneity of the melange raise the problem of the origin of the accumulated material.

The blocks of basic and ultrabasic rocks, often bulky and very frequent, doubtlessly belong to the ophiolitic complex and it is sometimes difficult to distinguish them from the ophiolitic nappe itself when the slabs are big.

As for the limy xenolithes, some of them of Triassic age come from series outcropping in inner zones. But the origin of numerous chips (non dated cristalline limestones, radiolarites, sandstones, greywackes) is still unknown, but probally situated mainly in inner zones.

The origin of the metamorphic elements is much more difficult to determine ; and they can have been torn from the paleozoic basement or from series metamorphised during jurassico-cretaceous tectonic phases.

The setting of such a material must take into account the data of the sedimentology about the relations between the wrapped blocks and their matrix. The blunt aspect of the elements, indeed dissociated and fractured previously to their deposit, their surface coated with pelitic and clayey material, the close union of the interfaces, the scarceness of shearing contacts and the absence of coherent net of fractures being generalised into the whole of the formation lead us to discard the hypothesis of the merely tectonic origin of a series previously stratified or of a tectonic mélange (DIMITRIJEVIC 1973, GANSSER 1974, HSÜ 1974). This induces us to consider the deposit of this exogenous blocks in a basin with a pelitic sedimentation and several supplies to that basin, itself submitted to a subsequent tectonization. The phenomena of resedimentation would have occured following early deformations phases having affected the Dinaric and Hellenic ranges at the upper Jurassic and early Cretaceous either under the shape of olistostrome, olistolith and xenolith (CELET et al 1976).

Considering the narrow association of the ophiolites and the volcano-detritic mélange, the explanatory schemas must account for their origin and situation. If they represent a slab of oceanic crust tectonically settled during upper Jurassic (cf DERCOURT 1970, BERNOUILLI and LAUBSCHER 1972 etc), several genetic models can be proposed.

- The mélange could result from the transportation and accumulation in a subsident basin of a material originating in a subduction trench on the edge of an island-arc or a marginal sea (Fig. 5 A).

- The volcano-detritic mélange could have been formed following obduction with dilaceration of two crusts and redeposit of the products in a basin and or upon platform (Fig. 5 B) situated at the front of the system.

- The two mechanisms, obduction and subduction, can be combined in a more complex model (Fig. 5 C) enabling us to explain the coexistence of such varied elements inside the terrigenous matrix.

So the volcano-detritic mélange reflects the instability of the sedimentary environment.The hypothesis of a sedimentary complex with olistostromes already considered for Hellenids (CELET 1976, CELET et al 1976, FERRIERE 1976, Othrys, NAYLOR et al 1976 Vourinos), and Dinarids (DIMITRIJEVIC 1973) fits better in the paleogeographic and tectonic frame of the Dinaric branch of the alpine range.

Conclusions

These triassic and jurassic events are highlights in the history of the chain. With the volcano-sedimentary formation of the middle Trias, we witness a revolution whose consequences are fundamental with the birth of the genuine mesozoic alpine paleogeography which will progressively be diversified. This stage, contemporary of the opening period of the tethysian ocean, directly leads to the late Jurassic upsetting. A progressive strain dynamics is developped and creates the first nappes, notably the ophiolites overthrust. The volcano-detritic deposits are the advanced witnesses and testify of the violence of the phenomenon. The following flyschs are a more remote consequence.

AUBOUIN J., BLANCHET R., CADET J.P., CELET P., CHARVET J., CHOROWICZ J., COUSIN M. et RAMPNOUX J.P., 1970. Essai sur la Géologie des Dinarides. Bull.Soc. Géol. Fr., 7, XII, 1060-1095.

BAILEY E.B. et Mc CALLIEN, 1950. The Ankara mélange and the Anatolian thrust. Nature, 166, 938-940.

BEBIEN J., BLANCHET R., CADET J.P., CHARVET J., CHOROWICZ J., LAPIERRE H. et RAMPNOUX J.P., 1975. La série "Porphyrit-Radiolarite" élément de reconstitution du cadre géotectonique des Dinarides au Trias. 3ème réunion des Sc. de la Terre, Montpellier (ronéot.), 31.

BERNOUILLI D. and LAUBSCHER H., 1972. The palinspastic problem of the Hellenides. Eclogae geol. Helv., 65, 1, 107-118.

BLANCHET R., 1974. De l'Adriatique au Bassin pannonique. Essai d'un modèle de chaîne alpine. Mém. Soc. Géol. Fr., LIII, 120, 171 p

BLANCHET R., DURAND-DELGA M., MOULLADE M. et SIGAL J., 1970. Contribution à l'étude du Crétacé des Dinarides internes : la région de Maglaj, Bosnie (Yougoslavie). Bull. Soc. Géol. Fr., 7, XII, 1003-1009.

CADET J.P., 1976. Contribution à l'étude géologique des Dinarides : les confins de la Bosnie-Herzégovine et du Montenegro. Essai sur l'évolution alpine d'une paléomarge continentale. Thèse, ronéot. Orléans, 450 p.

CELET P., 1976. A propos du mélange de type "volcano-sédimentaire" de l'Iti (Grèce méridionale). Bull. Soc. Géol. Fr., XVIII, (7), 299-307 (Col. ORSAY, 1975).

CELET P., CLEMENT B. et FERRIERE J., 1976. La zone béotienne en Grèce. Implications paléogéographiques et structurales. Eclogae géol. Helv., 69/3, 577-599.

CELET P., FERRIERE J. et WIGNIOLLE E., 1976. Le problème de l'origine des blocs exogènes du mélange de type volcano-sédimentaire au Sud du Sperchios et dans le Massif de l'Othrys (Grèce). Bull. Soc. Géol. Fr. (colloque de la Méditerranée, Montpellier), sous presse.

CHARVET J., 1970. Aperçu géologique des Dinarides au méridien de Sarajevo (Bosnie). Bull. Soc. Géol. Fr., XII, (7), 968-1002.

CHARVET J., 1973. Sur les mouvements orogéniques du Jurassique-Crétacé dans les Dinarides de Bosnie Orientale. C.R. Acad. Sc. Paris, 276 (D), 257-259.

CIRIC B., 1954. Einige Betrachtungen über die Diabas-Hornstein Formation der Dinariden Bull. Serv. Geol. Serbie, XI, 31-88.

CIRIC B., 1975. The geological composition and tectonic structure of the outer Dinaric Alps in the Montenegro. Acta geologica (res. anglais), VIII, 41, 317-345, Zagreb (en Serbo-croate).

CLEMENT B., 1968. Observations sur le Trias du Patseras et du Parnis en Attique C.R. somm. S.G. Fr., 9, 332-334.

COLEMAN R.G., 1971. Origin and emplacement of the Ophiolite suite : Appalachian ophiolites in New-foundland. Journ. Géophys.Res., 5,14, 1212-1222.

DERCOURT J., 1970. L'expansion océanique actuelle et fossile ; ses implications géotectoniques. Bull. Soc. Geol. Fr., XII, (7), 261-309.

DERCOURT J., CELET P., COTTIN J.Y., DEWEVER P., FERRIERE J., GRANDJACQUET C., HACCARD D. et WIGNIOLLE E., 1976. Importance d'une tectonique Jurassique supérieur sur les marges de la plaque d'Apulie (Hellenides et Apennins ligures). Bull. Soc. Geol. Fr. (Colloque de la Méditerranée, Montpellier), sous presse.

DEWEY J.F., PITMAN W.E., RYAN W.B. and BONNIN J., 1973. Plate tectonics and the evolution of the Alpine system. Geol. Soc. Ann. Bull., 84 (10), 3137-3180.

DIMITRIJEVIC M.D. and DIMITRIJEVIC M.N.,1973. Olistostrome mélange in the Yougoslavian Dinarides and late mesozoïc plate tectonics. Journ. Geol. 81, 328-340.

FERRIERE J., 1974. Etude géologique d'un secteur des zones helleniques internes subpélagonienne et pélagonienne (massif de l'Othrys - Grèce continentale). Importance et signification de la période orogénique anté-crétacé supérieur. Bull. Soc. Geol. Fr. XVI, (7), 543-562.

FERRIERE J., 1976. Sur la signification des séries du Massif de l'Othrys (Grèce continentale orientale) : la zone isopique maliaque. Ann. Soc. Geol. Nord, XCVI, 121-134.

FLEURY J.J., 1976. Unité paléogéographique originale sous le front de la nappe du Pinde Olonos : l'Unité du Megdovas (Grèce continentale), C.R. Ac. Sc. Paris, 282, D, 25-28.

GANSSER A., 1974. The Ophiolis mélange, a world-wide Problem on Tethyan examples. Eclog. Geol. Helv. 67 (3), 479-507.

GUERNET Cl., 1971. Etudes géologiques en Eubée et dans les régions voisines (Grèce). Mém. Lab. Geol. 1, Fac. Sc. Paris, 395 p.

HSÜ J.K., 1971. Franciscan mélange as a model for eugeosynclinal sedimentation and underthrusting tectonics. Journ. Geophys. Res., 76, 5, 1162-1170.

HSÜ J.K., 1974. Melanges and their distinction from olistostromes. Soc. Eco. Paleon. Mineral., spec. pub. (M.Kay), 19, 321-333.

HYNES A.J., 1974. Igneous activity at the birth of an ocean in Eastern Greece, Canad. Journ. Earth Sc., 11, 842-853.

KARIG D.E. and SHARMAN G.F., 1975. Subduction and accretion in trenches. Geol. Soc. Amer. Bull., 86, 377-389.

KATSIKATSOS G., 1971. Les formations triasiques de l'Eubée centrale. An. Geol. Pays Hellen., 22, 62-76.

LAPIERRE H. et ROCCI G., 1976. Le volcanisme alcalin du Sud-Ouest de Chypre et le problème de l'ouverture des régions téthysiennes au Trias. Tectonophysics, 30, 299-313.

MAXWELL J.C., 1974. Anatomy of an Orogen. Geol. Soc. Amer. Bull. 85, 1195-1204.

MERCIER J. et VERGELY P., 1972. Les mélanges Ophiolitiques de Macédoine (Grèce) : décrochements d'âge anté-crétacé supérieur. z. deutsch. Geol. Ges., 123, 469-489.

NAYLOR M.A. and HARLE T.J., 1976. Paleogeographic significance of rocks and structures beneath the Vourinos ophiolite, Northern Greece. Journ. Geol. Soc. Lond., 132, 667-675.

PAPA A., 1970. Conceptions nouvelles sur la structure des Albanides (présentation de la carte tectonique de l'Albanie au 500.000e). Bull. Soc. Geol. Fr., XII, (7), 1096-1109.

RAMPNOUX J.P., 1970. Regards sur les Dinarides internes yougoslaves (Serbie méridionale et Monténégro oriental) : stratigraphie, évolution paléogéographique et magmatique. Bull. Soc. Geol. Fr., XII, (7), 948-966.

RAMPNOUX J.P., 1973. Contribution à l'étude géologique des Dinarides : un secteur de la Serbie méridionale et du Montenegro oriental (Yougoslavie). Mém. Soc. Geol. Fr., LII, 119 100 p.

TERRY J., 1971. Sur l'âge triasique de laves associées à la nappe ophiolitique du Pinde Septentrional (Epire et Macédoine, Grèce). C.R. Somm. S.G. Fr., 7, 384-385.

TERRY J., 1975. Echo d'une tectonique jurassique : les phénomènes de resédimentation dans le secteur de la nappe des ophiolites du Pinde septentrional (Grèce). CR. Somm. S.G. Fr., 2, 49-51.

TURKU I. et NDOJAJ I. GJ., 1973. Mbi disa karakteristika gjeologo-petrgrafike te vulkanizmit mesozoik ne Shqiperi. Permbledhye Studimesh, 2, 23-33 (résumé en français).

VERRIEZ J.J., 1976. Sur les formations volcaniques basiques d'Atalanti (Locride Grèce). Bull. Soc. Geol. Fr., XVIII, (7), 293-298.

WIGNIOLLE E., 1975. Contribution à l'étude géologique de la région centrale du Massif de l'Iti (Grèce continentale). Diplôme Et. Approf. Lille, 94 p.

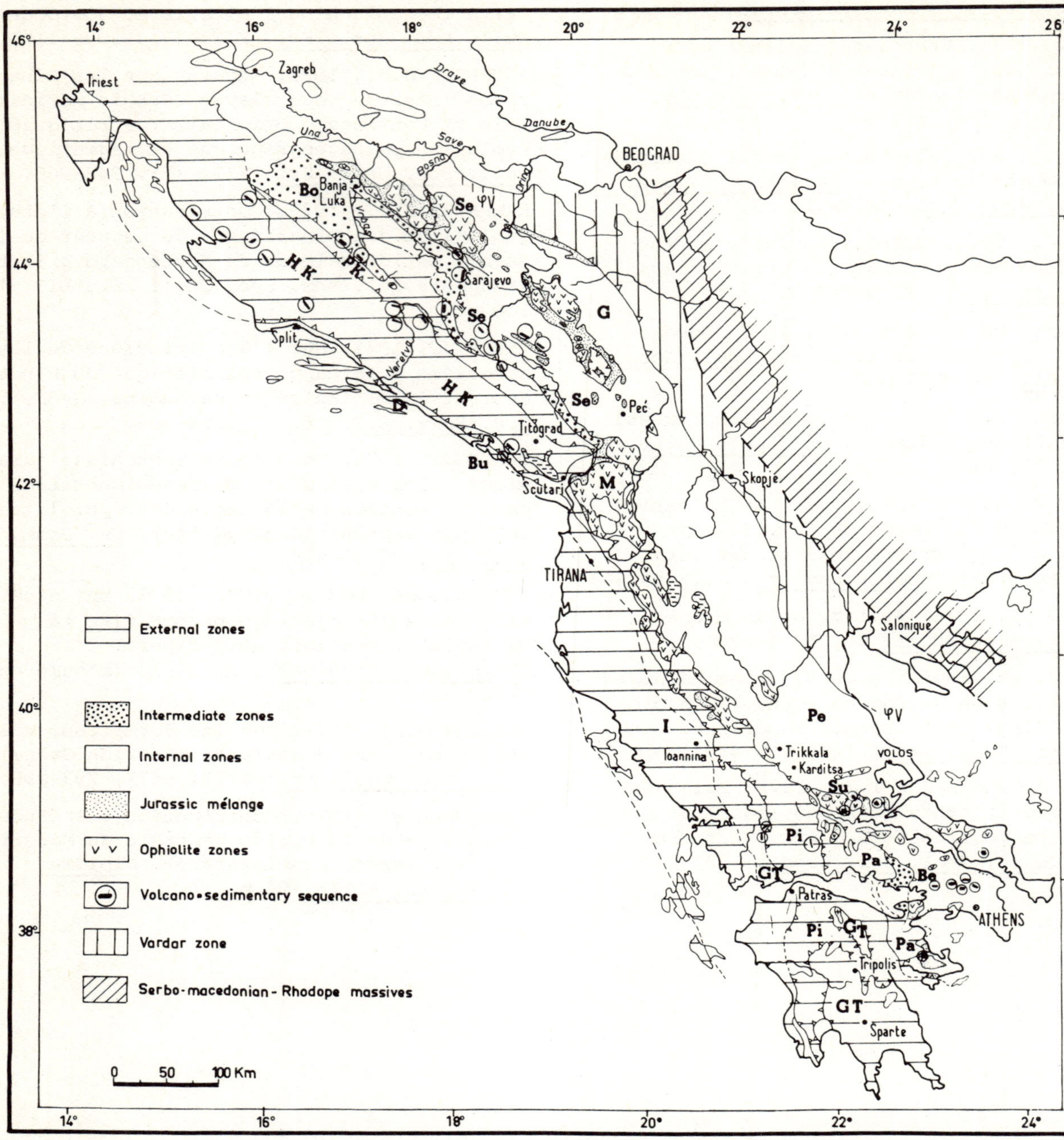

Fig. 1 - Place and distribution of volcano-sedimentary triasic outcrops and of volcano-detritic Jurassic mélange in Dinarides and Hellenides.
External zones. I : Ionian zone ; D : Dalmatian zone ; GT : Gavrovo-Tripolitza zone ; Bu : Budva-Cukali zone ; Pi : Pindus zone ; HK : High-Karst zone ; Pa : Parnassian zone.
Intermediate zones. Bo : Bosnian zone ; Be : Beotian zone.
Internal zones. Se : Serbian zone ; M : Mirdita zone ; Su : "Subpelagonian" (Maliac) zone ; G : Golija zone ; Pe : Pelagonian zone.

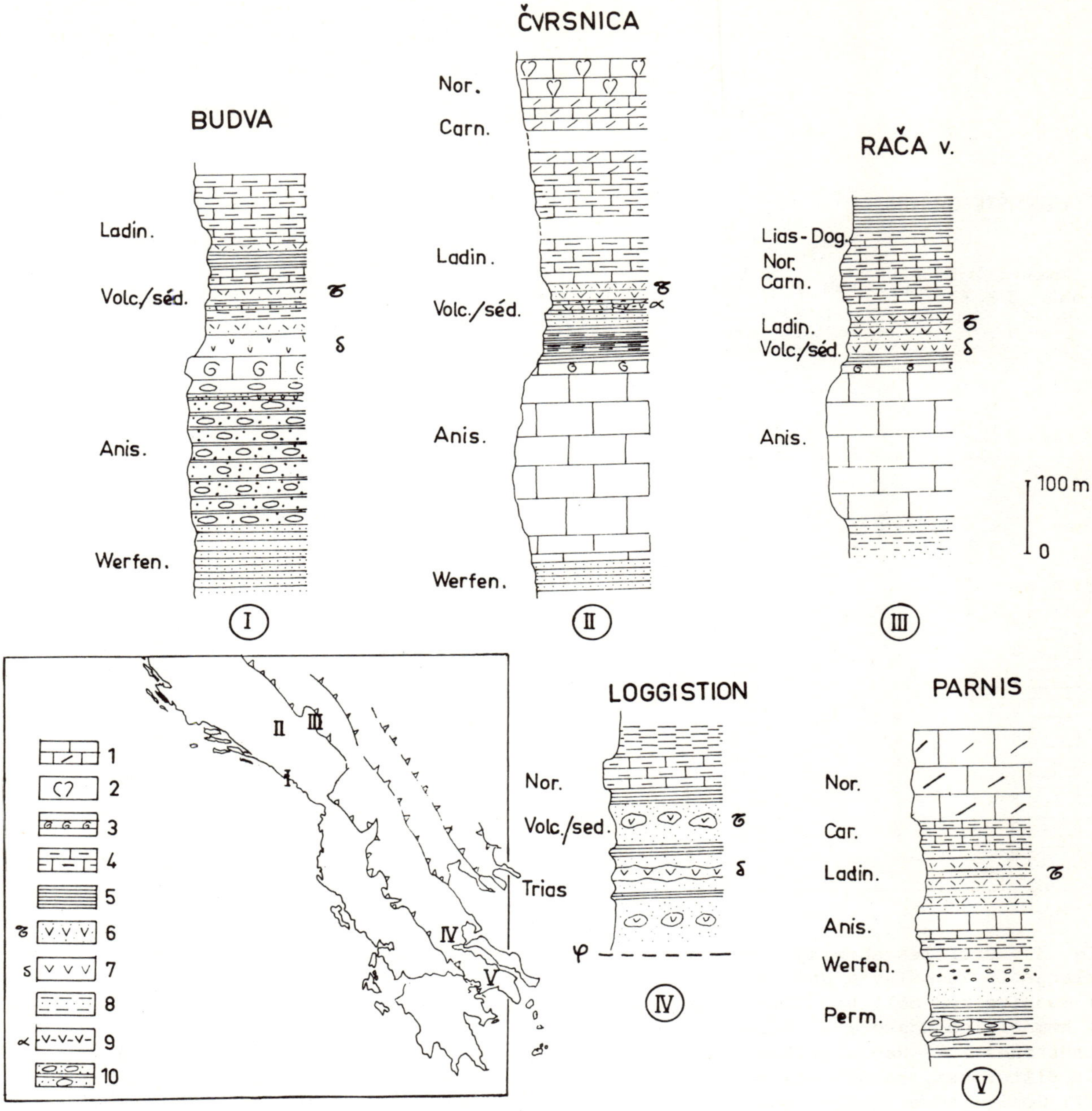

Fig. 2 - Diagrams of a few typical triasic series in Yugoslavia and Greece. 1) limestones and dolomitic limestones ; 2) Megalodonte limestones ; 3) Ammonitico-Rosso ; 4) Cherty limestones and cherts ; 5) radiolarites and siliceous pelites ; 6) pyroclastic rocks ; 7) dolerites ; 8) spilites ; 9) diabases ; 10) triasic detritic rocks (sandstones, conglomerates). Dinarides : I Budva zone ; II : High-Karst zone ; III : Internal bosnian zone. Hellenides : IV : Maliac zone ; V : Pelagonian zone.

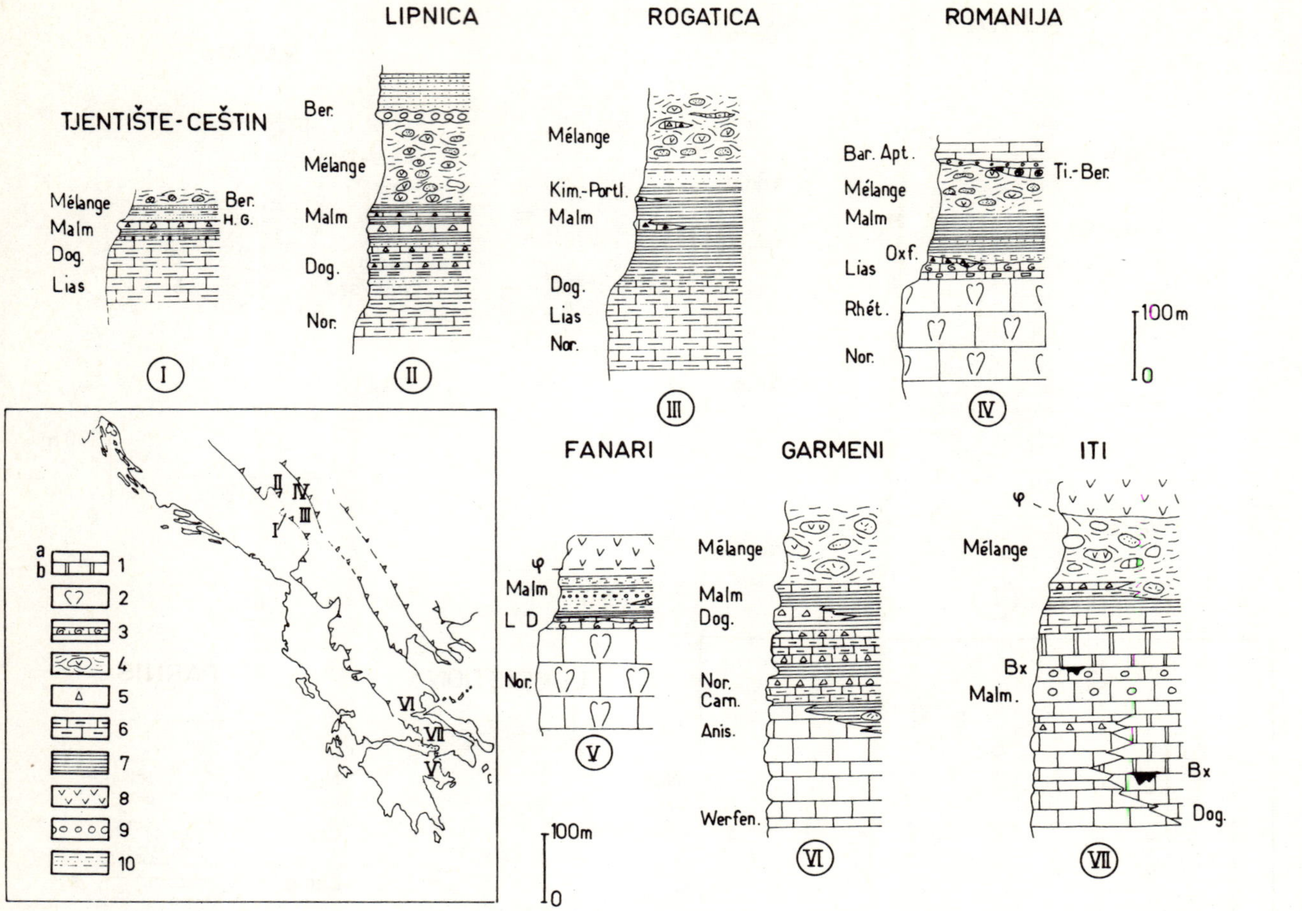

Fig. 3 - Profiles showing different stratigraphic environments of the volcano-detritic melange in Yugoslavia and in Greece.
1) Massive and oolithic limestones (a) ; reef limestones (b) ; 2) Megalodonte limestones ; 3) Ammonitico-Rosso ; 4) Mélange ; 5) Breccias and microbreccias ; 6) Pelagic cherty limestones ; 7) Radiolarites and siliceous pelites ; 8) Ophiolites ; 9) conglomerates ; 10) Siltstones, sandstones and associated rocks. Dinarides : I to III : Serbian zone, IV : Golija zone. Hellenides : V to VII : Internal zones (V : Argolis, VI : Othrys, VII : Locride).

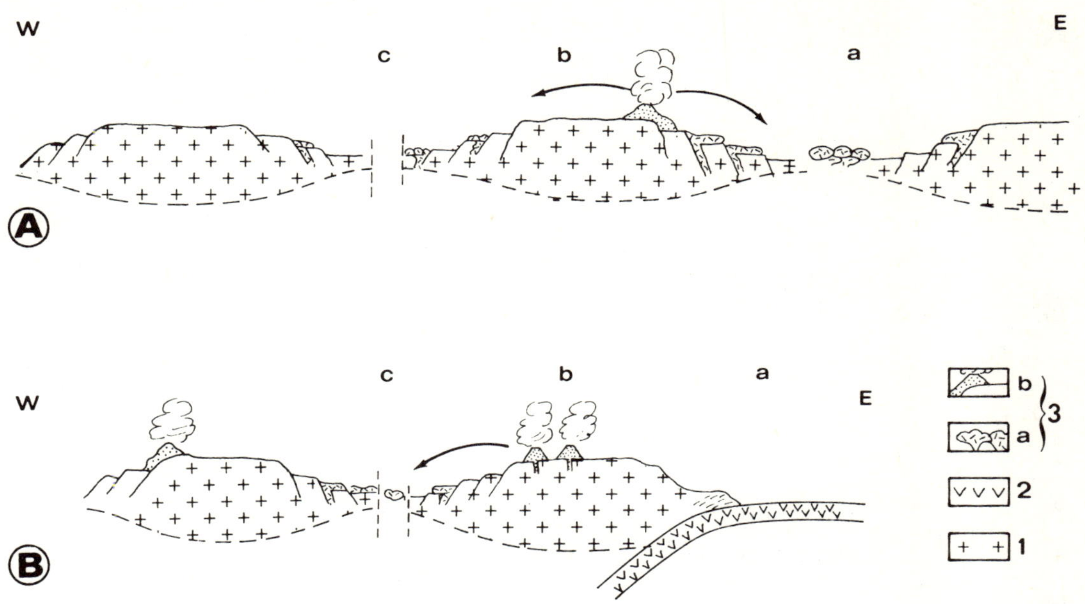

Fig. 4 - Hypothesis about the genesis of volcano-sedimentary triasic events of Dinarides and Hellenides.
1) Continental crust ; 2) Oceanic crust ; 3) effusive materials ; 3 a) mainly submarine ; 3 b) partly outward.

Remarks : The diagrams are not paleogeographic surveys on precisely determined zones but only illustrations of possible working of settlements. The correlations (a,b,c,) given as indications, are pure speculations and can be modified according to the chosen geotectonic patterns and the considered fields.

Possible equivalences : In a platform-basin or insular arc marginal sea system, b) could illustrate anyone of the neritic zones affected by triasic volcanism (Golija or High-Karst in Dinarides, Pelagonian or Parnassian in Hellenides). If b) corresponds to the Pelagonian zone or Golija, c) would be equivalent to Maliac zone and a) to Vardar zone in Greece.

A) Platform breakage that corresponds to a strong paleogeographic transformation after Anisian time and differentiation of the main deep basins ("troughs").

This hypothesis fits to the alkalin type of zone triasic Hellenide lavas.

B) Craton breakage and ocean crust subduction.

Besides necessary breakage to explain facies changes one or several subduction zones are thought over. If the process is able to explain the calcalkaline nature of some triasic lavas, more especially in Yugoslavia, there remains to solve the quick subduction disappearance.

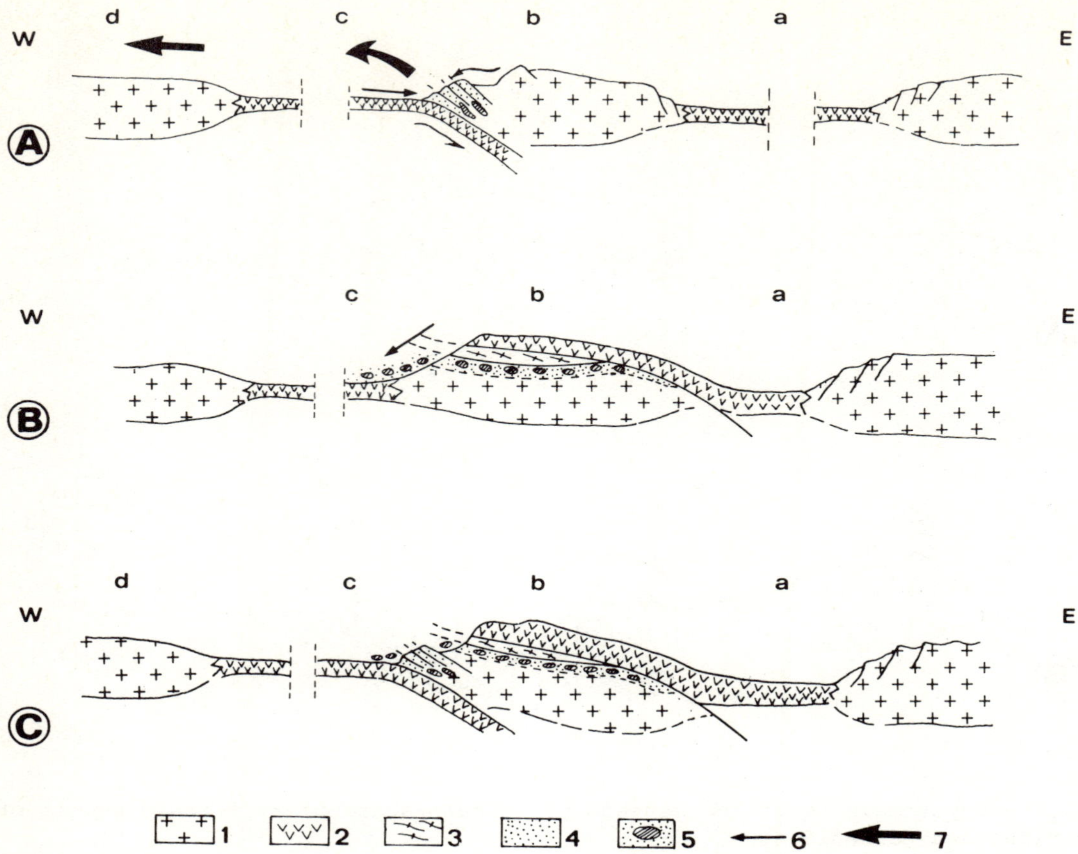

Fig. 5 - Different models of volcano-detritic upper Jurassic mélange working out in Dinarides and Hellenides.
1) Continental crust ; 2) Oceanic crust ; 3) Amphibolites ; 4) Detritic rocks ; 5) Blocks (peridotites and others in the terrigenous matrix) ; 6) Origin and transport of the detritic material ; 7) Tectonic reworking and resedimentation of the detritic formations.
N.B. : Same remarks as fig. 4 and some possible equivalences in the B or C hypothesis considered below.

A) A melange genesis at a subduction zone level (accretionary prisms).

This origin, frequently set forth leads to consider a melange formation without its own basement. Yet, such basements exist and make think of a tectonic reworking followed with resedimentation at least in certain places. In the Hellenides case, the process implies a melange deposit in the internal zones (a : Peonias subzone ; b : Païkon subzone ; c : Almopias subzone ?) and a westward conveying of the materials.

B) Formation of olistostrome melange through destruction of the ophiolitic nappe and the gathered sheets.

That is an intermediate stage where the front of the ophiolitic nappe is destroyed but yet it already covers a melange newly formed during the tectonic shifting of the nappe. There can be also set forth resedimentation of chaotic previously built formation and the "mélange" deposit over various substrata.

C) Simultaneous action of the two phenomena : obduction, subduction and redepositing.

That schema enables to explain "Mélange" without any own basement (accretionary prism) and others with one (olistostromes lying on triasico-jurassic unbroken series).

Nevertheless, such patterns can be seen through a tectonic "décollement".

INTERNATIONAL SYMPOSIUM ON THE STRUCTURAL HISTORY OF THE MEDITERRANEAN BASINS. SPLIT (YUGOSLAVIA) 25-29 OCTOBER 1976. B. BIJU-DUVAL AND L. MONTADERT, Eds. EDITIONS TECHNIP, PARIS 1977, pp. 47-72

BASSINS MARGINAUX ET TÉTHYS ALPINE : DE LA MARGE CONTINENTALE AU DOMAINE OCÉANIQUE DANS LES DINARIDES*

R. BLANCHET [(1)]

Abstract A comparison is done, between : 1) The paleo-oceanographic organisation and evolution, and the sedimentary sequences of the Dinarids during Mesozoic; and 2) The organisation, evolution and sedimentary process in marginal basins and small oceanic basins of Western Pacific, during upper Paleogene and Neogene.

Résumé L'organisation et l'évolution paléo-océanographique de la branche dinarique de la chaîne alpine au cours de l'ère secondaire sont comparées à l'organisation et à l'évolution des marges actives (systèmes bassins marginaux - arcs insulaires) de la bordure asiatique de l'océan Pacifique, du Paléogène supérieur à l'actuel.

Dans l'optique de la tectonique des plaques, plusieurs modèles d'évolution de la Téthys au Secondaire et au Tertiaire ont été proposés depuis 1970 (notamment Dewey et Bird, 1971; Dercourt, 1970, 1972; Bernoulli et Laubscher, 1972; Dewey et al., 1973; Biju-Duval, Dercourt et Le Pichon, 1976).

Pour les Dinarides, ces hypothèses cherchaient à intégrer au maximum les très nombreuses données de terrain connues et publiées alors. Si ces hypothèses simplificatrices n'ont pas tout expliqué ou ont été parfois même en désaccord avec certains faits, elles eurent le grand mérite de provoquer et ainsi de préparer des approches successives de plus en plus affinées. C'est ainsi que pour l'exemple des Dinarides dont nous traiterons essentiellement ici, des modèles d'évolution pour un cadre local ont été présentés (R. Blanchet, 1973, 1974; M. Dimitrijević, 1973; J.P. Rampnoux, 1974; J. Chorowicz et J. Geyssant, 1976; J.P. Cadet, 1976).

Dans cet essai de recherche du paléoocéan téthysien, les démarches suivantes sont adoptées :

1) Nous choisissons comme exemple de référence une portion de la chaîne alpine périméditerranéenne; il s'agit des Dinarides. Dans ce secteur, développé sur près de 2 000 km entre l'Autriche et la Turquie, par la Yougoslavie, l'Albanie et la Grèce, une nouvelle génération de travaux de terrain a apporté des faits géologiques nouveaux et fort importants.

2) Nous comparons l'organisation et l'évolution reconstituée pour l'ère secondaire d'une portion de chaîne alpine à un modèle océanique actuel. Il s'agit d'une portion du Pacifique; plus précisément du Pacifique occidental, en bordure du continent asiatique. Nous utilisons les données fournies par les récen-

(*) Une contribution voisine a été présentée oralement :
- à la Société Géologique de France (C.R. somm., B.S.G.F., séance du 3 mai 1976);
- au colloque final de l'ATP Géodynamique de la Méditerranée (CNRS-INAG, Montpellier, 14-15 oct. 1976).

(1) Département des Sciences de la Terre, Université de Bretagne Occidentale, 6 avenue Le Gorgeu, 29279 Brest Cédex, France

tes campagnes océanographiques effectuées dans le Pacifique, dans le cadre du Deep Sea Drilling Project (DSDP), et rassemblées dans les volumes des "Initial Reports of the Deep Sea Drilling Project", notamment lors des croisières 30 et 31.

Au total, nous avons cherché à comparer l'organisation et l'évolution paléo-océanographique d'une portion de chaîne alpine à l'organisation et à l'évolution d'une portion d'océan récent et actuel. Nous disposions des séries stratigraphiques connues sur le terrain; et de carottes de sédiments connues dans l'océan (matériel DSDP).

Nous avons voulu, autant que faire se pouvait, nous en tenir aux faits et aux objets. Nous pensons que la progression des connaissances sur l'histoire géologique d'une chaîne de type alpin, passe par la confrontation entre les résultats venus du terrain et ceux venus des océans actuels; cette étude comparée impliquant toutes les disciplines des Sciences de la Terre. Mais la géologie est une science historique qui, par conséquent, doit envisager l'évolution dans le temps et dans l'espace. C'est ce que nous allons tenter dans cet article.

I. RAPPEL DES GRANDES ETAPES DE L'EVOLUTION DES DINARIDES AU SECONDAIRE

Il s'agit de rappeler ici les étapes majeures de l'histoire de la chaîne, plusieurs articles permettant en effet au lecteur intéressé de retourner aux descriptions de détail [1].

L'organisation paléogéographique des Dinarides peut se ramener, au Jurassique moyen par exemple, aux subdivisions très générales suivantes, d'Ouest en Est :

1) Un domaine externe [2] en bordure du craton africain, ou plus précisément d'un lobe dit "apulien" de ce craton. En première approximation, ce domaine est comparable à l'époque considérée à une marge continentale ou précontinent de type atlantique. Cette paléomarge est dans le détail de sa topographie accidentée de sillons et de hauts fonds relatifs comme par exemple : la zone d'Apulie (ride), la zone ionienne (sillon), la zone du Gavroro-zone dalmate-zone du Haut Karst (ride). Une pente continentale permet de passer vers l'est à :

2) un domaine interne [2], que l'on a depuis longtemps assimilé à une zone océanique proprement dite. C'est le domaine des Ophiolites, des formations volcano-terrigènes, des Radiolarites (zone serbe, zone subpélagonienne, zone du Vardar).

3) Un domaine à croûte continentale qui borde le système vers l'Est; c'est le continent ou craton du Rhodope (massif Serbo-Macédonien), dépendance ou non, de l'Europe.

L'évolution dans le temps permet de retenir les grandes étapes suivantes :

1) L'individualisation des domaines de sédimentation a lieu dès le Trias.

2) Le développement des séries sédimentaires caractéristiques des zones isopiques se fait pendant le Trias supérieur-Jurassique. On peut opposer des sédiments de plateau continental (par exemple des calcaires néritiques) et des sédiments océaniques pélagiques (par exemple des calcaires pélagiques à silex et des radiolarites).

3) Au Jurassique supérieur-Eocrétacé, une période de crise tectonique pluriphasée affecte le domaine interne : les Ophiolites se mettent probablement en place par voie tectonique; plusieurs familles de structures (plis, nappes) prennent naissance, accompagnées de métamorphisme; on appelle, pour ce système alpin, crise paléotectonique, l'ensemble de ces phases Jurassique supérieur-Eocrétacé. A partir du Jurassique supérieur, s'organise une nouvelle paléogéographie crétacée notamment dans les zones internes; à cette époque débute le comblement par les puissants flyschs crétacés (flysch bosniaque-flysch béotien) à débris d'Ophiolites.

4) De la fin du Crétacé à l'Oligocène suivant les régions, une nouvelle crise tectonique reprend toutes les structures précédentes mais affecte de plus l'ensemble de la chaîne; des processus métamorphiques accompagnent cette phase Crétacé terminal-Tertiaire inférieur.

---

[1] On consultera notamment : Aubouin (1973); Aubouin et al. (1970); Bulletin de la Société géologique de France consacré aux Dinarides (1970); Blanchet (1974); Rampnoux (1974) et Cadet (1976).

[2] "Externe" et "interne" sont employés au sens des géologues.

Cette tectonique détermine une nouvelle paléogéographie oligo-miocène avec développement de bassins de sédimentation en bordure ou sur l'édifice né de la tectonique tertiaire; les molasses s'y accumulent.

5) Au Miocène, des bombements à grands rayons de courbure (tarditectonique) donnent aux chaînes leurs aspects orographiques d'ensemble, les voûtes culminantes correspondant à des vastes anticlinaux tarditectoniques.

6) La néotectonique mio-plio-quaternaire reprend alors l'édifice construit jusqu'au Miocène, par des réseaux de failles; elle détermine des reliefs (horsts) et des plaines (plaines balkaniques, mer Adriatique elle-même).

II. PRESENTATION DE LA BORDURE OUEST PACIFIQUE

A. Situation.

Dans ce vaste domaine sont installés plusieurs bassins marginaux. On doit distinguer deux grands groupes :

1) Une première famille de bassins marginaux est celle qui est actuellement sur la marge du continent asiatique entre ce dernier et l'océan Pacifique. Ces bassins marginaux sont installés entre le continent asiatique à l'ouest et un archipel de type arc insulaire, archipel à l'est duquel se développe l'océan Pacifique ou la mer des Philippines. On distingue notamment, du nord au sud, la mer d'Okhotsk, la mer du Japon, la mer de Chine méridionale.

2) Vers l'est, c'est-à-dire du côté Pacifique, se développe une autre série de bassins marginaux plus franchement en domaine océanique : les bassins de la mer des Philippines (Mariannes, Parece Vela, bassin ouest Philippines) et les bassins du Pacifique méridional (Nouvelles-Hébrides, Fiji, Lau).

On distingue, depuis Karig (1971) :

- des bassins marginaux actifs (bassins inter-arcs) comme ceux des Mariannes, des Nouvelles Hébrides, de Lau;

- des bassins marginaux inactifs parmi lesquels il oppose : des bassins inactifs à flux de chaleur élevé (mer d'Okhotsk, du Japon, de Chine Méridionale, bassin de Parece Vela) et des bassins inactifs à flux de chaleur normal (bassin ouest Philippine, sud Fiji, Nouvelle-Calédonie).

B. Caractères des bassins marginaux.

Plusieurs travaux ont traité des arcs insulaires et des bassins marginaux. Nous renvoyons donc le lecteur à quelques articles fondamentaux parmi lesquels ceux de Dewey et Bird (1970); Dickinson (1970, 1971, 1973); Karig (1970, 1972, 1973, 1974, 1976); Karig et Mammerickx (1972); Sigimura et Uyeda (1973); Coleman (1973, 1975), etc.

On peut retenir, à côté d'un plan d'organisation général décrit par Karig, des variations importantes, dans la position par rapport aux arcs insulaires et aux continents, dans la physiographie, dans la nature du fond océanique, dans le remplissage sédimentaire, dans l'activité tectonique, magmatique, le métamorphisme et le flux de chaleur.

C'est dire que la comparaison qui va suivre portera sur des exemples précis, localisés dont les caractères propres seront alors précisés.

III. COMPARAISON ENTRE UN SECTEUR OCEANIQUE (L'OUEST PACIFIQUE) ET UN SECTEUR PALEO-OCEANIQUE (LA CHAINE ALPINE DES DINARIDES)

La démarche est de comparer l'organisation et l'évolution telles qu'elles ont été reconstituées pour un domaine alpin (il s'agira des Dinarides) avec celles des bassins marginaux de l'ouest Pacifique. Diverses rubriques sont successivement envisagées.

A. Position par rapport aux continents.

1) Le paléo-océan mésogéen correspondant pour le Mésozoïque aux Dinarides, était situé à la périphérie d'un craton apulien, appendice probable du continent africain (J. Aubouin, 1958; J. Dercourt, 1970, 1972; D. Bernoulli et H. Laubscher, 1972). De plus ce paléo-océan était situé entre ce craton apulien et le craton rhodopien (massif Serbo-Macédonien), appendice (ou non) du craton européen.

2) Les actuels bassins marginaux de l'ouest Pacifique, par exemple la mer du Japon ou la mer de Chine méridionale se développent à la périphérie du continent asiatique entre ce dernier et un archipel cratonisé représenté au Japon et dans les Philippines (on connaît dans certains de ces archipels plusieurs cycles orogéniques superposés : cycle néotectonique mio-plio-quaternaire, cycle "alpin" anté-tertiaire, cycles paléozoïque et même antécambrien).

3) La position du paléo-océan dinarique entre ses deux bordures continentales non symétriques et celle de la mer du Japon néogène et actuelle, ou de la mer de Chine méridionale, sont donc à ce titre assez comparables.

B. Echelle.

1) Il est difficile d'évaluer la largeur de la portion de paléo-océan mésogéen considérée ici. Dans un édifice où l'allochtonie est la règle, l'évaluation des raccourcissements tectoniques est toujours délicate; d'autant plus que les découvertes récentes montrent que l'allochtonie des Dinarides est généralement sous-estimée. Ainsi, au droit de la Crête, tout l'édifice dinarique est allochtone ou en fenêtre (J. Aubouin, 1974; M. Bonneau, 1975).

Les arguments les moins incertains proviennent des secteurs ou l'allochtonie est (relativement) plus faible et où des transitions existent entre les séries charriées. Cependant les zones internes, ophiolitiques, paléo-océaniques sont toujours très largement allochtones.

La tectonique des plaques à l'échelle du globe entier, impose pour la période jurassique, certaines limites aux dimensions de la Téthys, comme les divers schémas de reconstitution le montrent (Dewey et al., 1973; Biju-Duval et al., 1976). Il s'agissait d'un domaine à croûte océanique, relativement petit, comparé aux grands océans actuels.

2) Ainsi le modèle des bassins marginaux de l'ouest Pacifique, larges, en moyenne, de 1 500 km, entre le continent asiatique et les arcs insulaires, semble avoir l'échelle de ce qu'a pu être le domaine mésogéen au Jurassique entre le bord apulien et la zone du Vardar. Les comparaisons avec les grands océans Atlantique et Pacifique n'ont pas tenu compte en général de cette question d'échelle.

C. Organisation morphologique des fonds.

1. LES DINARIDES.

En partant de la bordure du continent africain (craton apulien) et en allant vers le Nord-Est, on peut définir, avant l'invasion par les flyschs :

1) Une paléo-marge continentale : vaste plate-forme néritique carbonatée présentant dans le détail et suivant les époques et les régions, une morphologie accidentée de hauts fonds et de sillons (D'Argenio et al., 1971; Blanchet, 1973; Rampnoux, 1974; Cadet, 1976).

Cette paléogéographie peut être déterminée ou accusée par des jeux de failles en extension. Telle est la signification : du sillon ionien, de la ride Gavrovo-Kruja-dalmate en Grèce et en Albanie. Telle est probablement la signification du sillon de Budva et de la ride du Haut Karst au Monténégro. Tel est peut-être le cas d'autres éléments paléogéographiques en position plus internes, comme le haut fond du Parnasse en Grèce, celui du Durmitor, de Mihajlovici voire de Golija en Yougoslavie. De récentes considérations tectoniques conduisent à attribuer la même signification au domaine pélagonien de Grèce (Olympe, Eubée méridionale).

2) Un paléo-bassin océanique, à sédiments pélagiques calcaires ou siliceux (radiolarites) depuis le Trias jusqu'au Jurassique supérieur, et dont le fond pouvait être de nature océanique (ophiolites).

Entre ces deux domaines paléo-marge et paléo-bassin océanique peut apparaître (Yougoslavie) une aire de transition pentée à l'est, avec des brèches à éléments carbonatés venus de la plate-forme apulienne, des séries à sédimentation condensée et hiatus, des séries dolomitiques riches en niveaux schisteux noirs et en lits de silice et silex noirs ([1]).

Ce domaine paléo-océanique interne, est toujours très largement allochtone dans la chaîne dinarique.

Des travaux récents semblent montrer que les unités à ophiolites paléo-océaniques s'enracinent dans la zone du Vardar, à l'est du domaine pélagonien. Mais la paléogéographie de la zone du Vardar au Jurassique est difficile

([1]) Ce dispositif, en pente continentale, ces séries bien particulières, ne sont pas sans évoquer ce qu'on connaît des marges continentales passives, par exemple, sur la pente et au pied de la marge marocaine (cf. Leg 14).

à retracer. On sait que la configuration du bassin n'était pas simple (J. Mercier, 1966, y a défini un sillon d'Almopias à ophiolites une ride du Païkon (sialique), un sillon de Péonias).

3) Un domaine continental bordait à l'est ce paléo-océan mésogéen : Le massif serbo-macédonien et le Rhodope représenteraient une marge continentale en bordure nord-est du domaine mésogéen.

Au delà du Rhodope, entre celui-ci et le continent eurasiatique, se trouvait le domaine de sédimentation des séries de l'arc carpatho-balkanique, domaine marin alpidique dont certains auteurs envisagent le bouclage en cul de sac au-delà de l'Elbourz (Enay, 1972).

L'invasion par les flyschs, conséquence d'une crise orogénique jurassique terminal oblitère cette paléogéographie.

2. LES BASSINS MARGINAUX DE L'OUEST PACIFIQUE.

Ceux-ci montrent une morphologie de leur fond avec tout un ensemble de hauts fonds et sillons, en gros parallèles à la bordure du continent asiatique et à la direction d'allongement des arcs insulaires.

Cette topographie change longitudinalement; des hauts fonds se bouclent; d'autres prennent le relais. Ainsi la mer du Japon se subdivise au sud du 40ème parallèle, et d'ouest en est, en : une marge continentale asiatique, une zone en dépression, un haut fond : le plateau de Corée, un sillon : le bassin de Tsushima, la ride de Yamato (profondeur 300 à 500 m) à substratum fait de granite et de séries anciennes, un sillon : bassin de Honshu, une marge japonaise avec plusieurs hauts fonds, et bassins d'extension limitée.

Cette morphologie de détail disparaît au nord du 40ème parallèle. On n'a plus qu'un seul et vaste bassin (bassin central du Japon) entre le continent asiatique et l'archipel du Japon avec une marge continentale asiatique étroite en bordure ouest.

La mer de Chine méridionale montrerait des faits semblables entre le Vietnam et les Philippines, c'est-à-dire une morphologie en relatifs hauts fonds et sillons d'une marge de type atlantique.

3. LA MORPHOLOGIE DES FONDS DES BASSINS MARGINAUX.

Ainsi la morphologie des fonds des bassins marginaux a en commun avec les objets paléogéographiques des Dinarides les propriétés de variation bathymétrique souvent rapide, de changements longitudinaux avec bouclages de zones et relais de zones à zones.

On pourrait par exemple comparer la paléogéographie des Dinarides au Jurassique (sillon ionien, ride dalmate, sillon de Budva, ride du haut-Karst, sillon bosniaque, etc.) avec ce qui a été dit plus haut de la mer du Japon méridionale.

D. Logique des séquences sédimentaires.

1. LES SEQUENCES SEDIMENTAIRES DES DINARIDES.

Elles ont enregistré l'histoire de ce domaine et indiquent trois grandes étapes :

a. L'individualisation des milieux de sédimentation.

Sur un domaine continental, se produit à partir du Permo-Werfénien, une transgression. On observe la succession de dépôts évaporitiques, puis de formations détritiques très peu profondes (pélites et grès du Werfénien), puis d'une séquence carbonatée néritique à l'Anisien; cette série transgressive est interrompue dès le sommet de l'Anisien et au Ladinien par une formation volcano-sédimentaire, la série porphyrite-radiolarite (porphyrit-hornstein).

Cet épisode volcanique est présent dans toutes les Dinarides et assez généralement dans tout le domaine mésogéen. Ce régime d'instabilité du Trias moyen modifie l'évolution sédimentaire.

A partir du Trias supérieur, on peut en effet reconnaître un domaine à sédimentation néritique carbonatée avec émersions et un domaine à sédimentation pélagique calcaro-siliceuse. On peut interpréter cette séquence comme celle qui correspond aux prémisses d'une ouverture océanique : à des dépôts de bassins fermés, succèdent ceux d'une plate-forme marine très peu profonde; des effondrements (rifting) soulignés par une activité magmatique générale et importante précèdent l'apparition vers l'est du secteur de sédiments pélagiques profonds.

Une étude préliminaire du volcanisme triasique des Dinarides de Yougoslavie a montré que, en plus d'une famille acide (tufs rhyolitiques, trachyte) ou intermédiaire (andésites) assez généralement répandue sur toute la chaîne, il existait un volcanisme basaltique cantonné dans les domaines internes (zone bosniaque; zone serbe au nord de la transversale de Sarajevo). Des analyses et des comparaisons géochimiques ont montré que ce volcanisme évoquait un des types de volcanisme connu dans les actuels arcs insulaires (Bébien et al., 1976).

b. Le développement des séquences sédimentaires caractéristiques des divers milieux.

On peut opposer dès le Trias moyen-supérieur :

1) Des séquences caractéristiques d'une paléomarge continentale, dans les Dinarides externes : les séries sont soit carbonatées néritiques et très épaisses, entrecoupées dans certains domaines de courtes périodes d'émersion avec genèse de bauxites; soit carbonatées et pélagiques et moins épaisses dans des sillons souvent induits par le jeu de failles contemporaines de la sédimentation jurassique : zone ionienne en Grèce et en Albanie; zone de Budva, bassin des couches de Lemes en Yougoslavie. Des hiatus de sédimentation peuvent être présents.

Des apports détritiques carbonatés provenant du remaniement des séries carbonatées occupant l'ensemble du craton apulien, sont fréquemment intercalés au pied des escarpements de failles. On peut comparer ces apports détritiques externes avec ceux fournis actuellement par le continent nord américain sur sa marge nord atlantique ou par le continent asiatique sur sa marge de la mer du Japon ou de la mer de Chine. Un craton fournit des éléments de ses matériaux constitutifs; il s'agit d'apports terrigènes pour les deux exemples actuels ci-dessus évoqués. Il s'agissait d'éléments carbonatés pour les Dinarides; d'où, dans ce dernier cas, l'accumulation de brèches et microbrèches granoclassées avec figures de courant, probablement mises en place par des courants de turbidité.

2) Dans les Dinarides internes, des séquences dont la logique est différente :

a) Il s'agit, pour la période antérieure aux flyschs, de séries pélagiques carbonatées et siliceuses. Ce sont fondamentalement des calcaires pélagiques à microfaunes planctoniques à silex et lits de silice, depuis le Trias supérieur jusqu'au Jurassique supérieur. Les épaisseurs sont faibles comparées aux séries carbonatées des Dinarides externes. Au Jurassique supérieur, une série de "radiolarites" est présente. Il s'agit de lits réguliers de jaspes à radiolaires, holosiliceux ou hypersiliceux, pouvant alterner avec des pélites fines également très riches en silice. L'épaisseur varie de la dizaine à la centaine de mètres. Mais les "radiolarites" telles que les appellent les géologues alpins sont une des caractéristiques des zones internes des chaînes (M. Steinberg et al., 1976).

b) Dans ces domaines internes, les relations entre ophiolites et formations sédimentaires sont essentielles à connaître.

Il faut rappeler la distinction nécessaire au sein de ce qui est appelé complexe ophiolitique, de deux ensembles distincts (R. Blanchet, 1973, 1974) :

- Une formation volcano-sédimentaire : c'est, reposant stratigraphiquement sur une semelle pélagique (calcaire et radiolarites), une formation terrigène de type flysch, à laquelle sont associées des venues volcaniques. L'ensemble est perturbé et présente souvent un aspect chaotique. Dans le secteur où elle est présente, elle a signification de premier faciès de comblement et j'en parlerai plus loin.

- Les massifs ophiolitiques ou ophiolites : roches éruptives basiques et ultra-basiques formant des entités bien individualisées; ces cortèges ophiolitiques présentant une grande analogie pétrographique et géochimique avec les cortèges océaniques (cf. notamment C.J. Allègre et al., 1973) :

  - le mur : les ophiolites sont en contact tectonique avec le substratum sur lequel elles reposent dans la chaîne dinarique (Moores, 1969; Bernoulli et Laubscher, 1972; Aubouin, 1973; Blanchet, 1973; Mercier et Vergely, 1975);

  - le toit : si les ophiolites représentent des lambeaux de la croûte du paléo-océan, le toit sédimentaire serait fait des premiers sédiments déposés sur le fond de ce paléo-océan.

C'est dire que son étude est essentielle; malheureusement les travaux de terrain, faits à ce jour, ayant nécessairement des objectifs plus généraux n'ont pas apporté à cette question, tout le soin qu'elle méritait. C'est pourquoi nous entreprenons une recherche systématique dans la chaîne alpine des sites où l'on aurait des ophiolites surmontées stratigraphiquement de leurs premiers sédiments pour en faire une étude sédimentologique et géochimique, pour les comparer aux sédiments actuels.

Les seules données pour les Dinarides, comme pour les chaînes alpines méditerranéennes en général, font ressortir que le toit sédimentaire des pillow lavas est très souvent siliceux, et de type radiolarite (exemples : Apennin septentrional, Steinmann, 1913; Abbate et Sagri, 1970; Grèce, Kossmat, 1924; Mercier, 1966; Moores, 1969).

Il faut souligner :

- que cette couverture sédimentaire est généralement celle de pillows-lavas dont il est nécessaire de savoir exactement les relations avec les ophiolites;

- que cette couverture, est souvent constituée de radiolarites et de cherts; ceci est une différence fondamentale avec ce qu'on connaît des grands océans où les premiers sédiments déposés sur les dorsales sont des carbonates, les dorsales étant actuellement à une profondeur moyenne de 2 700 m, nettement au-dessus de la profondeur de compensation pour la calcite (CCD).

c. Le comblement progressif par les flyschs.

L'invasion des milieux par les flyschs se déclenche au Jurassique supérieur puis se développe progressivement de l'intérieur vers l'extérieur, c'est-à-dire de l'est vers l'ouest. Les sources du matériel terrigène sont toujours à l'est, c'est-à-dire, du côté interne.

Le domaine interne connaît les premiers apports détritiques terrigènes avec quartz détritique dès le Jurassique supérieur. La formation volcano-sédimentaire est le premier résultat de cette sédimentation terrigène à laquelle s'ajoutent :

- des coulées volcaniques;
- des olistolithes et des olistostromes de nature variée, notamment des blocs de roches du complexe ophiolitique.

Les blocs, parfois de taille importante, de roches du complexe ophiolitique dans la formation diabase-radiolarite peuvent être regardés comme les conséquences d'un écaillage du fond paléo-océanique, ce matériel étant débité en blocs au front, blocs qui s'additionnent à la sédimentation pélagico-terrigène du milieu. Les ophiolites sont ainsi associées tectoniquement à la formation volcano-sédimentaire. Les deux ensembles forment, à partir du Jurassique terminal, un "complexe ophiolitique" source partielle de matériel terrigène pour le flysch bosniaque.

Avec la limite Jurassique-Crétacé se généralisent les flyschs (flysch bosniaque, R. Blanchet, 1966, 1973, 1975; J. Aubouin et al., 1970). Ces flyschs précoces contiennent des éléments remaniés de la formation volcano-sédimentaire et des ophiolites. Il s'agit dans ce dernier cas soit d'olistolites, soit de galets, soit d'éléments centimétriques ou millimétriques, c'est-à-dire d'éléments beaucoup plus petits que dans le cas de la formation volcano-sédimentaire qui est, elle, plus précoce et plus interne.

Le flysch bosniaque cachète dans les régions les plus internes, le contact tectonique entre la formation volcano-sédimentaire et les Ophiolites. Ainsi cette tectonique précoce mais essentielle puisqu'elle écaille le fond paléo-océanique a pu être précisément datée comme étant post kimméridgienne et anté berriasienne (J. Charvet, 1972; R. Blanchet, 1973) c'est-à-dire de l'extrême fin du Jurassique.

Les Dinarides externes ne sont atteintes par les flyschs qu'au Crétacé terminal et cette formation gagne progressivement vers l'ouest pendant l'Eocène.

2. LES SEQUENCES SEDIMENTAIRES DANS LES MERS MARGINALES DE L'OUEST PACIFIQUE.

Elles n'ont pas encore fait l'objet de travaux systématiques de synthèse. L'effort a été mis jusqu'à maintenant sur les problèmes géophysiques et structuraux. J'ai bénéficié à la *Scripps Institution of Oceanography*, des résultats de forages Joïdes dans l'ouest Pacifique, Legs 6,21,30 et surtout Leg 31 (1975).

a. Soit l'exemple de la mer du Japon.

1) Dans l'espace et selon la surface temps actuelle. Divers types de sédiments tapissent actuellement le fond de ce domaine (Skornyakova, 1960) : sédiments détritiques, silts, argiles, sédiments carbonatés, sédiments siliceux, sédiments riches en manganèse.

Une carte synthétique (Skornyakova, 1960) montre la répartition des sédiments sur le fond actuel de la mer du Japon.

2) On a pu évaluer approximativement l'épaisseur de la couverture de sédiments au-dessus du "basement", à l'aide de profils sismiques. On observe une grande épaisseur de sédiments (plus de 2 km) au nord-est du bassin central c'est-à-dire au pied de la marge japonaise et à l'ouest de ce bassin, au pied de la pente continentale asiatique ainsi que dans le sillon de Honshu (de 1,5 à 2 km). Par contre, sur les rides, l'épaisseur est beaucoup plus réduite : 0,1 km à 0,4 km sur la ride de Yamato, 0,750 à 1 km au droit du plateau de Corée.

3) Le développement dans le temps et dans l'espace des séquences sédimentaires est encore imparfaitement connu, pourtant cette connaissance serait essentielle à bien des égards.

Le Leg 31 du programme DSDP a implanté 4 forages (sites 299, 300, 301, 302) en mer du Japon. Les résultats ont été inégaux pour diverses raisons techniques. On peut retenir les résultats fondamentaux suivants :

On observe la succession dans le temps de deux ensembles avant l'arrivée des turbidites :

a) A la base une série volcano-sédimentaire; (site 302 du Leg 31), ce sont des tufs verts à fragments volcaniques dévitrifiés, feldspaths, pyrite diagénétique et des sables à débris volcaniques. Cette série dont seule la partie supérieure a été atteinte est considérée comme l'équivalent de l'épisode des tufs verts (green tuff) andésitiques connus au Japon (nord est de Honshu) à l'Oligocène supérieur-Miocène inférieur (15 M.A.).

b) Lui succède, à partir du Miocène supérieur, une série pélagique (site 302 du Leg 31), comprenant de bas en haut :

- des argiles, indurées à la base, à zéolites (environ 170 m) à Diatomées et Radiolaires du Miocène supérieur;
- des boues à Diatomées, Radiolaires, spicules de Spongiaires, Silicoflagellés, de plus en plus lithifiées avec la profondeur, avec des passées de cherts. L'épaisseur totale est d'environ 280 m. Les organismes siliceux confèrent un âge Miocène terminal à Pliocène inférieur à cette série;
- des argiles vertes ou brunes zéolitiques (environ 40 m) avec des passées avec apparition de composants détritiques soit carbonatés, soit terrigènes (feldspaths, amphibole, chlorite, micas). Les Diatomées donnent un âge Pliocène supérieur-Pleistocène;
- des boues à Diatomées et des argiles silteuses (traversées sur environ 30 m) avec minces lits à éléments calcaires et à éléments de cendres volcaniques remaniés. Les composants terrigènes sont plus abondants (Plagioclases : jusqu'à 15 %; quartz : jusqu'à 12 %; verres volcaniques). On y connaît : Diatomées et Foraminifères d'âge Pleistocène supérieur.

Ainsi, il s'agit d'une série pélagique argilo-siliceuse très riche en sédiments siliceux biogènes, avec présence de niveaux de cherts. On note que c'est seulement au Pleistocène récent (site 301 et site 302) qu'apparaissent les microfossiles calcaires, au milieu de populations toujours nettement prédominantes de micro-organismes siliceux.

L'invasion des divers milieux de sédimentation de la mer du Japon par les formations terrigènes, a pu être analysée. Certains secteurs sont recouverts d'une épaisse accumulation de dépôts détritiques terrigènes.

Il faut distinguer :

a) Ceux qui sont au pied de la marge asiatique, c'est-à-dire externes (au sens géologique) dans le système; leur alimentation est facilement expliquée par la présence même du continent asiatique.

b) Ceux qui sont dans les secteurs orientaux de la mer du Japon (bassin central et sillon de Honshu). Les forages ont permis de faire ressortir les faits suivants :

- les éléments (roches et fossiles remaniés ont permis la reconnaissance précise des sources du matériel détritique (exemple : site 299 du Leg 31, Karig, 1976). La source principale d'alimentation est l'Est-Nord-Est, c'est-à-dire du côté de l'archipel du Japon (Honshu) du côté interne au sens géologique du terme;
- la migration de l'invasion terrigène dans le temps et dans l'espace est particulièrement claire : en un même point (site 299) on observe la superposition de formations détritiques distales au corps sédimentaire pélagique riche en Diatomées. Ces formations distales et ces contourites s'intercalent et se terminent dans des séries pélagiques ou hémipélagiques pliocènes. Puis arrivent des

séries proximales, beaucoup plus grossières, au Pleistocène. Les séquences comparées des divers sites en mer du Japon, montrent :

- . que des apports terrigènes s'additionnent à une séquence sédimentaire pélagique;
- . que les plaines abyssales les plus éloignées de l'archipel du Japon sont, soit épargnées par les apports internes, soit seulement atteintes par des apports distaux pleistocènes qui s'additionnent aux sédiments pélagiques (site 301 du Leg 31); alors que les sillons des secteurs plus proches (site 299) sont envahis par des apports proximaux dès le Pliocène et que certains sont remplis;
- . que les apports terrigènes s'accumulent en premier dans les sillons bordant l'archipel (sillon de Honshu), les rides ou leur prolongement pouvant être épargnées : nous avons vu que la séquence forée en bout de la ride de Yamato au site 302 était exclusivement pélagique. Il semble que l'on retrouve là un rôle de "barrière en creux" analysé dans les séries anciennes par exemple dans le flysch du Pinde (J. Aubouin, 1959);
- . que le moment de l'arrivée des apports terrigènes en un point donné est fonction de la morphologie du fond. Ainsi, en mer du Japon, le site 301 dans la plaine abyssale, est déjà atteint par des séries terrigènes distales alors que le site 302, plus interne, mais sur la retombée de la ride de Yamato n'est pas encore atteint actuellement;

- bien des structures et textures sédimentaires classiques pour les formations détritiques ont été rencontrées dans les séries de la mer du Japon, et décrites en détails. Les interprétations génétiques sur ces ensembles terrigènes divergent; s'agit-il de turbidites ? s'agit-il de "deep sea fan" ? (A. Bouma, 1975);
- en conclusion, on retiendra la superposition dans le temps : série volcano-sédimentaire, surtout andésitique; série pélagique essentiellement siliceuse biogène à taux de sédimentation 10 m/M.A. à 75 m/M.A. (ce qui est un taux élevé pour ce type de sédiments et pouvant témoigner, par exemple, de la grande productivité organique, notamment au Miocène supérieur); invasion des milieux par des apports terrigènes avec accroissement caractéristique du taux de sédimentation d'un facteur 2 (jusqu'à 170 m/M.A.), que l'on ne manquera pas de paralléliser avec les trois étapes distinguées dans les Dinarides : individualisation des milieux; développement des séquences caractéristiques; comblement par les flyschs.

On notera l'absence de dépôts carbonatés et de microfossiles calcaires avant le Pleistocène supérieur. Or nous verrons que la croûte sous-jacente peut être une croûte océanique; on sait que dans les grands océans (Atlantique, Pacifique, Indien) les premiers sédiments déposés sur la croûte océanique nouvellement créée sont des carbonates; les dorsales étant à une élévation moyenne de 2 700 m, c'est-à-dire nettement au-dessus de l'actuelle surface de compensation de la calcite. Ce serait là une caractéristique importante des séquences sédimentaires des bassins marginaux par rapport à celles des grands océans.

En l'état actuel des connaissances, plusieurs hypothèses explicatives concernant cette importance de la sédimentation pélagique siliceuse et l'absence de sédimentation carbonatée en mer du Japon peuvent être avancées parmi lesquelles :

- la surface de compensation de la calcite serait beaucoup moins profonde dans les bassins marginaux que dans les océans : certains auteurs placent la CCD à une profondeur de 2 100 m en mer du Japon (Ujüe et Ichikura, 1973) alors que dans le Pacifique voisin elle est en moyenne à 4 500 m. A quoi est liée cette remontée de la CCD ? aux conditions climatiques ? à un volcanisme sous-marin intense ? à une retenue des carbonates sur les marges ? à des évènements tectoniques ?
- les milieux ont des températures trop basses pour permettre le développement des micro-organismes calcaires, pendant l'intense refroidissement plio-pleistocène;
- les milieux seraient dépourvus d'importantes circulations profondes et resteraient, sous conditions anaérobies, empêchant les microfossiles calcaires d'être présents. La présence quasi-générale de pyrite dans les sédiments est en faveur de cette explication.

J'ai indiqué que les premiers sédiments signalés au sommet des complexes ophiolitiques pouvaient être des séries pélagiques siliceuses (radiolarites et argilites associées). Nous cherchons maintenant à comparer ces premiers sédiments à ceux connus dans les bassins marginaux. Ces derniers peuvent constituer des milieux particuliers en égard au développement de séries hypersiliceuses.

b. Soit l'exemple de la mer des Philippines.

Ce système complexe de bassin marginal, compris entre la fosse des Mariannes et la fosse des Philippines a été décrit dans plusieurs publications, par D. Karig.

Le Glomar Challenger a effectué huit forages dans la mer ouest Philippine lors du Leg 31. Du point de vue séquence sédimentaire, je ne retiendrai, à titre d'exemple, qu'un seul site (291). Le forage fut implanté immédiatement à l'est de la fosse des Philippines, à plus de 5 000 m de profondeur. A 120 m au-dessous du fond, on a atteint un niveau de basalte.

Le basalte est tholéiitique, sa surface est vitreuse; des fragments de verre sont interprétés comme parties de carapace de pillow lavas. Ce basalte a été traversé sur 1,60 m. S'agit-il du sommet du réel "basement" océanique ? ou bien s'agit-il d'une simple coulée intercalée au sein d'un ensemble sédimentaire ?

Le contact basalte-sédiment a pu être observé : on trouve au-dessus de la lave, une accumulation pélagique inorganique (minéraux argileux, fragments de verre volcanique, zéolites) avec des concentrations d'oxydes de fer.

Cette série d'argile brune (jusqu'à 87 % de minéraux argileux), zéolitique, varie progressivement vers le haut sur environ 16 m d'épaisseur : les oxydes de fer, les fragments de verres volcaniques, les zéolites diminuent. Puis apparaissent des organismes siliceux avec des lits riches en radiolaires et des zones silicifiées riches en cristobalite et tridymite (radiolarian porcelanite, R.E. Garrison et al., 1975). Les fossiles indiquent un âge Eocène moyen.

Vers le haut cette série s'enrichit en organismes (surtout en radiolaires et aussi en nannofossiles) et sur 22 m d'épaisseur on trouve une boue argileuse brune à radiolaires (40 % de minéraux argileux, 5 % de silts, des traces de zéolites, spicules d'éponges et nannofossiles), avec localement des passées de boues argileuses à radiolaires et nannofossiles. L'âge est : Eocène supérieur-Oligocène inférieur.

A l'Oligocène correspondent 16 m d'une boue argileuse à nannofossiles (80 % de nannofossiles, 20 % de minéraux argileux).

Entre l'Oligocène et le Pliocène s'accumulent 60 m d'un silt argileux (minéraux argileux, zéolites, palagonite) à faible contenu biogénique et à importante participation (30 %) de débris volcaniques.

Ce site permet donc d'étudier en détail, le processus sédimentaire au-dessus d'un corps basaltique, sans qu'on sache toutefois si ce basalte est vraiment le toit du "basement" océanique. Mais il est clair qu'une telle série sédimentaire océanique doit être comparée dans ses paramètres, lithologiques, minéralogiques, et géochimiques avec des séries sédimentaires connues immédiatement sur les pillows lavas dans les chaînes alpines.

En pensant à cette comparaison, on retiendra du site 291 :

- le développement d'une série pélagique avec fond continu de matériaux inorganiques, coupés par des périodes de sédimentation biogène;
- l'absence de dépôts carbonatés dans les premiers sédiments au-dessus du basalte. Si ce basalte appartient au fond océanique, la croûte océanique néoformée de cette partie du bassin marginal a été créée au-dessous de la CCD;
- la présence de lits à potentialité hypersiliceuse (lits silicifiés, porcelanite à Radiolaires, boues à Radiolaires);
- l'apparition sporadique de boues carbonatées à nannofossiles. S'agit-il d'apports organiques soudains par des courants transportant des faunes carbonatées de milieux moins profonds ? s'agit-il au contraire de variations dans la position de la CCD par rapport au fond selon les facteurs préposés par divers auteurs (Bramlette, 1962; Berger, 1973; Winterer, 1974) ?

c. Soit l'exemple du bassin de Lau.

Il s'agit d'un petit bassin marginal, en arrière de l'Arc des Tonga, dans le sud-ouest Pacifique. Une récente synthèse sur cette mer marginale due à J.W. Hawkins Jr. (1974), permet de relever certains faits majeurs pour une étude comparée océans actuels-chaînes plissées.

1) Le fond du bassin est connu par dragages, comme formé de basaltes tholéiitiques et de gabbros. La géochimie et la pétrologie de ces constituants montrent qu'ils ressemblent aux produits magmatiques des rides médio-océaniques.

2) Le fond est tapissé par une couche de sédiments argileux et clastiques riches en silice (60 à 65 % de SiO_2) avec abondance de verres volcaniques et zéolites.

Pour J.W. Hawkins, qui s'appuie sur des travaux expérimentaux sur la genèse des cherts (Keene et Kastner, 1974), on peut voir dans ces sédiments, l'origine de schistes siliceux, pouvant par diagénèse donner des lits siliceux (ribbon cherts).

E. Propriétés crustales.

1. CROUTE OCEANIQUE ET OPHIOLITE.

1) Les Ophiolites des chaînes alpines périméditerranéennes et notamment celles des Dinarides sont interprétées comme des copeaux de la croûte d'un domaine paléo-océanique, mis en place tectoniquement (obductés) sur une marge continentale (J. Dercourt, 1970; Bernoulli et Laubscher, 1972; R. Blanchet, 1973, 1974; J. Aubouin, 1973; Dewey et al., 1973; B. Biju-Duval et al., 1976).

2) Dans certains bassins marginaux on a mis en évidence depuis longtemps l'existence d'une croûte de nature océanique typique ou intermédiaire sur des aires plus ou moins étendues de ces bassins, dont c'est d'ailleurs l'une des caractéristiques.

Ainsi, c'est le cas dans la mer du Japon ou dans le bassin sud chinois, pour nous en tenir aux exemples pris plus haut.

Les bassins marginaux semblent se caractériser par des variations très rapides de l'épaisseur de leur croûte, montrant des passages relativement brusques, d'aire à croûte continentale à des secteurs à croûte océanique ou subocéanique.

Dans la mer du Japon partant de la marge continentale asiatique à croûte continentale, on passe au bassin du Japon à croûte océanique (largeur 200 km environ) puis à la ride de Yamato, à substratum continental, puis au bassin de Yamato (largeur 150 km environ) à croûte océanique. Vers le nord, la ride de Yamato se boucle "géographiquement" laissant place à un vaste domaine océanique (Japan Basin); enfin, les îles du Japon, sont typiquement de nature continentale.

Dans le bassin sud chinois, la partie orientale de la mer de Chine, a une structure identique à celle d'une croûte océanique aux différences suivantes près (Ludwig, 1971) : le niveau 3 aurait une épaisseur moyenne inférieure de moitié à son épaisseur dans une croûte océanique typique; le moho est moins profond (de 2 à 3 km).

On observe un passage très rapide de cette croûte océanique à la croûte continentale correspondante à Luzon (Philippines).

Ces différences entre croûte océanique d'un bassin marginal et croûte océanique d'un grand océan, pour relativement mineures qu'elles puissent apparaître, sont peut-être significatives.

3) Une croûte de nature océanique est bien développée dans certains secteurs des bassins marginaux. On doit donc envisager que les ophiolites des chaînes alpines puissent être des fragments de nature paléoocéanique mais provenant de la tectonisation de bassins marginaux ou de petits bassins océaniques.

Dans cette hypothèse, le paléo domaine océanique téthysien ou mésogéen, dans le secteur considéré (les Dinarides) serait à comparer, pour partie au moins, au domaine de nature océanique d'un bassin marginal, ou d'un petit bassin océanique.

Un modèle actualiste pour une partie au moins de l'ancien domaine océanique correspondant aux Dinarides serait celui d'un bassin marginal ou d'un petit bassin océanique : c'est vers cette hypothèse de travail que convergent les données exposées dans ces pages.

Du point de vue pétrologique et géochimique, rien ne s'oppose en l'état actuel des connaissances à ce que les ophiolites puissent avoir leur origine dans la croûte océanique de bassins marginaux (C.J. Allegre et al., 1973).

On peut aussi penser que les ophiolites sont un exemple de "phénomène de convergence", et qu'il puisse exister, plusieurs familles d'ophiolites, les unes provenant de bassins marginaux, les autres provenant de grands océans (de type Pacifique, par exemple). Sera-t-il possible de distinguer par des paramètres géochimiques, croûte océanique des grands océans et croûte océanique des bassins marginaux ?

On doit cependant remarquer, du point de vue structural, que les ophiolites sont toujours en relation tectonique, avec les terrains encaissants d'une paléo-marge continentale de type passif ayant une remarquable continuité et servant ainsi de fil directeur aux analyses structurales : les alignements ophiolitiques sont (en position) marginaux et tectoniquement superposés à des marges continentales ayant une continuité dans l'espace.

2.

Trois hypothèses ont été avancées pour expliquer la présence de croûte océanique dans les bassins marginaux :

1) Subsidence avec océanisation d'une portion de croûte continentale (V.V. Beloussov, 1961, 1970; Y. Kaseno, 1972).

2) Reliques des grands océans tels que le Pacifique (Vasikovsky et al., 1971).

3) Accrétion et expansion du fond dans les bassins marginaux actifs par un processus dont le résultat peut-être comparé à ce qu'on observe dans les grands océans.

Cette dernière solution, à la suite de travaux nombreux, notamment ceux du DSDP, a la faveur d'une majorité d'auteurs (Karig notamment, 1971-1974; Isezaki et Uyeda, 1973; Karig, Mammerik et al., 1973).

L'étude des anomalies magnétiques dans les bassins marginaux ne montrent pas la belle régularité dans la symétrie de part et d'autre d'une éventuelle dorsale. Cependant des alignements d'anomalies existent, comme par exemple dans la mer du Japon (Isezaki et Uyeda, 1973), où ils sont cependant plus difficiles à cerner que dans le Pacifique voisin.

Il existe donc des différences dans les résultats de l'accrétion crustale dans les bassins marginaux et dans les grands océans. D'où des hypothèses concernant le mécanisme profond de l'ouverture et de l'expansion des bassins marginaux actifs; la plus souvent émise étant la montée d'un dome de matériel "chaud" du manteau ("notion de diapir thermique") (Karig, 1971, 1974; Oxburgh et Turcotte, 1971; Hasebe et al., 1970). Des auteurs discutent pour savoir si cette montée thermique est la cause (diapirisme actif) ou la conséquence de l'expansion.

La zone d'expansion océanique dans les bassins marginaux ne devait pas nécessairement présenter les caractères typiques d'une dorsale-médio océanique. Mais bien des recherches sur le terrain sont encore à faire pour éclairer définitivement ce problème de l'accrétion dans les bassins marginaux.

Enfin, on peut souligner, à la suite des travaux de Karig (1971) :

- l'évolution depuis un bassin marginal actif vers un bassin marginal inactif (cf. <u>supra</u>, p. 7);
- l'existence d'un stade de vacuité maximale dans les bassins marginaux inactifs à flux élevé, c'est-à-dire au moment où déferle l'invasion par les apports terrigènes;
- l'existence d'une étape avec flux de chaleur encore élevé au stade bassin marginal inactif.

IV. L'ORGANISATION ET L'EVOLUTION DANS LE TEMPS

A. Les bassins marginaux.

Concernant les bassins marginaux envisagés, on peut reconnaître la succession dans le temps de différents stades :

<u>Stade 1</u>. C'est la fragmentation tectonique (rifting) en bordure d'un continent (pour l'exemple ouest Pacifique) mais sur celui-ci, ou au sein même d'une plaque en partie continentale. Un volcanisme de type intermédiaire souvent acide au début (ignimbrites et tufs ignimbritiques) accompagne ces effondrements.

<u>Stade 2</u>. Création d'une croûte océanique et accrétion du bassin marginal actif. Le bassin marginal peut être compartimenté en rides et sillons; des dénivellations de plus de 1 km sont fréquentes : elles peuvent être dues à des failles normales. Ce bassin est moins profond que les grands océans, moins profond aussi que les bassins inactifs.

<u>Stade 3</u>. Arrêt de l'accrétion et passage au stade bassin marginal inactif à flux de chaleur élevé puis à flux normal, ce bassin devenant plus profond que le bassin actif.

Le début du comblement par les apports terrigènes internes (au sens géologique) peut être la conséquence d'un processus d'activation tectonique des sources (orogénèse ? tectogenèse ?). On peut y voir le début <u>de phénomène de blocage</u> du système "ouverture arrière arc-subduction sous l'arc" (c'est-à-dire du début d'un processus de collision), pouvant, dans certaines conditions, conduire au débitage et à l'empilement tectonique d'écailles dans le matériel compris entre le continent et l'arc insulaire.

Ainsi le domaine marginal s'organise en partant du continent asiatique et en allant vers l'océan pacifique selon le schéma suivant :

- sur le parallèle du Japon : continent asiatique; marge continentale stable (de type atlantique) avec sillons et hauts fonds; domaine à croûte océanique (subdivisé ou non par des archipels); arc insulaire à croûte continentale; fosse avec subduction; océan;
- sur le parallèle des Philippines : continent; marge stable; domaine océanique; fosse avec subduction; arc insulaire à croûte continentale.

Il est important de noter qu'entre le bassin océanique et l'arc insulaire peut s'intercaler une fosse : c'est le cas de la fosse de Manille entre le bassin sud chinois et les Philippines.

Ainsi, le système est dissymétrique de part et d'autre du bassin marginal océanique : on a, d'un côté une marge continentale passive, et de l'autre un domaine de type arc insulaire avec fonctionnement d'un plan de Benioff. La notion de marge continentale active recouvre souvent, en fait, tout l'espace compris entre le continent eurasiatique et le plan de Benioff ouest Pacifique, par opposition aux marges continentales passives de type Atlantique.

Il est à noter que le caractère actif d'une marge active est liée à l'existence et au fonctionnement d'un plan de Benioff; le prototype en étant la marge occidentale de l'Amérique du Sud. Dans cette optique, le profil ouest Pacifique montre l'existence d'un domaine original, celui des bassins marginaux entre une marge inactive de type atlantique (marge asiatique) et une marge active de type Japon ou de type Philippines.

Les bassins marginaux sont ainsi des entités qui peuvent être "actives" pour leur propre compte, lorsqu'elles sont le siège d'accrétion.

B. Le secteur dinarique.

L'organisation et l'évolution dans le temps du secteur dinarique peuvent être mises en parallèle avec celles des bassins marginaux de l'ouest Pacifique.

On peut voir dans l'histoire triasique inférieure des Dinarides, le stade 1, dont j'ai parlé ci-dessus (rifting). Depuis J. Dercourt (1970), les auteurs sont d'accord sur ce point. Des précisions récentes concernant la nature et la répartition du volcanisme triasique des Dinarides soulignent encore les ressemblances (Bébien et al., 1975, 1976).

Le Trias supérieur-Jurassique correspondrait au stade 2 (bassin marginal actif), avec création d'une croûte océanique. C'est aussi l'époque du dépôt des séries pélagiques et des formations radiolaritiques si typiques des chaînes alpines. Or, dans la mer du Japon, dans la mer des Philippines, comme dans le bassin de Lau, j'ai insisté sur l'existence d'une série pélagique siliceuse biogène litée au-dessus des séries volcaniques ([1]), et j'ai discuté plus haut, de la position de la CCD dans les bassins marginaux.

La crise majeure du Jurassique supérieur-Eocrétacé dans les Dinarides pourrait correspondre au développement du stade 3, dont le début pourrait s'observer en mer du Japon par exemple.

C. Conclusion.

On peut ainsi comparer et mettre en parallèle :

- l'organisation et l'évolution de bassins marginaux de l'ouest Pacifique de l'Oligocène supérieur à l'actuel;
- l'organisation et l'évolution du domaine dinarique à partir du Trias et dans certaines de ses étapes.

Il faut insister sur le fait que le paléo-océan dinarique s'installe au sein d'un pays né de l'orogenèse hercynienne. De la même manière, les bassins marginaux de l'ouest Pacifique s'installent à partir du Paléogène supérieur sur un édifice structural né des orogenèses du Mésozoïque et du Paléogène inférieur.

([1]) Remarque : Dans les Dinarides, il existe deux séries siliceuses associées à un volcanisme. L'une est Triasique supérieur (porphyrite, radiolarites), l'autre est Jurassique (diabase, radiolarite). On doit donc étudier en détail ces épisodes volcaniques, pour faire des comparaisons valables. Des précisions sur la nature du volcanisme triasique des Dinarides ont été récemment apportées (J. Bébien et al., 1975, 1976) soulignant sa parenté avec un volcanisme de marge active. Les séries siliceuses de la mer du Japon seraient-elles comparables aux radiolarites triasiques ? ou bien auraient-elles déjà la signification des radiolarites jurassiques ?

Il est fondamental de bien mettre en évidence ces superpositions d'édifices dans le temps, avec leurs paléogéographies et leurs tectoniques propres; c'est la seule bonne manière de placer des limites aux tranches de temps, afin de comparer des éléments qui soit effectivement comparables. Trop de comparaisons entre les édifices anciens et les modèles actuels, mettent en jeu des termes appartenant à des édifices, à des histoires géologiques différents.

Un exemple concernant l'ouest Pacifique : le bassin marginal du Japon est un système né à la fin du Paléogène et continuant à évoluer à l'actuel. La polarité générale (au sens géologique) est de l'est vers l'ouest avec des zones «internes» à l'est, et «externes» à l'ouest (des apports terrigènes ont leur source à l'est; les résultats d'une tectonique pleistocène en compression sont connues sur la côte ouest de l'archipel japonnais). Il est clair que ce système est superposé à une chaîne née au cours du secondaire, achevée à la fin de l'Eocène, et dont la vergence n'était pas nécessairement de même polarité. Il n'est donc pas correct d'attribuer des faits appartenants à l'édifice ancien à des causes qui agissent dans le système actuel post-Oligocène.

La même superposition d'édifices tectoniques est claire au nord de la mer du Japon (mer d'Okotsk par rapport aux chaînes du nord-est de l'URSS) comme au sud (bassin sud chinois, par rapport aux chaînes des Philippines) mais demande à être structuralement débrouillée.

V. L'ORGANISATION ET L'EVOLUTION DANS L'ESPACE

En considérant des surfaces-temps données, on peut décrire la conformation dans l'espace du bassin marin (ou du paléo-bassin) considéré.

A. S'agissant des domaines marins actuels.

1) Le modèle fournit par la Californie (Etats-Unis-Mexique) montre un processus d'ouverture d'un domaine océanique en marge d'un continent.

La morphologie actuelle montre, le long d'un profil au Mexique, orienté est-ouest : un continent et sa marge inactive, un étroit domaine à fond océanique en expansion depuis le Pliocène : le golfe de Californie, une péninsule continentale bordée à l'ouest par une marge présentement inactive : la péninsule de basse Californie, un océan : le Pacifique.

Dans cet exemple, on retiendra l'existence claire de terminaison, de bouclages de zones en doigt de gant. La péninsule de Californie se boucle au sud où le domaine à croûte océanique devient alors unique. Le bassin océanique du golfe de Californie, marginal par rapport au continent américain, se boucle au nord où le domaine continental devient unique.

2) L'Amérique centrale voit le développement d'un rift volcano-tectonique où des formations ignimbritiques ont précédé le puissant volcanisme andésitique quaternaire.

On note que ce rift occupe, par rapport au continent et à la fosse d'Amérique centrale, la position d'un bassin marginal actif qui, dans ce cas, montrerait la première étape de la genèse.

3) On connaît de plus des exemples où s'observe dans l'espace le passage du rift au bassin marginal actif (D. Karig, 1971) : dans le sud Pacifique, le rift de Tampo (Nouvelle-Zélande) dépression tectonique avec ignimbrites puis andésites, passe longitudinalement au bassin marginal actif (inter arc bassin) de Tonga-Kermadec (sillon de Lau-Havre).

4) L'évolution dans l'espace de l'ensemble mer de Chine méridionale-Taïwan-archipel des Philippines, mérite une attention particulière :

a) Le bassin marginal sud chinois est du type inactif à flux de chaleur élevé (type mer du Japon). De l'ouest à l'est, un profil schématique montre :

- la marge continentale asiatique (marge passive) largement développée avec dans le détail une série de hauts fonds et de sillons;
- un bassin océanique dans la partie orientale (au moins) montre une structure crustale de type océanique avec une épaisseur réduite de la couche 3 et un Moho plus proche de la surface (de 2 à 3 km) que la normale.
  Les sédiments de ce bassin océanique disposés à plat sont accoustiquement transparents. Leur épaisseur de 500 m environ en moyenne est variable; mais elle augmente lorsqu'on approche de la pente continentale de la marge asiatique;
- une fosse, la fosse de Manille, placée au pied de l'archipel des Philippines (Luzon)

et à l'ouest de celui-ci; elle contient des sédiments (turbidites présumées, jusqu'à 2 000 m d'épaisseur, et sédiments transparents);

- une marge continentale étroite, montrant dans le détail :
  - . une pente bordant à l'est la fosse de Manille;
  - . une ride ayant un rôle clair de barrière pour les sédiments qu'elle piège entre elle et l'archipel;
  - . un sillon sédimentaire dit sillon W. Luzon de 55 km de large, 220 km de long; le remplissage sédimentaire en arrière de la ride est de 1 000 à 1 500 m de sédiments en alternance accoustiquement transparents et opaques;
  - . une pente raccorde ce sillon à un étroit plateau continental;
- l'île de Luzon, l'archipel des Philippines.

Le géologue de terrain retiendra, pour une comparaison avec les Dinarides, les faits suivants d'ouest en est :

- une marge continentale passive, avec des variations topographiques; un bassin océanique au large de cette marge; un archipel continental au pied duquel se développe une fosse;
- une dissymétrie du système telle que le domaine océanique soit bordé par une marge passive largement étalée du côté externe, et par une marge étroite avec des unités physiographiques de détail entre une fosse et l'archipel du côté interne;
- le dispositif avec une fosse et un plan de subduction penté à l'est, entre le bassin océanique et l'arc insulaire, remarquable dans cet exemple de bassin marginal. A l'est de Luzon, on passe directement, sans fosse à la mer des Philippines. Ce n'est que plus au sud qu'apparaît la fosse des Philippines.

On a donc un système actuel à polarité vers l'ouest entre Luzon et le continent asiatique. A ce propos, il convient de souligner une indépendance des fosses par rapport aux archipels :

- au niveau du Japon, une fosse borde à l'est le nord de l'archipel (fosse du Japon), puis vers le sud, la même fosse s'écarte très nettement de Shikoku et de Kyushu (Izu trench) pour se prolonger dans la fosse des Mariannes;
- au niveau des Philippines, l'archipel est bordé par une fosse tantôt sur son bord ouest (au nord, Luzon), tantôt sur son bord est (au sud, Mindanao).

b) Le dispositif présenté ci-dessus évolue longitudinalement vers le nord, et passe par exemple, au système de la transversale de Taïwan. La structure de l'île de Taïwan est le résultat d'une tectonique très récente. On peut retenir :

- l'empilement d'unités charriées à vergence ouest, débitées dans des formations néogènes;
- l'existence d'une unité de type "mélange" ophiolitique formant un étroit liseré sur la côte orientale de l'île.

Il serait donc extrêmement intéressant de savoir si l'unité dite ophiolitique de Taïwan est de même nature que les unités ophiolitiques des chaînes alpines et si cette unité appartient à un cycle antérieur au cycle actuel, ou si cette unité a une liaison avec le dispositif actuel.

c) Au sud et au sud-est du bassin sud chinois, l'évolution de la bordure Pacifique est complexe dans le détail. Il n'en sera pas question ici. On retiendra cependant l'existence d'une vaste unité "ophiolitique" (Papouasie-Nouvelle-Calédonie-Nouvelle-Zélande), ou "couronne ophiolitique péri-australienne" (J. Aubouin et al., 1976).

B. La comparaison de l'organisation spatiale des systèmes précédents avec l'organisation paléogéographique des Dinarides à différentes surfaces-temps, à partir du Trias.

Pour l'organisation paléo-océanique des Dinarides, on se reportera aux travaux nombreux depuis 1970 (J. Dercourt, 1970; D. Bernoulli et H. Laubscher, 1972; J. Dewey et al., 1973; J. Aubouin, 1973; R. Blanchet, 1973, 1974; J.P. Rampnoux, 1974; J.P. Cadet, 1976; B. Biju-Duval et al., 1976).

Quelques points méritent d'être soulignés dans la comparaison que j'ai proposée ci-dessus :

- l'échelle des deux termes comparés est du même ordre;
- l'organisation générale est comparable :

  marge continentale passive (précontinent asiatique);

. bassin croûte océanique;

. archipel interne, source du matériel terrigène;

. polarité de l'archipel vers la marge;

. dissymétrie dans les marges bordant le bassin océanique;

- à ce thème général relativement simple, s'ajoutent de grandes et rapides variations "géographiques" aussi bien transversalement que longitudinalement :

. système de rides et de sillons dans les marges;

. présence de petits bassins à croûte océanique entre des rides sialiques (cf. mer du Japon);

. bouclage des zones physiographiques et relais de zones entre elles.

Cette complexité des détails d'organisation est le propre des bassins marginaux et des petits bassins océaniques (actuels), comme des paléo-océans reconstitués par les géologues dans les Dinarides.

VI. BASSIN MARGINAL, PETIT BASSIN OCEANIQUE ET OCEAN

Dans la comparaison développée ci-dessus le domaine océanique duquel seraient issues les Dinarides est comparable, par son échelle, sa position, sa logique sédimentaire, ses propriétés crustales, à un bassin marginal ou à un petit bassin océanique plutôt qu'à un grand océan de type Atlantique.

Les modèles retenus suivant le stade du développement de l'édifice dinarique sont géographiquement situés sur la bordure occidentale du Pacifique.

Plusieurs auteurs envisageant les dispositifs ouest Pacifique, ont recherché à placer dans leurs reconstitutions des Dinarides un grand océan au large du bassin marginal, soit à l'ouest (Pinde), soit à l'est (océan péonien) de celui-ci (J. Mercier, P. Vergely, 1975).

1) Une première remarque concerne les ophiolites dinariques : sont-elles des témoins du fond océanique d'un bassin marginal ? Des présomptions de divers ordres exposées dans les pages précédentes sont en faveur de cette hypothèse. Cependant on doit se demander et chercher à savoir si les ophiolites dinariques sont homogènes avec une origine unique ou si, au contraire, certains massifs sont des témoins de bassins marginaux, d'autres d'océans de type Pacifique ou Atlantique. Il est clair que la géochimie appliquée parallèlement aux ophiolites et aux fonds océaniques est l'outil pour une éventuelle possibilité de distinction.

2) Il faut noter en second lieu que les bassins marginaux de l'ouest Pacifique sont fondamentalement en marge d'un continent, comme en témoignent leur origine et leur genèse et qu'ils sont (toutes proportions gardées) relativement indépendants de l'énorme machine océanique pacifique.

3) Il faut ajouter qu'en bien des régions, l'océan Pacifique apparaît entièrement indépendant de ses marges dont il recoupe souvent les structures même récentes (paléogènes) comme sur la côte ouest du Mexique (M. Tardy, 1974) en Californie, au Japon, dans les Philippines.

4) Un bassin marginal ouest Pacifique présente un domaine océanique compris entre deux domaines continentaux :

- l'un occidental de grandes dimensions, bordé par une marge de type atlantique;
- l'autre oriental dont l'importance dimensionnelle est souvent mésestimée du fait de sa morphologie en archipel.

Si l'on prend l'exemple de l'ouest Pacifique, il ne faut pas perdre de vue, qu'outre les grandes îles de l'archipel japonnais, une croûte continentale (gneiss, schistes, granite, calcaires éocènes à Nummulites) semble exister dans tout le secteur nord de la mer des Philippines (Daito Ridge) et jusqu'à Okinawa.

Au total, la domaine océanique mésozoïque duquel sont issues les Dinarides peut être comparé à un domaine de type bassin marginal avec fond océanique développé au large de la marge du craton apulien entre celle-ci et le craton européen (ou un "appendice", ou un archipel de celui-ci, le craton rhodopien, R. Blanchet, 1973-1974).

La nature, l'origine, l'évolution de ce paléodomaine océanique permettent de comparer certains stades aux bassins marginaux de l'ouest Pacifique, développés en bordure du continent asiatique. Sa position entre deux domaines cratoniques, évoque aussi celles de la mer des Caraïbes, ou de la mer Méditerranée. Or on sait que ces "petits bassins océaniques" ont des caractères de stades d'ouverture avec possibilité de croûte océanique en création, comme dans les bassins marginaux. Sa configuration de domaine à croûte océanique souvent "en coin" entre des continents, peut évoquer aussi certains domaines comme le golfe de Gascogne.

Ainsi, on peut regarder le domaine téthysien à paléo-croûte océanique, comme un unique bassin océanique de petite taille, ayant des propriétés représentées aujourd'hui dans certains bassins marginaux, bref comme un petit bassin océanique entre Afrique et Europe, pouvant s'ouvrir à l'est sur un océan plus important, à certaines périodes.

CONCLUSION : ESQUISSE D'UNE RECONSTITUTION PALEO-OCEANOGRAPHIQUE

Les Dinarides peuvent se résoudre d'un point de vue paléo-océanographique en considérant la période Trias supérieur-Jurassique, en :

- une paléomarge continentale à l'ouest : de type "atlantique", elle montre une suite de hauts fonds et de sillons, dont le nombre peut varier selon la transversale considérée. Ce dispositif est vraisemblablement guidé pendant le Jurassique par des failles en extension (zone ionienne en Grèce (*IFP-IGG*, 1966) en Albanie (A. Papa, 1970));
- un paléodomaine à croûte océanique, situé à l'est du précédent; les Ophiolites en seraient le témoin actuellement visible. Partout charriées et souvent très précocement (dès le Jurassique supérieur), les Ophiolites dinariques et leur genèse sont un élément fondamental de toute reconstitution paléo-océanographique. Leur mise en place tectonique peut s'expliquer par un rabotage de croûte océanique par suite d'un blocage dans un système en subduction;
- un paléodomaine continental, archipel (ou péninsule du craton européen ?), source du matériel détritique terrigène (R. Blanchet, 1973).

Ces trois éléments majeurs peuvent être comparés, respectivement :

- à la marge continentale asiatique;
- au domaine océanique d'un bassin marginal ouest Pacifique (mer du Japon, mer de Chine méridionale);
- au domaine continental bordant ces bassins marginaux (archipel du Japon, des Philippines).

Les bassins marginaux actuels peuvent ainsi, dans l'état actuel des connaissances, servir de modèles d'organisation et d'évolution pour certaines étapes de l'histoire du domaine à croûte océanique Téthysien qui, dans son stade d'état devait représenter un petit bassin océanique plutôt qu'un océan à l'échelle Atlantique ou Pacifique. Pour aller plus avant dans la comparaison, il est nécessaire de connaître de façon beaucoup plus précise, l'histoire sédimentaire et l'évolution structurale des bassins marginaux et des petits bassins océaniques dépendants ou non d'un des grands océans actuels.

Il convient en effet de déterminer l'âge et la nature des séries, l'âge des tectoniques, l'âge et la nature de la croûte, pour fixer aux reconstitutions et aux recherches de modèles, un cadre bâti sur des faits. Le programme IPOD, qui portera en 1977-1978 sur les marges actives, devrait permettre de faire un pas important et décisif dans cette voie.

Nous tenons à remercier les organismes qui ont permis la poursuite de nos travaux et toutes les personnes qui nous ont aidé :

- Le *Centre National de la Recherche Scientifique* et l'*Institut National d'Astronomie et de Géophysique* (ATP Géodynamique, groupe sédimentologie) qui ont subventionné un stage de post doctoral fellow à la *Scripps Institution of Oceanography, Université de Californie* à San Diego.
- Les responsables et chefs d'équipe du Deep Sea Drilling Project et la *National Science Foundation*, qui ont mis à ma disposition l'ensemble des carottes de l'océan Pacifique et des documents correspondants.
- Le Professeur E.L. Winterer, *Université de Californie* à San Diego, qui m'a accueilli à la *Scripps* dans son service, et avec lequel j'eus des discussions très stimulantes, bénéficiant ainsi de sa grande expérience des sédiments et de la structure du Pacifique.
- Les Professeurs W.H. Berger, M. Bramlette, J. Curray, M. Kastner, W.R. Riedel, de la division de la recherche géologique à la

Scripps avec lesquels j'eus des échanges très fructueux. Le Docteur A. Bosellini, de l'*Université de Ferrara* avec lequel j'eus de fréquentes discussions sur les Alpes méridionales.

- Messieurs J. Aubouin, B. Biju-Duval, J. Dercourt, L. Montadert, X. Le Pichon, E.L. Winterer, dont les critiques et les conseils après présentation orale et sur lecture du manuscrit, ont contribué largement à la mise au point de cet article.

ORIENTATION BIBLIOGRAPHIQUE

Il est impossible, dans la place impartie, de citer tous les travaux qui ont servi à l'élaboration de ce texte. L'orientation bibliographique proposée retient pour les divers objets les publications récentes et synthétiques. Chacune comporte une importante bibliographie auquel le lecteur pourra se reporter.

[1] ALLEGRE (C.J.), MONTIGNY (R.), BOTTINGA (Y.).- (1973). Cortège ophiolitique et cortège océanique, géochimie comparée et mode de genèse. Bull. Soc. géol. Fr. (7), XV, p. 461-477.

[2] ANDREWS (J.E.), PACKAM (G.) et al.- Initial Reports of the Deep Sea Drilling Project, vol. 30, Washington, US Gov. Printing Office.

[3] ASANO (Sh.), UDINTSEV (G.B.).- (1971). Island Arc and Marginal Sea. Proceed. 1st Symp. on Solid Earth Sc., Japan-USSR, Tokai University Press, 319 p.

[4] AUBOUIN (J.).- (1973). Des tectoniques superposées et de leur signification par rapport aux modèles géophysiques : l'exemple des Dinarides; paléotectonique, tectonique, tarditectonique, néotectonique. Bull. Soc. Géol. Fr. (7), XV, pp. 426-460.

[5] AUBOUIN (J.), BLANCHET (R.), CADET (J.P.), CELET (P.), CHARVET (J.), CHOROWICZ (J.), COUSIN (M.), RAMPNOUX (J.P.).- (1970). Essai sur la géologie des Dinarides. Bull. Soc. Géol. Fr. (7), XII, pp. 941-1129.

[6] BEBIEN (J.), BLANCHET (R.), CADET (J.P.), CHARVET (J.), CHOROWICZ (J.), LAPIERRE (H.), RAMPNOUX (J.P.).- (1975). La série "Porphyrite Radiolarite" élément de reconstitution du cadre géotectonique des Dinarides au Trias. 3ème réunion annuelle des Sci de la Terre, Montpellier, p. 31.

[7] BELOUSSOV (V.V.).- (1968). Some problems of Development of the Earth Crust and upper Mantle of Oceans. In : The Crust and upper Mantle of the Pacific Area. Am. Geoph. Union. Mon., 12, pp. 449-459.

[8] BERGER (W.H.) et WINTERER (E.L.).- (1974). Plate Stratigraphy and the fluctuating Carbonate Line. Spec. Pub. Int. Ass. Sed., 1, pp. 11-48.

[9] BERNOULLI (D.) et LAUBSCHER (H.).- (1972). The Palinspatic Problem of the Hellenides. Eclog. Geol. Helv., 65, (1), pp. 107-118.

[10] BIJU-DUVAL (B.), DERCOURT (J.), LE PICHON (X.).- (1976). La genèse de la Méditerranée, La Recherche, n° 71, pp. 811-822.

[11] BIQ (C.C.).- (1972). Dual Trench Structure in the Taiwan-Luzon Region. Geol. Soc. China Proc., 15, pp. 65-75.

[12] BLANCHET (R.).- (1973). Contribution à l'étude géologique des Dinarides. Un profil en Yougoslavie : de la Dalmatie au Bassin pannonique. Thèse Doct. Etat, Paris, 2 vol., 727 p., 1 carte geol. 1/200 000, Soc. Géol. France.

[13] BLANCHET (R.).- (1974). De l'Adriatique au bassin pannonique. Essai d'un modèle de chaîne alpine. Mémoire Soc. Géol. France, n° 120, pp. 1-172.

[14] BLANCHET (R.).- (1976). Essai d'un modèle pour la mésogée alpine : Bassins marginaux de l'ouest Pacifique et chaîne alpine des Dinarides. Résum. In : Colloque final A.T.P. Géodynamique de la Méditerranée, CNRS/INAG, Montpellier, p. 43.

[15] BOUMA (A.H.).- (1975). Sedimentary Structures of Philippine Sea and Sea of Japan. DSDP, Leg 31. In : Initial Reports of the DSDP, 31, US Gov. Printing Office, pp. 489-498.

[16] BRAMLETTE (M.N.).- (1961). Pelagic sediments. Ocean. Amer. Ass. for Adv. of Sci., pp. 345-366.

[17] BURK (C.A.) et DRAKE (C.L.).- (1974). The Geology of Continental Margins. Springer-Verlag, 1009 p.

[18] CADET (J.P.).- (1976). Contribution à l'étude géologique des Dinarides : les confins de la Bosnie-Herzégovine et du Monténégro. Essai sur l'évolution alpine d'une palécmarge continentale. Thèse Doct. Etat, Paris, 4 vol., 450 p.

[19] CHAI (B.H.T.).- (1972). Structure and Tectonic Evolution of Taiwan. Am. J. of Sci., 271, pp. 389-422.

[20] CHARVET (J.).- (1973). Sur les mouvements orogéniques du Jurassique-Crétacé dans les Dinarides de Bosnie orientale. C.-Rendu Acad. Sc., 276 (D), pp. 257-259.

[21] CHOROWICZ (J.) et GEYSSANT (J.).- (1976). La paléofaille transformante Split-Karlovac-Vienne; Introduction à un modèle d'évolution des chaînes alpines de l'Europe moyenne. Rev. Géol. Dyn. et Géographie Physique, XVIII, fasc. 1, pp. 127-142.

[22] COLEMAN (P.J.).- (1973). The Western Pacific. University of Western Australia Press, 675 p.

[23] DERCOURT (J.).- (1970). L'expansion océanique actuelle et fossile : ses implications géotectoniques. Bull. Soc. Géol. France, (7), XII, Paris, pp. 261-317.

[24] DERCOURT (J.).- (1972). The Canadian Cordillera, The Hellenides and the Sea-floor Spreading Theory. Canad. J. Earth. Sci., 9 (6), pp. 709-743.

[25] DEWEY (J.F.) et BIRD (J.M.).- (1970). Mountain Belts and the new global Tectonics. J. Geophys. Res., 75, pp. 2625-2647.

[26] DEWEY (J.F.), PITMAN (W.C.), RYAN (W.B.F.) and BONNIN (J.).- (1973). Plate Tectonics and the Evolution of the Alpine System. Bull. Geol. Soc. Amer., 84, pp. 3137-3180.

[27] DICKINSON (W.R.).- (1974). Sedimentation within and beside ancient and modern magmatic Arcs. SEPM, special publ., n° 22, pp. 1-27.

[28] DIMITRIJEVIĆ (M.D.) et DIMITRIJEVIC (M.N.).- (1973). Olistostrome Melange in the Yugoslavian Dinarides and late Mesozoic Plate Tectonics. Journ. Geol., 81, pp. 328-340.

[29] EMERY (K.O.), BEN-AVRAHAM (Z.).- (1972). Structure and Stratigraphy of China Sea. AAPG Bull., 56, n° 5, pp. 839-859.

[30] ENAY (R.).- (1972). Paléobiogéographie des Ammonites du Jurassique terminal (Tithonique/Volgien/Postlandien s.l.) et mobilité continentale. Geobios, 5, pp. 355-407.

[31] GARRISON (R.E.), ROWLAND (S.M.), HORAN (L.J.), MOORE (J.C.).- (1975). Petrology of siliceous Rocks recovered from marginal Seas of the Western Pacific, Leg 31, DSDP. In : Init. Rep., vol. 31, pp. 519-530.

[32] HAWKINS (J.W.) Jr.- (1974). Geology of the Lau Basin : A marginal sea behind the Tonga Arc. In : C.A. Burk et C.L. Drake, 1974, pp. 505-520.

[33] HAYES (D.E.), LUDWIG (W.J.).- (1967). The Manila Trench and West Luzon Trough. Gravity and Magnetics Measurment. Deep Sea Res., 14, pp. 545-560.

[34] HO (C.S.).- (1967). Structural Evolution of Taiwan. Tectonophysics, 4, pp. 367-378.

[35] IGRA-IFP.- (1966). Etude géologique de l'Epire (Grèce nord occidentale). Paris, Editions Technip, 306 p.

[36] ISEZAKI (N.).- (1975). Possible spreading Centers in the Japan Sea. Mar. Geophys. res. Netherl., 2, n° 3, pp. 265-277.

[37] ISEZAKI (N.), UYEDA (S.).- (1973). Geomagnetic anomaly Pattern of the Japan Sea. Marine Geoph. Res., 2, pp. 51-59.

[38] KARIG (D.E.).- (1973). Plate Convergence between the Philippines and the Ryukyw Islands. Marine Geol., 14, pp. 153-168.

[39] KARIG (D.E.).- (1974). Evolution of Arc Systems in the Western Pacific. Ann. Rev. Earth Planet. Sci., 2, (1974), pp. 51-75.

[40] KARIG (D.E.).- (1975). Basin Genesis in the Philippine Sea. Initial Rep. of DSDP, US Gov. Printing Office, vol. 31, pp. 857-880.

[41] KARIG (D.E.), MOORE (G.F.).- (1975). Tectonically controlled Sedimentation in marginal Basins. Earth and Planet. Sci. Letters, 26, pp. 233-238.

[42] KARIG (D.E.), INGLE (J.C.) Jr. et al.- (1975). Initial Reports of the Deep Sea Drilling Project, vol. 31, Washington, US Gov. Printing Office, 927 p.

[43] KASENO (Y.).- (1971). Geological Features of the Japan Sea Floor : a Review of recent Studies. Pacific Geology, n° 4, pp. 91-111.

[44] KASENO (Y.).- (1972). On the Origin of the Japan Sea Basin. 24th Int. Geol. Congr., 8, pp. 37-42.

[45] LE PICHON (X.), FRANCHETEAU (J.), BONNIN (J.).- (1973). Plate Tectonics. Elsevier Pub., 300 p.

[46] LUDWIG (W.J.).- (1970). The Manila Trench and West Luzen Trough; Seismic Refraction Measurments. Deep Sea Res., 17, pp. 553-571.

[47] LUDWIG (W.J.), HAYES (D.E.), EWING (J.I.).- (1967). The Manila Trench and West Luzon Trough; Bathymetry and Sediment Distribution. Deep Sea Res., 14, pp. 533-544.

[48] LUDWIG (W.J.), MURAUCHI (S.), HOUTZ (R.E.).- (1975). Sediments and Structure of the Japan Sea. Bull. Geol. Soc. Am., vol. 86, pp. 651-664.

[49] MERCIER (J.L.), VERGELY (P.), BEBIEN (J.).- (1975). Les Ophiolites helléniques "obductées" au Jurassique supérieur sont-elles les vestiges d'un océan Téthysien ou d'une mer marginale péri-européenne ? C.R. Somm. Soc. Geol. France, 4, pp. 108-111.

[50] MOORES (G.M.).- (1969). Petrology and Structure of the Vourinos Ophiolitic Complex of Northern Greece. Bull. Geol. Soc. Am., 80, pp. 3-74.

[51] PAPA (A.).- (1970). Conceptions nouvelles sur la structure des Albanides (présentation de la carte tectonique de l'Albanie). Bull. Soc. Géol. Fr. (7), XII, Paris, pp. 1096-1109.

[52] RAMPNOUX (J.P.).- (1974). Contribution à l'étude géologique des Dinarides : un secteur de la Serbie méridionale et du Monténégro oriental (Yougoslavie). Mémoire Soc. Geol. France, n° 119, pp. 1-100.

[53] SKORNYAKOVA (N.S.).- (1961). Bottom Sediments of the Japan Sea. In : Stepanov P.N. (Ed.), Inst. Okeanol., Akad. Sci., USSR, pp. 23-34.

[54] SUGIMURA (A.) et UYEDA (S.).- (1973). Island Arcs; Japan and its Environs. Dev. in Geotectonics, 3, Elsevier, Ed., 247 p.

[55] TARDY (M.).- (1974). Les phases tectoniques du secteur transverse de Parras, Sierra Madre Orientale (Mexique). Bull. Soc. Géol. France, (7), XV, n° 3-4, pp. 362-366.

[56] VASILKOVSKY (N.P.), UDINTSEV (G.B.), KULP (B.G.), MOURAVOVA (E.A.).- (1971). Japan Sea, Relict. of Ocean. In : Island Arc and marginal Sea, Tokai Univ. Press, pp. 58-64.

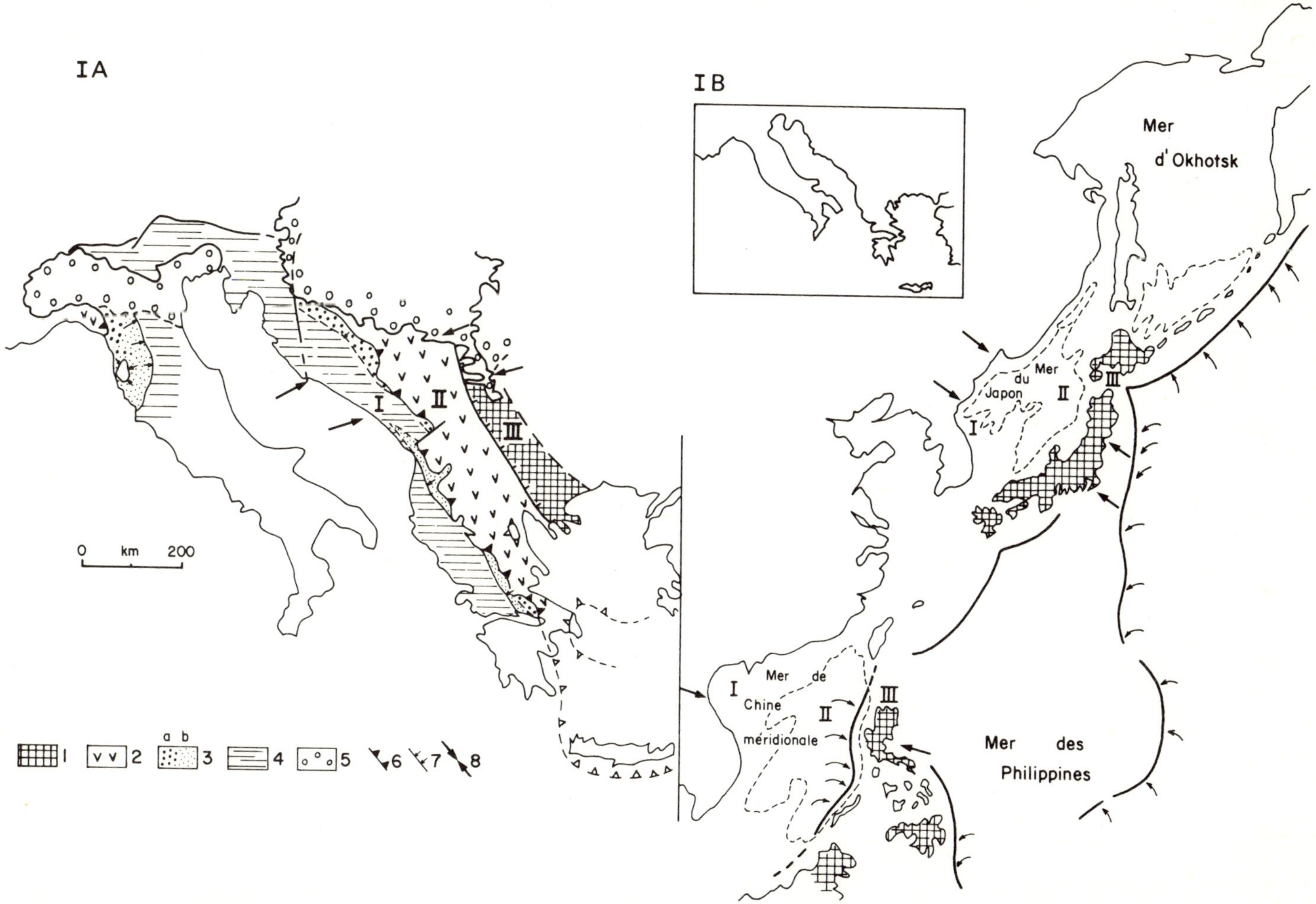

Fig. I.- Schéma général de présentation sur lequel les grands ensembles comparés sont mis en place, dans leur cadre structural.

IA.- Le secteur dinarique des chaînes alpines méditerranéennes. I : domaine externe (bordure du craton apulien) ; II : domaine interne, ophiolitique, largement allochtone (témoins paléo-océaniques). Entre I et II : nappes de flysch (Tithonique-Crétacé, Crétacé supérieur-Paléogène). 1. Croûte continentale du massif serbo-macédonien ; 2 : nappes internes, pour parties ophiolitiques ; 3 : a - nappe bosniaque (flysch tithonique-crétacé) ; b - nappe du Pinde (flysch Crétacé supérieur-Paléogène) ; 4 : bordure du craton apulien et paléomarge continentale ; 5 : bassins molassiques ; 6 : front de charriage des nappes internes ; 7 : front de charriage des nappes de flyschs ; 8 : situation des profils utilisés préférentiellement dans le texte.

IB.- Les marges actives sur la bordure Ouest Pacifique : on peut distinguer sur les sections, la suite : bordure continentale asiatique (I), bassin marginal (petit bassin océanique, bassin arrière arc) (II), arc insulaire (archipel) (III), fosse. Mais on notera la position de la fosse de Manille, entre la mer de Chine méridionale et l'archipel des Philippines avec subduction vers l'Est. De plus, le dispositif à l'Est des archipels est variable ; on a soit le Pacifique, soit l'une des unités de la mer des Philippines, soit le système complexe de la mer de Sulu, mer des Célèbes.

Enfin, le carton représente à même échelle, le secteur alpin présenté en IA.

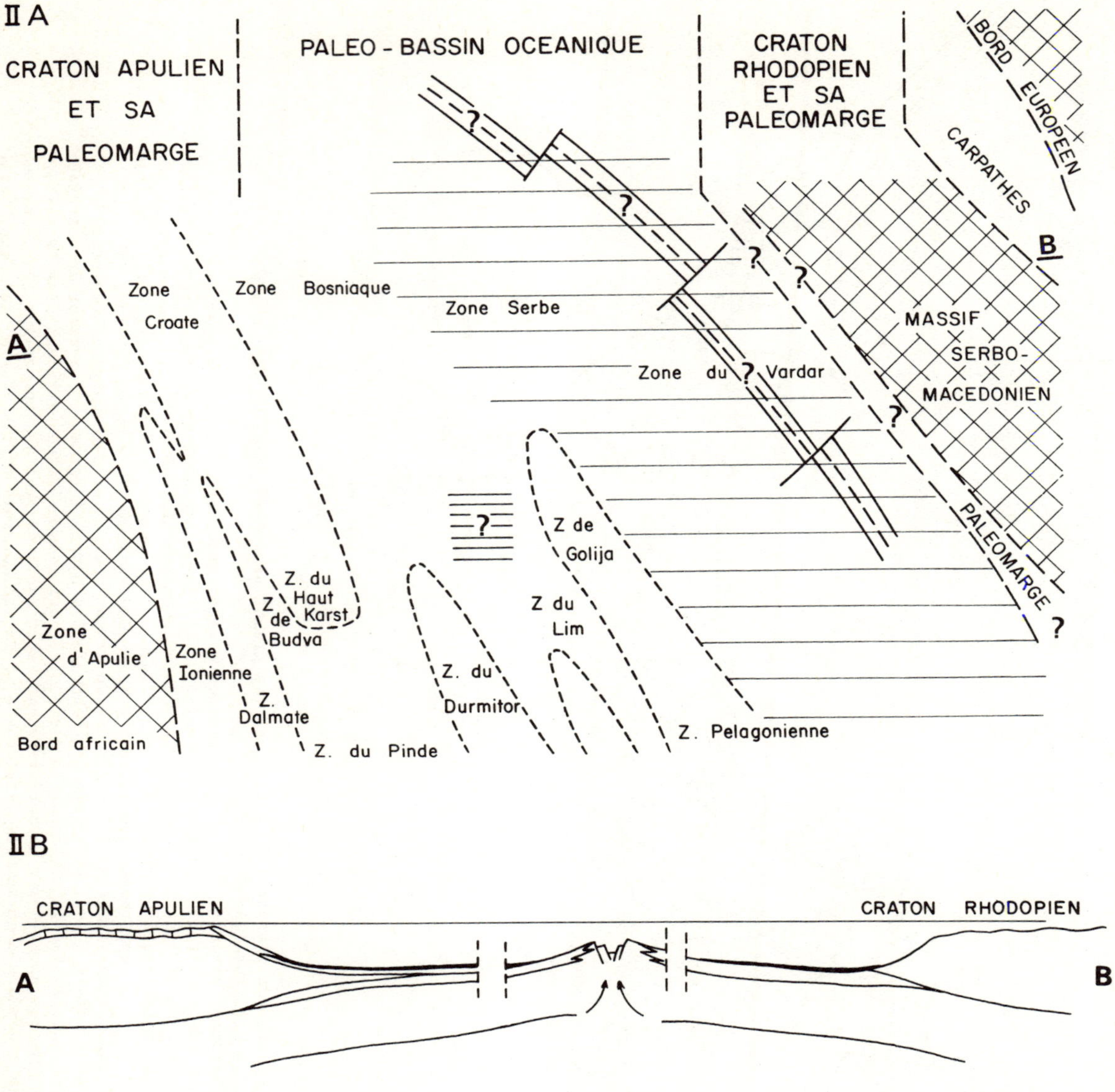

Fig. II.-

IIA.- Esquisse cartographique du dispositif paléo-océanique des Dinarides au Jurassique, reconstitué à partir des données de terrain (d'après les auteurs cités dans le texte). Le secteur considéré est encadré de quatre flèches sur la figure IA.
IIB.- Reconstitution très schématique selon le profil AB, mettant en place les grands ensembles (d'après R. Blanchet, 1974).

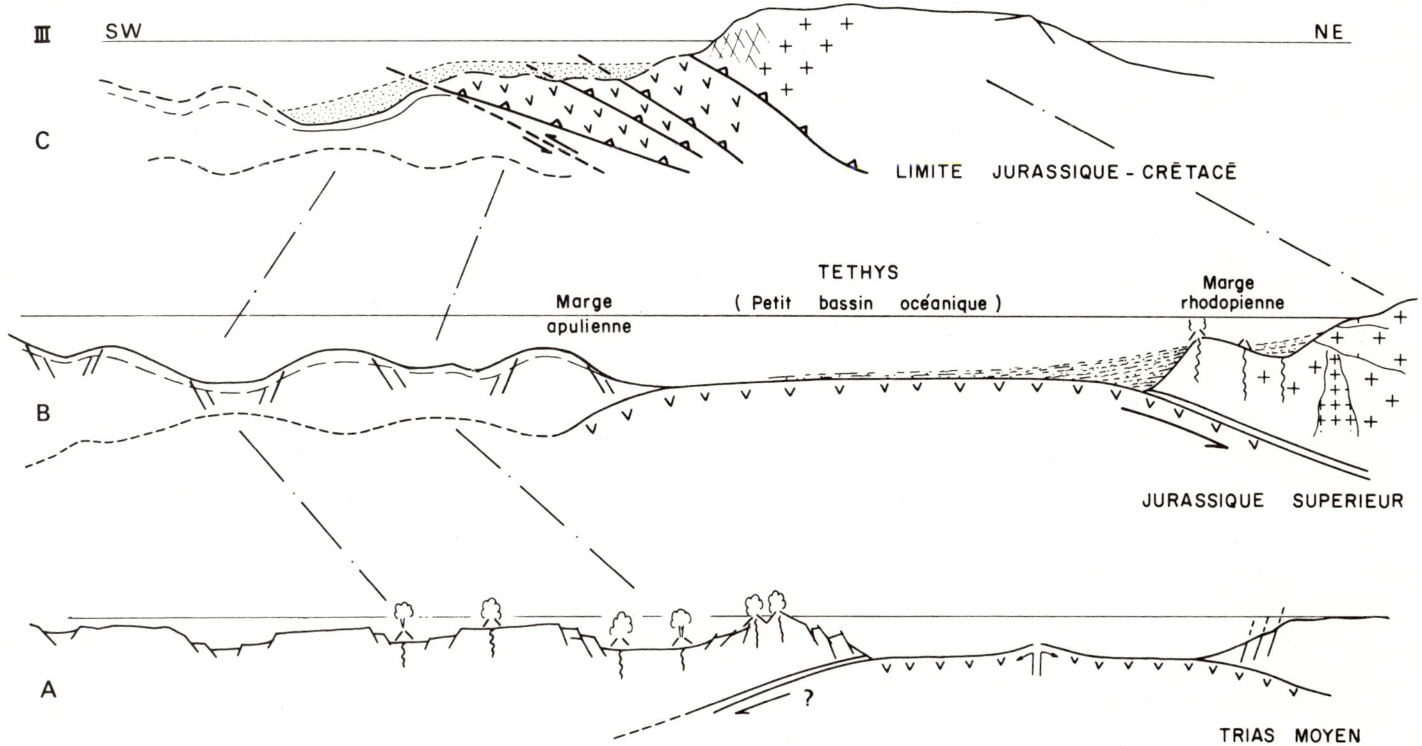

Fig. III.- Quelques grandes étapes de l'évolution paléo-océanique du secteur alpin considéré.
Trias moyen : période d'effondrement sur et en bordure du domaine apulien (africain). Le volcanisme aniso-ladinien des Dinarides pourrait s'interpréter comme conséquence de mouvements de convergence de deux plaques (Bébien, Blanchet et al., 1975 ; Cadet, 1976).
Jurassique supérieur : le domaine océanique, du type petit bassin océanique connait un début de subduction vers l'Est. Le dispositif général pourrait alors être comparé, par exemple, à celui de la mer de Chine méridionale avec, d'Ouest en Est : bordure continentale avec sillons et hauts fonds, petit bassin océanique, zone de subduction, archipel. Des sédiments volcano-détritiques recouvrent les séries siliceuses (formation volcano-terrigène ou "diabase-radiolarite").
Limite Jurassique-Crétacé : crise tectonique (paléotectonique) majeure ; la croûte océanique du petit bassin océanique de type bassin marginal est tectoniquement mise en place (charriée ; obductée) sur la bordure continentale apulienne.Plusieurs processus sont possibles, l'étape IIc ne représentant que le résultat. Les flyschs précoces à débris d'ophiolites commencent à se former. Dans cette optique, on peut réfléchir à la position structurale du complexe "ophiolitique" et volcano-sédimentaire à l'Est de Taïwan (cf. Fig. VI).

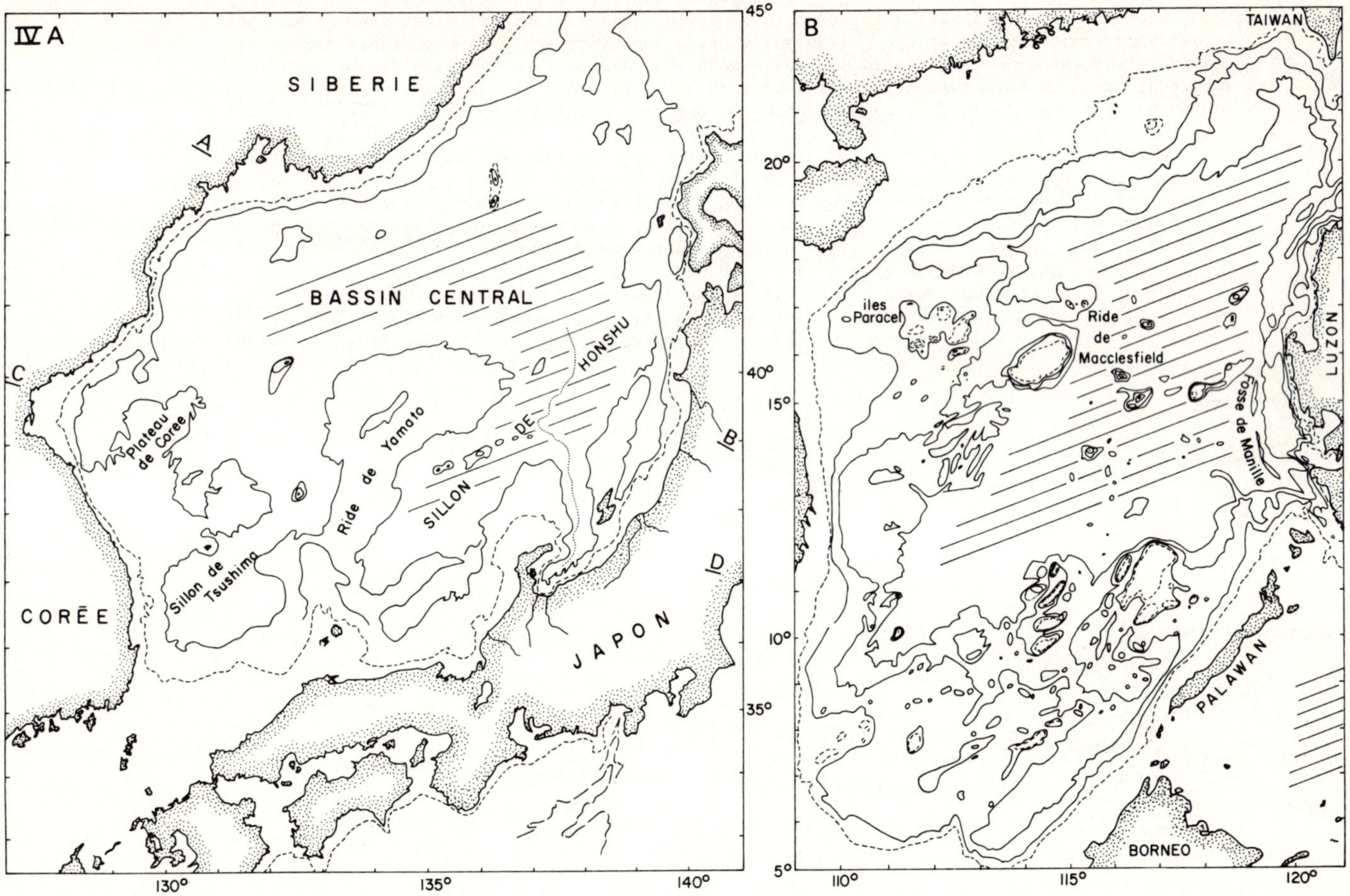

Fig. IV.-

IVA.- Esquisse de la mer du Japon : on notera la morphologie des fonds en "rides" et "sillons" et l'existence de croûte océanique-subocéanique (hachures) (d'après W.L. LUDWIG et al., 1975, complété).

IVB.- Esquisse de la mer de Chine méridionale mettant en évidence la variabilité de la morphologie des fonds, la localisation de la croûte océanique-subocéanique (hachures), et de la zone de subduction (fosse de Manille) (d'après Emery et Ben-Avraham, 1972, complété).

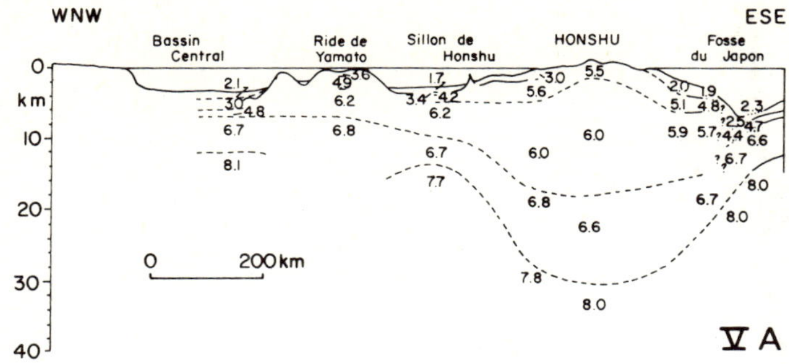

Fig. V.-

VA.- Coupe de la croûte dans le système mer-arc-fosse au niveau du Japon (d'après Murauchi et Yasui, 1968). Comparer à la figure IVA.

VB.- Coupe de la croûte dans la partie orientale de la mer de Chine méridionale (d'après W.J. Ludwig, 1970). Comparer aux figures IVB et IIIB.

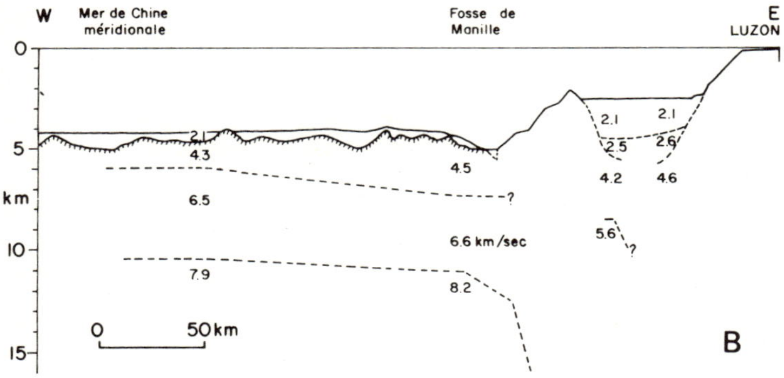

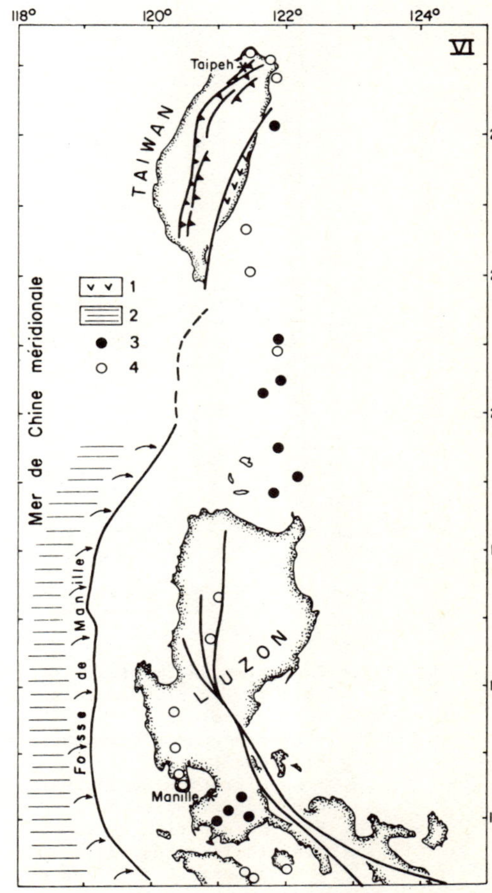

Fig. VI.- Schéma du dispositif structural dans le secteur Taïwan-Philippines-mer de Chine méridionale (d'après Karig, 1973, Chai, 1972, Ludwig, 1970). 1 : nappe volcano-sédimentaire et "ophiolitique" (?) de la chaîne côtière orientale de Taïwan ; 2 : croûte océanique de la mer de Chine méridionale ; 3 : volcanisme actuel ; 4 : volcanisme néogène à récent. On pourra comparer ce dispositif aux schémas B et C de la figure III et B de la figure V.

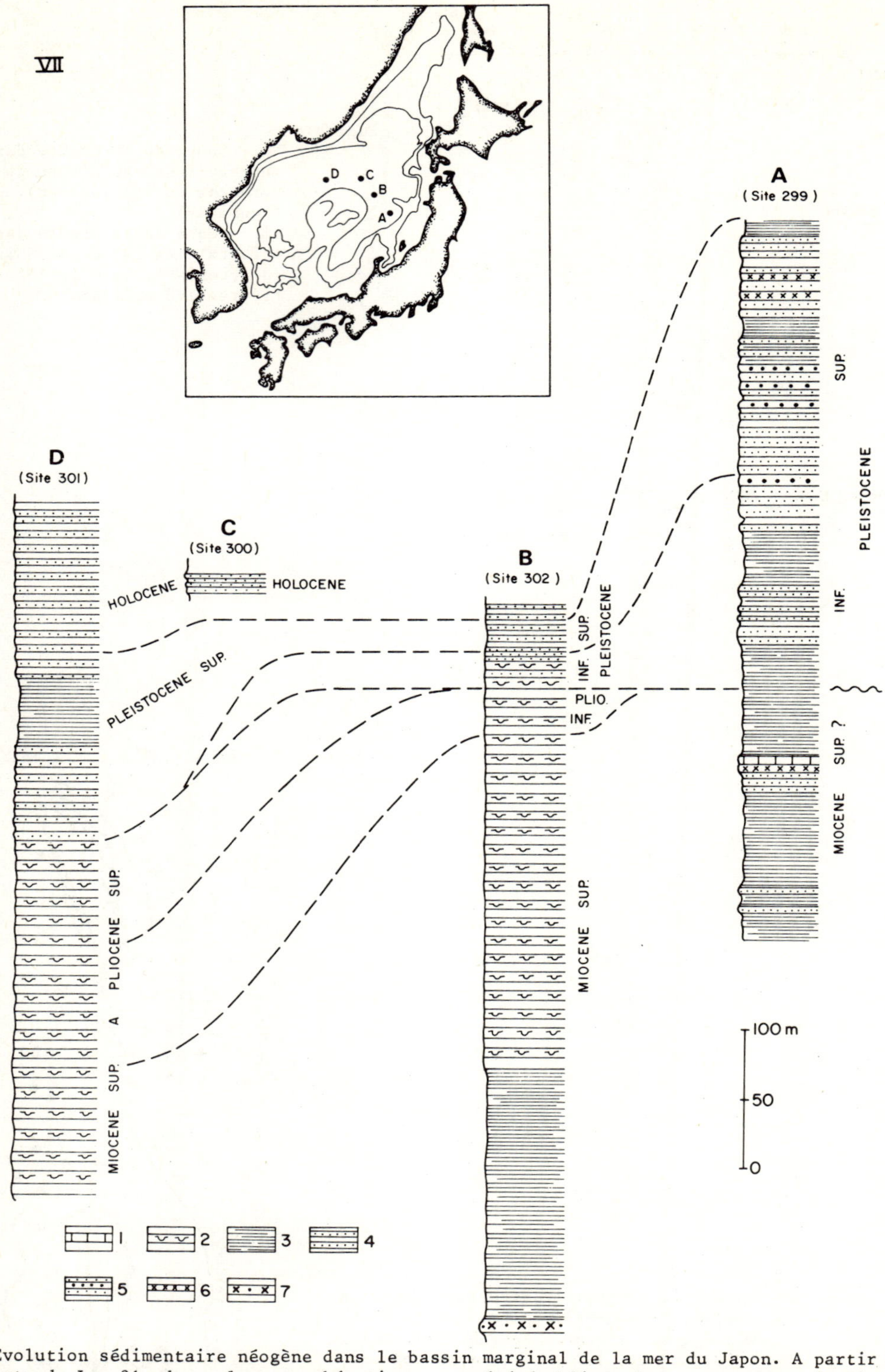

Fig. VII.- Evolution sédimentaire néogène dans le bassin marginal de la mer du Japon. A partir des résultats du Leg 31, des colonnes schématiques ont été établies. On notera en particulier (cf. texte) :
- La présence d'une séquence de boues pélagiques à diatomées et radiolaires, Miocène supérieur-Pliocène ;
- La migration de l'invasion terrigène dans l'espace et le temps (Miocène supérieur à Pleistocène supérieur).
- La distinction de turbidites proximales d'âge pleistocène sur le site 299 et de turbidites distales sur les sites 302 et 301.
1 : carbonates ; 2 : boues siliceuses à Diatomées Radiolaires ; 3 : argiles et argiles silteuses ; 4 : silts ; 5 : sables et sables silteux ; 6 : volcano-sédimentaire ; 7 : volcano-terrigène.

INTERNATIONAL SYMPOSIUM ON THE STRUCTURAL HISTORY OF THE MEDITERRANEAN BASINS. SPLIT (YUGOSLAVIA) 25-29 OCTOBER 1976. B. BIJU-DUVAL AND L. MONTADERT, Eds. EDITIONS TECHNIP, PARIS 1977, pp. 73-78

K-AR AGES FROM THE KIZIL DAGH OPHIOLITIC COMPLEX (HATAY, TURKEY) AND THEIR INTERPRETATION

M. DELALOYE, M. VUAGNAT and J.-J. WAGNER [1]

RESUME

Dans leur tentative de dater par la méthode géochronométrique K-Ar les ophiolites du Hatay, les auteurs ont reconnu la possibilité de réponses multiples. D'une part : l'âge vrai ou génétique, l'âge de la mise en place des ophiolites sur le plancher océanique, compte tenu du métamorphisme faible propre à cette situation ; d'autre part l'âge tectonique ou âge de mise en place, en compagnie des formations sédimentaires, dans la position géologique actuelle. Nos résultats montrent que les âges mesurés sur les pillow lavas correspondent très probablement à l'âge tectonique tandis que les âges obtenus sur les diabases du complexe filonien sont très voisins de l'âge vrai. Les recherches en cours portent plus particulièrement sur les gabbros - roches totales et minéraux - pour tenter d'apporter un début de solution aux problèmes soulevés.

1. Scope of this paper

Isotopic age dating of ophiolites, particularly by the potassium - argon method leads to several problems of interpretation. One is the problem to ascertain the cooling age of these rocks in the accreting ocean ridge, the other is to evaluate the effect of metamorphic overprints often related to processes of obduction and subduction along destructive plate margins.

The aim of this paper is to present geochronological data on rocks of the Kızıl Dagh ophiolite in a attempt to unravel the geological history of the Eastern Mediterranean region (Vuagnat et al. 1967). The significance of the radiometric ages is also critically discussed.

2. Introduction

In the Eastern Mediterranean, several ophiolitic complexes have been studied during the last decade and interpreted in terms of plate tectonics. From west to east a long chain of massifs, for example Vourinos (Greece), Troodos (Cyprus), Kızıl Dagh (Hatay) are regarded as fragments of the oceanic lithosphere but the original location and age of this sequence is still a matter of controversy. For example the Troodos was considered by Gass (1968) as a slice of oceanic crust and mantle formed in an old Tethyan ridge whilst Miyashiro (1973) interpreted the same massif as part of an island arc. Dewey (1973) compares the Troodos with other western Pacific features and reaches the conclusion that this massif was formed in a marginal or back arc basin.

(1) Department of Mineralogy, The University, CH-1211 Geneva 4, Switzerland

Similar arguments could be applied to the Kızıl Dagh.

It should also be mentioned that Lapierre and Parrot (1972) compared the Paphos ophiolite (near Troodos) and the Baer-Bassit ophiolite (located in Syria to the south of Kızıl Dagh complex) and found strong similarities between them.

It is evident that geochronology alone cannot decide between these various hypotheses. Several types of evidence should be considered in particular the geochemistry (Pearce, 1975). This will be the subject of a forthcoming article. Here we will only show how our geochronological data fit within the stratigraphic evidence and complement it.

3. Stratigraphic evidence

Dubertret (1953) has described limestones of Vindobonian age (15 m.y.) which cover the Sheeted Complex of Kızıl Dagh at Djebel Moussa. The sedimentary sequence begins with a basal conglomerate containing a large amount of ophiolitic pebbles. On the north side of the massif near Uluçinar (formerly named Arsus), he has also described Vindobonian limestones and conglomerates bearing ophiolitic material. In the middle part of the Kızıl Dagh range near Uçoluk, Maestrichtian limestone transgress over pillow lavas. In the vicinity of Uluçinar Dubertret has also described Maestrichtian limestone dipping under the basic-ultrabasic sequence. From these observations Dubertret interpreted the Kızıl Dagh basic-ultrabasic body as a huge flow in the Maestrichtian. Unfortunately we were not able to find the lower contact near Uluçinar. From other observations we think that there is a strong probability that the lower contact (Maestrichtian limestone under igeous rocks) is tectonic. Thus from field evidence and geochronology, Dubertret's model appears to be no longer valid.

4. K-Ar age determination

As the stratigraphical evidence is inconclusive and does not give the true age of the ophiolitic complex, an attempt has been made using the potassium-argon dating method. For the technical details of our method see Delaloye et al. (1974). Table 1 gives the analytical data.

5. Discussion of the results

It is evident that the different rocks of the ophiolitic complex have undergone physical and chemical changes during and after their emplacement. Because of the very low potassium content of most of the ultrabasic rocks of the sequence, age determinations were not possible for the dunites and harzburgites. On the other hand, gabbros, dolerites and pillow lavas contain enough potassium for potassium-argon dating.

A preliminary determination on a gabbro indicates a Maestrichtian age. It is evident from microscopic studies that the gabbros are in the range of low grade metamorphism. In consequence this rock was rejuvenated and its age is probably too young but it is not possible to give a figure for the argon loss.

The sheeted complex is 5 km wide in the Kızıl Dagh with almost 100% dolerite dykes. The dated section represents some 500 meters of the section. The chilled margins of the dykes are so numerous that a low rate of accretion (in the range of 1 cm per year) is suggested. The results on dolerites lead to an age of 94 m.y. (Turonian-Cenomanian limit) on an isochron plot $^{40}Ar/^{36}Ar$ versus $^{40}K/^{36}Ar$ (fig. 1). To draw this isochron we have assumed that all dykes were cogenetic and contemporaneous. The initial ratio $^{40}K/^{36}Ar$ is close to 334 and does not seem to be significantly different from the atmospheric ratio 295.5 so that argon overpressure cannot be held responsible for the spread of the ages. Traces of low grade type metamorphism (prehnite, laumontite, thompsonite) have been observed under the microscope so that, here also, the calculated age is probably to young. According to our petrophysical study (Wagner et al., 1976), the sheeted complex has a predominantly normal magnetic polarity. If we accept the isotopic age results, the dolerites could have crystallised during the stable Cretaceous normal Interval (Irving et al., 1973 ; Larson et al., 1975).

Pillow lavas give potassium ages from 44 to 74 m.y.. These ages are significantly different from those of the sheeted complex. Deep alteration of the pillow lavas either in situ on the sea floor or later on, during and after their tectonic emplacement, and atmospheric weathering can cause a rejuvenation. We made measurements across several pillows. Both of the ratios $^{40}Ar/^{36}Ar$ and $^{40}K/^{36}Ar$ do not decrease proportionally from the center to the rim which indicates some argon loss by alteration. On the other hand, the potassium content increases so that there is as well some indication of addition of potassium due to hydrothermal alteration on the ocean floor or through meteoric waters in the continental environment.

6. Comparison with other ophiolitic massifs from the same area

The neighbouring ophiolitic complexes such as the Troodos in Cyprus and the Vourinos in Greece have been extensively studied petrographically and structurally ; however only a few age determinations are available. Hynes et al. (1972) in their study of the Othris ophiolites from the subpelagonian zone in Greece, have obtained ages of 125 and 127 m.y. interpreted as minimum primary crystallisation ages. This is evidence for an earlier spreading in the Eastern Mediterranean area. Rex and Dodson (1973) reported ages of 56 to 88 m.y. from the Troodos which have probably to be interpreted as alteration ages. A systematic study of this massif is now being carried out by A. Desmet in Nancy and will certainly be of great help in this matter. Gabbros from Kızıl Dagh and the Troodos give similar ages but on the other hand the pillow lavas from Kızıl Dagh seem to be somewhat younger. However it is not yet possible to conclude on the basis of these more or less similar alteration ages that the emplacement of both complexes was contemporaneous.

In the Taurus range, to the north of the Troodos-Hatay "axis", many ophiolitic outcrops occur. For example in the province of Burdur, Sarp (1976) has reported ages of 93 to 98 m.y. for basic dykes and has interpreted these results as representative of the age of the metamorphism accompanying the emplacement of the Lycian Units. It is interesting to note that the ages for the tectonic event in the Taurus and for the emplacement of the ophiolites in the Troodos as well as in the Kızıl Dagh are almost the same. A relationship must exist between the opening of the marginal Troodos-Kızıl Dagh basin in the south and the displacement of the Taurus. This movement was apparently directed southward but it may also have resulted from the northerly underthrusting of the ocean floor.

7. Conclusion

Potassium-argon age determinations on ophiolitic rocks can be interpreted in several ways : as the age of cooling on the ocean floor or as the age of a metamorphic event more or less contemporaneous with their tectonic emplacement on to the continental margin.

For the age of crystallisation only an upper limit is given by the Maestrichtian transgression on the pillow lavas. Neglecting a possible but minor rejuvenation of the doleritic dykes in the sheeted complex, the age of about 100 m.y. (Albian-Cenomanian limit) can be assumed to be a cooling age. The younger ages on a gabbro and on pillow basalts are probably related to sea floor or tectonic metamorphic overprints.

We suggest that very interesting ages will be found for the gabbros in using both whole rocks and separated minerals. In the delicate task of dating ophiolites, agreement between geochronological, paleomagnetic and stratigraphic data is a prerequisite. As long as we are uncertain of the age of formation of these rocks, no valid reconstruction of the geological history of the region can be proposed.

ACKNOWLEDGMENTS

We thank Prof. R. Laurent and Dr. I. Hedley for their constructive reviews of this text. This work was supported by the grant No. 2.322-075 from the Swiss National Science Foundation. Technical assistance was supplied by M.T.A. Ankara for which we are thankful.

BIBLIOGRAPHY

DELALOYE, M. and J.-J. WAGNER. 1974 : Potassium-Argon Dating : an automatic equipment with digital output for computer processing. C.R. Soc. Phys. Hist. nat. Genève, 9, 66-74.

DEWEY, J.F., PITMAN, W.C., RYAN, W.B.F. and T. BONIN. 1973 : Plate tectonics and the evolution of the Alpine System. Geol. Soc. Am. Bull. 84, 3137-3180.

DUBERTRET, L. 1953 : Géologie des roches vertes du N.W. de la Syrie et du Hatay (Turquie). Notes et Mém. sur le Moyen-Orient, Tome VI. 227 pages.

GASS, I.G. 1968 : Is the Troodos Massif of Cyprus a fragment of Mesozoïc ocean floor ? Nature, 220, 39-42.

HYNES, A.J., NISBET, E.G., GILBERT SMITH, A., WELLAND, M.J.P. and D.C. REX. 1972 : Spreading and emplacement ages of some ophiolites in the Othris region (eastern central Greece), Z. Deutsch. Geol. Ges., 123, 455-468.

IRVING, E. and R.W. COUILLARD. 1973 : Cretaceous Normal Polarity Interval. Nature, Physical Science, 244/131, 10-11.

LAPIERRE, H. et J.F. PARROT. 1972 : Identité géologique des régions de Paphos (Chypre) et du Baer-Bassit (Syrie). C.R. Acad. Sc., Paris, 274, 1999-2002.

LARSON, R.L. and T.W.C. HILDE. 1975 : A Revised Time Scale of Magnetic Reversals for the Early Cretaceous and Late Jurassic. J. Geoph. Res. 80/17, 2586-2594.

MIYASHIRO, A. 1973 : The Troodos ophiolite complex was probably formed in an island arc. E.P.S.L. 19, 218-224.

PEARCE, J.A. 1975 : Basalt geochemistry used to investigate past tectonic environment on Cyprus. Tectonophysics 25, 41-67.

REX, D.C. and M.H. DODSON. 1973, in VINE, F.J., POSTER, C.K. and I.G. GASS : Aeromagnetic survey of the Troodos igneous massif (Cyprus), Nature, physical Science, 224, 34-38.

SARP, H. 1976 : Etude géologique et pétrographique du cortège ophiolitique de la région située au N-W de Yesilova Burdur-Turquie. Thèse, Université de Genève, no. 1731, 408 pages.

VUAGNAT, M. et E. COGULU. 1967 : Quelques réflexions sur le massif basique ultrabasique du Kızıl Dagh, Hatay, Turquie. C.R. Soc. Phys. Hist. Nat. Genève, 2, 210-216.

WAGNER, J.-J., DELALOYE, M., HEDLEY, I., VUAGNAT, M. and E. COGULU. 1976 : Magnetic properties of a fragment of Tethyan lithosphere. EGS abstract in EOS Am. Geophys. Union Trans., 57, 655.

TABLE I

No	% K	^{40}Ar rad 10^{-10} moles/g	% $^{40}Ar_r$	$^{40}Ar/^{36}Ar$	$^{40}K/^{36}Ar$	Age
Pillow lavas						
KA-914	0.34	0.597	65.0	$0.929 \cdot 10^3$	$0.103 \cdot 10^6$	100.7 ± 6.0
915	0.46	0.374	78.2	$0.135 \cdot 10^4$	$0.387 \cdot 10^6$	45.5 ± 6.3
916	1.46	1.510	88.7	$0.263 \cdot 10^4$	$0.674 \cdot 10^6$	57.2 ± 1.8
917	0.53	0.481	52.7	$0.623 \cdot 10^3$	$0.108 \cdot 10^6$	50.2 ± 2.6
918	0.63	0.854	69.1	$0.957 \cdot 10^3$	$0.146 \cdot 10^6$	74.5 ± 3.1
919	0.57	0.458	75.4	$0.121 \cdot 10^4$	$0.340 \cdot 10^6$	44.4 ± 2.9
920	0.49	0.405	40.3	$0.494 \cdot 10^3$	$0.718 \cdot 10^5$	45.9 ± 2.5
921	0.29	0.252	51.7	$0.612 \cdot 10^3$	$0.109 \cdot 10^6$	47.9 ± 5.0
922	0.79	0.503	38.0	$0.475 \cdot 10^3$	$0.846 \cdot 10^5$	35.5 ± 1.8
923	0.49	0.444	63.7	$0.811 \cdot 10^3$	$0.170 \cdot 10^6$	50.2 ± 1.9
924	0.49	0.436	30.9	$0.427 \cdot 10^3$	$0.443 \cdot 10^5$	49.3 ± 2.4
925	0.33	0.365	45.7	$0.541 \cdot 10^3$	$0.669 \cdot 10^5$	60.8 ± 2.9
926	0.54	0.481	54.5	$0.645 \cdot 10^3$	$0.118 \cdot 10^6$	49.4 ± 2.7
927	0.65	0.704	30.5	$0.425 \cdot 10^3$	$0.355 \cdot 10^5$	60.1 ± 4.4
Dolerites from the sheeted complex						
KA-930	0.23	0.579	28.4	$0.411 \cdot 10^3$	$0.136 \cdot 10^5$	135.3 ± 9.6
933	0.23	0.465	20.4	$0.371 \cdot 10^3$	$0.112 \cdot 10^5$	109.4 ± 9.2
934	0.32	0.720	32.6	$0.434 \cdot 10^3$	$0.181 \cdot 10^5$	124.2 ± 9.9
935	0.32	0.727	32.5	$0.440 \cdot 10^3$	$0.188 \cdot 10^5$	122.4 ± 9.0
936	0.27	0.710	31.0	$0.431 \cdot 10^3$	$0.150 \cdot 10^5$	144.4 ± 8.8
938	0.41	0.790	41.3	$0.502 \cdot 10^3$	$0.318 \cdot 10^5$	106.0 ± 6.9
939	0.36	0.900	36.2	$0.462 \cdot 10^3$	$0.202 \cdot 10^5$	133.7 ± 6.9
940	0.46	0.925	35.5	$0.457 \cdot 10^3$	$0.243 \cdot 10^5$	108.6 ± 5.8
942	0.43	0.856	41.1	$0.500 \cdot 10^3$	$0.309 \cdot 10^5$	108.1 ± 6.0
943	0.39	0.920	38.7	$0.481 \cdot 10^3$	$0.235 \cdot 10^5$	128.1 ± 5.9
Gabbro						
KA-713	0.54	0.675	47.4	---	---	69.0 ± 2.7

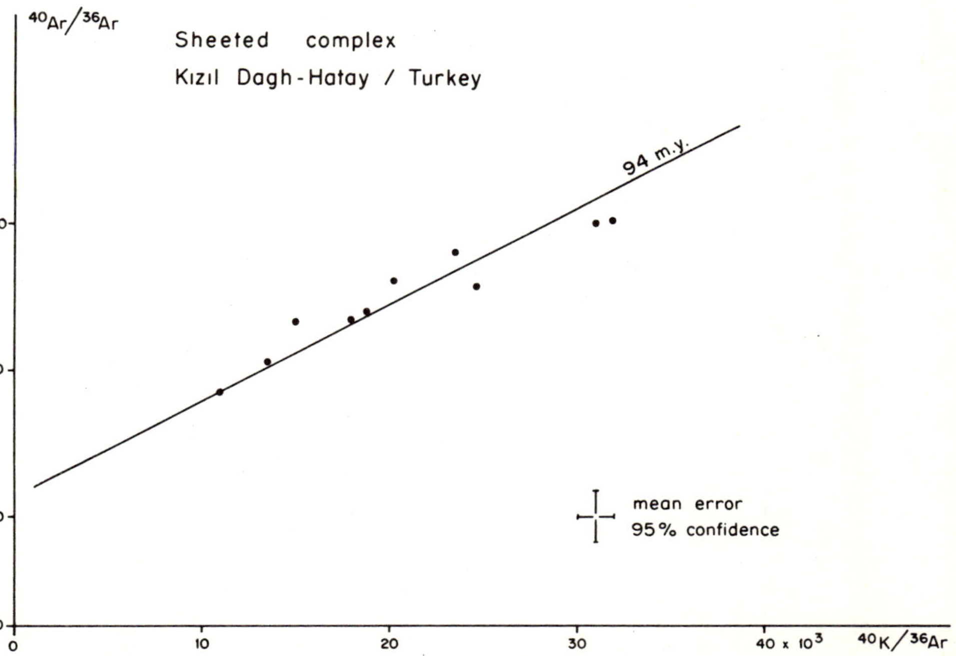

Figure 1 : Isochron plot $^{40}Ar/^{36}Ar$ versus $^{40}K/^{36}Ar$ from the Kizil Dagh dolerites.

INTERNATIONAL SYMPOSIUM ON THE STRUCTURAL HISTORY OF THE MEDITERRANEAN BASINS. SPLIT (YUGOSLAVIA) 25-29 OCTOBER 1976. B. BIJU-DUVAL AND L. MONTADERT, Eds. EDITIONS TECHNIP, PARIS 1977, pp. 79-94

MODÈLE D'ÉVOLUTION MÉSOZOÏQUE DE LA PALÉO-MARGE TÉTHYSIENNE AU NIVEAU DES NAPPES RADIOLARITIQUES ET OPHIOLITIQUES DU TAURUS LYCIEN, D'ANTALYA ET DU BAËR-BASSIT

M. DELAUNE-MAYERE [1], J. MARCOUX [2]*, J.-F. PARROT [1] et A. POISSON [3]*

RÉSUMÉ

Dans les trois secteurs étudiés, Taurus Lycien, nappes d'Antalya et Baër-Bassit, le matériel radiolaritique et ophiolitique impliqué dans les dispositifs structuraux est fondamentalement le même. La formation de ce matériel débute au Trias supérieur par suite de l'ouverture de l'océan téthysien dont l'extension se poursuit pendant le Jurassique et la majeure partie du Crétacé. L'extension vers le sud du domaine pélagique se produit aux dépens de la bordure méridionale du bassin qui joue pendant toute cette période le rôle de marge inactive. A eux trois, les secteurs étudiés permettent de reconstituer l'histoire sédimentologique d'un domaine qui comprend le plateau continental, le bassin et une zone intermédiaire.

Au Crétacé supérieur, une phase tectonique majeure marque la fin de l'extension du bassin. Elle provoque le charriage nord-sud du matériel radiolaritique et ophiolitique sur la plate-forme arabo-africaine. De plus, dans les secteurs du Taurus Lycien et d'Antalya, les phases tectoniques ultérieures ont disséqué le paléo-autochtone et la zone intermédiaire qui apparaissent ainsi en position frontale.

ABSTRACT

In the three studied sectors, Lycian Taurus, Antalya nappes and Baer-Bassit area, the radiolaritic and ophiolitic material comprised in the structural arrangement is more or less the same. The formation of this material began during upper triasic age, and it was provoqued by the opening of the tethyan basin and its extension during the jurassic and almost all the cretaceous periods. The pelagic area was encroaching on the northern continental border which played at mesozoic time the role of an inactive margin. The three studied sectors allowed us to reconstitute the sedimentological story of an area comprising the continental margin, the basin and an intermediate zone.

At upper cretaceous the radiolaritic and ophiolitic material is thrusted on the african platform. Furthermore, in the Lycian Taurus and Antalya, the following tectonic phases have dissected the paleoautochtonus and the intermediate zone which appear now in a frontal position.

* Travail réalisé dans le cadre des accords CNRS-MTA Enstitüsü (Ankara, Turquie)

(1) Laboratoire de Géologie, Office de la Recherche Scientifique et Technique Outre-Mer, 70 route d'Aulnay, 93140 Bondy, France

(2) Laboratoire de Géologie Historique, Université de Paris-Sud, 91405 Orsay, France

(3) Laboratoire de Pétrographie-Volcanologie, Université de Paris-Sud, 91405 Orsay, France

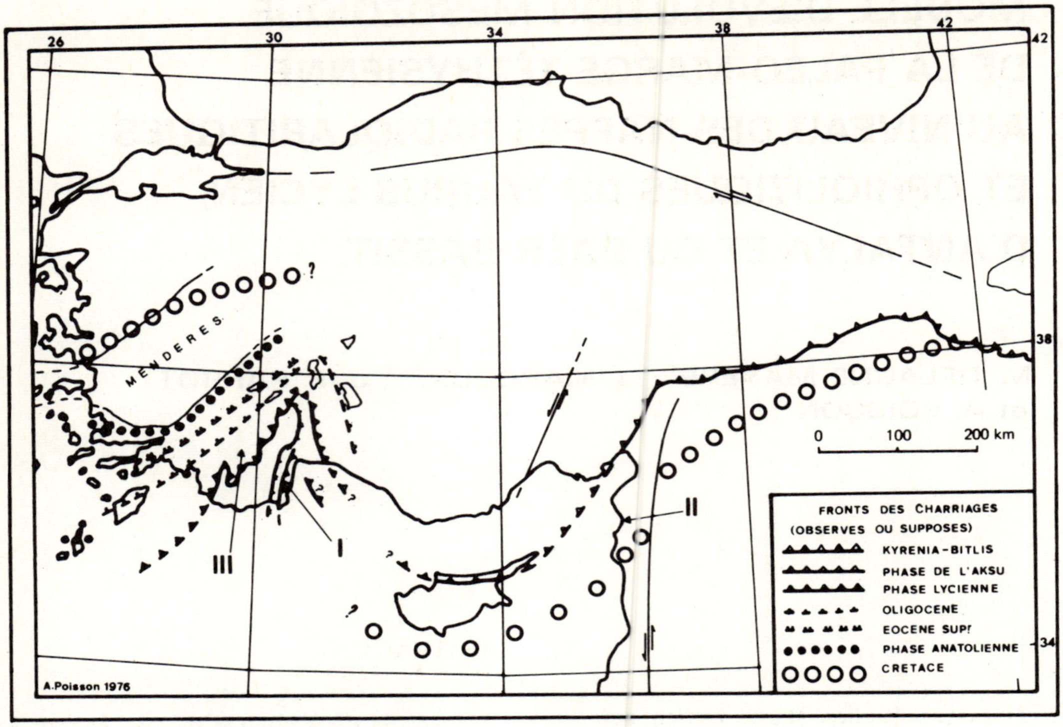

Fig. 1 - Carte de situation des trois secteurs étudiés

INTRODUCTION

Dans les trois secteurs étudiés, le matériel radiolaritique et ophiolitique impliqué dans les dispositifs structuraux est fondamentalement le même. De plus, lors de la première phase tectonique du Crétacé supérieur, l'autochtone relatif des nappes radiolaritiques et ophiolitiques est dans tous les cas la plate-forme arabo-africaine ; cette plate-forme peut en outre être reprise lors des mouvements ultérieurs qui se développent surtout dans le secteur occidental de la zone étudiée. En effet, dans le Baër-Bassit qui correspond à la partie occidentale du croissant ophiolitique péri-arabe, l'histoire tectonique tangentielle s'achève après les grands charriages du Crétacé supérieur ; en revanche, dans le Taurus Lycien, terminaison orientale de l'arc égéen, et le secteur d'Antalya, situé dans la zone charnière entre les arcs taurique et égéen, la tectonique tangentielle se poursuit pendant le Tertiaire.

L'étude des faciès observés et de leur relation avec le substratum de la première phase tectonique nous permet d'envisager une organisation paléogéographique originelle identique.

Les trois secteurs seront successivement décrits en détaillant pour chacun d'eux les traits particuliers qu'ils présentent ; ceux-ci correspondent notamment à des stades différents de l'évolution de la paléo-marge.

NAPPES D'ANTALYA
par J. Marcoux

Le secteur des nappes d'Antalya décrit ici correspont à leur segment sud occidental (I de la figure 1). On trouvera une information sur la géologie régionale de cette partie de la chaîne taurique dans des publications récentes (Brunn, 1976 - Marcoux, 1976a et b). L'évolution stratigraphique des séries antalyotes étant présentée de manière synthétique, la biostratigraphie ne sera pas détaillée ici (Marcoux, inédit).

DESCRIPTION DES SÉRIES ET FRACTURATION TRIASIQUE

Les tectoniques fini-crétacé et tertiaires ont dilacéré l'ensemble des formations de la paléo-marge amenant même, parfois, le paléoautochtone à chevaucher les séries allochtones de la phase crétacée. Cette histoire tectonique aboutit à l'édifice de nappes observables sur le terrain que nous avons schématisé sur la figure 2. La figure 3 indique quelle devrait être la disposition des unités tectoniques dans le cadre paléographique de la marge mésozoïque. Cette reconstitution palinspastique est fondée essentiellement sur les enchaînements de faciès observables tout au long de l'histoire secondaire et sur l'âge d'apparition des détritiques à éléments d'ophiolites.

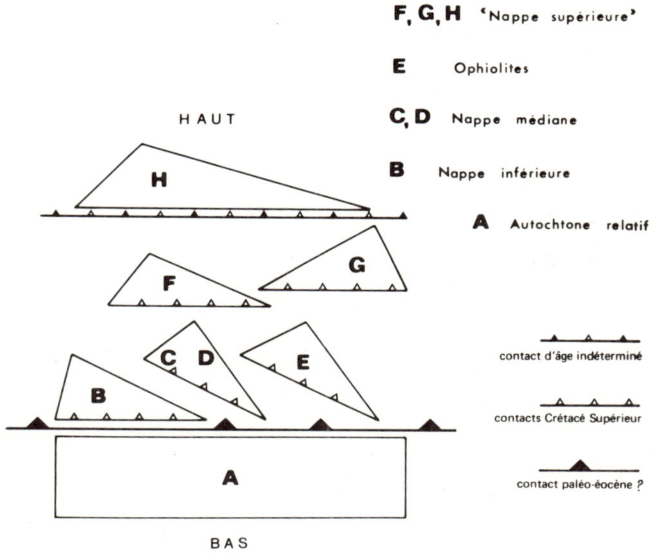

Fig. 2 - Édifice structural des nappes d'Antalya

Les séries de type «plate-forme carbonatée» (figure 4) se retrouvent aussi bien dans l'autochtone relatif : Bey Dagları orientaux (A) situé à la base des nappes d'Antalya que dans l'unité du Tahtalı Dag *sensu stricto* - Çam Dag (H) située tout à fait au sommet de l'édifice des nappes. Ces séries appartiennent au domaine de la plate-forme arabo-africaine.

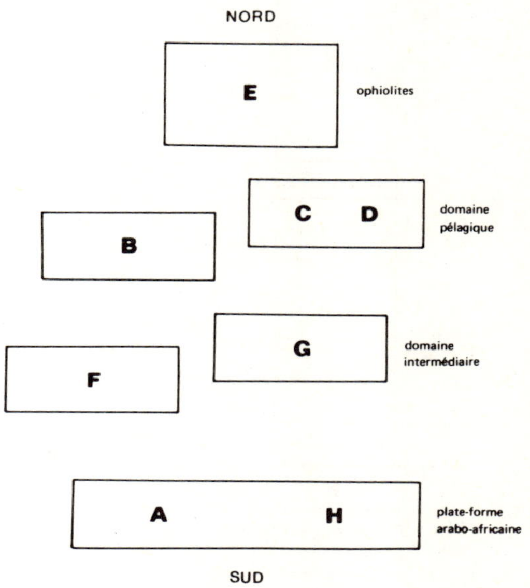

Fig. 3 - Reconstruction palinspastique des nappes d'Antalya

Les nappes situées en position intermédiaire comprennent des unités formées par des séries sans substratum anté-triasique (B C D) et une nappe ophiolitique (E). Ces séries sont représentées dans l'unité B (Dereköy) par des marnes à blocs récifaux, grès calcaires à Ammonites et Halobies (Trias supérieur) ; elles passent à une série à dominante radiolaritique allant du Lias au Maestrichtien. Dans les unités C et D (Alakırçay et Kara Dere-Çalbalı Dag) nous voyons s'amorcer, dès le Trias moyen et se généraliser au Trias supérieur une sédimentation à caractère pélagique (calcaires à Halobies, Ammonitico Rosso, radiolarites). Simultanément un important volcanisme sous-marin se manifeste ; d'une manière subordonnée des turbidites carbonatées et terrigènes témoignent de l'existence d'une topographie accidentée, elle-même contrôlée par une active tectonique synsédimentaire.

Malgré le peu de données de terrain à notre disposition, l'histoire sédimentaire de ces séries au Jurassique et au Crétacé reste fondamentalement pélagique (radiolarites et pélites siliceuses), quelques niveaux de calcarênites viennent sporadiquement s'y intercaler. De telles

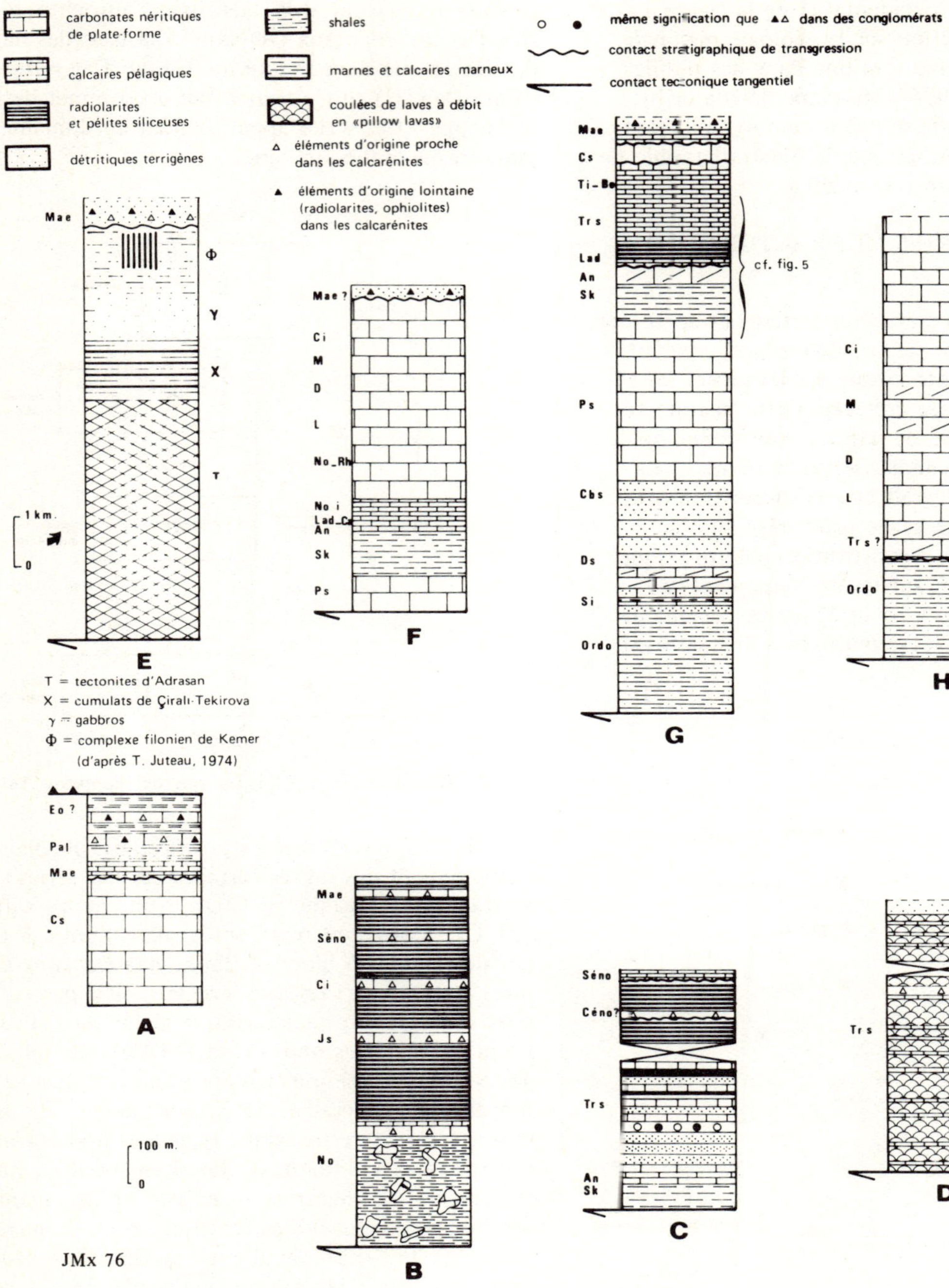

Fig. 4 - Schéma lithostratigraphique des principales unités des nappes d'Antalya

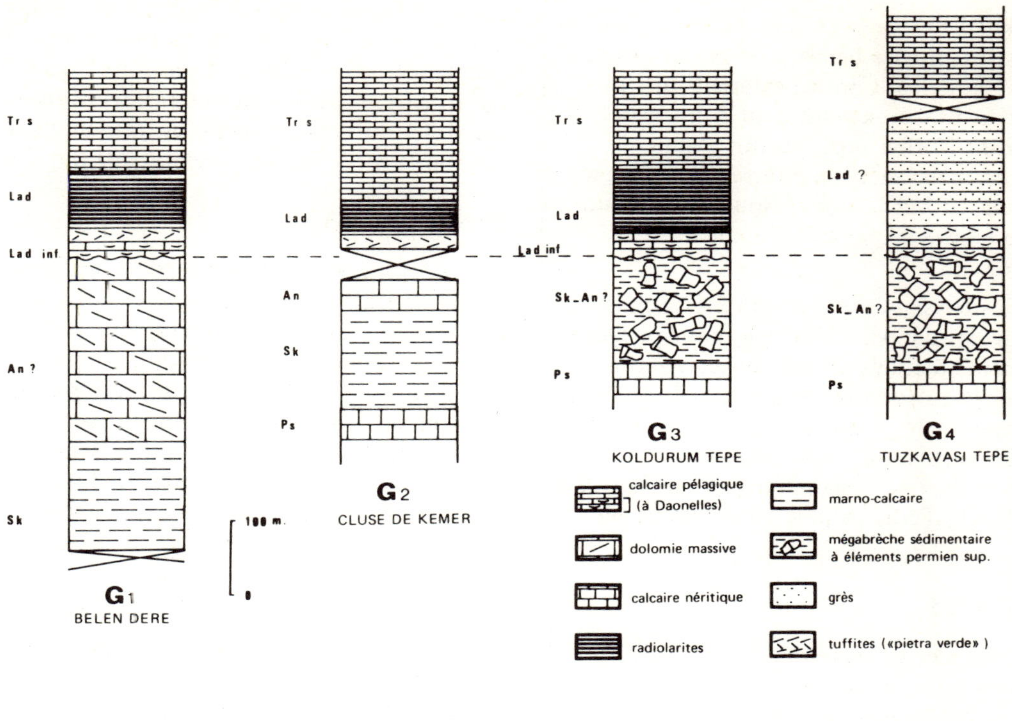

Fig. 5 - Exemples de séries triasiques de transition dans les unités de Kemer

séries, qui reflètent des conditions de dépôt de type «bassin» sont tout à fait analogues aux séries radiolaritiques pindiques décrites dans la littérature de la Yougoslavie à l'Oman.

Les séries B C et D et leurs sédiments pélagiques correspondent, indépendamment de la nappe ophiolitique proprement dite, aux formations les plus distales (septentrionales) de la paléomarge dans ce secteur. Nous devons d'ailleurs noter que ces séries, notamment dans leurs termes inférieurs (triasiques) traduisent la proximité de zones «hautes» à caractère souvent récifal dont la destruction fournit des matériaux de toutes tailles.

Cette instabilité du milieu au Trias moyen et supérieur est particulièrement nette dans les unités de type G (figure 4) qui font partie de l'ensemble structural de la «nappe supérieure d'Antalya». En effet, dans ce groupe d'unités (unités de Kemer) qui, toutes, possèdent un substratum paléozoïque de type plateforme épi-baikalienne arabe, connu depuis d'Ordovicien inférieur, nous observons (figure 5) une diversification et une évolution des faciès très remarquables tout au long du Trias. Ces phénomènes sédimentaires sont étroitement contrôlés par une tectonique distensive intratriasique directement observable sur le terrain. Au moins deux générations de failles normales ont pu être datées grâce au cachetage de niveaux fossilifères susjacents (antéladinien inférieur et intranorien). Les séries G_3 et G_4 (figure 5) montrent au Skythoanisien une spectaculaire mégabrèche à éléments de Permien terminal dont l'origine tectono-sédimentaire est claire.

Dans l'ensemble de ces séries G_1, G_2, G_3, G_4, le Skythien, lorsqu'il est présent, reflète par ses faciès des conditions de sédimentation de mer très peu profonde («tidal flat») ; ce type de sédimentation se poursuit à l'Anisien (dolomies et calcaires vermiculés). L'apparition des premiers faciès pélagiques francs (calcaires à Daonnelles du Ladinien inférieur) succède à une première activité tectonique distensive matérialisée par la mégabrèche décrite plus haut. A partir du Ladinien inférieur un régime de sédimentation pélagique s'instaure sous des faciès variés (radiolarites rouges manganésifères, calcaires à accidents siliceux, turbidites terrigènes, etc..). Ces faciès que nous avons décrits dans les unités B, C et D y présentent cependant un développement beaucoup plus important. Signalons enfin que dans les unités de la figure 5 (G) apparaissent au Trias moyen les témoins d'une activité volcanique sous la forme de tuffites vertes, acides, de type «pietra verde».

L'histoire sédimentaire des unités de Kemer (G) au Jurassico-Crétacé reste, dans certaines coupes, franchement pélagique au moins jusqu'au Tithonique-Berriasien (G de la figure 4) et dans d'autres coupes, reflète sous forme de brèches une grande instabilité du milieu (travaux en cours)

L'unité F (Bakul Dag) présente une histoire triasique semblable dans ses grandes lignes à celles des unités G, avec au Ladinien supérieur-Carnien inférieur une formation d'Ammonitico Rosso extrêmement condensée ; par contre, à partir du Norien supérieur un régime de type plate-forme carbonatée néritique se réinstaure et se poursuit tout au long du Mésozoïque. Nous avons là un exemple de série se trouvant, par la suite, annexée au domaine de plate-forme.

Grâce aux séries triasiques que nous venons d'analyser dans le secteur des nappes d'Antalya, nous pouvons apporter des arguments factuels décisifs concernant la chronologie et le cadre paléogéographique du phénomène distensif majeur responsable de la fracturation du domaine épicontinental arabo-africain. Cette fracturation qui débute dès le Trias moyen correspond aux premiers stades du développement de la Téthys alpine en Méditerranée orientale.

PHASES TECTONIQUES

Aucun cachetage n'étant observable dans l'édifice des nappes d'Antalya nous ne pouvons que nous appuyer sur des arguments indirects pour préciser les phases tectoniques les ayant affectées. Nous notons que la plupart de ces séries (E F G) possèdent des formations détritiques à ophiolites au Crétacé supérieur qui enregistrent les importants mouvements tectoniques se déroulant à cette époque.

Comme nous l'avons indiqué plus haut, les phases tertiaires ont à leur tour affecté le domaine tectonisé au Crétacé supérieur, ce qui explique la position allochtone actuelle des compartiments de plate-forme qui auraient joué le rôle d'autochtone transitoire à la fin du Crétacé. Enfin, ce sont ces mêmes phases tertiaires qui sont responsables du repos anormal de l'ensemble des nappes d'Antalya sur le versant oriental des Bey Dagları.

CONCLUSIONS

Le secteur d'Antalya apporte des preuves du processus de fracturation triasique et des informations sur le domaine affecté par la tectonique distensive. On peut mettre en évidence l'instauration progressive d'un régime pélagique dans ce domaine qui correspondra à une zone «océanique» fonctionnant pendant la quasi-totalité du Mésozoïque avant que ne se produisent, au Crétacé supérieur, les charriages nord-sud des formations héritées de l'expansion sur la plate-forme arabo-africaine.

Sur le plan structural, rien ne s'oppose (Ricou *et al.,* 1975) à une origine interne des nappes d'Antalya. Les séries des nappes d'Antalya qui débutent au Trias sont analogues à celles de même âge de la zone interne du Taurus occidental, par exemple dans certaines unités de Bozkir (Ozgül, 1976 - Marcoux, observ. person.). Ces observations montrent que toutes ces formations faisaient en fait partie intégrante d'un même domaine paléogéographique, ce qui exclut toute hypothèse s'appuyant sur la présence d'un bassin méridional fonctionnant indépendamment du bassin téthysien.

BAER-BASSIT
par M. Delaune-Mayere et J.-F. Parrot

Comme le pensaient Kober dès 1915 et Dubertret dans ses premiers articles (1937 et 1939), le secteur du Baër-Bassit (cf. figure 1) se présente comme un ensemble de nappes ophiolitiques et radiolaritiques reposant sur les formations carbonatées de la plate-forme syrienne. Pour Blanckenhorn (1891), les formations à radiolaires et serpentines correspondaient à la limite nord de cette plate-forme. Cette limite est en fait considérée comme plus septentrionale par Dubertret en 1953 qui postule à cette époque que ces formations résultent de venues fissurales recoupant le nord de la plate-forme arabe. Elle est aussi évidemment plus septentrionale dans les plus récents travaux allochtonistes (Kazmin et Kulakov, 1968- Parrot, 1976).

Les séries radiolaritiques ont été interprétées par Dubertret (1953) comme partie intégrante et supérieure de l'appareil ophiolitique. Ayant indiqué que l'âge de la formation des ophiolites est intra-maestrichtien, cet auteur estimait que ces séries étaient nécessairement de même âge. Dans cette optique, les formations triasiques rencontrées étaient interprétées comme des lambeaux de substratum arrachés lors de l'épanchement ophiolitique. Définissant également de leur côté les séries radiolaritiques comme niveaux supérieurs de l'appareil ophiolitique, Kazmin et Kulakov (1968) se sont précisément appuyés sur la présence de calcaires à Halobies à la base des formations radiolaritiques qu'ils estimaient ne pas pouvoir dépasser le Jurassique supérieur, pour attribuer un âge infra-triasique aux ophiolites mises en place tectoniquement pendant le Maestrichtien.

Lapierre et Parrot (1972) ont esquissé le style tectectonique en écailles des formations du Baër-Bassit. De plus, Parrot (1976) a montré que les séries radiolaritiques du Baër-Bassit ne correspondent en aucune manière aux termes supérieurs de l'assemblage ophiolitique, et sont, en fait, l'équivalent de séries pélagiques poussées

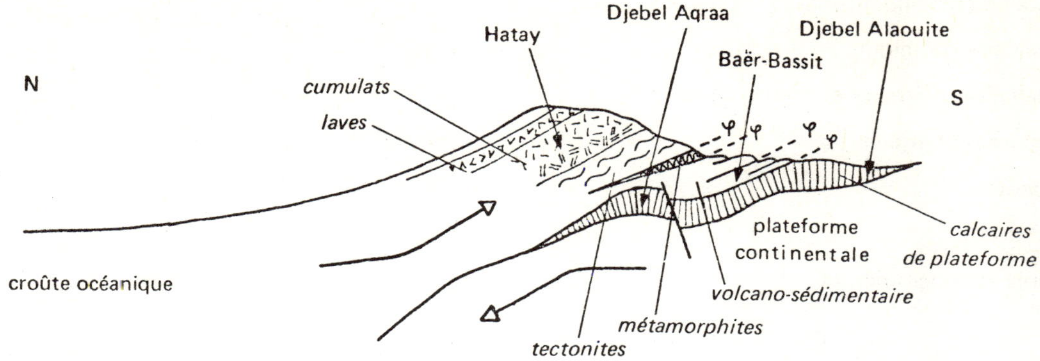

Fig. 6 - Rapport entre les nappes et le substratum au cours du charriage maestrichtien

et chevauchées par le matériel ultrabasique et basique lors de la phase tangentielle du Crétacé supérieur qui a provoqué le charriage nord-sud de ces formations sur la plate-forme arabo-africaine (cf. figure 6). Le secteur du Baër-Bassit qui représente l'extrémité occidentale du croissant ophiolitique péri-arabe défini par Ricou (1971), serait alors le front méridional d'une grande nappe qui pourrait s'enraciner au nord de l'axe calcaire du Taurus (Ricou *et al.*, 1975). Ainsi, les séries pélagiques à radiolaires correspondraient à des sédiments supra-crustaux océaniques. Par ailleurs, Delaune-Mayere et Parrot (1976) ont montré que l'âge de ces séries s'échelonne dans le temps du Trias supérieur au Cénomano-Turonien.

DESCRIPTION DES SÉRIES

Les sédiments pélagiques entraînés avec les nappes ophiolitiques sont peu épais ; les niveaux les plus anciens sont ici datés du Trias supérieur (calcaires à Halobies). Leur fait suite une sédimentation pélitique qui, dans les parties sud du secteur étudié, est plus épaisse et riche en niveaux gréseux (figure 7 B) ; cette séquence méridionale traduit vraisemblablement une plus grande proximité d'une plate-forme. Un niveau de calcarénites remaniant des calcaires à Calpionnelles termine fréquemment cette séquence pélitique dont le dépôt se serait fait pendant le Jurassique et, peut-être, le début du Crétacé. Les radiolarites qui terminent la séquence pélagique sont d'âge crétacé (niveaux de calcarénites à Orbitolines, Hedbergelles du Crétacé moyen, Préalvéolines et radiolaires (1) du Cénomanien). Elles sont surmontées par des brèches du Sénonien inférieur remaniant, outre des calcaires de plateforme, des éléments de la série pélagique décrite ci-dessus.

On n'observe pas, dans le Baër-Bassit, les termes de la zone de transition qui, nous le verrons, traduisent dans le Taurus Lycien l'existence d'une zone de passage entre celle où se produit la sédimentation pélagique et les dépôts carbonatés de plate-forme.

Deux coupes de référence illustrent la sédimentation sur la plate-forme au cours du Mésozoïque.

La coupe du Djebel Aqraa (figure 7 C) correspond au substratum des nappes radiolaritiques et ophiolitiques qui apparaît à la faveur de mouvements de surrection post-miocènes. Les termes infra-bajociens n'y sont pas visibles et la série s'arrête au Maestrichtien inférieur.

La coupe du Djebel Alaouite (figure 7 B) situé à l'avant des nappes radiolaritiques et ophiolitiques correspond à une zone de la plate-forme que n'auraient pas dépassée ces nappes. La série débute au Trias supérieur (Mouthy, 1976) et se poursuit de façon continue jusqu'au Sénonien compris, exception faite d'une lacune entre le Callovo-Oxfordien et l'Aptien, que l'on retrouve dans le Djebel Aqraa.

Dans l'ensemble, la sédimentation est monotone : calcaires et calcaires marneux à faune néritique. Les termes de même âge sont deux fois plus épais dans le Djebel Aqraa, ce qui traduit bien l'approfondissement du bassin vers le nord.

ÉVOLUTION DE LA SÉDIMENTATION AU MÉSOZOÏQUE

Au Trias supérieur, la sédimentation des termes de base de la série radiolaritique est d'emblée pélagique. Elle s'accompagne de volcanisme tholéitique à tendance alcaline présentant avec l'environnement sédimentaire des rapports assez comparables à ceux qui ont été précédemment décrits dans le secteur des nappes d'Antalya. Ici encore, quoique de façon moins spectaculaire, on enregistre des phénomènes qui s'inscrivent dans une phase d'ouverture affectant la plate-forme arabo-africaine.

(1) Nous remercions A. SCHAAF du Centre de Sédimentologie de l'Université de Strasbourg qui a déterminé trois espèces du Cénomanien : *Hemicryptocapsa prepolyhedra,* DUMITRICA, *Holocryptocanium barbui,* DUMITRICA et *Cryptamphorella conara,* FOREMAN.

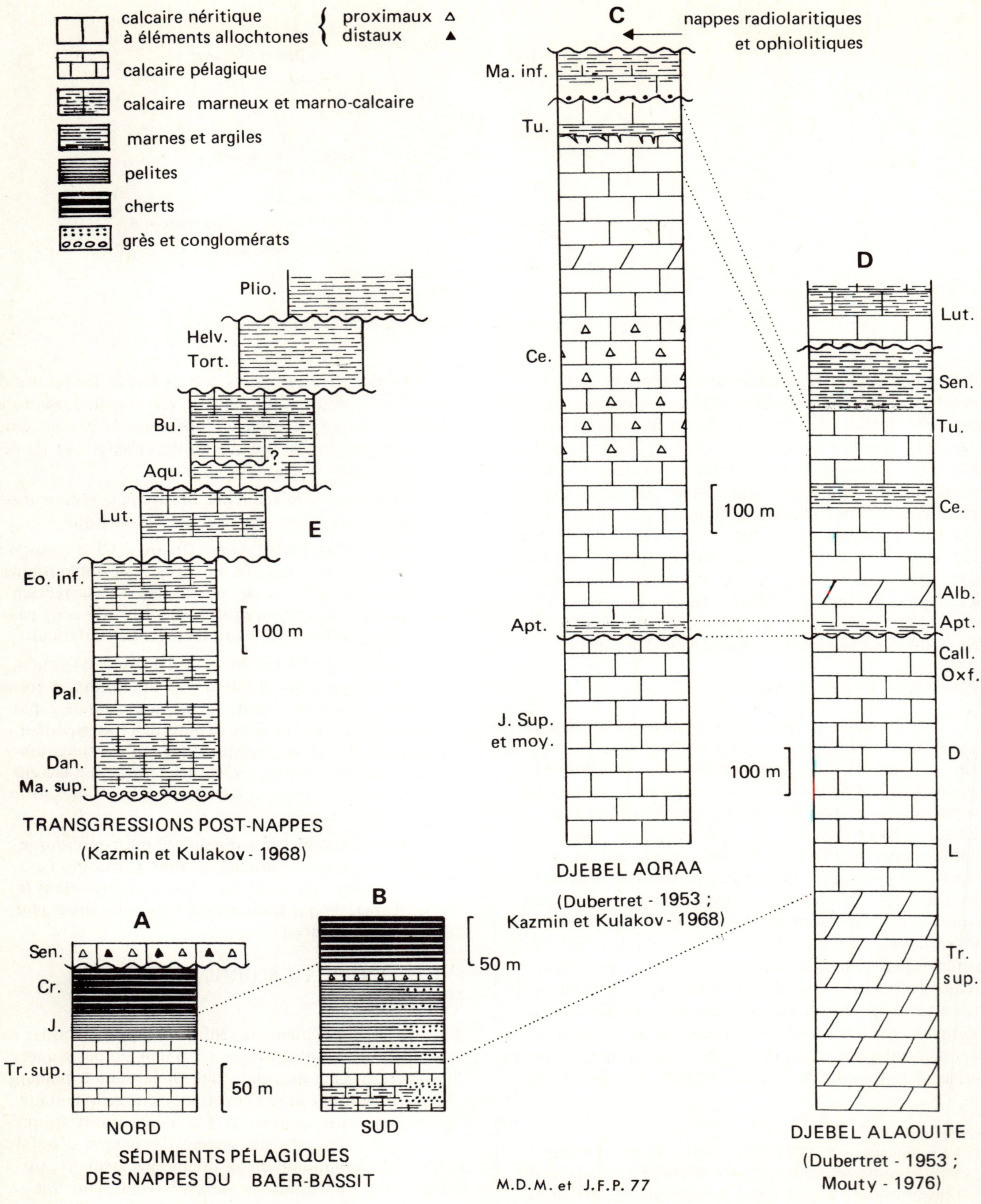

Fig. 7 - Stratigraphie schématique des nappes radiolaritiques, de leur substratum et de leur couverture dans le secteur du Baër-Bassit

Alors que le domaine épicontinental s'approfondit vers le nord, les sédiments qui se déposent dans le bassin deviennent franchement siliceux. Jusqu'à la fin du Jurassique, ils enregistrent des apports détritiques indiquant que le bassin n'est pas encore très éloigné de la plate-forme (éléments allochtones à caractères de turbidites distales).

Au Crétacé, la sédimentation devient uniquement pélagique et se poursuit au moins jusqu'à la base du Turonien qui est localement surmonté par une brèche du Sénonien inférieur (figure 7 A) renfermant des éléments de série pélagique et des fragments de plate-forme carbonatée. La formation de cette brèche est peut-être en relation avec les mouvements de surrection qui affectent la plate-forme dès le Turonien supérieur comme l'a montré Al Maleh (1975) dans la région voisine du Kurd Dagh (cf. également le hard-ground de la coupe 7 C). On peut aussi raisonnablement penser que ces mouvements ont atteint la zone de transition entre les domaines pélagique et épicontinental, et que cette zone pourrait être en partie à l'origine du matériel que l'on trouve dans la brèche sénonienne.

FERMETURE DU BASSIN ET PHASES POST-NAPPES

Les nappes radiolaritiques et ophiolitiques reposent sur le Maestrichtien inférieur calcaro-marneux du Djebel Aqraa (figure 7 C). Elles sont recouvertes par une série de transgressions dont la première débute au Maestrichtien supérieur avec, parfois, un poudingue de base à galets d'ophiolites.

Dans le massif calcaire du Djebel Alaouite qui correspond vraisemblablement à un élément de plate-forme syrienne non chevauché par les nappes, on ne peut mettre en évidence, en raison de l'érosion profonde qui l'affecte jusqu'à ses termes sénoniens, la présence d'une continuité de la sédimentation à ce niveau pendant tout le Maestrichtien qui montrerait que les nappes n'ont pas dépassé cette latitude. Toutefois, cette continuité se manifeste dans la vallée de l'Aafrine, Kurd Dagh, située à une trentaine de kilomètres au nord-est de l'extrémité septentrionale du Djebel Alaouite. L'épaisse sédimentation marneuse de la vallée de l'Aafrine qui, dès le Sénonien, forme une zone basse dans le domaine épicontinental, reste en effet continue du Crétacé supérieur à la base du Lutétien (Dubertret, 1953 - Wolfart, 1963). Le Lutétien débute par une transgression qui coïncide avec la phase anatolienne du secteur plus occidental du Taurus Lycien. Dans le Baër-Bassit (figure 7 E), elle recouvre, soit les sédiments paléocènes, soit les nappes radiolaritiques et ophiolitiques. Dans le Djebel Alaouite, comme il a déjà été dit, elle recouvre directement les marnes sénoniennes (figure 7 D).

Les phases ultérieures décrites dans le Taurus Lycien et les nappes d'Antalya sont marquées dans le Baër-Bassit par une série de transgressions esquissées dans la coupe E de la figure 7.

CONCLUSIONS

Les séries radiolaritiques des nappes du Baër-Bassit comprennent des faciès essentiellement pélagiques qui se seraient déposés dans une zone du bassin plus éloignée de la plate-forme qu'elle ne l'était dans le cas des séries équivalentes du secteur d'Antalya.

De plus, on constate la présence, au Jurassique supérieur, d'un volcanisme per-alcalin (Parrot, 1974) qui pourrait correspondre à un accident purement local affectant la croûte océanique (Parrot, 1976). Ce volcanisme est associé à des radiolarites rouge foncé dont la position par rapport aux différentes zones du bassin reste à préciser.

Le charriage nord-sud des nappes radiolaritiques et ophiolitique sur la plate-forme arabe est bien visible dans le secteur du Baër-Bassit, et les phases postérieures qui seront détaillées dans le secteur suivant, s'y manifestent par une suite de transgressions contemporaines des phases tangentielles que l'on observe dans le Taurus.

TAURUS LYCIEN

par A. Poisson

Le Taurus Lycien est situé à l'extrémité occidentale des chaînes tauriques ; il représente la terminaison de l'arc égéen. De nombreux travaux récents en ont fait connaître par le détail des aspects variés (Richard, 1967 - de Graciansky, 1972 - Boray *et al.*, 1973 - Sarp, 1976 - Poisson, 1976). Du point de vue synthétique divers auteurs se sont attachés au problème des relations avec l'arc égéen, et, par delà, avec les Héllénides (Bernouilli *et al.*, 1973 - Dürr, 1975 - Brunn *et al.*, 1976). Les relations avec le reste des Taurides a été également abordé (Ozgül et Arpat, 1973 - Ozgül, 1976), ainsi qu'avec les chaînes alpines en général (Brunn, 1976).

Outre le fait que le Taurus Lycien renferme également les termes de la série pélagique décrits dans les deux autres secteurs, il se prête bien à l'étude des phases tectoniques tertiaires qui suivent la mise en place des formations du domaine téthysien et de sa marge sur la plate-forme arabo-africaine.

DESCRIPTION DES SÉRIES

Le schéma proposé (et toute l'évolution structurale qui en découle) repose sur une hypothèse fondamentale qui est celle de l'origine très interne des ophiolites. Dans cette hypothèse, il est logique d'admettre l'existence d'un seul domaine ophiolitique situé quelque part au N du massif (au sens large) du Menderes, et également au N des domaines d'origine des unités sédimentaires des nappes lyciennes (Poisson, 1976).

Les figures 8 et 9 rendent compte, dans cette optique, de l'organisation générale de cette marge.

Les études de détail effectuées dans le Taurus Lycien ont permis d'y mettre en évidence une grande variété de séries sédimentaires que l'on peut ranger en trois grands groupes :

1 - Des séries actuellement très dilacérées qui ont été tectonisées en même temps que les ophiolites et qui étaient peut-être, originellement, étroitement associées avec elles (sédiments supra crustaux). Il s'agit essentiellement de radiolarites (figure 9 AI).

2 - Des séries que je qualifie d'intermédiaires (figure 8 et figure 9 A2), dont la base est quelquefois de faciès néritique, et le sommet toujours de faciès pélagique. Ces séries sont à dominance carbonatée (Domuz Dag, Gülbahar), elles sont peu épaisses et très souvent lacunaires.

3 - Des séries de plate-forme (figure 8 et figure 9 A3), carbonatées, épaisses et de faciès néritique jusqu'au Jurassique (Barla Dag) ou jusqu'au Cénomanien (Bey Dagları), pélagiques ensuite jusqu'à l'arrivée des sédiments détritiques terrigènes (flyschs).

ÉVOLUTION DES SÉRIES AU COURS DU MÉSOZOÏQUE

En l'absence d'arguments directs, dans les nappes lyciennes en faveur d'une ouverture au Trias, la figure 9 (ABCD) résume l'organisation originelle au Trias supérieur (A) de la marge sud-téthysienne en trois domaines fondamentaux :

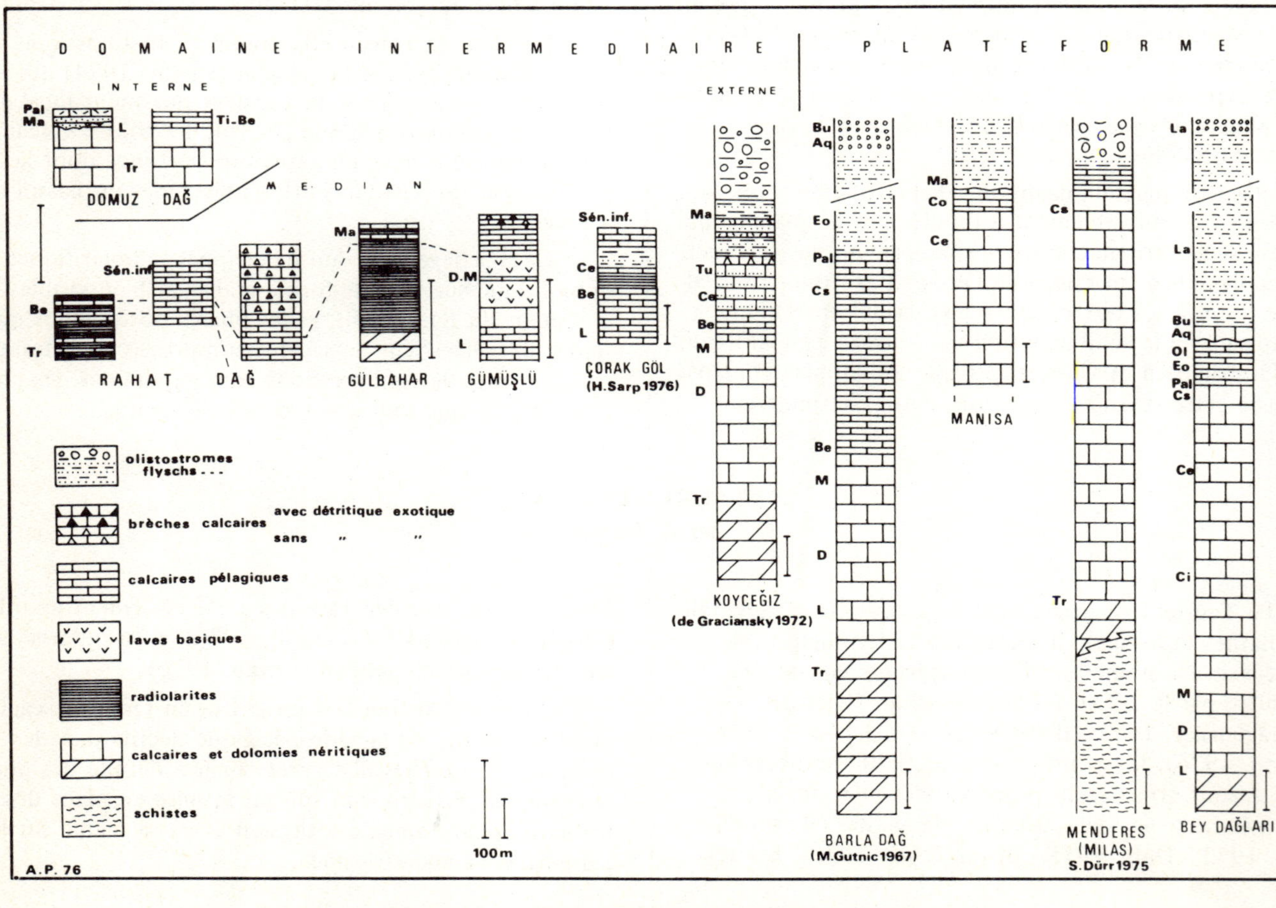

Fig. 8 - Tableau synoptique des principales unités du Taurus Lycien

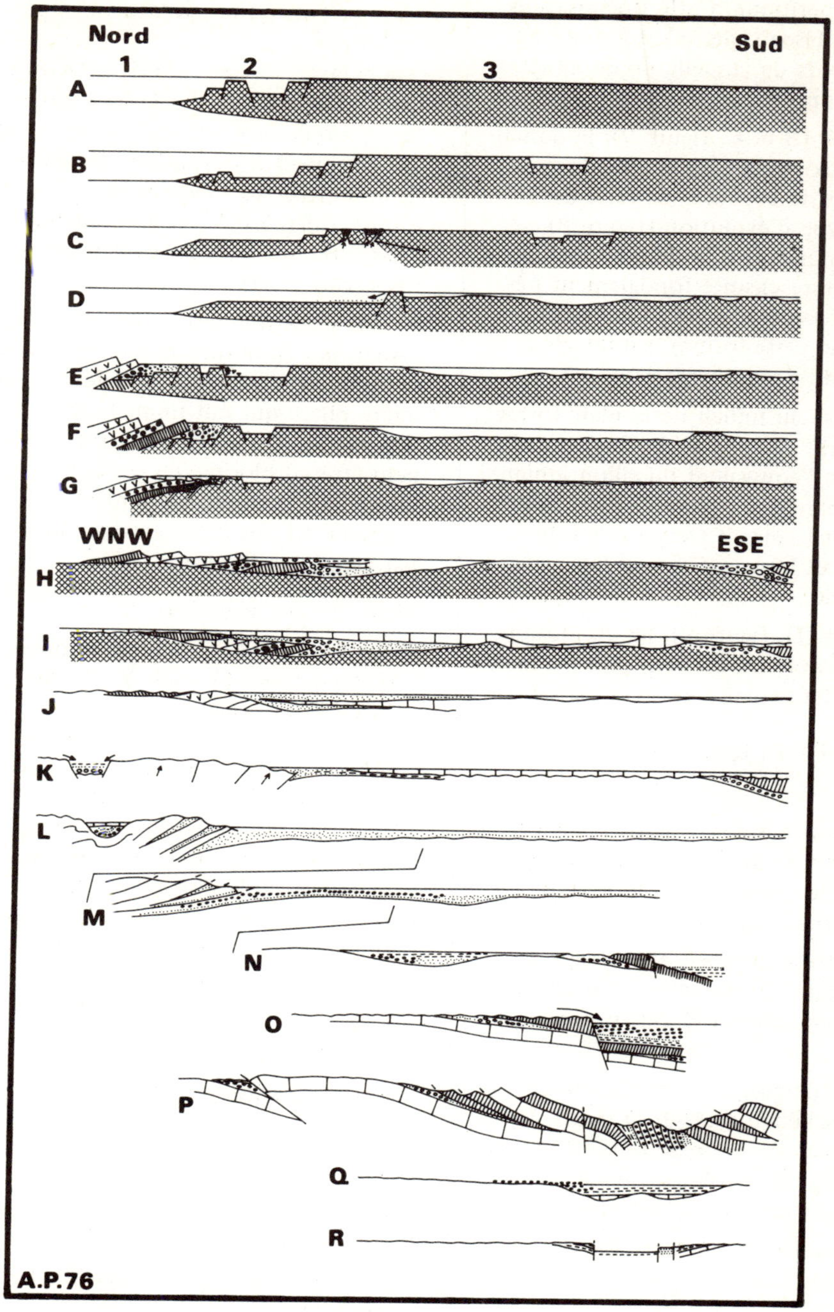

Fig. 9 - Évolution mésozoïque de la marge africaine
et tectogénèse fini-crétacé et tertiaire dans le secteur du Taurus Lycien

A à D : Évolution de la marge africaine au cours du Mésozoïque - Période à dominante distensive. A : Trias supérieur, B : Lias, C : Dogger-Malm, D : Sénonien inférieur.

E à P : Évolution au cours de la période à dominante compressive. E : Campanien, F : Maestrichtien, F : Maestrichtien, G : Paléocène moyen, H : Paléocène supérieur - Éocène inférieur, I : Lutétien, J : Éocène supérieur, K : Oligocène-Aquitanien, L : Burdigalien, M : Langhien, N : Serravalien, O : Tortonien, P : Miocène terminal, Q : Pliocène inférieur, R : actuel.

1 : domaine ophiolitique, 2 : domaine intermédiaire, 3 : plate-forme carbonatée néritique. Cette organisation subsistera tout au long du Mésozoïque, c'est-à-dire pendant toute la période au cours de laquelle prédomineront les mouvements de distension.

Au cours du Jurassique (B et C, figure 9), le domaine intermédiaire s'élargit aux dépens de la plate-forme par effondrement des zones de bordure. La série de Koycegiz illustre bien ce type d'évolution (figure 8) : les faciès pélagiques y apparaissent à la limite Jurassique-Crétacé, la sédimentation reste ensuite fondalement pélagique mais des turbidites à matériel néritique (témoignant de la proximité d'une plate-forme), s'intercalent dans la série tout au long du Crétacé inférieur.

Un sillon s'individualise au milieu de la plate-forme au cours du Jurassique (série de type Barla Dag). Ce sillon a été interprété comme l'équivalent du sillon ionien (Brunn *et.al.*, 1976). Il sépare deux plates-formes qui deviendront respectivement le massif (*sensu lato*) du Menderes du côté interne, et le massif (*sensu lato*) des Bey Daglari du côté externe.

Au Sénonien inférieur (D, figure 9), les conditions deviennent plus uniformes avec immersion, plus ou moins prononcée, du domaine intermédiaire et de la plate-forme.

PHASES TECTONIQUES ET FERMETURE DU BASSIN

L'extension de chaque phase a été schématisée sur la figure 1 qui montre la position des fronts de charriage successifs. On remarquera le décalage très important du front crétacé entre le Taurus Lycien et Chypre.

La figure 9 (E à P) présente quelques aspects de l'évolution paléogéographique du domaine de plate-forme au cours de la période compressive (Sénonien supérieur à Miocène terminal).

Les principales séries détritiques accompagnant les phases tangentielles sont schématisées sur la figure 10. En B, on voit le «cachetage» de la phase sénonienne par des calcaires néritiques d'âge paléocène et, en A, celui de la phase anatolienne par des calcaires d'âge lutétien.

En résumé, les caractéristiques essentielles de chaque phase sont les suivantes :

Phase sénonienne

- âge du premier détritique terrigène connu : Santonien-Campanien (Yeleme) ;
- cachetage : Paléocène (moyen) (calcaires néritiques à Polypiers) (Kızılcadag, Domuz Dag (Söbüce), Yeleme) ;
- extension : générale en Turquie, Chypre, croissant ophiolitique (Ricou, 1971) ;
- zone affectée : domaine ophiolitique et bordure interne du domaine intermédiaire ;
- direction supposée : du nord vers le sud.

Phase anatolienne (Parejas, 1943)

- âge du détritique : Paléocène supérieur (?) à Éocène inférieur ;
- cachetage : Éocène inférieur (Kızılhisar-Akhisar), Lutétien inférieur (Kızılcadag), à moyen (Inceler) ;
- extension : générale en Turquie (Parejas, 1943), semble se retrouver à Karpathos (Davidson, 1974) ;
- zones affectées : nappes sénoniennes et leur autochtone relatif (c'est-à-dire le domaine intermédiaire dans son ensemble) et, probablement, la bordure de la plate-forme interne.

Il s'agit d'une des phases les plus spectaculaires de l'Anatolie. Sa nature exacte reste à replacer dans une optique de tectonique globale. Elle correspond à la première phase qui fait immédiatement suite à la collision des plaques eura-asiatique et africaine. Elle correspond peut-être au blocage de ces deux plaques. Les phases ultérieures correspondraient alors à l'affrontement proprement dit des blocs avec début du glissement de l'un par rapport à l'autre.

Phase Éocène supérieur

- âge du détritique : Lutétien moyen (Kızılcadag), Lutétien supérieur (Varsak Y., Yavuz), à Éocène supérieur (Varsak Y., Yavuz) ;
- cachetage : aucun n'a été observé ;
- extension : reconnue comme phase tangentielle dans le nord de la courbure d'Isparta par Monod (1976). Dans les nappes lyciennes elle ne se traduit que par des flyschs, et elle correspond vraisemblablement alors au début de la phase suivante ;
- zones affectées : domaine des nappes précédentes. Les translations tangentielles semblent peu importantes (dans le Taurus Lycien) ;
- direction supposée : du nord vers le sud (?).

Phase Oligocène

- âge du détritique : Oligocène s.l. (Yavuz, N de la courbure d'Isparta), Oligocène à Aquitanien (Tavas-Kale) ;
- cachetage : par du Burdigalien inférieur à Kale (calcaires néritiques) ;
- extension : phase majeure prolongeant la précédente dans tout le Taurus occidental (Monod, 1976). Reprend les nappes antérieures dans le Taurus Lycien et à Rhodes.

Phase lycienne

- âge du détritique : débute au Burdigalien inférieur et semble s'achever au Langhien inférieur avec des faciès mollassiques (précédant l'émersion générale ?) ;
- cachetage : par des formations lacustres (Burdur, Tefenni, Çavdir) d'âge Plio-quaternaire ;
- extension : semble exclusivement limitée au Taurus Lycien, et probablement à ses prolongements égéens. La flèche du charriage est d'au moins 75 km au front

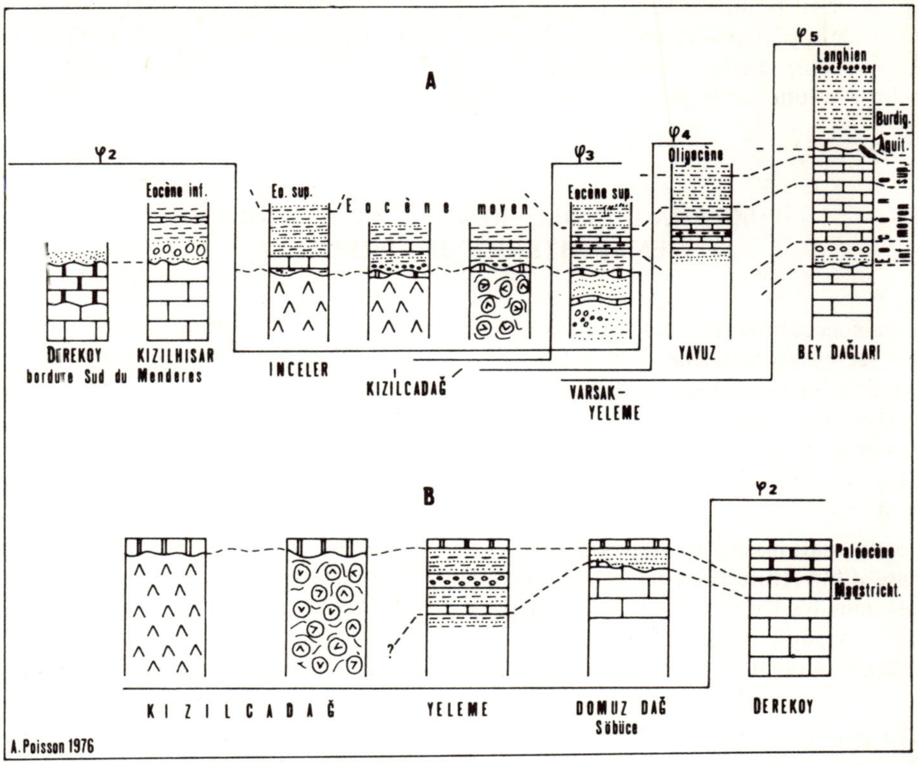

Fig. 10 - Formations détritiques fini-crétacée et tertiaires du Taurus Lycien

des nappes lyciennes orientales. N'affecte pas le Taurus occidental ;
- direction : du secteur N à NW vers le S-SE.

Phase de l'Aksu (ou phase pamphylienne)

- âge du détritique : Tortonien ? (vallée de l'Aksu) ;
- cachetage : Messinien (?) ou Pliocène inférieur (calcaires néritiques et marnes de Gebiz) ;
- extension : bordure orientale du Taurus Lycien, bassin de l'Aksu (au NE d'Antalya) jusqu'à la plaine d'Isparta (Davras Dag) et bordure ouest du golfe d'Antalya, y compris le massif des Bey Daglan qui acquiert sa structure anticlinale à axe N-S ;
- direction : de l'est vers l'ouest.

Cette phase se prolonge en mer dans le golfe d'Antalya mais elle ne semble pas se confondre avec l'arc de Kyrenia à Chypre, qui a rejoué au cours du Pliocène.

Phase de Kyrenia-Bitlis

Cette phase affecte du Pliocène dans le bassin d'Adana. Elle est donc plus récente que la phase de l'Aksu qui, elle, est cachetée par du Pliocène inférieur. Cette phase n'a pas été mise en évidence jusqu'à présent dans le bassin d'Antalya.

A ces phases tangentielles se superposent probablement plusieurs phases de distension correspondant au réseau des grandes failles de directions principales N-S et NW-SE (Bey Daglan) ou bien SW-NE (Susuz Dag au N de Kaş). Ces grands accidents ont joué au départ en failles normales (au cours du Miocène probablement). Certains par la suite ont pu jouer en décrochement et d'autres se transformer en failles inverses au cours de la dernière phase en compression (phase de l'Aksu).

CONCLUSION

On peut clairement mettre en évidence dans le Taurus Lycien plusieurs exemples d'unités particulières intermédiaires entre celles du domaine océanique et celles de la plate-forme. Ces unités sont caractérisées par leur faible épaisseur, par les lacunes importantes de leur série mésozoïque et par leurs faciès pélagique (carbonatés et siliceux). Ces derniers apparaissent plus ou moins tardivement et brutalement, au dessus de couches néritiques correspondant à la bordure effondrée de la plate-forme. Il en résulte un élargissement vers le sud du bassin pélagique téthysien aux dépens du domaine de la plate-forme.

L'existence de séries détritiques fini-crétacées et tertiaires dans le Taurus Lycien permet de suivre l'évolution tectonique au cours de la période tertiaire à dominante compressive. On suit ainsi le cheminement progressif des unités allochtones loin sur la plate-forme.

La dernière phase tangentielle (phase de l'Aksu), de direction E-W, provoque dans le secteur d'Antalya une inversion structurale qui amène le dernier autochtone relatif des nappes en position haute (Poisson, 1974).

SCHÉMA D'ÉVOLUTION DE LA PALÉO–MARGE EN MÉDITERRANÉE ORIENTALE

Au point de vue paléogéographique, les formations observées dans les trois secteurs permettent d'envisager la succession des faciès lithologiques dans un domaine comprenant du sud vers le nord, la plate-forme, une zone intermédiaire, le bassin et la croûte océanique dont les massifs ophiolitiques qui accompagnent les sédiments pélagiques charriés, sont les témoins.

Dans le temps, les formations mésozoïques se développent entre une phase d'ouverture dont les traces sont particulièrement nettes dans les nappes d'Antalya, et une phase de fermeture qui se traduit par le charriage nord-sud des formations téthysiennes sur la plate-forme arabo-africaine.

L'ouverture se fait par fracturation d'un domaine épicontinental qui peut appartenir, soit à la zone de transition entre l'Afrique et l'Europe soudées depuis le choc hercynien, soit au domaine africain proprement dit (Argyriadis, 1975).

L'évolution ultérieure des trois secteurs étudiés appartenant à la paléo-marge téthysienne méridionale ressort déjà, en partie, des schémas A, B, C et D de la figure 9 auxquels il faudrait en fait adjoindre un schéma correspondant au stade initial représentant le domaine épicontinental en voie de fracturation.

Dans l'ensemble, l'ouverture s'est faite aux dépens de séries épicontinentales en amenant au moins à ce stade une relative symétrie entre les bordures nord et sud du bassin naissant. En revanche, au cours de l'extension du bassin, conjuguée à l'expansion océanique, celui-ci s'est très rapidement développé de façon assymétrique; en effet, dès le Lias (Fourquin, 1975), sa bordure nord joue le rôle d'une marge active à l'inverse de ce que nous observons sur la bordure sud.

Dans les trois secteurs, les sédiments pélagiques jurassico-crétacés sont essentiellement siliceux. Ils s'accompagnent en outre, dans le Baër-Bassit qui correspond au domaine le plus éloigné de la marge, d'un volcanisme per-alcalin fini-jurassique ; le caractère ponctuel de cet évènement effusif est interprété par Parrot (1976) comme le résultat d'une venue volcanique intra-plaque océanique.

Lors de la phase sénonienne, une partie de la croûte océanique formée au cours du Mésozoïque pousserait vers le sud les sédiments supra-crustaux océaniques et ceux de la zone intermédiaire méridionale sur la plate-forme arabo-africaine, toutes ces formations pouvant être ultérieurement reprises lors des mouvements tectoniques tertiaires.

L'ensemble des affleurements radiolaritiques et ophiolitiques de Méditerranée orientale, qu'il s'agisse des trois secteurs étudiés, de Chypre, voire même des zones internes situées derrière l'axe calcaire du Taurus, ont appartenu à une seule et même nappe charriée à la fin du Crétacé sur la plate-forme arabo-africaine. Cette assertion repose notamment sur l'identité du matériel composant partout l'allochtone sédimentaire et sur celle de l'autochtone au Crétacé supérieur ; de plus, la stratigraphie de l'axe calcaire du Taurus montre que celui-ci ne saurait constituer une barrière faisant obstacle au passage des nappes. Ainsi, la reconstitution paléogéographique de la Téthys dans cette région au Mésozoïque peut facilement faire l'économie d'un bassin marginal méridional.

CONCLUSIONS

1 - L'étude du secteur d'Antalya met en évidence l'existence d'une période distensive débutant au Trias moyen et se poursuivant au moins jusqu'à la fin du Trias. Cette fracturation affecte aussi bien des séries antétriasiques correspondant au substratum que des termes de transition entre ces séries et celles du Trias supérieur pélagique sans substratum connu. Elle se développe tout d'abord aux dépens des séries continentales paléozoïques de type arabo-africain. Des tuffites apparaissent au Ladinien, puis au Trias supérieur, ce sont d'importantes venues volcaniques sous-marines à tendance alcaline (Juteau et Marcoux, 1973 - Juteau, 1974). L'arrivée de matériel détritique à tous niveaux dans les sédiments pélagiques de zones basses créées par l'effondrement de portions de la plate-forme, traduit bien l'instabilité du milieu.

Dans le Taurus Lycien et le Baër-Bassit, les niveaux infra-noriens sont inconnus. Un volcanisme alcalin et/ou tholéitique à tendance alcaline accompagne fréquemment des formations calcaires rouges à filaments (ammonites ou halobies) (Sarp, 1976 - Parrot, 1976) ; il en est d'ailleurs de même dans la région de Mamonia, Chypre (Henson *et al.*, 1949 - Lapierre, 1975). Dans le Taurus Lycien, les séries pélagiques observées se sont formées dans une zone intermédiaire entre bassin et plate-forme. Dans le Baër-Bassit, ces mêmes séries se sont déposées dans une zone nettement plus éloignée de la plate-forme.

La fracturation de la plate-forme et le développement du volcanisme triasique sont interprétés comme les témoins de l'ouverture du bassin téthysien où se forme ultérieurement une croûte océanique. L'existence de cette ouverture dès le Trias supérieur est en outre accréditée par la coupure faunistique qui se manifeste dès cette époque (Hirsch, 1976) et se confirme au Lias (Enay, 1976) entre les domaines nord et sud téthysien. Par ailleurs, la formation d'une croûte océanique dès la phase distensive marquée par la fracturation de la plate-forme arabo-africaine semble confirmée par la présence, dans la région d'Ankara (Karagöl) d'un détritique à ophiolites et d'un volcanisme triasique supérieur cacheté par du Lias (Fourquin, 1975).

2 - Les trois secteurs étudiés montrent qu'il existe, au Mésozoïque, des séries pélagiques pérennes et des séries devenant pélagiques au cours du temps. Ceci traduit la présence d'un domaine pélagique jusqu'au Crétacé supérieur et son extension sur la bordure méridionale du bassin téthysien, notamment par suite d'effondrements successifs de la marge continentale, le style tectonique observé au Trias supérieur semblant se poursuivre au delà. Si ces phases distensives entraînent localement des émissions volcaniques et la présence de hauts-fonds, il n'en demeure pas moins que contrairement à ce qui se passe au nord du bassin téthysien, la bordure sud joue le rôle de marge inactive pendant tout le Mésozoïque.

3 - La phase tangentielle du Crétacé supérieur marque la fin de l'extension du bassin téthysien. S'il n'est pas possible en l'état actuel des connaissances de dire si cette phase entraîne une fermeture complète du bassin, on peut en tout cas noter que dans les trois secteurs étudiés, le cachetage de cette phase se fait toujours par l'intermédiaire de faciès peu profonds.

De plus, dans les deux secteurs occidentaux, en disséquant le paléo-autochtone et les séries intermédiaires, les phases tectoniques ultérieures, à l'inverse de ce que l'on observe plus à l'est, ont dépassé vers le sud le domaine atteint par les nappes au cours de la phase fini-crétacé.

RÉFÉRENCES BIBLIOGRAPHIQUES

AL MALEH, A.K. - 1975 - Évolution sédimentologique et paléographique du NW syrien (Kurd Dagh) pendant le Crétacé - *9e Congr. int. sédimentol. Nice sédim.,*

ARGYRIADIS, I. - 1975 - Mésogée permienne, chaîne hercynienne et cassure téthysienne. *Bull. Soc. Géol. Fr.*, 7, 17, 1, 56-57.

BERNOUILLI, D., de GRACIANSKY, P.C., MONOD, O. - 1974 - The extension of the Lycian nappes (SW Turkey) into the Southeastern Aegean Islands. *Eclogae Geol. Helv.*, 67, 39-90.

BLANCKENHORN, M. - 1891 - *Grundzüge der Geologie und physikalischen Geographie von Nordsyrien.* 102 p., 2 cartes, Berlin.

BORAY, A., AKAT, U., ANDENIZ, N., AKÇÖREN, Z., CAGLAYAN, A., GÜNEY, E., KORKMAZER, B., ÖZTÜRK, E.M., SAV, H. - 1973 - Some geological problems and their possible solutions along the southern border of Menderes massif. *Pub. 50th Ann. Turkish Repub., M.T.A. Ankara.*

BRUNN, J.H. - 1976 - L'arc concave zagro-taurique et les arcs convexes taurique et égéen : collision et arcs induits. *Bull. Soc. Géol. Fr.*, (7), 18, 2, 553-567.

BRUNN, J.H., ARGYRIADIS, I., RICOU, L.E., POISSON, A., MARCOUX, J., de GRACIANSKY, P.C. - 1976 - Éléments majeurs de liaison entre Taurides et Hellénides. *Bull. Soc. Géol. Fr.*, (7), 18, 2, 481-498.

DAVIDSON-MONETT, J. - 1974 - Contribution à l'étude géologique de l'arc égéen : l'île de Karpathos. Thèse 3e cycle, Université Paris VI.

DELAUNE-MAYERE, M., PARROT, J.F. - 1976 - Évolution du Mésozoïque de la marge continentale méridionale du bassin téthysien oriental d'après l'étude des séries sédimentaires de la région ophiolitique du NW syrien. *Cah. ORSTOM, sér. Géol.*, VIII, 2 (sous presse).

DUBERTRET, L. - 1937 - Sur la constitution et la genèse des roches vertes syriennes. *C.R. Ac. Sc.*, 204, 1663.

DUBERTRET, L. - 1939 - Sur la genèse et l'âge des roches vertes syriennes. *C.R. Ac. Sc.*, 209, 763.

DUBERTRET, L. - 1953 - Géologie des roches vertes au nord-ouest de la Syrie et du Hatay (Turquie). *Notes, Mém. Moy. Orient*, VI, 1-179.

DÜRR, S. - 1975 - *Uber Alter und geotektonische Stellung des Menderes-Kristallins/SW-Anatolien und seine Aequivalente in der mittleren Aegaeis.* Habil. -Schr., 106s, Marburg.

ENAY, R. - 1976 - Faunes anatoliennes (*Ammonitina,* Jurassique) et domaines biogéographiques nord et sud téthysiens. *Bull. Soc. Géol. Fr.,* (7), XVIII, 2, 533-541.

FOURQUIN, C. - 1975 - L'Anatolie du nord-ouest, marge méridionale du continent européen. Histoire paléogéographique, tectonique et magmatique durant le Secondaire et le Tertiaire. *Bull. Soc. Géol. Fr.,* (7), XVII, 1058-1069.

GRACIANSKY, P.C. de - 1972 - Recherches géologiques dans le Taurus lycien occidental. Thèse Université Paris XI, Orsay.

GUTNIC, M., MOULLADE, M. - 1967 - Données nouvelles sur le Jurassique et le Crétacé inférieur du Barla Dag au S de Semirkent (Taurus de Pisidie, Turquie). *Bull. Min. Res. Explor. Inst., Turkey,* 69, 60-78.

HENSON, F.R.S., BROWNE, R.V., Mc GINTY, J. - 1949 - A synopsis of the stratigraphy and geological history of Cyprus. *Quart. J. Geol. Soc., London,* 105, 1-41.

HIRSCH, F. - 1976 - Sur l'origine des particularismes de la faune du Trias et du Jurassique de la plate-forme africano-arabe. *Bull. Soc. Géol. Fr.,* (7), XVIII, 2, 347-356.

JUTEAU, T. - 1974 - Les ophiolites des nappes d'Antalya (Taurides occidentales, Turquie). Pétrologie d'un fragment de l'ancienne croûte téthysienne. Thèse Doc. État, Université de Nancy, 420 p. multigr.

JUTEAU, T., MARCOUX, J. - 1973 - Un exemple de volcanisme sous-marin au Trias supérieur : le strato-volcan de Kara Dere - Çalbali Dag (Ophiolites des nappes d'Antalya, Taurides occidentales, Turquie). Étude des manifestations éruptives et de leur contexte sédimentaire. *Réun. Ann. Sc. Terre,* I, Paris, p.258 (résumé).

KAZMIN, V.G., KULAKOV, V.V. - 1968 - The geological map of Syria, 1:50 000. *Explanatory notes. Technoexport, contract 944, Moscou.*

KOBER, L. - 1915 - Geologischen Forschungen in Vorderasien. 1 - Zur Tektonik des Libanons. *Denkschr. Akad. Wiss. Wien. math.-naturwiss.,* 9, 379-427.

LAPIERRE, H. - 1975 - Les formations sédimentaires et éruptives des nappes de Mamonia et leur relation avec le massif du Troodos (Chypre occidentale). *Mém. Soc. Géol. Fr.,* LIV, 123, 322 p.

LAPIERRE, H., PARROT, J.F. - 1972 - Identité géologique des régions de Paphos (Chypre) et du Baër-Bassit (Syrie). *C.R. Acad. Sc., Paris,* D, 274, 1999-2002.

MARCOUX, J. - 1976 a - Les séries triasiques des nappes à radiolarites et ophiolites d'Antalya (Turquie) : homologies et signification probable. *Bull. Soc. Géol. Fr.,* (7), 18,2, 315-316.

MARCOUX, J. - 1976 b - La fracturation de la plate-forme scythienne et les stades initiaux du développement de la Téthys alpine en Méditerranée orientale. *4e Réun. Ann. Sc. Terre,* Paris, p. 285.

MONOD, O. - 1976 - La courbure d'Isparta : une mosaïque de blocs autochtones surmontés de nappes composites à la jonction de l'arc hellénique et de l'arc taurique. *Bull. Soc. Géol. Fr.,* (7), 18, 2, 521-532.

MOUTY, M. - 1976 - Présence du Lias dans le massif Alaouite (Syrie). *C.R. somm. Soc. Géol. Fr.,* 3, 104-105.

ÖZGÜL, N. - 1976 - Some geological aspects of the Taurus orogenic belt. Turkey (en turc, résumé en anglais). *Bull. Geol. Soc. Turkey,* 19, 65-78.

ÖZGÜL, N., ARPAT, E. - 1973 - Structural units of the Taurus orogenic belt and their continuation in neighbouring regions. *Bull. Geol. Soc. Greece,* 10, 156-164.

PAREJAS, E. - 1943 - Une phase orogénique d'âge yprésien en Anatolie. *Arch. Sc., Genève,* 25, 89-92.

PARROT, J.F. - 1974 - Le secteur de la Tamimah (Tourkmannli) : étude d'une séquence volcano-sédimentaire de la région ophiolitique du Baër-Bassit (NW de la Syrie). *Cah. ORSTOM, sér. Géol.,* 6, 127-146.

PARROT, J.F. - 1976 - Assemblage ophiolitique du Baër-Bassit et termes effusifs du volcano-sédimentaire. Pétrologie d'un fragment de la croûte océanique téthysienne charriée sur la plate-forme syrienne. Thèse Doc. État, Université de Nancy, 333 p. multigr.

POISSON, A. - 1974 - Chronologie des évènements tectoniques depuis le Crétacé sur la bordure NW du golfe d'Antalya. CIESM, 24e Congrès Monaco.

POISSON, A. - 1976 - Essai d'interprétation d'une transversale Korkuteli-Denizli (Taurus ouest-Anatolien, Turquie). *Bull. Soc. Géol. Fr.,*

RICHARD, F. - 1967 - Étude géologique de la fenêtre de Göcek-Aygir Dag (Taurus lycien occidental, Turquie). Thèse 3e cycle, Grenoble.

RICOU, L.E. - 1971 - Le croissant ophiolitique péri-arabe : une ceinture de nappes mises en place au Crétacé supérieur. *Rev. Géogr. Phys. Géol. Dyn., Paris,* 13, 327-349.

RICOU, L.E., ARGYRIADIS, L., MARCOUX, J. - 1975 - L'axe calcaire du Taurus : un alignement de fenêtres arabo-africaines sous les nappes radiolaritiques, ophiolitiques et métamorphiques. *Bull. Soc. Géol. Fr.,* (7), 17, 1024-1044.

SARP, H. - 1976 - Étude géologique et pétrographique du cortège ophiolitique de la région située au nord-ouest de Yeşilova (Burdur, Turquie). Thèse Doc. État, Université de Genève, 378 p. multigr.

WOLFART, R. - 1963 - Untersuchungen zur Grundwasserschliessung in West-Syrien. 1. Geologie und Hydrogeologie. *Bundesanst. Bodenforsch.,* 26u, 203 S, Hanovre.

INTERNATIONAL SYMPOSIUM ON THE STRUCTURAL HISTORY OF THE MEDITERRANEAN BASINS. SPLIT (YUGOSLAVIA) 25-29 OCTOBER 1976. B. BIJU-DUVAL AND L. MONTADERT, Eds. EDITIONS TECHNIP, PARIS 1977, pp. 95-110

A MESOZOIC FOSSIL EDGE OF THE ARABIAN PLATE ALONG THE LEVANT COASTLINE AND ITS BEARING ON THE EVOLUTION OF THE EASTERN MEDITERRANEAN

A. BEIN and G. GVIRTZMAN [(1)]

ABSTRACT

A paleodepositional hinge belt during Cretaceous, Jurassic and probably Triassic times along the present Levant coastline has been postulated from data accumulated from drillholes and seismic surveys carried out in the Coastal Plain and Continental Shelf of Israel. This belt separated between shallow-marine and continental sediments deposited on the Arabian Platform to the east and the continental slope and rise sediments in the Levant Basin to the west. The sharp facies boundary, accompanied by deep canyon incision and the accumulation of deep water sediments in the form of sedimentary prisms indicates that this hinge belt coincides with the Mesozoic continental margins of the Arabian plate. Ultrabasic rocks of mantle origin, which are surrounded by sheeted intrusions and pillow lavas, outcropping in the Troodos massif in Cyprus and the massifs of Baer-Bassit in northwest Syria, are interpreted as a mid-Tethyan Ridge and spreading center. This center is assumed to have been active in the Late Triassic to Late Cretaceous time interval (200-70 my BP) between the Arabian and Turkish plates. During the Paleozoic, the Tethys Ocean extended to the north of Iran and Turkey only. These two continents were split away from Africa-Arabia by a Late Triassic rifting. The new Troodos-Baer-Bassit Mesozoic spreading center accreted a new ocean floor in the new southern Mesozoic branch of the Tethys. The activity along the spreading center ceased at Late Cretaceous times, and in the process of collision and subduction between Turkey-Iran and Africa-Arabia along the Zagros-Taurus-Mediterranean Ridge since the Tertiary, the Mesozoic ocean floor between the Turkish plate and the Troodos-Baer-Bassit spreading center was consumed. The Levant Basin is therefore only a remnant of the Mesozoic ocean floor.

RESUME

Les renseignements fournis par les sondages et levés sismiques effectués dans la plaine côtière et sur le précontinent d'Israël, permettent de mettre en évidence une zone bordière paléosédimentaire pendant le Crétacé, Jurassique et probablement Trias, qui longe l'actuelle côte du Levant. Cette zone sépare l'aire de sédimentation néritique et continentale de la plateforme arabe à l'Est, de celle du talus continental et de son relèvement dans le bassin du Levant à l'Ouest. La limite abrupte de faciès, accompagnée de profonds canyons sous-marins fossils et les prismes sédimentaires, formés de dépôts accumulés en eau profonde, suggèrent que la dite zone représente la marge continentale mésozoïque de la plaque arabe. Des ultrabasites en provenance du manteau, et leur couverture de roches intrusives et de laves en coussins, qui affleurent dans les massifs du Troodos (Chypre) et de

(1) Oil Research Division, Geological Survey of Israel, 30 Malkhei Israel Street, 95501 Jerusalem, Israel

Baër-Bassit (Syrie nord-occidentale), sont interprétées comme les témoins d'une dorsale et d'un centre d'expansion océanique téthysiens, actifs du Trias terminal jusqu'au Crétacé terminal (-200 à -70 m.a.) et séparant la plaque turque de la plaque africano-arabe. En effet, au Sud de la Téthys océanique paléozoïque unique deux micro-continents, l'Iran et la Turquie, se sont détachés de la plaque africano-arabe dans le courant du Trias supérieur. Le néocentre d'expansion océanique mésozoïque du Troodos-Baër-Bassit a produit l'accrétion du fond océanique de la nouvelle branche méridionale de la Téthys ainsi formée, dont l'activité expansive prit fin vers la fin du Crétacé. A partir du Tertiaire, le fond océanique situé entre la plaque turque et le centre d'expansion océanique Troodos-Baër-Bassit fut consumé dans le processus collisionaire des éléments eurasiatiques et africano-arabes et par la subduction en bordure de la ride méditerranéenne et des chaînes Zagro-Tauriques, en Turquie et en Iran. Ainsi, le Bassin du Levant représente l'unique vestige du fond océanique méridional de la Téthys mésozoïque.

I. INTRODUCTION

The origin and crustal nature of the eastern Mediterranean and of the Levant basin in particular (Fig. 1) has been discussed in many studies. However, because no deep drillholes have penetrated the thick Tertiary-Recent sedimentary cover, the problem is still controversial. Biju-Duval et al. (1974) suggested that the younger sediments of the Levant Basin may overlie the remainder of a former oceanic crust (Mesozoic Tethys) or that the whole basin may be a prolongation of the Arabian Platform. The second possibilty implies that the Mesozoic deep Tethys belt had been located farther north, beyond the present position of the Mediterranean Ridge. Rabinowitz and Ryan (1970), who assumed an oceanic crust below the eastern Mediterranean, have suggested that the Levant Basin is a fragment of a younger ocean which originated in the separation of microcontinents from North Africa rather than a remnant of the ancient Tethys which separated Africa from Eurasia prior to the birth of the North Atlantic Ocean. Neev (1975) and Neev et al. (1976) suggested that the contact of the oceanic crust of the Levant Basin with that of the continental crust of the Arabian Platform occurs some 60 km off the Israel coast along the "Pelusium Line" and is, in their opinion, the present edge of the Sinai plate. Smith (1971) pointed out the fit between Turkey and the southeastern corner of the Mediterranean, and Dewey et al. (1973) consequently suggested, in their plate tectonic model for the evolution of the Alpine system, that the Turkish plate slid northward from the Levant corner during Late Triassic times along a transform fault. Freund et al. (1975) suggested that the Early Mesozoic orogenic deformation and magmatism in Iran, Turkey and Greece occurred in island arcs on the western side of the Tethys Ocean, behind which the eastern Mediterranean expanded as a marginal basin. They interpreted a few analyses of potassium rich alkaline igneous rocks from Israel and Lebanon as an indication of a Benioff zone which descended from Iran-Turkey beneath the Levant.

The significance of the Tethys ophiolites in general, and those of Cyprus and northwest Syria in particular, are of crucial importance in the interpretation of the origin of the Levant Basin. In the present study, sedimentological data from the eastern Levant margin are linked together with data on these ophiolites and a model explaining the origin and structure of the Levant Basin is proposed.

II. SEDIMENTOLOGICAL CONSIDERATIONS

A. THE MESOZOIC DEPOSITIONAL HINGE-BELT ALONG THE LEVANT COAST

A paleodepositional hinge-belt during Cretaceous, Jurassic and probably Triassic times along the present Levant coastline has been postulated (Bein, 1971, 1976; Cohen, 1971; Gvirtzman and Klang, 1972; Gvirtzman et al., 1973; Druckman et al., 1975) from the data accumulated through drilling and seismic surveying in the Coastal Plain and Continental Shelf of Israel (Fig. 2). This belt separates between shallow-marine and continental sediments deposited on the Arabian Platform to the east and continental slope and rise sediments in the Levant Basin to the west.

Since most of the Mesozoic sequence in this area is found in the subsurface, the data available decrease with increasing age of the strata. Therefore, the following description of the stratigraphic-sedimentological sequence is given from top to bottom, giving the better known sequence first, so that the lesser known ones may be presented in reference to it.

B. CRETACEOUS

The facies changes from the platform in the east, through the hinge-belt, to the basin in the west during the Cretaceous are presented in Figure 3. The sequence, which overlies an unconformable surface, may be divided into two main parts: the Berriasian-Aptian strata, which consists mostly of terrigenous clastics; and the Albian-Turonian strata, which consists mostly of carbonates. The Cretaceous sequence may be further subdivided into five sedimentary cycles (two in the lower part and three in the upper), each deposited during a transgression and regression which gradually advanced over larger parts of the platform. The change from terrigenous clastic predominance in the lower part of the sequence to carbonate predominance in the upper part reflects the first large transgression during which shallow water carbonate deposition extended over large parts of the platform. From that time on the contribution of terrigenous material from the mainland became negligible, probably as a result of the very low relief of the shallow shelf.

During the first two cycles, which comprise the lower part of the sequence, the basin sedimentation is represented by the black shales and silts of the Gevar'am Formation which were deposited in a relatively deep and open marine environment (Cohen, 1971; Shenhav, 1971). It appears that the Gevar'am Formation accumulated in the form of a sedimentary prism whose axis of maximum thickness (1500 m in the Delta 1 borehole, Fig. 2) parallels the hinge-belt and is located about 15-20 km west of it. The thinning mode of the prism further to the west has not yet been established due to insufficient data. The sediments deposited on the platform during the first cycle are included in the Heletz Formation and those of the second cycle in the Telamim Formation. The Heletz Formation is composed of beach sands, eolian sands, estuarine silts, shales and dolomitized reefs (Shenhav, 1971). The Telamim Formation, which overlies the Heletz Formation in the Coastal Plain (Fig. 3), consists mostly of oolitic limestones, some sands and shales and a few coral-stromatoporid-rudist reefs. Further to the east and southeast (beyond the eastern limit given in Fig. 3), both formations change and pass into the Hatira Formation, which consists of continental sands, silts and shales of "Nubian facies".

Deep submarine canyons were cut perpendicular to the platform edge during Early Cretaceous times. Cohen (1976) described a deep canyon in the Heletz area in the central Coastal Plain of Israel, which had been cut almost 1000 m into Jurassic limestones and was subsequently filled by the black shales of the Gevar'am Formation.

During the three subsequent sedimentary cycles, which comprise the upper part of the Cretaceous sequence, the basin sedimentation is represented by the Talme Yafe Formation. This formation consists mostly of lutite to arenite calcareous detritus derived from the shallow shelf (Bein and Weiler, 1976).

The Talme Yafe Formation also accumulated in the form of a sedimentary prism with its axis of maximum thickness (over 3000 m in the Item 1 borehole, Fig. 2) paralleling the hinge-belt and located about 7-10 km west of it. Further to the west, at a distance of 15-20 km from the hinge belt, the formation thins to about 350 m (Bein and Weiler, 1976). The entire suit of the Talme Yafe Formation has been interpreted by Bein and Weiler (1976) as a contour current shaped sedimentary prism which accumulated at a high rate of deposition on the continental slope and rise off the Arabian Platform.

The platform sediments deposited simultaneously with the Talme Yafe Formation make up the Yachini-Yagur, the Negba and the Bina-Dalya formations. These formations (known as the Judea Group) were deposited through three sedimentary cycles on a carbonate shallow shelf over large areas of the Arabian Platform, attaining thicknesses of 600-900 m. During the first cycle rudistid reefs fringed the shelf edge; in the following cycles reefs developed mainly along shelf basin margins (Freund, 1965; Bein, 1976).

Several hiatuses have been suspected in the Cretaceous sequence at, and west of, the hinge-belt (schematically presented by a gap between the Gevar'am and the Talme Yafe formations, Fig. 3). In some places, in the vicinity of the hinge-belt, the Talme Yafe unconformably overlies Aptian (Telamim Fm.) to Jurassic sediments. The hiatuses and unconformities are interpreted as non-deposition resulting from slumps and submarine erosion in areas of maximum bottom inclination - the continental slope. Similar features have been recently reported from passive continental margins in which huge erosional escarpments and missing sequences in the continental slope and rise were observed (the Blake Escarpment off the coast of Florida, Heezen *et al.*, 1966; the missing sequence off the coast of northwest Africa, Ryan and Von Rad *et al.*, 1976).

C. JURASSIC

The facies changes from the platform in the east to the basin in the west are presented in Figure 4. The Jurassic sequence may also be subdivided into four sedimentary cycles of transgressions and regressions. Each cycle is terminated by an influx of terrigenous clastics which were deposited during the regression. These cycles are expressed in and span the following formations: 1) Ardon-Lower Inmar; 2) Qeren-Upper Inmar-Rosh Pinna; 3) Daya-Sherif-Karmon; 4) Zohar-Kidod-Be'er Sheva'-Halutza (Fig. 4).

The information on the Jurassic sequence of the Coastal Plain and Continental Shelf was obtained from boreholes only and is very sparse as very few wells penetrated either the Lower Jurassic sequence on the platform or the Upper Jurassic sequence in the western basin (Fig. 2). Although the sedimentology of the Jurassic sequence is less known than that of the Cretaceous, the facies change on both sides of the hinge-belt is evident. Friedman *et al.* (1971) were the first to distinguish a major facies change along an E-W section in northern Israel. They suggested that during the Upper Jurassic, platform carbonates were deposited in the Asher 1 and Haifa 1 wells (see Fig. 2), reefs developed in Caesarea 1 and open sea, basinal sediments accumulated in Delta 1 to the west. Using criteria of energy of deposition, Barzel and Friedman (1970) and Derin and Gerry (1972) noted the changes in facies from the center to the edge of the platform. They established a belt of low energy environment over most of the platform, and a belt of higher energy, characterized by calcarenites, biosparites, reefs and oolites, at the platform edge. Derin (1974) presented a more detailed scheme, in which he divided the platform paleoenvironment into a shelf lagoon (including an inner basin) and a shelf margin. On the latter he distinguished barrier reef, threshold and shoal belts. The shelf margin is characterized according to Derin (1974) by high to intermediate energy sediments. These facies changes were observed in the Lower, Middle and Upper Jurassic sequences. The deposition in the area west of the hinge-belt is represented only by the Delta Formation in the Delta 1 well in the north, with Upper Jurassic shale, and by the Barnea Formation of Middle Jurassic limestone, which was penetrated in several wells in the central Coastal Plain.

D. TRIASSIC

Paleogeographic reconstruction of the area during the Triassic is postulated along general lines only since no borehole has penetrated the Triassic sequence in the vicinity of the hinge-belt area. Belts of depositional environments distinguished in the Upper Triassic sediments of the northern margins of the Arabian Platform are presented in Figure 5. Druckman *et al.* (1975) suggested that sea marginal flats, hypersaline lagoons and sabkha environments prevailed throughout the margins of the Arabian Platform and North Africa. The nearest points of information to the hinge-belt area, in which Upper Triassic dolomites and evaporites were found, is about 40-70 km east of the hinge-belt (northwestern Negev, Ramalla 1 well near Jerusalem and Baflioun 1 well in northwest Syria - Druckman *et al.*, 1975, Fig. 3). To the west of the hinge-belt area, allochthonous blocks of Upper Triassic-Jurassic-Cretaceous oceanic sediments, composed of radiolarites and nodular limestones with *Daonella* and *Halobia* associated with pillow lavas and ophiolites, were found in the Mamonia Formation in Cyprus (Ingham, 1960; Turner, 1968a, 1968b; Lapierre and Parrot, 1972; Robertson and Hudson, 1974), in the Nahar Kandil, Baer-Bassit regions of northwest Syria (Dubertret, 1963; Shatsky *et al.*, 1966; Lapierre and Parrot, 1972; Parrot, 1974a, 1974b; Delaune-Mayere, 1974) and also in the Antalya region of Turkey (Dumont *et al.*, 1972; Monod *et al.*, 1974). Druckman *et al.* (1975) suggested that the Levant hinge-belt, along the coastline of Israel, may have been the boundary between the platform and the continental slope in Late Triassic times, as in the Jurassic and Cretaceous models.

E. THE LEVANT HINGE-BELT: A MESOZOIC CONTINENTAL MARGIN OF THE ARABIAN PLATFORM

The sharp facies boundary around the hinge-belt during the Cretaceous, the Jurassic and probably the Upper Triassic, combined with the deep canyon incised at the shelf edge in the Early Cretaceous and the accumulation of at least part of the deep water sediments in the form of a sedimentary prism west of the belt, appear to be characteristic features of continental margin sedimentation. The Mesozoic sedimentation along the hinge-belt of the Levant closely resembles sedimentation of Atlantic type continental margins such as that off the eastern coast of North America. The Mesozoic sediments of the Arabian Platform "thickens out" towards the platform edge in the same manner as the Lower Cretaceous to Recent sediments off Cape Hatteras (Dietz and Holden, 1966). The deep water sediments, especially the Cretaceous sedimentary prism of the Levant margin, is analogous to the American rise prism (Dietz and Holden, 1966) and its recent counterpart which has been shaped by contour currents (Heezen et al., 1966).

From this we conclude that the Levant hinge-belt is a fossil Mesozoic continental margin of the Arabian Platform. This belt probably trends N-S from the northern coastline of Syria, through Lebanon and Israel, changing to an E-W direction in northern Sinai and continuing below the Nile cone.

Information from boreholes indicate that from the Senonian to Middle Eocene, the significance of this depositional hinge-belt ceased and sediments of the same facies, composed of open marine chalks and marls of the Hashefela Group, were deposited on both sides of the hinge-belt (Gvirtzman, 1970). The new Tertiary tectonic and sedimentological regime in the Levant Basin is discussed later.

III. REGIONAL CONSIDERATIONS

A. MID-TETHYAN MESOZOIC SPREADING CENTER IN CYPRUS AND NORTHWEST SYRIA

The ophiolites of the Troodos in Cyprus and the Baer-Bassit massifs in northwestern Syria have been interpreted by many investigators as remnants of oceanic crust which originated at a mid-Tethyan spreading center. Gass (1968) was the first to call attention to the similarity between the magmatic sequence characterizing recent mid-oceanic spreading centers and the "stratigraphic" order of the Troodos ophiolites, in which a plutonic complex is followed by a sheeted intrusive complex which in turn is covered by a pillow lava series, suggesting that the Troodos originated as a "mid-Tethyan rise". Moores and Vine (1971) provided additional petrographic evidence supporting Gass' interpretation and proposed a correlation between the various layered magmatic complexes in the Troodos and the different geophysical layers defined in oceanic crusts. They recognized a fossil transform fault along which dislocation of the spreading center took place over a lateral distance of about 30 km. Moores and Vine (1971) also suggested an approximately 90° counterclockwise rotation of the entire Troodos since the time of its formation, which implies formation at a ridge crest trending east-west. The slow rate of spreading proposed for the Troodos (1-2 cm per year by Moores and Vine, 1971, and less than 1 cm per year by Smewing et al., 1975) means that the 110 km across strike extent of the outcrop area represent 5 to 11 million years of spreading history. The absence of normal and reversed magnetic lineations, usually found on both sides of spreading centers, are explained by the generation of the Troodos massif at a mid-oceanic ridge during a long period of normal polarity (Moores and Vine, 1971), such as that recognized during the Aptian-Coniacian period (Larsen and Pitman, 1972). Many recent studies on petrography, mineralogy, bulk chemical composition and rare elements of the Troodos rocks provide much evidence supporting the suggested spreading center origin for the massif (Greenbaum, 1972; Gass and Smewing, 1973; Pearce and Cann, 1973; Kidd and Cann, 1974; Smewing and Potts, 1976; Church and Coish, 1976). The presence of umbers on top of the Troodos pillow lavas also reinforces this theory. These umbers are interpreted by Robertson and Hudson

(1973) and Robertson (1975) to be iron manganese and trace metal enriched mudstones of underwater volcanic exhalant origin. Similar material is reported, according to these authors, from the East Pacific Rise.

Late Triassic allochthonous blocks of oceanic sediments associated with volcanics around the Troodos massif (Robertson and Hudson, 1974), radiometric datings of the massif rocks (Gass, in Moores and Vine, 1971) and autochthonous Campanian pelagic sediments deposited on top of the pillow lavas (Robertson and Hudson, 1974) may indicate that the mid-Tethyan spreading center was active at least during Late Triassic-Late Cretaceous times (about 200-70 my BP). The scattered occurrences of umbers overlain by Campanian radiolarites and radiolarian mudstones, overlain in turn by Maastrichtian pelagic ooze deposited close to the carbonate compensation depth wherever the upper pillow lavas crop out (Robertson and Hudson, 1974), appears to indicate that the Troodos massif represents a dying phase in the history of spreading axes (Smewing et al., 1975).

Miyashiro (1973) challenged Gass' almost overall accepted interpretation and suggested that the Troodos and other ophiolitic complexes were created as basaltic volcanos in an island arc with a relatively thin oceanic type crust. His conclusion is based on the relative abundance of SiO_2 and the FeO/MgO ratio which is similar to that found in island arc rocks. He suggested that the composition of about one-third of the volcanic Troodos rocks analysed follows a calc-alkalic trend which is characteristic of island arc volcanism. Miyashiro's conclusion raised a controversial discussion and many investigators opposed this conclusion. Using field observations as well as petrographic data, they claimed that the use of bulk chemical analysis to determine the chemical affinity of the Troodos ophiolites is unjustified since these rocks exhibit conspicuous alteration in thin sections. (See discussion between Miyashiro, 1975a, 1975b, on the one hand, and Hynes, 1975, Moores, 1975, and Gass et al., 1975, on the other).

The similarity between the ophiolites and the associated sediments of northwestern Syria and those of the Troodos in Cyprus has been discussed by Lapierre and Parrot (1972). Field observation, petrographic characteristics and petrological considerations based on bulk chemical analysis and rare element composition combined with the presence of umbers on top of the pillow lavas led many investigators to conclude that the Baer-Bassit ophiolites, like those of the Troodos, represent an uplifted fragment of a Mesozoic mid-Tethyan spreading center as schematically represented in Figure 6 (Parrot, 1974b; Delaune-Mayere, 1974; Whitechurch and Parrot, 1974).

It appears that the ophiolites of the Hatay complex in southern Turkey are analogous to the Baer-Bassit ophiolites (Parrot, 1974b), but data on this complex are still insufficient.

B. THE SEPARATION OF TURKEY AND IRAN FROM THE ARABIAN-AFRICAN PLATE AND THE CREATION OF THE MESOZOIC TETHYS

We propose that the hinge-belt of the Levant was part of the continental margin which delimited the Arabian-African continent to the northwest during Early Mesozoic (Fig. 6). The differentiation between platform and oceanic sediments started at the latest in Late Triassic. Dumont et al. (1972) presented a stratigrapic scheme in which Late Triassic platform sediments in Turkey change southward toward Antalya into deep water oceanic sediments (the "Pamphylien Basin"). We interpret these data as indicating the southern continental margins of the Turkish plate (Fig. 7).

Monod et al. (1974), suggested that the shallow-marine Late Triassic sequence of Tahatali Dag, found in the "nappe superieur" in the Antalya Basin, might represent a fragment of the Arabian-African platform, which originally formed the northern extension of the Arabian-African platform south of the Pamphylien Basin. This hypothesis does not necessarily follow from the field data. It is more likely that the "nappe superieur", as well as the other Antalyan nappes, the Mamonia nappe of Cyprus and the nappes of northwest Syria, were emplaced during the Maastrichtian orogeny. It seems unreasonable that the "nappe superieur" of Antalya could represent the Late Triassic Arabian platform facies if the nappes of Cyprus and northwest Syria, which are much farther south and east, represent the Late Triassic continental margin facies (see Fig. 5).

Smith (1971) suggested that Turkey was separated from the Levant and Dewey et al. (1973) proposed that this event occurred at Late Triassic. Lapierre and Rocci (1976) suggested that the splitting started with a Late Triassic rifting phase along the northern edge of the Arabian-African plate, associated with the alkali basalt suite found in the Mamonia nappe in Cyprus, the Antalya nappes in Turkey and in the Baer-Bassit nappes in Syria. Late Maastrichtian southward emplacement of the Mamonia thrust sheets (Robertson and Hudson, 1974; Lapierre and Rocci, 1976) accounts for the close association of these alkali basalts with the ophiolites of the Troodos massif.

In a Middle Triassic (Ladinian) biofacies map of the Alpine-Mediterranean belt, Hirsch (1976) defined a "biofacies sepharade" province (in contrast to the "biofacies tethysien" province)which is distributed over the Arabian-African platform. This province, characterized by fauna from Israel, Jordan and the western Mediterranean, is found also in the Seydisehir section of Antalya, a location which, according to Dumont et al. (1972), represents the shallow-marine Turkish platform north of the Pamphylien Basin (Hirsch, 1976, Fig. 3). The biofacies similarities between the Turkish and Arabian platforms in Middle Triassic times, in our opinion,reinforces the hypothesis that the rifting and spreading between Turkey and the Arabian-African plate postdates the Middle Triassic (Ladinian).

It appears that the separation of the Iranian plate along the Zagros took place at about the same period (Stocklin, 1974). We accept the hypothesis that the Troodos and Baer-Bassit ophiolites represent a Tethyan ridge and we suggest that this ridge was active during Late Triassic to Campanian (from about 200 to 70 my BP). The material accreted along the mid-Tethyan spreading center (represented by the Troodos and Baer-Bassit massifs) subsequent to the Late Triassic rifting phase formed an oceanic crust in the newly developing ocean located between the African-Arabian plate in the south and the Iranian and Turkish plates in the north (Fig. 7).

In conclusion, it appears to us that during the Paleozoic, Iran and Turkey were the southern border of the Tethys Ocean (the "Paleozoic Tethys" in Fig. 7). Following a Late Triassic rifting phase, Iran and Turkey separated from Africa-Arabia and started to move northward. This process, combined with the new oceanic crust accreted in the expanding sea located between Turkey-Iran and Africa-Arabia, resulted in the creation of the Mesozoic Tethys as a southern oceanic branch of the Paleozoic Tethys. Previous to, and simultaneous with, the early rifting and expansion of the Mesozoic Tethys, the Paleozoic Tethys shrunk through the collision and consumption process between the Turkish-Iranian and Eurasian plates, marked by Early Mesozoic orogenic deformation along the Elburz in northern Iran and the Pontids in northern Turkey (Stocklin, 1974). A recent analogue to the entire process is probably the collision between the Turkish-Iranian and Arabian plates, marked by the orogenic deformation along the Zagros-Taurus range in the north and the simultaneous spreading of the Red Sea as a new ocean in the south.

The absence of the northern counterpart of the Mesozoic ocean floor between the Troodos-Baer-Bassit massifs and the Turkish margins is a result of a later tectonic regime. The Tethyan spreading was terminated and a collision between the African-Arabian and the Turkish-Iranian plates along the Zagros-Taurus-Mediterranean Ridge subduction zone became the dominant tectonic process at Late Cretaceous times. Most of the oceanic crust which was formed during the Mesozoic between the mid-Tethyan ridge and the Turkish-Iranian plates was probably consumed during this process. The Levant Basin located further southeast of the subduction zone has thus far escaped the consumption process. A thick sedimentary cover (up to 12 km, Morelli, 1976) was deposited in the eastern Mediterranean. It appears that most of this sequence was deposited from Tertiary to Recent times over a thin Mesozoic sediment veneer.

Freund et al. (1975) suggested that the eastern Mediterranean expanded during Triassic-Late Cretaceous times as a marginal basin behind island arcs above a subduction zone descending from the Tethys under Africa in a similar process to that occurring recently beyond the Japanese island arc. They consider the high potassium content of the igneous Mesozoic rocks of Israel and Lebanon and the increase of the K_2O/SiO_2 ratio of these rocks from Lebanon to southern Israel as an indication of island arc volcanism and a southward descending Benioff zone. They also use Miyashiro's theory that the Troodos ophiolites are island arc products to reinforce their hypothesis.

In principle, the Freund et al. (1975) model is possible and does not even necessarily contradict the model presented here, since spreading centers probably also develop in marginal seas, beyond island arcs (cf. Hawkins, 1974). However, in our opinion, the island arc-marginal sea model is not in line with the data from Iran, Turkey and the Troodos and Baer-Bassit massifs and the significance of the chemical composition of the Mesozoic igneous rocks from Israel and Lebanon presented in Figure 9, p. 18, of Freund et al. (1975) is not unequivocal and may be subject to different interpretations. Furthermore, if Turkey and Iran had evolved as island arcs during the Early Mesozoic, more andesites should have been found in these land masses.

C. THE POSSIBILITY OF A TRANSFORM FAULT ALONG THE LEVANT COAST

Dewey et al. (1973) proposed a transform fault along the Levant coast to match the apparent discrepancy between the assumed spreading centers separating Iran from Arabia and Turkey from the Levant.

The N-S trend of the hinge-belt, which is almost perpendicular to the general E-W trend of the North African coast, and the rapid facies changes across the hinge-belt within a distance of about 10 km may be better explained by assuming continental margins along a transform fault. In such a case, an E-W section normal to the hinge-belt will not appear as schematically presented in Figure 6, since it would run parallel to the assumed strike of the accretion axis and the age of the oceanic crust would be uniform. But in any case, the schematic presentation in Figure 6 is valid for a N-S cross-section across the Sinai coast. The sedimentological data on which our Mesozoic continental margin theory of the Levant has been postulated are insufficient, however, to indicate whether these margins originated along a transform fault or along normal rifting.

IV. CONCLUSIONS

1) The depositional hinge-belt of the Levant differentiated between platform sediments in the east and deep water sediments in the west during Late Triassic to Late Cretaceous times (about 130 my).

2) The hinge-belt of the Levant evolved as Mesozoic continental margins delimiting the Arabian plate to the northwest.

3) The Troodos massif in Cyprus and the Baer-Bassit massifs in northwest Syria are remnants of oceanic crust which originated at a Mesozoic mid-Tethyan spreading center.

4) The splitting of Turkey and Iran away from Africa-Arabia together with the activity along the spreading centers between them resulted in the formation of a new southern oceanic branch, the Mesozoic Tethys.

5) During the Paleozoic, the Tethys extended to the north of Iran and Turkey only. Prior to and simultaneous with the expansion of the Mesozoic Tethys, the Paleozoic Tethys shrunk through the collision and consumption process between the Turkish-Iranian and Eurasian plates, marked by Early Mesozoic orogenic deformation along the Elburz in northern Iran and the Pontids in northern Turkey.

6) The Mesozoic continental margins of the Levant originated either as a transform fault connecting the spreading center separating Iran from Arabia with that between Turkey and Africa, or as normal rifting.

7) Since Late Cretaceous times, subduction and collision of the African-Arabian and Turkish-Iranian plates along the Zagros-Taurus-Mediterranean Ridge became the dominant tectonic process in the Tethys.

8) The Levant Basin consists of an oceanic crust, which is a remnant of the Mesozoic Tethys, overlain by a thick sedimentary cover (probably exceeding 12 km) mostly of Tertiary to Recent age.

ACKNOWLEDGMENTS

This study presents some of the results of Research Project No. 20230 of the Geological Survey of Israel. The authors would like to express their appreciation to Dr. Zvi Garfunkel of the Department of Geology, The Hebrew University in Jerusalem, to Dr. Francis Hirsch of the Paleontology Division, Geological Survey of Israel, and to Dr. Yehezkeel Druckman of the Oil Research Division, Geological Survey of Israel, for reviewing the manuscript. They would also like to thank Mrs. Beverly Katz for editing the manuscript, Mrs. Laura Buchbinder for editing the abstract presented at the C. I. E. S. M. Assembly and Mrs. Carmella Hadar for drafting the figures.

REFERENCES

BARZEL, A.; FRIEDMAN, G. M., 1970, The Zohar Formation (Jurassic) in souther Israel: A model of shallow water marine carbonate sedimentation. Israel J. Earth-Sci., v. 19, pp. 183-207.

BEIN, A., 1971, Rudistid reef complexes (Albian to Cenomanina) in the Carmel and the Coastal Plain, Israel. Institute for Petroleum Research and Geophysics, Report 1051; Geol. Surv. of Israel, Report OD/2/71, 87 p.

BEIN, A., 1976, Rudistid fringing reefs of Cretaceous shallow carbonate platform of Israel. AAPG Bull., v. 60, pp. 258-272.

BEIN, A.; WEILER, Y., 1976, The Cretaceous Talme Yafe Formation: A contour current shaped sedimentary prism of calcareous detritus at the continental margin of the Arabian Craton. Sedimentology, v. 23, pp. 511-532.

BIJU-DUVAL, B.; LETOUZEY, J.; MONTADERT, L.; COURRIER, P.; MUGNIOT, J. F.; SANCHO, J., 1974, Geology of the Mediterranean Sea basins, pp. 695-721 in: Burk, C. A. and Drake, C. (eds.), The geology of continental margins. Springer-Verlag, Berlin, Heidelberg, New York, 1009 p.

BOSELLINI, A.; HSU, K. J., 1973, Mediterranean plate tectonics and Triassic paleogeography. Nature, v. 244, no. 5412, pp. 144-146.

CHURCH, W. R.; COISH, R. A., 1976, Oceanic versus island arc origin of ophiolites. Earth Planet. Sci. Lett., v. 31, pp. 8-14.

COHEN, Z., 1971, The subsurface geology of the southern Coastal Plain and petroleum research. Ph. D. thesis, The Hebrew University, Jerusalem, 67 p.

COHEN, Z., 1976, Early Cretaceous buried canyon: Influence on accumulation of hydrocarbons in Helez Oil Field, Israel. AAPG Bull., v. 60, pp. 108-114.

DELAUNE-MAYERE, M., 1974, Donnees mineralogiqueset sedimentologiques sur le "volcano-sedimentaire" du Bassit (Syrie). (English summary) Cahiers ORSTOM-Serie Geologie, v. 6, no. 2, pp. 161-172.

DERIN, B., 1974, The Jurassic of central and northern Israel. Ph. D. thesis, The Hebrew University, Jerusalem (in Hebrew, English abstract), 152 p.

DERIN, B.; GERRY, E., 1972, Jurassic biostratigraphy and environments of deposition in Israel. Israel Institute of Petroleum Report 2/72, 20 p.

DEWEY, J. F.; PITMAN, W. C.; RYAN, W. B. F.; BONNIN, J., 1973, Plate tectonics and the evolution of the Alpine system. Geol. Soc. of America Bull., v. 84, pp. 3137-3180.

DIETZ, R. S.; HOLDEN, J. C., 1966, Miogeoclines (miogeosynclines) in space and time. Jour. Geology, v. 74, pp. 566-583.

DRUCKMAN, Y.; GVIRTZMAN, G.; KASHAI, E., 1975, Distribution and environments of deposition of Upper Triassic on the northern margins of the Arabian Shield and around the Mediterranean Sea. Proc. IX Intern. Congress of Sedimentology, Nice, Theme 5, v. 1, pp. 183-192.

DUBERTRET, L., 1963, Liban, Syrie, part A, pp. 1-153 in: Dubertret, L. (ed.), Lexique Stratigraphique International, v. III, Asie, Fascicule 10 c 1, Liban, Syrie, Jordanie. Centre National de la Recherche Scientifique, Paris, 437 p.

DUMONT, J. F.; GUTNIC, M.; MARCOUX, J.; MONOD, O.; POISSON, A., 1972, Le Trias des Taurides occidentales (Turquie), Definition du bassin pamphylien: Un nouveau domaine a ophiolithes a la marge externe de la chaine taurique. Z. Deutsch. Geol. Ges., Band 123, pp. 385-409.

FREUND, R., 1975, Upper Cretaceous reefs in northern Israel. Israel J. Earth-Sci., v. 14, pp. 108-121.

FREUND, R.; GOLDBERG, M.; WEISSBROD, T.; DRUCKMAN, Y.; DERIN, B., 1975, The Triassic-Jurassic structure of Israel and its relation to the origin of the eastern Mediterranean.

Geol. Surv. of Israel Bull., 65, 26 p.

FRIEDMAN, G. M.; BARZEL, A.; DERIN, B., 1971, Paleoenvironments of the Jurassic in the Coastal belt of northern and central Israel and their significance in the search for petroleum reservoirs. Institute for Petroleum Research and Geophysics, Report 1050; Geol. Surv. of Israel, Report OD/1/71, 26 p.

GASS, I. G., 1968, The Troodos Massif, Cyprus: A fragment of Mesozoic ocean floor? Nature, v. 220, pp. 39-42.

GASS, I. G.; NEARY, C. R.; PLANT, J.; ROBERTSON, A. H. F.; SIMONIAN, K. O.; SMEWING, J. D.; SPOONER, E. T. C.; WILSON, R. A. M., 1975, Comments on "The Troodos ophiolitic complex was probably formed in an island arc", by A. Miyashiro and subsequent correspondence by A. Hynes and A. Miyashiro, Earth Planet. Sci. Lett., v. 25, pp. 236-238.

GASS, I. G.; SMEWING, J. D., 1973, Intrusion, extrusion and metamorphism at constructive margins: Evidence from the Troodos Massif, Cyprus. Nature, v. 242, pp. 26-29.

GOLDBERG, M., 1970, The lithostratigraphy of the Arad Group (Jurassic) in the northern Negev. Ph. D. thesis, The Hebrew University, Jerusalem, (Hebrew, with English summary), 137 p.

GOLDBERG, M.; FRIEDMAN, G. M., 1974, Paleoenvironments, paleogeography and evolution of the Jurassic system in southern Israel. Geol. Surv. of Israel Bull., 61, 44 p.

GREENBAUM, D., 1972, Magmatic processes at ocean ridges: Evidence from the Troodos Massif, Cyprus. Nature, v. 238, pp. 18-21.

GVIRTZMAN, G., 1970, The Saqiye Group (Late Eocene to Early Pleistocene) in the Coastal Plain and Hashephela regions, Israel. Geol. Surv. of Israel, Report OD/5/67, 180 p. (Hebrew, English summary).

GVIRTZMAN, G.; BEIN, A.; BUCHBINDER, B., 1973, Oil prospects of the Coastal Plain and the Continental Shelf of Israel, pp. 36-39 in: Gvirtzman, G. and Perath, I. (eds.), Oil Exploration in Israel in the Seventies, Symposium Abstracts, Jerusalem, April 1973, Ministry of Development, Geol. Surv. of Israel and Petrol. Commissioner's Office, 46 p. (English), 38 p. (Hebrew).

GVIRTZMAN, G.; KLANG, A., 1972, A structural and depositional hinge-line along the Coastal Plain of Israel evidenced by magneto-tellurics. Geol. Surv. of Israel Bull., 55, 18 p.

HAWKINS, J. W., Jr., 1974, Geology of the Lau Basin, a marginal sea behing the Tonga Arc, pp. 505-520 in: Burk, C.A. and Drake, C. (eds.), The geology of continental margins, Springer-Verlag, Berlin-Heidleberg-New York, 1009 p.

HEEZEN, B. C.; HOLLISTER, C. D.; RUDDIMAN, W. F., 1966, Shaping of the continental rise by deep geostrophic contour currents. Science, v. 152, pp. 502-508.

HIRSCH, F., 1976, Sur l'origine des particularismes de la fauna Trias et du Jurassique de la plateforme africano-arabe. Bull. Soc. Geol. France, v. (7) 18, pp. 543-552.

HYNES, A., 1975, Comment on "The Troodos ophiolitic complex was probably formed in an island arc", by A. Miyashiro. Earth Planet. Sci. Lett., v. 25, pp. 213-216.

INGHAM, F. T., 1960, Cyprus, in: Dubertret, L. (ed.), Lexique Stratigraphique International, v. III, Asie, Fascicule 9d. Centre National de la Recherche Scientifique, Paris, 47 p.

KIDD, R. G. W.; Cann, J. R., 1974, Chilling statistics indicate an ocean floor spreading origin for the Troodos complex, Cyprus. Earth Planet. Sci. Lett., v. 24, pp. 151-155.

LAPIERRE, H.; PARROT, J. F., 1972, Identite Geologique des regions de Paphos (Chypre) et du Baer-Bassit (Syrie). C. R. Acad. Sci. Paris no. 274, Ser. D., pp. 1999-2002.

LAPIERRE, H.; ROCCI, G., 1976, Le volcanisme alcaline du sud-ouest de chypre et le problem de l'ouverture des regions tethysiennes au Trias. Tectonophysics, v. 30, pp. 299-313.

LARSEN, R. L.; PITMAN, W. C., III, 1972, World wide correlation of Mesozoic magnetic anomalies and its implications. Geol. Soc. Amer. Bull., v. 83, pp. 3645-3662.

MIYASHIRO, A., 1973, The Troodos ophiolitic complex was probably formed in an island arc. Earth Planet. Sci.Lett., v. 19, pp. 218-224.

MIYASHIRO, A., 1975a, Origin of the Troodos and other ophiolites: A reply to Hynes. Earth Planet. Sci. Lett., v. 25, pp. 217-222.

MIYASHIRO, A., 1975b, Origin of the Troodos and other ophiolites: A reply to Moores. Earth Planet. Sci. Lett., v. 25, pp. 227-235.

MONOD, O.; MARCOUX, J.; POISSON, A.; DUMONT, J. F., 1974, Le domaine d'Antalya, temoin de la fracturation de la plateforme africaine au cours du Trias. Bull. Soc. Geol. France, v. (7) 16, pp. 116-127.

MOORES, E. M., 1975, Discussion of "Origin of Troodos and other ophiolites: A reply to Hynes", by Akiho Miyashiro. Earth Planet. Sci. Lett. v. 25, pp. 223-226.

MOORES, E.M.; VINE, F.J., 1971, The Troodos Massif, Cyprus, and other ophiolites as oceanic crust: Evaluation and implication. Phil. Trans. Roy. Soc. London, A, v. 268, pp. 443-466.

MORELLI, C., 1976, Advancement in the geophysical knowledge of the Mediterranean Sea. Preprint, presented in the 25th Congress and Plenary Assembly of ICSEM, Committee on Geology and Geophysics, 4 p.

NEEV, D., 1975, Tectonic evolution of the Middle East and the Levantine basin (easternmost Mediterranean). Geology, v. 3, no. 12, pp. 683-686.

NEEV, D.; ALMAGOR, G.; ARAD, A.; GINZBURG, A.; HALL, J.K., 1976, The geology of the southeastern Mediterranean. Geol. Surv. of Israel Bull., 68, 52 p.

PARROT, J.F., 1974a, Le secteur de Tamimah (Tourkmannli): etude l'une sequence volcano-sedimentaire de la region ophiolitiane du Baer-Bassit (Nord-ouest de la Syrie). (English summary). Cahiers ORSTOM - Serie Geologie, v. 6, no. 2, pp. 127-146.

PARROT, J.F., 1974b, L'assemblage ophiolitique du Baer-Bassit (Nord-ouest de la Syrie): etude petrographique et geochemique du complexe filonnien des laves en coussins qui lui sont associees, et d'un partie des formation effusives du volcanosedimentaire. (English summary). Cahiers ORSTOM - Serie Geologie, v. 6, no. 2, pp. 97-126.

PARROT, J.F.; DELAUNE-MAYERE, M., 1974, Les terres d'Ombre du Bassit (nord-ouest Syrien). Comparaison avec le termes similaires du Troodos (Chypre). (English summary). Cahiers ORSTROM - Serie Geologie, v. 6, no. 2, pp. 147-160.

PEARCE, J.A.; CANN, J.R., 1973, Tectonic setting of basic volcanic rocks determined using trace element analysis. Earth Planet. Sci. Lett., v. 19, pp. 290-300.

RABINOWITZ, P.D.; RYAN, W.B.F., 1970, Gravity anomalies and crustal shortening in the eastern Mediterranean. Tectonophysics, v.10, pp. 585-608.

ROBERTSON, A.H.F., 1975, Cyprus umbers: Basalt-sediment relationships on a Mesozoic ocean ridge. Jour. Geol. Soc. London, v. 131, pp. 511-531.

ROBERTSON, A.H.F.; HUDSON, J.D., 1973, Cyprus umbers: Chemical precipitates on a Tethyan ocean ridge. Earth Planet. Sci. Lett., v. 18, pp. 93-101.

ROBERTSON, A.H.F.; HUDSON, J.D., 1974, Pelagic sediments in the Cretaceous and Tertiary history of the Troodos Massif, Cyprus, in: Hsu, K.J. and Jenkyns, H.C. (eds.), Pelagic sediments: On land and under the sea. Spec. Publ. Internat. Assoc. Sediment., no. 1, pp. 403-436.

RYAN, B.F.; VON RAD, U. et al., 1976, Passive continental margin. Geotimes, v. 21, no. 10, pp. 21-24.

SHATSKY, V.N.; KAZMIN, V.G.; KULAKOV, V.V., 1966, The geological map of Syria, 1:200,000, Sheets I-37-XIX; I-36-XXIV, Explanatory notes (V.P. Ponikarov, ed.), Syrian Arab Republic, Ministry of Industry, Dept. of Geological and Mineral Research, 107 p.

SHENHAV, H., 1971, Lower Cretaceous sandstone reservoirs, Israel: Petrography, porosity, permeability. AAPG Bull., v. 55, pp. 2194-2224.

SMEWING, J.D.; POTTS, P.J., 1976, Rare earth abundance in basalts and metabasalts from the Troodos Massif, Cyprus. Contrib. Mineral. Petrol., v. 57, pp. 245-248.

SMEWING, J.D.; SIMONIAN, K.O.; GASS, I.G., 1975, Metabasalts from the Troodos Massif, Cyprus: Genetic implication deduced from petrography and trace element geochemistry. Contrib. Mineral. Petrol., v. 51, pp. 49-64.

SMITH, A.G., 1971, Alpine deformation and the oceanic areas of the Tethys, Mediterranean and Atlantic. Geol. Soc. Amer. Bull., v. 82, pp. 2039-2070.

STOCKLIN, J., 1974, Possible continental margins in Iran, pp. 873-887 in: Burk, C.A. and Drake, C. (eds.), The Geology of continental margins, Springer-Verlag, Berlin, Heidelberg, New York, 1009 p.

TURNER, W.M., 1968a, A progress report on the geology of western Cyprus including the Akamas Peninsula. Univ. of New Mexico, Albuquerque, 41 p.

TURNER, W.M., 1968b, The general stratigraphy of western Cyprus including the Akamas Peninsula. Univ. of New Mexico, Albuquerque, 14 p.

WHITECHURCH, H.; PARROT, J.F., 1974, Les ecailles metamorphiques infraperidotitique du Baer-Bassit (nord-ouest de la Syrie) (English summary). Cahiers ORSTOM - Serie Geologie, v. 6, no. 2, pp. 173-184.

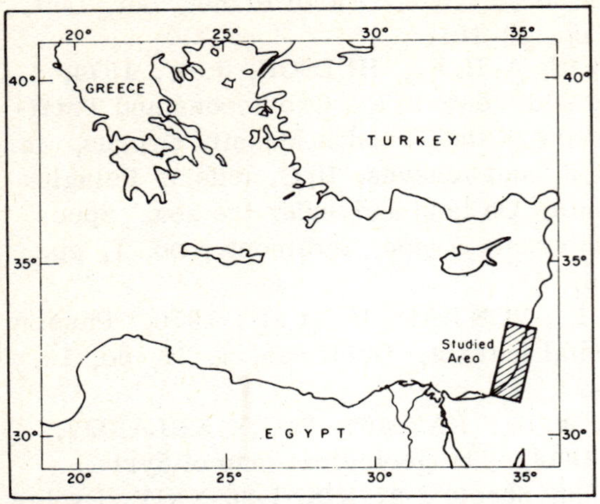

Figure 1 - The studied area and the eastern Mediterranean.

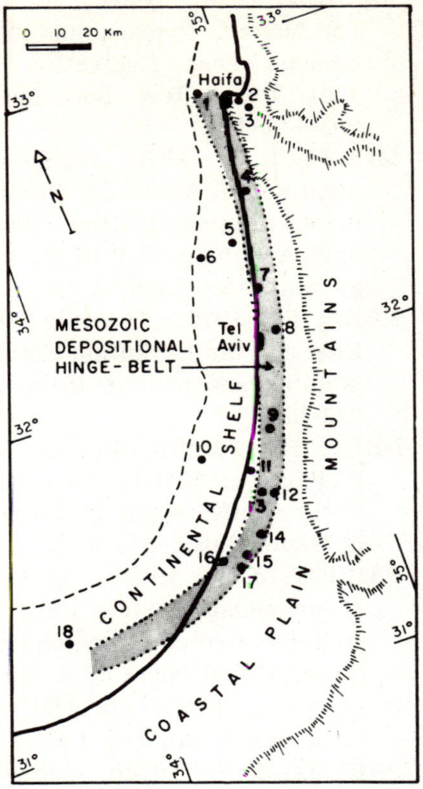

Figure 2 - The Israeli Coastal Plain and Continental Shelf and the inferred extension of the Mesozoic depositional hinge-belt. The subsurface information is mainly from the boreholes numbered on the map as follows: 1 - Foxtrot 1; 2 - Haifa 1; 3 - Asher 1; 4 - Caesarea 1; 5 - Item 1; 6 - Delta 1; 7 - Ga'ash 1; 8 - Petah Tiqwa 1; 9 - Gan Yavne 3; 10 - Bravo 1; 11 - Barne'a 1; 12 - Heletz 22; 13 - Talme Yafe 4; 14 - Sa'ad 1; 15 - Be'eri 1; 16 - Kefar Darom 1; 17 - Nirim 1; 18 - Gal 1.

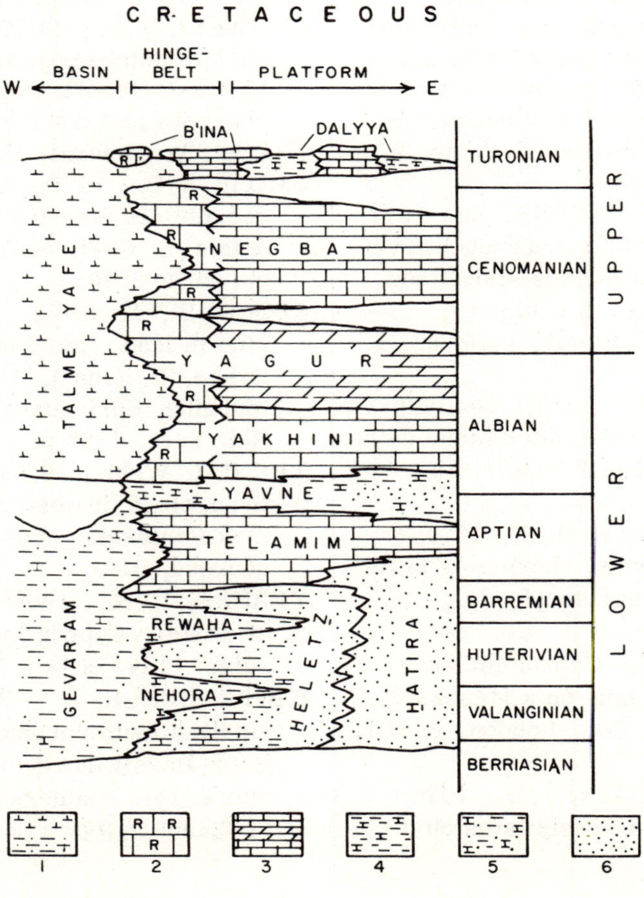

Fig. 3

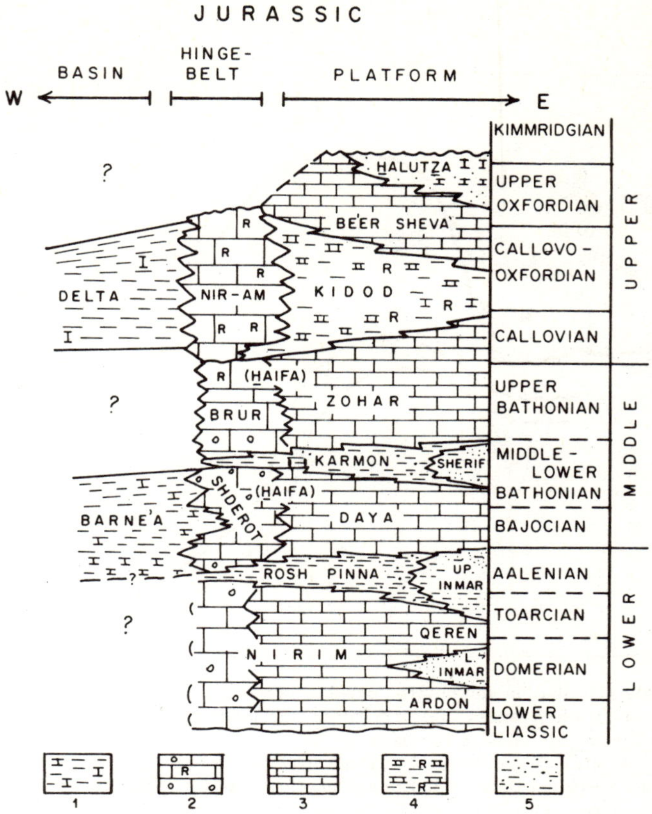

Figure 4 - A composite generalized scheme of the stratigraphic units and environments of deposition of Jurassic sediments in Israel, showing facies changes across the Arabian Platform, the hinge-belt and the continental slope. Compiled from data of Friedman, Barzel and Derin (1971), Derin and Gerry (1972), Derin (1974) and partly from Goldberg (1970), Goldberg and Friedman (1974). Since the shale break of the Karmon and Kidod formations is missing in northern Israel, the Daya, Zohar and Be'er Sheva' formations are included by Derin (1974) in one unit, called the Haifa Formation.

1 - Open marine, off-shelf, basin slope and basin marginal micrites and shales, with sponge spiculae (Delta and Barne'a).
2 - Shelf-edge high energy zone; coral and stromatoporid reefs, shoals, oolites, biosparites (Nir-Am, Brur, Shderot and western part of Nirim).
3 - Inner and outer shelf low energy zone; micrites, dolomicrites, oncolites, occasional patch reefs (Be'er Sheva', Zohar, Daya, Nirim, Qeren, Ardon).
4 - Shelf lagoon, low energy zone; shales and some micrites, occasional patch reefs (Kidod).
5 - Continental to shallow marine transitional zone, terrigenous influx, sands, silts, shales (Halutza, Sherif, Karmon, Rosh Pinna, Upper Inmar, Lower Inmar).

Figure 3 - A composite generalized scheme of the stratigraphic units and environments of deposition of Cretaceous sediments in the Coastal Plain and Continental Shelf of Israel, showing facies changes across the Arabian Platform, the hinge-belt and continental slope. Compiled from data of Cohen (1971), Gvirtzman (1970), Shenhav (1971) and Bein (1971, 1976).

1 - Continental slope and rise, open marine calcilutites, micrites and shales (Talme Yafe and Gevar'am).
2 - Shelf-edge rudist reefs (western facies of B'ina, Negba, Yagur and Yakhini).
3 - Shelf platform limestones and dolomites (B'ina, Negba, Yagur and Yakhini).
4 - Shelf-lagoon micrites and marls (Dalya).
5 - Continental to shallow marine transitional zone, terrigenous influx, sands, silts, limestones, dolomites and shales (Yavne, Heletz, Revaha and Nehora).
6 - Continental fluvial and eolian sands and silts (Hatira).

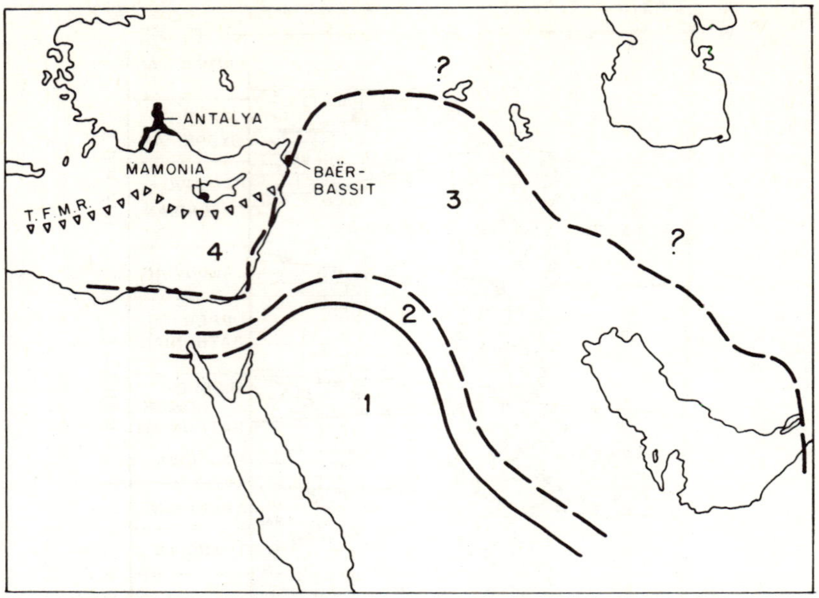

Figure 5 - Facies distribution and environments of deposition of Upper Triassic in the northern margins of the Arabian Platform (after Druckman et al., 1975).

1 - Continental: clastic red beds
2 - Transitional: continental and thin marine intercalation
3 - Sea marginal flats, sabkhas and shallow marine: dolomite, anhydrite, limestone
4 - Oceanic: micrites, radiolarites and nodular limestone associated with ophiolites (Baer-Bassit, Syria; Mamonia, Cyprus; Antalya, Turkey)
T. F. M. R. - Thrust front of the Mediterranean Ridge orogenic belt

The boundary between belts 3 and 4 in the eastern Mediterranean is highly hypothetical. Although the oceanic Upper Triassic outcrops are situated in the orogenic belt and are allochthonous, it seems that the facies distribution concept is still valid.

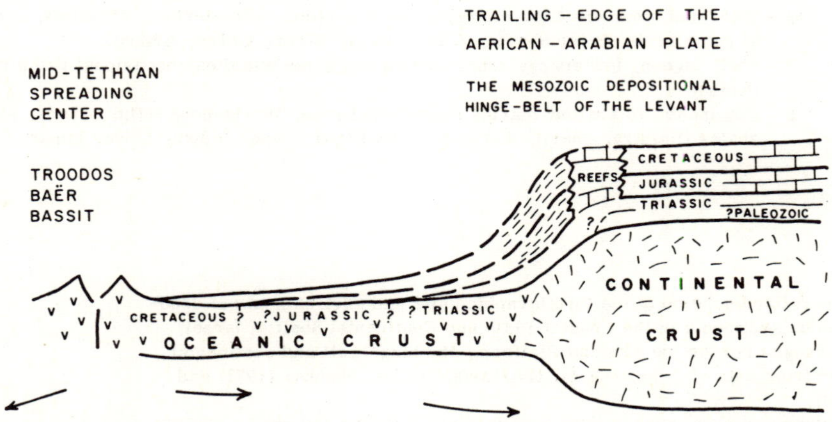

Figure 6 - Schematic representation of the hypothetical structure of the Levant Basin and the Arabian plate during the Mesozoic. Note that the diagram is not to scale. The width of the hinge-belt, which is about 10 km, is much exaggerated compared to the distance from the continental margin to the spreading center. For further clarification see text.

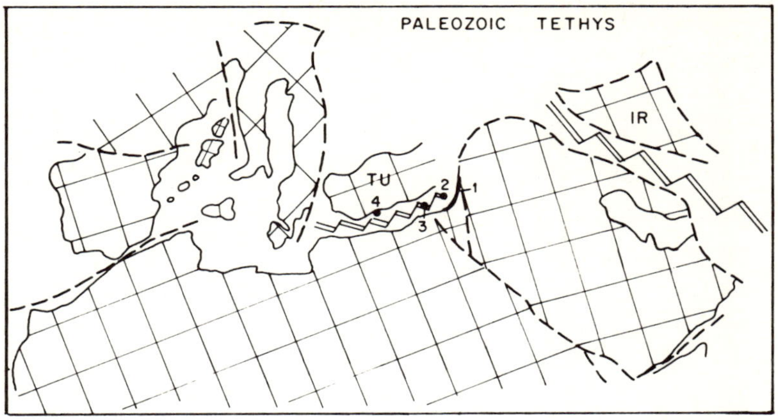

Figure 7 - Paleogeographic distribution of Upper Triassic micro-continents around the Mediterranean area. Modified from data and conceptual models of Smith (1971), Bosellini and Hsu (1973), Dewey et al. (1973) and Druckman et al. (1975).

1 - The Mesozoic depositional hinge-belt of the Levant coinciding with the Mesozoic continental margins of the Arabian plate.
2 - Baer-Bassit (northwest Syria) ophiolites, which indicate a spreading center.
3 - Troodos (Cyprus) ophiolites, which indicate a spreading center.
4 - Antalya (Turkey) ophiolites and Late Triassic oceanic sediments.
TU - Turkish plate
IR - Iranian plate

The Triassic splitting of Turkey-Iran away from Africa-Arabia and the spreading between them mark the embryonic phase of the southern Mesozoic Tethys. Dewey et al. (1973) suggested a transform zone, parallel to the Levant coast, which links these spreading centers.

INTERNATIONAL SYMPOSIUM ON THE STRUCTURAL HISTORY OF THE MEDITERRANEAN BASINS. SPLIT (YUGOSLAVIA) 25-29 OCTOBER 1976. B. BIJU-DUVAL AND L. MONTADERT, Eds. EDITIONS TECHNIP, PARIS 1977, pp. 111-128

CARTES SÉDIMENTOLOGIQUES ET PALÉOGÉOGRAPHIQUES DES RÉGIONS PÉRIMÉDITERRANÉENNES OCCIDENTALES

P.F. BUROLLET [1] et E. WINNOCK [2]

Résumé

Pour évaluer l'évolution paléogéographique du hiatus alpin et des cratons qui l'entourent, les auteurs ont essayé une reconstitution par cartes paléosédimentologiques.

L'aire étudiée comprend les pays suivants : France, Suisse, Italie, Malte, Espagne, Maroc, Algérie, Tunisie et Libye occidentale ainsi que les mers qui les bordent.

Le premier type de carte est une représentation de la lithologie par lithofaciès ; certaines informations sédimentologiques sont reportées en surcharges telles que les oolites, récifs, brèches, volcanisme, etc.. Les isopaques de l'étage figurent chaque fois qu'elles sont disponibles.
Le second type est une interprétation bathymétrique et paléogéographique des cartes précédentes.

Les différents secteurs ont été figurés dans leur position géographique actuelle. Les nappes de charriage sont dans leur cadre régional actuel mais décomposées en unités classées d'après leur polarité probable. Cette représentation a l'avantage de ne pas émettre d'hypothèse palinpastique mais elle a l'inconvénient d'écraser les régions d'origine des chaînes de montagne et d'étirer probablement des zones médianes maintenant occupées par des bassins méditerranéens profonds.

Ces cartes montrent les sillons profonds alpins, apenniniques, bétiques et telliens, les zones intermédiaires et les plateformes ; elles soulignent les anomalies sédimentaires.
On pressent sur ces documents l'antériorité de la Tethys avec son prolongement à l'Ouest du Maroc, confrontée au Crétacé avec les processus d'ouverture de l'Atlantique.

Abstract

As a contribution to the study of the evolution of the western Alpine System, paleosedimentologic maps are presented here and discussed.

Described area includes France, Switzerland, Italy , Malta, Spain, Morocco, Algeria, Tunisia and Western Libya and the surrounding seas.

First types of plates are isopach and lithofacies maps with indication of some sedimentological features as oolites, reefs, breccia volcanism, etc.. The second kind is a tentative palaeophysiographic map with marine bathymetric representation ; it is an interpretation of the first maps.

Sedimentary features and areas are located on the maps in their present geographical position ; nappes and large overthrusts are included in their present general

(1) Compagnie Française des Pétroles, Tour Mirabeau, 39-43 quai André Citroën, 75739 Paris Cédex 15, France

(2) Société Nationale Elf Aquitaine (Production), 26 avenue des Lilas, 64001 Pau, France

setting unless they have been reconstituted in tectonic units classified after their polarity.

This representation avoids premature palinspastic reconstructions ; on the other hand it has the defect of squeezing geosynclinal zones and of stretching internal areas which are presently replaced by recent deep marine basins.

These maps emphasize deep furrows (alpine, apenninic, tellian, and betic), intermediate areas, and platforms. They underline sedimentologic anomalies.

They suggest anteriority of the Tethys, with its possible extension West of Morocco, confronted during the Cretaceous with the Atlantic Ocean opening process.

De la connaissance des relations tectoniques et structurales entre les bassins actuels de la Méditerranée et le Système Alpin peuvent découler des applications économiques pour la prospection des hydrocarbures en mer. C'est pourquoi un programme d'études paléogéographiques a été mis en oeuvre en association, par des géologues de la C.F.P. et de la S.N.E.A.(P). A l'inverse des multiples hypothèses, souvent contradictoires, de reconstruction de la mosaïque méditerranéennes présentées par divers auteurs, nous avons envisagé de commencer par une évaluation des caractères sédimentologiques et paléogéographiques à différents moments représentatifs de l'histoire géologique en établissant des cartes où des données soient, autant que possible, représentées dans leur position géographique actuelle, la seule dont on soit sûr.

La collection et l'examen critique des faits connus sur une région aussi vaste se sont révélés plus longs et plus délicats que nous ne le pensions au départ et ont limité un programme initialement plus ambitieux. Les documents présentés ici concernent juste l'étage choisi au sein du Jurassique: le Bajocien. Les cartes du Cénomanien sont en cours de rédaction mais ne peuvent pas encore être présentées.

Le travail initial de compilation et d'interprétation a été fait au cours des stages industriels à la C.F.P. et à la SNEA par MM. J.C. CALTERO, D. GARDETTE, C. MARTIN, D. SOREL. Outre les auteurs de cet article, les collaborateurs suivants ont participé à l'établissement des cartes d'isopaques et de lithofaciès : MMes M.C. BERNET-ROLLANDE, G. COCHET et S. GAUCHERY pour C.F.P., MM. M. SURIN et L. ESCARTIN pour SNEA (P.). Notons la contribution importante représentée par l'Atlas du bassin d'Aquitaine publié par le BRGM avec le concours des Sociétés SNPA-ELF (= SNEA.P.) et ESSOREP.

Au cours de la conception de ce travail nous avons bénéficié des conseils de D. CONTINI et P. RAT. Nous avons ensuite discuté certains secteurs avec MMes C. RUGET et C. VIRGILI et avec MM. J. DEBELMAS, J. DELFAUD, R. du DRESNAY, N. KASI-TANI, M. LEMOINE, R. MOUTERDE et R. TRUMPY. Nous leur exprimons ici toute notre reconnaissance pour les avis et les critiques reçus.

Nous remercions enfin les dessinateurs et photographes de la Direction Exploration de la CFP pour leur précieuse collaboration : J. MOINE, M. CHAMBOLLE, L. SOLER.

ETABLISSEMENT DES CARTES -

Les données de base ont été reportées sur des cartes à l'échelle du 1/1.000.000: épaisseurs, lithologie, divers caractères sédimentologiques tels que brèches, conglomérats, nodules de manganèse, récifs, oolites, volcanisme. D'autres informations sont consignées sur des fiches pour chaque point de contrôle : faune, figures sédimentaires, stratification, discontinuités, couleur, etc.., de manière à aider l'interprétation de la carte lithologique en vue de définir les faciès.

Les cartes que l'on présente dans les expositions sont établies à l'échelle du 1/3.000.000°, ce qui donne l'ensemble des dimensions acceptables pour les moyens de reproduction, pour le transport et pour la manipulation.

Deux cartes ont été établies pour le Bajocien. La première regroupe les épaisseurs, les lithologies et certains critères sédimentologiques; nous dirons que c'est un document essentiellement fondé sur des faits.

Le second document est une interprétation faciologique de la première, utilisant toutes les données acquises pour imaginer la paléobathymétrie ou paléophysiographie des régions considérées.

Dans cet article, ne pouvant utiliser les couleurs et réduisant fortement l'échelle, nous avons dû simplifier les cartes et diviser l'information en trois documents : 1°/ épaisseur et données sédimentologiques diverses, 2°/ lithofaciès, 3°/ Interprétation paléophysiographique.

La première exposition publique a eu lieu à la Convention annuelle de l'American Association of Petroleum Geologists (Poster Session), Nouvelle Orléans 1976. La présentation officielle, avec les commentaires, s'est produite au cours du Symposium de Split (CIESM 1976).

<u>DIFFICULTES RENCONTREES</u> -

La confection de ces cartes s'est heurtée à de nombreuses difficultés dont les trois principales doivent être mentionnées ici :

a - <u>La disparité de l'information</u>: il y a des régions bien connues, où les données de terrain sont clairement publiées et les forages nombreux et accessibles au public; c'est le cas de la France en général, du Sahara algérien et, en certaine mesure, de la Tunisie et de la Tripolitaine. D'autres régions sont moins connues et il est difficile de réunir certaines catégories de renseignements, les épaisseurs en particulier (Maroc, Espagne).

Enfin, il y a les secteurs mobiles, très tectonisés où les informations sont multiples mais fragmentaires, souvent délicates à interpréter : Alpes en général, Atlas, Chaînes Bétiques et surtout Italie où, en outre, les données de forages ne sont pas dans le domaine public.

La disparité des connaissances complique le choix de la tranche de temps géologique choisie. En Franche Comté où dans le Bassin de Paris, il est dommage de regrouper tout le Bajocien alors qu'une subdivision détaillée, par zones, ou tout au moins en deux, permettrait une interprétation plus fine de la sédimentation et de son cadre. Par contre, en Tunisie, en Libye ou en Aquitaine on ne pourrait que difficilement subdiviser le Bajocien. Au Sahara algérien et dans les zones de nappes il est souvent impossible d'identifier l'étage et les auteurs auraient préféré s'en tenir au Dogger "sensu lato"; et il y a même certains secteurs, en Italie en particulier, où nous avons arbitrairement découpé une tranche pouvant correspondre à l'étage dans des ensembles lithologiques compréhensifs regroupant une part importante du Jurassique, voire du Secondaire.

b - <u>La représentation des lithofaciès</u> a été tentée à l'aide de triangles distinguant des pourcentages entre trois constituants principaux. Mais la nature est infiniment complexe et la région étudiée très vaste, ce qui a nécessité l'introduction de 6 constituants principaux: sables, argiles, calcaires, dolomies, évaporites et roches siliceuses. Un triangle n'ayant que trois sommets, nous avons utilisé différents types de triangles selon les régions considérées; en principe tous les figurés présents sur la carte devraient exister dans l'un des quatre triangles que l'on a situés en marge de la carte à proximité de la région où ils sont le plus utiles. Les pôles: 80 à 100% de calcaire et 50 à 100% d'argiles sont communs à trois de ces triangles alors que le pôle des sables n'est présent que dans un triangle et concerne principalement la marge saharienne; de même le pôle des roches siliceuses n'apparaît qu'une fois et sert pour les faciès rencontrés dans les sillons de type alpin.

c - La dernière difficulté principale est dûe au principe même de la carte, c'est à dire la <u>représentation en position géographique actuelle</u>. Si cette conception est favorable sur les plate-formes continentales et même en ce qui concerne leurs relations réciproques, son application aux zones mobiles est difficile. Dans les régions de nappes superposées il a fallu éviter le "pointillisme" et l'on a donc essayé de regrouper les faciès correspondant aux différentes unités (ou à certains groupes d'unités) en les classant selon la polarité le plus généralement admise mais en gardant approximativement pour l'ensemble de ces unités la place qu'elles occupent actuellement, ce qui entraîne une compression considérable de ces domaines sédimentaires par rapport à leur extension à l'époque. A l'inverse, au travers des bassins récents de la Méditerranée, dûs principalement à des distensions et des effondrements de la fin du Néogène, on a une dilatation considérable de zones sédimentaires certainement beaucoup moins vastes. A titre d'exemple, il n'est pas certain que le bloc kabylo-bétique ait été plus étendu en largeur au Bajocien que la dorsale du Briançonnais. Cette difficulté ne nous a pas échappé, et elle s'accentue lorsqu'on aborde des périodes plus récentes, mais ses inconvénients, une fois connus, nous ont paru inférieurs à ceux de reconstruction palinpastiques prématurées.

<u>Le choix de l'étage</u>, le Bajocien "sensu stricto" pourrait être critiqué, il n'est pas spécialement représentatif des évènements du Jurassique supérieur. Mais, justement, nous lui avons trouvé les avantages suivants :

- Transition entre la paléogéographie stable du Lias et les distensions du Jurassique supérieur.

- Prédominance des facteurs eustatiques sur les plate-formes.

- Apparition de fossés de distension dans les zones alpines avec caractérisation de sillons profonds mais pratiquement sans écartement de domaine à croûte simique, ce qui aurait accentué les difficultés de représentation en position géographique actuelle.

- Existence de vastes aires à sédimentation évaporitique sur la plate-forme saharienne.

<u>PRINCIPAUX RESULTATS</u> -

Ces cartes, dira-t-on, n'apportent rien de bien nouveau. C'est vrai, vu leur origine essentiellement bibliographique, quoiqu'elles intègrent un certain nombre de travaux originaux des deux compagnies et en particulier des données de forages auxquelles il est souvent assez difficile d'accéder.
Bien qu'elles puissent paraître sommaires à chaque spécialiste d'un secteur, elles essaient néanmoins de regrouper différentes régions. Nous rêvons du jour où nous aurions la possibilité d'étendre la zone d'étude, vers le Nord et vers l'Est.

La confrontation des faciès dans de nombreux secteurs permet de limiter les hypothèses tectoniques. Nous ne pouvons pas, par exemple, croire à un décrochement de centaines de kilomètres le long des Pyrénées et dans ce domaine nous aurions donc tendance à nous rallier à une répartition des contraintes en plusieurs décrochements plus modestes comme cela a été proposé par B. PEYBERNES (1976) et B. PEYBERNES & P. SOUQUET (1975).

De même une comparaison générale entre l'Espagne et le Maroc montre une symétrie dans les zones rifaine et bétique avec par contre un caractère assez différent des secteurs de plate-formes instables : Chaînes Cantabriques comparées à l'Atlas marocain. Dans la Péninsule Ibérique nous avons peu de faciès côtiers et l'on peut considérer que la répartition actuelle des affleurements est en grande partie dûe à l'érosion postérieure; au Maroc, si ce facteur n'est pas à négliger, il semble q'une large partie du Sud et même de la Meseta était émergée et représentait une source de matériel gréseux.

REMARQUES PALEOGEOGRAPHIQUES -

Le Bassin de Paris était une plate-forme stable où s'opposaient des faciès marneux infranéritiques et des plate-formes carbonatées (Bourgogne par exemple).

Dans les Alpes on voit, à partir des plate-formes carbonatées du Jura ou du Languedoc une évolution progressive vers des sillons plus profonds, accidentés d'amygdales ou de lèvres soulevées au cours de la distension; celle-ci est particulièrement marquée en Ligurie où la présence d'ophiolites est possible.

Les sillons profonds italiens se développaient entre les blocs cratoniques corso-sardes d'une part et le promontoire adriatique (Pr. Africain d'E ARGAND) avec des faciès détritiques épais de margue connus du Frioul au Nord, à la Calabre au Sud.

L'Aquitaine semble faire partie d'un domaine unique s'étendant du seuil du Poitou au bassin de l'Ebre et caractérisé par une sédimentation confinée marquée par des dolomies. Localement la sédimentation évolue vers le dessalement (lignites). De part et d'autre des Pyrénées centrales on reconnaît des lagunes à gypse.

Il est difficile de dire si la sédimentation recouvrait largement le Massif Central. En l'absence d'éléments détritiques on peut supposer que les Pyrénées n'existaient pas et le Massif Central se marquait peu. La plate-forme plus ou moins confinée s'étendait du seuil des Cévennes jusqu'à une barrière de dunes oolitiques et calcaires biodétritiques qui courrait de l'Angoumois au Béarn.

A l'Ouest s'étendait une mer plus ouverte avec des faciès nicritiques à microfilaments.

Il est délicat de tirer des conclusions sur la profondeur des dépôts du bassin Ouest Aquitaine, profondeur réputée faible, et sur son extension vers l'Ouest.

Sur la péninsule Ibérique l'importance de l'érosion rend les reconstructions hasardeuses. Une plate-forme émergée ou non (elle sera émergée à partir de l'Oxfordien comme le montre les grès dits "Purbeckien" qui apparaissent alors) occupe l'emplacement de la Meseta centrale.

Autour de la Meseta beaucoup des dépôts de plate-forme interne ont pu être décapés ultérieurement. Il reste les dolomies de la région d'Albacete, de Valence et des parties externes du Bassin de Lisbonne au Portugal.

Sur leur périphérie on trouve des formations carbonatées diverses. La mer ouverte se fait sentir vers l'Ouest au coeur du Bassin de Lisbonne et surtout vers le Sud dans le domaine bétique.

Venant du domaine ibérique (dolomies) on rencontre schématiquement en se dirigeant vers le domaine bétique :

- une zone prébétique autochtone de type plate-forme externe carbonatée.
- une zone subbétique allochtone présentant des faciès pélagiques argileux (à radiolarites, ammonitico-rosso, pillowlavas.
- la dorsale calcaire située entre zone bétique et zone subbétique montre une étroite bande de faciès carbonatés de plate-forme avec un approfondissement vers le Nord. Cette disposition permettrait de concevoir la zone subbétique comme un sillon profond entre une zone bétique émergée et la plate-forme prébétique et ibérique.

Il faut pourtant savoir qu'une autre hypothèse à la faveur de certains tectoniciens : le sillon subbétique récelerait le matériel sédimentaire d'un sillon méridional charrié par dessus la zone bétique

La position d'une prolongation éventuelle du Sillon bétique vers l'Est est discutée par différents auteurs.

Au Maroc on note successivement :

- le secteur du Rif symétrique du domaine bétique, dont le sillon sédimentaire se poursuit presque sans interruption jusqu' en Sicile et en Italie par le sillon tellien.
- une meseta marocaine sans doute partiellement émergée car elle comporte à plusieurs endroits de sa périphérie des dépôts sableux.
- un bassin océanique ouvert vers l'Atlantique: bassin de Tarfaya.
- des sillons du Moyen et du Haut Atlas.

Dans le Rif la situation est parfaitement symétrique de celle que nous avons reconnue en domaine bétique : le sillon dont les dépôts se rencontrent entre dorsale rifaine et plate-forme du récif est, soit subautochtone et largement dilacéré en nappes superposées, soit entièrement allochtone et provenant d'un sillon entièrement situé à l'origine au Nord de la dorsale calcaire actuelle.

Que devenaient vers l'Ouest les fosses bétiques et rifaines ? Le couple se perd dans l'Atlantique.

En fait rien n'empêche de faire de la fosse de Tarfaya l'équivalent et la suite du sillon qui s'ouvre au Nord du Récif. La plate-forme continentale située au large de Rabat et Casablanca n'en a pas gardé de traces mais des travaux ultérieurs pourraient le déceler dans les grands fonds du large.

Au Tarfaya et au Rio de Oro on connaît en affleurement ou par forages et sismique de l'ESE vers l'WNW:

une frange de sables continentaux, des depôts internes de plate-forme plus ou moins confinés, des dépôts externes carbonatés avec éventuellement des épisodes récifaux (mieux connus au Malm), enfin un talus ouvert sur des dépôts plus profonds.

Si l'on veut bien reconnaître ici la suite probable du sillon rifain, ne faut-il pas aller chercher en Atlantique ou même outre Atlantique, la suite du sillon bétique déplacé par l'ouverture de l'océan.

Du Maroc en Tunisie se tordent obliquement entre Sahara et géosynclinal mésogéens des sillons moins profonds, longs et étroits, laissant entre eux des domaines de plate-forme dont - à la différence de la Meseta marocaine - nous savons qu'ils furent émergés au Dogger (Oranie).

Le Haut Atlas marocain était le siège d'un sillon bordé de platiers récifaux avec, en pérophérie et surtout à l'W, des formations détritiques ou confinées.

Il se poursuit en Algérie par le sillon Atlas saharien qui reçoit une impressionnante quantité de matériel terrigène en provenance directe du Sahara. Sous l'afflux des sédiments le sillon devait être comblé très tôt dans l'W tandis qu'il passait à des dépôts beaucoup plus profonds à l'ENE.

Nous ne connaissons pas dans le Sahara occidental l'équivalent du bassin évaporitique du Sahara oriental. La présence de sables en bordures W du bassin évaporitique indique que le Sahara occidental était probablement émergé et explique l'abondance des sédiments sableux ou argileux qui aboutissaient dans la fosse de l'Atlas Saharien

Entre Atlas Saharien d'une part, et sillon tellien d'autre part, on note la présence d'une plate-forme carbonatée peu subsidente dans l'ensemble mais disloquée en compartiments d'inégales épaisseurs de sédimentation. Cette plate-forme oranaise et marocaine était bordée de marges plus subsidentes donnant naissance à d'impressionnantes falaises calcaires en direction de la fosse de l'Atlas Saharien.

En Tunisie, une suture majeure dite Axe Nord-Sud limite la zone instable atlasique vers l'Est; au-delà la plate-forme pélagienne est plus stable avant de s'abaisser vers les faciès plus profonds de la Sicile orientale.

En Tripolitaine, comme dans l'Est du Sahara algérien et dans l'Extrême Sud tunisien, le Bajocien correspond à une vaste lagune à sédimentation évaporitique; seuls les alentours du môle émergé dit de Médenine sont soulignés par des réductions d'épaisseurs et des faciès dolomitiques.

Au Fezzan les détritiques d'origine continentale s'intercalent dans les évaporites et finissent par dominer.

CONCLUSIONS -

1 - De nombreux secteurs où n'existent pas actuellement de dépôts attribuables au Bajocien ont dû être le siège d'une sédimentation néritique puis d'érosions ultérieures (Massif Central, Espagne, etc..) et l'on ne connaît pas de faciès côtiers aux alentours.

Les cartes présentées soulignent les anomalies locales de la sédimentation; bien souvent une zone de faible épaisseur, voire d'absence, ne correspond pas à un haut fond mais soit à une pente où les sédiments ne pouvaient pas s'arrêter, soit à un bassin profond où il y avait un déficit des apports en regard de la subsidence.

2 - Les faciès internes bético-rifains, ceux de la Dorsale calcaire nord-africaine et les séries décrites aux Baléares, appuient l'hypothèse d'un bloc central regroupant ces éléments et probablement en partie émergé au Dogger. Ses relations avec la Sardaigne, ainsi que les rapports entre cette île et la Calabre cristalline sont difficiles à préciser et il faut bien considérer que l'hypothèse choisie est très risquée.

3 - On voit s'esquisser nettement les bassins de type périalpin caractéristiques des "chaînes intermédiaires" : Atlas saharien algéro-marocain, Fosse vocontienne, etc.. Même dans une période aussi calme que le Jurassique moyen l'instabilité relative des zones périalpines et périatlasiques apparaît bien avec ces sillons subsidents à sédimentation néritique et les hauts fonds qui leur sont parfois opposés (Môles de Médenine en Tunisie, de Constantine en Algérie, etc..) A ces zones périalpines s'opposent déjà les sillons profonds de distension, bordés par des lèvres soulevées comme le Briançonnais et caractérisés généralement par des sédiments peu épais de type bathyal ou abyssal, et des formations de pente parfois grossièrement détritiques d'épaisseurs très inégales.

4 - On trouve presque sur le même méridien, de part et d'autre des Pyrénées, des faciès identiques et pourtant très particuliers (dolomie et anhydrite).

Cette observation s'oppose à l'existence d'un décrochement important d'une plaque ibérique par rapport à l'Europe le long d'un accident Nord-Pyrénéen depuis le Dogger.

Adhérer à l'hypothèse de mouvements relatifs de grande amplitude de blocs situés de part et d'autre des Pyrénées, depuis le Dogger, c'est admettre que l'hypothèse trouve dans d'autres disciplines considérées comme plus séduisantes le support que la géologie régionale ne peut lui apporter.

5 - Les sillons profonds de la Téthys sont parfaitement perçus avec leur symétrie très caractéristiques jusqu'à Gibraltar vers l'W. Que devenaient-ils au delà?

Concernant leur marge Sud l'hypothèse la plus simple consiste à la rechercher dans le bassin de Tarfaya. Il n'y a interruption d'observation qu'à hauteur de Rabat et Casablanca où la trace du bassin doit être recherchée en eau profonde au-delà de la plate-forme continentale.

L'autre marge de la Téthys serait à rechercher au sein ou au delà de l'Atlantique actuel où elle a été déplacée par l'ouverture océanique ultérieure.

L'hypothèse la plus simple concernant les relations entre Téthys et Atlantique consiste à donner l'antériorité à la Téthys (s'étendant de la Méditerranée aux Caraïbes). Ultérieurement - au cours du Crétacé - l'ouverture Atlantique séparera en deux les blocs Laurasien d'une part et Gondwanien d'autre part en empruntant localement le tracé de la Téthys dont elle éloignera les deux marges. Ainsi sur les marges de l'Atlantique moyen l'histoire atlantique succède à l'histoire mésogéenne suivant le processus pressenti par beaucoup d'auteurs mais dont l'étude n'a pas été abordée sous cet éclairage. En Espagne on assisterait à une polarité NS s'intégrant dans le système alpin, dans les provinces orientales en particulier, et à un basculement Est-Ouest avec soulèvement de la Meseta et enfoncement d'un gradin portugais, annonce du "Graben" atlantique.

BIBLIOGRAPHIE

(1) - ARNAUD M., MONLEAU C., 1975 - Le Lias et le Dogger inférieur du bord de mer Varois - Géologie méditerranéenne Tome II, n° 4, pp. 147 à 158.

(2) - AYME J.M., GUILLAUME R., 1957 - Rapport géopétrolier sur le Jurassique du Bassin de Paris - C.F.P. (non publié).

(3) - AZEMA J., 1966 - Sur l'existence des phénomènes de ravinement, de condensation et de remaniement au sein des formations secondaires de la Sierra de Grevillende (Prov. d'Alicante, Espagne), Bull. Soc. Géol. de France (7) VIII, 1966, pp. 518-520.

(4) - BARFETY J.C., DEBELMAS J., MOUTERDE R. 1972 - Caractères stratigraphiques, paléontologiques et structuraux du Jurassique inférieur et moyen des bordures W et SE du Massif de Belledonne - Géologie alpine, T. 48, pp. 61-86.

(5) - BEAUDOIN B., 1975 - Le Bassin subalpin au Jurassique - IX Congrès International de Sédimentologie, Nice, Thème 5, T. 1, pp. 51-64.

(6) - BENZAQUEN M., 1964 - Puissance des Formations Jurassiques du Djbel Ouarit (Moyen Atlas septentrional) - C.R. SGF n° 5, pp. 190-191.

(7) - BERTRAND J.P., 1975 - Données sur la sédimentologie du Jurassique français. Faciès de bassin et plateforme S.E. de la France - IX Congr. International de Sédimentologie, Nice, Thème 5, T. 1, pp. 73-75.

(8) - BIELY A., 1969 - Bajocien sous faciès "Ammonitico rosso" au Jebel Zaghouan - Notes Serv. Géol. Tun. 30. 1, pp. 11-16, Tunis.

(9) - BISMUTH H., BONNEFOUS J., DUFAURE Ph. 1967 - Mesozoic microfacies of Tunisia - Guidebook to the geology and history of Tunisia - Petr. Expl. Soc. of Libya, 1967, pp. 159-173, Tripoli.

(10) - BONNEFOUS J., 1972 - Thèse Paris VI-Contribution à l'étude stratigraphique et micropaléontologique du Jurassique de Tunisie (Tunisie septentrionale et centrale, Sahel, zone des Chotts).

(11) - BONNEFOUS J., RAKUS M., 1965 - Précisions nouvelles sur le Jurassique du Djebel Bou Kornine d'Hamman Lif (Tunisie) - Bull. Soc. Géol. Fr., 7 sér., T. 7, n° 6, pp. 855-858, Paris.

(12) - BORDALO DA ROCHA R., MOUTERDE R., TINTANT E., 1972 - Formations récifales dans le Jurassique moyen d'Algarve occidental - Revista da Faculdade de Ciencias de Lisboa, 2 ser., C, v. 17, 1., pp. 45-58.

(13) - BRUN, 1969 - Le Dogger des Pyrénées ariégeoises.

(14) - BULARD P.F., 1974 - Sur les faciès et l'extension du Jurassique au sud de Saragosse - C.R. Acad. Sci. France T. 263 D, n° 3, pp. 220-223.

(15) - BUROLLET P.F., 1960 - "Names and Nomenclature Committee de la Petrol. Expl. Soc. of Libya" - Lexique stratigraphique international : Libye - CNRS Paris, 4 a, 62 pp.

(16) - BUROLLET P.F., 1966 - Remarques sur la stratigraphie du Jebel Nefusa (Tripolitaine - Libye) - Riv. Ital. Paleont. e Stratigr., vol. 72, 4, pp. 1313-1317.

(17) - BUROLLET P.F., 1971 - Tunisia - Geologia - Enciclopedia del petrolio e del gas naturale - Pubbl. Ente naz. Idrocarb., Colombo ed., Rome, vol. 8, pp. 1246-1274.

(18) - BUSSON G., 1967 - Le Mésozoïque saharien - 1ère partie - L'Extrême Sud Tunisien - Publ. Centre Rech. Zones arides (CNRS), Paris, Géol. n° 8, 194 pp.

(19) - BUSSON G., 1970 - Le Mésozoïque saharien - 2ème partie - Essai de synthèse des données des sondages algéro-tunisiens - Publ. Centre Rech. zones arides (CNRS) Paris, Géol. n° 11, 2 vol., 810 pp.

(20) - CANERAT J.M., 1974 - Recherches géologiques aux confins des chaînes ibérique et catalane (Espagne) - Thèse, Juin 1974.

(21) - CHAMPETIER Y., 1966 - Etude du Jurassique et du Crétacé de la Sierra Fontanello - Province Valence (Espagne) - Mém. Fac. Sci. Univ. de Nancy - DES Sci. Nat.

(22) - CHOUBERT G., 1971 - Domaine rifain Atlasique - Unesco 1971 - Tectonique de l'Afrique (Sciences de la Terre,6).

(23) - CONTINI D., 1970 - L'Aalénien et le Bajocien du Jura franc-comtois - Annales scientifiques Université de Besançon - 3è série - Géol., fasc. 11.

(24) - DUBAR G., 1952 - Variation de faciès et lacunes dans le Bajocien inf. du Haut Atlas marocain - 19è Congr. Géol. Intern., Alger, Fasc. 15 - pp. 17-22.

(25) - DU DRESNAY R., 1962 - Les discontinuités de sédimentation pendant le Jurassique dans le domaine oriental de l'Atlas marocain - Colloque du Jurassique.

(26) - DU DRESNAY R., 1963 - Interprétations stratigraphique et paléogéographique du Jurassique moyen Jbel Klakh (Haut Atlas marocain oriental) - C.R. Acad. des Sci., 1er Avril 1963, pp. 3157-3159.

(27) - DU DRESNAY R., 1971 - Extension et développement des phénomènes récifaux jurassiques dans le domaine atlasique marocain - Extrait du Bull. de la S.G.F. - 7è série, T. XIII, pp. 46-56.

(28) - DU DRESNAY R., 1971 - Relations paléogéographiques entre sillons atlasiques et plate-forme des Hauts Plateaux pendant l'Aalénien et le Bajocien - Extrait Coll. du Jurassique Luxembourg 1967 - Mém. BRGM, Fr. n° 75. 1971.

(29) - DURAND J.G., 1961 - Le Dogger dans les sondages de la C.E.P. - Ouest Bassin de Paris - Revue IFP, Vol. XVI, n° 5, pp. 511-537.

(30) - DURAND DELGA M., 1969 - Mise au point sur la structure du Nord-Est de la Berberie - Bull. Serv. Géol. Algérie, n. sér., 39, pp. 89-131.

(31) - FLEURY J.P., 1968 - La marge orientale du Causse Méjean et du Causse Noir - Aalénien-Bajocien dans la région de

Meyrueis - Bull. Soc. Géol. de France (7) X, pp. 645-652.

(32) - GLAÇON G. et ROUVIER H., 1970 - Découverte de Jurassique dans les monts de la moyenne Medjerda (Tunisie septentrionale) - C.R. Acad. Sc. Paris, t. 270 sér. D, 25, pp. 3007-3009.

(33) - GOUDARZI GUS H., 1970 - Geology and mineral resources of Libya - A reconnaissance - US Geol. Surv. Prof. Paper, n° 660, Washington, 104 pp.

(34) - GROUPE FRANCAIS D'ETUDES DU JURASSIQUE, 1973 - Stratigraphie et sédimentologie du Jurassique en Côte d'Or - 6è excursion annuelle - Institut des Sciences de la Terre, Dijon, pp. 15-19.

(35) - GWINNER M.P., 1971 - Geologie der Alpen - E. Schweizerbart'sche Verlagsbuchhandlung (Nägele u. Obermiller) Stuttgart, 477 pp.

(36) - KIEKEN M., 1962 - Les traits essentiels de la géologie algérienne dans "Livre à la mémoire du Professeur Paul Fallot" Soc. Géol. Fr., pp. 545-614 - 7 fig., 3 pl.

(37) - KIEKEN M., 1975 - Etude géologique du Hodna, du Titteri et de la partie occidentale des Biban (Dept. d'Alger - Algérie) - Serv. Géol. d'Alger, Bull. n° 46, T. II, 269 pp.

(38) - LORENCHET M., 1964 - Limite du Bajocien et du Bathonien dans le moyen Atlas oriental - C.R. SGF, Fasc. 3, 1964, pp. 122-124.
Le Bajocien du moyen Atlas oriental pp. 110-111.
Bathonien et Baj. sup. de la plaine de Guercif et de ses bordures - pp. 102-104.

(39) - LUCAS G., 1942 - Description géologique et pétrographique des monts Ghar Rouban et du Sidi el Abed (Frontière algéro-marocaine) - Bull. Serv. Carte Géologique Algérie - 2è sér. Strat.

(40) - LUCAS G., 1952 - Bordure Nord des Hautes Plaines dans l'Algérie occidentale - XIX Congrès Géologique Intern., Alger - 1ère série : Algérie n° 21, pp. 57-113.

(41) - MAGNIER Ph., 1963 - Etude stratigraphique dans le Gebel Nefousa et le Gebel Garian (Tripolitaine Libye) - Bull. Soc. Géol. Fr., 7è sér., E. 5, 1(1964), pp. 89-94, Paris.

(42) - MOUTERDE R., KERRIEN Y., LABOURGUIGNE J., MANIVIT J., 1966 - Le Lias et le Bajocien de La Javie (B.A.) - Bull. Soc. Géol. de France (7), VIII, pp. 347-352.

(43) - MOUTERDE R., RUGET Ch., 1975 - Esquisse de la paléogéographie du Jurassique inférieur et moyen au Portugal - Bull. Soc. Géol. Fr., 7è sér., 17, 5, pp. 779-786.

(44) - PEYBERNES B., 1976 - Le Jurassique et le Crétacé inférieur des Pyrénées franco-espagnoles entre la Garonne et la Méditerranée - Thèse doctorat, Toulouse.

(45) - PURSER B.H., 1975 - Le Jurassique moyen de la Bourgogne - IX Congr. Intern. de Nice - pp. 341-346.

(46) - RAKUS M., 1968 - Recherches stratigraphiques dans la formation Nara (Jurassique) - Axe Nord-Sud de Tunisie - Notice préliminaire - Bratislava.

(47) - RIOULT M., FILY G., 1975 - Discontinuité de sédimentation et unités lithostratigraphiques dans le Jurassique de Normandie - IX Congr. Intern. de Nice, thème 5, pp. 353-360.

(48) - RUGET Ch. 1961 - Etudes stratigraphiques du Dogger et du Malm inf. du Portugal au bord du Tage.

(49) - SUAREZ VEGA L.C., 1974 - Estratigrafia del Jurasico en Asturias - Publ. del Departamento de Geologia economica - Cuadernos de geologia iberica 74, 3, T. 1 et 2, Madrid, 369 pp.

(50) - SAPIN S., 1967 - Principaux résultats géologiques des travaux réalisés par SNPA dans le S.O. du Bassin de Paris - Bull. Soc. Géol. de France, 7è série, T. IX, pp. 327-354.

(51) - TEMPLIER C., 1972 - Les faciès calcaires du Jurassique provençal - Thèse Sciences Marseille.

(52) - VIALLARD P., 1973 - Recherches sur le cycle alpin dans la chaîne ibérique sud occidentale - Thèse Sci. Nat., 1973.

(53) - VIRGILI C., 1968 - Quelques résultats sur le Jurassique des Asturies (Espagne) - C.R. som. des séances de la SGF - fasc. 7, 1968, p. 226.

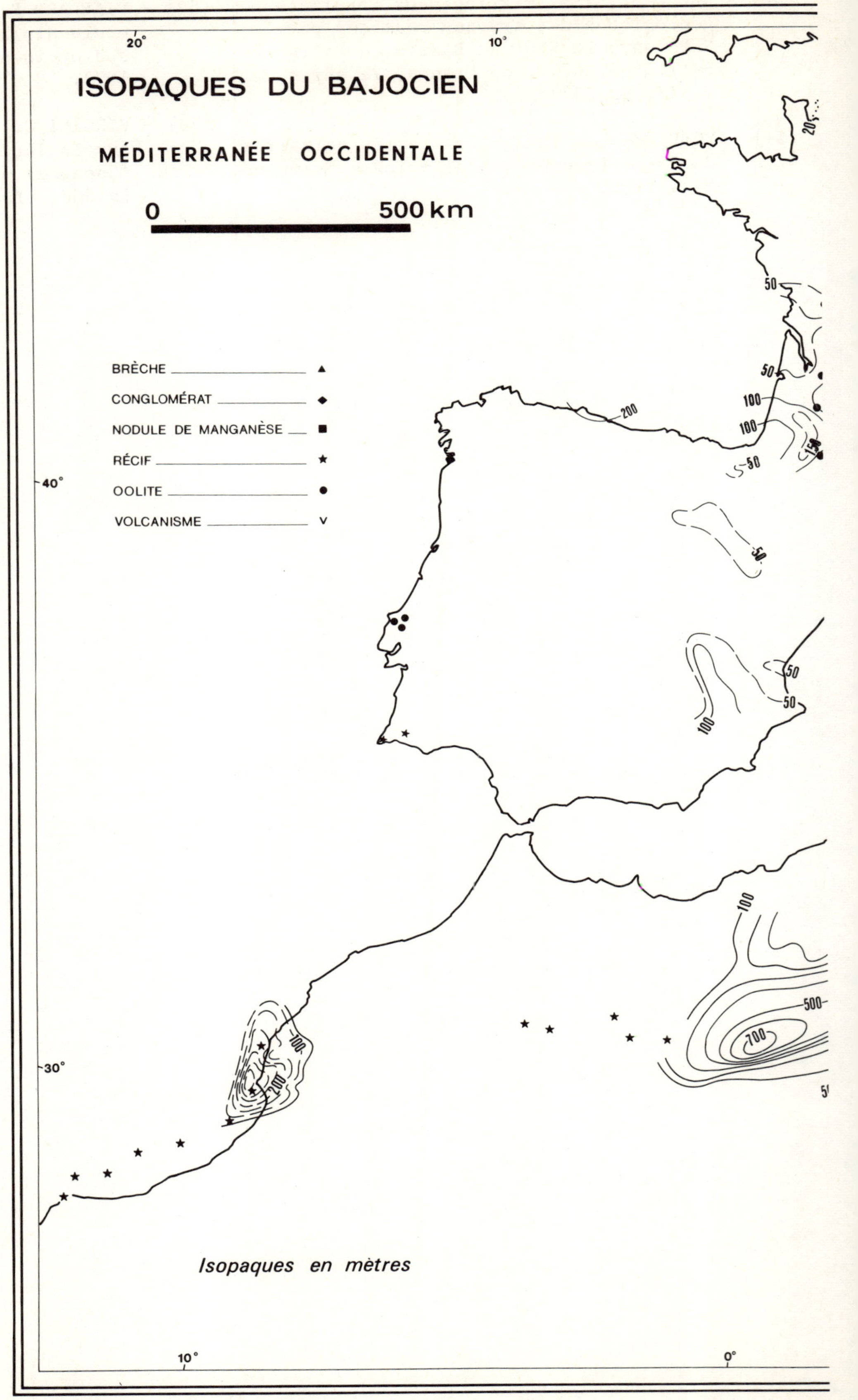

ISOPAQUES DU BAJOCIEN
MÉDITERRANÉE OCCIDENTALE
0 500 km
BRÈCHE ▲
CONGLOMÉRAT ◆
NODULE DE MANGANÈSE ■
RÉCIF ★
OOLITE ●
VOLCANISME V
Isopaques en mètres
20°
10°
40°
30°
10°
0°

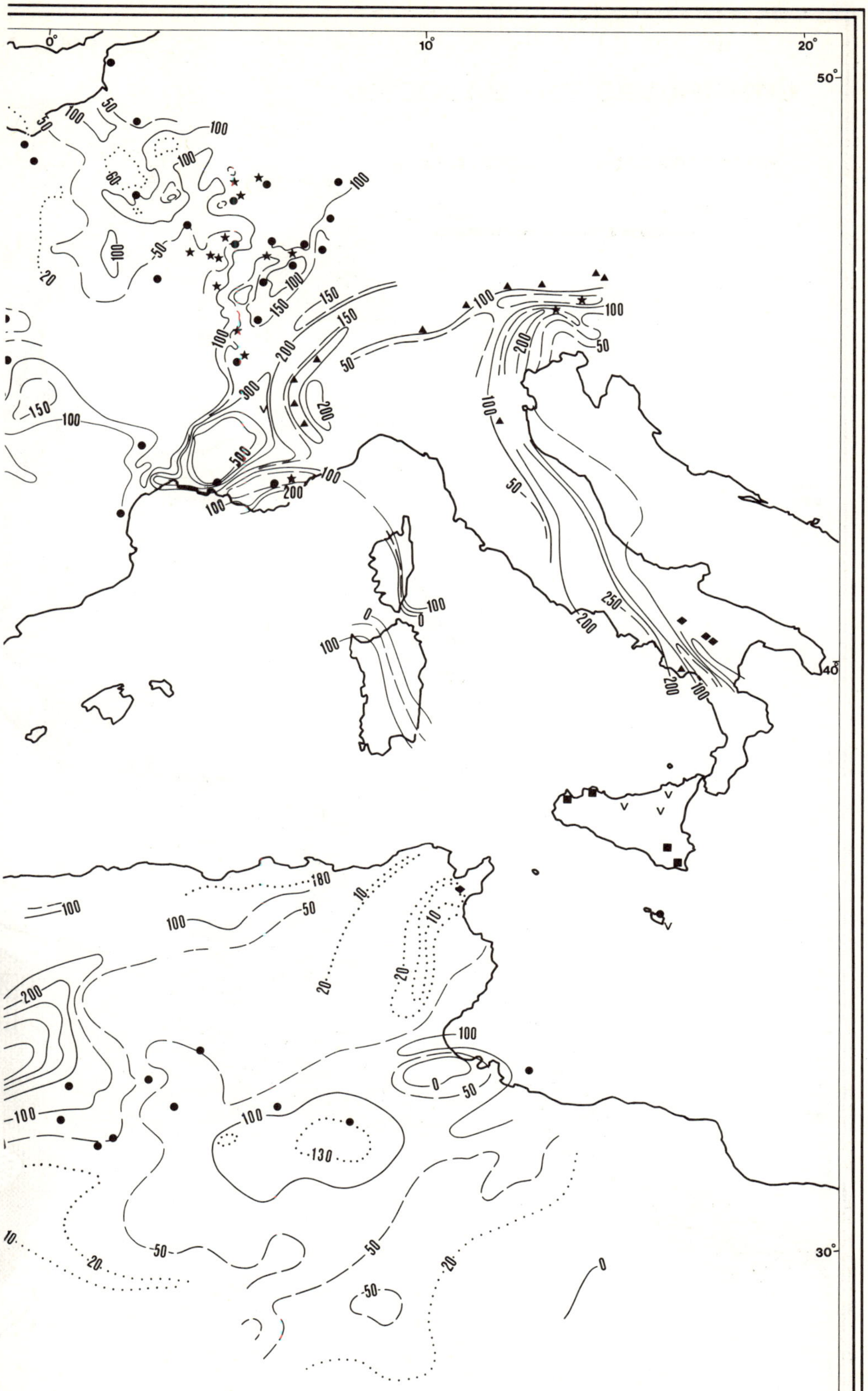
0°
10°
20°
50°
40°
30°
10°

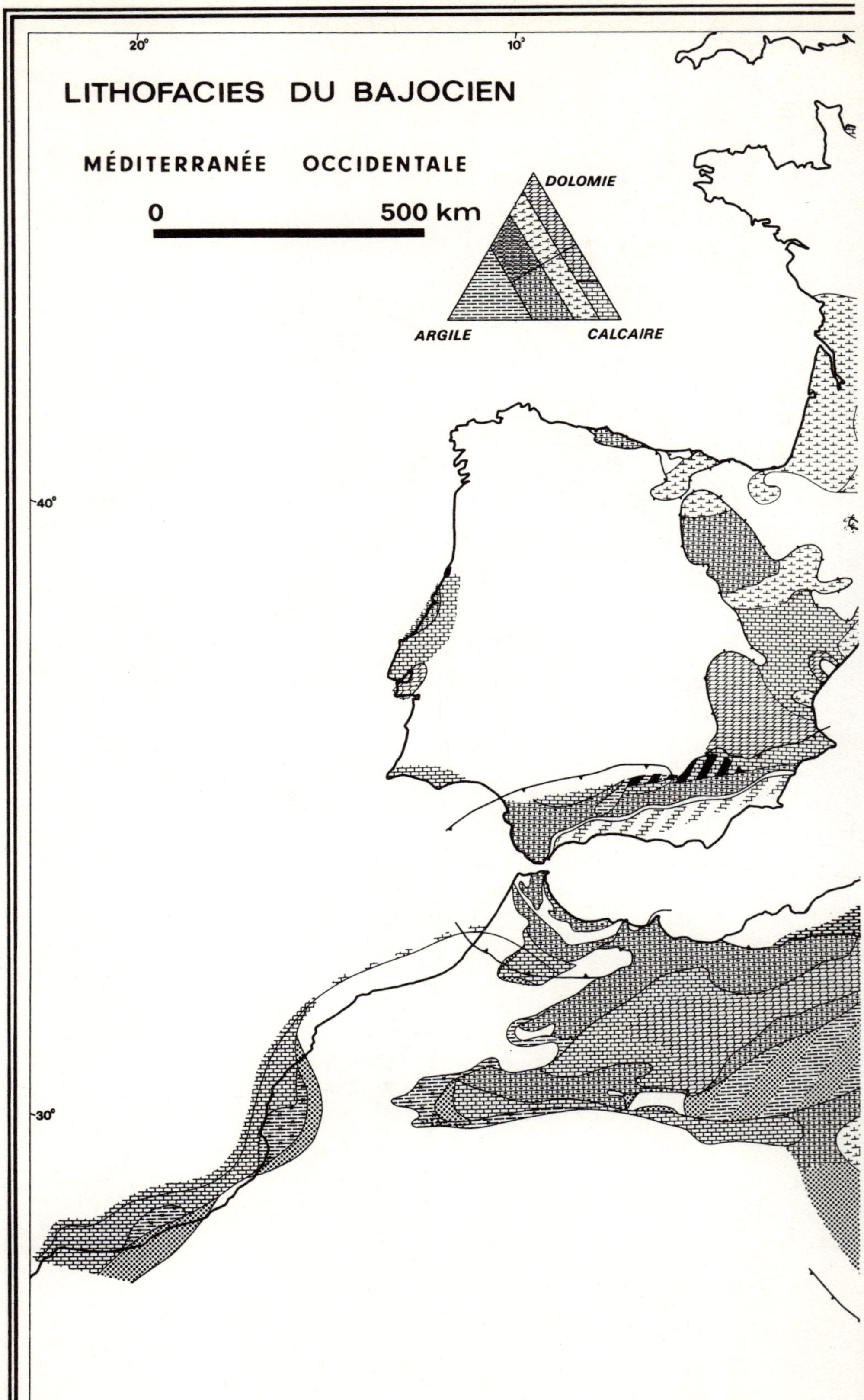
20°
10°
LITHOFACIES DU BAJOCIEN
MÉDITERRANÉE OCCIDENTALE
0
500 km
DOLOMIE
ARGILE
CALCAIRE
40°
30°
10°
0°

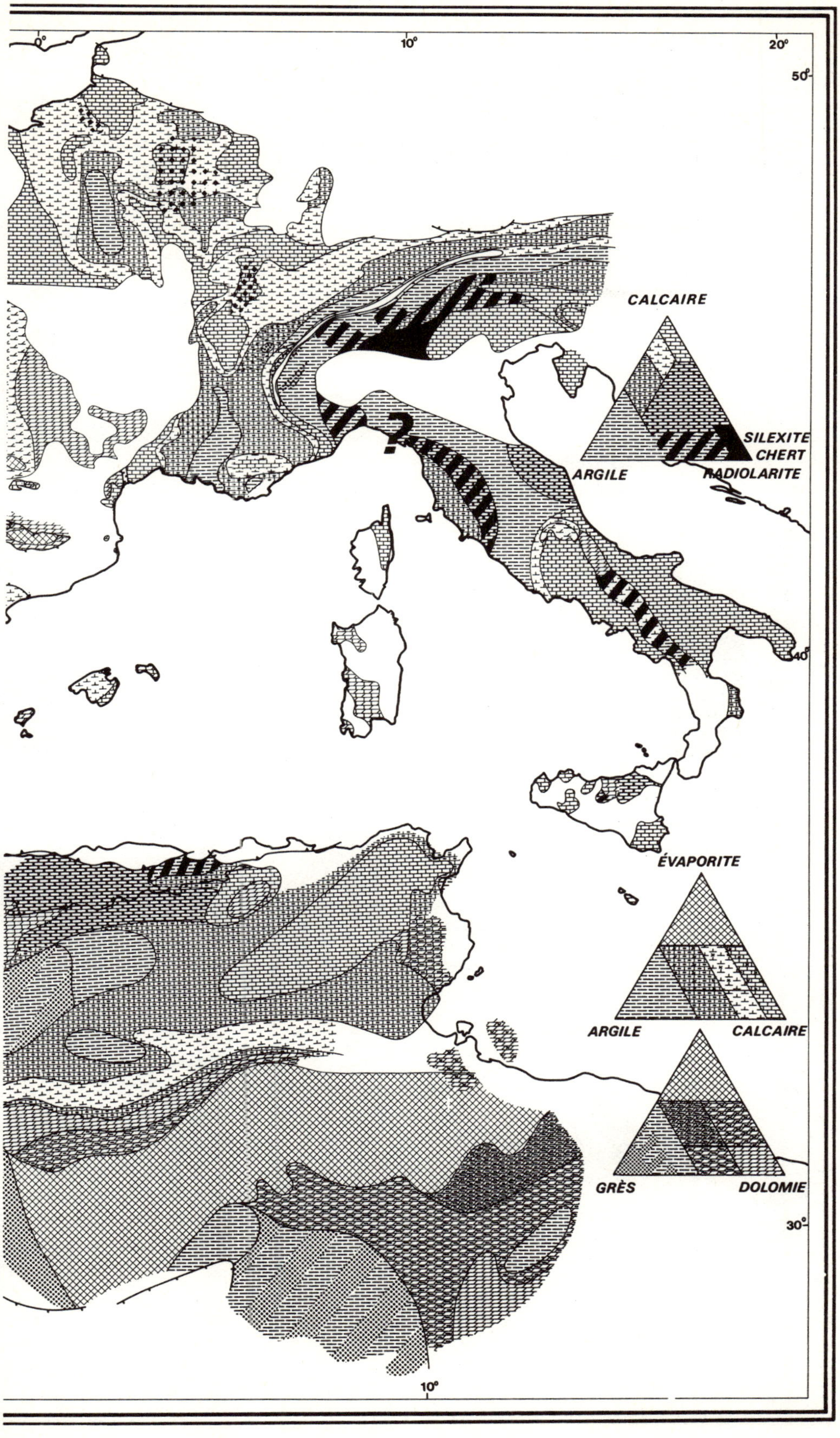
0°
10°
20°
50°
CALCAIRE
SILEXITE
CHERT
ARGILE
RADIOLARITE
?
40°
ÉVAPORITE
ARGILE
CALCAIRE
GRÈS
DOLOMIE
30°
10°

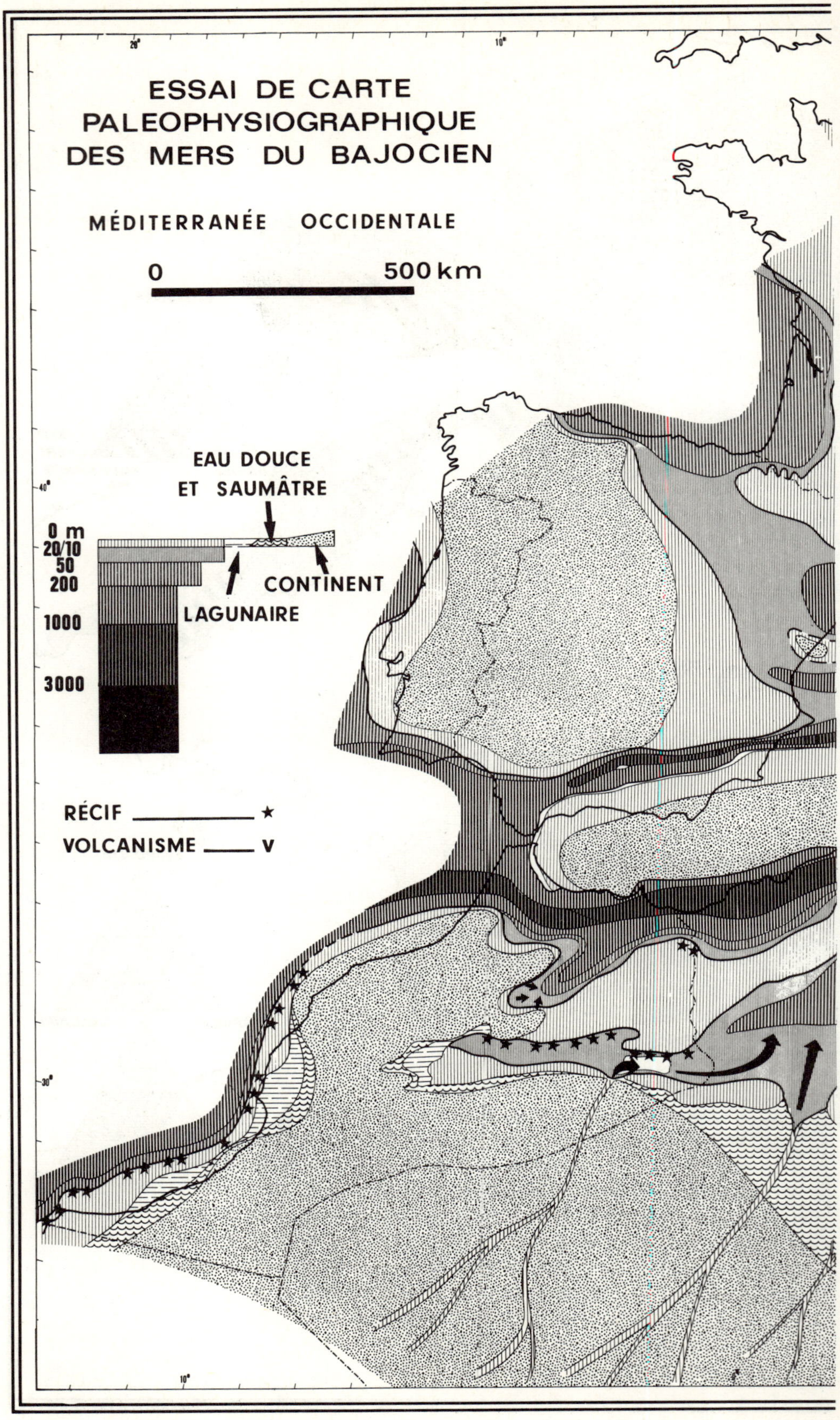
ESSAI DE CARTE
PALEOPHYSIOGRAPHIQUE
DES MERS DU BAJOCIEN
MÉDITERRANÉE OCCIDENTALE
0
500 km
EAU DOUCE
ET SAUMÂTRE
0 m
20/10
50
200
1000
3000
CONTINENT
LAGUNAIRE
RÉCIF
VOLCANISME

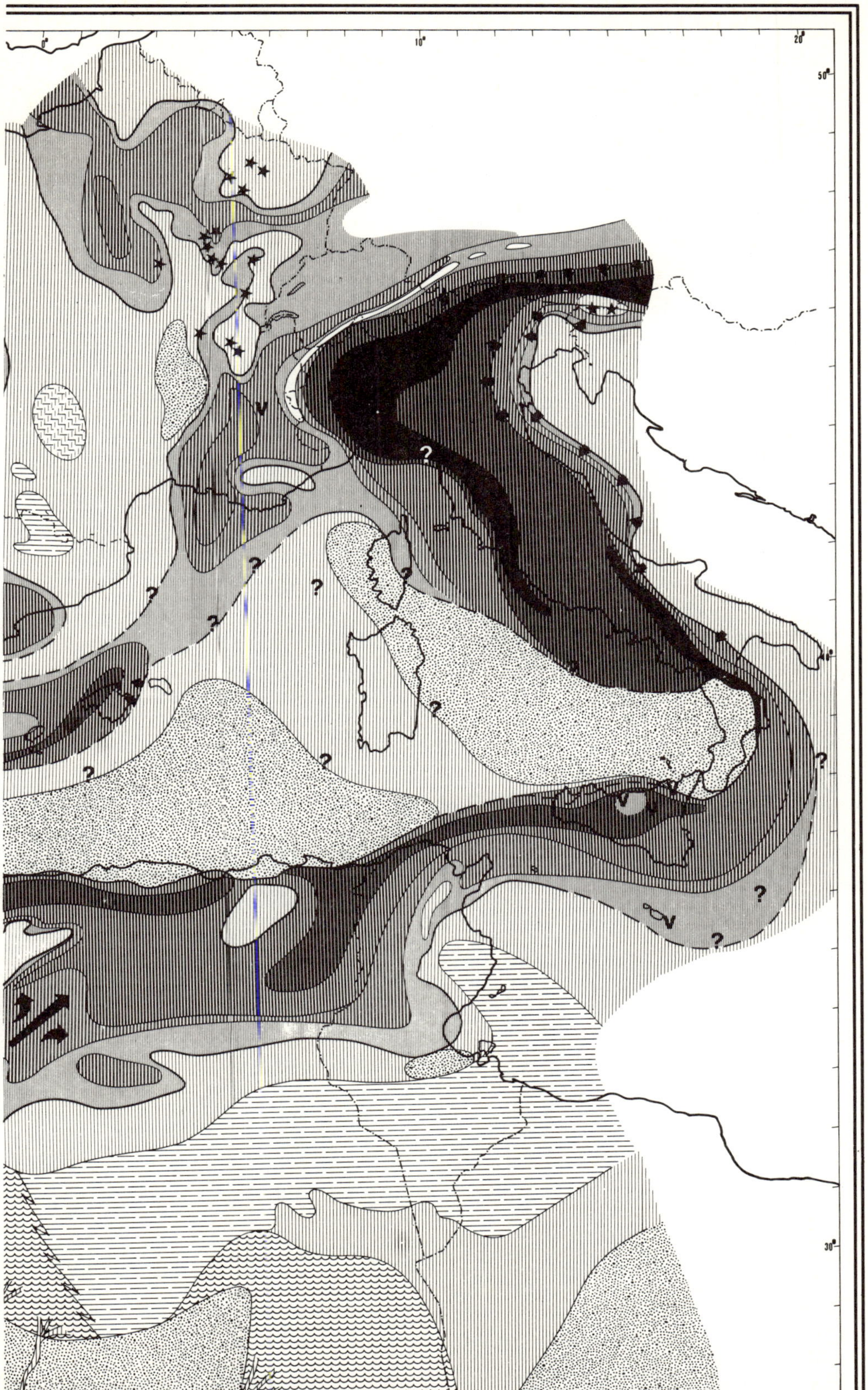
10°
20°
50°
40°
30°

INTERNATIONAL SYMPOSIUM ON THE STRUCTURAL HISTORY OF THE MEDITERRANEAN BASINS. SPLIT (YUGOSLAVIA) 25-29 OCTOBER 1976. B. BIJU-DUVAL AND L. MONTADERT, Eds. EDITIONS TECHNIP, PARIS 1977, pp. 129-132

MEDITERRANEAN AND TETHYS

H. LAUBSCHER and D. BERNOULLI [1]

SUMMARY

"Tethys" and "Mediterranean" are marine systems at the Eurasia-Africa and Eurasia-India plate boundaries. These boundaries had a dynamic history from the Triassic on and the boundary seas can only be understood in the light of this history, which may be viewed from the vantage-point of a number of guiding concepts. These are (1) a pre-Jurassic assembly of the continental masses into a Pangaea as proposed e.g. by Dietz and Holden (1970) and modified for the Atlantic by Dewey et al.,(1973)(fig.1);(2) an opening history for the Atlantic as proposed by Pitman and Talwani (1972) and modified by Dewey et al. (1973);(3) a distribution of oceanic and continental margin belts in the Mesozoic boundary zone as suggested by surface geology, i.e. a central oceanic domain flanked by continental margins of the Atlantic type (compare figs. 2a, 2b, 4); (4) consumption of the Mesozoic ocean and a large part of its margins by Cretaceous-Tertiary orogeny as plate boundaries become compressive, and foundering of the mountain system in a series of successor basins due to new lithospheric instabilities in connection with subduction. It is proposed to call the pre-Mesozoic embayment of Panthalassa that reached as far west as Sicily the "Paleotethys", whereas the term "Tethys" should be reserved for the marine systems associated with Mesozoic rifting, and the term "Mediterranean" for those Tertiary boundary seas dominated by successor basins.

After the Variscan orogeny the central Atlantic and western Mediterranean areas were part of one large continental mass; no remnants of pre-Jurassic oceanic crust are known from this area and Permo-Triassic deposits are predominantly continental, with episodic marine incursions (fig. 1). However, in the east the pre-drift configuration shows a wedge-shaped space between Africa and Eurasia. This probably once oceanic area - called here Paleotethys - has completely fallen victim to subduction in Mesozoic-Tertiary times and only extensive deposits of its shallow continental margins are conserved in the nappes of continental origin in the Alpine-Himalayan system.

In the Triassic early rifting was associated with widespread volcanism and marine incursions as far west as the Betic Cordillera. Also pelagic domains were established as far west as Sicily though it is doubtful that their basement had been oceanic (figs.1, 3).

In the Jurassic major rifting began in the Central Atlantic, and dominated in the Tethyan domain as far east as the Balkan triple point where independent rifting in the Paleotethys domain began to dictate plate tectonic development. This rifting in places was discordant with respect to Triassic paleotectonics as suggested in figs. 1 and 3. Oceanic areas of Tethys today are recognized in the relics of intricately contorted ophiolite nappes which are flanked by nappes derived

(1) Geological Institute at the University, Bernoullistrasse 32, 4056 Basel, Switzerland

from Atlantic type continental margins on both sides. From kinematic considerations it is plausible that these oceanic areas belonged to the Jurassic-Early Cretaceous world rift system and accreted along a spreading ridge rather than in small basins behind island arcs. Arguments for this may be seen in the close affinities in the sedimentary and most probably also bathymetric evolution of the central Atlantic and Liguria-Piemont ocean basin and the evolution of typical, subsiding continental margins of Atlantic-type. The southern continental margin in particular was characterized by extensive Bahamas-like carbonate platforms large portions of which sank into sub-photic depths with pelagic sedimentation at one time or another (fig. 4).

Farther east, there must have been compressional plate boundaries at this time, and particularly in the area of the triple point between the African, Eurasian and Paleotethys plates extensional and compressional movements were closely juxtaposed in space and time (fig. 2).

There is a possibility that the present eastern Mediterranean is a relic branch of Tethys (fig. 2b).

With the opening of the North Atlantic in the Late Cretaceous the Tethyan scenery was subject to radical change. The sinistral and dilational movements of the Jurassic and Early Cretaceous were substituted by dextral and compressional ones, and the former oceanic areas were consumed in orogeny. By Late Eocene times the entire central Tethys north of the south-Tethyan carbonate belt had been consumed, and only the Eastern Mediterranean, south of that belt, is a potential relic of the Paleotethys or a separate Jurassic-Cretaceous ocean basin. At the same time, successor basins began to founder discordantly on the orogenic belt, and have continued to do so up to Recent times, with some of the earlier ones being destroyed by subsequent compression. The rapid change in configuration of these successor basins eventually lead to repeated separation from the open oceans which precipitated the Messinian salinity crises. These were a cluster of pan-Mediterranean events which affected a number of basins of different origin and depth.

The full extent of this paper will be published in: Laubscher, H. & Bernoulli, D. (1977): Mediterranean and Tethys. In: Nairn, A.E.M., Stehli, F.G., & Kanes, W. (Eds.): The Ocean Basins and Margins. IV.: Mediterranean. -Plenum Publ., Corp. New York. This also contains the bibliography of the references in the present abridged version.

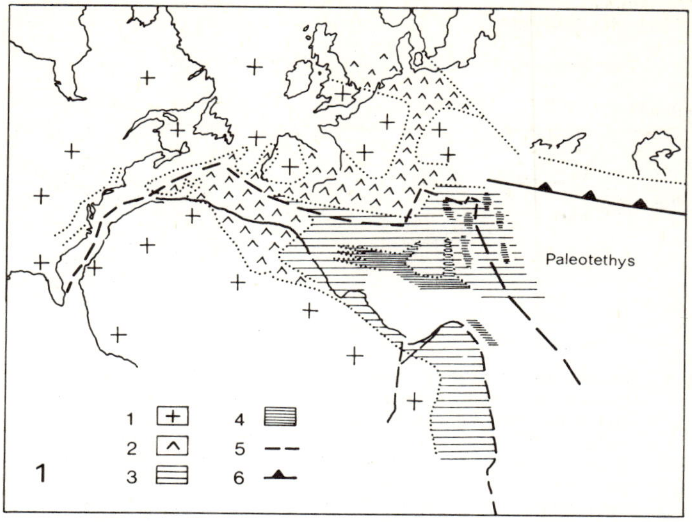

Figure 1 : Palinspastic restoration of the Late Triassic paleogeography of the Mediterranean and central Atlantic area. Reassembly of the principal continental masses in the Triassic, about 220 ± 20 m.y.b.p., after Dewey et al. (1973). The dashed heavy line indicates the later E-W-oriented Mesozoic break-up of Pangaea and barbed lines approximate locations of Mesozoic subduction zones.

1 - Continental areas with erosion or local continental sediments.
2 - Basins occasionally flooded by marine waters including important evaporite deposits.
3 - Strongly subsiding basins with thick carbonate platform (and evaporite) deposits.
4 - Major basins with deep-water facies.
5 - Jurassic break-up of Pangaea.
6 - Mesozoic subduction zones.

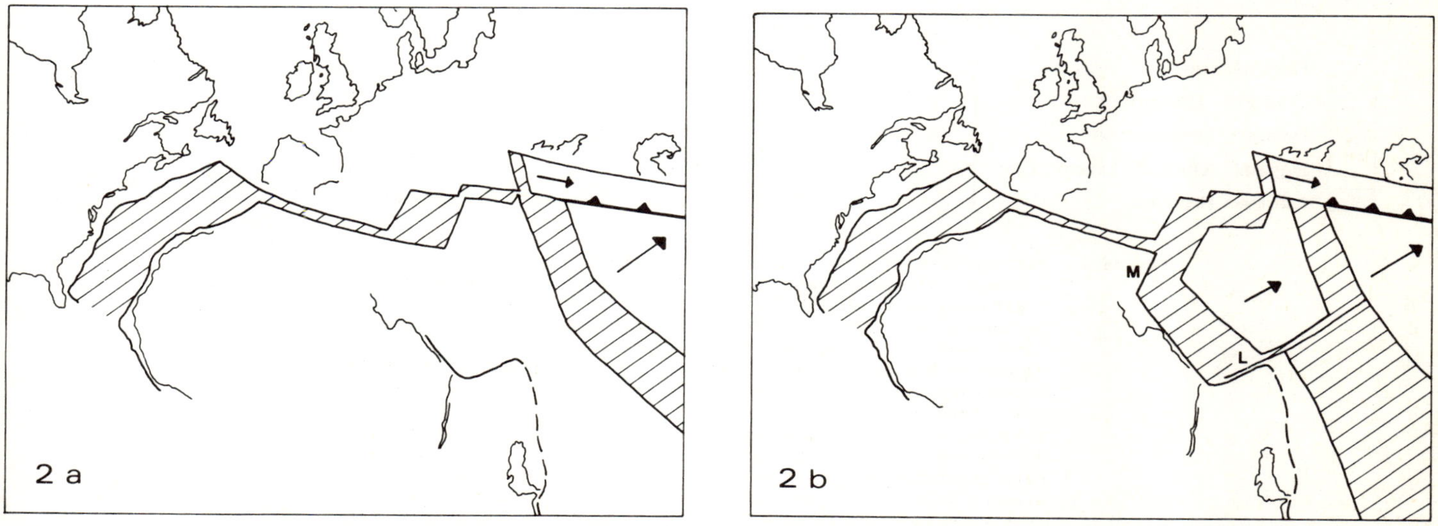

Figure 2 : Alternative palinspastic restorations for the Late Jurassic (Kimmeridgian) Tethys, a. assuming a northern origin of the Antalya-Cyprus-Hatay ophiolites and a young age of the Eastern Mediterranean basin, b. assuming a southern origin of the Antalya-Cyprus-Hatay ophiolites and a Mesozoic ocean in the eastern Mediterranean.

M = Malta escarpment. L = Levantine escarpement.

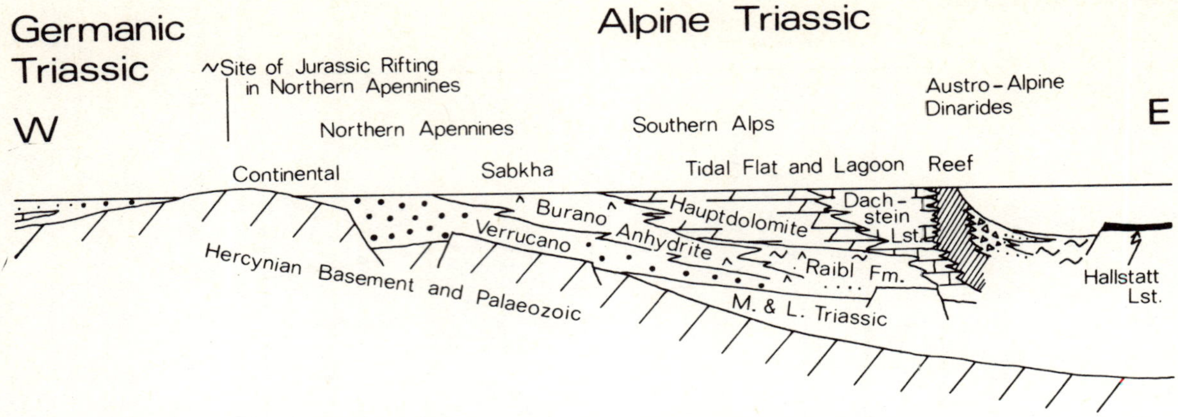

Figure 3 : Paleographic profile through the Late Triassic deposits of the central Mediterranean area, after Bosellini (1973), modified. The section combines depositional areas now occuring in different tectonic units into one palinspastic profile, in order to show the general arrangement of facies types (not to scale). The main feature is a Late Triassic sediment wedge with a westward transgression on the Variscan basement. Note that the later, Jurassic, rifting is discordant with respect to the Triassic facies distribution: In the transection between the central Apennines and Corso-Sardinia it coincides with the boundary between the Germanic Muschelkalk province and the "Alpine" Triassic, whereas in the central Alps the later suture lies within the "Alpine" Triassic limestone belt (cf.figures 1 and 4).

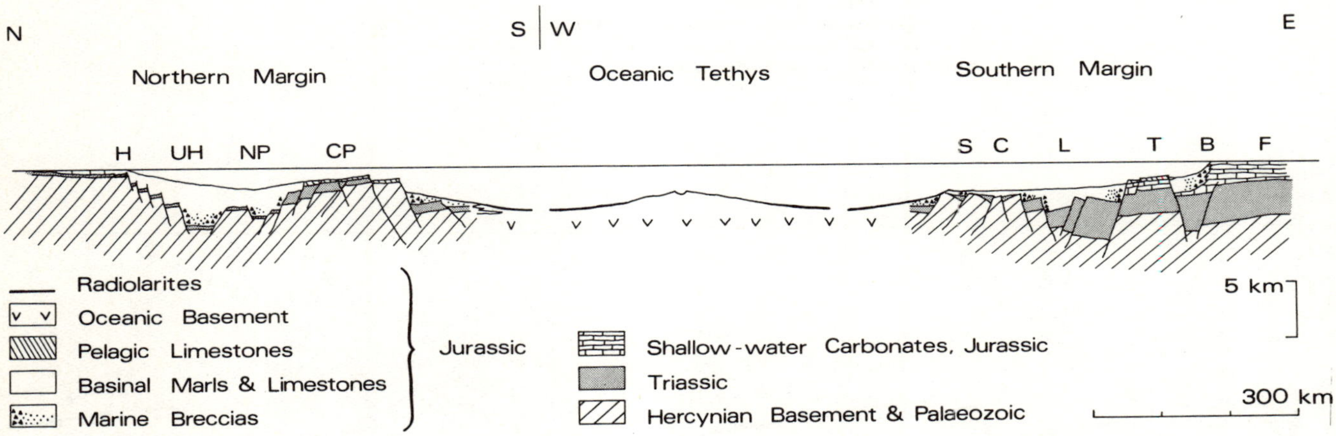

Figure 4 : Palinspastic section through the Liguria-Piemont ocean in the Late Jurassic (Kimmeridgian). The main features are: 1. Thick "Alpine" shallow-water sediments in the Triassic of the central Pennine realm on the northern continental margin. 2. The Early Jurassic phase of block-faulting in the embryonic continental margins, well documented by marine breccia formations. 3. The onlap of pelagic deep-sea sediments over the deeply submerged continental margins following the initial phase of block-faulting. These sediments are characterized by a sequence of solution facies indicating prolonged subsidence of the distal continental margins. 4. The development of Bahamian-type carbonate platforms along the southern continental margin with adjacent troughs with pelagic and turbiditic carbonate sedimentation. 5. The development of the oceanic Tethys, documented by the oceanic ophiolite suite. Its width and shape are somewhat arbitrary, but 1000 km as a rough figure are consonant with the history of spreading in the Atlantic.

H = Helvetic
UH = Ultrahelvetic
NP = North Pennine
CP = Central Pennine
S = Sesia Zone-Dent-Blanche nappe

C = Canavese Zone
L = Lombardia Zone
T = Trento Zone
B = Belluno Zone
F = Friuli Zone

INTERNATIONAL SYMPOSIUM ON THE STRUCTURAL HISTORY OF THE MEDITERRANEAN BASINS. SPLIT (YUGOSLAVIA) 25-29 OCTOBER 1976. B. BIJU-DUVAL AND L. MONTADERT, Eds. EDITIONS TECHNIP, PARIS 1977, pp. 133-142

FURTHER EVIDENCE RELEVANT TO THE AFRICAN/ADRIATIC PROMONTORY AS A PALAEOGEOGRAPHIC PREMISE FOR ALPINE OROGENY

F. HORVATH [(1)] and J.E.T. CHANNELL [(2)]

SUMMARY

It has been implied that the Adriatic region represents a promontory of the African plate. Recent evidence is concordant with this hypothesis and it is shown that the phases of motion of an Adriatic promontory since the Early Mesozoic, as determined from the Atlantic magnetic anomaly data, can account for the space/time relationships of Periadriatic orogeny.

Recent geological and geophysical data provide new constraints on the definition and behaviour of microplates in the Carpatho-Pannonian region. The existence of an African/Adriatic promontory is consistent with these data.

I. INTRODUCTION

One of the most important factors which helps to explain mountain building is the kinematic history of interacting plates. If at least one margin of a plate has been accretional (oceanic ridge) the relative movement history can be deduced on the basis of marine magnetic anomalies. The relative movement of the African and European plates was determined in this way (Pitman and Talwani, 1972; Dewey et al., 1973). In the region between Europe and Africa, however, the Tethys sea was destroyed by subduction and hence the kinematics of the independent microplates can be monitored only by detailed palaeomagnetic measurements. It is therefore a great advantage if a permanent or temporary connection of smaller blocks with the African of European plates can be established. Over fifty years ago Argand (1924) suggested that the Adriatic region formed a promontory of Africa and interpreted the Alpine orogeny in terms of collision of this promontory with Europe.

In the following, we show that recent geological and geophysical data are concordant with Argand's suggestion. We define the regions which have constituted the Adriatic promontory, and the effect of the motion of this African element on Periadriatic orogeny is investigated.

II. THE AFRICAN/ADRIATIC PROMONTORY

The area now submerged beneath the Adriatic Sea represented a stable foreland during Alpine orogeny. The emergent fragments of this foreland such as Iblei and Apulia are relatively undeformed. However, the surrounding continental margin has been severely

(1) Geophysical Department, Eötvos University, H-1083 Budapest, Kun B.t.2, Hungary

(2) Institut für Geophysik, ETH-Hönggerberg, CH-8093 Zürich, Switzerland

deformed to produce the Apennines, Southern Alps, Dinarides and Hellenides. In these mountains a more-or-less continuous arrangement of isopic zones may be observed characteristic of a continent/ocean transition (Fig. 1). These zones are made up mainly by Mesozoic carbonate rocks which differ markedly from another continental margin facies found in Europe. The carbonate rocks of the Periadriatic region (from the Atlas Mts. to southern Turkey) had low rates of detrital sedimentation (e.g. Ammonitico rosso) while the "European" carbonate rocks represent higher rates of terrigenous influx (e.g. Fleckenmergel). Laubscher (1970) and Bernoulli and Jenkyns (1974) suggested that the two carbonate complexes constituted the northern and southern continental margin facies of Mesozoic Tethys (Fig. 2) and interpreted their differences in terms of different climatic condition on the northern and southern landmasses. The Adriatic region exhibits carbonate rocks deposited on the southern (African) margin of the Tethys. The two continental margin facies became juxtaposed by closing of the Tethys. The obducted remnants of the deep-sea region (pelagic sediments and ophiolites) make up the internal zones and encircle the Adriatic region (Fig. 1 and 2).

Palaeomagnetism should be able to give evidence for or against the African affinity of the Adriatic region. The palaeomagnetic data are, however, not fully conclusive. One of the main reasons is that there are very few reliable Mesozoic palaeomagnetic data from stable Europe and Africa with which to compare the Periadriatic data. The inferred pole positions determined by Van der Voo and French (1974) on the basis of the available palaeomagnetic data and the Atlantic seafloor spreading model can be compared with the Jurassic and Cretaceous palaeomagnetic data from the Southern Alps, Gargano, Apulia and the Iblei zone of Sicily (Schult, 1973; Barberi et al., 1974; Gregor et al., 1975; Channell and Tarling, 1975; Van den Berg and Wonders, 1976; Channell, unpublished data). The magnetic directions from these four autochthonous units are concordant with the inferred African directions (Fig. 2). Much of the palaeomagnetic work in Italy has been focused on the Mesozoic Umbrian sequence which forms a curved belt in the southeastern part of the Northern Apennines. An African "pattern" has been observed by Van den Berg and Wonders (1976) and Van den Berg et al. (1977) in Jurassic-Cretaceous data for sites in northern Umbria (Cagli), however this pattern appears to have been rotated 25° counterclockwise. They concluded that Italy was coupled with the African plate from at least Late Early Jurassic until Early Eocene, but subsequently a 25° counterclockwise rotation of Northern Apennine autochthon occurred relative to Africa. This "surplus" rotation of the Northern Apennine autochthon is valid however only if the Umbrian sequence is truly autochthonous.

Does the hypothesis of an Adriatic/African promontory conflict with seismic data for the Eastern Mediterranean, and is there any evidence for continuous connection of the Adriatic region and African platform in post-Paleozoic times ? The available seismic data for the Ionian and Levantine seas indicate 20-30 km thick crust (Fig. 2) which is usually supposed to be a reduced continental crust. From our point of view, however, it is not critical whether the Eastern Mediterranean crust is modified continental or oceanic in nature. The vital point is that the Mesozoic and perhaps even the Paleozoic layers of the African platform extend without major interruption under the young sedimentary cover of the Eastern Mediterranean (Malovitskiy et al., 1975; Morelli, 1976). There is some recent rifting in the Strait of Sicily but the amount of separation is negligible. Present day seismicity appears to follow a periadriatic line (McKenzie, 1972).

In summary it may be concluded that recent evidence provided by facies considerations, palaeomagnetism and crustal structure investigations tend to suggest that, on a large scale, the assumption of the African/Adriatic promontory is valid.

III. CARPATHO-PANNONIAN REGION

The Carpatho-Pannonian region forms the northeastern boundary of the Adriatic promontory (Fig. 2). Both the external and internal facies which are characteristic of the margin of the Adriatic promontory turn to the northeast at the eastern boundary of the Austroalpine nappes.

Indeed, facies and faunal considerations led Géczy (1973 a,b) to suggest that the NW part of the Carpatho-Pannonian region originated from the southern margin of the Mesozoic Tethys (i.e. it belonged to the African/ Adriatic promontory). On the other hand the SE part of the region was associated with the European marginal complex of the Tethys. On the basis of Lower Jurassic brachiopod fauna, Vörös (1977) has also provided evidence for the Mediterranean and European affinity, respectively, of the NW and SE parts of the Carpatho-Pannonian region. Having studied some thousand drill cores from the basement of the Pannonian region, Szepeshazy (1975) demonstrated the marked difference in the Paleozoic history of the two regions. The two contrasting regions were probably microplates as they are surrounded by ophiolitic suture zones. The ophiolites of the Carpathians, Apuseni Mts. and Dinarides are well known, but the ophiolite zone bisecting the Pannonian basin has only recently been recognized. It has been pointed out that this NE-SW striking belt characterized by deep-water sedimentation in Jurassic time, constitutes the border zone of the two contrasting regions (Wein, 1969). In this belt and a parallel belt to the SE, typical ophiolitic complexes (radiolarites-pillow lavas- gabbros-peridotites) of Late Jurassic-Early Cretaceous age were found by deep drilling through 1-2000 m of mainly Neogene sediments (Szepeshazy, 1975; Juhasz and Vass, 1974). In the Bükk Mts. outcrops of ophiolites can be found (Földessy, 1975).

IV. TECTONIC EVOLUTION OF THE PERIADRIATIC REGION

The movements of an African/Adriatic plate relative to Europe, as predicted by the Atlantic magnetic anomaly data (Fig. 3) can be correlated with the most important tectonic events of the Periadriatic region (Fig.4). As the Central Atlantic began to open, Africa moved southeastwards away from Europe and North America (Fig. 3, Phase I). The initial continental split both in the Atlantic and Tethyan regions is marked by widespread Middle to Upper Triassic volcanism (Fig. 4). The geological record indicates that the Jurassic was a period of distension in the Tethyan region. Ophiolitic rocks with recurrent Upper Jurassic spreading ages occur throughout the Periadriatic mountain belt, in the Apennines, Alps, Dinarides and Hellenides (Fig. 4). The development of a Late Jurassic-Early Cretaceous oceanic area at the eastern margin of the Adriatic promontory is interpreted here as marginal basin development induced by the subduction of Paleo-Tethys at the eastern margin of the Adriatic promontory (Fig. 3, Phase I). The Pelagonian microplate and the African Pannonian microplate separated from the Adriatic promontory as an "island-arc" during this process.

The marginal basin began to close during the Latest Jurassic, and the Early Cretaceous is the first main compressional period in the internal zones of Dinarides and Hellenides (Fig. 4). There is no clear evidence for compression to the north or west of the Adriatic promontory during the Early Cretaceous and it may be supposed that curstal accretion continued into the Cretaceous (Fig. 3, Phase II).

This orogenic event in the Dinarides and Hellenides subsided in the Late Cretaceous and was superceded by platform carbonate deposition in quiescent tectonic conditions. Dimitrejevic (1974) and Dercourt (pers. comm.) have recognised evidence for strike-slip movement along the Vardar zone during the Late Cretaceous and Early Tertiary. The first recognisable phase of Alpine deformation and regional metamorphism in the Alps occurred during the Late Cretaceous (Fig. 4) as the oceanic region to the north of the Adriatic promontory began to close (Fig. 3, Phase III). The first recognisable phases of compression to the west of the Adriatic promontory occurred during the Late Cretaceous and affected the rocks of the Northern Apennines.

The tectonic quiescence in the Dinarides and Hellenides which began in the Late Cretaceous continued into the Palaeocene. The Palaeocene is also a period of tectonic quiescence in the Alps. This "Palaeocene restoration" (Trümpy, 1973) appears to coincide with the Phase IV motion (Fig. 3). The compressional tectonics to the west of the Adriatic promontory which began in the Late Cretaceous continued into the Tertiary (Fig. 4). During the Eocene and Miocene, tectonic compression and regional metamorphism recured in the Dinarides, Hellenides and in the Alps (s.s.). Corsica, Sardinia and Calabria began to drift eastward (Alvarez et al., 1974) during the Eocene causing the thrusting of allochthonous units onto the western margin of the Adriatic promontory.

The concordance between the predicted movements of an African/Adriatic promontory and geologic evidence for phases of crustal accretion and subduction in the Periadriatic region suggests that the concept of an Adriatic promontory provides a useful paleogeographic framework for Alpine orogeny.

The authors are grateful to Professor W. Lowrie and Dr. F. Heller for commenting on the manuscript.

REFERENCES

ARGAND, E. (1924): La Tectonique de l'Asie. Proc. Int. Geol. Congr. XIII, 171.

AUBOUIN, J. (1973): Des tectoniques superposées et de leur signification par rapport aux modèles géographiques: l'exemple des Dinarides; paléotectonique, tectonique, tarditectonique, néotectonique. Bull. Soc. Geol. Fr., XV, 426.

ALVAREZ, W., COCOZZA, T. & WEZEL, F.C. (1974): Fragmentation of the Alpine Orogenic belt by microplate dispersal. Nature 248, 309.

BARBERI, F., GIVETTA, L., GASPARINI, P., INNOCENTI, F., SCANDONE, P. & VILLARDI, L. (1974): Evolution of a section of the Africa-Europe plate boundary: paleomagnetic and volcanological evidence from Sicily. Earth Planet. Sci. Lett., 22, 123.

BERNOULLI, D. & JENKYNS, H.C. (1974): Alpine, Mediterranean, and Central Atlantic Mesozoic Facies in Relation to the Early Evolution of the Tethys. In: DOTT, R.H. & SHAVER, R.H. (eds.): Modern and Ancient Geosynclinal Sedimentation. - Soc. of Economic Paleontologists Mineralogists, Special Publication no. 19.

CHANNELL, J.E.T. & TARLING, D.H. (1975): Palaeomagnetism and the Rotation of Italy. Earth Planet. Sci. Lett. 25, 177.

CHANNELL, J.E.T. & HORVATH, F. (1976): The African/Adriatic Promontory as a palaeogeographical premise for Alpine Orogeny and Plate Movements in the Carpatho-Balkan Region. Tectonophysics 35, 71.

D'ARGENIO, B. (1974): Le piattaforme Carbonatiche Periadriatiche una Rassegna di Problemi nel Quadro Geodinamico Mesozoico dell'Area Mediterranea. Mem. Soc. Geol. It. 13/2, 1.

DEWEY, J.F., PITMAN III, W.C., RYAN, W.B.F. & BONNIN, J. (1973): Plate Tectonics and the Evolution of the Alpine System. Geol. Soc. America Bull. 84, 3/37.

DIMITRIJEVIC, M.D. (1974): The Dinarides: A Model Based on the New Global Tectonics. Metallogeny and concepts of the geotectonic development of Yugoslavia, 141, Belgrade.

FOLDESSY, J. (1975): Petrological study of a diabase-spite magmatic rock suite, Darno-hegy (Hungary). Proc. Xth Cong. CBGA, 1973, Sect. VI, 55, Geol. Ust. D. Stura, Bratislava.

GAETANI, M. (1975): Jurassic Stratigraphy of the Southern Alps. In: Geology of Italy. C. Squyres (ed.). Earth Sciences Soc. Libyan Arab Republic.

GECZY, B.(1973a): Plate tectonics and paleogeography in the East-Mediterranean Mesozoic. Acta Geol. Acad. Sci. Hung., 17, 421.

GECZY, B. (1973 b): The origin of the Jurassic faunal provinces and the Mediterranean plate tectonics. Ann. Univ. Sci. Budap. Rolando Eötvös Nom., Sect. Geol., 16, 19.

GREGOR, C.B., NAIRN, A.E.M. & NEGENDANK, J.F.W. (1975): Paleomagnetic Investigations of the Tertiary and Quaternary Rocks: IX The Pliocene of Southeast Sicily and some Cretaceous Rocks from Capo Passero. Geol. Rundschau 64(3), 948.

HAILWOOD, E.A. & MITCHELL, J.G. (1971): Paleomagnetic and radiometric dating results from Jurassic intrusions in Southern Morocco. Geophys. J. Roy. astr. Soc. 24, 351.

HELLER, F. (1977): Palaeomagnetism of Jurassic limestones from Franconia, Southern Germany. J. Geophys. (in press).

JUHASZ, A. & VASS, G. (1974): Mesozoische Ophiolite im Beckengrund der grossen ungarischen Tiefebene. Acta Geol. Acad. Sci. Hung., 18, 3-4, 349.

LAUBSCHER, H.P. (1970): Das Alpen-Dinariden-Problem und die Palinspastik der südlichen Tethys. Geol. Rundschau 60, 813.

LOWRIE, W. & ALVAREZ, W. (1975): Paleomagnetic evidence for the Rotation of the Italian peninsula. J. Geophys. Res. 80, 1579.

MALOVITSKIY, Ya.P., EMELYANOV, E.M., KAZAKOV, O.V., MOSKALENKO, V.N., OSIPOV, G.V., SHIMKUS, K.M. & CHUMAKOV, I.S. (1975): Geological structure of the Mediterranean Sea Floor (Based on Geological-Geophysical Data). Marine Geology 18, 231.

McKENZIE, D.P. (1972): Active tectonics in the Mediterranean region. Geophys. J. Roy. astr. Soc. 30, 109.

MORELLI, C. (1976): Advancement in the geophysical knowledge of the Mediterranean sea. Paper presented at the XXVth Cong. of ICSEM, Split.

MORELLI, C., PISANI, M. & GANTAR, C. (1975): Geophysical studies in the Aegean Sea and in the Eastern Mediterranean. Boll. Geofis. teor. appl., 18/66, 127.

PITMAN, W.C. & TALWANI, M. (1972): Sea floor spreading in the North Atlantic. Geol. Soc. Am. Bull. 83, 619.

SCHULT, A. (1973): Palaeomagnetism of Upper Cretaceous Volcanic rocks in Sicily. Earth Planet. Sci. Lett. 19, 97.

SPENCER, A.M. (ed.) (1974): Mesozoic-Cenozoic Orogenic Belts. Scottish Acad. Press.

TRUMPY, R. (1973): The timing of Orogenic Events in the Central Alps. In: DE JONG, K.A. & SCHOLTEN, R. (eds.): Gravity and Tectonics. - J. Wiley & Sons, London.

VAN DEN BERG, J. & WONDERS, A.A.H. (1976): Paleomagnetic evidence of large fault displacement around the Po-basin. Tectonophysics, 33, 301.

VAN DEN BERG, J., KLOOTWIJK, C.T. & WONDERS, A.A.H. (1977): The Late Mesozoic and Cenozoic movements of the Italian Peninsula. - Further paleomagnetic data from the Umbrian sequence. Geol. Soc. Am. Bull. (in press).

VAN DER VOO, R. & FRENCH, R.B. (1974): Apparent polar wandering for the Atlantic bordering continents: Late Carboniferous to Eocene. Earth Sci. Revs. 10, 99.

VOROS, A. (1977): Proviciality of the Mediterranean Lower Jurassic Brachipod fauna: Causes and plate tectonic implications. Palaeogeogr. Palaeoclim. Palaeoecol., 21, 1.

WEIN, G. (1969): Tectonic review of the Neogene-covered areas of Hungary. Acta Geol. Acad. Sci. Hung., 13, 399.

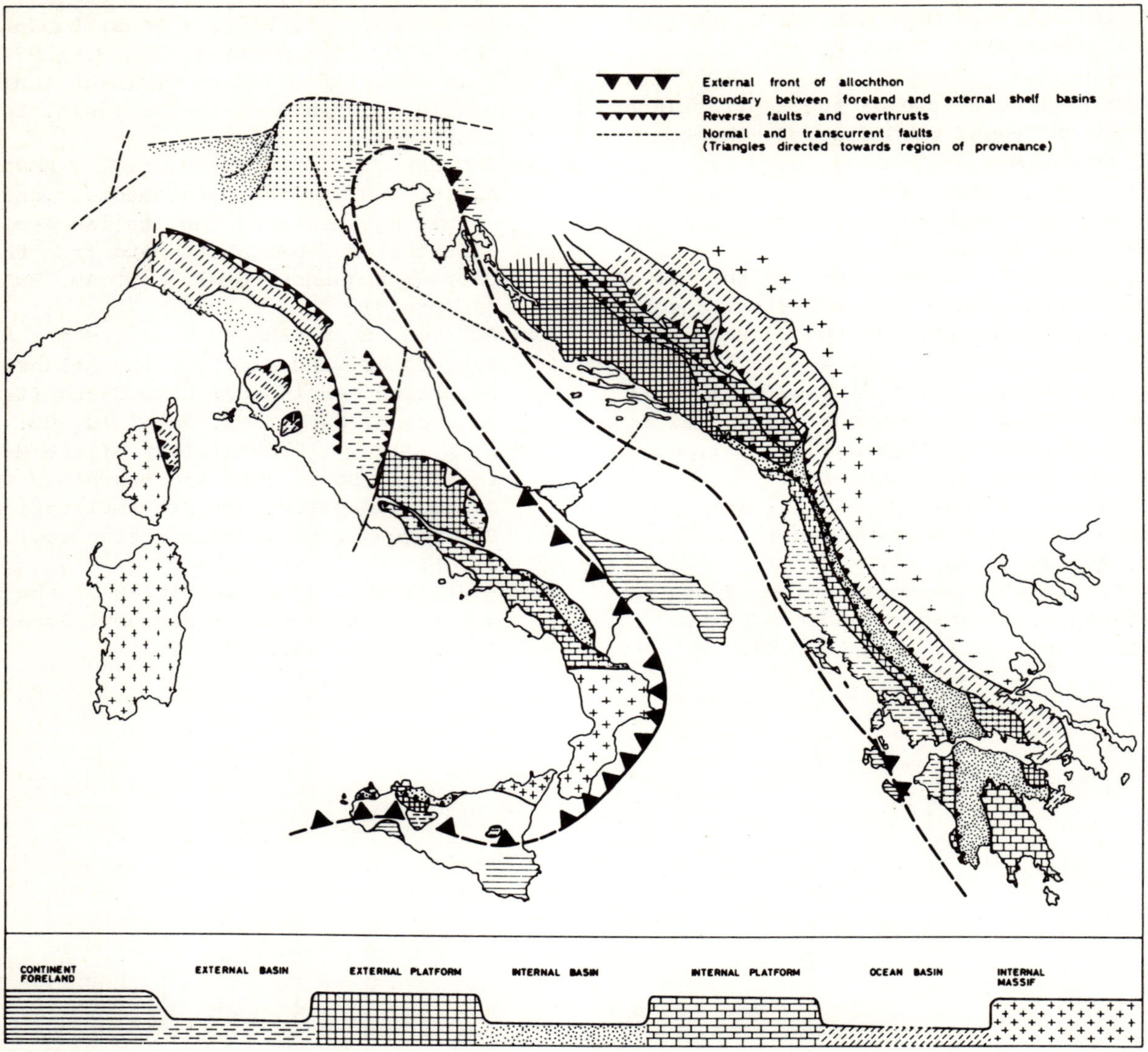

Fig. 1 : Sketch map of the present location of the Jurassic palaeogeographic zones which formed the Periadriatic continental margin facies (after D'Argenio, 1974; Gaetani, 1975; Aubouin, 1973; and many others).

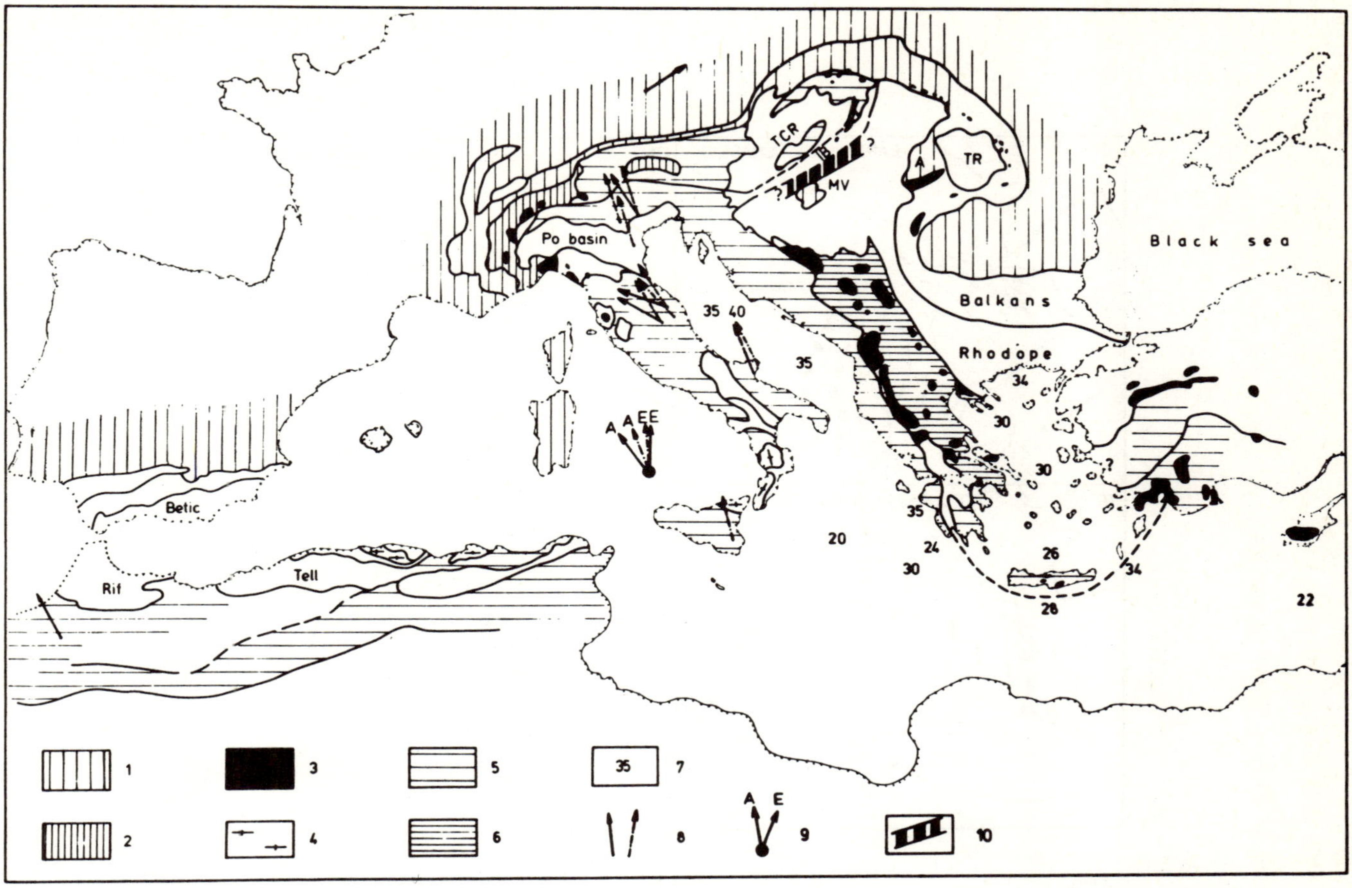

Fig. 2 :

Some recent data relevant to the hypothesis that the Adriatic region has represented part of the African plate.

The base map is modified after Bernoulli and Jenkyns (1974); crustal thicknesses after the summary by Morelli et al. (1975); palaeomagnetic data after Hailwood and Mitchell (1971); Schult (1973); Barberi et al. (1974); Van der Voo and French (1974); Gregor et al. (1975); Lowrie and Alvarez (1975); Channell and Tarling, (1975); Van den Berg and Wonders (1976); Van den Berg et al. (1977); Heller (1977); Channell (unpublished data).

Symbols:

1. Tectonic units of the northern continental margin of the Tethys.
2. Tectonic units of the northern continental margin and emplaced oceanic sediments.
3. Ophiolites
4. Massifs
5. Tectonic units of the Southern continental margin of the Tethys.
6. Tectonic units of the Southern continental margin and emplaced oceanic sediments.
7. The depth of the Moho.
8. Jurassic (solid lines) and Upper Cretaceous (dotted lines) palaeodeclinations.
9. Inferred Jurassic (solid lines) and Upper Cretaceous (dotted) palaeodeclinations for Italy which correspond to the palaeomagnetic pole positions of Van der Voo and French (1974) for Africa (A) and Europe (E).
10. The distribution of ophiolitic rocks under the post-Lower Cretaceous sediments of the Pannonian basin.

IB : Igal-Bükk zone

MV : Mecsek-Villany Belt

TR : Transylvanian Basin

TCR : Hungarian Central Mountains

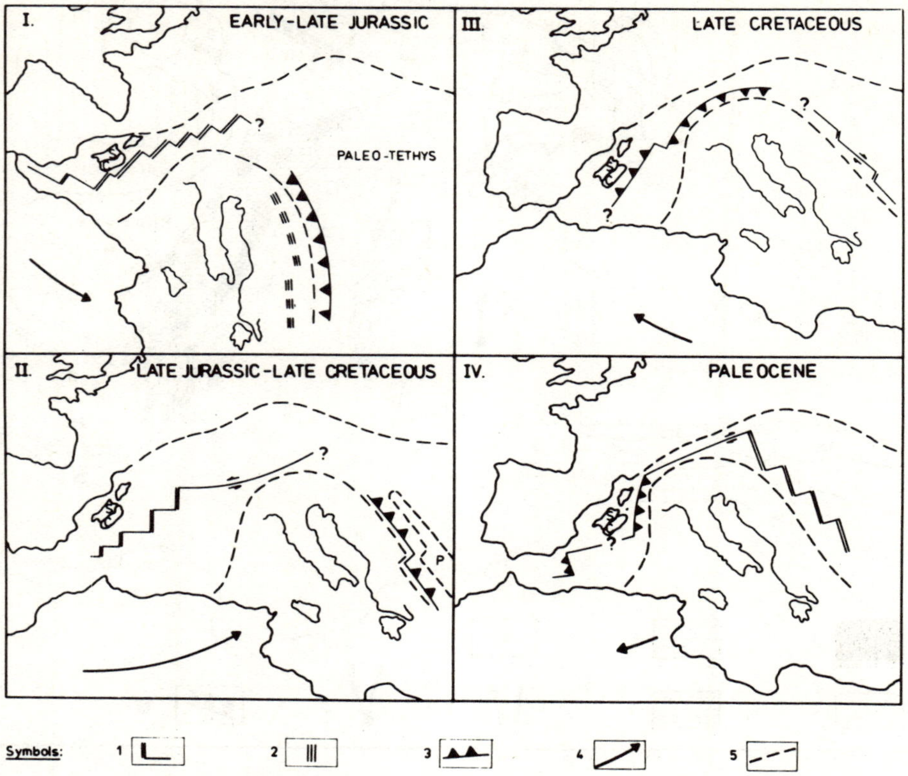

Fig. 3 : The position of Africa with its Adriatic promontory relative to Europe at the end of Phases I - IV as delineated by the Atlantic opening model of Pitman and Talwani (1972).

<u>Symbols:</u>

1. Accreting and transform plate boundaries.
2. Zone of "multispreading" (opening of back-arc basins).
3. Zone of subduction.
4. Path taken by the African plate.
5. Hypothetical margin of Europe and the Adriatic promontory.

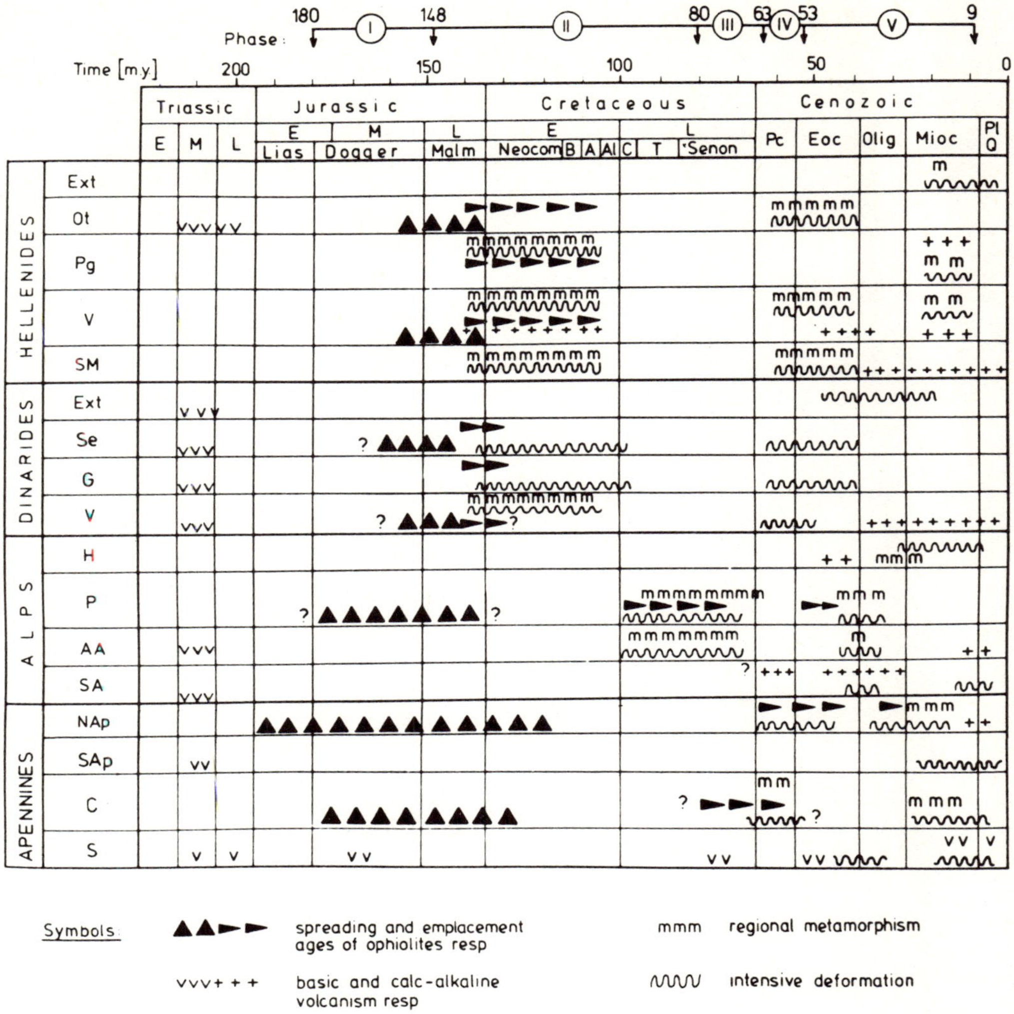

Fig. 4 : Main Alpine tectonic events in the Periadriatic region (mainly after Spencer (1974) and personal communications from Cadet, Catalano, Dercourt, Dietrich, Dimitrijevic and Scandone).

Abbreviations:

Hellenides: Ext=Ionian, Gavrovo-Tripolitza, Pindos, Parnassos zones; Ot=Othrys (Subpelagonian) zone; Pg=Pelagonian zone; V=Vardar zone; SM=Serbo-Macedonian zone.

Dinarides: Ext=Dalmatian-Herzegovinian, Budva, High Karst, Bosnian zones; Se=Serbian zone (Durmitor nappes, Ophiolite belt); G=Golija zone (Drina-Ivanjica element); V=Vardar zone.

Alps: H=Flysch zone, Helvetic domain; P=Penninic zone; AA=Austro-alpine unit; SA=Southern Alps.

Apennines: NAp=Northern Apennines; SAp=Southern Apennines; C=Calabrian zone; S=Sicilian zone.

INTERNATIONAL SYMPOSIUM ON THE STRUCTURAL HISTORY OF THE MEDITERRANEAN BASINS. SPLIT (YUGOSLAVIA) 25-29 OCTOBER 1976. B. BIJU-DUVAL AND L. MONTADERT, Eds. EDITIONS TECHNIP, PARIS 1977, pp. 143-164

FROM THE TETHYS OCEAN TO THE MEDITERRANEAN SEAS : A PLATE TECTONIC MODEL OF THE EVOLUTION OF THE WESTERN ALPINE SYSTEM

B. BIJU-DUVAL[1], J. DERCOURT[2] and X. LE PICHON[3]

ABSTRACT

We propose a series of paleogeographic maps of the western alpine system, from Triassic to Present, based on the geologic structure and evolution of the Mediterranean, its adjacent folded belts and the associated basins, and on the evolution of the North Atlantic ocean according to the magnetic anomaly informations. The kinematic framework is relatively precise for the Cenozoic but the accuracy is quite poorer for the Mesozoic. Between Triassic and Upper Cretaceous, the reconstructed continental margins can be compared to present passive or active margins. The geologic evolution can be described in terms of plate tectonics. After Upper Cretaceous, the oceanic areas are quite limited and the geologic evolution is dominated by the continental collision processes. For each of eight different key geologic periods (including Present), Africa and Europe are positionned according to the kinematics deduced from the opening of the Atlantic ocean. Three major intermediate plates are identified : Iberia, Apulia and Anatolia. The dimensions of the initial cratons are obtained through a palinspastic restoration taking into account the basement nappes and the effect of superposed tectonics. It is shown that, prior to Lower Jurassic, an ocean (the Tethys) separated Eastern Europe from Africa-Arabia. It is being consumed during Mesozoic because of the left lateral motion of Africa with respect to Europe, which induces the separation of Apulia and Anatolia from Africa and their accretion to Europe. Distension prevails in the wake of the motion of Apulia-Anatolia, especially between Tunisia and Arabia, which is related to the formation of a new oceanic surface, called the Mesogea. The collision of Apulia occurs at different times depending on the morphology of the European southern margin. It is initiated in Uppermost Jurassic, with a resulting fracturation of Apulia in three blocks ; and ends during Uppermost Cretaceous to the east and west of Rhodope in the median portion. The consumption of the Tethys is accompanied by the creation of marginal basins : the Carpathian flyschs trough and the Black Sea. The motion of Moesia toward the southwest, related to the opening of the Black Sea, increases the Carpathian arcuation. After Upper Cretaceous, the change in the motion of Africa will lead to the following facts. New strike-slip faults appear in the Apulian plate. The Mesogea is being consumed as well as its different narrow troughs which extended toward the Tethys and the Atlantic. This consumption is accompanied by the formation of marginal basins. In the Western Mediterranean, where the consumption is complete, the marginal seas are well developed. In the Eastern Mediterranean, where the consumption is still only partial, they are barely initiated.

(1) Institut Français du Pétrole, 1-4 avenue de Bois-Préau, 92506 Rueil-Malmaison, France

(2) Université des Sciences et Techniques de Lille, B.P. 36, 59650 Villeneuve d'Ascq, France

(3) Centre National d'Exploitation des Océans, 39 avenue d'Iéna , 75116 Paris ; Centre Océanologique de Bretagne, B.P. 307, 29273 Brest, France

FROM THE TETHYS OCEAN TO THE MEDITERRANEAN SEAS : A PLATE TECTONIC MODEL OF THE EVOLUTION OF THE WESTERN ALPINE SYSTEM.

I. INTRODUCTION

This paper is an intermediate stage in a systematic attempt at reconstructing the evolution from the early Mesozoic Tethys ocean to the present Mediterranean seas, within the framework of plate tectonics. In an earlier publication (Biju-Duval et al, 1976a), we outlined the way in which we approach this reconstruction, which incorporates, in a synthetic fashion, the very large amount of data now available on the continental margins as well as on the folded belts surrounding the Mediterranean area (Biju-Duval et al. 1977). The main result of this effort is a series of paleogeographic maps at key geologic periods which were prepared with the collaboration of V. Apostolescu, for the distribution of the main geologic facies, and of J.C. Sibuet, for the plate kinematic reconstructions (Biju-Duval et al., 1976 b). These maps are presented in such a way that they can be easily used, for addition or modification, by other scientists.

The purpose of this paper is to give to the reader the necessary elements to have a clear knowledge of the basic methodology followed by us and to outline the main conclusions, the major problems and the directions along which we intend to proceed in future stages.

We have limited, rather arbitrarily, the present reconstruction to the western part of the alpine system which surrounds the Mediterranean seas, Black sea included. It will be obvious to the reader that a full understanding of the eastern Mediterranean area would require a broader framework including the Caspian sea area and extending to the Oman sea to the south-east.

Starting from the present structure of the Mediterranean region, (Ryan et al, 1971, Biju-Duval, 1974, Biju-Duval et al, 1974), we tried to investigate how the geologic evidence for past distension, for the distribution of epicontinental and pelagic facies, and for the superposition of tectonic deformations in the folded belts fit within the pattern of relative motion between Africa and Eurasia since early Mesozoic time, as given by plate kinematics.

In the last few years the structure of the Mediterranean seas has been intensively studied (Auzende et al., 1972 ; Rehault et al., 1974 ; Finetti and Morelli, 1973, 1974 ; Biju-Duval et al., 1977 ; etc ...). Plate I[x] shows that it is characterized by narrow zones of thin crust inserted within the alpine folded belt, along the fairly broad Africa-Eurasia plate boundary which is outlined by a diffuse but intense seismicity (e.g. Mc Kenzie, 1972). The limited size of these oceanic basins has been the source of considerable controversy. Actually, only a small part of the surface of the Mediterranean basins are located more than 100 km away from the continental shelfs. The continental margins, which mark the transition from continental to oceanic crust have an average width of 100 km or more (e.g. Worzel, 1965) and their deep structure is characterized by a rapid evolution in space and time. It cannot be easily related to either oceanic or continental crust. As a result, the nature of the crust, in the Mediterranean basins, where opposite continental margins can be considered to be contiguous, is necessarily complex and difficult to characterize clearly as either oceanic or continental.

However, we consider that the oceanic nature of the crust underlying the abyssal plains of the western Mediterranean and the southern Tyrrhenian seas is well established. The very large accumulations of sediments over the deepest parts of the eastern Mediterranean and the Black seas indicate that the igneous crust is very thin (Biju-Duval et al, 1976 a). In addition, the existence of deep seismic zones under the Tyrrhenian and Hellenic arcs (e.g. McKenzie, 1972) indicates that active subduction of crust has proceeded there for the last few million years. We consequently also assume an oceanic nature for these portions of crust. The age of formation of the basins can be estimated on the basis of a study of the continental margins which surround them. Although they have often been affected by tectonics, it is possible to recognize that some of them are quite old (Mesozoic) whereas others are much more recent. We estimate on this basis that the eastern Mediterranean basins were formed in Jurassic time, the Black sea in Cretaceous-Eocene, the Western Mediterranean in Oligo-Miocene (Le Pichon et al., 1971 ; Hsü, Montadert et al., 1977), the Tyrrhenian since Upper Miocene (e.g. Morelli, 1975). The present Pannonian basin (see Channell and Horwath, 1976), although not of an oceanic nature, was also formed by distension in Neogene (Boccaletti et al., 1976, Biju-Duval et al, 1976 a).

[x] Plates I to VIII are inserted in a paper pocket placed at the end of the volume.

The notion of resorbable oceanic lithosphere is fundamental to any reconstruction of folded belts. It requires however to be able to clearly identify traces of former continental margins, whether active or passive. Unfortunately, our ability in this domain is still very limited as shown by the numerous differences between the interpretations made by different authors.

In addition, it becomes ever clearer that considerable shortening may also occur through folding and thrusting of continental crust. For example, in the Alps (Trumpy, 1975) and in the Hellenides (Dercourt et al. 1976), the basement nappes obviously originated from a much wider continental platform than the present folded belt. The numerous studies made within the perimediterranean basins have shown that, to restore the original basement platform, it is necessary to unfold the present tectonic structures, due to a superposition of tectonics, over distances which often exceed 500 km. Thus, the reconstruction of the original cratonic plates requires such a palinspastic restoration.

Finally, as vividly illustrated by Mc Kenzie (1972), the present pattern of the Africa-Eurasia collision is dominated by the fragmentation of the continental crust through a wide belt of deformation. This state of fragmentation through intracontinental collision processes was probably initiated in Upper Cretaceous. It is the present complex pattern of fragmentation which probably led to the identification of numerous (20) plates which evolved independently since early Mesozoic in the Dewey et al. (1973) reconstruction. This reconstruction of the western alpine system is probably the most widely used. But we prefer to consider that the present complexity is mostly the result of the late Mesozoic and Cenozoic intracontinental collision : the main independent plates should be identified only on the basis of an association of unambigous evidences for the existence of former plate margins.

As wasdone in the pioneering work of Dewey et al (1973), we have placed our reconstruction within the framework of plate kinematics. The solution adopted for the Africa-Eurasia relative motion was essentially derived from the work of Pitman and Talwani (1972) from anomaly 31 (68 M.y.) on, whereas the earlier kinematics is obtained from the work of Le Pichon et al (1977). It will not be discussed in detail here.

Similarly, the bases for our choice on the present structure of the Mediterranean basins have been discussed by Biju-Duval et al (1977) and will not be presented again here.

Finally, due to space limitations, we will not comment in detail the geological data which are used in the paleogeographic maps of (PL. II to VIII). We will only give brief general comments and will postpone a more elaborate discussion to a later paper.

Many of the earlier attempts at reconstructing the alpine evolution are partial attempts which only consider portions of the Mediterranean system (Laubscher, 1969, 1971 ; Dercourt, 1970 ; Alvarez 1972, 1974 ; Boccaletti et al. 1974, etc ...). Others ignore a considerable part of the available information in one or several of the three major domains : plate kinematics, present geophysical structure, geological structure (1) (Smith, 1971 ; Bosellini and Hsü, 1973; Channell and Horwath, 1976, etc ...). Our attempt, on the contrary, is built on a broad base in each of these three domains.

But there is one major set of data which we have not used for the reconstruction. These are the so-called "absolute" position data which relate the position of a plate to the axis of rotation of the earth. We feel that there is still too much uncertainty to use these data to position the intermediate plates. We did, however, position the paleolatitudes on our paleogeographic maps modified from the paleomagnetic work of Van Der Voo and French (1974).

A last word should be said about the nomenclature adopted. We call Tethys the ocean which existed between Europe and Africa during early Mesozoic. We call Mesogea the fairly broad ocean which was created during Mesozoic, and we reserve the name of Mediterranean to the present basins, parts of which are relics of the Mesogea.

(1) Numerous reconstructions have been recently proposed : Herz and Savu, 1974 ; Dimetrijevic and Dimetrijevic, 1973 ; Géczy, 1973 ; Hawkesworth et al, 1972 ; Bernoulli and Laubscher, 1972 ; Boccaletti et al. 1974 a, 1974 b; Dimetrijevic, 1974 ; Hadzi et al.1974 ; Radulescu and Sandulescu, 1973 ; Dal Piaz, 1974 ; Dal Piaz et al, 1972 ; Hsü, 1971 ; Boccaletti et al., 1971, 1972 ; Elter and Pertusati, 1973; Mattauer and Proust, 1976 ; etc ...

II. METHODOLOGY

1. The framework of our attempt is the plate kinematic evolution of the Africa/Europe plate system obtained from an identification of the Vine and Matthews (1963) isochrons in the Atlantic ocean between Africa and North America and between Europe and North America. The first systematic attempt at identifying the magnetic anomalies on a large scale over these two areas is the one by Pitman and Talwani (1972) which was the basis of the Dewey et al (1973) kinematic solution. Unfortunately, mostly because of fairly slow spreading rates, the anomalies are difficult to identify and the Pitman and Talwani (1972) identifications cannot be considered as final. Several modifications or additions have been proposed since, which include, Williams (1975), Kristoffersen and Talwani (1977) and Kristoffersen (1977) for the Europe-America system, Hayes and Rabinowitz (1975), Barret and Keen (1976) for the Africa-America system and others will undoubtedly come in the near future.

In addition, the kinematic solution is obtained by fitting together corresponding anomalies from both sides of the ridge crest. However, in general, we do not have a good definition of the isochrons over a sufficient length to obtain a very precise fit. As a result, the Africa-Europe relative motions which are obtained by composition of the Africa/America and America/Europe motions cannot be expected to be known with a good precision. For example, Francheteau (1973), using the anomalies identified by Pitman and Talwani (1972) as well as informations on fracture zones trends, has shown that a somewhat different kinematic solution could be obtained for the Africa/America motion. Kristoffersen and Talwani (1977) and Kristoffersen (1977) have obtained a different kinematic solution for the America/Europe motion mostly because of additional anomaly identifications to the north of 50°N.

The complexity of the kinematic problem is increased by the existence of two relatively mobile zones on each side of the Açores-Gibraltar line. Iberia has had an independent motion from Europe in Late Mesozoic and part of Early Cenozoic and the anomalies between Iberia and America cannot be used to obtain the America/Europe motion without allowing for this independent displacement. Similarly there is some indication that Morocco has been affected by limited displacement with respect to Africa during the opening (e.g. Le Pichon et al., 1977) and the anomalies between Morocco and America should only be used with caution to obtain the Africa/America motion.

Finally, there is some significant uncertainty on the absolute ages of these Vine and Matthews anomalies. The most recent Cenozoic polarity time scale by Tarling and Mitchell (1976), which we have adopted, differs considerably from the Heirtzler et al (1968) time scale or even the one proposed by Larson and Pitman in 1972. For example, according to Larson and Pitman (1972), anomaly 24 is 60 M.y. old (Middle Paleocene) whereas it is 49 M.y. old (Lower Eocene) for Tarling and Mitchell (1976). For the Mesozoic, we have adopted the time scale proposed by Thierstein (1977). The maximum difference between the time scales may reach 11 M.y. for a given anomaly, as shown in the example above.

However, the greatest difficulty in the reconstruction of the Africa/Europe kinematics is related to the early stages for which we do not have Vine and Matthews isochrons but rather have to rely on a reassembly of the continents prior to the break-up. Yet this reassembly is difficult to obtain on purely objective ground. It becomes abundantly clear that no given isobath can be considered to reflect accurately the edge of the original continental crust break-up (Le Pichon et al, 1977). As the break-up between Europe and America most probably did not occur before 95 - 90 My (Cenomanian), all reconstructions prior to that time are plagued by the fairly large uncertainty in the Europe-America reconstruction. This is illustrated in figure 1 which shows the large differences produced by the different Europe/America solutions prior to 90 M.y. At one extreme, there is the solution of Bullard et al. (1965) which fit at the 1000 m isobath. It is not acceptable as it is now known that large portions of the sea-floor, down to a water depth of 4000 m in places, are continental. For example, between Europe and North America, this is true for the Rockall plateau in its entirety, the Orphan Knoll area, the area just south of the Porcupine sea-bight, the Galicia bank etc .. (e.g. Le Pichon et al., 1977). At the other extreme there is the most recent solution by Kristoffersen (1977). Actually, the initial fits obtained from a study of the ocean floor features and continental margins have regularly tended to get looser from Bullard et al (1965) to Dewey et al (1973) to Le Pichon et al (1977) and Kristoffersen (1977). We feel that the actual position is probably within 200 km of the Le Pichon et al (1977) solution for Europe-North America and that the resulting position for Africa - Europe may be eventually in error by as much as 300 km to the west and 500 km to the east.

The precision, however, gets considerably better posterior to 90 M.y., as shown in figure 1. It is of the order of 200 km to the west and somewhat more to the east. It may be fairly poor for the Campanian-Maestrichtian as Kristoffersen (1977) has suggested that Pitman and Talwani (1972) generally identified anomaly 31 (which is Maestrichtian, 68 M.y.) where it should be 33 (which is Campanian, 76 M.y.).

In this paper, we do not propose any new magnetic anomaly identifications or kinematic solutions. Rather, we have chosen among the solutions proposed what seems to us to be the most probable, recognizing that it may be illusory at the present time to expect a precision in the position of Africa with respect to Europe better than 300 - 500 km prior to 90 M.y. ago and 200 - 300 km later than that. Although these possible errors are large, they only represent about 20 % of the motion of Africa with respect to Europe since the corresponding time.

Table I indicates the solution adopted and figure 2 illustrates the resulting kinematics. The solution chosen is essentially based on the work of Pitman and Talwani (1972) from anomaly 31-33 on. However, between Africa and North America, we have used the slightly different finite rotation angles proposed by Francheteau (1973). Between Europe and North America, we have also used the parameters proposed by Kristoffersen (1977) for anomaly 24 (49 M.y., Ypresian) and anomaly 31-33. Prior to anomaly 31-33, we adopt the solution proposed by Le Pichon et al (1977) for the initial reconstruction and the Oxfordian reconstruction. The Barremian position (anomaly M2) is not very precise and the Cenomanian position which is an interpolation, even less precise. The latter one has not been used in this paper.

2. Within this kinematic framework, we have chosen seven key geologic periods for which there was a fair control of the relative position of Africa and we have established a paleogeographic map for each of these periods (figure 3 and plates II to VIII). Starting from the present time, these are :

- Tortonian, 9 M.y., anomaly 5 (plate VIII) : just before the Messinian event
- Stampian, 35 M.y., anomaly 13 (plate VII) : just before the opening of the western basin.
- Lutetian, 44 M.y., anomaly 21 (plate VI), which corresponds to an important tectonic event in the whole Mediterranean area (Pyrenean phase)
- Campanian - Maestrichtian, 76-68 M.y., anomaly 33-31 (plate V), which represents the main period of evolution of the future eastern Mediterranean area
- Tithonian-Berriasian, 140 M.y., interpolated between anomaly M 25 and M 2 (plate IV) because of the very important sedimentary and tectogenetic processes (first obduction of ophiolites)
- Dogger, 165 M.y., interpolated between anomaly M 25 and the initial reconstruction (Plate III) during which the main isopic zones are defined as a result of the liassic distension
- Upper Triassic, about 200 M.y., initial fit (plate II) corresponding to the first distentional phases between the Tethys and the future Atlantic ocean.

In addition, plate I gives the present situation. We have not made the Cenomanian reconstruction (112 M.y.), although it is a key geologic period, as we felt that the relative position of Africa was poorly known.

However, it is obvious that, for a given geologic period, the major events that should be illustrated are not synchronous over the whole area. It should be stressed, further, that the dating of geologic facies or events may still be quite imprecise. For example, plate IV shows data concerning the Campanian as well as the Maestrichtian era. Actually, each map shows the general geological pattern in the vicinity of the chosen period and it may include events which occurred immediately before or after the date chosen.

Thus, we often find the same range of possible errors in the chronology of geologic data as in the kinematic data, which confirms that it is probably illusory at the present time to look for much better accuracy in such a reconstruction. It is quite clear that the precision of our paleogeographic reconstructions gets worse with increasing age. The following features have been shown on each map :

The repartition of continental and oceanic areas with a transitional zone which corresponds to the continental margins. This continental margin area distribution has been determined in a very schematic fashion. We must keep in mind, as previously said, that their deep structure is characterized by a rapid evolution in space and time, especially in the case of the active margins. So the size and the shape of such margins have been drawn with some modifications from one map to the other. Moreover, we have not attempted to illustrate systematically the possible existence of marginal basins which have since disappeared. Although we do not exclude this possibility, we have only shown on the plates those for which the evidence seems to us to be unambiguous.

The repartition of major facies in each zone based on both our field experience and a detailed bibliographic study which cannot be exposed in full in this paper for lack of space. Taking in account the previous remarks about the precision of dating and the fact that many rapid changes of facies are known to occur in space and time, we have only drawn in a schematic fashion the main facies, many fine variations being impossible to show at the scale of the maps.

To show the lateral variations we had to adopt a certain number of hypotheses on the initial positions of the different paleogeographic and (or) structural units within the different belts. For example the Rif, Kabyle and Calabrian basement units have been considered to be parts of the southern margin of Iberia to the north of the "Mauretanian-Massylian" trough. Our choice is illustrated in figure 4 which should be used as a reference map for plates II to VIII. To establish these reconstructions one needs to reposition at each stage the paleogeographic zones ; this requires analysing each part of the Alpine belt in which superimposed tectonics are known. Because of lack of space it is not possible to mention here the data we have used on the different branches of the Alpine belt, data which are necessary to understand, and to explain the paleogeographic or tectonic hypotheses chosen. This will be done in a future paper. Nevertheless figure 4 localizes the key paleogeographic zones which existed after the liassic distension. It shows also the main bibliographic references in a condensed form (author, year) even if we have not adopted the conclusions of the mentioned authors. The figuration of zones within stable areas is not difficult but to represent the former margins now included in "internal zones" is not easy : they correspond to the most tectonized, polyphased, granitized and metamorphosed areas. The study of present margins show that they are formed by horsts and grabens which size and shape are difficult to define in these former margins. It is also known that the type of deposits varies enormously over short distances on present continental margins (closeness of clastics sources, relative bathymetry-carbonate compensation depth, currents, ..).

Moreover we have used a study of the major lineaments revealed by the Landsat I satellite to define major fractures related to continental collision (Biju-Duval et al, 1976).

A possible or probable kinematic pattern, although in many cases it would be possible to reconstruct fairly different kinematic patterns. The main rationale was one of minimum complexity. It is clear that onland geological data have been used in priority. For example the relation between facies, tectonic development, overthrusting of ophiolites, calcoalcaline volcanism in north Anatolia during Late Cretaceous leads us to suggest the existence of a subduction zone. In a few cases clear geological data do not exist but the kinematic evolution obliges us to infer the existence of a plate boundary. The minimum complexity criterion has led us to identify only three main plates between Africa and Europe, which are Iberia, Apulia and Anatolia. The size and shape of these intermediate plates is obtained from a rough palinspatic restoration for Apulia and Iberia. For Anatolia the geological data are often so imprecise that we simply chose a size roughly compatible with the inferred kinematic evolution ; it should in particular result in continental collision by Late Cretaceous.

The major volcanic manifestations and the main phases of tectonic emplacement of ophiolites. Because the knowledge of the geochronology, petrography and geochemistry is still quite variable from one area to another a synthetic view of the volcanic activity all around the Mediterranean area is not yet well established (see Bellon and Letouzey, this volume). Nevertheless we have mentioned the main occurrences of widespread volcanism (including resedimented volcanic deposits). We have assumed in this paper that the obduction of ophiolites was related to the last stage of disparition of an oceanic area. This is not the only possibility (see for example Dewey, 1976) and we actually chose a different one in our preliminary publication (Biju Duval et al, 1976 a).

Finally, we note that these maps are prepared in such a way that they can be easily adapted or modified by others in order to obtain a better fit to other sets of data. The solution chosen is only one among several possible at the present time and we are still seriously considering an alternative solution in which the continental collision between Apulia and Europe only occurs in Late Cretaceous. All the maps are at the same scale in an oblique Mercator projection (pole at 52° N, 150° E) with Europe in its present position. The scale of the map is given by the latitude marks on the side. The paleolatitudes are figured modified from Van Der Voo and French (1974).

III. PROPOSED MODEL OF EVOLUTION

Figures 2 and 3 illustrate schematically the displacements of Africa with respect to Europe (arbitrarily kept fixed in its present position) since the beginning of the opening of the Atlantic ocean. Three main phases can be recognized.

a)From Mesozoic to Late Cretaceous, Africa is moving left laterally at a fairly high rate (2 to 4 cm/yr).

b)In Upper Cretaceous, the motion is slowed, most probably because of the first major intracontinental collision between Arabia, Anatolia and Eurasia. A relatively modest right lateral motion of Africa follows between Upper Cretaceous and Ypresian (49 M.y.), being related to the initiation of spreading between Europe and North America.

c)Finally, from Ypresian on, there is a general state of intracontinental collision which is characterized by a trigonometric rotation of Africa around a eulerian pole situated in the vicinity of Morocco. Thus the motion is very limited to the west but increases rapidly to the east where it is essentially north-south. Note that the pole of rotation tends to migrate with time to the west in the Atlantic ocean, which results in increasing compression in the whole western Mediterranean area.

The corresponding paleogeographic evolution is shown in figure 3 and plates II to VIII. During the first phase (prior to Upper Cretaceous), the geologic structure is mostly the result of the plate tectonic evolution of the three main intermediate plates, Iberia, Apulia and Anatolia. From Early Cretaceous on, the geologic structure corresponds to the fragmentation of the intermediate plates, with progressive disparition of the remaining oceanic areas. It is a process of continental collision where the concept of individual plates is difficult to use any more.

In this paper, the motion of Iberia has been obtained from kinematic criteria in the Bay of Biscay, as proposed by Le Pichon et al (1970, 1971), Le Pichon and Sibuet (1971) and Choukroune et al (1973). We do not have such criteria to obtain the motion of Apulia. We chose a solution which is compatible with the Africa-Europe motion and results in collision with Europe in Late Jurassic at the time of emplacement of ophiolites in the Hellenides (Dercourt et al, 1976). We still consider as a viable alternative a solution where collision only occurs in Upper Cretaceous and ophiolites obduction is not related to this continental collision. Note that the formation of the Bay of Biscay and of the Mesogea is not of a "marginal sea" type but rather is due to the displacements of intermediate plates induced by the motion of the main plates. On the other hand, the Black Sea and the Western Mediterranean probably represent fossil marginal basins.

The paleogeographic map will be commented starting from Early Mesozoic. However, their reconstruction was done starting from the present situation (plate I*) and moving back and forth in time.

1. <u>Upper Triassic</u> (<u>plate II</u>)*

The opening fit, which is pre-middle-Jurassic, is most probably valid for the whole Triassic time. We have chosen to draw the paleogeography of the Upper Triassic (200 M.y.). We previously discussed the

* Plates I to VIII are inserted in a paper pocket placed at the end of the volume.

fairly large inaccuracy of this reconstruction, especially in the Iberia-Morocco area.

a) Plate II shows that a broad Tethys ocean exists to the northeast although. Argyriadis (1975) has argued that there is no geologic evidence for oceanic facies in this area. Nevertheless, the distribution of fauna may suggest the existence of such an oceanic hiatus (Enay, 1972 ; Hirsch, 1976). It could be argued further that the distribution of the epicontinental facies is in fair agreement with the existence of an eastward Tethys. For example, north of Apulia, platform carbonate facies merged into deep water deposits (Bernoulli and Jenkyns, 1974). Sediments associated will pillow-lavas have been described but their present tectonic setting is complicated ; so it is still difficult to define exactly the relation between oceanic and marginal facies (Terry, 1971 ; Juteau, 1970). Evaporite germano - andalousian facies extend of Europe and over North Africa ; Nubia west facies are found to the south east ; and carbonate platform deposits are found all along the margins of the Tethys and along the areas of future rifting (Bernoulli and Jenkyns, 1974).

b) Upper Triassic is the period when the first distension occurred (Fallot, 1945, 1948 ; Jenkyns, 1970 ; Ginzburg et al, 1975 etc ...) along the future continental margins of the Mesogea and the future Atlantic ocean. However, the distension on the eastern margins of Apulia, which already existed at that time, is not properly explained by this model, unless one relates it to the formation of marginal seas (Blanchet, this volume). It would accordingly be interesting to be able to distinguish between the different types of volcano-sedimentary deposits in the now tectonized Dinarides - Hellenides folded belts (Ciric, 1966 ; see Cadet et al, this volume)

c) In plate II, the gap between Apulia (still connected to Africa) and the African margin is too large. It could be reduced in part with a different more compact fit of the now dissociated parts of Iberia (including Corsica-Sardinia, Calabria, East end West Kabylias, Rif etc ...) such as the one proposed by Bayer et al. (1973) and Le Pichon et al (1977) or with a different position of Northern Africa. In this paper, we have adopted the more conservative reconstruction of Corsica and Sardinia proposed by Biju-Duval et al (1977). The gap would be greatly increased, in any case, if the position for Iberia is the one proposed by Bullard et al (1965). We feel that this gap may not be significant in view of the large inaccuracies involved in such a reconstruction.

d) Another problem is the interpretation of the Isparta line of Turkey (Brunn et al., 1971) as corresponding to an ancient Transform Fault separating Apulia from Arabia - Anatolia, in spite of very little field evidence for it.

e) A better knowledge of the Cimmerian folding phase in Dobroudgea, Crimea, Caucasus would probably lead to a modification of the scheme proposed in the north-eastern area where an active margin already existed. Continuous deformations and thrusting occurring NW of the Pontides until Triassic time and including so-called "green rocks" (Fourquin, 1975) could indicate the existence of a subduction zone south of Europe.

2. Dogger (pl. III)

The distension gets generalized in Liassic time during which the left-lateral rotation of Africa, related to the opening of the Central Atlantic, has begun

a) We have chosen a position close to the position corresponding to M 25 (150 M.y., Oxfordian) in the absence of other magnetic anomaly data.

b) The left-lateral motion of Africa results in distension in Southern Iberia ("dorsale") (Fonbote et Quintero, 1960 ; Busnardo et Chenevoy, 1962 ; Paquet, 1974), in the Alps (Briançonnais) (Bourbon et al, this volume), in East Sicily and along the Levantine margin (Gvirstman , this volume) which can be considered as passive continental margins. These correspond to the formation of zones of Transform Faulting where the oceanic space is small (Alps) or even absent (Pyrénées, South Iberia) or to the formation of the future Mesogea (Monod, 1975), in the wake of the motion of Apulia away from Africa.

c) The left - lateral motion of Africa also results in the beginning of consumption of the Tethys with corresponding formation of marginal seas (Eastern Black Sea - Caspian, Dobroudgea sphenochasm). Important volcanism and granitisations known in the Pontides and Caucasus (Bergougnan, 1975 ;

Khain, 1963) could be related to a such phenomenon.

d) As a result of this evolution, the sedimentary facies get greatly differentiated around Apulia and in areas of subsidence near the margins of Iberia, Africa and Europe ("Citrabetic and Betic troughs" in Iberia, Azema et al. 1974, Atlas trough, Busson 1972, etc ..). Broad epicontinental seas extend on the cratons.

e) As discussed previously, the positions of Apulia and Anatolia are fairly arbitrary. The size of the Mesogea depends of course on the positions adopted. The connection between Apulia and Anatolia is also quite speculative. The absence of known oceanic facies between the two plates leads us to consider that they were separated by a Transform continental Fault with left- lateral motion at that time.

3. Tithonian - Berriasian (pl. IV)

The position of Africa was obtained by interpolation between the position corresponding to anomaly M 25 (Oxfordian) and anomaly M2 (Barremian).

a) We consider that this is the time of a first limited closure of the Tethys between Apulia and the Moesian area because this is the main time of tectonic emplacement of ophiolites in the Hellenides (Dercourt, 1970 ; Aubouin et al, 1970 ; Juteau et al, 1973 ; Terry, 1974) which we interpret here as due to incipient continental collision (Maxwell, 1970 ; Hynes et al, 1972). We previously noted that we consider as possible another solution where collision only occurs in Upper Cretaceous time in this area (Biju-Duval et al, 1976 a). In our present scheme, this collision will initiate the fragmentation of Apulia.

b) The oceanic realms, to the east (North Anatolian trough) and west (Alpine Margins) (Bezzi et Picardo, 1971 ; Decandia and Elter, 1972 ; Dietrich and Scandone, 1972 ; Contesogno et al., 1975) of the continental collision (Rhodopian), are fairly restricted. Two marginal basins have been created, the eastern Black Sea and the Carpathian flysch trough ("Siret Ocean" of Herz and Savu, 1974 ; Boccaletti et al. 1974 b) which are affected by important volcanic activity.

c) This very important event is followed by the wide-spread deposition of flyschs during Lower Cretaceous. We note the occurrence of such clastic turbiditic deposits not only along the Hellenides - Dinarides tectonized margins (Blanchet et al. 1969 ; Charvet, 1973 ; Terry et Mercier, 1971 ; Celet et al. 1976) but also in other areas (Durand-Delga et Lambert, 1955) extending to the Atlantic ocean. The flysch deposition may be, at least partly, related to a major change in the paleogeography of the continental areas (Purbeckian facies in England, France, Spain, etc ...) accompanied by a general regression (see Histoire structurale du golfe de Gascogne, Technip edit.). This is also the time during which the Bay of Biscay and the North African trough (Bouillin et al., 1970 ; Bouillin et Kornprobst, 1974 ; Obert, 1974) were initiated.

d) The size of the Mesogea depends on the position chosen for Apulia. It is possible that an active margin started at this time along the southern border of Anatolia, inducing the formation of a marginal sea (the future Troodos - Kizil Dag).

4. Campanian - Maestrichtian (pl. V)

The Barremian (anomaly M 2) and Campanian - Maestrichtian (anomaly 31 - 33) positions being fairly inaccurate, we have not used the interpolated Cenomanian position, in spite of the important paleogeographic changes occurring at that time. The Cenomanian saw an early tectonic event in the Alps, the initiation of new flyschs and a broad epicontinental transgression.

However, the Uppermost Cretaceous is a major period in the evolution leading from the Tethys to the Mediterranean, as it is the time of the final consumption of the Tethys. The left-lateral motion of Africa is stopped due to continental collision between Europe-Anatolia and Africa - Arabia. This continental collision, which is the first true collision between the Africa - Arabia and Europe plates, will continue from then on.

a) The consumption of the Tethys is marked by a broad phase of obduction of ophiolites and mélanges (accretionary prisms) south of the Tethys (Pontides, Bergougnan, 1975 ; Apuseni , Lupu, 1974) and north of Africa - Arabia (from Cyprus to Oman; Gass, 1968 ; Lapierre, 1976, Ricou, 1971). An important calco-alcaline volcanism from the Balkans to the Lesser Caucasus (Adamia, 1975 ; Boccaletti et al., 1974) accompanies the final oceanic subduction in this area ; the completion of the opening of the Black Sea is probably related to this event (see Letouzey et al., this volume). The Carpathian Sea continues to evolve as a marginal basin.

b) The narrow oceanic trough which ran from the Tethys to the Atlantic north of Apulia has completely disappeared along the Alps (See Tollman, 1969 ; Haccard et al., 1972). To the south, on the contrary, the Mesogea, which is broadly opened toward the Atlantic, has reached its maximum size. Flysch continues to be deposited on both margins of the North African trough (from Lower Cretaceous to Eocene) (see Paquet, 1974). The tectogenesis of the Pyrénées, related to the final stage of opening of the Bay of Biscay, is now engaged (Choukroune et al, 1973 ; Mattauer et Henry, 1974).

c) This general collision to the east results in the fragmentation of Apulia and Anatolia.

The motion of Africa will now change and, until Ypresian (49 M.y.), will correspond to right lateral strike slip, because of the faster opening between Europe and North America than between Africa and North America. After this limited right-lateral motion, Africa will start rotating trigonometrically around a pole near Morocco, resulting in pure compression to the east (fig. 3).

5. Lutetian (plate IV)

a) To the west, this is the time of consumption of the narrow western extension of the Mesogea between Africa and Iberia, attested by overthrusts which include fragments of ophiolites (Durand-Delga, 1971 ; Andrieux et Mattauer, 1963 ; Bouillin et Kornprobst, 1974). Paquet (1974) has suggested that transverse wrench faults cut the southern margin of Iberia at this time. The small maximum size of this oceanic area and the slow rate of consumption might explain the absence of calco-alcaline volcanism.

b) This is also the time of the main phase of tectogenesis in the Pyrénées (Mattauer et Henry, 1974) and the Alps (see Debelmas et Lemoine, 1970 ; Caby 1973). The fragmentation of Apulia may help to explain the arcuate development of the Alps, as proposed by Laubscher (1969, 1971).

c) To the east, the oceanic area gets smaller, mostly by Transform faulting along the northern border of the Mesogea. Anatolia is now fully accreted to Arabia but a new active margin is initiated south of the Caucasus (see Khain, this volume).

d) The collision in the Rhodopian - Moesian area generates underthrusting associated with volcanism (Boccaletti et al., 1974) and new continental fragmentation with movement toward the west, to the west, and toward the east, to the east.

6. Stampian (plate VII)

a) The process of continental collision with progressive consumption of the oceanic areas continues to the east. In particular, the old marginal Carpathian basin (initiated in Jurassic) begins to get reduced in size (Contescu, 1974).

b) The present Western Mediterranean still does not exist (Wezel, 1974). However, to the north, the rifting will take place during Upper Oligocene - Aquitanian (Le Pichon et al., 1971 ; Hsu, Montadert et al. 1977). To the south, we assume the formation from Gibraltar to Calabria of a marginal basin located north of the continental fragments of Iberia (Kabylies, Peloritan and Calabria) along the North African consuming plate boundary. The Numidian flysches are deposited both in internal and external positions as proposed by Caire (1971), Wezel (1970), Hoyez (1974).

It is interesting to note that the fragmentation of Iberia corresponds to the migration of the Africa-Europe pole of rotation toward the west.

c) To the east, the Mesogea is being consumed from the west to the east. This will lead to the future development of the Tyrrhenian arc and the opening of the Western Mediterranean followed by the opening of the Tyrrhenian sea. It is the beginning of the Apennines orogenesis (Roman, 1970 ; Boccaletti et al., 1976 ; Haccard et al. 1972).

7. Tortonian (plate VIII)

The paleogeography gets very similar to the present one (Biju-Duval et al., 1977).

a) The Mesogea is being consumed along two arcs, the Calabrian and Aegean arcs and is progressively being reduced to the size of the present Eastern Mediterranean.

b) The opening of the Red Sea corresponds to a faster northward movement of Arabia. As a result, the fragmentation of Anatolia is accentuated. The North Anatolian fault is initiated and allows a migration of Anatolia to the west (e.g. McKenzie, 1972).

c) The Carpathian arc continues to evolve with the formation of the Pannonian basin, as a continental marginal basin (Stegena et al, 1975 ; Contescu, 1974).

d) The subduction ends to the west in North Africa (Magne et Raymond, 1974) and along Italy. It is the age of the main nappes emplacement. The western basin has essentially its present configuration although the Tyrrhenian is smaller than today. Its basins are deep, subsidence is continuous, and Messinian salt deposition will occur in the deepest area, the margins being then affected by erosion (Montadert et al. 1977).

IV. CONCLUSIONS

The evolution from the Tethys to the present Mediterranean occurred in two main stages

1. From Triassic to Uppermost Cretaceous

Africa is moving left laterally with respect to Europe. The paleogeography is essentially controlled by the plate tectonic evolution of three intermediate plates, Iberia, Apulia and Anatolia, with respect to the two major plates, Africa and Eurasia. We adopted the hypothesis of a marginal basin for the Carpathian flyschs basin and for the Black Sea. We did not however try to reconstruct the evolution of the Hellenides in terms of marginal basins, although we do not exclude this possibility which is compatible with our scheme.

In Early Mesozoic, the paleogeography is characterized by a wide Tethys ocean between Europe and Africa. Its exact surface is difficult to define precisely. Previous studies had minimized the extent of the basement overthrusts in the Alps, the Dinarides and the Hellenides etc... Recent studies tend to increase them considerably (see Aegean Congres, Paris ; Tollmann, 1969 ; Derycke et al., 1974 ; Ferrière, 1974). The analysis of the different superposed metamorphic units of alpine age leads to an increasing surface estimate for the now tectonized initial cratons. Is it possible to get further than we have done and to completely eliminate the oceanic surface extending from the Carpathians to the Alps ? In any case, the existence of the Eastern Tethys between the Apulian margin, now incorporated into the Dinarides and Hellenides, the Anatolian margin, now incorporated into the Pontides, and the Rhodope and European margins, is difficult to doubt. The presence of the Eastern Tethys may be inferred from the volcanic activity in Pontides caucasus which suggests its subduction under Europe since Triassic. The geometry of the Tethys is partly controlled by the position of North Africa with respect to the old African shield. We did not take into account a possible deformation due to intra-African fractures, in particular the so-called "South Atlas" fault system (Dubourdieu, 1962).

The kinematics of Apulia cannot be directly deduced from the kinematics of the Africa-Europe plates system. Does Apulia collide with Europe as early as Uppermost Jurassic or does the corresponding Jurassic ophiolites obduction occur on the margin of an oceanic basin ? We have chosen, in this paper, the first solution whereas we developed the second one in an earlier paper (Biju-Duval et al., 1976 a). The large amount of clastic turbiditic deposits in the alpine belts during Upper Jurassic - Berriasian seems to us to be easier to interpret if high reliefs have been produced by continental collision. The alternative is still open.

From Triassic to Jurassic, a period of distension, which manifests itself in particular in the formation of continental margins affects the whole area. This is the time during which the Mesogea was formed in the wake of Apulia. After the Upper Jurassic-Lower Cretaceous collision of Apulia with Europe, intracontinental fractures cut Apulia in three sections.

During Cretaceous, the portions of Apulia situated to the east and to the west of the Upper Jurassic collision zone also collide with Europe. Consequently, the initial Tethys has been completely consumed in Maestrichtian time whereas the Mesogea has reached its maximum size.

2. From Uppermost Cretaceous to Present

Africa ceases to move left laterally and starts a right-lateral motion of limited extent until Ypresian. Then it rotates trigonometrically about a pole of rotation close to Morocco, thus resulting in north-south compression to the east. This is a time of intracontinental collision during which the concept of plates is difficult to use any more. The collision of Apulia and Anatolia with Europe results in an intracontinental fragmentation accompanied by large strike-slip faulting which allows the formation of important overthrusts in the Alps.

The Mesogea starts being consumed ; this consumption is accompanied by the creation of heterochronous marginal basins (the Balearic and Tyrrhenian basins). The remnents of the Eastern Mesogea correspond to the present Eastern Mediterranean where the Aegean marginal basin is barely initiated. The consumption of the Carpathian flyschs marginal basin is closely related to the Pannonian basin distension which is marked by very active volcanicity.

We have not taken into account the paleomagnetic data for the reconstruction of the positions of the intermediate plates, as we feel that they still are inadequate for this purpose. Due to a lesser quality of geological data, the reconstructions of Anatolia are less precise than those of the other parts of the Mediterranean area.

We are aware that we did not have enough constraints to reconstruct unambignously the position of Apulia and Anatolia. As we pointed out earlier, we previously chose another solution in which Apulia collides later with Eurasia with a much smaller rotation during Jurassic. If this is so, we would have only to change the Dogger and Upper Jurassic paleogeographic maps. It is clear that the answer to this problem lies primarily in a better knowledge of the evolution of the paleomagnetic poles for the intermediate plates.

However, although this model is far from being final, it suggests some interesting new geologic problems.

a) Great differences exist between the Mesozoic and Cenozoic sedimentary history and tectonic development. During the Mesozoic, in spite of great uncertainties in the plates positions, the size of the oceans and the exact locations of subduction zones, a comparison can be attempted with present oceans and the tectonic evolution of their associated marginal basins. But, in Cenozoic, after the continental collision occurring in Late Cretaceous between Africa and Europe, the tectonic development changes completely because of reduced oceanic areas and intracontinental fragmentation with large strike-slip faulting which makes comparison with present oceanic margins quite dubious.

b) The former Tethys ocean has completely disappeared. The present Mediterranean basins are either remnants of the Mesozoic Mesogea created by rifting in the wake of the motion of Apulia - or marginal type basins (whether oceanic or intracontinental) associated to a subduction zone.

c) The recent history of the Mediterranean basins (which includes subsidence, the Messinian evaporitic episode, the continuous Africa-Europe collision) masks differences.

d) The importance given to intracontinental strike-slip faulting suggests a new examination of different portions of the Alpine belt.

e) The presence of a Mesogea which is being consumed to the north and west suggests a revision of the schemes of evolution which have been proposed for the Hellenides, and of the relationship between Calabria and the Apennines.

f) The major roles given to the Dobroudgea and Isparta geologic features should be confirmed by more detailed studies.

ACKNOWLEDGEMENTS

We are much indebted to several friends and colleagues with which we had numerous discussions: V. Apostolescu who prepared palaeographic maps, J. Letouzey and L. Montadert with which an attempt of synthesis of Mediterranean Basins was established, J.C. Sibuet who worked on plate kinematics, J. Paquet and R. Blanchet which gave comments of this paper and C. Delaunay who drew the figures.

BIBLIOGRAPHY

ABBATE E., BORTOLOTTI V., PASSERINI P., SAGRI M. (1970).- Introduction to the Geology of the Northern Apennines. Sediment. Geol. 4, n° 3-4, 207-250

ADAMIA S.A. (1975).- Plate tectonic and the evolution of the Alpine System : discussion. Geol. Soc. Amer. Bull., 86, n° 5, 719-720

ALVAREZ W., GOHRBANDT K.H.A. (1970).- Geology and History of Sicily Tripoli. Petr. Explor. Soc. Libya, 291 p. (12th annual field conference)

ALVAREZ W. (1972).- Rotation of the Corsica Sardinia microplate Nature (Phys. Sci.), 235, 103-105

ALVAREZ W., COCOZZA T., and WEZEL F.C. (1974).- Fragmentation of the Alpine orogenic belt by microplate dispersal Nature (Phys. Sci.), 248, 309-314

ANDRIEUX J., MATTAUER M. (1963).- Sur la présence de roches vertes à la base des nappes ultra du Rif. Soc. Géol. France, Compt. Rend., 213-215

ANDRIEUX J. (1971).- La structure du Rif central. Etude des relations entre la tectonique de compression et les nappes de glissement dans un tronçon de la chaîne alpine. Notes Mem. Serv. Geol. Maroc, 235, 1-155

APOSTOLESCU V., BIJU-DUVAL B., GONNARD R., LETOUZEY J., DEVAUX E. (1975).- Synthèse des données géologiques et géophysiques de la Méditerranée occidentale. Internal Report Inst. Franc. Pétrol. Ref. 23458

ARGYRIADIS I. (1975).- Mésogée permienne, chaîne hercynienne et cassure téthysienne. Soc. Géol. France, Bull. (7), XVII, 56-70

AUBOUIN J. (1960).- Essai sur l'ensemble italo-dinarique et ses rapports avec l'arc alpin. Soc. Géol. France, Bull. (7), II, 487-526

AUBOUIN J. (1963).- Essai sur la paléogéographie post-triasique et l'évolution secondaire et tertiaire du versant Sud des Alpes orientales (Alpes méridionales ; Lombardie et Venétie, Italie ; Slovenie occidentale, Yougoslavie). Soc. Géol. France, Bull. (7), V, 730-766

AUBOUIN J., NDOJAJ I. (1964).- Regards sur la Géologie de l'Albanie et sa place dans la géologie des Dinarides. Soc. Géol. France, Bull. (7), VI, 593-625

AUBOUIN J., BLANCHET R., CADET J.P., CELET P., CHARVET J. CHOROWICZ J., COUSIN M., RAMPNOUX J.P. (1970).- Essai sur la géologie des Dinarides. Soc. Géol. France, Bull. (7), XII, 1060-1095

AUZENDE J.M., OLIVET J.L., BONNIN J. (1972).- Une structure compressive au Nord de l'Algérie : Deep-Sea Res., 19, n° 2, 149-155

AZEMA J., BOURROUILH R., CHAMPETIER Y., FOURCADE E., RANGHEARD X. (1974).- Rapports stratigraphiques, paléogéographiques et structuraux entre la Chaîne ibérique, les Cordillères bétiques et les Baléares. Soc. Géol. France, Bull. (7), XVI, 140-160

BARRETT D.L., KEEN C.E. (1976).- Mesozoic Magnetic lineations, the Magnetic Quiet zone, and Sea Floor Spreading in the North west Atlantic. J. Geophys. Res., 81, n° 26, 4875-4884

BAYER R. LE-MOUEL J.L., LE PICHON X. (1973).- Magnetic anomaly pattern in the western Mediterranean : Earth Planetary Sci. Lett., 19, 168-176

BEIN A. (1976).- Rudistic Fringing Reefs of Cretaceous shallow Carbonate Platform of Israël. Amer. Ass. Petrol. Geologists, Bull., 60, n° 2, 258-272

BERGOUGNAN H. (1975).- Relations entre les édifices pontique et taurique dans le Nord-Est de l'Anatolie. Soc. Géol. France, Bull. (7), XVII, 1045-1057

BERNOULLI D., LAUBSCHER H. (1972).- The palinspatic problem of the Hellenides. Eclogae Geol. Helv., 65, 107-118

BERNOULLI D., JENKYNS H.C. (1974).- Alpine, Mediterranean and central Atlantic Mesozoic facies in relation to the early evolution of the Tethys. In : Modern and Ancient Geosynclinal sedimentation, R.H. Dott and R.H. Shaver, edit. Tulsa, Soc. Econ. Pal. Mineral., 129-160 (Spec. Bull. n° 19)

BEZZI A., PICCARDO G.B. (1971).- Structural features of the ligurian ophiolites : Petrologic evidence for the "oceanic" floor of the Northern-Apennines geosyncline : A contribution to the problem of the Alpine type gabbro-peridotite association. Mem. Soc. Geol. Ital., 10, 53-64

BIJU-DUVAL B. (1974).- Commentaires de la carte géologique et structurale des bassins tertiaires du domaine Méditerranéen. Rev. Inst. Franc. Pétrol., 29, n° 5, 607-639

BIJU-DUVAL B., LETOUZEY J., MONTADERT L., COURRIER P., MUGNIOT J.F. SANCHO J. (1974).- Geology of the Mediterranean Sea Basins. In : The Geology of continental Margins Drake and Burk Edit. New-York, Heidelberg, Berlin, Springer Verlag, 695-721

BIJU-DUVAL B., DERCOURT J., LE PICHON X. (1976a).- La genèse de la Méditerranée. La Recherche, 7, n° 71, 811-822

BIJU-DUVAL B., DERCOURT J., LE PICHON X., APOSTOLESCU V. (1976b).- Evolution from Tethys to Mediterranean Basins. Abstract In XXV° Congrès CIESM Split in press

BIJU-DUVAL B., LAMPEREIN C., RIVEREAU J.C., LOPEZ N. (1976c).- Commentaires de l'Esquisse Photogéologique du Domaine Méditerranéen grands traits structuraux à partir des images du satellite Landsat 1. Rev. Inst. Franc. Pétrol. 31, 365-400

BIJU-DUVAL B., LETOUZEY J., MONTADERT L. (1977).- Structure and Evolution of the Mediterranean Sea Basins. In : Initial Report of DSDP, vol. XXXXII, Washington (U.S. Government Printing Office) in press

BLANCHET R., CADET J.P., CHARVET J., RAMPNOUX J.P. (1969).- Sur l' existence d'un important domaine de flysch tithonique-crétacé inférieur en Yougoslavie : l'unité du flysch bosniaque. Soc. Géol. France, Bull. (7), XI, 871-880

BOCCALETTI M., ELTER P., GUAZZONE G. (1971).- Plate tectonic models for the development of the western Alps and Northern Apennines, Nature (Phys. Sci.), 234, n° 49, 108-111

BOCCALETTI M., GUAZZONE G. (1972).- Remnant arcs and marginal basins in the Cainozoic development of the Mediterranean, Nature, (Phys. Sci.) 242, 18-21

BOCCALETTI M., GUAZZONE G. (1974a).- Remnant arcs and marginal basins in the Cainozoic development of the Mediterranean, Nature, 252, 18-21

BOCCALETTI M., MANETTI P., PECCERILLO A. (1974b).- Hypotheses on the plate tectonic evolution of the Carpatho-Balkan arcs. Earth Planetary Sci. Lett., 23, 193-198

BOCCALETTI M., HORVATH F., LODDO M. MONGELLI F., STEGENA L. (1976).- The Ty-rhenian and Pannonian Basins : A comparison of two Mediterranean interarcs bassins. Tectonophysics, 35, 45-69

BOGDANOFF A.A., MOURATOV M.V., SCHATSKY N.S. (1964).- Tectonique de l'Europe ; notice explicative pour la carte tectonique internationale de l'Europe au 1:2 500 000 Nanka, Moscou, Nedra ... Congrès géol. Intern.) 360 p.

BOSELLINI A., HSU K.J. (1973).- Mediterranean Plate tectonics and Triassic Paleogeography, Nature, 244, n° 5412, 144-146

BOUILLIN J.P., DURAND DELGA M., GELARD J.P., LEIKINE M., RAOULT J.F., RAYMOND D., TEFIANI M., VILA J.M. (1970).- Définition d'un flysch massylien et d'un flysch mauretanien au sein des flyschs allochtones d'Algérie. Compt. Rend. Acad. Sci. Fr., 270, 2249-2252

BOUILLIN J.P., KORNPROBST J. (1974).- Associations ultrabasiques de petite Kabylie : péridotites de type alpin et complexe stratifié comparaison avec les zones internes bético-rifaines. Soc. Géol. France, Bull. (7), XVI, 183-194

BOUSQUET J.C. (1973).- La tectonique récente de l'Apennin calabro-lucanien dans son cadre géologique et géophysique. Geol. Romana, 22, 104 p.

BRINKMANN R. (1972).- Mesozoic troughs and crustal structure in Anatolia. Geol. Soc. Amer. Bull., 83, n° 3, 819-826

BRUNN J.H., DUMONT J.F., GRACIANSKY P. ch. de., GUTNIC M., JUTEAU Th., MARCOUX J., MONOD O., POISSON A. (1971).- Outline of the geology of the western Taurids. In Geology and History of Turkey A.S. Cambell Edit. Tripoli, Petr. Explor. Soc. Libya, 225-255 (13th annual field conference)

BULLARD E.C., EVERETT J.E., SMITH A.G. (1965).- The fit of the continents around the Atlantic. In Symposium continental drift, P. M.S. Blackett. E. Bullard and S.K. Runcorn, edit., Phil. Trans. R. Soc. London, A. 1088, 1-51

BUREAU DE RECHERCHES GEOLOGIQUES ET MINIERES, SOCIETE ELF RECHERCHE, SOCIETE ESSO DE RECHERCHE ET D'EXPLOITATION DU PETROLE, SOCIETE NATIONALE DES PETROLES D'AQUITAINE (1974).- Géologie du Bassin d'Aquitaine, Paris, B.R.G.M., 27 pls.

BUSNARDO R., CHENEVOY M. (1962).- Dolérites intrusives dans le lias et de dogger d'Andalousie. Leurs différenciations pegmatitiques alcalines et auréoles de Métamorphisme. Soc. Géol. France, Bull. (7), IV, 461-470

BUSSON G. (1972).- Principes, méthodes et résultats d'une étude stratigraphique du Mésozoïque saharien. Mém. Mus. Nat. Hist. Nat., Nouv. sér., C, XXVI, 441 p.

CABY R. (1973).- Les plis transversaux dans les Alpes occidentales, implications pour la genèse de la chaîne alpine. Soc. Géol. France, Bull. (7), XV, 624-634

CAIRE A. (1971).- Chaînes alpines de la Méditerranée centrale. In : Tectonique de l'Afrique, Paris, UNESCO, "Sciences de la Terre n° 6, 61-90

CAMPBELL A.S. ed. (1971).- Geology and History of Turkey, Tripoli. Petr. Expl. Soc. Libya, 511 p. (13th annual field conf.)

CELET P., CLEMENT B., FERRIERE J. (1976).- La zone béotienne en Grèce : implications paléogéographiques et structurales. Eclogae. Geol. Helv., 69, n° 3, 577-599

CHANNEL J.E.T., HORVATH F. (1976).- The African - Adriatic Promontory as a palaeogeographical premise for alpine orogeny and plate movements in the Carpatho-Balkan region. Tectonophysics, 35, 71-101

CHARVET J. (1973).- Sur les mouvements orogéniques du Jurassique - Crétacé dans les Dinarides de Bosnie orientale. Compt. Rend. Acad. Sci. Fr., 276, sér. D, 257-259

CHOUKROUNE P., LE PICHON X., SEGURET M. SIBUET J.C. (1973).- Bay of Biscay and Pyrénées, Earth Planetary Sci. Lett., 18, 110-118

CIRIC B. (1966).- The magmatism of Montenegro, a geological Survey Bull. Inst. Geol. Geophys. Res., Beograd, 22-23, 17-33

CONANT L.C., GOUDARZI G.H. (1967).- Stratigraphic and tectonic framework of Libya : Amer. Assoc. Petrol. Geologists Bull., 51, n° 5, 719-730

CONTESCU L.R. (1974).- Geologic history and paleogeography of Eastern Carpathians : example of Alpine Evolution. Amer. Assoc. Petrol. Geologists Bull., 58, n° 12, 2436-2476

CORTESOGNO P., GIANELLI G., PICCARDO G.B. (1975).- Preorogenic metamorphic and tectonic evolution of the ophiolite mafic rocks (northern Apennine and Tuscany), Boll. Soc. Geol. Ital., 94, 291-327

DAL PIAZ V., HUNZIKER J.C., MARTINOTTI G. (1972).- La zona Sesia-Lanzo e l'evoluzione tettonico metamorfica delle alpi nord occidentali interne. Mem. Soc. Geol. Ital., XI, 433-460

DAL PIAZ V. (1974).- Le métamorphisme de haute pression et basse température dans l'évolution structurale du bassin ophiolitique alpino-apenninique (1° partie) : considérations paléogéographiques. Boll. Soc. Geol. Ital., 93, 437-468

DEBELMAS J., LEMOINE M. (1970).- The Western Alps : palaegeography and structure. Earth. Sci. Rev., 6, n° 4, 221-256

DEBELMAS J. (1972).- A propos de quelques synthèses récentes sur la genèse de l'arc alpino-apenninique. Rev. Géogr. Phys. Géol. dyn. XIV, 3, 229-241

DECANDIA F.A., ELTER P. (1972).- La zona ofiolitifera del Bracco nel Settore compresso fra Levanto e lo Val Graveglia (Apennino ligure). Mem. Soc. Geol. Ital., XI, 503-530

DERCOURT J. (1970).- L'expansion océanique actuelle et fossile ; ses implication géotectoniques. Soc. Géol. France, Bull. (7), XII, 261-317

DERCOURT J., CELET P., COTTIN J.Y., DE WEVER P., FERRIERE J., GRANDJACQUET C., HACCARD D., TERRY J., WIGNIOLLE E. (1976).- Importance d'une tectonique jurassique supérieur sur les marges de la plaque d'Apulie (Hellenides et Apennin ligure). Coll. ATP Montpellier, Soc. Géol. France in press.

DERYCKE F., GODFRIAUX I., ROBASZYNSKY F. (1974).- Sur quelques paragenèses du métamorphisme de haute pression-basse température dans l'axe et le pourtour de la fenêtre de l'Olympe (Grèce). Compt. Rend. Acad. Sci. Fr., 279, sér. D, 227-230

DEWEY J.F., PITMAN W.C. III, RYAN W.B.F., BONNIN J. (1973).- Plate Tectonics and the evolution of the Alpine system. Geol. Soc. Amer. Bull., 84, n° 10, 3137-3180

DEWEY J.F. (1976).- Ophiolite obduction. Tectonophysics, 31, 93-120

DIETRICH D., SCANDONE P. (1972).- The position of the basic and ultrabasic rocks in the tectonic units of the Southern Apennines. Att. della Accad. Pont., XXI, 1-15

DIMETRIJEVIC M.D., DIMETRIJEVIC M.N. (1973).- Olistostrome melange in the Yugoslavian Dinarides and Late Mesozoic Plate tectonics. J. Geol., 81, 328-340

DIMETRIJEVIC M.D. (1974).- The Dinarids : a model based on the new global tectonics. In : Jankovic S., Metallogeny and concepts of the geotectonic development of Yugoslavia. Belgrade. 141-178

DUBOURDIEU M. (1962).- Déplacement et déformation du pourtour de la Méditerranée occidentale depuis la fin du Miocène. Compt. Rend. Acad. Sci. Fr., 254, 2029-2031

DURAND DELGA M., LAMBERT A. (1955).- Existence d'un flysch schisto-gréseux tithonique aux abords des massifs anciens kabyles. Soc. Géol. France, Compt. Rend., 200-203

DURAND DELGA M. (1969).- Mise au point sur la structure du Nord-Est de la Berbérie. Publ. Serv. Geol. Algérie (N.S) 39, 89-131

DURAND DELGA M. (1971).- Les unités à mésozoïque métamorphique d'El Milia à Texenna (Algérie) et leur cadre structural. Soc. Géol. France, Bull. (7), XIII, 328-337

DZOTSENIDZE A.D. (1968).- Essai of comparison of the Meso-cenozoic magmatism of the caucasus, Crimea, Balkans and the Carpathians. Int. Geol. Congr., 23d, Prague, Proc. Sect. 2, 81-98

ELTER P., PERTUSATI P. (1973).- Considerazioni sul limite alpi-appennino sur relazioni con l'arco delle alpi occidentali. Mem. Soc. Geol. Ital., 12, 359-375

ENAY R. (1972).- Paléobiogéographie des Ammonites du Jurassique terminal (Tithonique-Volgien-Portlandien l.s.) et mobilité continentale. Soc. Géol. France, Compt. Rend., 163-168

ENTE NAZIONALE IDROCARBURI (1962-1971).- Colombo, Carlo, ed., Enciclopedia del Petrolio e del gaz naturale, v. 1-8

FALLOT P. (1945).- Estudios geologicos en la zona subbetica entre Alicante y el Rio Guadiana Menor ; Madrid ; Publ. Inst. "Lucas Malladas", C.S.I.C., 719 p.

FALLOT P. (1948).- Les cordillères bétiques. Est. Geol., 8, 83-172 Madrid (Cons. Sup. Invest. Cient. - Inst. Invest geol "Lucas Mallada")

FERRIERE J. (1974).- Etude géologique d'un secteur des zones helleniques internes subpélagonienne et pélagonienne (massif de l'Othrys - Grèce continentale). Importance et signification de la période orogénique anté Crétacé supérieur. Soc. Géol. France, Bull. (7), XVI, 543-562

FINETTI I., MORELLI C. (1973).- Geophysical exploration of the Mediterranean Sea. Boll. Geofis. Teorica. Appl., 15, n° 60, 263-341

FONBOTE J.M., PUINTERO L. (1960).- Lavas almohadillas (pillow-lavas) en los afloramientos volcanicos de la transversal Iznallos-Jaen (Cordilleras beticas). Notas comun. Inst. Geol. Miner. Espan., n° 60, 85-90

FOURQUIN C. (1975).- L'anatolie du Nord-Ouest, marge méridionale du continent européen, histoire paléogéographique, tectonique et magmatique durant le secondaire et le tertiaire. Soc. Geol. France, Bull. (7), XVII, 1058-1070

FRANCHETEAU J. (1973).- Plate tectonic model of the opening of the Atlantic ocean south of the Azores. In : Implication of Continental Drift to the Earth Sciences, ed. D.H. Tarling and S.K. Runcorn, 1, London, Academic Press, 197-202

GASS I.G. (1968).- Is the Troodos massif of Cyprus a fragment of Mesozoïc ocean floor ? Nature, 220, n° 5162, 39-42

GECZY B. (1973).- Plate tectonics and paleogeography in the east mediterranean mesozoic. Acta. Geol. Acad. Sci. Hung., 17, 421-428

GEYSSANT J., TOLLMANN A. (1966).- Réunion extraordinaire de la S.G.F. Alpes autrichiennes. Soc. Géol. France, Compt. Ren., 413-472

GINZBURG A., COHEN S.S., HAY-ROE H., ROSENZWEIG A. (1975).- Geology of Mediterranean shelf of Israël. Amer. Ass. Petrol. Geologists, 59, n° 11, 2142-2160

GRAY C. ed. (1971).- Symposium on the geology of Libya. Tripoli, Faculty of Science, Univ. Libya, 522 p.

HACCARD D., LORENZ C., GRANDJACQUET C. (1972).- Essai sur l'évolution tectogénétique de la liaison Alpes-Apennins (de la Ligurie à la Calabre). Mem. Soc. Geol. Ital., XI, 309-341

HADZI E., PANTIC N., ALEKSIC V., KALENIC M. (1974).- The Alpides of Southeastern Europe in the light of plate tectonics. In : Jankovic op. cité, 275-310

HAWKESWORTH C.J., WATERS D.J., BICKLE M.J. (1975).- Plate tectonics in the Eastern Alps, Earth Planetary Sci. Lett., 24, n° 3, 405-413

HAYES D.E., RABINOWITZ P.D. (1975).- Mesozoic magmatic lineations and the magmatic quiet zone of the eastern North Atlantic, Earth Planetary Sci. Lett., 28, 105-115

HEIRTZLER J.R., DICKSON G.O., HERROW E.M., PITMAN W.C., LE PICHON X. (1968).- Marine magmatic anomalies, geomagmatic field reversals, and notions of the ocean floor and continents. J. Geophys. Res., 73, 2119-2136

HERZ N., SAVU H. (1974).- Plate tectonic history of Romanie. Geol. Soc. Amer. Bull., 85, n° 9, 1429-1440

HIRSCH F. (1976).- Sur l'origine des particularismes de la faune du Trias et du Jurassique de la plateforme africano-arabe. Soc. Geol. France, Bull. (7), XVIII, n° 2, 543-552

HORSTINK J. (1971).- The late Cretaceous and Tertiary Geological Evolution of Eastern Turkey. In : First Petrol. Cong. Turkey, Keskin C. and Demirmen F. Edit. Ankara, Ass. Turkish Petr. geologists, 25-41

HOYEZ B. (1974).- Cadre et évolution du bassin numidien sicilien. Compt. Rend. Acad. Sci. Fr., 278, sér. D, 1007-1010

HSU K.J. (1971).- Origin of the Alps and Western Mediterranean. Nature, 233, n° 5314, 44-48

HSU K.J., MONTADERT L. et al. (1977).- Initials Reports of the Deep Sea Drilling Project, vol. XXXXII, Washington (US Government Printing office) in press

HYNES A.J., NISBET E.G., SMITH A.G., WELLAND M.J.P., REX D.C. (1972).- Spreading and emplacement ages of some ophiolites in the Othrys region (Eastern Central Greece). Z. Deut. Geol. Ges. 123, n° 2, 455-468

INSTITUT FRANCAIS DU PETROLE ET INSTITUT DE GEOLOGIE DE GRECE (1966).- Etude géologique de l'Epire, Grèce nord-occidentale. Paris, Technip ed., 306 p.

JENKYNS H.C. (1970).- Submarine volcanism and the Toarcian pisolites et Western Sicily. Eclogae Geol. Helv., 63, 2, 549-572

JUTEAU T. (1970).- Pétrogenèse des ophiolites des nappes d'Antalya (Taurus Lycien oriental, Turquie). Leur liaison avec une phase d'expansion océanique active au Trias supérieur. Sci. Terre, 15, 3, 265-288

JUTEAU T., LAPIERRE H., NICOLAS A., PARROT J.F., RICOU L.E., ROCCI G., ROLLET M. (1973).- Les idées actuelles sur la constitution, l'origine et l'évolution des assemblages ophiolitiques mésogéens. Soc. Geol. France, Bull. (7), XV, 478-493

KESKIN C., DEMIRMEN F. Edit. (1971).- First Petroleum Congress of Turkey - Proceedings Ankara. Ass. Turkish Petrol. Geologists, 175 p.

KHAIN V.E., MILANOVSKY E.E. (1963).- Structure tectonique du Caucase d'après les données modernes. In : Livre à la Mémoire de P. Fallot, Soc. Géol. France, Mém. H. ser. 663-703

KLEMME H.D. (1958).- Regional geology of circum-Mediterranean region : Amer. Assoc. Petrol. Geologists Bull., 42, n° 3, 477-512

KRISTOFFERSEN Y. (1977).- Sea-floor spreading and the early opening of the North Atlantic, Earth Planetary Sci. Lett., in press.

KRISTOFFERSEN Y., TALWANI M. (1977).- The extinct triple junction south of Greenland and Tertiary motion of Greenland relation to North America. Geol. Soc. Amer. Bull., in press.

LAPIERRE H. (1975).- Les formations sédimentaires éruptives des nappes de Mamonia et leurs relations avec le massif du Troodos (Chypre occidental). Mém. Soc. Géol. France, N.S., t. LIV, 123, 1-132

LARSON R.L., PITMAN W.C. III (1972).- Worldwide correlation of Mesozoic magnetic anomalies and its implications. Geol. Soc. Amer. Bull., 83, 3645-3666

LAUBSCHER H.P. (1969).- Mountain Building. Tectonophysics, 7, n° 5-6, 551-563

LAUBSCHER H.P. (1971a).- Das Alpen Dinariden problem und die Palinspatik der Süduchen Tethys. Geol. Rundsch., 60, 819-833

LAUBSCHER H.P. (1971b).- The large scale kinematics of the Western Alps and the northern Apennines and its palinspatic implications. Amer. J. Sci., 271, 193-226

LE PICHON X., BONNIN J., SIBUET J.C. (1970).- La faille nord-pyrénéenne : faille transformante liée à l'ouverture du golfe de Gascogne. Compt. Rend. Acad. Sci. Fr., 271, ser. D, 1041-1044

LE PICHON X., SIBUET J.C. (1971).- Comments on the evolution of the Northeast Atlantic, Nature, 233, 257-258

LE PICHON X., BONNIN J., FRANCHETEAU J., SIBUET J.C. (1971).- Une hypothèse d'évolution tectonique du golfe de Gascogne. In Histoire structurale du golfe de Gascogne, Paris. Edit. Technip, "Coll. et séminaires" n° 22, 11-44

LE PICHON X., PAUTOT G., AUZENDE J.M., OLIVET J.L. (1971).- La méditerranée Occidentale depuis l'Oligocène. Schéma d'évolution : Earth and Planetary Sci. Lett., 13, n° 1, 145-152

LE PICHON X., SIBUET J.C., FRANCHETEAU J. (1977).- The fit of the continents around the North Atlantic ocean, Tectonophysics, in press

LIVRE JUBILAIRE M. SOLIGNAC (1973).- Ann. Mines et Geol. Tunis, n° 26, 643 p.

LUPU M. (1974).- The Southern Apuseni Zone of metaliferous Mountains. In : Tectonics of the Carpathian - Balkan Regions. Bratislava. Inst. Geol. Dionyz Stur. éd., 234-239

McKENZIE D. (1972).- Active tectonics of the Mediterranean region. Geophys. J., 30, n° 2, 109-185

MAGNE J., RAYMOND D. (1974).- Le Néogène "post-nappes" de la région de Dellys-Tizi Ouzou (Algérie), un enregistrement de l'évolution dynamique du NW de la Grande Kabylie après le Burdigalien. Soc. Géol. France, Bull. (7), XVI, 537-542

MAHEL M. édit. (1973).- Tectonic problems of the Alpine system Bratislava. Veda - Slovak Acad. Sci.

MAHEL M. édit. (1975).- Carte tectonique du système Carpathes - Balkans. Notice explicative. Paris UNESCO - Bratislava inst. Geol. Dionyz Stur., 9-44 (Sciences de la Terre n° 14)

MARTIN L. édit. (1967).- Guide-book to the Geology and history of Tunisia. Petr. Exploi. Soc. Libya. 9th annual field Conference, 293 p.

MATTAUER M., HENRY H. (1974).- Pyrénées in : Spencer A.M. ed. op. cit.

MATTAUER M., PROUST F. (1975).- La Corse alpine : un modèle de genèse du métamorphisme haute pression par subduction de croûte continentale sous du matériel océanique. Compt. Rend. Acad. Sci. Fr., 282, sér. D, 1249-1252

MAXWELL J.C. (1970).- The Mediterranean ophiolites and continental drift. In : Johnson H. edit. The megatectonics of continents and oceans, New Brunswick N.J. Rudgers Univ., 167-193

MONOD O. (1976).- La "courbure d'Isparta" : une mosaïque de blocs autochtones surmontés de nappes composites à la jonction de l'arc hellenique et de l'arc taurique. Soc. Geol. France, Bull. (7), XVIII, 521-532

MONTADERT L., MAUFFRET A., LETOUZEY J. (1977).- Messinian Event : seismic contribution. In : Initial Reports of the Deep Sea Drilling Project, vol. XXXXII, Washington op.cit.

MORELLI C. (1975).- Geophysics of the Mediterranean. Rapp. Comm. Int. Mer. Medit. Sp. Issue 7, 29-111

NAIRN A.E.M. edit. (1977).- The Mediterranean, vol. IV, New-York. Plenum, in press.

NEGRONI P. (1971).- Bassin Pannonien. Intern. Report IFP. Réf. 20492

OBERT D. (1974).- Phases tectoniques mésozoïques d'âge anté cénomanien dans les Bâbors (Tell nord-setifien, Algérie). Soc. Geol. France, Bull. (7), XVI, 171-176

OKAY H., DILEKÖZ E. edit. (1974).- Second Petroleum Congress of Turkey. Proceedings. Ass. Turk. Petrol. Geol., Ankara, 310 p.

OZGUL N. (1976).- Some Geological aspect of the Taurus orogenic belt (Turkey). Bull. Soc. Geol. Turkey, 19, 65-78

PAQUET J. (1974).- Tectonique éocène dans les Cordillères Bétiques : vers une nouvelle conception de la paléogéographie en Méditerranée occidentale. Soc. Géol. France, Bull. (7), XVI, 58-73

PITMAN W.C., TALWANI M. (1972).- Sea Floor spreading in the North Atlantic. Geol. Soc. Amer. Bull., 83, 619-649

RADULESCU D.P., SANDULESCU M. (1973).- The plate tectonic concept and the geological structure of the Carpathians. Tectonophysics, 16, 155-161

RAOULT J.F. (1974).- Géologie du centre de la chaîne numidique (N. Constantinois, Algérie). Soc. Géol. France, Mém. Nelle série, n° 121, 163 p.

REHAULT J.P., OLIVET J.L., AUZENDE J.M. (1974).- Le Bassin nord-occidental méditerranéen : structure et évolution. Soc. Géol. France, Bull. (7), XVI, 281-294

RICOU L.E. (1971).- Le croissant ophiolitique péri-arabe. UNe ceinture de nappes mises en place au Crétacé supérieur. Rev. Géogr. Phys. Géol. Dyn., (2), 13, 4, 327-350

ROMAN C. (1970).- Seismicity in Romania - Evidence for the sinking lithosphere. Nature, 228, 1176-1178

RYAN W.B.F., DANIEL J., STANLEY J., HERSEY J.B., FAHLQUIST D.A., ALLAN J.D. (1971).- The Tectonics and Geology of the Mediterranean Sea. In : The Sea, vol. IV, (II), 387-492, New York, Wiley

SAID R. (1962).- The Geology of Egypt, Amsterdam. Elsevier, 377 p.

SALEM R. (1976).- Evolution of Eocene - Miocene sedimentation Patterns in Parts of Northern Egypt. Amer. Ass. Petrol. Geologists Bull., 60, n° 1, 34-64

SANDER N.J. (1970).- In : Geology and History of Sicily, 43-132 op. cit.

SANDULESCU M. (1975).- In : Les Carpathes roumaines. Soc. Géol.

France, Bull. (7), XVII, 299-358

SOCIETE GEOLOGIQUE DE FRANCE (1962-1963).- Livre Mémoire Prof. P. Fallot

SOCIETE GEOLOGIQUE DE FRANCE (1970).- Géologie des Dinarides. Soc. Géol. France, Bull. (7), XII, 941-1124

SOCIETE GEOLOGIQUE DE FRANCE (1973).- L'arc de Gibraltar. Soc. Géol. France, Bull. (7), XV, 77-189

SOCIETE GEOLOGIQUE DE FRANCE (1974).- Débat sur la structure des Taurides occidentales. Soc. Géol. France, Bull. (7), XVI, 101-127

SOCIETE GEOLOGIQUE DE FRANCE (1975a).- Alpes maritimes franco-italiennes et Apennin septentrional. Soc. Géol. France, Bull. (7), XVII, 901-1014

SOCIETE GEOLOGIQUE DE FRANCE (1975b).- Anatolie. Soc. Géol. France, Bull. (7), XVII, 1015-1083

SOCIETE GEOLOGIQUE DE FRANCE (1976a).- Colloque sur la géologie des régions égéennes. Soc. Géol. France, Bull. (7), XVIII, 199-567. and Coll. intern. CNRS n° 244

SOCIETE GEOLOGIQUE DE FRANCE (1976b).- Réunion Extraordinaire en Corse. Soc. Géol. France, Bull. (7), XVIII, in press.

SOCIETE GEOLOGIQUE DU NORD (1970).- Mémoires du Centenaire.

SMITH A.G. (1971).- Alpine deformation and the oceanic areas of the Tethys, Mediterranean, and Atlantic : Geol. Soc. Amer. Bull., 82, n° 8, 2039-2070

SPENCER édit. (1974).- Mesozoic-Cenozoic orogenic belts data for orogenic studies - Edinburgh, Scottish Acad. Press, 809 p.(The Geol. Soc. Spec. Publ., n° 4)

STEGENA L., GECZY B., HORVATH F. (1975).- Late Cenozoic Evolution of the Pannonian Basin. Tectonophysics, 26, 71-90

SUBBOTIN S.I., SOLLOGUB V.B., PROSEN D., DRAGASEVIC T., MITUCH E. POSGAY R. (1969).- Junction of deep structure of the Carpatho-Balkan region with those of the Black and Adriatic Seas. Can. J. Earth Sci., 5, 1027-1035

TAPPONNIER P. (1976).- Essai d'interprétation globale de l'évolution tectonique du système alpin en Europe. Coll. Geodyn. Méditer. occid. Montpellier. CNRS, INAG, p. 176

TARLING D.H., MITCHELL J.G. (1976).- Revised Cenozoic polarity time scale - Geology, 4, n° 3, 133-136

TECHNIP Edit. (1971).- Histoire structurale du Golfe de Gascogne. t. 1 et t. 2. Inst. Fr. Pétrole. Collection "Colloques et séminaires" n° 22, 944 p., 315 fig.

TERRY J., MERCIER M. (1971).- Sur l'existence d'une série détritique beriasienne intercalée entre la nappe des ophiolites et le flysch éocène de la nappe du Pinde (Pinde septentrional, Grèce) Soc. Géol. France, Compt. Rend, 71-73

TERRY J. (1974).- Ensembles lithologiques et structures internes du Cortège ophiolitique du Pinde septentrional (Grèce). Construction d'un modèle pétrogénétique. Soc. Géol. France, Bull. (7), XVI, 204-213

THIERSTEIN H.R. (1977).- Biostratigraphy of marine Mesozoic sediments by calcareous nannoplankton, Third planktonic Conference Proceedings, in press.

TOLLMANN A. (1969).- Die tektonische Guederung des Alpen-Karpaten Bogens. Geol. Jahrfuch, 18, n° 10, 1131-1158

TRUILLET R. (1970).- In : Geology and History of Sicily. Petr. Expl. Soc. Libya op. cit.

TRUMPY R., HACCARD D. (1969).- Réunion extraordinaire de la Société géologique de France - Les Grisons. Soc. Géol. France, Compt. Rend., 329-396

TRUMPY R. (1971).- Stratigraphy in mountain belts. Quart. Geol. Soc London, 126, 293-318

TRUMPY R. (1975).- Penninic-Austroalpine boundary in the Swiss Alps. A presumed former continental margin and its problems. Amer. J. Sci., 275A, 209-238

TRUMPY R. (1976).- Du pelerin aux Pyrénées. Eclogae Geol. Helv., 29, n° 2, 249-264

VAN DER VOO R., FRENCH R.B. (1974).- Apparent polar Wandering for the Atlantic-Bordering Continents : late Carboniferous to Eocene. Earth. Sci. Rev., 10, 99-119

VINE F.J., MATTHEWS D.H. (1963).- Magnetic anomalies over oceanic ridges. Nature, 199, 947-949

VOGT P.R., ANDERSON C.L., BRACEY D.R. (1971).- Mesozoic magnetic anomalies, sea floor spreading and geomagnetic reversals in the southwest north Atlantic. Journal Geophys. Res., 66, 4796-4823

WEZEL F.C. (1970).- Numidian Flysch : an Oligocene early Miocene continental rise deposit off the African platform. Nature, 228, 275-276

WEZEL F.C. (1974).- Flyschs successions and the tectonic evolution of Sicily during the oligocene and early miocene. In : Guide book of the Geology of Italy. Petr. Expl. Soc. Libya

WILLIAMS C.A. (1975).- Sea floor spreading in the Bay of Biscay and its relationship to the North Atlantic. Earth Planetary Sci. Lett. 24, 440-456

WORZEL J.L. (1965).- Deep structure of coostal margins and mid-oceanic ridges. In : Submarine Geology and Geophysics, W.H. Whittard and R. Bradshaw éd., London. Butterworths, 335-361

YOUSSEF M.I. (1968).- Structural pattern of Egypt and its interpretation. Amer. Ass. Petrol. Geologists Bull., 52, 601-614

ZIEGLER W.H. (1975).- Outline of the Geological History of the West Europe. vol. 1. Wooland A.W. edit., 165-187. Barking, Essex. Applied. Science Publ.

TABLE I

PARAMETERS USED FOR RECONSTRUCTION

Anomaly	Age in M.y.	Geologic Age	Rotation applied to Africa			Movement AF/NA	Movement NA/EU
			Latitude	Longitude	Angle		
	Initial	-	52.	- 0.4	- 49.6	Initial fit (1)	Initial fit (1)
M 25	148	Oxfordian	50.	- 2.1	- 41.4	Quiet zone fit(1)	Initial fit (1)
M 2	112	Barremien	45.7	2.6	- 31.5	(2)	Initial fit (1)
-	90	Cenomanian-Turonian	30.3	- 3.6	- 18.4	Interpolation	Initial fit (1)
31-33	76-68	Campanian Maestrichtian	20.1	- 11.3	- 16.0	(3)	(4)
24	49	Ypresian	31.9	- 16.6	- 14.8	(3)	(4)
21	44	Lutetian	33.7	- 18.8	- 12.9	(3)	(5)
13	35	Stampian	29.3	- 26.6	- 7.8	(3)	(5)
5	9	Tortonian	27.1	- 27.9	- 2.4	(3)	(5)

(1) Le Pichon et al (1977)
(2) Hayes and Rabinowitz (1975), Vogt et al (1971), Larson and Pitman (1972)
(3) Francheteau (1973)
(4) Kristoffersen (1977)
(5) Pitman and Talwani (1972)

Ages according to Tarling and Mitchell (1976) for anomaly 31 on, according to Thierstein (1977) prior to that. Positive latitude for north, positive longitude for east, convention for sign is negative when seen clockwise from pole of rotation.

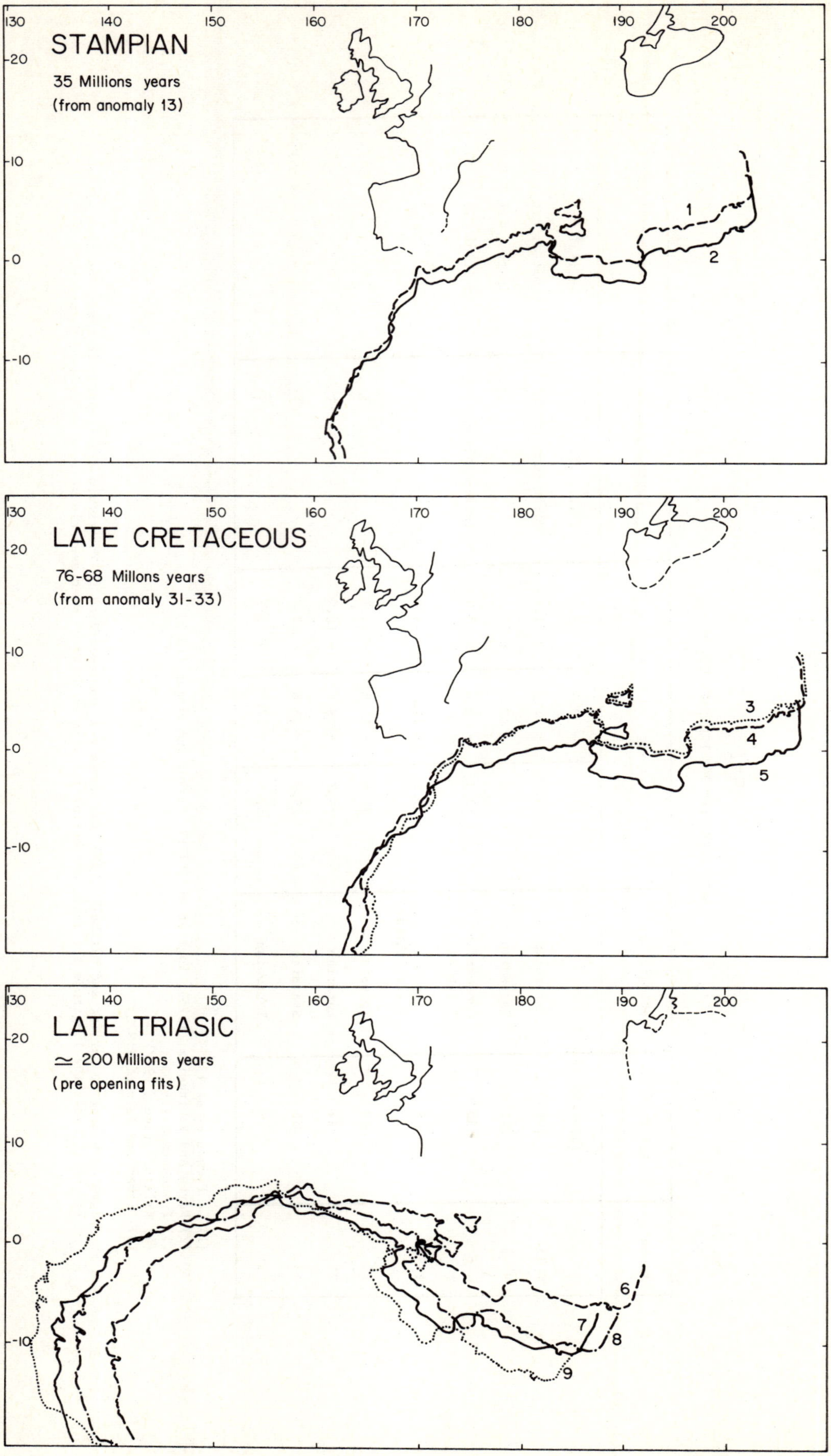
STAMPIAN
35 Millions years
(from anomaly 13)
LATE CRETACEOUS
76-68 Millons years
(from anomaly 31-33)
LATE TRIASIC
≃ 200 Millions years
(pre opening fits)
130
140
150
160
170
180
190
200
20
10
0
-10
1
2
3
4
5
6
7
8
9

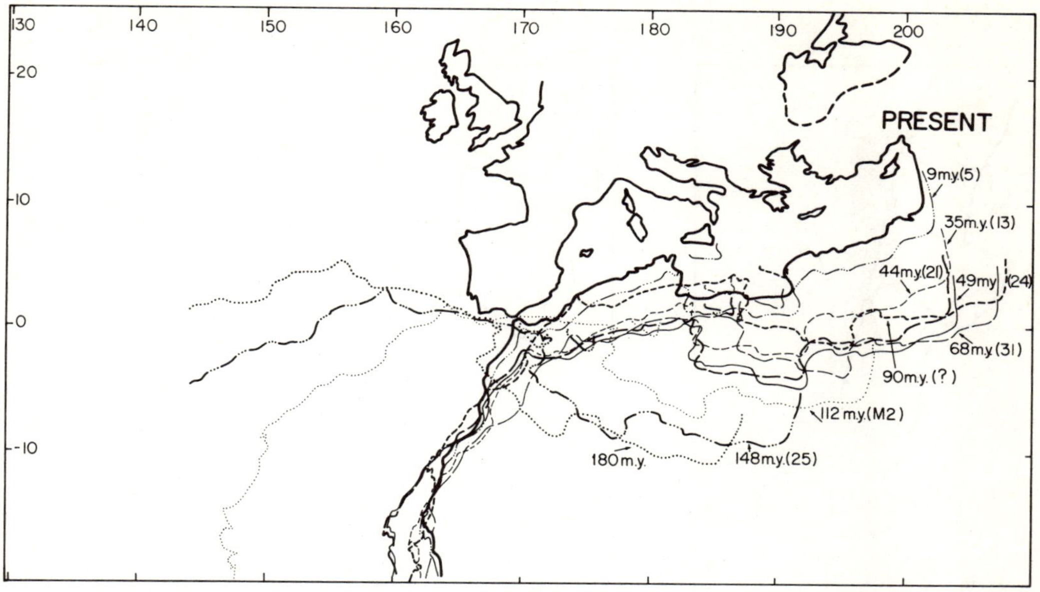

Figure 2 : Plate Kinematics of the Africa-Europe plate system. Parameters are given in table I.

← Figure 1 - Comparison of different kinematic models for the Africa displacements with respect to Europe in its present position. Oblique Mercator projection with pole at 52° N and 150° E as in all the maps of this paper. Numbers refer to references of table I : 1) - ref. 5 ; 2) - ref. 3 for Africa-America and ref. 5 for Europe-America ; 3) - ref. 3 for Africa-America and ref. 5 for Europe-America ; 4) - réf. 5 for both ; 5) - ref. 5 for Africa-America and ref. 4 for Europe-America ; 6) - Bullard et al. 1965 ; 7) - Dewey et al., 1973 ; 8) - ref. 1 for both ; 9) - ref. 1 for Africa-America and ref. 4 for Europe-America.

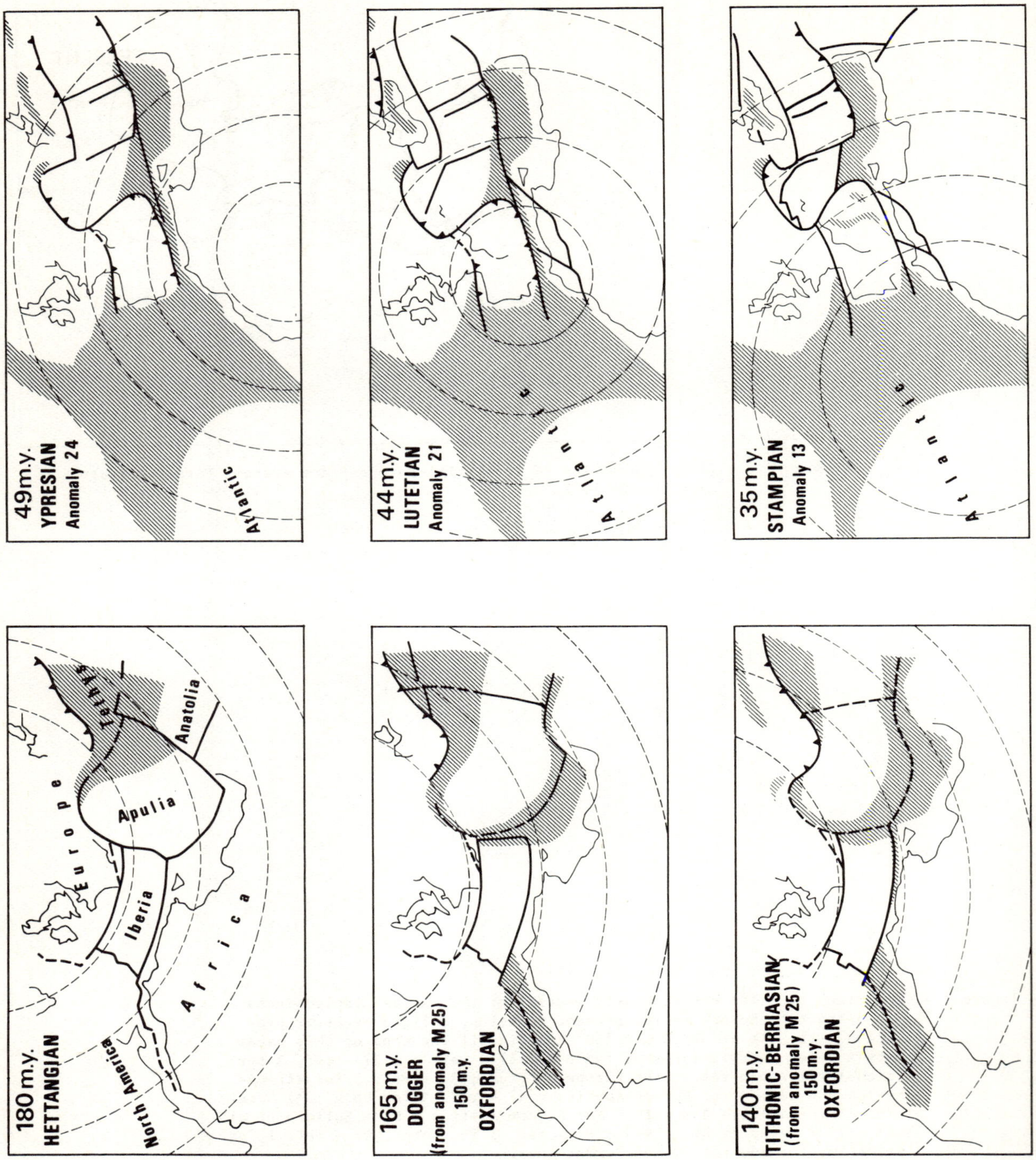
49 m.y.
YPRESIAN
Anomaly 24
Atlantic
44 m.y.
LUTETIAN
Anomaly 21
A t l a n t i c
35 m.y.
STAMPIAN
Anomaly 13
A t l a n t i c
180 m.y.
HETTANGIAN
North America
Europe
Tethys
Anatolia
Apulia
Iberia
A f r i c a
165 m.y.
DOGGER
(from anomaly M25)
150 m.y.
OXFORDIAN
140 m.y.
TITHONIC-BERRIASIAN
(from anomaly M 25)
150 m.y.
OXFORDIAN

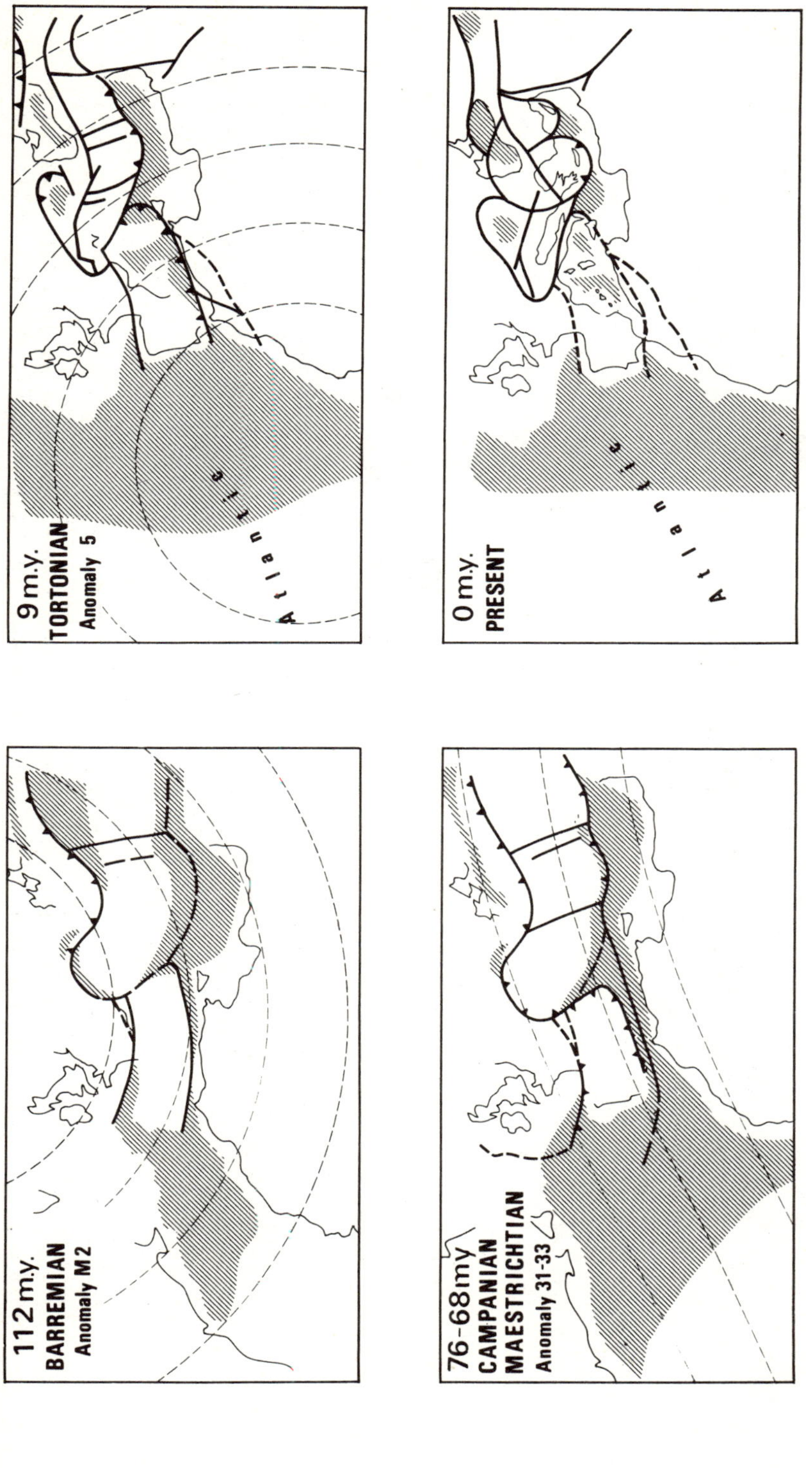

Figure 3 : Proposed evolution from the Tethys to the Mediterranean basins. Parameters are given in table I for positions of Africa. For positions of intermediate plates, see text. Small datted circles are about pole of rotation of Africa with respect to Europe during the immediatly following period.

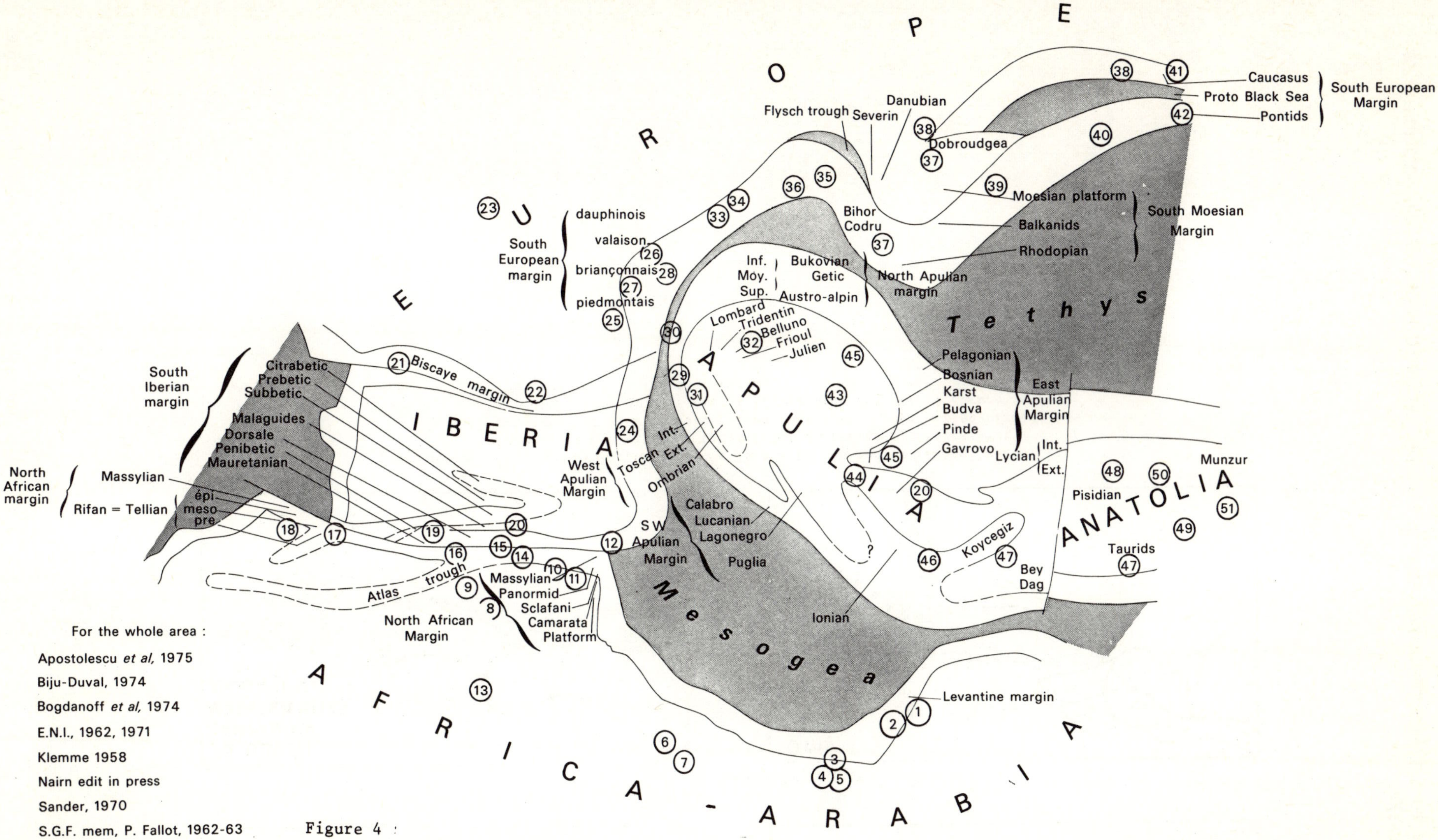

For the whole area :

Apostolescu *et al*, 1975

Biju-Duval, 1974

Bogdanoff *et al*, 1974

E.N.I., 1962, 1971

Klemme 1958

Nairn edit in press

Sander, 1970

S.G.F. mem, P. Fallot, 1962-63

S.G.F., 1974

Spencer edit, 1974

Figure 4 :

Main paleogeographic zones and general geological bibliography used in this paper illustreted on the Dogger reconstruction (see plate III). Numbers refer to following references (see detailed bibliography) : 1 - Ginzburg et al., 1975 ; 2 - Bein, 1976 ; 3 - Salem, 1976 ; 4 - Saïd, 1962 ; 5 - Youssef, 1968 ; 6 - Carlyle Gray edit, 1971 ; 7 - Conant and Goudarzi, 1967 ; 8 - Livre Jub. M. Solignac, 1973 ; 9 - PESL, Martin edit., 1967 ; 10 - Caire, 1971 ; 11 - PESL, Alvarez and Gahrbandt edit., 1970 ; 12 - Bousquet, 1973 ; 13 - Busson, 1972 ; 14 - Raoult, 1975 ; 15 - Bouillin et al., 1970 ; 16 - Durand Delga, 1969 ; 17 - S.G.F., 1973 ; 18 - Andrieux, 1971 ; 19 - Paquet, 1974 ; 20 - S.G.N., Mem. centenaire, 1970 ; 21 - Hist. Golfe Gascogne, Technip edit., 1971 ; 22 - BRGM, 1974 ; 23 - Ziegler, 1975 ; 24 - S.G.F., 1976 b ; 25 - Debelmas et Lemoine, 1970; 26 - Trumpy, 1975 ; 27 - Caby, 1973 ; 28 - Trumpy et Haccard, 1969 ; 29 - Haccard et al., 1972 ; 30 - S.G.F., 1975 ; 31 - Abbate et al., 1970 ; 32 - Aubouin, 1963 ; 33 - Tollmann, 1969 ; 34 - Geyssant et Tollmann, 1966 ; 35 - Sandulescu, 1975 ; 36 - Boccaletti et al., 1976 ; 37 - Mahel edit., 1975 ; 38 - Dzotsenide, 1968 ; 39 - Fourquin, 1975 ; 40 - Bergougnan, 1975 ; 41 - Khain, 1963 ; 42 - Adamia, 1975 ; 43 - Aubouin, 1960 ; 44 - Aubouin et Ndojaj, 1964 ; 45 - S.G.F., 1970 ; 46 - S.G.F., 1976 a ; 47 - PESL, Campbell edit., 1971 ; 48 - Brinkmann, 1972 ; 49 - Ozgül, 1976 ; 50 - Okay and Dileköz edit., 1974 ; 51 - Kestin and Demisman, 1971.

INTERNATIONAL SYMPOSIUM ON THE STRUCTURAL HISTORY OF THE MEDITERRANEAN BASINS. SPLIT (YUGOSLAVIA) 25-29 OCTOBER 1976. B. BIJU-DUVAL AND L. MONTADERT, Eds. EDITIONS TECHNIP, PARIS 1977, pp. 165-184

VOLCANISM RELATED TO PLATE-TECTONICS IN THE WESTERN AND EASTERN MEDITERRANEAN

H. BELLON [(1)] and J. LETOUZEY [(2)]

Data on Mesozoic and Cenozoic magmatism in the Mediterranean region are numerous. An attempt at synthesis has recently been done from new results (K-Ar ages and chemical considerations) (BELLON, 1976) on the western domain; on the contrary, no critical review has recently been published about the magmatism outcropping in the eastern domain.

In outcrops, these domains are contrasting: magmatism is largely distributed in the eastern domain;outcrops are discontinuous and smaller in the western one. The age spectra stretches itself out in a large time (since Liassic) in the eastern area. It limits itself to the Neogene and Quaternary, with a few exceptions recently discovered, in the western area.

Scientific works of the last years have demonstrated that magmatism might be a tracer of geodynamical processes, if its nature and distribution in space and time are well known. Our attempt is to use this marker as an element of critical value in relation to three main structural and paleogeographical problems, choosen to clarify our report:

- The Tethys closing and the Pannonian Basin and Black Sea opening
- The possible opening and closing of a proto Eastern Mediterranean
- The building of the present Western Mediterranean.

I - TETHYS CLOSING (TABLE I) - (MAP I)

In the beginning of Mesozoic, an ocean, the Tethys, separated Europe from Africa. According to kinematic reconstructions from the magnetic anomaly pattern in the Atlantic, this ocean was widely opened to the east. We find its trace in the facies of ancient margins and in obducted and transported ophiolites in the Apennines, the Western Alps and in the Carpathian Chains. The width of this "Western Tethys" has not been important. On the contrary, the "Eastern Tethys" sublined now by ophiolitic massives, which separated the southern district of the European plate (Rhodop, Pontids, Adzhar-Trialet area, Lesser Caucasus) from the intermediate sialic blocks (Apulia, Aegea, Anatolia, Armenia) was much wider. The closeness of Europe and Africa during Mesozoic led to the closing of this ocean marked by the tectonic emplacement of ophiolites (obduction). During the stage of continental collision, the obduction of oceanic material: the ophiolites, occurred. This collision is of Jurassic age in the Vardar zone, and was achieved at the end of Cretaceous to the South of Pontids.

I-1 - Closing of the "Western Tethys" and opening of the Pannonian Basin.

In the whole folded area, from the Western to the Eastern Alps, the elements of a Mesozoic and Cenozoic magmatism are rare (Table 1). However, some words about the Alpine chains must point out the existence of volcanic pebbles in some flysch formations.

(1) Laboratoire de Pétrologie, Université de Paris-Sud, 91405 Orsay, France

(2) Institut Français du Pétrole, 1-4 avenue de Bois-Préau, 92506 Rueil-Malmaison, France

(Taveyanne sandstones and Gets flysch (BERTRAND et al., 1976)). Their calc-alkaline nature is doubtless. Age data obtained on these samples (DELALOYE et al., 1975) are around 60 MY and 26-18 MY. These apparent ages, due to a probable metamorphic rejuvenation are the clues of at least two magmatic stages during the Alpine story (Mesozoic and Cenozoic). A tectonic feature, in Southern Alps, is sublined by intrusive bodies of granodioritic composition: so, in Adamello, one of these bodies has been dated at 33 MY (BORSI et al., 1966). Near Nice, esterellite intrusions of same chemical nature are dated at 30 MY. The Veneto volcanism (DE VECCHI et al., 1976) dated at 35 MY must be related to the last Alpine magmatic activity. The existence of a stable margin on the edge of the European continent and the narrowness of the oceanic area before the closure could perhaps explain the lack of volcanism.

In Pannonian region and the Carpathians, the problem offers some complexity. The geochronology of magmatism allows us to distinguish three stages.

The first stage is basic and ultrabasic.

Rocks are located in the Metaliferi Mountains and along the eastern edge of the Transylvanian basin in the East Carpathians. The activity began in Liassic times in the Metaliferi Mountains (SAVU, 1968) and later in the East Carpathians where volcanism is interbedded in the Sinaia Beds zone and was initiated probably during middle Jurassic (BLEAHU, 1962). In age, this magmatic stage can be compared with the magmatism present in the Dinarids and Hellenids chains (BOCCALETTI et al., 1974). A question arises from these facts: does this magmatism represent either an oceanic floor of great extension, or two distinct oceanic regions (one of which being a marginal basin)?

The second stage of Senonian -Paleocene age (RADULESCU et al., 1973) and geochronologically dated (K-Ar method) between 75 and 61 MY (PAVELESCU et al., 1975) is of calc-alkaline affinity The rocks are known as Banatitic rocks. They are located in the Apuseni Mountains, and also in the southern extremity of the South Carpathians. Along the Balkan Chains, the Vistosha pluton, near Sofia, is ante-Maestrichtian (GOCEV, 1973). This magmatism cuts the basic rocks of the Metaliferi Mountains, and also the continental crust. So, the authors (RADULESCU et al., 1973) consider that this magmatism, because of its calc-alkaline affinity and its structural emplacement, is connected to subduction processes.

The geographical distribution of these rocks, between the Pannonian basin and the Transylvanian one, and along the Balkan Chains indicated a structural alignment and probably a zone where subduction processes have been very active at the end of Cretaceous. It extends eastward (Srednogoria, Pontids, see later...).

The third stage of general calc-alkaline affinity, starts at the beginning of Miocene; this fact is suggested by geochronological dates (K-Ar) obtained by KONECNY et al., (1969).

Some new results, not yet published, obtained by one of us (B) give identical data. A diachronism appears, in the light of these data. First products (21 MY) are located in Czecho-Slovakia. Identical products appear only around 14 MY in Hungary and later in Romania (BOCCALETTI et al., 1973). It seems that there is a displacement of the activity from the west to the east. We must notice the abundance of rhyolitic products in Czecho-Slovakia and Hungary due to a probable anatexis of continental crust, and the scarcity of rhyolites in Romania. Several questions arise: is this fact linked to the heterogeneity of the basement ? Is the displacement of activity process due to the activity of two well distinguished subduction zones, or to the displacements of two sialic zones upon one subduction zone ? Some authors (STEGENA et al., 1975) have proposed the existence of a fossil subduction zone dipping from the north east to the south west under the Carpathian Chains. The chronology of the volcanism does not allow us to accept this idea, because of the diachronism.

A basaltic activity terminates this magmatic stage. It begins at 10 MY near the Tojak Massive (drilling hole of Sarospatak) in Hungary. An other K-Ar date proves the continuity of basaltic activity until 3 MY (BELLON, unp. data) at least.

Some basaltic flows are spreading around the Ciuc Basin since 1 MY until 0.5 MY in Romania, and close the magmatic activity.

Many young basaltic outcrops are aligned along a north 340° direction, the "Balaton line".

From our point of view, the three volcanic stages are a large subduction process with the creation of a marginal basin in correlation with the displacement of sialic blocks upon the subduction zone.

This magmatism in three stages is located inside the Carpathian system. The existence of a rift volcanism, outside the Carpathian system, which started around 60 MY until subpresent times, must be pointed out (BELLON, KOPECKY, in the press) in the Bohemian Massive.

I-2 - Eastern Tethys closing and Black Sea opening.

The present mark of the eastern part of the Tethys ocean is located, according to most recent authors, in the south of the Rhodopian massive, of Pontids, of Adzhar Trialet zone and Lesser Caucasus, and in the north of Aegean Sea, of Anatolia and Armenia. The closing of this ocean took place at the end of the Cretaceous. Obducted oceanic crust delineates the suture. Before the continental collision stage, it seems possible to compare the magmatic provinces located on the northern and southern edges.

I-2a - Along the southern European margin.

Volcanic evidences appear since the Jurassic: spilits of Tithonian age, in Bulgaria, (Strandja zone) and to the east of Marmara Sea (FOURQUIN, 1975); thick Liassic volcano sedimentary series (2000m) with albitophyrs and trachyandesites near Kelkit in the south west from Trabzon (BERGOUGNAN, 1976). The region located to the east of the Black Sea is the seat of an important volcanism, extending from Lesser Caucasus to the south of the Greater Caucasus and Crimea:

Jurassic

- 160 MY - Andesites and dacites in Lesser Caucasus to the south of the Greater Caucasus in north west of the Sevan Lake (BAGDASARJAN, 1974).
- Bajocian - thick series (1.5 to 3 km) of andesites and basalts laid upon Transcaucasian massives (ADAMIYA et al., 1974). This volcanism extends to the Pontids and the southern margin of th Greater Caucasus (KHAIN, 1974).
- Bathonian - to the north, the volcanic activity ceases, but is still present to the south.
- End of Jurassic - Neocomian - the southern volcanism becomes andesitic (KHAIN, 1974).

Middle and Upper Cretaceous

A flysch type basin accompanied by an intense magmatism (volcanism and intrusive bodies) is present to the north of the ophiolitic belt and to the south of the European continent. It is continued, very often, until

Cenozoic

In Srednogoria, a Campanian-Maestrichtian volcanism, perhaps Paleocene, exhibits three cycles in 100 m of thickness. At the base, medium basic andesites exist; the second cycle is represented by dacites and rhyolites and the third by basic andesites and trachyandesites. To the east, volcanism seems to be of alkaline type (olivine-basalt) (CANKOV, 1974).

These volcanic units are known in the Pontids: an offshore deep hole cuts Cretaceous volcanics in the east of Strandja ridge; between the Marmara Sea and Ankara, a large volcanic complex (7000 sq. km), the Galatic Massive, was emplaced from Maestrichtian until Eocene. The series is essentially of rhyolitic and trachytic nature. During the Oligocene-Miocene, a new acidic volcanism occurred.

Near Trabzon, on the southern coast of Black Sea, a tholeiitic series is followed by basaltic andesites, then by high potassic rhyo-dacites. The last products are basic and probably contemporaneous with the post upper-Cretaceous tectonics (STOJANOV, 1970).

In the Adzhar-Trialet zone, the volcanic complex startsin the Paleocene. Until the middle-Eocene, a space and time association of alkaline, strongly alkaline and calc-alkaline rocks exists. The series has a thickness of 4 or 5 km. For the authors (ADAMIYA et al. 1974), this partition seems to be characteristic of a rift zone active in Eocene times. In the upper-Eocene, intermediate lavas with shoshonitic affinities would characterize a part of an island arc. Grained intrusive bodies would represent the roots of Paleogene volcanoes.

In the Greater Caucasus, no volcanic activity is reported during the Cretaceous. On the contrary, to the south, in Trans-Caucasia and in the Lesser Caucasus, pyroclastics and lavas of andesitic nature may attain a thickness of 3 to 5 km (KHAIN, 1974).

I-2b - Along the ophiolitic suture

To the south of it, one can oppose the Armenia which, since the Kimmeridgian (150 MY) until Campanian (75 MY), is the seat of an intense calc-alkaline volcanism (BAGDASARJAN, 1974) to the north. In Anatolian domain and north Aegean one, where the volcanism begins during Eocene (trachytes and syenites of Sivas - Malatya, LEO et al., 1974) or during Miocene, this activity is very difficult to relate to the Tethys consumption, because it is very late after the continental collision.

I-3 - In conclusion, it seems that during the Tethys closing, during the Mesozoic, Armenia excepted, the south European margin shows a major volcanic activity. Thus, this north Tethysian margin would have been, during the closing, an active margin. Structural informations about the Black Sea taken into account, it seems that the basin may be interpreted in terms of a marginal basin (LETOUZEY et al., 1977, and this volume) with a southern drift of the Pontids domain upon a subduction zone. The basement of the Black Sea is prior, at least, to the Oligocene and since that time subsidence was very active.

The problem of a hypothetical subduction zone to the south of the Greater Caucasus since the Jurassic (KHAIN, 1974) remains unsolved. One may remark that after the colliding stage, volcanism has been widely spreading on each side of the suture. The breaking down, during the collision, of a lithospheric slab and the slow consumption of it in the atmosphere may constitute an explanation. Such a situation exists in the Fidji plateau, where the seismic activity shows the presence of a slab descending through the asthenosphere since several tens of M.Y.

II - POSSIBLE OPENING AND CLOSING OF A PROTO EASTERN MEDITERRANEAN

II-1 - Opening.

The possible dispersal of the Apulian intermediate plate from the African plate at the beginning of the Mesozoic would lead to the opening of a proto Eastern Mediterranean located between the Arabo-African block and the Aegean-Anatolian and Armenian block (BIJU - DUVAL et al., 1977). Consecutive to a coming together of the Arabo-African and the European plates, it would be represented today only by the abyssal part of the present Eastern Mediterranean, overthrusted during Cenozoic times by the Aegean plate. The part that was situated to the north of the Levantine Basin and the Arabic platform, has disappeared during the oceanic consumption then during the continental collision with obduction of ophiolites. We may consider as the evidence of the Apulo-African distension the Triassic pillow-lavas overthrusted during the Maestrichtian and outcropping in Antalya tectonic sheets (MARCOUX, 1970), in Mamounia tectonic sheets, in Cyprus (Carnian-Norian age) (LAPIERRE, 1976), in Hatay (Syria). Contrary to this hypothesis, RICOU (1974) and BRUNN et al., (1976) proposed that the pillow-lavas located to the south today were issued from the Tethys, to the north of the intermediate plates.

II-2 - Volcanism in the present Eastern Mediterranean.

According to seismic reflection and magnetism in this area (GONNARD et al., 1975), volcanism is scarce with few exceptions since Tertiary in deep zones of the Eastern Mediterranean.

- In the offshore basin of Antalya and in the basins located to the east - north east from Cyprus, magnetic anomalies may be related to pillow-lavas and ophiolites settled under these basins during the late Cretaceous, overthrusting contemporarily with Troodos (BIJU - DUVAL et al., 1977)

- The magnetic anomaly of the south eastern flank of Mount Erathostenes, in the Levantine Sea, is due probably to Mesozoic magmatic bodies, because seismic profiles do not show any Tertiary volcanism..

- In the Ionian Basin, a volcanism related to a large NE-SW fracturation network is interbedded in Miocene levels but may occur earlier.

- A large volcanism marks out the escarpment between the Ionian Basin and Malta platform. Let us remember that to the south east of Sicily, an alkaline magmatism extends since the Liassic to subpresent times with some breaks (KAFKA et al., 1959).

- In all the submerged continental domain located to the north of the Syrtan basin, magnetic anomalies mark out volcanism presumed of Tertiary age.

- On the Siculo-Tunisian platform, recent volcanism of Pantelleria (BARBERI et al., 1973) seems to be related with graben structures of NW-SE direction, of late Miocene-Pliocene and Quaternary ages. Perhaps, these grabens indicate the initiation of rifting.. We must note that the most important volcanoes are located on the graben extremities and may be in relation with transverse faults of NE-SW direction.

II-3 - The Arabian and African platform.

Data from litterature show that alkaline volcanism is scattered all along Mesozoic times and Cenozoic times.

- During Mesozoic:
  - magmatism of upper Jurassic and lower Cretaceous in the south of Israel. In Galilea, volcanism is active until Senonian (FREUND et al., 1975)
  - calk-alkaline olivine basalts and alkaline

rhyolites are dated at 120 MY (MENEISY et al. 1974) in the eastern Egyptian desert and mark a lower Cretaceous distensive phenomenon.
- in Sicily, basalts of Capo Passero are 71 MY old (BARBERI et al., 1974). Let us remember the long alkaline activity in south east Sicily since Liassic.

- During Cenozoic:

  - volcanism spreads largely in Syria, Lebanon, Israel, Egypt and Libya. The geochronological details are given in Table 2. A part of the volcanism may be related to the opening of the Red Sea.
  - In Sicily (Mt Iblei), where volcanic products are either of tholeiitic affinity or of alkaline one (BARBERI et al., 1973), volcanism is dated (K-Ar) between 5 and 2 MY. Present activity of Etna Volcano (TANGUY , 1967) is of alkaline and tholeiitic tendency.

II-4 - Active margins to the north of the proto-Eastern Mediterranean and the present Eastern Mediterranean

In Turkey, volcanism is active during the Cenozoic in Taurus-Anatolia and in the north of Cyprus. In the north of Taurus from the Eskisehir to the Sivas Malatya, magmatism begins in the Paleocene:

- Sivas: Syenites and trachytes 75 to 65 MY (LEO et al., 1974).
- South of Tuz Golü basin: andesites (ARIKAN, 197).
  During the Oligo-Miocene: south east of Tuz Golü, andesites.
  During the Miocene: andesites of Sivas 18 to 14 MY (LEO et al., 1974).
  During the Plio-Quaternary: basalts and rhyolites are widely spreading in Capadoccia (JUNG et al., 1972) and around Tuz Golü basin (ARIKAN, 197)..

To the north of Cyprus, in Kyrenia range, volcanism supposedly lasted from the Maestrichtian until the Paleocene (BAROZ et al., 1976).

In the Aegean arc:

- In Creta, diorites and basalts are dated at 77 to 66 MY (BARANIY et al., 1976). This fact confirms the Paleocene activity, also known in the south of Pindus (78 to 61 MY) (HYNES, 1972).
- In north of the Aegean Sea, from the south of Marmara Sea until Eubea in Greece, a large magmatism is known on land and delineated offshore by magnetic anomalies and disturbances of seismic reflectors.

In Turkey, to the north of Izmir, a calc-alkaline magmatism active since the Oligocene is dated between 28 and 11 MY (BORSI et al., 1972). The crustal magmatism due to anatexis does not seem to be negligible. An alkaline activity comes later (9 to 1 MY).

To the south of the Aegean Sea, in the Sporades and Cyclades archipelago, there is a recent volcanism (2.7 to 0.2 MY) (FYTIKAS et al., 1976).

II-5 - In conclusion.

During the creation of the proto Eastern Mediterranean and the present Eastern Mediterranean, a contrast appears between the magmatic provinces located to the north and to the south. The southern province is of alkaline type during the Cenozoic and the Mesozoic. Around the northern province, one may see the calc-alkaline tendency of magmatism. The crustal anatexis played a non negligible role in magma generation.

A delay appears between the stage of the formation of the eastern part of proto Mediterranean (late Cretaceous) and the beginning of calc-alkaline magmatism (Paleocene).

This fact may reflect - either a low width of the proto-Eastern Mediterranean in this region and therefore no calc-alkaline magmatism generated during its closing, the depth attained by the subducted slab being too low, or a time delay between the beginning of the subduction and the beginning of an associated calc-alkaline magmatism, or perhaps the two mechanisms are linked.

III - THE WESTERN MEDITERRANEAN BASIN. (TABLE III) - (MAP II)

Until recently, very few results were reported about the existence of an old magmatic activity. Yet some data on basalts from Morocco (HERNANDEZ et al., 1976) and andesites from Morocco (BELLON, 1976) gave results between 80 and 60 MY.

New geochro nometrical data obtained by K-Ar method (BELLON, 1976) lead to give a Neogene and Quaternary age to the magmatism outcropping along the North-African coast, in South-Eastern Spain, in France and in Sardinia, almost on the western coast.

Volcanic structures are known but only a few geochronological data have been obtained on the volcanism dredged and drilled, until recently.

III-1 - Chemical affinities of magmatism.

It consists of the three main types of chemical and petrological associations spreading during a short time in a narrow area:

- calc-alkaline with, in some cases, influences by crustal anatexis
- alkaline
- plutonic, a part of which undoubtedly having a calc-alkaline affinity; the other part must be related, in our opinion, to a crustal anatexis: this fact is supported by the Sr isotopic composition of the material.

III-2 - Ages and partition of magmatism.

150 results of absolute dating, carried on the magmatism, disclose two major conclusions.

All the magmatic events are subsequent to the Aquitanian tectonic event.

Two large magmatic processes exist, well distinguished from each other in time.

The first one, with a calc-alkaline affinity, the second showing an alkaline trend.

III-2-1 - Calc-alkaline magmatism.

A diachronism appears on the North African margin: magmatism is old (22 to 15 MY) in the central part (Algiers to Bougaroun Cape) and younger in the eastern part (12 to 8 MY), in Tunisia, and still younger (9 to 5 MY) in the western part (Algiers to Morocco). A problem arises about the presence of an acid plutonism with some tendency to a calc-alkaline association, between Algiers and Tunisia. In this area, an anatexis process has been demonstrated recently by isotopic investigation (METRICH, 1976).

In Morocco also, volcanism of shoshonitic affinity (especially highly potassic) (HERNANDEZ, 1975) is younger from north to south (since 10 to 5 MY in Gourougou, and only since 7 - 5 to 5 MY in Guilliz).

In the Alboran Sea, the submarine volcanism is very important as shown by the magnetic studies and the dredges. An age result obtained on an "alboranite" (poorly potassic lava from this island with two chemical affinities: tholeiitic and calc-alkaline) is in the order of 20 MY.

In South-Eastern Spain, the first products emitted on the present coast are andesites and dacites between 15 and 8 MY in age. The last ones (6 MY) known as lamproitic rocks (Verite or Fortunite), with a general hyperpotassic and hypermagnesian tendency, are located in land around Murcia, in an external position.

A problem arises about the explanation of this highly potassic volcanism located on both sides, or a poorly potassic volcanism. In the whole zone, here considered, the first products appear in Spain, and with an exception in regards to Ras Tarf volcanoe (13 to 12 MY) in Morocco, the Moroccan volcanism appears later. ARANA et al. (1974) suggest one subduction zone dipping from Africa beneath Spain; a detailed analysis of age and nature of magmatism (BELLON, 1976) shows that the process is not so simple.

In Sardinia, the volcanic period of calc-alkaline type lasted from 30 MY (BELLON, 1976; SAVELLI, 1975) to 13.5 MY (COULON et al., 1974). In comparison with other Western Mediterranean regions, it is a long period of calc-alkaline volcanicity.

In France, in the "Cap d'Ail" region, a calc-alkaline volcanism and plutonism is 26 MY old (BELLON et al., 1971).

III-2-2 - Alkaline volcanism.

In Tunisia, its age is from 8 to 6 MY. In Sardinia, the first products are 3 MY old, the last being subpresent. Near Cartagena (S-E Spain), basalts are 2.8 MY old. In Morocco, calc-alkaline volcanoes of Gourougou and Guilliz are covered by basaltic flows younger than 2.8 MY (BELLON, 1976). In Algeria, near Oran, the three basaltic volcanoes are younger than 2 MY.

In France, near Montpellier, basaltic outcrops (alkaline affinity) have been dated between 30 and 25 MY (BELLON, 1976). In the Massif Central, in France, there has been a long alkaline volcanic activity with several differentiated cycles (BELLON et al., 1975) since 20 MY. The volcanism linked to the initiation of small rifts (Forez - Limagne) may be older likewise the Eocene activity in the Charollais rift (BELLON et al., 1974).

Around the Valencia trough, volcanism is a major event without any calc-alkaline affinity. Here, the oldest products found today are 21 MY old (DSDP 123) (FERRARA et al., 1973). The end of activity is very recent.

On the Spanish coast, alkaline volcanism (basalts) exists since 10 MY until today (Olot zone) (DONVILLE, 1973).

III-2-3 - Offshore data.

Offshore in the Valencia trough, some dredged basalts gave a 4.5 MY age near DSDP hole 122 (MAUFFRET, 1976), another one having given an age of 0.3 MY.

In the Valencia trough, seismic reflection profiles and magnetism show the presence of a large interbedded volcanism in the sedimentary cover (GONNARD et al., 1975) of mainly late

Miocene age. This volcanism lays in the basin axis. A more recent volcanism exists in the western part of the trough, it outcrops in Columbretes islands where it is of alkaline tendency. A thin crust exists in the basin axis (HINZ, 1973). It seems that this trough might be interpreted as an aborted rifted area.

Elsewhere in the western basin: Balearic-Provençal Basin, South Balearic - North Algerian Basin, outside of a limited zone located in the south-west of Sardinia, one does not observe well defined linear magnetic anomalies indicating a single spreading ridge. Most of the prominent magnetic anomalies are linked with acoustic basement heights (volcanoes) or with interbedded volcanism younger than the basin itself. An opening of this basin by a mechanism comparable with the one proposed for the Tyrrhenian Basin, could be proposed.

III-3 - The Tyrrhenian Basin.

Volcanism and plutonism located in Italy along the north western - south eastern edge of the basin, belong to three magmatic associations.

The first one is found in the Oligocene-Aquitanian Tusa flysch in Sicily as relics of volcanism (WEZEL, 1973) and in a deep hole near Naples. It presents some affinity with a calc-alkaline sequence. One may ask if this volcanism is similar to the calc-alkaline volcanism of Sardinia; from this point of view, it will represent the calc-alkaline volcanism, before the Tyrrhenian opening, upon a subduction zone.

The second one present in Tuscany is acidic.Strong petrographical and isotopic clues (METRICH, 1976) lead to conclude that the magmatism is a consequence of sialic anatexis. In age, this latter goes from 7 MY (Elba island) to 2 MY (BORSI et al., 1967).

The Roman volcanism is highly potassic and very recent: 1-2 MY to the present (EVERNDEN et al., 1965) (CIVETTA et al., 1969).

The third one, of calc-alkaline and shoshonitic tendency (KELLER, 1974), is very young (less than 1 MY). It outcrops in the Aeolian arc. It is the only recent calc-alkaline volcanism in the Western Mediterranean area and it can be related to the only active seismic Benioff zone of the whole area. Transition between calc-alkaline and shoshonitic rocks has been interpreted (BARBERI et al., 1974) in terms of end of magmatic evolution in an arc.

In abyssal plain and on the margins exists a close relation between magnetic anomalies and topographical structures, now interpreted as submarine volcanoes.

Seismic reflection profiles show that volcanism is often interbedded in recent sediments:

- some dredged basalts are alkaline, the other being tholeiitic (SELLI, 1974).

The DSDP hole 143 located in the abyssal plain has drilled tholeiitic basalts dated by K-Ar at 6-5 MY (BARBERI et al., 1977). The lack of magnetic lineations of Vine and Matthews type, and the presence of a scattered volcanism, show that the creation of the abyssal plain may result of an opening related to a multi-spreading process. Such a mechanism might have begun in the Oligocene, resulting in an oceanic type crust. According to several authors (BOCCALETTI et al., 1972) this basin is interpreted as a marginal basin, with travels of sialic blocks (Calabria ...), created upon a subduction zone presently located to the east and the south-east of Calabria and Northern Sicily.

III-4 - Conclusions: Western Mediterranean.

In the Western Mediterranean area, outcrops of volcanism prior to 50 MY are very scarce; it is due either to the consumption of a great part of volcanism, probably tectonically, or to a real scarcity of activity before 50 MY, according to the narrowness of the oceanic zone at this time, and a shearing movement of the two plates (European and African) before 50 MY. Yet, new results obtained by BOUILLIN et al. (1974) argue for a tholeiitic activity before Miocene (near Cap Bougaroun in Algeria).

If some magmatic activities, in some areas of Western Mediterranean are in agreement with other structural and paleographical arguments (i.e. in Tyrrhenian area), some others, despite new data, remain very gloomy.

For instance, the pene simultaneous presence of volcanism on each margin of the Alboran Sea, where calc-alkaline is much younger as it is located further from the present shore line, has not received satisfactory explanation with subduction processes. New results clearly show the complete magmatic change from calc-alkaline to alkaline since 3 MY in the Western Mediterranean area: nothing let suppose in the precedent history of this zone this phenomenon except, as proposed (BELLON, 1976), a complete blocked system continent - continent, realized since some millions of years.

Cenozoic volcanism is much more important offshore than inland. Unfortunately, its nature (often described as alkaline) and its age are badly known.

IV - CONCLUSION

This short review of the Mesozoic and Cenozoic volcanism in the Mediterranean area shows the large differences in age, nature and importance of magmatism around the Western Mediterranean, Eastern Mediterranean and around the old Tethysian area now consummed.

The difference in the importance can be partly related to the initial size of the adjacent ocean which had never been broad in the western part of the Tethys and in the proto Mediterranean, on the contrary of the Eastern Tethys which was very large.

It seems that age and opening mechanisms are not the same for the oceanic zones of the different present Mediterranean Basins. Because the nature of magmatism is one of the bases for determining fossil subduction zones and rift zones, many new geochronometrical and geochemical results would be necessary to clarify these ages and mechanisms and thus confirm or infirm the different hypotheses on the structural evolution of the Mediterranean area.

REFERENCES

. ADAMIYA SH.A, GANKRELIDZE I.P., ZAKARIDZE G.S., LORDKIPANIDZEN M.B, 1974
Adzhar-Trialet trough and the problem of the Black Sea deep water trough Geotektonik, n° 1 - p. 39-47

. ARANA , VEGAS R. 1974
Plate tectonics and volcanism in the Gibraltar arc
Tectonophysics 24 - p. 197-212

. ARIKAN Y. 1975
The geology and petroleum prospects of the Tuz Golü Basin
Bull. M.T.A. Ankara - T. 85 - p. 17-37

. ATANASSOV G., GORANOV A. 1961
Bentonic clays from the region of the town of Kardzali
Anc. Univ. Sofia - Vol. 56 - Livre 2 Géologie - p. 149-188

. BAGDASARJAN G-P. 1974
Datation absolue des processus tectoniques et magmatiques du cycle alpin d'après l'exemple de l'Arménie
in: Novye dannye absoljutnos Geokhronology (in Russian)

. BARANYI I., LIPPOLT H.J, TODT W. 1975
Kalium Argon Datierungen an zwei Magmatiten von Kalo Chorio, Nord-Ost Kreta
N. Jahrbuch für Geologie und Paläontologie H.5, p. 257-262

. BARBERI F., CIVETTA L., GASPARINI P. INNOCENTI F., SCANDONE R., VILLARI L. 1974
Evolution of a section of the Africa-Europe plate boundary : paleomagnetic and volcanological evidence from Sicily
E. Pl. Sc. L. 22 - p. 123 - 132

. BARBERI F., INNOCENTI F., FERRARA G., KELLER J., VILLARI L. 1974
Evolution of Aeolian arc volcanism (Southern Tyrrhenian Sea)
E. Pl. Sc. L. 21 - p. 269-276

. BARBERI F., BIZOUARD H., CAPALDI G., FERRARA G., GASPARINI P., INNOCENTI F., JORON J.L., LAMBRET B., TREUIL M., ALLEGRE C. 1977
Age and nature of basalts from the Tyrrhenian abyssal plain
Report of D.S.D.P. - Vol. 42 A (in Press)

. BAROZ F., DESMET A., LAPIERRE H. 1976
Les traits dominants de la géologie de Chypre
B.S.G.F. (7) - T. XVIII - n° 2 - p. 419 - 427

. BELLON H., BROUSSE R. 1971
L'âge oligo-miocène du volcanisme ligure
C.R.A.S. - Sie D - T. 272 - p. 3109-3111

. BELLON H., GILLOT P.Y, NATIVEL P., 1974
Eocene volcanic activity in Bourgogne-Charollais, Massif Central (France)
E. Pl. Sc. L 23-1 - p. 53-58

. BELLON H., BROUSSE R., GILLOT P.Y, 1975
Recognition of Mio-Pliocene distensional phases in the North Mediterranean region by analysis of Cantalian volcanism, France
Ist. Meeting of European geological Societies - Oxford - Septembre

. BELLON H. 1976
Séries magmatiques néogènes et quaternaires du pourtour de la Mediterranée occidentale comparées dans leur cadre géochronométrique : implications géodynamiques
Thèse Univ. Paris XI - Orsay n° 1750

. BELLON H., KOPECKY L. 1977
Absolute age datation of the rift volcanism of the Bohemian Massif (to be published)

. BELLON H. Unplished data

. BENTOR Y.K. 1957
Cenozoic volcanism of Israel
Cong. Geol. Mexico - XXème Session p. 121 - 126

. BERGOUGNAN H. 1975
Relations entre les édifices pontique et taurique dans le Nord-Est de l'Anatolie
B.S.G.F. - 7ème sie - T. XVII - n° 6 - p. 1045 - 1057

. BERGOUGNAN H. 1976
Structure de la chaîne pontique dans le Haut-Kelkit - (N.E. de l'Anatolie)
B.S.G.F. (7) - T. XVIII - n° 3 - p. 675 -686

. BERTRAND J., DELALOYE M. 1976
Datation par la méthode K-Ar de diverses ophiolites du flysch des Gets (Haute-Savoie - France)
Ecl. Geol. Helv. - Vol. 69 - n° 2 p. 335

. BIJU-DUVAL B., LETOUZEY J., MONTADERT L. 1977
Structure and evolution of the Mediterranean basins
to be published in BLUE BOOK

. BLEAHU P. 1962
In Romanian with French summary recherches géologiques dans le bassin supérieur de Valaa Ruscova
Inst. Geol. Rom - Dari Seama Sedin 45 - p. 298-308

. BOCCALETTI M., GUAZZONE G., 1972
Gli archi appenninici, il Mar Ligure ed il Tirreno nel quadro della tettonica dei Bacini marginali retroarco
Mem. Soc. Geol. Ital. - 11 - p. 201 216

. BOCCALETTI M., MANETTI P., PECCERILLO A., PELTZ S. 1973
Young volcanism in the Calimani - Harghita Mountains (east Carpathians) evidence of a paleoseismic zone
Tectonophysics, 19, p. 299-313

. BOCCALETTI M., MANETTI P., PECCERILLO A. 1974
Hypothesis on the plate tectonic evolution of the Carpatho-Balkan arcs
E. Pl. Sc. L. - 23 - p. 193-198

. BORSI S., FERRARA G., TONGIORGI E. 1966
Rb-Sr and K-Ar age of intrusive rocks of Adamello and M. Sabion
E. Pl. Sc. L. - 1 - p. 55 - 57

. BORSI S., FERRARA G., TONGIORGI E. 1967
Determinazione con metodo K-Ar della rocce magmatiche della Toscana
Boll. Soc. Geol. 1 T. Vol. 86 p. 403 - 410

. BORSI S., FERRARA G. INNOCENTI F., MAZZUOLI R.B. 1972
Petrology and geochronology of recent volcanics of eastern Aegean Sea (West Anatolia and Les vos Island)
Bull. Volcanol. - V. 36 - n° 3 - p. 473 - 496

. BOUILLIN J.P., KORNPROBST J. 1974
Associations ultrabasiques de Petite Kabylie : péridotites de type alpin et complexe stratifié : comparaison avec les zones internes Bético-Rifaines
B.S.G.F. (7) - T. XVI - n° 1 - p. 101 - 234

. BRUNN J.H., ARGYRIADIS J., RICOU L.E., POISSON A., MARCOUX J., J. GRACIANSKY P. Ch. de 1976
Eléments majeurs entre les Taurides et les Hellénides
B.S.G.F. (7) - T. XVIII n° 2 - p. 481 - 498

. CANKOV C. 1974
The Bulgarian - Carpathian Balkan area in : tectonics of the Carpathian-Balkan region
Geol. Inst. of Dionz stur. Bratislava p. 303

. CIVETTA L., GASPARINI P., ADAMS J.A.S. 1970
Geochronology and geochemical trends of volcanic rocks from Campania (S. Italy)
Ed. Geol. Helv. - Vol. 63-1 - p.57 68

. CONANT L.C., GONDARZI G.H 1967
Stratigraphic and tectonic framework of Libya
Am. Ass. of Petr. Geol. Bul - p. 719 - 730

. COULON C., DEMANT A., BELLON H. 1974
Premières datations par la méthode K-Ar de quelques laves cénozoïques et quaternaires de la Sardaigne nord-occidentale
Tectonophysics - 22 - p. 41 - 57

. DELALOYE M., COGULU E., CHESSEX R. 1972
Etude géochronométrique des massifs cristallins de Rize et de Gümüsane-Pontides orientales
C.R. Arch. Sci. Genève - Vol. 7/2

. DELALOYE M., SAWATZKI G. 1975
Potassium-argon ages on volcanic elements from the Helvetic flysch in the area of Thônes (Haute-Savoie)
Arch. Sc. Genève - Vol. 28/1

. de VECCHI G., GREGNANIN A., PICCIRILLO E.M., 1976
Tertiary volcanism in the Veneto : magmatology, petrogenesis and geodynamic implications
Geol. Rundschau - Vol. 64-2 - p. 701 - 709

. DONVILLE B. 1973
Géologie néogène et ages des éruptions volcaniques de la Catalogne Orientale
Thèse Univ. de Toulouse n° 567

. DUBERTRET L. 1962
Carte géologique Liban-Syrie et bordure des Pays voisins 1/100 000ème in Lexique Stratigraphique international III 10

. EVERDEN J.,CURTIS G.H. 1965
Potassium argon dating of late Cenozoic rocks in East Africa and Italy
Current anthropology - p. 343-345

. FERRARA G., BIGGAZI G., BONADONNA F.P., GIULANI O. 1973
Radiometric dating of the Valencia volcanic rocks
In. Rep. D.S.D.P. - Vol. XIII 1447p

. FOURQUIN C.
L'Anatolie du Nord-Ouest, marge méridionale du continent européen, histoire paléogéographique, tectonique et magmatique durant le Secondaire et le Tertiaire
B.S.G.F. - 7ème Série - T. XVII - n° 6 - p. 1058 - 1070

. FREUND R., GOLDBERG M., WEISSBROD T., DRUCKMAN Y., DERIN B. 1975
The Triassic-Jurassic structure of Israel and its relation to the origin of the eastern Mediterranean
Geol. Surv. of Israel - Bul. n°65 p. 1-26

. FYTIKAS M., GIULANI O., INNOCENTI F., MARINELLI G., MAZZUOLI R., 1976
Geochronological data on recent magmatism of the Aegean sea
Tectonophysics 31 - T. 29 - T. 34

. GOCEV P.M. 1973
Structure of the Vitosha Pluton
Bulgarian Ac. of Sc. - Geol. Inst. Bul. - Vol. XXI - XXII p. 105-136

. GONNARD R., LETOUZEY J., BIJU-DUVAL B. MONTADERT L. 1975
Apport de la sismique - réflexion aux problèmes du volcanisme en Méditerranée et à l'interprétation des données magnétiques
3è Réunion - An. Sc. de la Terre Montpellier - p. 171

. GVIRTZMAN G. 1969
Outcrops of neogene formation in the central and southern coastal

plain
Hashephela and Be'er sheva regions
Israel
Geol. Surv. - Bul. 50 - p. 1.74

. HERNANDEZ J. 1975
Sur le caractère shoshonitique des andésites du Gourougou, Rif Oriental (Maroc)
C.R.A.S. - Sie D - T. 180 - p. 233-236

. HERNANDEZ J., LEBLANC D., MARCAIS J. 1976
Une manifestation volcanique d'âge paléocène dans le Rif : les laves basaltiques de Sidi Maatoug
B.S.G.F. (7) - T. XVIII - n° 3 - p. 697 - 706

. HINZ K. 1973
Crustal structure of the Balearic sea
Tectonophysics 20 (1-4) - p. 295-302

. HYNES A.J., NISBET E.G., SMITH A.G., WELLAND M.J.P., REX D.C., 1972
Spreading and emplacement ages of some ophiolites in the Othris region (eastern central Greece)
Z. Dt. Geol. Ges. 123 - p. 455-468

. ILIE M. 1935
Recherches géologiques dans les Monts de Trascau et dans le bassin de l'Aries
Inst. Géol. Rom. Anu. 17 - p. 1-329

. JUNG D., KELLER J., 1972
Die Jungen Vulkanite in Raum Zwischen Konya und Kayseri (Zentral Anatolien)
z. Deutsch. Geol. Ges. B. 123 - p. 503-512

. KAFKA F.T., KIRKBRIDE R.K. 1959
The Ragusa oil Field Sicily
5th World Pet. Congr. I (1959)

. KELLER J. 1974
Petrology of some volcanic rock series of the Aeolian Arc. Southern Tyrrhènian sea : calc-alkaline and shoshonitic associations
Contr. Mines - Petr. 46 - p. 19-47

. KHAIN V.Y. 1975
Principle stages of tectonic magmatic development of the Caucasus and experiment in geodynamical interpretation
Geotectonics n° 1 - p. 6-13

. KLERKX J., RUNDLE C.C. 1975
Preliminary K.-Ar. results on different igneous rock formations from the Jebel Uweinat region (S.E. Lybia)
Rap. ann. 1975 - Dept. Geol. Min. Mus. Roy. Afr. Centr.

. KONEČNY V., BAGDASARJAN G.P., VASS D., 1969
Evolution of neogene volcanism in Central Slowakia and its confrontation with absolute ages
Acta. Geol. Ac. Sc. Hungaricae - T. 13 - p. 245-258

. LAHNER L.
Geologische Untersuch ungen an der Ostflanke des mittleren Amanos (S.E. Türkei)
Geotekt. Forsch - B. 42 - p. 64-96

. LANPHERE M.A., COLEMAN R.G., KARAMATA S., PAMIC J. 1975
Age of amphi bolites associated with Alpine peridotites in the Dinarides ophiolite zone, Yugoslavia
E. Pl. Sc. Let. 26 - p. 271-276

. LAPIERRE H. 1972
Les formations sédimentaires et éruptives des nappes de Mamounia et leurs relations avec le Massif de Troodos (Chypre)
Thèse Nancy I. - Aq. n° 5153

. LAPIERRE H., ROCCI G. 1976
Le volcanisme du Sud-Ouest de Chypre et le problème de l'ouverture des régions téthysiennes au Trias
B.S.G.F. (7) - T. XVIII - n° 2 - p. 428

. LEO G.W., MARVIN R.F., MEHNERT H.H. 1974
Geologia Framework of the Kuluncak Solufar area, East Central Turkey and K-Ar ages of igneous rocks
Geol. Soc. of America Bull. - Vol. 85 - p. 1785 - 1788

. LETOUZEY J., KHRISTCHEV K., GONNARD R., MONTADERT L., DORKEL A., 1977
Black Sea : geological and recent deposits deposition - from seismic reflection data

Blue book - Vol. 42 B - Initial report of the deep sea drilling project (in Press)

. LETOUZEY J., BIJU-DUVAL B., DORKEL A., GONNARD R., KHRISTCHEV K., MONTADERT L., SUNGURLU O., 1977
The black Sea : a marginal basin - Geophysical and geological data
This volume

. MARCOUX J. 1970
Age carnien de termes effusifs du cortège ophiolitique des nappes d'Antalya (Taurus lycien oriental - Turquie)
C.R.A.S. - Paris - T. 271 - p. 285
187

. MAUFFRET A. 1976
Etude géodynamique de la marge des Iles Baléares
Thèse - Univ. Paris VI - 137 pages

. MENEISY M.Y., KREUZERH 1974
Potassium-argon ages of Egyptian basaltic rocks
Geol. Jb, B 9 - p. 21-31

. METRICH N., 1976
Fusion partielle dans les enclaves gneissiques - Applications au volcanisme de Toscane et de Tunisie
Thèse 3ème cycle - Univ. Paris XI - Centre d'Orsay

. PAL' SHIN I.G., SIMOV S.D., ARAKELYANTS M.M., CHERNYSHEV I.V, 1974
Absolute age of Alpine activations in Rhodope median massif, Bulgaria
Internat. Geology. Rev. - Vol. 17 n° 10

. PARROT J.F 1974
Les différentes manifestations effusives de la région du Baär Bassit - (N.O. de la Syrie)
C.R.A.S. Paris - Sie D. - T. 279 - p. 627 - 680

. PAVALESCU L., POP. G., AILENEI G., ENE I., SOROIU M., POPESCU G., 1975
K-Ar age determinations from the Apuseni and the Banat Mountains
Rev. Roumaine Geol. Geophs. Geogr. 19

. PICCOLI G. 1971
Outlines of volcanism in Northern Tripolitania
Symp. Geol. Libya Tripoli - p. 323-339

. RADULESCU D.P., SANDULESCU M. 1973
The plate-tectonics concept and the geological structure of the Carpathians
Tectonophysics - 16 - p. 155 - 161

. RICOU L.E. 1974
L'étude géologique de la région de Neyris-Zagros iranien-et l'évolution structurale des Zagrides
Thèse Univ. Paris-Sud - Orsay - n° 1269

. ROSSET J., CHAROLLAIS J., TOUMARKINE M., MANIVIT H., CHATEAUNEUF J.J., SCHAUB H. 1976
Présentation des différentes variétés de flyschs du synclinal de Thônes (Haute Savoie) (France)
Ec. Geol. Helv. - Vol. 69 - n° 2 - p. 335

. SAVELLI C. 1975
Datazioni preliminari con metodo K-Ar di vulcaniti della Sardegna Sud-Occidentale
Soc. It. Min. e Petr. Milano - Vol. XXXI (1) - p. 191 - 198

. SAVU H. 1968
Considérations concernant les relations stratigraphiques et la Pétrologie des ophiolites mésozoïques de Roumanie
Com. Geol. Rom. Anu. 36 - p. 143 - 175

. SELLI R. 1974
Appunti sulla geologia del Mar Tirreno
in Paleogeografia del Terziaro nell'ambito del Mediterraneo occidentale
Seminario Fac. Sc. Universit. di Cagliari - Sup. 43 - p. 327 - 351

. STEGENA L., GECZY B., HORVATHH F. 1975
Late Cenozoïc evolution of the Pannonian basin
Tectonophysics, 26 - p. 71-90

. STOJANOV R. 1973
Petrology of the Pontides volcanic rocks in the River Aera Harsit (Turkey)
Congress of earth Sciences - 50 th anniversary of Republic - Ankara
Editors M.T.A. - D.S.I.

. TANGUY J.C. 1967
Presence de basaltes à caractère tholéiitique dans la zone de l'Etna (Sicile)
C.R.A.S. - Paris - T. 254 - p. 21-24

. TERRY J. 1971
Sur l'âge triasique des laves associées à la nappe ophiolitique du Pinde septentrional (Epire et Macédoine - Grèce)
C.R. somm. Soc. Géol. Fr. - p. 384

. VERRIEZ J.J. 1976
Sur les formations volcaniques basiques d'Atalanti (Locride - Grèce)
B.S.G.F. (7) - T. XVIII- n° 2 - p. 293-298

. WEISSBROD T. 1969
The paleozoic of Israel and adjacent countries
Geol. Surv. Israel - Bull. 48 - 32 p.

. WESTERVELD J. 1957
Phases of Neogene and quaternary volcanism in Asia Minor
Cong. Geol. Intern. XXth Ses. Mexico - Vulcanologia del Cenozoico T. 1 - p. 103 - 119

. WEZEL F-C., GUERRERA F. 1973
Nuovi dati sulla eta e posizione strutturale del flysch di Tusa in Sicilia
Boll. S.G.I - Vol. 92 (2) p. 193-211

TABLE 1. TETHYS CLOSING

M.Y		GREATER CAUCASUS	LESSER CAUCASUS	ARMENIA	PONTIDS	SREDNOGORIA	DINARIDS HELLENIDS	PANNONIAN BASIN	WESTERN ALPS
0–2	PLE.	(Alk) β, ρ (Khain 1974)		Mt ARARAT (Alk) (Westerveld 1957)				BALATON β (Bellon)	
2–5	PL.								
5–21	MIOCENE	NO ACTIVITY (Khain 1974) [5–100 m.y.]		ARMENIA (Bagdasarjan 1974) (Alk) [0–c.35 m.y.]				5 α β ROMANIA CZECHO.SL. (Konečny 1969 Bellon); ρ, α CZECHO-SLOVAQUIA 21	
	OLIG.					33 (ρ, D, α) (γ) (Pal Shin 1974) 45 RHODOP			($γ_D$) ADAMELLO 33 (Ferrara 1966)
50	EOC.		ADZHAR-TRIALET (Sh. δ) (Adamiya et al. 1974); Alk, α (Adamiya et al 1974)		55 (γ) RIZE (Delaloye 1972)	(α, γ) (Atanassov 1963)			(β) VICENZO; (ρ, α) EUGANEI (De Vecchi et al. 1976)
	P.EOC.						61 OTHRYS (ρ, β) (Hynes 1972) 78	61 (Pavalescu et al. 1973) Banatitic rocks δ, γ, δ (Radulescu 1973) METALIFERI 75	
	UPPER CRET.		TRANSCAUCASUS and LITTLE CAUCASUS (Khain 1974)	75 (α) ARMENIA (Bagdasarjan 1974)	GALAT (τα) (Fourquin 1975); TRABZON (βα) (Stojanov) 1973 (ρα) (βα) th	(βα) (α, γ) (Cankov 1974) α SREDNOGORIA			
100	LOWER CRET.								
150	UP. JUR.	(α) (Khain 1974)	(α, γ)?	(α, γ) 150 (γ-γδ)	E. MARMARA Sea (α?)	STRANDJA (sp)			GETS flysch (op) (Bertrand et al. 1976)
	M. JUR	CRIMEA (Alk)?	SEVAN 160 AKERA (α) (Bagdasarjan 1974)				157 Amphibolites DINARIDS (Lanphere et al. 1975) 174	SINAIA BEDS (Bas) (Bleahu, 1962)	
	L. JUR.	(Sp. D) (Khain 1974) SVANETO DAGESTAN			? KELKIT (τα, alb) (Bergougnan 1976) ?			METALIFERI (Bas) (Ilie, 1935)	
200	TRIASSIC						PINDUS (β, dol, ρ) (Terry 1971)		

(β) Basalt (α) Calc. alkaline rocks (τα) Trachyandesite (ρ) Rhyolite (γ) Granite (γδ) Granodiorite (dol) Dolerite (alb) albitophyr (op) Ophiolites (Sp) Spilite (D) Diabase (Bas) Basic (Alk) Alk tendency (Sh) Shoshonitic

TABLE 2 POSSIBLE OPENING AND CLOSING OF A PROTO EASTERN MEDITERRANEAN

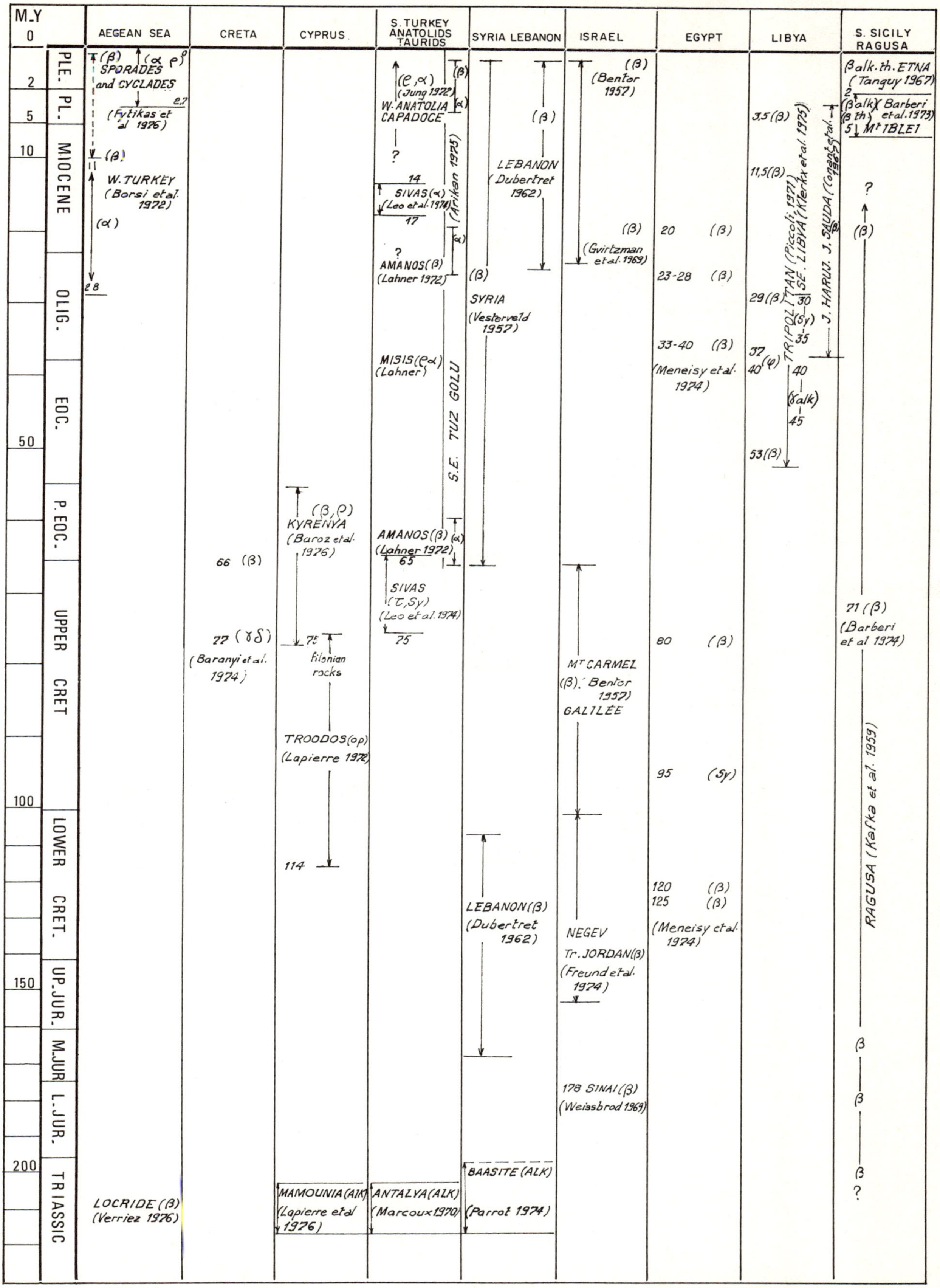

(β) Basalt (α) Calc. alkaline rocks (ρ) Rhyolite (τ) Trachyte (Sy) Syenite (γ) granite (γδ) Granodiorite (op) Ophiolites (Alk) alkaline tendency (th) tholeiitic tendency

TABLE III

PETROLOGICAL AND GEOCHRONOMETRIC DATAS
ON WESTERN MEDITERRANEAN MAGMATISM

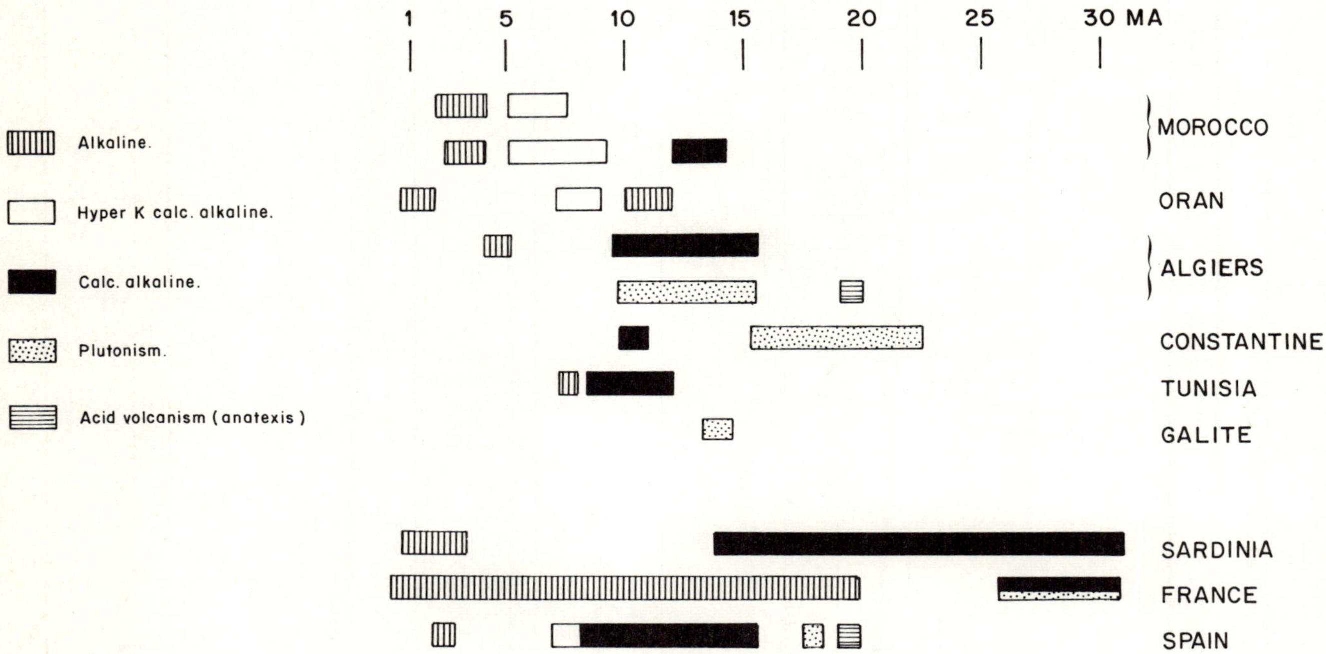

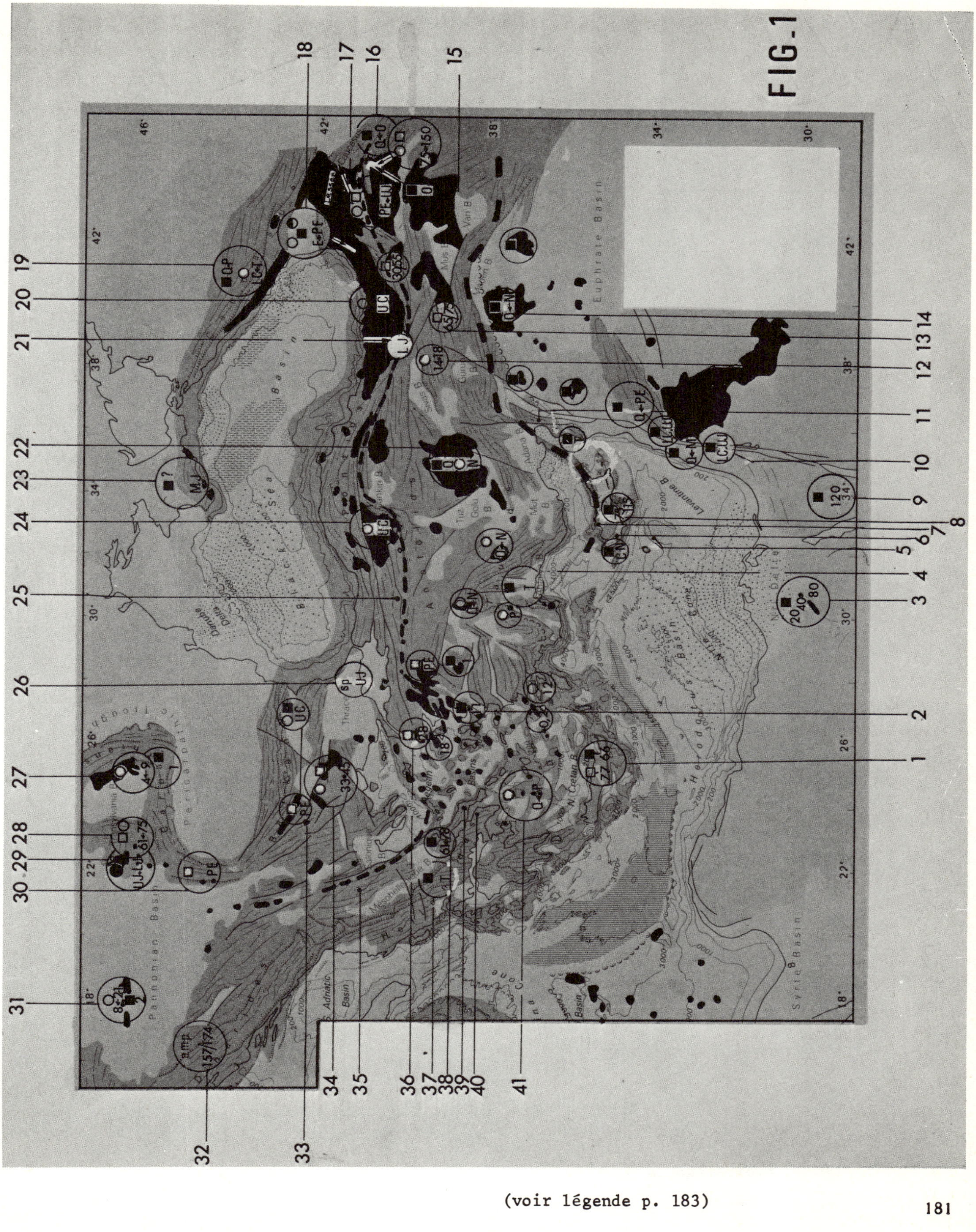

FIG_1

(voir légende p. 183)

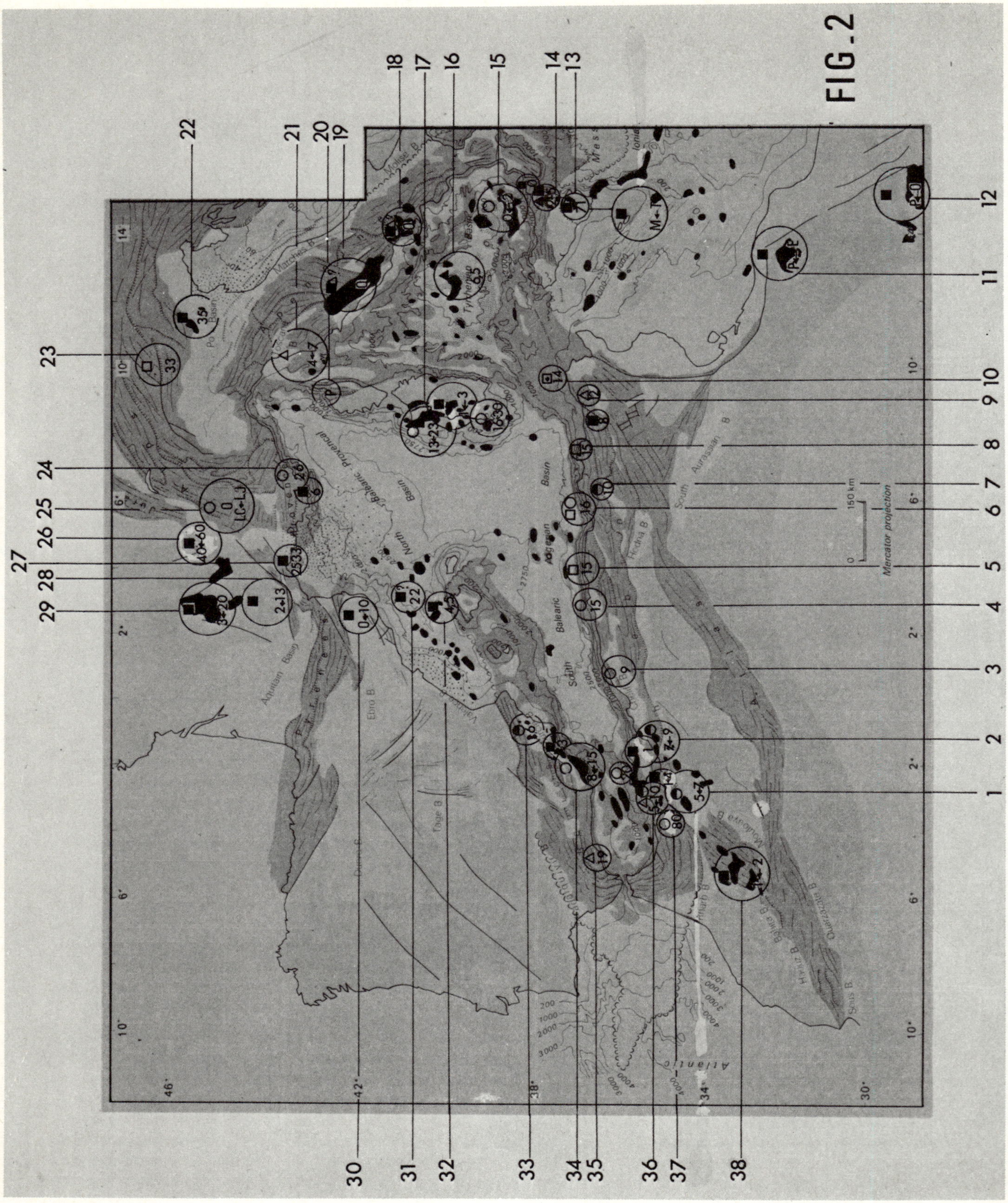

FIG. 2

(voir légende p. 183)

Fig. I (MAP I)

Number	in the circle
1	Creta
2	Western Turkey
3	Egypt
4	Antalya
5	Eratosthene Seamount
6	Mamonia
7	Kyrenia
8	Troodos
9	Sinaï
10	Negev
11	Hatay - Amanos - Baasite
12, 13	Sivas - Malatya
14	Diyarbakir
15	Ararat Mountain
16	Armenia
17	Lesser Caucasus
18	Adzhar-Trialet
19	Greater Caucasus
20	Trabzon
21	Kelkit
22	Tuz Golü
23	Crimea
24	Galat
25	Eskisehir
26	Strandja
27	East Carpathians (Romania)
28	Banatitic rocks Metaliferi
29	Metaliferi - Sinaï Beds
30	Banatitic rocks
31	Czechoslovakia - Balaton
32	Dinarids
33	Srednogoria - Vitosha pluton
34	Rhodop
35	Vardar
36	Western Turkey
37	Pindus
38	Othrys
39	Eubea
40	Locride
41	Cyclades - Sporades

Fig. 2 (MAP II)

Number	in the circle
1	Guilliz
2	Oran
3	Tenes
4	Algiers
5	Dellys volcanism
6	Cap Bougaroun
7	Kef Hahouner (near Constantine)
8	Herbillon
9	Tunisia
10	Galite
11	Tripolitan
12	Dj-Haruj
13	Capo Passero Ragusa
14	Mont Iblei - Etna
15	Aeolian islands
16	Tyrrhenian
17	Sardinia
18	Naples
19	Romanian province
20	Corsica
21	Tuscanian
22	Veneto
23	Adamello
24	"Cap d'Ail"
25	Alpine flysch volcanism
26	Charollais volcanism
27	Montpe llier volcanism
28	Causses volcanism
29	Cantalian volcanism
30	Catalunya volcanism
31	Joïdes D.SD.P. 123
32	North Balearic escarpment
33	Murcia
34	Cabo de Gata
35	Near Malaga
36	Gourougou
37	Old andesite (Bab Tafrant)
38	Central Morocco

The circle must be reported to the neigh bouring volcanic massive. For another precision, one may consult the litterature

▬ ▬ ▬	Suture
▲	Tholeiitic lavas
●	Basic magmatism
Amp	Amphibolites
■	Alkaline lavas
○ ◓	Calc-alkaline lavas High potassic calc-alkaline lavas
□	Plutonism
△	Anatectic volcanism Roma volcanism ◣

MAP I and MAP II

Stratigraphic data abreviations

Q	Quaternary
P	Pliocene
M	Miocene
O	Oligocene
E	Eocene
P.E	Paleocene
U.C	Upper Cretaceous
L.C	Lower Cretaceous
U.J	Upper Jurassic
M.J	Middle Jurassic
L.J	Lower Jurassic
T	Triassic {N. Norian, C. Carnian}

Geochronometrical data

0 ← 2	Activity between 2 and 0 MY
25 33	Activity at 25 and 33 MY

INTERNATIONAL SYMPOSIUM ON THE STRUCTURAL HISTORY OF THE MEDITERRANEAN BASINS. SPLIT (YUGOSLAVIA) 25-29 OCTOBER 1976. B. BIJU-DUVAL AND L. MONTADERT, Eds. EDITIONS TECHNIP, PARIS 1977, pp. 185-188

second part

recent tectonics in the mediterranean area

EDITORS' COMMENTS

Clearly, great progress in Mediterranean geology has been possible because the observations have been interpretated in the light of plate tectonic concepts. Nevertheless, as appears on general geological cross-sections of the different Mediterranean basins (Biju-Duval and Montadert, this volume), the interpretation of the tectonics of the whole area must take into account the fact that since the late Cretaceous there have been more and more areas where plates with a continental crust are colliding, while the oceanic areas are being consummed or severely restricted (Biju-Duval et al., part one, this volume). It appears that, in this state of continental collision (McKenzie, 1972; Molnar and Tapponnier, 1975; Tapponnier, 1977), plate tectonics concepts are not strictly applicable as shown by the broad distribution of the stresses within the colliding continental plates. We observe, in this case, a fragmentation into small blocks with limited areas of compression, distension, or strike-slip faulting. This new approach seems to be very fruitful, especially for interpreting the relatively young tectonic features of the Mediterranean (McKenzie, in this volume; Tapponnier, abstract, in this volume). Data are based : 1) basically on seismicity (Papazachos and Comninakis, McKenzie, in this volume; see also descriptions in central Mediterranean by Schick*[1] or in Gibraltar area by Hatzfeld*); 2) on studies of seismic surface waves, as presented by Mueller* for the zone of contact between the Eurasian and African plates; 3) and on mi-

---

([1]) Names with the symbol* correspond to papers given during the Symposium, for which an abstract will be published in the "Rapports et Procès-Verbaux des Réunions" de la <u>Commission Internationale pour l'Exploration Scientifique de la mer Méditerranée</u> (<u>CIESM</u>) (in press).

crotectonic studies in the field. On this last domain, great progress has recently been made on active margins : e. g. Bousquet (abstract, in this volume) on the southern part of the Western Basin from Gibraltar to Calabria ; Schroeder* on the sedimentary history and neotectonic development of the northern Peloponnesus (Greece). Such microtectonic studies on the adjacent plates (see for example de Charpal et al., 1975) enable the chronology of the constraints to be defined as was proposed during the Symposium by Schaffer* who presented a paper on the consistent crustal stresses around the western Mediterranean during the African and European approach, and by Letouzey et al. (in this volume) also using an approach of the structure of the Mediterranean area from a satellite photogeological study. Indeed, satellite imagery is now used as a tool to define major lineaments whose significance has to be checked in the field (Biju-Duval et al., 1975, 1976; Molnar and Tapponnier, 1975; Foose, in this volume).

A different technique, but making possible the same type of observations on the bottom of the sea, is the long-range side-scan sonar which enabled Kenyon and Belderson (this volume) to define structural alignments in the eastern Mediterranean, interpreted as compressional features.

Recent tectonics can also be studied by using conventional shallow penetration seismics as exemplified by Nesteroff et al. (in this volume) between the Aegean and the Cyprus Arc. Such shallow seismic recordings used by Kouprin* and Bobier* revealed the recent faulting and volcanism in the Tunisian Strait between the Pelagonian platform and Sicily. In the eastern Mediterranean, Goedicke* has also shown that the submarine canyons crossing the upper continental shelf of the Lebanon is linked to the faults structures of the " Pelusium line" (see Neev et al., 1976).

BIBLIOGRAPHY

BIJU-DUVAL (B.), DEVAUX (E.), GONNARD (R.), LATARCHE (G.), RIVEREAU (J.C.), LAMPEREIN (C.), LOPEZ (N.).- (1975). Apports de l'étude des images du satellite Landsat-1 à la connaissance de la structure du domaine méditerranéen. Rev. Inst. Franç. du Pétrole, XXX-6, 841-853.

BIJU-DUVAL (B.), LAMPEREIN (C.), RIVEREAU (J.C.), LOPEZ (N.).- (1976 c). Commentaires de l'esquisse photogéologique du domaine Méditerranéen. Grands traits structuraux à partir des images du satellite Landsat-1. Rev. Inst. Franç. du Pétrole, XXXI-3, 365-400.

DE CHARPAL (O.), TREMOLIERES (P.), JEAN (F.), MASSE (P.).- (1974). Un exemple de tectonique de plate-forme : les Causses majeurs (sud du massif Central, France). Rev. Inst. Franç. du Pétrole, XXIX-5, 641-659.

McKENZIE (D.).- (1972). Active Tectonics of the Mediterranean Region. Geophys. Jour., vol. 30, n° 2, 109-185.

MOLNAR (P.), TAPPONNIER (P.).- (1975). Cenozoic Tectonics of Asia : Effects of a Continental Collision. Science, 189, 419-426.

NEEV (D.), ALMAGOR (G.), ARAD (A.), GINZBURG (A.), HALL (J.K.).- (1976). The Geology of the Southeastern Mediterranean. Geol. Surv. Israel, 68, 1-51.

TAPPONNIER (P.).- (1977). Poinçonnement rigide-plastique et évolution tectonique du système alpin en Méditerranée. Soc. Géol. France Bull., 7ème sér., XIX, in press.

INTERNATIONAL SYMPOSIUM ON THE STRUCTURAL HISTORY
OF THE MEDITERRANEAN BASINS. SPLIT (YUGOSLAVIA)
25-29 OCTOBER 1976. B. BIJU-DUVAL AND L. MONTADERT, Eds.
EDITIONS TECHNIP, PARIS 1977, pp. 189-196

CAN PLATE TECTONICS DESCRIBE CONTINENTAL DEFORMATION ?

D. MCKENZIE [1]

1 ABSTRACT

A detailed investigation of the seismicity and focal mechanisms of shallow earthquakes in Western Turkey and the Aegean Sea shows that the deformation is not confined to narrow linear regions as it is in oceanic regions. This behaviour is probably the result of interactions between old tectonic trends and present motions. There is no reason to believe that other active continental regions or older orogenic movements were any simpler than those now active in the Mediterranean. Because of these complications paleogeographic reconstructions produced by using rigid plate tectonics are unlikely to be even approximately correct.

2. INTRODUCTION

The general acceptance of the ideas of plate tectonics is now almost complete, and it is beginning to be used to obscure the shortcomings of the original simple ideas. Such use is to be expected: similar inglorious examples can be quoted from any branch of the Natural Sciences. The purpose of this paper is to expose one major gap in our present knowledge, to attempt to counteract the recent and enthusiastic application of the basic ideas to continental deformation.

The difficulty is clearly illustrated by figure 1, which shows the seismicity of the western part of the Alpide-Himalayan belt and part of the North Atlantic. The distribution of earthquakes along the Mid-Atlantic Ridge where the oceanic lithosphere is being deformed is quite different from that of the Mediterranean region where the deformation is mostly taken up by continental lithosphere. This behaviour could equally well be illustrated from any other continental active belt, such as the eastern end of the Alpide-Himalayan belt, the Southwestern United States, or Siberia. Wherever a plate boundary crosses from an oceanic to a continental region the deformation ceases to be confined to a single fault and the observed motions can only with difficulty be related to the relative motion between the plates on either side of the active zone. This behaviour occurs whatever the nature of the relative motion. In the region shown in fig. 1 the activity results from the northward motion of Africa and Arabia towards Eurasia, but in California, Nevada, and Central and Southern Africa the deformation is the result of strike slip or

(1) Department of Geodesy and Geophysics, University of Cambridge, Madingley Rise, Madingley Road, Cambridge CB3 OEZ, Great Britain

extension and is equally complicated. This difference between oceanic and continental deformation was clearly recognised at the time plate tectonics was put forward (see Isacks et al. 1968), and has been amplified by later work. Until the origin of this difference is properly understood plate tectonic reconstructions of continental orogenic belts cannot be carried out with any confidence. For the same reason the deformation of continental regions mapped by tectonic geologists cannot be unambiguously related to the history of large scale plate motions.

In 1967 it seemed reasonable to suppose that a detailed study of the present deformation of the Mediterranean and Middle East might help make the connexion between oceanic and continental deformation. A detailed account of the first stage of this work (McKenzie 1972) showed that the deformation in figure 1 resulted from normal and strike slip motions as well as the thrusts expected from the large scale plate motions. This behaviour was unexpected and could only be reconciled with the oceanic magnetic anomaly patterns by postulating the existence of a number of small continental fragments moving independently of the major plates. The boundaries of several of these fragments were not clearly outlined by the seismicity, nor were their motions always well defined by the focal mechanisms. A different interpretation of some of the same data was suggested by Nowroozi (1972), and even some of the mechanisms were disputed. Others, especially Molnar and Tapponnier (1975) have questioned whether any interpretation in terms of plate motions is useful, and have preferred to use a continuum model for the deformation of the Eastern end of the Alpide-Himalayan belt. Clearly the problem of continental deformation can not be considered to be solved! For these and other reasons more information was required and therefore the project was continued. About a hundred new focal mechanisms have accumulated, together with five more years of activity. In addition satellite photographs have become available and clearly show several of the active faults. A full account of this new information is in preparation, and the account below contains only a summary.

Figure 2 shows a sketch of the present plate motions, corresponding to figure 2(b) of McKenzie (1972). The principal changes required by the new information involve Eastern Iran and the Western Aegean. In Eastern Iran both the seismicity (fig. 1) and the focal mechanisms indicate that the deformation does not continue eastward into Afghanistan but is taken up by North-South structures which connect the Kopet Dag to the Makran Coast. But the Aegean is obviously of greater interest to symposium on the Mediterranean, and for this reason is discussed in more detail.

3. WESTERN TURKEY AND THE AEGEAN

A plot of epicentres of earthquakes since 1961 in figure 3 is dominated by the foreshock and aftershock sequences of two large earthquakes which have occurred since 1969. The available fault plane solutions in figure 4 show that the region is being extended by motion on a number of normal faults, and the satellite photographs (Allen 1975) clearly show a number of dominantly east-west grabens bounded by active normal faults. This information is summarized in figure 5 which shows the strike and sense of the fault motion, together with the horizontal projection of the slip vector obtained from the fault plane solutions. The solid lines mark surface breaks which have been mapped in the field, whereas faults visible on the satellite mozaics on which individual earthquakes in figure 4 are presumed to have occurred are shown by dashed lines. Many more normal faults are clearly visible on the mozaics, but these have not been shown in figure 4 because they have either not moved seismically since 1963 or have not produced a known surface break. Most, if not all, of these faults will probably turn out to be active, and should obviously be so considered when seismic risk, rather than tectonics, is of concern.

Figure 5 shows various important structures. The large strike slip fault entering the figure in the northeast is the North Anatolian Fault, and forms the northern boundary of a relatively aseismic region (fig. 1). It cannot be followed west of about 30.5°E, where the graben system and the normal faulting begin (fig. 4). The two strike slip faults further west have a different strike and are not directly connected to the North Anatolian Fault. The northerly fault has only recently been recognised by Allen (1975): it is clearly visible on the satellite photographs and broke in 1912. Unfortunately the sense of motion has not been determined though it is almost certainly right handed. This fault is alligned along the northern boundary of a deep trough in the northern Aegean (see below). The major east-west normal faults have produced a series of shocks in the last seven years whose mechanisms (fig. 4) are strikingly consistent. Several surface breaks have been described and agree excellently with the teleseismic focal mechanisms. The mechanisms also suggest the active grabens extend beneath the Aegean to form the intricate coastline of Western Turkey, though the active faults have not been mapped on the ground. In Southwestern Turkey the direction of extension changes slightly but otherwise the pattern of normal faulting remains the same.

The effect of the motions in figure 5 is to stretch the continental crust of Western Turkey. The directions of extension correspond to a fan being opened, with all the opening being taken up by the southward movement of the southern edge of the fan relative to Eurasia. Such motions will reduce the thickness of the continental crust and hence produce subsidence. The allignment of the submarine features (fig. 6) with similar features on land and the extensive normal faulting still active beneath parts of the Aegean Sea (fig. 7) strongly suggest that the same process has produced most of the sea floor which is now relatively aseismic.

This discussion shows that all the new observations agree with the deformation model proposed earlier (McKenzie 1972). The main difficulty, then as now, is to understand why the deformation is so complicated when a single north-south strike slip fault could take up the observed motions. The only obvious explanation is that the pre-existing structures trended east west, and could not be easily cut to form a north-south fault.

The bathymetry of the Aegean is now much better known, and figures 6 and 7 show the seismicity and focal mechanisms together with the 600 m isobath (Morelli et al. 1975). In the Northern Aegean a large trough extends from Northwestern Turkey southwestward and is truncated by Greece. This structure is known as the North Aegean or Anatolian Trough, and a variety of explanations have been offered for its existence. It is associated with strike slip and normal faulting and its eastern part forms a continuation of a strike slip fault on land. One of the submarine shocks in figure 7 has been extensively studied by North (1973) who showed that the aftershock distribution agreed with the fault plane striking NE-SW, or parallel to the Anatolian Trough. The sense of motion is then right lateral. There is therefore little doubt that this feature is basically due to strike slip movement on a right lateral fault system. Similar bathymetric features are commonly found associated with oceanic transform faults. It is just possible that the large magnetic anomalies associated with the Anatolia Trough are produced by a short ridge segment at right angles to its trend, as has been demonstrated for a similar structure called the Cayman Trench in the Caribbean.

The bathymetric trends beneath the Aegean which extend westwards from Turkey are truncated against features parallel to the NW-SE trends in Greece, and no strike slip fault plane solutions are found southwest of this line. Once again the motion must be taken up by structures at right angles to the slip vector and, as in Western Turkey, this produces a number of parallel grabens with normal faulting, two of which are visible near the western edge of figures 6 and 7. These faults have not been as active recently as those in Western Turkey, though several normal faults of this type have been active in other parts of Greece (see McKenzie 1972 fig. 17). In the nineteenth century, however, two large shocks occurred one of which was probably and the other certainly the result of slip on a normal fault (Richter 1958). The fault trace produced by the second of these earthquakes is shown in figure 6.

4. DISCUSSION

Both McKenzie (1972) and Molnar and Tapponnier (1975) attempted to explain the difference between continental and oceanic deformation by using the bouyancy forces produced by continental shortening. Such forces only exist because thrust faults thicken the continental crust. Though this explanation can plausibly explain the general motions in figure 2 and those north of the Himalaya, it cannot explain the present deformation of Western Turkey, the Aegean Sea and Greece. In these areas the observed complications could be the result of interference between existing structures and present motions. Such an explanation would also account for the simple structure of the North Anatolian Fault which is parallel to the older faults. Whatever the explanation the deformation in this region is not confined to a few simple plate boundaries. Those drawn in figure 2 represent zones of complicated deformation and do not correspond to single faults. It is clear that no simple plate tectonic model can reconstruct the earlier geometry. This difficulty is particularly important if the regions of deformation change frequently with time. In the Aegean this has almost certainly happened, because the bathymetry shows block faulted structures in regions which are relatively aseismic. At present there seems no reason to believe that the Aegean region is anomalously complicated: most of the earlier motions in the Western Mediterranean seem equally so, and no-one who has studied the Alps before plate tectonics was proposed thought that they could be explained by simple models. Nor do the well exposed parts of older mountain belts which have been mapped in detail suggest that the movements involved were any simpler than those in the Aegean. For these reasons I doubt that the types of reconstructions of the Mediterranean which are now so popular can produce realistic geographies for earlier periods. The difficulty is not connected with our ignorance of mantle circulations but of the large scale material science of continental deformation. It is this which controls the way in which the deformation is expressed and about which so little is known.

The answer to the question 'Can Plate Tectonics describe Continental Deformation?' appears to be 'Yes, but not in detail'. Molnar and Tapponnier (1975) may be correct in their view that simpler descriptions exist, but their model cannot account for the observations discussed above.

5. ACKNOWLEDGMENTS

I am grateful to N. N. Ambraseys for his help in producing figure 5.

ALLEN, C. R. 1975. Geological Criteria for Evaluating Seismicity Bull. geol. Soc. Amer. 86 1041.

ISACKS, B. L., OLIVER, J. and SYKES, L. R. 1968. Seismology and the new global tectonics J. geophys. Res. 73 5855.

MCKENZIE, D. P. 1972. Active Tectonics of the Mediterranean Region Geophys. J. R. astr. Soc. 30 109.

MOLNAR, P. and TAPPONNIER, P. 1975 Cenozoic tectonics of Asia: effects of a Continental collision Science 189 419.

MORELLI, C., GANTAR, C. and PISANI, M. 1975. Geophysical Studies in the Aegean Sea and in the Eastern Mediterranean Boll. Geof. teor. e appl. 17 66.

NORTH, R.G. 1973. Seismic Source Parameters, Ph.D. thesis Cambridge University.

NOWROOZI, A.A. 1972. Focal Mechanisms of Earthquakes of the Persian Plateau, Eastern Turkey, Caucasus and the Hindu-Kush regions. Bull. seism. Soc. Amer. 61 317.

RICHTER, C.F. 1958. Elementary Seismology W.H. Freeman & Co.

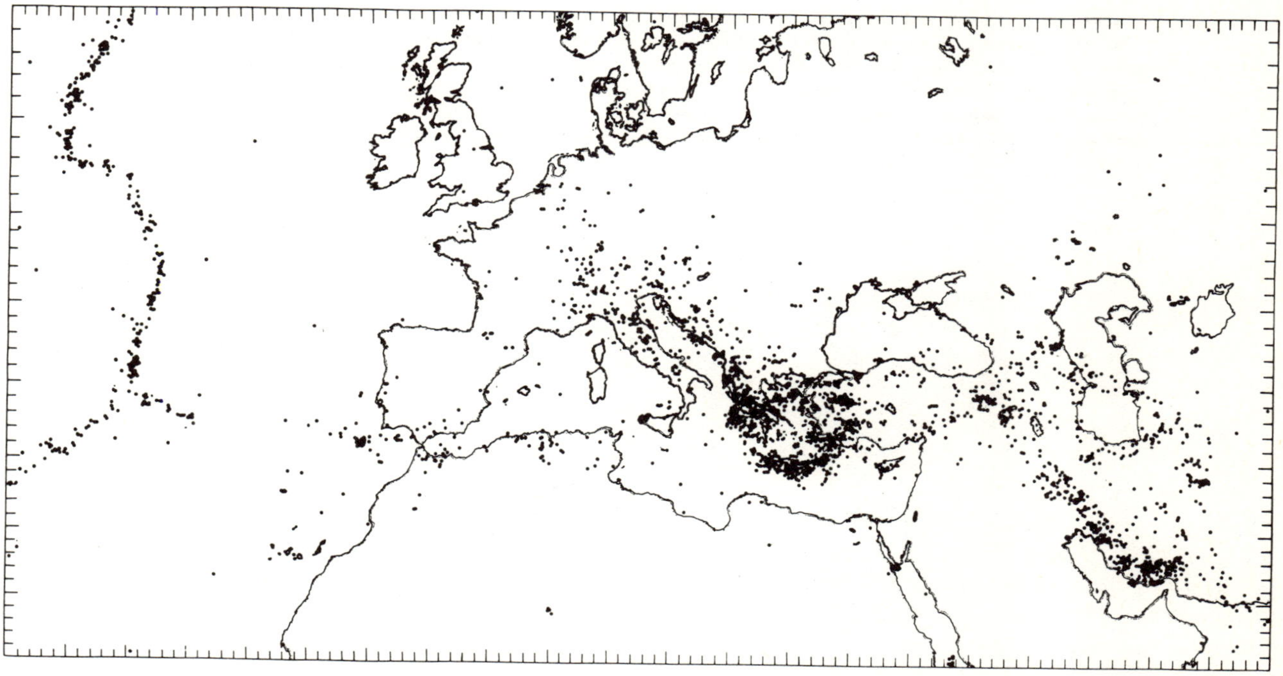

Fig. 1 Epicentres of those earthquakes shallower than 50 km located by the U.S. Coast and Geodetic Survey and its successors between January 1 1961 and April 30 1975.

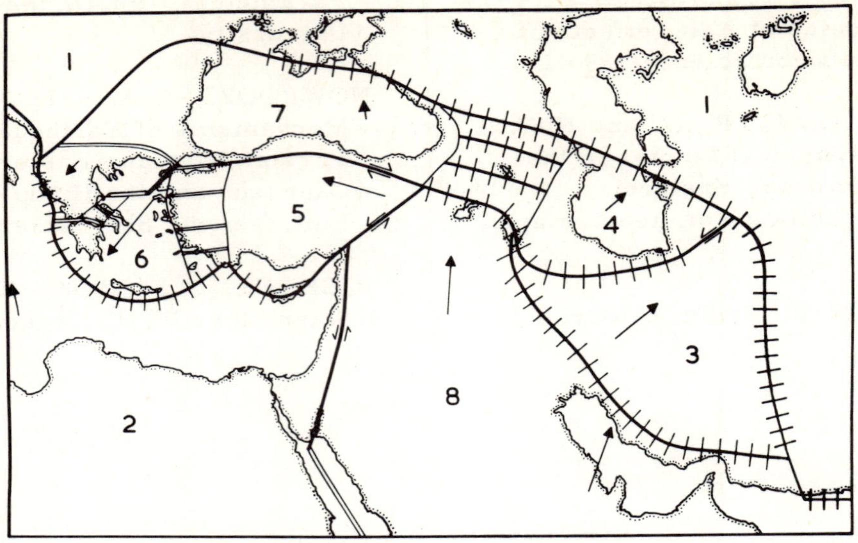

Fig. 2 Sketch of plate boundaries and motions. The arrows show the directions of motion relative to Eurasia and their lengths are approximately proportional to the magnitude of the relative velocity. Except for the North Anatolian Fault in Northern Turkey, the plate boundaries are diffuse active zones and only the general nature of the deformation is represented. The plate numbers are: 1 Eurasian, 2 African, 3 Iranian, 4 South Caspian, 5 Turkish, 6 Aegean, 7 Black Sea and 8 Arabian. The two changes to the earlier diagram (McKenzie 1972) required by new data are in the Western Aegean and in Eastern Iran.

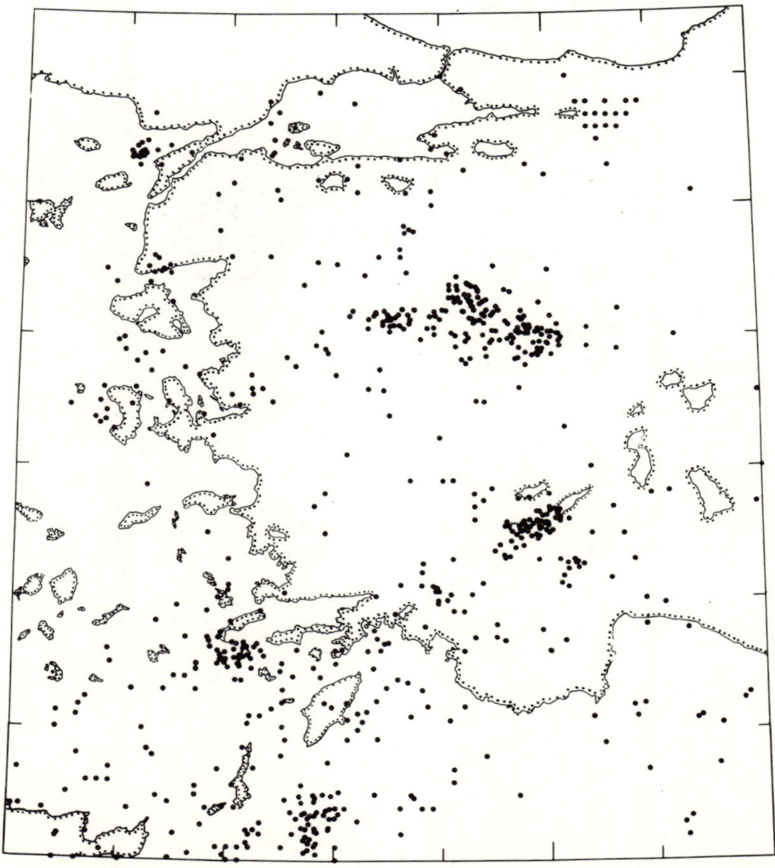

Fig. 3 Epicentres of shocks between 0 and 150 km which occurred between January 1 1961 and April 30 1975.

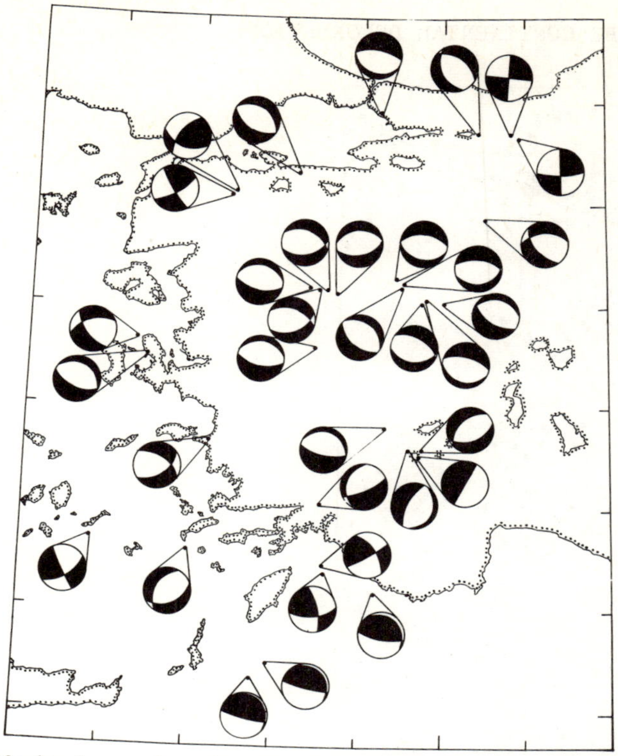

Fig. 4 Mechanisms of earthquakes whose depths were shallower than 50 km in Western Turkey. The symbols represent equal area projections of the lower focal sphere, and the compressional quadrants are shaded.

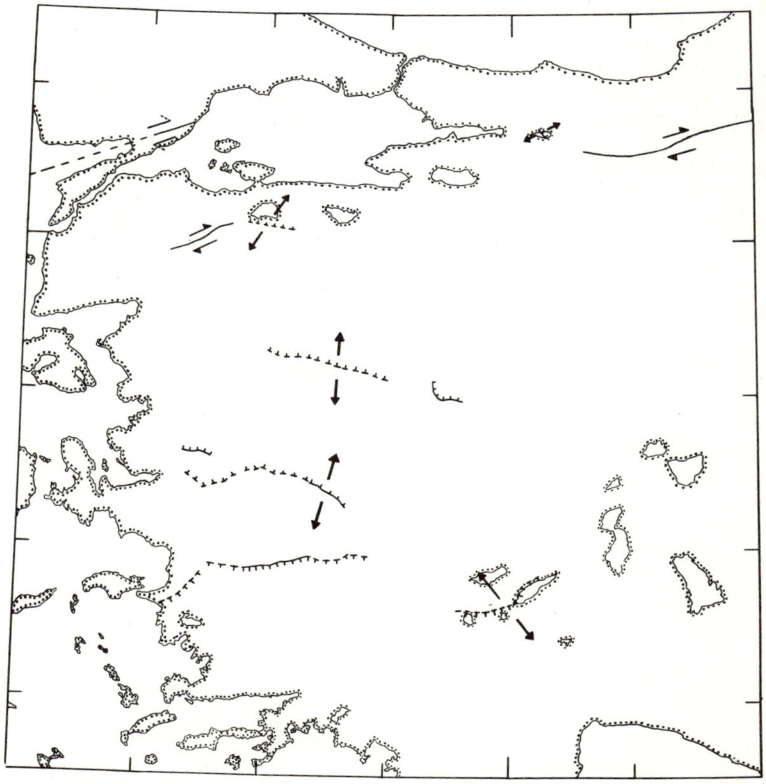

Fig. 5 Active fault systems in Western Turkey. Surface breaks produced by earthquakes are shown with a continuous line, those visible on satellite photographs and seismically active but not yet mapped in the field, are shown by dashed lines. The arrows show the relative motion between the two sides of the relevant faults obtained from the mechanisms in figure 4 and the strike of the faults.

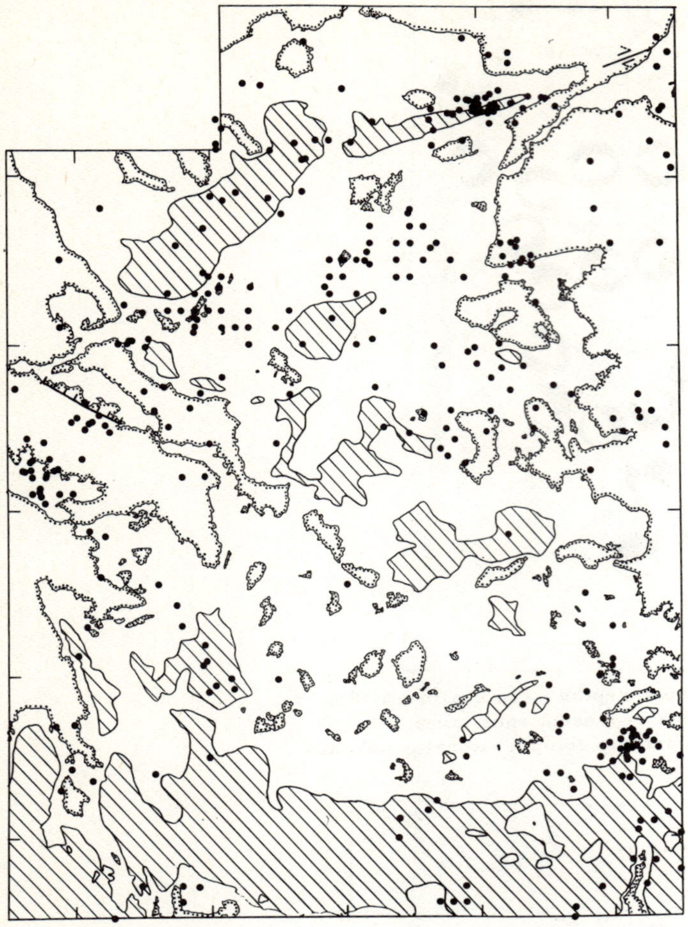

Fig. 6 The distribution of shocks shallower than 50 km in the Aegean between January 1 1961 and December 31 1975. Shaded regions are deeper than 600 m (Morelli et al. 1975).

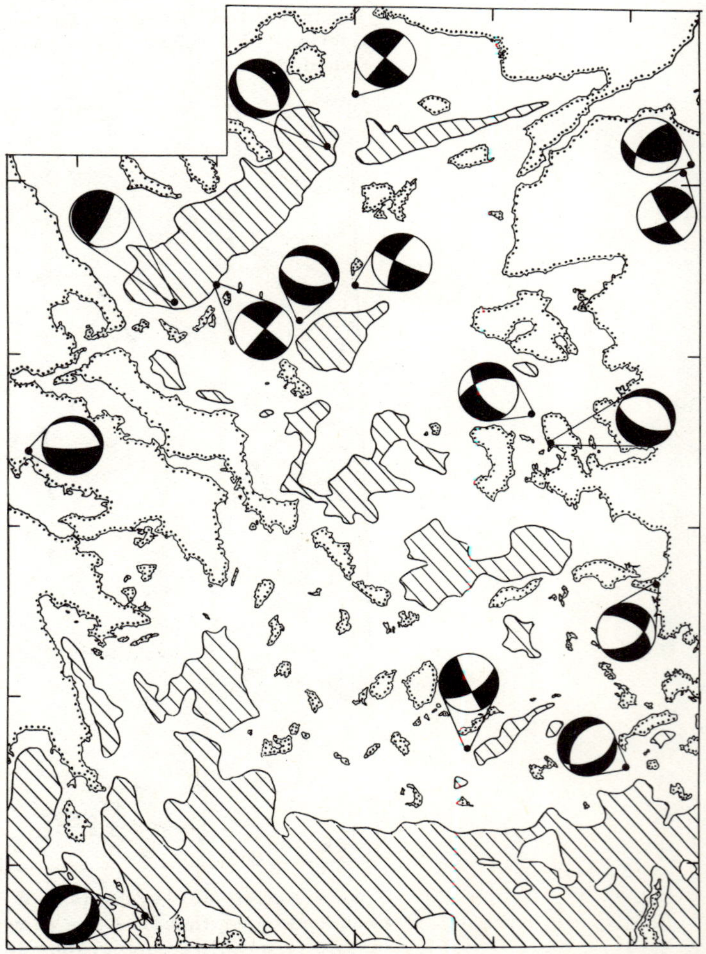

Fig. 7 Mechanisms of Earthquakes whose depths were shallower than 50 km.

INTERNATIONAL SYMPOSIUM ON THE STRUCTURAL HISTORY OF THE MEDITERRANEAN BASINS. SPLIT (YUGOSLAVIA) 25-29 OCTOBER 1976. B. BIJU-DUVAL AND L. MONTADERT, Eds. EDITIONS TECHNIP, PARIS 1977, pp. 197-198

RIGID-PLASTIC INDENTATION AND TECTONIC EVOLUTION OF THE ALPINE SYSTEM IN EUROPE

P. TAPPONNIER [(1)]

Abstract The distribution in space and succession in time of the major tectonic events in the Western Alpine system since the Jurassic is interpreted as the result of progressive oblique indentation of Europe by peninsulas of the African continent.

Résumé La répartition géographique et l'enchaînement dans le temps des différents épisodes tectoniques majeurs en Méditerranée (des Balkans à Gibraltar) sont interprétés à l'aide d'un formalisme nouveau (analogie rigide plastique) développé pour l'étude des déformations intraplaques qui résultent des collisions continentales, comme par exemple en Asie nord himalayenne. La cause principale des déformations alpines en Méditerranée depuis le Jurassique est le poinçonnement oblique et progressif de l'Europe par divers promontoires de l'Afrique.

The tectonics of the alpine chains in the Western Mediterranean are related to the slow convergence between the African and European continents since the Jurassic. In this context situations of continental interactions and collisions have arisen which show many similarities with situations related to the India-Eurasia collision and the large scale continental tectonics it has caused in Asia (1). We therefore suggest that new methods which are being developped for the analysis of intraplate tectonics in Asia (2) are also adapted to the study of the Alpine system in the Mediterranean. These methods imply a rigid-plastic behavior of the continental lithosphere and use slip-line field analysis to estimate the stress field in the lithosphere. We believe that :

1) The existence of ocean basins, which is the main difference between the mediterranean and continental Asia, may not give rise to major difficulties. Because these basins are small and have a complex shape, subduction of oceanic lithosphere at their margins may be difficult. They may therefore behave like rigid blocks (such as the shields in Asia) and transmit the forces that are exerted at their boundaries.

2) A most important requirement is to explain in a coherent way the distribution in space and the succession in time of the different tectonic events and phases. From the Balkans to Gibraltar, these events should be interrelated.

We suggest that most of the tectonics of the Western Mediterranean since the late Jurassic can be analysed as a combination of three basic processes :

1) The first one is an oblique and progressive "indentation" of Europe by peninsulas of the African continent. The main indenters are limited by important strike slip fault systems. The major system, oriented NE-SW extends from the Rif mountains to the Carpathians, over a total length of more than

(1) Laboratoire de Géologie Structurale, Université des Sciences et Techniques du Languedoc, Montpellier, place Eugène Bataillon, 34060 Montpellier Cédex, France

2 000 km. It "isolates" to the West the French-Austrian Alps and the Betic-Gibraltar arc. Segments of it have been activated, mostly in a left lateral sense, at differents epochs since the Jurassic.

A smaller, right-lateral strike-slip fault-system separates to the East the Balkan mountains and the Aegean arc from the Dinarids and the Carpathians. The maximum shortening in the whole Western Alps (from 500 km to 1 000 km since the Jurassic) has occurred in the Dinarids, Carpathians and Apennines between these two fault systems.

2) The second process is a consequence of indentation : it is a "lateral flow" of the continental lithosphere along large strike-slip faults, which prevents excessive crustal thickening in front of the indenter. This flow enhances subduction of ocean basins adjacent to the indenter, and thus causes the formation of small lateral subduction arcs. We believe that the formation at different epochs of the four major syntaxis in the Western Alpine system (French Alps and Gibraltar arc to the West, Transylvanian and Tyrrhenian arc to the East) may be related to this process.

3) The third process is a "cracking" of the lithosphere (formation of graben, and alcaline basalt volcanism) which can originate at major strike slip faults. The cause of this crustal thinning and break up is a stress state analogous to "secondary tension" (known in mechanical engineering) which develops in rigid-plastic media for some indentation geometries and boundary conditions. We think that the Oligocene rifting in France and Germany, the quaternary rifting and alcaline basalt volcanics (Etna) in Calabria and Sicily, and the opening of the Provencal ocean basin may be due to this process. At a larger scale, the opening of the North Atlantic in the Eocene might also have a similar origin, and would then appear as a direct consequence of the progressive indentation of Africa into Europe.

BIBLIOGRAPHY

(1) MOLNAR (P.) and TAPPONNIER (P.).- (1975). Cenozoic Tectonics of Asia : Effects of a Continental Collision. Science, 189, 419-426.

(2) TAPPONNIER (P.) and MOLNAR (P.).- (1976). Slip line Field Theory as a quantitative method for understanding large scale continental Tectonics. Nature, in press.

INTERNATIONAL SYMPOSIUM ON THE STRUCTURAL HISTORY OF THE MEDITERRANEAN BASINS. SPLIT (YUGOSLAVIA) 25-29 OCTOBER 1976. B. BIJU-DUVAL AND L. MONTADERT, Eds. EDITIONS TECHNIP, PARIS 1977, pp. 199-214

CONTRIBUTION A L'ÉTUDE DE LA TECTONIQUE RÉCENTE EN MÉDITERRANÉE OCCIDENTALE : LES DONNÉES DE LA NÉOTECTONIQUE DANS L'ARC DE GIBRALTAR ET DANS L'ARC TYRRHÉNIEN

J.-Cl. BOUSQUET [(1)]

- ABSTRACT.

The structural analysis of recent deformations (Tortonian to Quaternary) in the Gibraltar arc and Tyrrhenian arc shows that the recent history of these two regions is different, particularly during the Quaternary, when the Gibraltar arc undergoes compressional tectonics, while in the Tyrrhenian arc the normal faulting tectonic is very important. The general meaning of these phenomena is explaned within the frame work of the geodynamic evolution of the western Mediterranean sea.

Our main conclusions are the following:

- Along the Gibraltar arc : after a first period of extension during the upper Neogene and earlier Quaternary, collision of the European an African plate occured from the Early Quaternary onward and most of the deformation was absorbed in a zone that coincides approximately with the Alpine belt; therefore it is not possible to draw a clear cut boundary beetween both plates. At present the same phenomena continues as shown by the seismicity and the computed convergence vector beetween both plates.

- Along the Tyrrhenian arc the Plio-Quaternary extension may be related to an oceanic extension restricted to the Tyrrhenian sea, following the marginal sea-model.

Thus from the early Quaternary onward, the evolution in these two areas was fundamentally different, so that the southern part of the western Mediterranean sea has to be divided into two domains, i.e. 1) the Gibraltar arc and north Africa which were subjected to compression and 2) the Tyrrhenian arc subjected principaly to distension. The limit beetween both domains is most probably the NW-SE trending fracture zone of Sicilia-Tunisia strait which would then work as a righ-handed transform zone. Nevertheless, in both these domains, the extension however diachronous (Pliocene to earlier Quaternary in the first domain, late Pliocene to recent in the second one) is sufficient to cause the deepening of the western Mediterranean after the Messinian.

- RESUME.

L'analyse structurale des déformations récentes (Tortonien à Quaternaire) dans l'Arc de Gibraltar et dans l'Arc tyrrhénien montre que l'histoire récente de ces deux régions est différente. En particulier au Quaternaire, l'Arc de Gibraltar est soumis à un régime de compression, alors que l'Arc tyrrhénien est le siège d'une tectonique de distension particulièrement importante. La signification générale de ces phénomènes est précisée dans le cadre de l'évolution géodynamique de la Méditerranée occidentale. Les conclusions principales sont les suivantes :

a- Sur l'Arc de Gibraltar : le rapprochement entre les plaques européenne et africaine se

(1) Laboratoire de Géologie Structurale, Université des Sciences et Techniques du Languedoc, Montpellier, place Eugène Bataillon, 34060 Montpellier Cédex, France

fait à partir du Quaternaire ancien sous forme d'une tectonique de collision, et la déformation est absorbée principalement dans une zone qui se superpose à peu près à la chaîne alpine. Il n'est pas possible de tracer de limite précise entre les deux plaques. Actuellement le même phénomène se poursuit comme l'indique la séismicité et le vecteur de rapprochement des deux plaques.

b- Sur l'Arc tyrrhénien : La distension plio-quaternaire peut être reliée à une extension océanique limitée de la mer tyrrhénienne, sur le modèle des mers marginales.

c- Sur la Méditerranée occidentale: l'évolution des deux arcs est fondamentalement différente à partir du Quaternaire ancien. La partie méridionale de la Méditerranée occidentale doit être divisée en deux domaines (l'Arc de Gibraltar et l'Afrique du Nord soumis à un régime de compression, et l'Arc tyrrhénien, siège d'une tectonique de distension) séparés vraisemblablement par la zone des fractures NW-SE du détroit siculo-tunisien qui jouerait alors le rôle d'une zone transformante dextre. Dans ces deux domaines, tout en étant décalés dans le temps (Pliocène à IVre ancien à l'W, et Pliocène supérieur à IVre récent à l'E), les phénomènes de distension sont suffisamment importants pour avoir provoqué l'approfondissement de la Méditerranée occidentale après l'épisode messinien.

INTRODUCTION.

Depuis quelques années, en Méditerranée, nos connaissances géologiques et géophysiques se sont considérablement enrichies. Les données recueillies éclairant en premier lieu l'histoire récente de la Méditerranée, il en est résulté un nouvel intérêt pour les événements qui se sont déroulés depuis une dizaine de millions d'années et en particulier pour ceux qui eurent lieu pendant le Quaternaire.

Des tectoniciens se sont alors tournés vers l'analyse des déformations quaternaires et un progrès considérable a pu être fait dans ce domaine grâce à l'utilisation des méthodes de la microtectonique cassante (1). Celles-ci permettent de caractériser sans trop d'ambiguité si les déformations observées sont le fruit d'un régime de compression ou de distension, et dans certains cas de déceler le caractère polyphasé des déformations et d'en établir une chronologie relative. En outre, les directions d'allongement ou de raccourcissement des déformations quaternaires peuvent être comparées avec les mécanismes au foyer des séismes (Mercier et al. 1973) ou avec certains paramètres généraux tels les vecteurs de déplacement des plaques.

Bien que l'étude néotectonique du domaine méditerranéen est encore incomplète, des résultats importants ont déjà été recueillis, qui éclairent de façon nouvelle nos concepts sur les phénomènes géodynamiques quaternaires et actuels. Nous avons choisi ici, pour illustrer la diversité et la signification de ceux-ci, de les comparer dans deux secteurs clefs de la Méditerranée occidentale : l'"Arc de Gibraltar" et l'"Arc tyrrhénien" (2).

Il faut toutefois souligner tout de suite, que dans ces deux régions, nos connaissances de la néotectonique et de l'évolution tectonique récente ne sont pas comparables. Dans l'Arc de Gibraltar et sur les bordures de la mer d'Alboran, une étude générale (3) a été entreprise de 1973 à 1976 par un groupe pluridisciplinaire d'une quinzaine de chercheurs (cf. liste en annexe). Par contre, dans l'Arc tyrrhénien, aucune étude de même envergure, c'est-à-dire concernant l'ensemble de l'arc, n'a été jusqu'à maintenant réalisée, et nous exposerons ici principalement les données que nous avons personnellement recueillies dans le Nord de l'Arc tyrrhénien, auxquelles nous adjoindrons quelques compléments tirés de la bibliographie.

(1) Les méthodes utilisées ont été développées par les laboratoires de Géologie Dynamique interne d'Orsay et de Géologie Structurale de Montpellier (cf. Pegoaro 1972, Philip 1974, Sorel 1976).

(2) Le terme d'"arc" est utilisé ici sans aucune signification géologique précise et ne fait qu'indiquer la courbure de ces deux régions.

(3) Etude réalisée dans le cadre de l'Action thématique programmée (A.T.P.) du Centre NaNational de la Recherche Scientifique : "Géodynamique de la Méditerranée occidentale".

LA NEOTECTONIQUE DANS L'ARC DE GIBRALTAR ET DANS L'ARC TYRRHENIEN.

I. LES DONNEES DE LA NEOTECTONIQUE DANS L'ARC DE GIBRALTAR ET LES BORDURES DE LA MER D'ALBORAN.

Les recherches ont été divisées en deux volets complémentaires :

- L'étude des dépôts littoraux quaternaires sur la façade atlantique de l'arc et sur le pourtour de la mer d'Alboran, le long des côtes marocaines et espagnoles, afin d'établir un bilan général des variations altimétriques des lignes de rivage quaternaires.

- L'étude structurale des déformations quaternaires, dans le cadre de l'évolution récente (à partir du Tortonien) de l'Arc de Gibraltar, réalisée en Espagne au sud d'une ligne Séville-Alicante, au Maroc dans le Rif, les bassins du Guercif et du Rharb, et en Algérie nord-occidentale (Oranie).

Les résultats obtenus sont ici brièvement résumés, ayant déjà fait l'objet de plusieurs publications citées dans la bibliographie. Pour un aperçu général, on se reportera en particulier à la synthèse présentée au colloque final de l'A.T.P. "Géodynamique de la Méditerranée occidentale" (Groupe de recherche néotectonique de l'Arc de Gibraltar).

A. Les grands traits de l'histoire tectonique récente (Tortonien à Quaternaire).

La signification et l'importance de la néotectonique doivent être précisées par rapport aux phénomènes qui l'ont immédiatement précédé. D'autre part les déformations quaternaires si elles doivent être recherchées en premier lieu dans les terrains quaternaires, intéressent également les terrains néogènes, et au sein de ceux-ci il est nécessaire de faire la part de ce qui est déformation quaternaire (néotectonique) de ce qui est antérieur. Aussi dans l'Arc de Gibraltar, on a pris en compte les événements qui se produisent après la dernière phase importante de compression. Cette phase située à la fin du Serravallien intéresse aussi bien les zones internes que les zones externes de l'édifice alpin. Par la suite nous distinguerons trois périodes d'une durée très inégale.

1. LA PERIODE DE DISTENSION DU TORTONIEN AU QUATERNAIRE ANCIEN. (soit approximativement entre 10 millions d'années et 1,5 million d'années).

Cette période est marquée par le déroulement de phénomènes distensifs, à l'exception d'une phase de compression, brève et très localisée, reconnue entre le Tortonien et le Messinien dans la région de Melilla au Maroc (Guillemin 1976).

Des structures en horst et en graben vont se réaliser par étapes au cours de cette période, aussi bien dans les zones internes et externes de l'édifice alpin que dans son avant pays (Oranie, bassin de Guercif, Rharb, bassin du Guadalquivir). En général ces structures ne seront que peu modifiées par la phase de compression ultérieure et elles sont en grande partie responsables de la disposition actuelle du relief. Du Tortonien au Pliocène y compris, la sédimentation et l'évolution des bassins est en rapport assez étroit avec la tectonique du fait de la prépondérance des mouvements à composante verticale. Des mouvements synsédimentaires (subsidence, olistolithes, failles...) et des discordances en général faibles (sauf parfois en bordure de bassin) sont les reflets de cette activité tectonique.

Pendant cette période, des manifestations volcaniques notables sont connues en mer d'Alboran et à terre dans les Cordillères bétiques orientales, le Maroc nord-oriental et l'Oranie. Un volcanisme de type "calco-alcalin" (Tortonien à Messinien) précède un volcanisme de type "alcalin" (Pliocène supérieur en Espagne et au Maroc, Pliocène supérieur et Quaternaire ancien en Oranie).

A ces deux types de volcanisme, il est possible que correspondent deux épisodes distincts de distension. En effet des mesures microtectoniques encore assez rares (la phase de compression quaternaire ayant très souvent effacé les microstructures de distension) semblent indiquer que l'Arc de Gibraltar a été tout d'abord soumis à une extension N-S (à NE-SW au Maroc), responsable de la formation de nombreux bassins, bordés par des failles de direction E-W. Par contre, des mesures effectuées dans des couches pliocènes en Espagne (Bousquet et Philip 1976, Benkhelil 1976) révèlent une extension E-W, ayant donné naissance à des failles de direction N-S. Il faut souligner qu'outre ces failles assez rares, et celles plus courantes d'orientation E-W, créées vraisemblablement par les deux épisodes distensifs envisagés, des failles antérieures au Tortonien jouent également un rôle notable. Ces accidents fonctionnent en décrochement lors de la dernière phase "alpine" de serrage. Les plus importants constituent une ligne de fractures de direction NE-SW, qui à partir du Maroc, traverse la mer d'Alboran et la sépare en deux bassins profonds (cf. Olivet, Pautot et Auzende 1972) et prend ensuite en écharpe la partie orientale des Cordillère bétiques. Une des failles majeures après avoir traversé le golfe d'Almeria (Rios 1976) se poursuit à terre sur près de 40 km (accident de Carboneras), avant d'être décalée vers le Nord par un accident de direction NNE-SSW (accident de Palomares) et de longer ensuite la côte près d'Aguilas. Egalement de direction NE-SW, l'accident d'Alhama de Murcia a été reconnu sur près de 100 km entre Huercal-Overa et Murcia. Ensuite sous une couverture d'alluvions récentes, son tracé a été révélé par une étude géophysique (Gauyau et al. 1976) et par le biais de plusieurs failles NE-SW, parfois décalées par des failles NW-SE, il arrive à la côte à proximité d'Alicante. Le prolongement de cette ligne de fractures a même été envisagé audelà d'Alicante jusqu'au niveau de la marge septentrionale d'Ibiza et de Majorque (Bousquet et Montenat 1974). A ces failles NE-SW, sont associées d'autres de moindre importance de direction NW-SE, qui apparaissent donc comme conjuguées des précédentes.

Tous ces accidents anté-tortoniens rejouent en général lors de la période de distension (4) et il est remarquable de constater qu'à terre dans l'Arc de Gibraltar, le volcanisme précédemment mentionné se trouve presque exclusivement à l'Est de la grande ligne de fractures NE-SW.

2. LA PHASE DE COMPRESSION DU QUATERNAIRE ANCIEN.

L'étude de vastes secteurs et l'utilisation de la microtectonique a permis de caractériser cette phase dans la plus grande partie de l'Arc de Gibraltar. En raison de l'imprécision des datations des formations continentales qui succèdent plus ou moins précocement aux niveaux marins du Pliocène, le début de cet épisode compressif est assez difficile à situer. En se fiant à certaines datations (en particulier du Quaternaire d'Oranie), et du fait de la permanence des mouvements distensifs jusqu'au début du Quaternaire, cette phase pourrait commencer au cours du Quaternaire ancien (entre 1,5 et 1 million d'années ?). On peut d'autre part observer que d'une manière générale les glacis et les terrasses fluviales et marines rapportés habituellement au Quaternaire moyen sont postérieures aux déformations majeures des terrains du Pliocène et du Quaternaire ancien. L'ampleur et la répartition des déformations résultant de cette phase de compression quaternaire sont assez inégales. Des plis d'échelle kilométrique sont présents dans les secteurs les plus déformés (Oranie, bassin de Guercif, bordure sud du Rharb, Cordillères bétiques occidentales et orientales) et une partie des failles ayant joué précédemment en distension, fonctionnent alors en décrochement (ou plus rarement en faille inverse) comme cela est particulièrement net dans les Cordillères bétiques orientales et en Oranie. Par contre dans d'autres secteurs où les terrains néogènes et quaternaires sont restés tabulaires, les microstructures (galets striés ou fracturés, microfailles et diaclases) sont les seuls témoins de cette phase. Il semble que la partie centrale de l'Arc de Gibraltar de part et d'autre du détroit, ait été une des zones les moins touchées par la compression quaternaire, et il est à noter, qu'alors que certaines zones de l'avant-pays de l'édifice alpin présentent des déformations notables (Rharb, bassin de Guercif...), dans le bassin du Guadalquivir seules quelques ondulations anticlinales ont été signalées sur son bord méridional (Viguier 1975).

Les mesures microtectoniques réalisées dans une cinquantaine de sites sur toute l'étendue de l'Arc, les directions des axes des plis et le sens du rejeu en décrochement des failles sont cohérents entre eux et indiquent une direction de raccourcissement en général NNW-SSE, seuls quelques plis et quelques stations au Maroc nord-occidental donnant une direction NE-SW.

3. LA PERIODE DU QUATERNAIRE MOYEN ET RECENT.

D'une étude plus délicate, elle a été distinguée de la précédente période, car au contraire de celle-ci elle ne présente pas les mêmes caractéristiques dans toute l'étendue de l'Arc de Gibraltar.

Dans les Cordillères bétiques orientales, en plusieurs endroits des niveaux rattachés au Quaternaire moyen (Paléotyrrhénien) et plus rarement récent (Würm ou Tyrrhénien à Strombes), sont affectés par des déformations qui sans ambiguité se sont formées dans un régime de compression. Ce sont des plis, des failles inverses ou des décrochements qui en général sont de moindre ampleur que les struc-

(4) En fonction d'une extension N-S puis E-W, il est probable que de telles failles aient alors joué avec une certaine composante horizontale.

tures de la phase de compression du Quaternaire ancien.

Un des secteurs les plus remarquables est situé à proximité d'Almeria : sur une distance de près de 15 km, le réseau de failles d'El Alquian découpe des glacis et des terrasses marines plus anciennes que le Tyrrhénien à Strombes. Ces failles disposées en échelon suivant une orientation générale NW-SE peuvent être interprétées comme des déformations de surface, liées au rejeu en décrochement dextre d'un accident profond, conjugué avec le jeu sénestre de l'accident de Carboneras voisin. En dehors de quelques exceptions (dont le secteur de Campo de Dalias également proche d'Almeria - cf. Fourniguet 1976), les déformations les plus nettes ont été observées le long des grands accidents précédemment mentionnés (accidents de Carboneras, Palomares et Alhama de Murcia).

Dans le reste des Cordillères bétiques et en Oranie, on a pu observer également mais en quelques points seulement des déformations en compression de niveaux plus récents que le Quaternaire ancien.

Par contre localement dans le Rif oriental, après l'épisode compressif du Quaternaire ancien, les phénomènes de distension semblent reprendre, comme l'attestent des jeux en faille normale de certains accidents, l'individualisation ou le maintien de zones subsidentes, et quelques mesures microtectoniques dans des terrains du Quaternaire moyen qui indiquent une extension EW à NE-SW. De plus, contemporainement le Maroc oriental est la seule région de l'Arc de Gibraltar, qui soit alors le siège d'une activité volcanique (basaltes alcalins).

En plus des informations chronologiques qu'elle fournit, l'étude systématique des dépôts littoraux (Gigout et collaborateurs) a permis de connaître l'amplitude et la répartition des mouvements à grand rayon de courbure, subis par l'Arc de Gibraltar au cours du Quaternaire moyen et récent. En repérant les lignes de rivages correspondant aux différents maxima transgressifs du Quaternaire, on met en évidence les gauchissements de repères initialement horizontaux. Les deux niveaux ayant la meilleure répartition géographique correspondent au Tyrrhénien récent (Ouljien) et au Tyrrhénien ancien (= Anfatien ou Paléotyrrhénien). L'analyse de leurs lignes de rivage indique un soulèvement des deux rives du détroit de Gibraltar, et en mer d'Alboran une succession de secteurs affaissés et surélevés avec une disposition similaire en Espagne et au Maroc.

Sur le rivage espagnol, près de Carboneras, le soulèvement du Tyrrhénien récent peut s'expliquer par un rejeu récent de l'accident de Carboneras (Bousquet, Cadet, Montenat 1975), mais dans les autres secteurs l'interprétation de ces déformations à grand rayon de courbure est délicate : au Maroc, elles pourraient accompagner les mouvements distensifs comme cela semble être le cas au niveau du secteur du Bas Nekor, où ont été caractérisés les déformations en distension indiquées précédemment. Par contre en Espagne et en Oranie, où les structures en compression ont pu être observées, elles pourraient correspondre à une accentuation en compression de la déformation des mégastructures.

D'autre part si l'on se réfère aux maxima atteints, on constate qu'ils sont relativement faibles (22 m pour le Tyrrhénien récent et 95 m pour le Tyrrhénien ancien dans la zone du détroit de Gibraltar et 165 m pour le Sicilien au Ras Torf (Maroc)).

B.- Conclusions sur l'Arc de Gibraltar et la mer d'Alboran.

Certaines des conclusions tectoniques exposées précédemment apportent de nouveaux éléments pour comprendre l'évolution récente de l'Arc de Gibraltar et de la mer d'Alboran. Nous n'aborderons pas ici le problème de la formation de cette mer, car la discussion de celui-ci nécessiterait de prendre en considération un grand nombre de données géologiques et géophysiques et nous ferait dévier de notre propos. Signalons seulement que, par rapport aux hypothèses récemment proposées qui font jouer un rôle majeur aux décrochements (Le Pichon et al., 1972, Olivet et al. 1974), les données qui viennent d'être recueillies confirment l'importance dans ce secteur de ce type d'accidents, et tout particulièrement de la grande ligne de fractures NE-SW qui traverse obliquement l'Arc de Gibraltar et la mer d'Alboran.

Quel que soit le mécanisme retenu pour expliquer la formation de celle-ci, et le rôle tenu par les phénomènes distensifs précédemment analysés, il apparaît maintenant qu'à partir du Quaternaire ancien, l'évolution dynamique de la région change considérablement. Il est à noter toutefois, que d'après des résultats préliminaires qui demanderaient à être confirmés, un changement se produirait déjà au cours du Pliocène, l'extension N-S étant remplacée par une extension E-W. Ceci pourrait correspondre à une modification du champ des contraintes d'une signification majeure étant donné que parallèlement un volcanisme de type "alcalin" (Pliocène supérieur et parfois également Quaternaire ancien) remplace celui de type "calco-alcalin" (Tortonien et Messinien).

Quoiqu'il en soit à partir du Quaternaire ancien, l'ensemble de l'Arc de Gibraltar et vraisemblablement de la mer d'Alboran (vu la présence des mêmes systèmes de failles à terre et en mer) sont soumis à un régime de compression qui conduit à un raccourcissement général NNW-SSE, y compris dans la zone de courbure de l'Arc, à l'exception du Rif sud-occidental. Le secteur déformé se superpose approximativement à la chaîne "alpine", mais déborde parfois sur son avant-pays (au Maroc et en Oranie). L'orientation relativement constante des directions de la déformation, la largeur de la zone déformée (5 à 600 km du Nord au Sud), et le rôle tenu par les décrochements, permettent de penser qu'à partir du Quaternaire ancien, le rapprochement entre les plaques européennes et africaines se fait au niveau de l'Arc de Gibraltar avec les caractéristiques d'une tectonique de collision, continent contre continent (5) suivant un modèle comparable toutes proportions gardées à ce qui s'est produit en Iran ou en Asie centrale (Molnar et Tapponnier 1975).

Au Quaternaire moyen et récent, cette tectonique se poursuit comme cela est assez net dans les Cordillères bétiques orientales, mais la réapparition des phénomènes distensifs dans le Rif oriental et d'un volcanisme de type alcalin pourrait indiquer un ralentissement dans la convergence des deux plaques dans ce secteur de la Méditerranée.

Quant à la période actuelle, elle présente les mêmes caractéristiques : on constate en effet que le vecteur de rapprochement actuel des deux plaques (Mac Kenzie 1972, Minster et al. 1974) a la même orientation NNW-SSE que les directions de raccourcissement mesurées pour la compression quaternaire; de plus la séismicité (séismes superficiels et normaux)(6) est répartie sur l'ensemble du secteur déformé au Quaternaire, et d'après Hatzfeld (1975) elle témoignerait également en faveur d'une collision continentale.

Les mécanismes au foyer publiés sur ce secteur sont peu nombreux, mais si l'on se réfère à Ritsema (1975), d'après ses propres calculs ils indiqueraient une compression dans les Cordillères bétiques et une distension au Maroc, ce qui serait donc tout à fait comparable avec ce qui a été observé pour la période du Quaternaire moyen et récent.

Au niveau de l'Arc de Gibraltar, nous estimons donc que la convergence des plaques européenne et africaine se traduit depuis le Quaternaire ancien par une tectonique de collision. De ce fait les données néotectoniques et séismiques ne permettent pas de situer dans cette région la limite entre ces deux plaques de façon précise, et de faire un choix entre les différents tracés proposés. (cf. Mac Kenzie 1970, Udias 1972, Dewey et al. 1973, Arana et Vegas 1974).

II.- LES DONNEES DE LA NEOTECTONIQUE DANS l'ARC TYRRHENIEN.

Comme nous l'avons précédemment souligné, cet arc n'a pas encore fait l'objet d'une étude néotectonique générale, comme dans l'Arc de Gibraltar ou l'Arc égéen. De nombreux problèmes subsistent aussi en ce qui concerne la tectonique récente dont certaines données apparaissent pour le moment contradictoires.

A. Les problèmes tectoniques posés par l'histoire néogène de l'Arc tyrrhénien.

Dans l'état actuel des connaissances, ce qui surprend principalement c'est que l'évolution tectonique néogène de la Calabre cristalline qui forme la partie centrale de l'Arc tyrrhénien, et celle des régions voisines (Apennin méridional et Sicile nord-occidentale) semblent différer sensiblement.

---

(5) En mer d'Arboran suivant les travaux récents, les zones à croûte de type océanique ou intermédiaire semblent très limitées, et ne sont donc pas susceptibles de modifier les caractéristiques générales de cette tectonique.

(6) Deux séismes profonds ont été enregistrés (1954 et 1973). Le mécanisme au foyer du séisme de 1954 indique une compression E-W (Hodgson et Cock 1956, Filippo et Peronaci 1959). Ces séismes peuvent être liés à des reliques lithosphériques, vestiges d'une subduction (Payo 1975).

En Calabre cristalline les phénomènes de distension paraissent prépondérants depuis l'Oligocène (Guérémy 1972, Dubois 1976). Des chevauchements du socle cristallin sur le Messinien ont toutefois été signalés (Di Nocera et al. 1974), et les travaux d'Ogniben (1955-1962) et de ses collaborateurs dans les bassins du bord oriental de la Calabre centrale, ont révélé l'existence d'un plissement modéré intra-pliocène.

Par contre dans l'Apennin méridional comme en Sicile nord-occidentale, à partir du Miocène des "phases" nombreuses se succèdent (cf. Scandone 1975, Ogniben 1975, Vezzani 1975, ...) et en particulier, celles qui sont responsables de la mise en place des nappes rigides formées de matériel carbonaté mésozoïque et qui sont dans l'Apennin méridional placées au Miocène moyen. Même si, comme cela a été souligné à de nombreuses reprises, on fait la part de ce qui peut être dû à la gravité, et qui peut donc être lié à des soulèvements ou des affaissements engendrés par des phénomènes distensifs, il n'en reste pas moins que certains épisodes semblent ne pouvoir être interprétés que dans un régime de compression (en particulier dans l'Apennin calabro-lucanien, des plis isoclinaux liés à un métamorphisme léger ont été observés dans des niveaux datés du Miocène inférieur) (Grandjacquet 1969). Il serait donc logique de retrouver en Calabre cristalline la trace même minime de ces épisodes compressifs, et dans ce domaine des études microtectoniques seraient précieuses. Elles permettraient également d'étayer les conclusions de Nocera et al. (1975) sur deux "phases" qui selon ces auteurs intéresseraient à la fois l'Apennin méridional et la Calabre cristalline, à savoir les phases intra-messinienne et intra-pliocène.

B. La néotectonique dans l'Arc tyrrhénien.

Pour important qu'aient pu être les phénomènes tant compressifs que distensifs pendant le Néogène, la morphologie actuelle de l'Arc tyrrhénien est surtout le fait de la tectonique distensive plio-quaternaire.

1. D'APRES L'EXEMPLE DE L'APENNIN CALABRO-LUCANIEN (cf. Bousquet 1973).

L'image qui nous est fournie par l'Apennin calabro-lucanien est celle d'un pays de nappes, découpé par un réseau assez serré de failles qui présentent plusieurs directions (dans l'ordre de fréquence : NNW-SSE, NE-SW, WNW-ESE, NNE-SSW) et qui peuvent délimiter de vastes fossés séparés par des horsts. Il est vraisemblable que les mouvements verticaux dus à ces failles furent pour l'essentiel assez tardifs, sinon ils auraient rendu impossible le glissement des nappes de "flysch" qui, au niveau de l'Apennin calabro-lucanien se mettent en place au Messinien puis progressent vers le fossé bradanique au cours du Pliocène (Mostardini et al. 1966, Lentini 1967). D'autre part, les dépôts du Pliocène supérieur (pro-parte) et du Calabrien s'effectuent en continuité dans des fossés bien délimités, et qui, au Calabrien, pouvaient être entourés par des reliefs atteignant dans l'Apennin calabro-lucanien 1000 m (au maximum). Aussi la tectonique distensive post-nappe n'apparaît ici avec une certaine importance qu'à partir du Pliocène supérieur et l'on peut juger de son ampleur grâce à quelques témoins des dépôts messiniens, ou par rapport à certains repères structuraux, comme la base des nappes de "flysch". On constate ainsi qu'entre l'Apennin calabro-lucanien et la fosse bradanique, pour des secteurs distants seulement de 45 km, la dénivellation maximum atteinte, due à la tectonique plio-quaternaire, est de l'ordre de 5500 à 6000 m. A l'intérieur même de l'Apennin calabro-lucanien, on arrive par des estimations à un chiffre de 4000 m, entre le horst de la chaîne du Pollino et le fossé de Scalea en bordure de la mer tyrrhénienne. Ces chiffres fournissent une idée globale de l'importance de la tectonique plio-quaternaire. On peut de plus distinguer dans le déroulement de celle-ci les étapes suivantes :

- en premier lieu, au Pliocène supérieur les mouvements distensifs qui guident l'installation des bassins plio-calabriens, et que nous avons précédemment mentionnés.

- ensuite la subsidence plio-calabrienne, accompagnée du jeu de failles synsédimentaires (les épaisseurs atteintes sont considérables et peuvent dépasser 1500 m, on peut calculer alors un taux moyen maximum de l'ordre de 2 mm/an).

- puis la distension post-calabrienne, qui intéresse tous les dépôts quaternaires, des formations rattachées au Riss et plus rarement au Würm, étant parfois affectées (cf. Bousquet et Gueremy 1967 et 1968). Les rejets verticaux mesurés par rapport aux terrains calabriens ou postérieurs sont encore notables : ainsi dans l'Apennin calabro-lucanien, la faille du Pollino a un rejet vertical post-calabrien de l'ordre de 500 m (dont 100 m post-Sicilien et anté-Riss).

- enfin un soulèvement général, qui commence à la fin du Calabrien et se poursuit tout au long du Quaternaire, comme l'attestent les altitudes des différentes terrasses marines. Les dépôts calabriens ont été portés à des altitudes considérables (altitude maximum observée : 835 m ; altitude maximum déduite : entre 1000 et 1200 m).

La tectonique plio-quaternaire apparaît donc pour l'essentiel de type distensif, mais d'après quelques indices, un épisode de compression vraisemblablement assez bref, a intéressé également cette région au cours du Quaternaire. En effet, dans le réseau de failles postérieures aux nappes, on remarque à côté des failles à disposition de faille normale, d'autres à disposition de faille inverse. Les mesures microtectoniques, si elles indiquent qu'environ trois quart des failles ont un pitch > 45° et rentrent ainsi dans la catégorie des failles normales ou dans celle des failles intermédiaires entre les failles normales et les décrochements, révèlent aussi que pour un quart, des failles ont fonctionné avec un mouvement à composante horizontale prépondérant (pitch < 45°). De plus, on constate que certains miroirs de faille normale ont été déformés en compression (microfailles inverses, petits plis, etc.) comme cela a été observé dans l'Arc Egéen (Philip 1974).

Cet épisode compressif est assez difficile à situer chronologiquement dans le déroulement de la tectonique plio-quaternaire: il pourrait s'agir d'un épisode post-calabrien et anté-sicilien, compte tenu de la discordance assez nette des dépôts siciliens sur ceux du Calabrien, du caractère tardif des déformations en compression (faille normale réutilisée en décrochement, ou à miroir déformé) et de quelques observations néotectoniques (brèches du Quaternaire ancien ployé en anticlinal par le rejeu d'un contact anormal au Mte Raparo, crochon de faille constitué de Calabrien renversé à 80° près de la faille du Pollino, signalé par Quitzow 1935).

2. D'APRES LES AUTRES REGIONS DE L'ARC TYRRHENIEN.

Ne pouvant passer ici en revue toutes les autres régions de l'Arc tyrrhénien, nous renvoyons à l'aperçu général précédemment publié (Bousquet 1973). De celui-ci, nous retiendrons que, dans les grandes lignes, en l'état actuel des recherches, on retrouve dans tout l'Arc des manifestations nombreuses de la tectonique de distension plio-quaternaire.

Si suivant les régions le réseau de failles normales peut présenter d'autres directions que dans l'Apennin calabro-lucanien (en particulier EW et NS comme en Calabre cristalline), le bilan global des mouvements verticaux est tout aussi considérable et seules les zones les plus externes de l'Arc tyrrhénien (Pouilles et Plateau de Raguse) sont les moins touchées par la néotectonique. Les terrains du Pliocène supérieur et du Calabrien sont très souvent affectés par des failles normales, et l'on peut citer comme exemple extrême, la faille de Scandale, dans le bassin de Crotone dont le rejet vertical post-calabrien est de 800 m (Signorini 1947, Selli 1962). Le soulèvement post-calabrien est général et les altitudes atteintes par les dépôts calcabriens sont maxima (entre 800 et 1000 m) en bordure de la mer tyrrhénienne et décroissent vers l'extérieur de l'arc. Le phénomène est particulièrement net de l'Apennin aux Pouilles, soulevées seulement de 300 à 400 m. Un tel soulèvement différentiel n'est pas dû à un réajustement isostatique, et sa cause doit être recherchée dans la dynamique générale de l'Arc tyrrhénien (Bousquet 1973).

Enfin mentionnons que les plis post-calabriens d'axe EW à ENE-WSW décrits par Mascle (1973) en Sicile occidentale, pourraient être une manifestation plus vigoureuse de la phase de compression envisagée dans l'Apennin calabro-lucanien entre le Calabrien et le Sicilien.

C. Conclusions sur l'histoire récente de l'Arc tyrrhénien.

Il est difficile de tracer un bilan général de l'évolution récente de l'Arc tyrrhénien en l'état actuel des connaissances. Au coeur de l'arc, la Calabre cristalline a gardé la trace de mouvements en distension antérieurs au Miocène, mais par contre ne semble pas avoir été touchée par les épisodes compressifs miocènes des régions voisines (Apennin méridional et Sicile), ce qui paraît assez difficilement explicable. Les "phases" intra-messinienne et intra-pliocène sont peut-être plus générales, et elles mériteraient d'être étudiées dans le détail avec l'aide de la microtectonique. Par la suite, les caractéristiques des phénomènes tectoniques paraissent plus clairs, étant donné l'ampleur de la tectonique de distension plio-quaternaire. Toutefois l'âge du début de cette tectonique pourrait être mieux connu et les effets d'une phase de compression post-calabrienne recherchés dans tout l'arc. Bien qu'il est probable que cette phase ait joué un rôle mineur dans l'élaboration de la structure et du relief de l'Arc tyrrhénien, par rapport aux phénomènes distensifs, son étude minutieuse ne serait pas sans intérêt d'autant qu'elle semble se situer à la même époque que celle décrite dans l'Arc égéen (Philip 1974, 1976).

Si l'on cherche maintenant à replacer ces phénomènes tectoniques dans le cadre d'une explication globale de l'Arc tyrrhénien et de la mer tyrrhénienne, il faut prendre en considération d'autres données géologiques et géophysiques. Nous indiquons seulement ici à nouveau (cf. Bousquet 1973) que l'hypothèse qui intègre le mieux toutes ces données, est à notre avis, celle de la formation de la mer tyrrhénienne sur le modèle des mers marginales à partir du Messinien. Cette océanisation tardive a pu toutefois être précédée par une période d'ouverture et de distension dès l'Oligocène.

Cette hypothèse est en grande partie confirmée par les nombreuses études récentes géochimiques et géochronologiques réalisées sur le volcanisme plio-quaternaire de ces régions, qui à l'intérieur de l'arc est de type "calco-alcalin" (Iles Eoliennes) ou de type tholéitique (DSDP 373). Par contre le volcanisme de type "alcalin" situé à l'extérieur de l'Arc (Etna et détroit siculo-tunisien), doit être relié à d'autres phénomènes géodynamiques.

III.- CONCLUSIONS GENERALES.

Pour établir des interprétations de l'évolution récente de la Méditerranée occidentale, on se heurte actuellement à deux difficultés majeures. Tout d'abord un manque de connaissances sur le déroulement précis des événements tectoniques au Pliocène et au Quaternaire. Du fait de l'importance des phénomènes distensifs, ceux-ci sont le plus souvent les seuls mentionnés, et des études longues et minutieuses seront nécessaires pour apporter des éléments nouveaux capables de compléter le bilan structural des nombreuses régions.

Ensuite, en supposant connu le déroulement des épisodes tectoniques récents et leurs caractéristiques, on doit rechercher des modèles généraux qui relient entre eux les phénomènes observés par une certaine logique. La subdivision de la Méditerranée actuelle en un certain nombre de petites plaques en fonction de la seule séismicité (Mac Kenzie 1970, 1972, Dewey et al. 1973), n'est pas un modèle satisfaisant pour les géologues, car il ne tient pas compte du fait que les failles actives (telle la faille nord-anatolienne), ne constituent pas nécessairement des limites de plaque, et que les déformations intra-plaques existent elles aussi. Une meilleure connaissance des caractéristiques des déformations intracontinentales provoquées par une collision continent contre continent entre l'Inde et l'Asie (Molnar et Tapponnier 1975, Tapponnier et Molnar 1976) a conduit Tapponnier à présenter (Montpellier et Split 1976) un modèle d'évolution générale de la Méditerranée depuis le Trias, faisant appel à un formalisme nouveau et dans lequel l'accent est mis sur trois processus :

- le "poinçonnement" de l'Europe par des promontoires du continent africain, ou l'"écrasement" de ceux-ci contre l'Europe.
- le "fluage latéral" de la lithosphère le long de grands décrochements; il provoque la formation d'arcs de subduction latéraux au niveau des bassins océaniques adjacents aux promontoires continentaux.
- la fracturation de la lithosphère avec émission de volcanisme de type "alcalin", due à des contraintes analogues à celles décrites comme "tension secondaire" dans les déformations des métaux. Les fractures qui apparaissent alors, et qui provoquent la formation de grabens, sont orientées suivant la direction de la contrainte horizontale maximum.

Dans ce modèle, l'évolution récente de la Méditerranée s'effectue seulement en fonction des deux grandes plaques européenne et africaine et de leur convergence suivant une direction N-S à NNW-SSE. En dehors des zones de subduction, cette convergence se traduit par une tectonique de collision avec des caractéristiques similaires à celles mises en évidence en Asie, et en particulier avec un rôle important tenu par les décrochements. De plus, le dernier des trois processus ci-dessus énumérés est particulièrement intéressant car il permet de saisir que la collision des plaques, peut se traduire non seulement par de la compression, mais aussi par de la distension suivant des failles orientées avec une direction voisine de celle du vecteur de déplacement global. Cela n'exclut pas que des phénomènes distensifs d'un autre ordre peuvent exister : ce sont ceux qui sont liés à l'ouverture de bassins marginaux, phénomènes eux-mêmes en liaison avec une subduction. Il faut toutefois bien préciser à propos des zones de subduction et des bassins marginaux, que la répartition générale des contraintes ne paraît pas obéir à des lois simples, de nombreux facteurs pouvant entrer en ligne de compte, dont nous ne citerons que la vitesse de subduction et l'inclinaison suivant laquelle elle s'effectue (cf. Mégard et Philip 1976); ceci permet d'ailleurs d'expliquer pourquoi dans ces zones se succèdent des phases de compression et de distension.

Si avec l'optique d'un tel modèle général, nous retournons maintenant vers l'interprétation des données recueillies, nous pouvons faire les remarques et les hypothèses suivantes, sur l'évolution récente de la Méditerranée occidentale.

Si nous prenons comme point de départ la période miocène, pour comprendre la signification de ce qui s'est passé ensuite en Méditerranée occidentale, nous constatons tout d'abord, qu'au Miocène les zones de subduction étaient beaucoup plus largement répandues qu'à l'heure actuelle. Ceci ressort de l'examen pour cette période des différents modèles explicatifs des chaînes alpines périméditerranéennes (cf. par exemple Fig. 18 de Dewey et al. 1973), de l'importance du volcanisme de type "calco-alcalin" (cf. Girod 1976) et de l'étendue des phénomènes distensifs à cette époque accompagnant la formation (pro-parte) de nouveaux bassins marginaux.

L'évolution ultérieure va correspondre au passage entre une période où la convergence entre les plaques européenne et africaine s'effectuait donc principalement par subduction, à une autre où la tectonique de collision devient prépondérante.

Le premier changement important par rapport à la période précédente, semble se produire au cours du Pliocène, avec le remplacement dans de nombreuses régions du volcanisme de type "calco-alcalin", encore important au Messinien, par un volcanisme de type "alcalin". Les données encore fragmentaires, recueillies dans l'Arc de Gibraltar, indiquent en même temps une extension E-W, donnant naissance à des failles N-S, et il est remarquable de constater qu'à la même époque dans de nombreuses régions (par ex.: volcanisme plio-quaternaire du Sud de la France de l'Escandorgue au Cap d'Agde, ou de Ciudad Real dans le Sud de la Meseta ibérique), le volcanisme de type "alcalin" est aligné suivant des fractures N-S.

L'arrêt des subductions miocènes, sauf dans l'Arc tyrrhénien,se traduit donc dans un premier temps par des phénomènes d'extension E-W, accompagnés par un volcanisme de type "alcalin", qui peut être interprété comme de la "tension secondaire", conséquence de la collision des plaques européenne et africaine.

Au cours du Quaternaire ancien, l'état des contraintes se modifie et la collision se traduit alors par des phénomènes principalement compressifs, dont l'apparition semble s'être fait sentir sur une grande étendue de la Méditerranée occidentale : on les connaît dans l'Arc de Gibraltar, mais également dans toute l'Afrique du Nord où ils ont été signalés depuis longtemps (Glangeaud 1927, 1955) et leur existence confirmée par nombre de travaux plus récents. On les retrouve, bien que de façon en général plus discrète, dans le Sud-Ouest de la France (Languedoc, Barrière et al. 1973), où l'on constate que le volcanisme s'arrête juste avant la phase compressive. De plus dans l'Arc tyrrhénien en Sicile sud-occidentale, et sans doute dans l'Apennin calabro-lucanien , cet épisode de compression quaternaire existe également. Dans cet arc, l'apparition brève de phénomènes compressifs, au milieu d'une tectonique plio-quaternaire en distension, paraît être comparable à ce qui s'observe précisément à la même époque dans l'Arc égéen voisin (cf. Mercier et al. 1976).

Pour la période du Quaternaire moyen et récent, la néotectonique est encore très mal connue en Méditerranée occidentale, et nous nous contenterons d'indiquer que dans l'Arc de Gibraltar, on voit réapparaître très localement au Maroc le volcanisme de type "alcalin" associé à une extension E-W, ce qui peut être interprété comme une nouvelle manifestation des phénomènes de "tension secondaire", alors que dans le reste de l'arc des déformations en compression sont localisées près des grands accidents.

En fonction de cette évolution récente de la Méditerranée occidentale et des connaissances séismiques actuelles nous pensons que la période présente doit donc être interprétée comme une période où la convergence N-S à NNW-SSE entre les plaques européenne et africaine se traduit essentiellement par une tectonique de collision se manifestant principalement par des déformations en compression. En conséquence, la séismicité est principalement associée aux jeux en décrochement (ou en faille inverse, suivant l'orientation par rapport aux contraintes) des grands accidents des domaines à croûte de type continental.

Des phénomènes distensifs sont encore présents dans les zones où se manifeste de la "tension secondaire", et dans l'Arc tyrrhénien et les régions qui lui sont associées. Mais, l'on remarquera que la subduction est peut-être en train de se terminer dans cet arc, car l'on constate une absence de séismes intermédiaires entre 50 et 200 km de profondeur et la présence d'un volcanisme "alcalin" dans le Sud de l'arc.

A ce cadre général de l'évolution récente de la Méditerranée occidentale, qui demanderait à être complété et en quelque sorte testé sur de nombreux points, nous ajouterons deux remarques complémentaires :

- l'évolution particulière de l'Arc tyrrhénien par rapport à l'ensemble Arc de Gibraltar-Afrique du Nord, ne peut se comprendre que si ces deux domaines sont séparés par une faille transformante. Ce rôle pourrait être tenu par la ligne de fracture NW-SE du détroit siculo-tunisien, qui aurait alors joué en zone transformante dextre.

- dans ces deux domaines, les phénomènes de distension sont suffisamment importants pour avoir provoqué un approfondissement notable de la Méditerranée après le Messinien, bien que chronologiquement on puisse constater qu'ils ne soient pas synchrones (Pliocène à Quaternaire ancien dans l'Arc de Gibraltar et l'Afrique du Nord, Pliocène supérieur et Quaternaire dans l'Arc tyrrhénien).

D'autre part nous voudrions ajouter une remarque d'ordre plus général en ce qui concerne le terme "néotectonique". Nous avons constamment utilisé ce terme comme synonyme de "tectonique de la période quaternaire". Il nous apparaît en effet difficile d'associer à ce terme une signification structurale (néotectonique = tectonique récente de distension) comme cela a été développé récemment. En Méditerranée, l'importance au Quaternaire des phénomènes compressifs a été sousestimée, au profit des phénomènes distensifs plus immédiatement visibles. Ces derniers par les dénivellations verticales qu'ils provoquent sont plus faciles à mettre en évidence tant sur le terrain que sur les profils géophysiques, alors que les mouvements de coulissage horizontal passent plus facilement inaperçus. La tectonique quaternaire peut être distensive ou compressive, ou même successivement l'une et l'autre. Si l'on examine par ailleurs son caractère de nouveauté, celui-ci tient essentiellement au fait qu'elle est la tectonique la plus proche de notre époque. Nous avons vu au contraire dans l'Arc de Gibraltar combien la néotectonique est guidée par les tectoniques antérieures, car la tectonique de compression quaternaire s'exerce sur un pays découpé par un réseau de fractures héritées d'une part de la période anté-tortonien et d'autre part de celle post-tortonien. L'héritage de la première période joue un rôle majeur : les accidents qui ont joué en décrochement lors de la dernière phase "alpine", sont réutilisés d'une manière identique et avec la même direction de raccourcissement lors de la phase quaternaire. De plus on peut constater que ces accidents ont au préalable rejoué en distension entre le Tortonien et le Quaternaire, et donc que les phénomènes distensifs eux-mêmes ne sont pas complètement indépendants des phénomènes qui les précèdent.

D'autre part dans l'optique du modèle général que nous avons choisi pour comprendre l'évolution récente de la Méditerranée occidentale, les différentes tectoniques de compression et de distension, n'apparaissent que comme des manifestations d'un même phénomène général, qui se déroule d'une façon continue suivant des modalités variables, qui sont les seules qu'il est important de définir. Aussi, pour toutes ces raisons, nous estimons qu'il faut garder au terme "néotectonique", sa seule signification chronologique.

REMERCIEMENTS.

L'auteur remercie vivement P. Tapponnier et H. Philip avec lesquels il a pu discuter de nombreuses fois des problèmes soulevés par l'évolution récente de la Méditerranée. Il remercie également M. Girod qui l'a fait profiter de sa connaissance du volcanisme récent en Méditerranée occidentale. Enfin, il associe à ce travail, tous les chercheurs ayant participé à l'étude néotectonique de l'Arc de Gibraltar et grâce auxquels de nombreux résultats nouveaux ont pu être acquis.

ANNEXE : LISTE DES CHERCHEURS, du Groupe de recherche néotectonique de l'Arc de Gibraltar 1972-1976.

ARMIJO R., Laboratoire de Dynamique de la Lithosphère, Université de Paris VII, 75221 Paris Cedex 05.

BENKHELIL J., GUIRAUD R., Laboratoire de Géologie, Faculté des Sciences Abidjan (Côte d'Ivoire).

BOUSQUET J.C., PHILIP H., Laboratoire de Géologie structurale, U.S.T.L. 34060 Montpellier-Cédex.

ESTEVEZ A., SANZ DE GALDEANO C., Laboratoire de Géologie structurale, Faculté des Sciences, Grenade (Espagne).

MONTENAT C., I.G.A.L., 21 ru d'Assas, 75270 Paris Cedex 06.

PAVILLON M.J., Laboratoire de Géologie appliquée, Université de Paris VI, 75230 Paris Cedex 05.

VIGUIER C., I.U.T., Génie Civil, Université de Bordeaux I, 33405 Talence.

RAMPNOUX J.P., COLETTA B., FUDRAL S., Laboratoire de Géologie, Centre Universitaire de Savoie, Chambéry.

GIGOUT M., CADET J.P., FOURNIGUET J., GUILLEMIN M., PIERRE G., Laboratoire de Géologie, Université d'Orléans, 45045 Orléans.

THOMAS G., Laboratoire de Géologie, Faculté des Sciences, 64016 Pau.

BIBLIOGRAPHIE

ARANA V. et VEGAS R., 1974.- Plate tectonics and volcanism in the Gibraltar arc. Tectonophysics 24, p. 197-212.

AUBOUIN J., 1971.- Réflexion sur la tectonique de faille plio-quaternaire. Geol. Rundschau, 60, 3, p. 833-848.

AUBOUIN J., 1973.- Paléotectonique, tectonique, tarditectonique et néotectonique en Méditerranée moyenne... C.R. Acad. Sc. Paris, t. 276, p. 457-460.

AUBOUIN J., 1974.- Des tectoniques superposées et de leur signification par rapport aux modèles géophysiques : l'exemple des Dinarides; paléotectonique, tectonique, tarditectonique, néotectonique. Bull. Soc. Géol. Fr., 7° s., t. 16, p. 426-460.

AUZENDE J.M., BONNIN J. et OLIVET J.L., 1973.- The origin of the Western Mediterranean basin. Bull. geol. Soc. London, vol. 129, p. 607-620.

AUZENDE J.M., OLIVET J.L. et PAUTOT G., 1972.- Balearic islands : southern prolongation. Init. Rep. Deep sea drilling project. XIII, p. 1441-1447.

BARRIERE J., BOUSQUET J.C. et TONI C., 1973.- Données nouvelles sur la néotectonique des Costières du Gard. C.R. Acad. Sc. Paris, t. 277, p. 285-288.

BELLON H., BORDET P., BOUSQUET J.C., MONTENAT C., 1976.- Principaux résultats d'une étude chronométrique du volcanisme néogène des Cordillères bétiques (Espagne méridionale). 4e Réunion Annuelle des Sciences de la Terre, Paris, p. 43.

BENKHELIL J.,1976.- Etude néotectonique de la terminaison occidentale des Cordillères bétiques (Espagne). Thèse 3e cycle, Nice.

BENKHELIL J.et GUIRAUD R., 1976.- A propos du style et de l'âge des déformations récentes de la terminaison occidentale des Cordillères bétiques. C.R. Ac. Sc. Paris, 282, pp. 1339-1342.

BIJU-DUVAL B., 1974.- Commentaires de la carte géologique et structurale des bassins tertiaires du domaine méditerranéen. Revue Inst. Fr. Pétrole, vol. 29, n° 5, p. 607-639.

BISCHKE R.C. 1974.- A model of convergent plate margins based on the recent tectonics of Schikoku. Japan. J. Geoph. Res. 79, 32, p. 4845-4857.

BOCCALETTI M. et GUAZZONE G, 1974.- Remnant arcs and marginal basins in the Cenozoïc development of the Mediterranean. Nature, 252, p. 18-21.

BOUSQUET J.C. et GUEREMY P., 1967 et 1968.- Quelques phénomènes de néotectonique dans l'Apennin calabro-lucanien et leurs conséquences géomorphologiques. Rev. Géogr. Phys. Géol. Dyn., 10 (3), p. 225-238 et 11 (2), p. 223-236.

BOUSQUET J.C., 1973.- La tectonique récente de l'Apennin calabro-lucanien dans son cadre géologique et géophysique. Géologica Romana, vol. XII, p. 1-104.

BOUSQUET J.C. et MONTENAT C., 1974.- Présence et décrochements NE - SW plio-quaternaires dans les Cordillères bétiques orientales (Espagne). Extension et signification. C.R. Ac. Sc. Paris, t. 278, p. 2617-2620.

BOUSQUET J.C. et PHILIP H., 1975.- Observations microtectoniques sur la compression Nord-Sud quaternaire des Cordillères bétiques orientales (Espagne méridionale, Arc de Gibraltar). Bull. Soc. Géol. de France, 1976, n° 3, p. 711-724.

BOUSQUET J.C., CADET J.P. et MONTENAT C., 1975.- Quelques observations sur le jeu quaternaire de l'accident de Carboneras (décrochement majeur sénestre NE-SW des Cordillères bétiques orientales). 3e Réunion Annuelle des Sciences dela Terre, Montpellier, p. 73.

BOUSQUET J.C., DUMAS B. et MONTENAT C., 1975.- L'accident de Palomares : décrochement sénestre du bassin de Vera (Cordillères bétiques orientales, Espagne). Cuad. Geol. Granada (sous presse).

BOUSQUET J.C., CADET J.P. et MONTENAT C., 1976.- Histoire néogène et quaternaire de l'accident de Carboneras, Cordillères bétiques orientales. Espagne (à paraître).

BOUSQUET J.C., MONTENAT C. et PHILIP H.,1976.- La evolution tectonica reciente de las Cordilleras beticas orientales. Reunion sobre la Geodinamica de la Cordillera betica y mar de Alboran, Mayo 1976, Granada (à paraître).

BOUSQUET J.C. et PHILIP H., 1976.- Observations tectoniques et microtectoniques sur la distension plio-pléistocène ancien dans l'Est des Cordillères bétiques (Espagne méridionale). Cuad. Geol. Granada (à paraître).

DEWEY J.F., PITMAN III W.C., RYAN W. et BONNIN J., 1973.- Plate tectonics and the evolution of the Alpine system. Geol. Soc. Amer. Bull., vol. 84, n° 10, p. 3137-3180.

DI NOCERA S., ORTOLANI F. et TORRE M., 1975.- Fase tettonica messiniana nell'Appennino meridionale. Boll. Soc. Nat. in Napoli, vol. 84, p. 1-17.

DUBOIS R., 1976.- La suture calabro-apennique crétacé-éocène et l'ouverture tyrrhénienne néogène. Thèse Paris, 567 p.

FONTBOTE J.M., y SANZ DE GALDEANO C., 1976.- Sobre las caracteristicas de la neotectonica de la region central de la Cordillera Betica. Reunion sobre la Geodinamica de la Cordillera Betica y mar de Alboran. Mayo 1976, Granada (à paraître).

FOURNIGUET J., 1975.- Néotectonique et Quaternaire marin sur le littoral de la Sierra Nevada, Andalousie (Espagne). Thèse 3e cycle, Univ. d'Orléans, 234 p.

ESTEVEZ A., LOPEZ-GARRIDO A., y SANZ DE GALDEANO C., 1976 - "Estudio de las deformaciones recientes en el sector del Negratin (Depresion Guadix-Baza". Reunion sobre la Geodinamica de la Cordillera Betica y mar de Alboran, Mayo 1976, Granada (à paraître).

FILIPPO (Di) B. et PERONACI F., 1959 - Natura fisica della scossa all'ipocentro di alcuni terremoti profundi nel Mediterraneo, Ann. Geof. p. 12-549.

GAUYAU F., BAYER R., BOUSQUET J.C., LACHAUD J.C., LESQUER A. et MONTENAT C., 1976 - La prolongation de l'accident d'Alhama de Murcia, entre Murcia et Alicante (Espagne méridionale). Résultats d'une étude géophysique. Bull. Soc. Geol. France (à paraître).

GIROD M. et GIROD N., 1976 - Contribution de la pétrologie à la connaissance de l'évolution de la Méditerranée occidentale depuis l'Oligocène. Bull. Soc. Geol. France (à paraître).

GLANGEAUD L., 1927 - Sur les plissements post-astiens dans le Nord de la province d'Alger. Bull. Soc. Geol. France, 4, t. 17, p. 239-246.

GLANGEAUD L., 1955 - Les déformations plioquaternaires de l'Afrique du Nord. Geol. Rundschau, 43, 1, p. 181-196.

GRANDJACQUET C., 1969 - Les phases tectoniques et le métamorphisme tertiaire de la Calabre du Nord et de la Campanie du Sud (Italie). C.R. Acad. Sc. Paris, 269, p. 1819-1822.

GROUPE DE RECHERCHE NEOTECTONIQUE DE L'ARC DE GIBRALTAR, 1976 - L'Histoire tectonique récente (Tortonien à Quaternaire) de l'Arc de Gibraltar et des bordures de la mer d'Alboran. Bull. Soc. Geol. Fr. (à paraître)

GUEREMY P., 1972 - La Calabre centrale et septentrionale. Guide d'excursion géomorphologique. Tr. Inst. Geog. Reims, n° 10, 128 p.

GUILLEMIN M. 1976 - Les formations néogènes et quaternaires des régions de Melilla-Nador et leurs déformations (Maroc nord-oriental). Thèse Orléans 220 p.

HATZFELD D., 1975 - Etude de sismicité dans la région de l'Arc de Gibraltar (à paraître).

HODGSON et COCK 1956 - Direction of faulting in the deep focus Spanish earthquake of the 1954. Tellus, 8, p. 321.

KAMPSHUUR W. et RONDEEL H.E., 1974 - The origin of the bètic orogen, southern Spain. Tectonophysics, 27, p. 39-56.

KARIG D.E., 1971 - Origin and development of marginal basins in the western Pacific. J. Geophys. Res., 76, p. 25-42.

KARNIK V., 1969-1971 - Seismicity of the European area (part. I et II). Reidel Edit. Dordrecht - Hollande.

LENTINI F., 1967 - Le successioni stratigrafiche plio-pleistoceniche sui due lati della dorsale Nocara-Colobraro (Matera). Atti Acc. Gioena Sc. Nat. Catania, S. 6, 18, 181-206.

LE PICHON X., 1968 - Sea-Flor spreading and continental drift. J. Geophys. Res., 73, p. 3661.

LE PICHON X., PAUTOT G. et WEILL J.P., 1972 - Openning of the Alboran sea. Nature V. 236, 67, p. 83-85.

MAC KENZIE D.P., 1970 - Plate tectonics of the Mediterranean region. Nature, V. 226, p. 230-243.

MAC KENZIE D.P., 1972 - Active tectonics of the Mediterranean region. Geophys. J.R. ast. Soc., 30, p. 109-185.

MASCLE G., 1973 - Etude géologique des Monts Sicani. Thèse Paris, 622 p.

MAUFFRET A., 1976 - Etude géodynamique de la Marge des Iles Baléares. Thèse Paris, 137 p.

MEGARD F. et PHILIP H., 1976 - Plio-quaternary tectono-magmatic zonation and plate tectonics zonation in the central Andes. Earth and Plan. Sc. Letters, 33, p. 231-237.

MERCIER J.L., BOUSQUET B., DELIBASSIS N., DRAKOPOULOS I., KERAUDREN B., LEMEILLE F. et SOREL D., 1972 - Déformations en compression dans le Quaternaire des rivages ioniens (Céphalonie, Grèce). Données néotectoniques et séismiques. C.R. Ac. Sc. Paris, 275,p. 2307-2310.

MERCIER J. et VERGELY P., 1973 - Comparison between deformation deduced from the analysis of recent faults and from focal mechanisms of Earth quakes. Tectonophysics, 19, p. 315-332.

MERCIER J., CAREY E., PHILIP H. et SOREL D., 1976 - La néotectonique plio-quaternaire de l'Arc égéen externe et de la mer Egée et ses relations avec la séismicité. Bull. Soc. Geol. Fr., 7, t. 18, p. 355-372.

MINSTER J.B., JORDAN T.H., MOLNAR P. et HAINES E., 1974 - Numerical modeling of instantaneous plate tectonics. Geophys. J.R. ast. Soc. 36, p. 541-576.

MOLNAR P. et TAPPONNIER P. - Cenozoïc tectonics of Asia : effects of continental collision. Science, 8 Aug. 1975, vol. 189, n° 4201, p. 419-426.

MONTENAT C., 1970 - Sur l'importance des mouvements orogéniques récents dans le Sud-Est de l'Espagne. C.R.Acad. Sc. Paris, 270, p. 3194-3197.

MONTENAT C., 1973a - Les formations néogènes et quaternaires du Levant espagnol. Thèse Sciences, Orsay, p. 1170.

MONTENAT C., 1973b - Le Miocène terminal des chaînes bétiques (Espagne méridionale) in "Messinian events in the Mediterranean". Konink Nedesl. Akad. Wetur. Amsterdam, p. 180-187.

MONTENAT C., 1975 - Tectonique et sédimentation pliocène dans les cordillères bétiques (Espagne méridionale). Giornale di Geologia (Bologna) XXXIX, 2, 469-480.

MONTENAT C. et BIZON C., 1976 - A propos de l'évolution géodynamique mio-pliocène en Méditerranée occidentale. L'exemple du bassin de Vera (Cordillères bétiques). C.R. Somm. S.G.F. 1, p. 15-16.

MOSTARDINI F., PIERI M. et PIRINI C., 1966 - Stratigrafia del foglio 212, Montalbano ionico. Boll. Serv. Geol. It., 87, p. 57-143.

MULDER C.J., 1973 - Tectonic framework and distribution of Miocene evaporites in the Mediterranean"in Messinian events in the Mediterranean," p. 44-59.

OGNIBEN L., 1955 - Le argille scagliose del Crotonese. Mem. Note Ist. Geol. Appl. Napoli, 6, 72 p.

OGNIBEN L., 1962 - Le argille scagliose ed i sedimenti mesiniani a sinistra del Trionto (Rossano, Cosenza). Geologica Romana, 1, p. 255-282.

OGNIBEN L., 1975 - Lithostratigraphic complexes and incidence for tectonic phases in Sicily and Calabria in structural model of Italy. Quad. Ric. Scientifica, 90, p. 365-408.

OLIVET J.L., PAUTOT G. et AUZENDE J.M., 1972 - Alboran sea. Init. Rep. Deep sea drilling proj. XIII, p. 1417.

OLIVET J.L., AUZENDE J.M. et BONNIN J., 1973 - Structure et évolution tectonique du bassin d'Alboran, Bull. Soc. Geol. Fr. (7), XV, p. 108-112.

PANNEKOEK A.J., 1969 - Uplift and subsidence in and around the western mediterranean since the Oligocene : a review. Verh. Kon. Ned. Geol. Mijnb. Gen. Vol. 26, 77 p.

PAUTOT G., AUZENDE J.M., OLIVET J.L., et MAUFFRET A., 1972 - Valencia basin. Init. Rep. Deep sea drilling proj., XIII, p. 1430-1447.

PAYO G., 1975 - Estructuras simicidad y tectonica del mar Mediterraneo. Ist. Geog. y Cadastral. Madrid 39 p.

PEGOARO O., 1972 - Application de la microtectonique à une étude de néotectonique : le Golfe Maliaque (Grèce centrale). Montpellier, 41 p.

PHILIP H., 1974 - Etude néotectonique des rivages égéens en Locride et en Eubée nord-occidentale (Grèce). Thèse, Montpellier. 78 p.

PHILIP H., 1976 - Un épisode de déformation en compression à la base du Quaternaire en Grèce centrale (Locride et Eubée nord-occidentale). Bull. Soc. Geol. France (7), t. XVIII, n° 2, p. 287-292 et coll. intern CNRS, Paris, n° 244, p. 91-96.

PHILIP H. et THOMAS G., 1976 - Détermination de la direction de raccourcissement de la phase de compression quaternaire en Oranie (Algérie). Rev. Geog. Phys. Geol. Dyn. (à paraître).

QUITZOW H.W., 1935 - Der Deckenbau des Kalabrischen Massivs und seiner Randgebiete Abh. Wiss. Gott. Math. Phys., 3, 13, p. 63-179.

RIOS J.M., 1975 - "El mar Mediterraneo occidental y sus costas Ibericas". Madrid ENADIMSA, 473 p.

RITSEMA A.R., 1970 - Notes on plate tectonics and arc movements in the mediterranean region. XIIe Ass. Gen. Comm. Seismologique Europ. Luxembourg 5 p.

RITSEMA A.R., 1975 - The contribution of the study of seismicity and earthquake mechanisme to the knowledge of Mediterranean geodynamic processes. Proc. of the nat. symp. Geodyn. Amsterdam, 12 p.

SANZ DE GALDEANO C. (sous presse). "Datos sobre las deformaciones neogenas y cuaternarias del sector del Padul (Granada). Reunion sobre le geodinamica de la Cordillera Betica y mar de Alboran. Mayo 1976, A Publ. Bol. Geol. y Min. Madrid.

SCANDONE P., D'ARGENIO et PESCATORE T., 1975 Structural pattern of the Campania-Lucania Apennines in structural model of Italy. Quad. Ric. Scientifica 90, p. 313 327.

SELLI R., 1962 - Le Quaternaire marin du versant adriatique ionien de la péninsule italienne. Quaternaria, 6, p. 391-413, Roma.

SIGNORINI R., 1942 - Cenni sulla formazione gesoso salifera del Bacino del Neto in Calabria. Rend. 2, Acc. Sci. fis. mat. nat. Soc. z. Napoli, 4, 12, p. 1-16.

SOREL D., 1976 - Etude néotectonique dans l'arc égéen externe occidental. Thèse Orsay, 196 p.

STANLEY D.J., GOT H., KENYON N.H., MONACO A. et WEILER Y., 1976 - Catalonian Eastern Betic and Balearic margins. Structural types and geologically recent foundering of the Western Medit. basin. Smithsonian contrib. Earth Sciences n° 20, 62 p.

TAPPONNIER P. et MOLNAR P., 1976 - Slipe-line field theory and large scale continental tectonics. Nature, vol. 264, p. 319-324.

TAPPONNIER P., 1976 - Evolution tectonique du système alpin en Méditerranée : poinçonnement et écrasement rigide-plastique. Bull. Soc. Geol. Fr. (à paraître).

THOMAS G., 1974 - La phase de compression pléistocène en Algérie nord-occidentale âge, premiers éléments cinématiques, relations avec les mouvements en distension. C.R. Acad. Sc. Paris, t. 279, p. 311-314.

THOMAS G., 1976 - Mise en évidence de décrochements dextres Est-Ouest d'âge quaternaire en Algérie nord-occidentale. C.R. Acad. Sc. Paris, t. 283, p. 893-896.

UDIAS A., 1972 - Plate tectonics and the Azore Gibraltar region. Nature, 237, p. 67-69.

VEZZANI L., 1975 - Lithostratigraphic complexes and evidence for tectonic phases in the Molise-Puglia-Lucania Apennines in structural model of Italy. Quad. Ric. Scientifica, 90, p. 329-363.

INTERNATIONAL SYMPOSIUM ON THE STRUCTURAL HISTORY OF THE MEDITERRANEAN BASINS. SPLIT (YUGOSLAVIA) 25-29 OCTOBER 1976. B. BIJU-DUVAL AND L. MONTADERT, Eds. EDITIONS TECHNIP, PARIS 1977, pp. 215-220

AN APPROACH TO THE STRUCTURE OF THE MEDITERRANEAN AREA : A SATELLITE PHOTOGEOLOGICAL STUDY

J. LETOUZEY, P. TREMOLIERES and B. BIJU-DUVAL [1]

Résumé L'étude des images du satellite Landsat-1 a permis de mettre en évidence de nombreux linéaments. Ceux-ci sont interprétés comme des décrochements récents liés à la collision continentale Afrique-Europe et à la fragmentation des masses intermédiaires.

The study of about 700 images taken by the Landsat-1 satellite has been used to draw a photogeological sketchmap of the Mediterranean realm. The interpretation of this imagery enables the major structural features of the margins of the Mediterranean deep basins to be outlined. A great many lineaments can be recognized; their trends and distribution in the different geological provinces are not haphazard (fig. 1) : a) Many of these lineaments cut through more recent tectonic units : they have to be put in relation to recent deformations; b) Nevertheless they are much denser in the areas where pre-Cenozoïc rocks outcrop than in the thick sedimentary basins. Thus they partly correspond to old (Paleozoïc or Mesozoïc) structures that have recently been reactivated, especially on the margins of the basins; c) They do not correspond to superficial structures (overthrusts, nappes) but probably to deep crustal structures. It is too early to conclude this study which needs additional work, *i.e.* comparison with detailed geological or geophysical maps, studies on the field, processing of satellite-recorded magnetic tapes. But we can emphasize the following points :

1. In spite of the difficulties of interpretation, the major systems of lineaments can be considered as wrench faults, and the importance of these horizontal shears is probably greater than generally admitted :

- some relationship exists with the morphology, *i.e.* the structure of the deep margins (Tyrrhenian, Siculo-Tunisian threshold, Valencia trough, etc.);
- the extension of structures from autochthonous to allochthonous areas raises the problem of the true size of the allochtonous domain under the tectonic pile in the different areas of the Alpine belt;
- especially the great arcuate systems of the Alpine belt in the Mediterranean seem to be related to such lineaments.

So we have to raise the following questions : what is the amplitude of horizontal displacement ? What are the ages of activity ? Are they responsible for sedimentary changes of facies ?

2) If we consider that the Mediterranean-Alpine area is in a state of continental collision since the Maestrichtian and if we agree with the idea of fragmentation of continents, the major lineaments can reflect the trace in the morphololögy of the major faults initiated by this fragmentation. Thus the maps which have been drawn can be used to help field studies; by their synthetic view they enable the major line of weakness to be situated.

(1) Institut Français du Pétrole, 1-4 avenue de Bois-Préau, 92506 Rueil-Malmaison, France

These lines express the result of different stresses that occurred in the geological history. The stress orientation is much more important to know than the orientation of lineaments. So further studies in the field are now necessary to understand the stress pattern which has caused the movement along these linear features. An example of such a field study made in southeast Turkey shows that a structural analysis based on the observation of minor features associated with the Landsat lineaments (shear planes, stylolites, associated folds, tensional joints) furnishes very useful data. It confirms, for example, that since the Maestrichtian, Anatolia and Arabia have been subjected to the same stresses. The stress orientation for different periods is in accordance with the general kinematics deduced from the study of Atlantic magnetic anomalies; but here the determination of the directions of shortening or lengthening and their chronology can be much more precise (fig. 2).

In conclusion, a photogeological sketchmap and field studies are complementary : onland work enable us to understand (and in some cases to predict for other areas) the structures associated with the major lineaments observed by satellite.

BIBLIOGRAPHY

BIJU-DUVAL (B.), LAMPEREIN (C.), RIVEREAU (J. C.), LOPEZ (N.).- (1976). Commentaires de l'esquisse photogéologique du domaine méditerranéen. *Rev. Inst. Franç. du Pétrole*, XXXI-3, pp. 365-400.

Succession des "phases tectoniques" - Essai de datation.

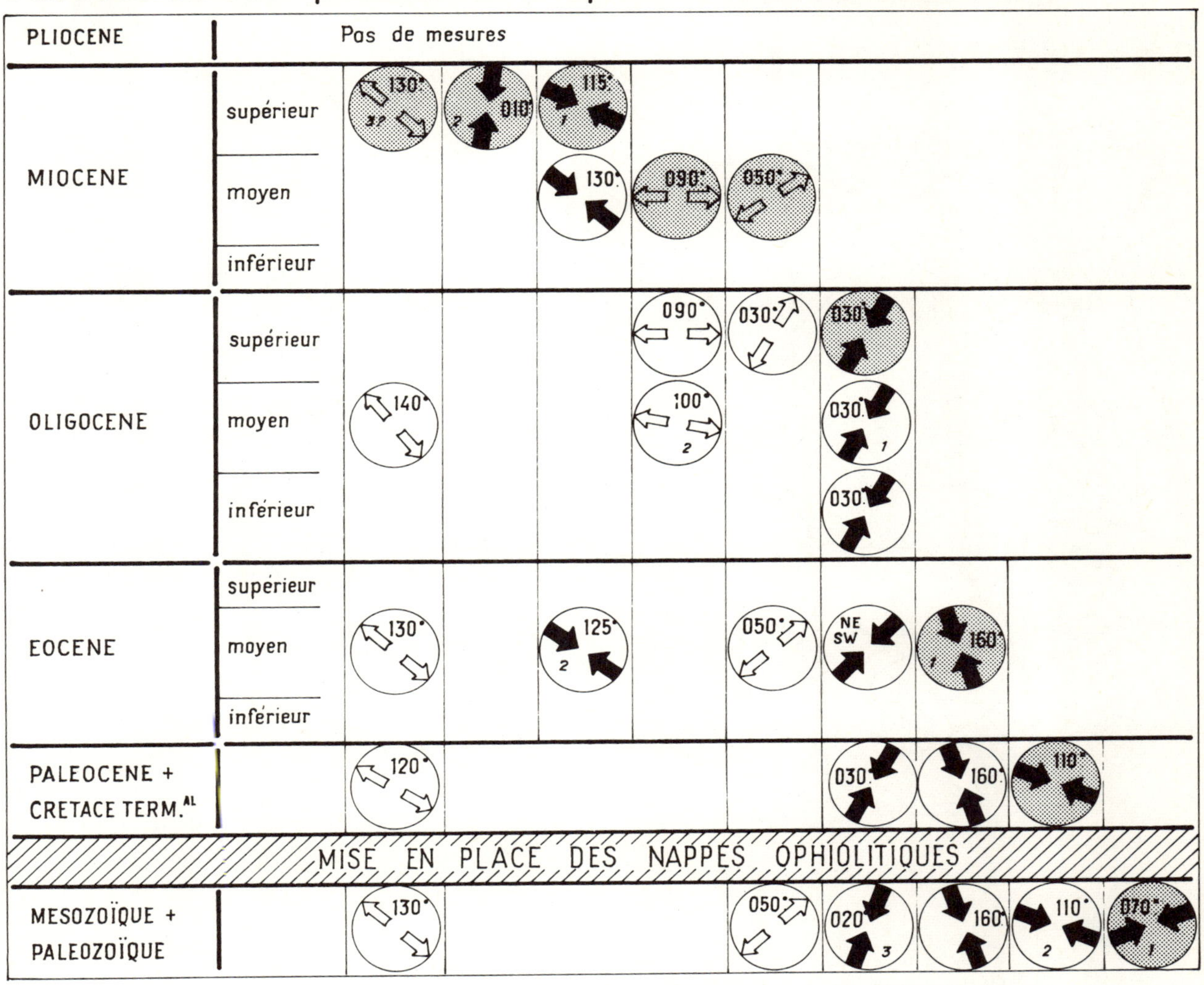

Fig. 2.- Various stress orientations for different periods, deduced from structural analysis on the field, southeast of Turkey (Arrows indicate shortening or lenghtening corresponding to measurements in various stratigraphic layers; shortening or lenghtening are thus younger than the layers they affect; for example shortening at 160° observed in Eocene beds is probably of Upper Eocene or Oligocene age).

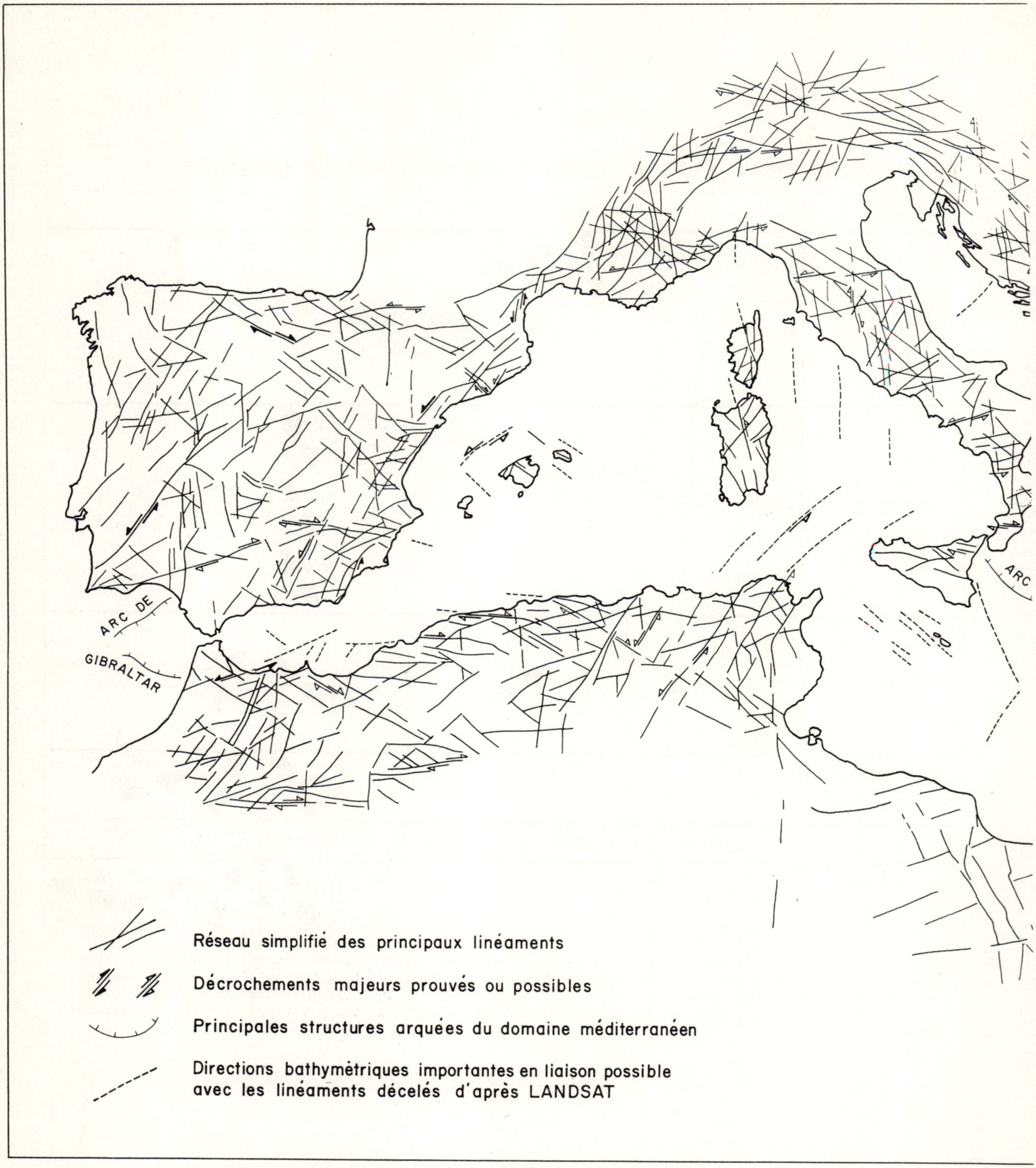

(From Biju-Duval et al. 1976, Rev. Inst. Fr. Pétrole)

Fig. 1.- Major lineaments of the Mediterranean

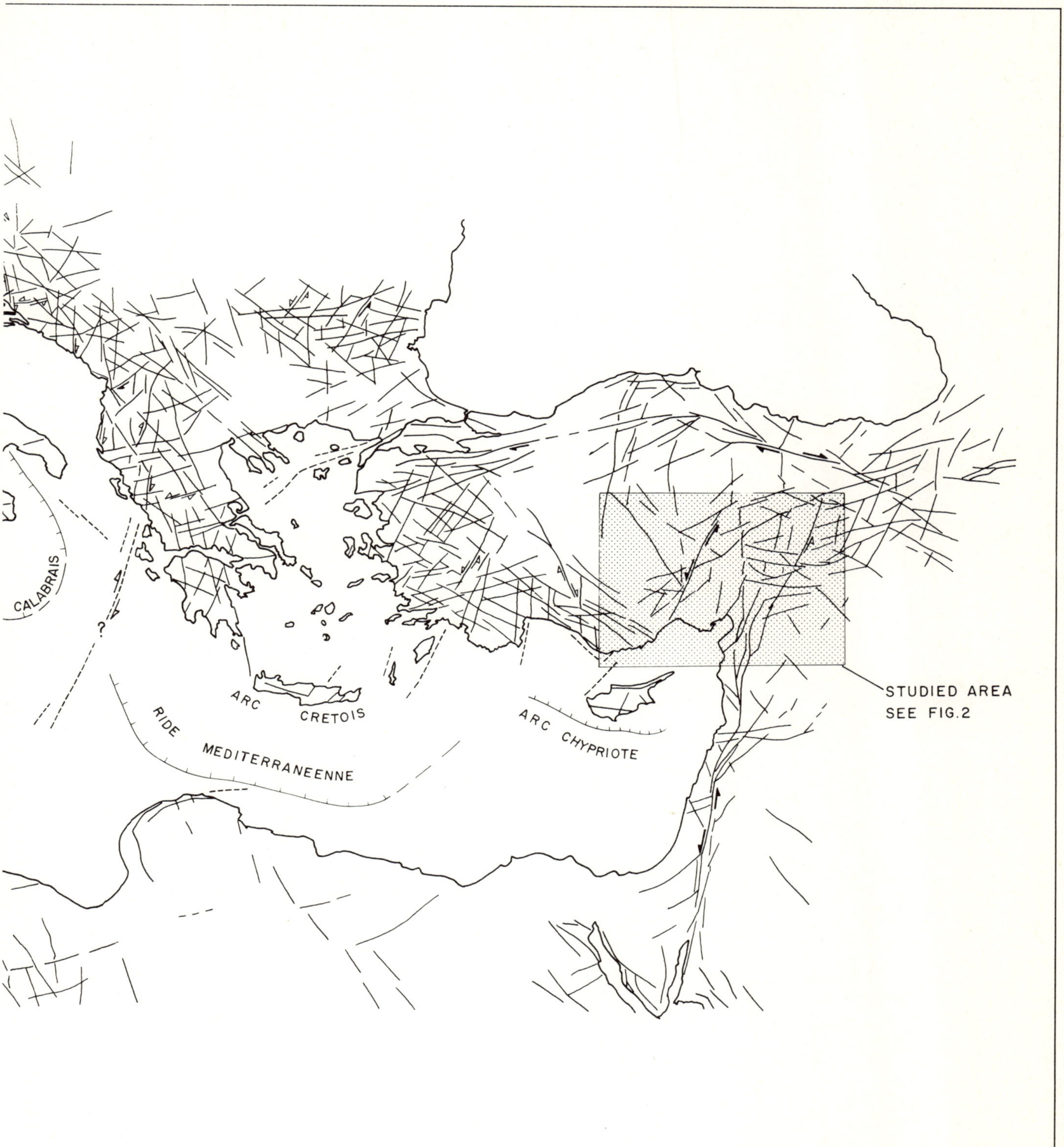

Figure 1

area (from BIJU-DUVAL et al., 1976).

INTERNATIONAL SYMPOSIUM ON THE STRUCTURAL HISTORY OF THE MEDITERRANEAN BASINS. SPLIT (YUGOSLAVIA) 25-29 OCTOBER 1976. B. BIJU-DUVAL AND L. MONTADERT, Eds. EDITIONS TECHNIP, PARIS 1977, pp. 221-232

STRUCTURAL LINEAMENTS AND TECTONICS OF THE MEDITERRANEAN BASIN

R.M. FOOSE [1]

I. INTRODUCTION

Since the launching of the original ERTS satellite in 1972 numerous investigators have demonstrated the value of studying the images for geologic interpretation. Most faults published on geologic maps are identifiable on the LANDSAT (ERTS) images (Isachsen, 1973). Many faults can be mapped for greater distances than shown on published maps (Short, N. and Lowman, P., 1973). Other structural features, not shown on published geological maps, are clearly identifiable within the same tectonic patterns (Goetz et al., 1973). Many other lineaments, not necessarily within the same tectonic pattern as shown on published geological maps, also are identifiable (Weidman, 1973).

However, where there has been careful field checking, it has been established that not all of the observed linear elements are caused by geological structures. Some are due to works of man, such as powerlines, property boundaries, highways, and reforested areas. Still others have not been identified as to their cause (Short, 1973. Isachsen, 1974). Furthermore, there is considerable variability in the number and length of lineaments identified on the same LANDSAT images by different observers, thus emphasizing the need for more objective analytical procedures (Podwysocki et al., 1975).

At the December, 1974 meetings of the C.I.E.S.M. in Monaco, Foose (1974) suggested that the lineament data extracted by observers from LANDSAT images might be classified according to length and shape, thereby providing the basis for interpreting the genetic character of the lineaments. This would enhance one's understanding of the structural style within an area and lead to an understanding of the origin of the regional tectonic framework. Specifically, he suggested that many, if not all, short straight lineaments would be caused by normal faults or prominent joint sets, these indicating tensional conditions in the crust. Longer, wavy-shaped lineaments would be created by erosion of the upper plate of thrust faults or of gravitationally detached crustal blocks, these indicating compressional conditions in the crust. Much longer, straight lineaments would be created by strike-slip (transcurrent) crustal movements, indicative of crustal shear. Even if these relationships were to be accurate for only half of the observed lineaments, this should prove to be a valuable tool in carrying out tectonic study of a large region, such as the Mediterranean Basin.

(1) Department of Geology, Amherst College, Amherst, Massachusetts 01002, U.S.A.

This paper presents some of the results of the extensive research that has been carried out since 1974 both on the analytical techniques for lineament observation, analysis, and interpretation and on the tectonic style and tectonic development of parts of the Mediterranean Basin.

I acknowledge the valuable contributions to this work of my student assistants, Robert H. Brigham, Richard E. Williams, Henry Shaw, and of other students who have worked on image-observation teams.* Many of the ideas and facts presented in this paper have been reviewed by staff members of the Goddard Space Flight Center, but responsibility for the interpretations belong solely to the author.

II. ANALYTICAL METHODS

A detailed description of the methods employed for image observation, lineament identification and mapping, and of the computer processing techniques for reducing the lineament data has been made by Foose and Brigham (1976) in the second conference on the New Basement Tectonics (Wilmington, Delaware, July 1976). The most important aspects are summarized here.

A. MORPHOTECTONIC UNITS

It is most important that all lineament data be studied and interpreted only within discrete morphotectonic provinces; that is, to analyze those lineaments within crustal units that have had a similar tectonic evolution and geologic history (i.e. Tyrrhenian Basin, Adriatic Basin, Aegean Basin, etc.). Within such crustal units the character and orientation of the lineaments generally are compatible throughout the entire unit. If not, the valid question may be asked: Has more than one morphotectonic unit been included? Thus, the analysis of lineament data in an area where very little prior geologic knowledge exists should prove to be a valuable tool in helping to define morphotectonic units and, indeed, the tectonic evolution of that area. The morphotectonic units for the northeastern Mediterranean Basin are shown in Figure 1.

B. LINEAMENT PARAMETERS

Three fundamental parameters of all lineaments have been observed and recorded: length, shape, and azimuth.

1. Lineament Length.

Based on the studies of Foose and Brigham (1976), the length of every lineament is further classified as short or long, with the critical length dividing them being 12 kilometers. Statistical analysis of thousands of measurements has shown that the surface expression on LANDSAT images of prominent joint sets and of normal faults produces lineaments with strike lengths less than 12 kilometers. Almost all low angle reverse or overthrust faults (including gravitational glide masses) have strike lengths greater than 12 kilometers. All strike-slip faults have strike lengths greatly exceeding 12 kilometers.

2. Lineament Shape.

The shape of each lineament is classified as straight or wavy. Most straight lineaments are created by joint sets, normal faults, or strike-slip faults and seldom include thrust (gravitational glide) faults. The statistical analyses of data by Foose and Brigham (1976) and the comparison by them of these data with published geologic maps support this conclusion.

3. Lineament Azimuth.

The azimuth of each lineament is measured as the bearing of the line that joins the end points of the lineament. All azimuth data are divided into 36 5-degree wide segments covering the 180-degree arc extending from East (N 90 E) to West (N 90 W).

C. METHODS OF ANALYSIS

Using the five parameters de-

*Tony Kuehn, Lawrence Garmezy, Michael Graham, Benny Loebner and Mark Heiligman.

scribed above (short and long lineaments, straight and wavy lineaments, and lineament azimuth), three methods of plotting the data have been employed:

- Frequency vs. Length histograms;
- Frequency vs. Azimuth histograms;
- Density (total lineament-kilometers) vs. Azimuth histograms.

1. Frequency vs. Length.

Frequency vs. Length histograms have been prepared for all lineaments within each morphotectonic unit to evaluate the nature of the correlation between lineament length and shape and also to establish the critical length (12 km.) which divides short from long lineaments. For each morphotectonic unit three Frequency vs. Length histograms have been plotted: one showing only straight lineaments, another showing only wavy lineaments, the third showing the combined total. These three histograms for Sicily are shown as an example on Figure 2. Note that the lefthand ordinate shows frequency (actual number) while the righthand ordinate shows the percent of the total number of all lineaments on Sicily. The abscissa shows individual lineament length in kilometers.

Four important facts have emerged from the analysis of these histograms from all morphotectonic units:

First, the mean kilometer length for straight lineaments is consistently and significantly smaller than for wavy lineaments from the same morphotectonic unit (Fig. 1).

Second, straight lineaments always are in the majority. This fact is clearly displayed by comparing Figure 2a with Figure 2b for the data from Sicily.

Third, the great majority of straight lineaments in all morphotectonic units are shorter than 12 kilometers.

Finally, as the mean lineament length for both total lineaments and wavy lineaments increases (see Table 1), the incidence of thrust faults (gravity glide) or strike-slip faults, or both, increases within that morphotectonic unit. For example, there are many thrust (gravity glide) faults in the Pindos Mountains and on Cyprus but few on Corsica and in Western Italy.

2. Frequency vs. Azimuth.

The obvious importance of lineament orientation (azimuth) to tectonic studies dictated that histogram plots of frequency to azimuth be constructed for each kind of lineament (i.e. short straight, long straight, short wavy, long wavy, etc.) and for each morphotectonic unit. The work of Foose and Brigham (1976) established several important facts that are summarized here:

- The sample size (number of actual measurements plotted as frequency) is very important. A sample of less than 50 measurements is poor and probably has little validity. A good sample should have more than 100 measurements. With increasing sample size the interpretation is easier to make and is more accurate.
- The number of lineaments within each 5-degree azimuth interval is counted but, in order to eliminate error, the histogram plot is "smoothed" by including half the number of lineaments in the two adjacent 5-degree intervals and averaging the total. Accurate interpretation of a histogram is much improved by making a "smoothed" plot.

3. Density vs. Azimuth.

Because the number of short lineaments always is greater than the number of long lineaments, regardless of the geologic and tectonic character of the morphotectonic unit, it is important to appropriately "weight" the longer lineaments. Therefore, histogram plots of lineament density (the total lineament kilometers) within each 5-degree azimuth interval were prepared. The work of Podwysocki (1974) and Foose and Brigham (1976) show that by this method the relative importance of the different lineament types (short straight, long straight, short wavy, and long wavy) is clearly revealed, whereas it is not shown by Frequency vs. Azimuth histograms. In Table 2 it may be seen that the lineament orientation N 5-30 W in the Pindos Mountains is comprised of nearly equal numbers of short straight and long wavy lineaments, as shown by the Frequency vs. Azimuth data. However, in the Density vs. Azimuth data it is shown that long wavy lineaments account for nearly two-thirds of the total lineament kilometers!

Reference to the published geologic details of the Pindos Mountains (Mercier, J., 1966; Brunn, J. and Mercier, J., 1971; Temple, P., 1968) fully confirms the more accurate depiction of the tectonic style by the Density vs. Azimuth data. The Pindos are dominated by a large number of gravity glide crustal masses transported southwesterly toward the Ionian Sea during the Miocene. Subsequent subsidence of the Adriatic, probably due to crustal loading from the gravity glide masses, produced a series of normal faults roughly parallel to the trend of the older, detached glide masses. Both fault types are identifiable in the Frequency and Density data, but it is the density plots that reflect the dominance of gravity glide tectonics. Furthermore, it confirms the hypothesis that long wavy lineaments are produced by low-angle reverse faulting and that short straight lineaments represent chiefly normal faulting.

Density vs. Azimuth data for the four key lineament types (short straight, long straight, short wavy, and long wavy) have been plotted and analyzed for each of the 14 morphotectonic units shown in Figure 1 and work on all remaining morphotectonic units of the entire Mediterranean Basin is now well advanced. Thus far, all lineament data from all units confirm the hypothesis discussed above, using the Pindos Mountains as an example.

III. LINEAMENT ANALYSIS WITHIN MORPHOTECTONIC UNITS

Analysis of the lineaments within a morphotectonic unit involves determination of:

- the azimuth of each preferred orientation <u>and</u> the width of the azimuth interval within which it lies;
- the relative percentage within each interval of the four lineament types (LW = long wavy; SW = short wavy; LS = long straight; SS = short straight). Greater than 25 percent is considered significant;
- the percentage of total lineament kilometers within each azimuth interval;
- the Dominance Factor.* This is a function of the percent of lineament kilometers related to the width of the azimuth interval in which they occur.

A. LINEAMENT ANALYSIS OF NORTHERN ITALY

As an example of the method described, Table 3 presents the summarized data for the morphotectonic province of northern Italy. Following is a brief analysis based both on the summarized data and the geologic work of other investigators.

Northern Italy exhibits a variety of tectonic styles. The first and most dominant is the gravity glide event with a northeast-southwest trend that was involved in the Dolomites (Van Bemmelen, 1966; Vink, 1968). This is reflected by the preferred orientation at N 55 E dominated by both long and short wavy lineaments. Subsequent subsidence of the Po River basin (Brinkmann, R., 1960; Rutten, M., 1969) is associated with the development of normal faulting in a <u>north-nort</u>heast - south-southwest

*From the work of Foose and Brigham (1976): "Given a perfectly uniform lineament distribution, each degree interval will contain 0.556 percent of the total. Multiplying the azimuth width of a preferred orientation by 0.556 yields a number whose units are also in percent. Subtracting this percentage from the <u>actual</u> percentage in a given preferred orientation compensates for differences in azimuth width. The resulting percentage is called simply the "Dominance Factor."

direction. Note the preferred orientation at N 27 E dominated by short lineaments. The third event was the development of the Insubric Zone (Van Bemmelen, 1966; Laubscher, 1971, Trumpy, 1975) characterized by east-west strike-slip faulting. This can be correlated with the orientation at N 88 W dominated by straight lineaments. The origin of the preferred orientation at N 15 W dominated by short straight lineaments is unknown.

IV. LINEAMENT ANALYSES OF MEDITERRANEAN BASINS

The same method is applicable to larger areas than single morphotectonic units although, obviously, complexities may ensue from processing and interpreting data within larger regions. Nevertheless, this is a significant tool for the geologist in terms of helping to understand tectonic details. Brief tectonic analysis are provided here, as examples, for the Tyrrhenian, Adriatic, and Aegean basins.

A. TYRRHENIAN BASIN

Lineament data from western Italy, Sicily, Sardinia, and Corsica have been combined in Table 4. These data show:

- short lineaments are in much greater abundance than long;
- straight lineaments are twice as abundant as wavy;
- the dominance factor of the four preferred orientations is small.

The obvious conclusions are that normal faults with relatively short strike-length are the dominant structures around the entire basin and that tensional conditions have prevailed in the crust during the time that these structures were formed.

The low value of the dominance factors for the four major tectonic orientations indicates that the Tyrrhenian Basin is a centrally stressed tectonic system with more or less equal distribution of strike directions of its structural features. A large number of radial and peripheral tensional faults within and around the edge of the Tyrrhenian Basin have produced many crustal blocks, many of them grabens with varying amounts of subsidence. With the knowledge that the Tyrrhenian Basin includes evaporites of Messinian age, crustal foundering must have been primarily in post-Miocene time. The deeper blocks have been subject to abundant sedimentary infilling since subsidence began.

One other significant conclusion may be made. The preferred orientation of N 55 E within the basin is manifested by thrust faults on Corsica but by normal faults on Sicily and in western Italy. This suggests that a pre-existing joint or fault system within the older crystalline crust (probably Hercynian) may have influenced or controlled late Tertiary tectonic activity.

B. AEGEAN BASIN

Lineament data from Crete, the Peloponessus, northern Greece, and western Turkey are combined in Table 5. These data show:

- long lineaments are slightly more numerous than short;
- straight lineaments are more abundant than wavy;
- the dominance factor of the five preferred orientations is small.

The larger number of long lineaments, many of them wavy, in the Aegean Basin is due to the presence of two extensions of the Alpine system into the region: one through the Peloponessus into Crete (with preferred orientations at N 17 W and N 90 W) and the other through northernmost Greece into the Pontic range in Turkey (with preferred orientation of N 70 W). These structures, which include many thrust faults as well as normal faults, are of early Tertiary age.

A second major tectonic event, that of subsidence of the Aegean Basin since the Pliocene, created a very large number of radial and peripheral normal faults that ring the basin and have broken it into many horsts and grabens. Once again, the low value of the dominance factors of all the preferred orientations indicates that the Aegean Basin also is a centrally stressed system dominated by crustal tension and geologically recent subsidence.

C. ADRIATIC BASIN

Lineament data from eastern Italy, northern Italy, the Dinaric Alps of Jugoslavia, and the Pindos Mountains of Albania and Greece are combined in Table 6. These data show:

- very strong preferred orientations at N 65 W and N 82 W, both with large dominance factors;
- a majority of short straight lineaments but a large number of thrusts.

Both of the preferred orientations are best seen in the Dinaric Alps. The N 82 W orientation is dominated along the northern Jugoslavian coast by gravity glide masses that have moved southwestward. In southern Jugoslavia and northern Greece the orientation is N 65 W but the earliest and dominant structures are also southwestward moving detached crustal masses (Mercier, 1966; Temple, 1968). The movement in a northwestward direction of similar crustal masses from the Appenines of central Italy has created a tectonic system in which the center of the Adriatic Basin is an axis of crustal foundering due to tectonic loading.

The tectonic loading and crustal foundering of the Adriatic Basin has occurred from early Miocene time to the present and has been accompanied by the progressive development of foredeeps in the subsiding basin with abundant normal faulting (thus accounting for the numerical majority of short straight lineaments).

Figures 3 and 4 show, respectively, the major lineaments of the Tyrrhenian-Adriatic basins and the Aegean basin as observed on LANDSAT images.

V. CONCLUSION

The Tyrrhenian and Aegean basins have had a similar tectonic evolution during the late Tertiary (Neogene), with the Aegean being more recent. Both basins exhibit dominantly tensional conditions in the crust with concurrent foundering. The lineament data derived from the LANDSAT images and analyzed by the methods described in this paper strongly support this conclusion. Furthermore, they provide the basis for a much more detailed structural analysis of both regions and for a better understanding of their tectonic evolution.

The Adriatic Basin has had an entirely different structural development and tectonic evolution. During the mid and late Tertiary the Adriatic crust has subsided in response to tectonic loading from gravitationally detached masses moving toward the basin axis from both Italy and from Jugoslavia. This tectonic evolution also is strongly suggested by the analysis of lineament data from that region.

The methods of lineament analysis described here provide a powerful tool for regional tectonic studies. Obviously it is possible to describe in much greater detail the structural relationships within any specific morphotectonic unit than has been provided in these short summary analyses. These have been provided to elucidate the method as well as to add to the understanding of the tectonic history of the individual basins. Work is now nearing completion on the entire Mediterranean Basin.

VI. REFERENCES

Bemmelen, R. W. van, 1966, Structural evolution of the southern Alps: Geol. Mijnb., v. 45, no. 12, pp. 405-444.

Brinkman, R., 1960, Geologic Evolution of Europe, Hafner Publ. Co., N. Y.

Brunn, J. and Mercier, J., 1971, Esquisse de la structure et de l'evolution geologique de la

Greece, in Tectonique de l'Afrique: UNESCO, pp. 103-111.

Foose, R. M., 1974; ERTS Satellite Photography as a tool for tectonic investigations in the Mediterranean Basin: XXIV Congres de C.I.E.S.M., Monaco, 3 pages.

Foose, R. M. and Brigham, R. H., 1976; Interpreting the genetic character of structural lineaments in the Mediterranean Basin: in press. Conference on New Basement Tectonics, Wilmington, Delaware, July 1976.

Goetz, A. F. H., Billingsley, F. C.; Elston, D.; Lucchitta, I., Shoemaker, E. M.; 1973, Preliminary geologic investigations in the Colorado plateau using enhanced ERTS images; in Symp. on Significant Results obtained from ERTS-1, Vol. I, paper G-19, pp. 403-411.

Isachsen, Y. W.; Fakundiny, R. H.; Forster, S. W.; 1973; Evaluation of ERTS-1 imagery for geological sensing over the diverse geological terranes of New York State; in Symps. on Significant Results obtained from ERTS-1, Vol. I, paper G-1, pp. 223-230.

Isachsen, Y. W.; Fakundiny, R. H.; Forster, S. W.; 1974; Assessment of ERTS-1 imagery as a tool for regional geological analysis in New York State: Geol. Survey - New York State Museum report to Goddard Space Flight Center, 52 pages.

Laubscher, H. P., 1971b; The large scale kinematics of the Western Alps and The Northern Appenines and its palinspastic implications: Am. Jour. Sci., v. 271, pp. 193-226.

Mercier, J., 1966; Paleogeographic, orogenese, metamorphisme et magmatisme des zones internes des Hellenides en Macedoine (Greece); vue d'ensemble: Soc. Geol. France, Bull. (7), 8, pp. 1020-1049.

Podwysocki, M. H.; 1974; An analysis of fracture trace patterns in areas of flat-lying sedimentary rocks for the detection of buried geologic structure; NASA - Goddard Space Flight Center Document X-923-74-200.

Podwysocki, M. H.; Moik, J. G.; Shoup, W. C.; 1975; Quantification of lineaments by manual and machine processing techniques: NASA - Goddard Space Flight Center Document X-923-75-183.

Rutten, M., 1969; Geology of Western Europe, Elsevier Publ. Co., N. Y. 520 pages.

Short, N. M.; Lowman, Jr. P. D.; 1973; Earth observations from space: outlook for the geologic sciences; NASA - Goddard Space Flight Center Document X-650-73-316.

Temple, P. G., 1968; Mechanics of large scale gravity gliding in the Greek Peloponessus: Geol. Soc. Amer. Bull., v. 79, pp. 687-700.

Trumpy, R., 1975; Penninic - Austroalpine boundary in the Swiss Alps: a presumed former continental margin and its problems: Am. Jour. Sci., v. 275-A, pp. 209-238.

Vink, B. W., 1968; Gravity tectonics in Eastern Cadore and Western Carnia, NE Italy, Geol. Ultraiectina, Geolog. Inst. Utrecht, No. 15, 64 pages.

Weidman, R.; in Short, N. M.: Lowman, P.D., 1973, page 14; Earth observations from space: outlook for the geological sciences; NASA - Goddard Space Flight Center Document X-650-73-316.

Table I. Summary of Mean Lineament Lengths for 14 Morphotectonic Units in the Northern and Eastern Mediterranean Region (in Kilometers).

Morphotectonic Unit	Total	Straight	Wavy
Sicily	6.26	5.48	9.43
Western Italy	5.92	5.45	6.49
Corsica	5.09	4.19	6.70
Sardinia	5.89	5.55	8.42
Northern Italy	6.67	6.28	7.19
Eastern Italy	9.34	8.37	10.59
Dinaric Alps	8.80	8.27	10.58
Pindos Mountains	10.58	7.29	14.98
Balkans	13.11	9.32	19.73
Crete	6.98	5.99	11.32
Aegean Basin	9.68	7.83	13.83
Turkey	10.66	9.60	12.53
Cyprus	10.21	8.05	15.17
Middle East	8.40	7.37	13.37
Averages	8.40	7.07	11.45

Table II. Percent of Lineament Types in the N 5-30 W Orientation in the Pindos Mountains

	FREQUENCY		DENSITY (Total Lineament Km.)	
	Number	Percent	Km.	Percent
Long Wavy	43	40	1124	63
Short Wavy	21	20	220	12
Long Straight	6	6	132	7
Short Straight	37	34	315	18
Total	100	100	1791	100

Table III. Lineament Analysis of Northern Italy

Total Lineament Kilometers: 2,738.8

Preferred Orientation	Degrees Width of Azimuth Interval	Lineament Type LW	SW	LS	SS	Percent of Total	Dominance Factor
N 55 E	30	38	27	10	25	31.5	14.8
N 27 E	25	9	37	12	42	24.8	10.9
N 88 W	15	00	16	48	36	9.2	0.9
N 15 W	20	9	17	15	59	9.2	-1.9

Table IV. Lineament Analysis of Tyrrhenian Basin
Total Lineament Kilometers: 14,210

Preferred Orientation	Degrees Width of Azimuth Interval	Lineament Type LW	SW	LS	SS	Percent of Total	Dominance Factor
N 88 W	15	14	36	12	38	13.0	4.7
N 55 E	20	21	23	7	49	15.0	3.9
N 60 W	20	6	34	11	49	12.4	1.3
N 12 W	15	21	23	6	50	6.5	-1.4

Table V. Lineament Analysis of the Aegean Basin
Total Lineament Kilometers: 20,637.7

Preferred Orientation	Degrees Width of Azimuth Interval	Lineament Type LW	SW	LS	SS	Percent of Total	Dominance Factor
N 90 W	20	41	9	28	22	16.2	5.1
N 70 W	20	31	12	29	28	15.4	4.3
N 25 E	20	17	9	25	49	11.1	0.0
N 17 W	15	51	11	18	20	6.4	-1.9
N 58 E	25	22	17	16	45	11.9	-2.0

Table VI. Lineament Analysis of the Adriatic Basin
Total Lineament Kilometers: 7,856.3

Preferred Orientation	Degrees Width of Azimuth Interval	Lineament Type LW	SW	LS	SS	Percent of Total	Dominance Factor
N 65 W	20	17	21	29	33	25.8	17.4
N 82 W	15	22	22	21	35	17.3	9.0
N 15 W	20	19	20	25	36	11.0	-0.1
N 28 E	15	28	20	9	43	7.4	-0.9

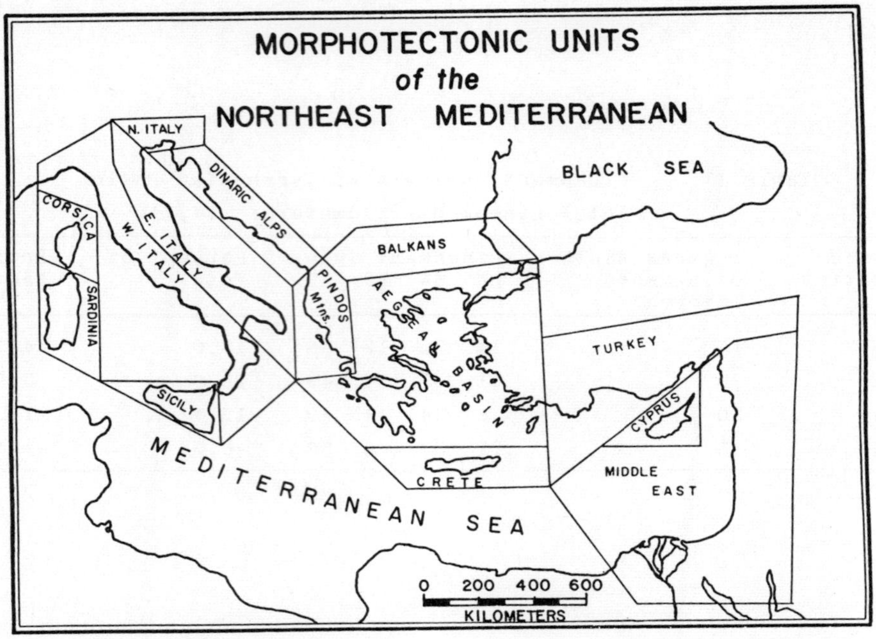

Figure 1. Morphotectonic Units of the Northeast Mediterranean.

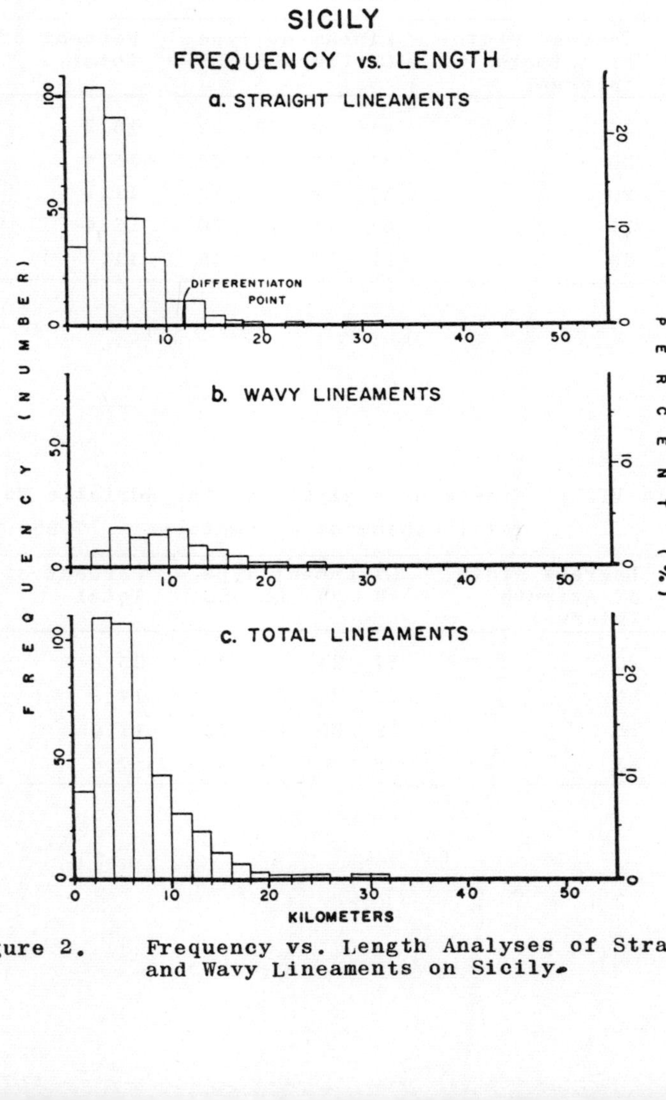

Figure 2. Frequency vs. Length Analyses of Straight and Wavy Lineaments on Sicily.

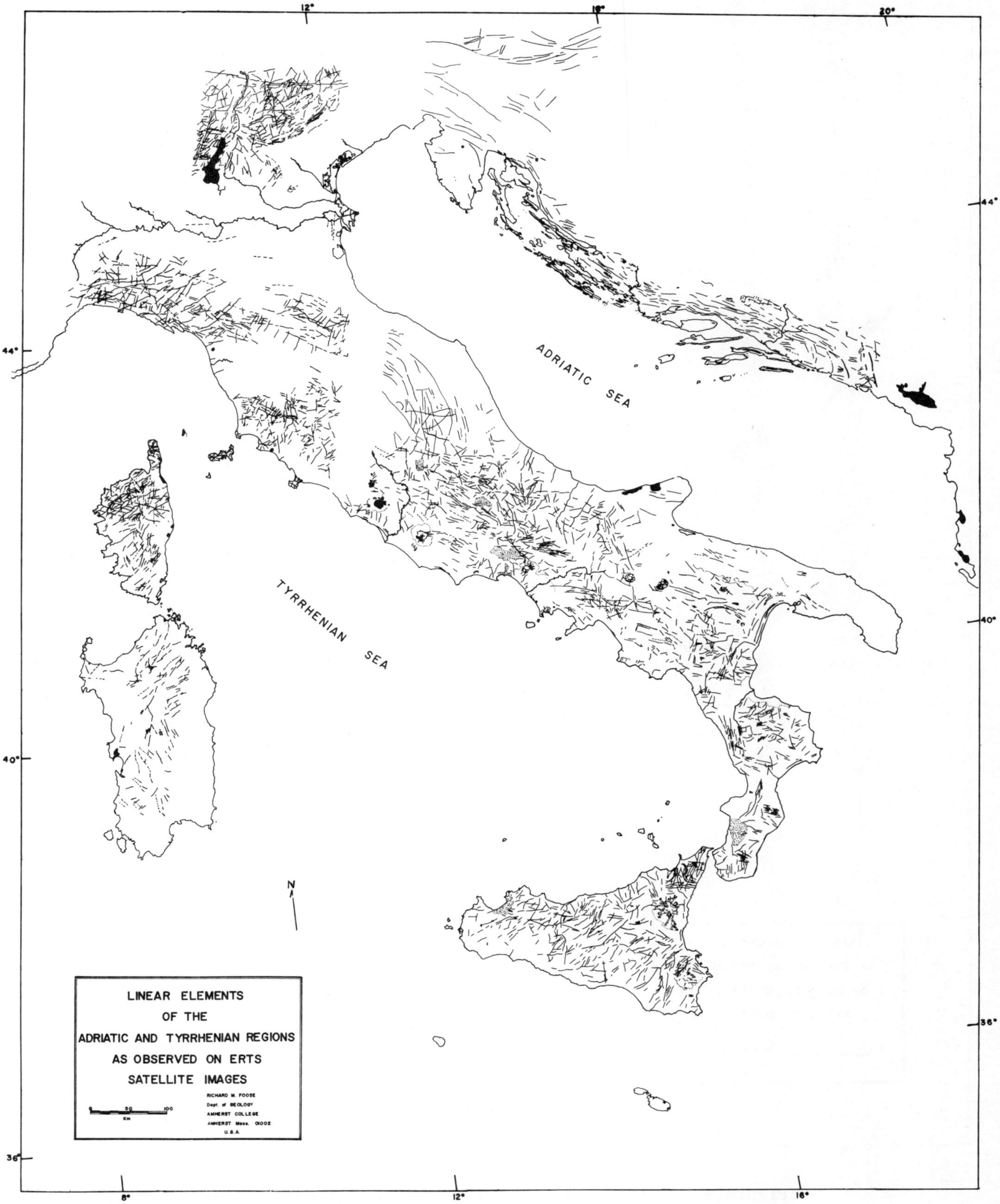

Figure 3. Structural Lineaments of the Tyrrhenian and Adriatic Basins.

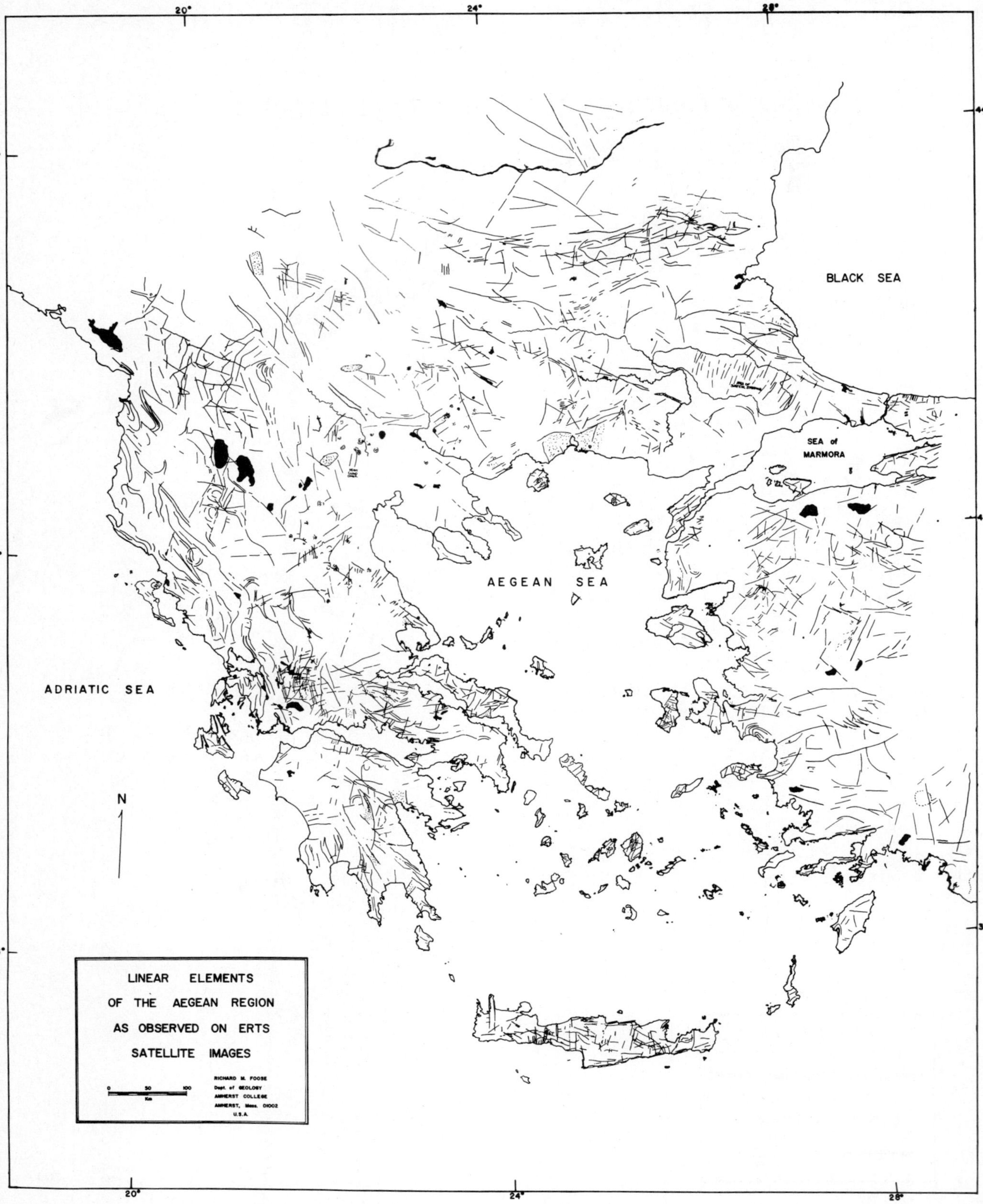

Figure 4. Structural Lineaments of the Aegean Basin.

INTERNATIONAL SYMPOSIUM ON THE STRUCTURAL HISTORY OF THE MEDITERRANEAN BASINS. SPLIT (YUGOSLAVIA) 25-29 OCTOBER 1976. B. BIJU-DUVAL AND L. MONTADERT. Eds. EDITIONS TECHNIP, PARIS 1977, pp. 233-240

YOUNG COMPRESSIONAL STRUCTURES OF THE CALABRIAN, HELLENIC AND CYPRUS OUTER RIDGES

N.H. KENYON and R.H. BELDERSON [1]

ABSTRACT

The small scale relief on the outer ridges of the East Mediterranean, mapped by long range side-scan sonar, is primarily compressional. Structures due to karstic conditions, tensional faulting, salt tectonics or olistostromes if present locally, are less important. Patterns of active folding and faulting are compared to structural patterns in foreland fold belts such as the Jura. Arguments are advanced for believing that emplacement of the suspected décollement sheets is by push from the rear, rather than by gravity gliding. Mantle diapirs (rather than plate subduction) are considered to be the cause of the compression, and the outer ridges to be miogeanticlines in the "European" rather than the "American" sense.

INTRODUCTION

There are three large swells on the floor of the East Mediterranean (rather than the single Mediterranean Ridge of other workers) that are formed in thick sedimentary rocks and Quaternary sediments. The three features (Fig. 1) being external and parallel to the active Calabrian, Hellenic and Cyprus arcs, are called the Calabrian, Hellenic and Cyprus Outer Ridges (Stride et al, 1977). The largest of them is the Hellenic Outer Ridge, 1300 km long and up to 2 km high, which lies to the south of a complex of deep trenches. At its western end it faces part of the smaller Calabrian Outer Ridge (Belderson et al, 1974) across a narrow trough (believed to be a thrust-bounded ramp valley). At its eastern end the Hellenic Outer Ridge appears to merge into the low, narrow Cyprus Outer Ridge (which is external to the Florence Rise and Cyprus block). A relatively small scale relief, up to 300 m high, previously called "cobblestone" topography, is characteristic of the outer ridges.

Seismic profiling results (reviewed by Biju-Duval et al, 1974) have been of only limited value in increasing our understanding of this extremely broken ground, whereas sonographs (plan views of the sea floor) obtained from R. R. S. Discovery by the long range side-scan sonar system, GLORIA, have enabled the relatively small scale surface topography to be described in detail as structural trends, and have greatly helped in choosing between some of the proposed origins. The present paper briefly summarises these side-scan results from the outer ridges and takes discussion of their origin further, particularly in the light of well documented structures from the Jura and other foreland fold belts.

PROPOSED ORIGINS OF THE OUTER RIDGES AND THEIR SURFACE RELIEF

One of the earlier views was that the Hellenic Outer Ridge is a currently active compressional ridge (analogous to earlier miogeanticlines of the Hellenides), and the site of tectonic thickening

(1) Institute of Oceanographic Sciences, Wormley, Godalming GU8 5UB, Surrey, Great Britain

(Ryan et al, 1971). However, interpretations of the first published deep seismic reflection profiles (Sancho et al, 1973) included one, extending south -wards from Crete, which appeared to show large normal faults bounding an "Upper Plateau", and a "Lower Plateau" underlain by disturbed sediments, thus implying a recent tensional history in that area. Furthermore a seismic refraction profile across the Ionian Sea, from Sicily to western Greece, showed little increase in the thickness of sedimentary layers (Hinz, 1974) required by the model of Ryan et al (1971).

The earlier proposals for the origin of the relatively small scale surface topography were related to the formation of the Hellenic Outer Ridge itself. Ryan et al (1971) believed it to be due to compressional deformation with some tensional deformation possible on the upbent surface of the ridge. Hieke et al (1973) emphasised possible tensional structures. More recently three origins have been proposed for the small scale relief which are to some extent independent of the origin of the outer ridges themselves:-

1. A karstic origin was proposed (Ryan, 1973) when deep drilling revealed the presence of evaporites of Upper Miocene age.
2. Salt tectonics was invoked as a possible origin for some of the folded structures on the southern part of the outer ridges (Biju-Duval, et al, 1974).
3. Gravitational gliding (Mulder, 1973; Biju-Duval et al, 1974) was credited as causing a mass of small allochthonous blankets (olistostromes, such as those described from Sicily and the Appenines) to move out from the uplifted Tyrrhenian and Aegean regions, followed by a collapse of the source areas to form the Tyrrhenian and Aegean basins and the Hellenic trough (a complex of trenches).

In practice the patterns on the sonographs bear little resemblance to what one would expect for karstic topography. For instance there are few signs of circular structures that might be interpretable as dolines, nor do the known trends follow expected drainage directions as they should if they were due to collapsed caverns. Circular features believed to be associated with solution of evaporites or piercement structures were seen by us on the Florence Rise, but not elsewhere. Neither the ridge-transverse trends nor the gross distribution of relief patterns fit in with an origin by karst or salt piercement.

An olistostrome origin for the superficial structures of the three outer ridges seems to be unlikely. As envisaged by Mulder et al.(1975) in discussion) there would be many separate, local and relatively small scale gravity slides. By comparison with examples studied on land these should have a chaotic internal structure. In practice DSDP Sites 125 and 377 on the Hellenic Outer Ridge showed a normal succession of marine faunas in both Plio-Quaternary and Middle Miocene deposits. Furthermore, sonographs show that orderly arc-parallel structures are present on most of the three outer ridges. The portion of the Hellenic Outer Ridge with the most confused pattern of surface relief (the "Upper Plateau", Fig. 2D) is better explained as due to a greater intensity of compressional deformation, related to its special situation where the Hellenic Outer Ridge is most constrained by the proximity of North Africa.

COMPRESSIONAL ORIGIN OF THE SURFACE RELIEF

A. Structural Patterns

Four patterns have been identified (Stride et al, 1977) on sonographs (Fig. 2), a common element of all the patterns being a ridge-parallel trend (Fig. 3). In addition, a transverse trend is found on the inner flank of the Hellenic Outer Ridge and on the western end of the Cyprus Outer Ridge. The greatest complexity of trends is on the highest part of the Hellenic Outer Ridge between Crete and Cyrenaica, in the region called the "Upper Plateau" by Sancho et al (1973). The four patterns are believed to be indicative of different degrees of deformation and this deformation is believed to be essentially compressional, with the intensity of deformation generally increasing towards the inside of the outer ridges but with the maximum intensity being found in the area of the "Upper Plateau". Simple folds are progressively destroyed by faulting and slumping parallel to the fold axes. The resulting faulted folds are further affected by strike-slip faulting of small displacement as well as suspected thrusting. The deformation may thus proceed through several stages. The transverse trend, for instance, may be younger than the ridge parallel trend. Comparable, perhaps, are the lineations seen on satellite images of Central Sicily, which are ascribed to very recent faults transverse to the main overthrusts (Cardamone et al, 1976).

B. Comparison with the Jura

An explanation for the structural patterns may be found by analogy with the external domains of other orogenic belts. Various structures are described from seismic reflection profiles across submarine outer boundaries of modern orogenic belts. However, as our data is of structures seen in plan view, a more useful comparison is with suitable regions that have been well studied by surface mapping on land. The similarity between the Jura fold belt and the outer ridges of the East Mediterranean was first proposed by Giermann (1966) and supported in more detail by Stride et al (1977).

The Jura has the same arcuate shape (Fig. 4) as the outer ridges of the East Mediterranean and is similar in size to the Calabrian Outer Ridge. Their respective forelands are continental and crossed by major graben systems (the Rhine-Bresse Graben in Europe, the Hon Graben of Libya and the grabens in the Strait of Sicily). The Lower Plateau, south of Crete, has been compared to an unfolded (tabular) area of the Jura (Stride et al, 1977), and the Hellenic Trough Complex may be in the same structural position with respect to the Hellenic Arc as the Molasse basin is to the Helvetic Alps. The boundaries of deformation on the outer margins of both the Jura and the outer ridges of the East Mediterranean are sharply defined compared to their inner margins. There are long foreland folds on both outer margins, which typically have wavelengths of 100 m to 1000 m and follow the arcuate shape of their respective fold belts. Thrusting accompanies the Jura folding, particularly in the form of large hidden outward directed overthrusts within the anticlines. Thrusting is suspected beneath the outer ridges of the East Mediterranean (Finetti, 1976; Stride et al, 1977; Mulder et al, 1975). Wrench faults in the Jura trend at a high angle to the fold arcs; larger wrench faults are accompanied by "countless smaller wrench faults" (Laubscher, 1972) and these latter may have their counterpart in the crossfault pattern found on the inner parts of the Hellenic and Cyprus Outer Ridges (Fig. 3). It is noteworthy that the main area of wrench faults in the Jura is limited to their inner part (Fig. 4). The sigmoidal curves or "pincées" in the bundles of folds in the Jura (Caire, 1975) may be matched by bends in the strike of the East Mediterranean folds (Fig. 2, C).

C. Décollement

If, as we believe for the reasons given above, the structures on the outer ridges of the East Mediterranean are frontal folds and faults of a still active orogenic belt then we should consider a choice between the mechanisms proposed for The emplacement of such structures. The main models that appear to be in vogue (Milici, 1975) are:

- movement of the "basement" (thick-skinned tectonics e. g. Caire, 1975).
- detachment from the "basement" at a regional décollement (thin-skinned tectonics). Deformation may be by compression alone (e. g. Laubscher, 1972), or by gravity gliding (e. g. Pierce, 1966).

It is noteworthy that there appears to be no consensus as to which model is best suited for the most studied land examples, such as the Western Alps and the Appalachians. Thus, perhaps no consensus should be expected for the submarine examples under consideration.

A model incorporating décollement has been proposed for the East Mediterranean Outer Ridges and the Barbados Outer Ridge by Belderson and Kenyon (1977) when comparing such ridges to the outer basement highs found in oceanic island arc systems. The interpretations of supposed tensional faulting within the outer basement highs can thus be reconciled with interpretations of compression in overlying sediments. Décollement is also proposed on the inner flanks of certain continental basement highs external to the major thrust belts. These include the triangular high beneath the Jura, lying between the Molasse Basin and the Rhine-Bresse Grabens (Laubscher, 1972) and the Kahta-Kastel Geanticline beneath the foothills structure belt of south-east Turkey (Rigo de Righi and Cortesini, 1964).

It is worth seeking out criteria relevant for testing the emplacement mechanisms proposed in thin-skin tectonics models. Those criteria related to the areal distribution of structures should be the most useful, as that is what we have attempted to map at sea. They should be applied with caution as none appear conclusive (e.g. Lemoine, 1973). However, pointers are found which favour a 'push' model:

1. Deformation by push should be greatest near the internal margin, the source area for the push. The push model invoked by Laubscher (1972) for the Jura (Fig. 4) can be roughly matched with that of the outer ridges of the East Mediterranean (Fig. 3). On the other hand, a foreland fold belt deformed by gravity sliding could be expected to be more compressed near the external margin, the "amortissement frontal" of Fallot and Faure-Muret (1949).

2. A further test used by Milici (1975) for the decollement sheets of the southern Appalachians is that deformation by push should start near the internal margin and progress outwards, whereas deformation by gravity sliding should start near the external margin and progress inwards. On the outer ridges of the East Mediterranean the youngest folding appears to be on the outer margin, from study of uplifted turbidites and from consideration of the way the Hellenic Outer Ridge is wrapped around the foot of the protruding African continental slope.

3. The youthfulness of some of the deformation is not in accord with gravity gliding. The need for a slope means that, for a source in the Hellenic Arc, any gravity gliding ought to predate collapse of the intervening Hellenic Trough Complex and the Cretan Trough.

DISCUSSION

The outer ridges are considered to be miogeanticlines, flanked by miogeosynclines and eugeosynclines, in the sense used by Aubouin (1965). These owe their origin, and develop within, an active orogenic setting, and must not be confused with the mio-eugeosynclines of Drake et al (1959) or the mio-eugeoclines of Dietz and Holden (1974), which develop on a passive continental margin and are only tectonised by a reversal in the direction of sea-floor spreading. Aubouin (1965) has already pointed out that the source area of the terrigenous material in the latter models lies in precisely the opposite direction to that observed in geosynclinal chains, and that there is a fundamental difference in their polarity, which is exactly the reverse of that found in orogenic arc systems.

It is now fashionable to explain the numerous belts of compressional deformation found throughout the Mediterranean region in terms of plate "consumption" at "subduction zones". In contrast to such an hypothesis, the driving force favoured by the authors is for each outward moving orogenic wave to be due to the outward spreading summit of a mantle diapir centred to the rear of the orogenic arc. Plate subduction, as currently envisaged in a variety of complicated and often bewildering scenarios involving numerous small "plates", is not favoured by us for explaining orogeny in and around the Mediterranean because, among other reasons, the arc systems face in so many different (and often directly opposing) directions. The plate gymnastics needed to fit them all into a satisfactory scheme seem to us to lack credibility.

ACKNOWLEDGEMENTS

The authors are grateful to many colleagues for their work at sea during two cruises of R. R. S. Discovery to the Eastern Mediterranean. Mr. C. D. Pelton assisted in the preparation of figures. Professor R. W. van Bemmelen is thanked for helpful comment. The manuscript was reviewed by Dr. A. H. Stride, who suggested several improvements.

REFERENCES

AUBOUIN, J., 1965. Geosynclines. Amsterdam: Elsevier Publishing Co.

BELDERSON, R. H., and KENYON, N. H., 1977. Outer ridges of orogenic arc systems. Deep-Sea Res.

BELDERSON, R. H., KENYON, N. H., and STRIDE, A. H., 1974. Calabrian Ridge, a newly discovered branch of the Mediterranean Ridge. Nature, 247, 453-454.

BIJU-DUVAL, B., LETOUZEY, J., MONTADERT, L., COURRIER, P., MUGNIOT, J. F., and SANCHO, J., 1974. Geology of the Mediterranean Sea basins. In The Geology of Continental Margins (ed. C. A. Burke and C. L. Drake), 695-721.

CAIRE, A., 1975. Les règles de la fracturation continentale et le rôle des géofractures dans l'évolution de l'écorce terrestre. Rev. Géog. Phys. et Géol. Dyn. 17, 319-354.

CARDAMONE, P., CASNEDI, R., CASSINIS, G., CASSINIS, R., MARCOLONGO, B., and TONELLI A. M., 1976. Study of regional linears in Central Sicily by satellite imagery. Tectonophysics, 33, 81-96.

DIETZ, R. S., and HOLDEN, J. C., 1974. Collapsing continental rises: actualistic concept of geosynclines - a review. In Modern and Ancient Geosynclinal Sedimentation (ed. R. H. Dott and R. H. Shaver), 14-25.

DRAKE, C., EWING, M., and SUTTON, G., 1959. Continental margins and geosynclines: the east coast of North America, north of Cape Hatteras. Physics and Chemistry of Earth, 3, 110-198.

FALLOT, P., and FAURE-MURET, A., 1949. Sur l'extension du décollement de la série de ouverture subalpine. Comptes Rendus Acad. Sci. Paris. 228, 616-619.

FINETTI, I., 1976. Mediterranean Ridge: a young submerged chain associated with the Hellenic Arc. Boll. Geofis. Teor. Appl. 19, 31-65.

GIERMANN, G., 1966. Gedanken zur Ostmediterranen Schwelle. Bull. Inst. Océanogr. Monaco. 66, 1-16.

HIEKE, W., SIGL, W., and FABRICIUS, F., 1973. Morphological and structural aspects of the Mediterranean Ridge SW of the Peloponnesus (Ionian Sea) Bull. Geol. Soc. Greece. 10, 109-126.

HINZ, K., 1974. Results of seismic refraction and seismic reflection measurements in the Ionian Sea. Geol. Jb. Reihe E2. 33-65.

LAUBSCHER, H. P., 1972. Some overall aspects of Jura dynamics. Am. J. Sci. 272, 293-304.

LEMOINE, M., 1973. About gravity gliding tectonics in the Western Alps. In Gravity and tectonics (eds. K. A. de Jong and R. Scholten). New York, John Wiley, pp. 201-216.

MILICI, R. C., 1975. Structural patterns in the southern Appalachians: evidence for a gravity slide mechanism for Alleghanian deformation. Bull. Geol. Soc. Am. 86, 1316-1320.

MULDER, C. J., 1973. Tectonic framework and distribution of Miocene evaporites in the eastern Mediterranean. In Messinian events in the Mediterranean (ed. C. W. Drooger), Amsterdam, North-Holland, 44-59.

MULDER, C. J., LEHNER, P., and ALLEN, D. C. K., 1975. Structural evolution of the Neogene saltbasins in the eastern Mediterranean. Rapp. Comm. int. Mer. Médit. 23, 141-146.

PIERCE, W. G., 1966. Jura tectonics as a décollement. Bull. Geol. Soc. Am. 77, 1265-1276.

RIGO DE RIGHI, M., and CORTESINI, A., 1964. Gravity tectonics in Foothills Structure Belt of south-east Turkey. Bull. Am. Ass. Petrol. Geol. 48, 1911-1937.

RYAN, W. B. F., 1973. Geodynamic implications of the Messinian crisis of salinity. In Messinian events in the Mediterranean (ed. C. W. Drooger), Amsterdam, North-Holland, pp. 26-38.

RYAN, W. B. F., STANLEY, D. J., HERSEY, J. B., FAHLQUIST, D. A., and ALLAN, T. D., 1971. The tectonics and geology of the Mediterranean Sea. In The Sea, Vol. 4 (ed. A. E. Maxwell), New York, John Wiley, 387-492.

SANCHO, J., LETOUZEY, J., BIJU-DUVAL, B., COURRIER, P., MONTADERT, L., and WINNOCK, E., 1973. New data on the structures of the eastern Mediterranean basin from seismic reflection. Earth Planet. Sci. Lett. 18, 189-204.

STRIDE, A. H., BELDERSON, R. H., and KENYON, N. H., 1977. Evolving miogeanticlines of the east Mediterranean (Hellenic, Calabrian and Cyprus Outer Ridges). Phil. Trans. R. Soc. Lond. (A).

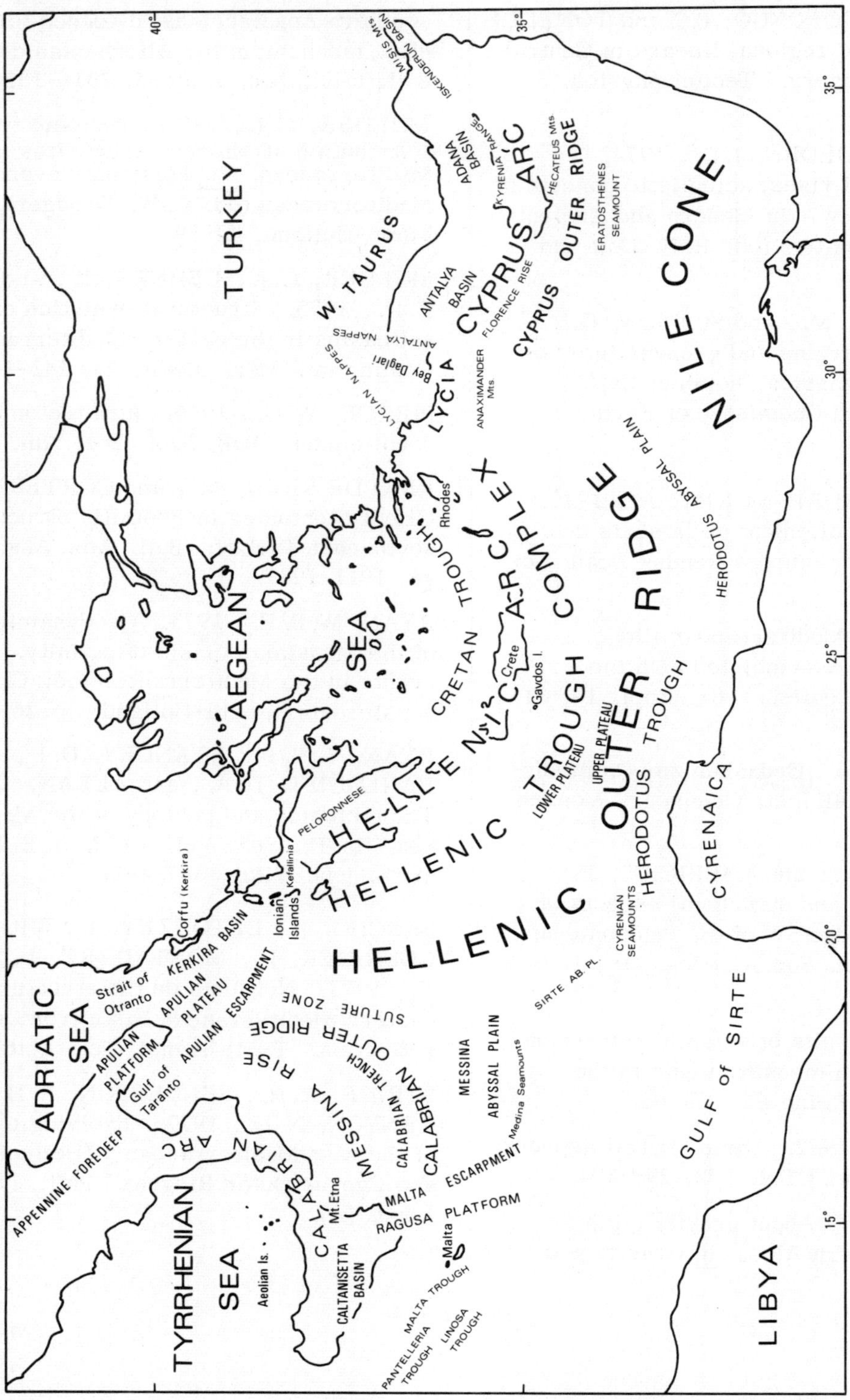

Figure 1 Location map for the eastern Mediterranean showing the geographical names used in this paper.

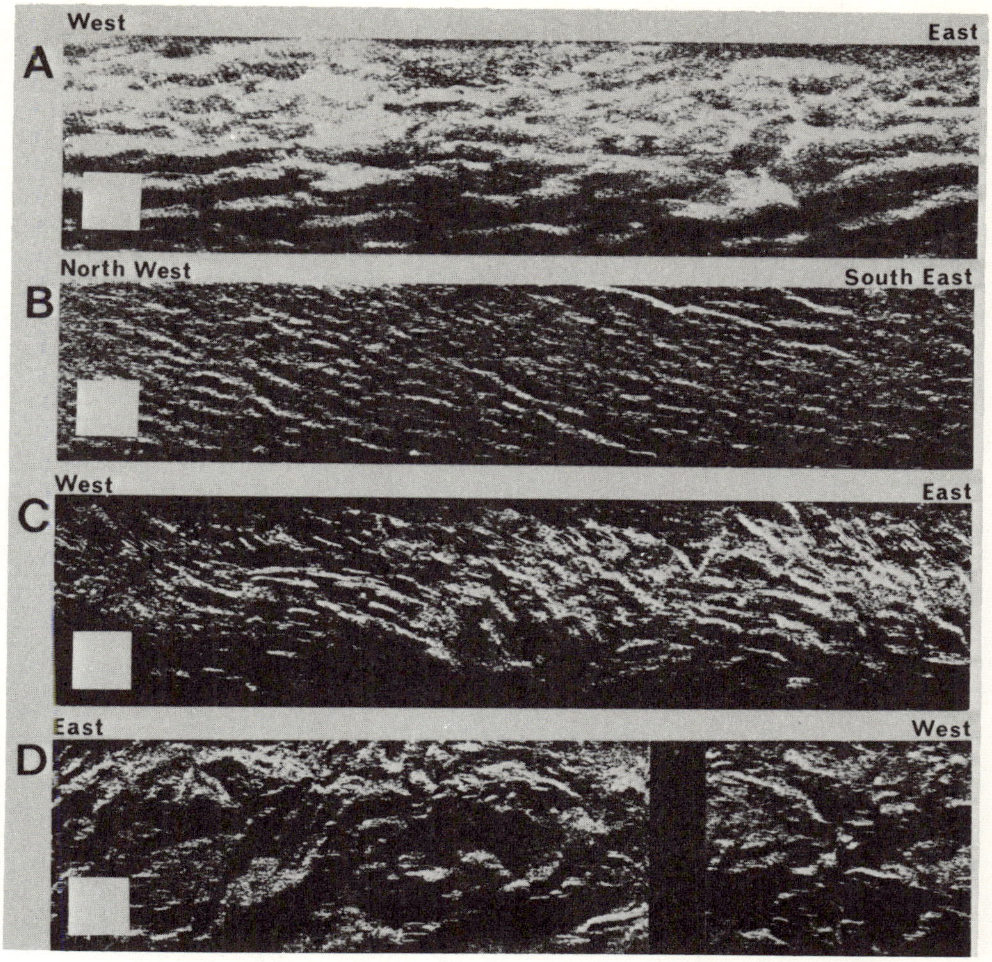

Figure 2 Sonographs of four characteristic styles of small scale tectonic relief recognised on the Hellenic Outer Ridge (and applicable to the other two Outer Ridges). These sonographs extend more or less along the line of the Ridge in order to show the relief to best advantage. The sides of the white squares represent 2 km.

Style A. Simple folds with an amplitude of about 270m and a wavelength of about 1-1.5 km. They are slightly sinuous in plan and up to about 20 km long. The interfingering and bifurcation of the folds is made more obvious if the sonograph is viewed along its length.

Style B. The folds are intensively faulted parallel to their strike, the resulting lineation having a smaller separation of about 0.7 km. The height and overall gradient of the original folds have been reduced by faulting and associated slumping to about 35-70 m.

Style C. On the north side of the Ridge cross faulting is as well developed as the longitudinal structures, but no sonograph of suitable quality is available to show this. The present sonograph illustrates a transition between tectonic styles B and C, in which Ridge-parallel structures are kinked and interrupted by incipient cross faults and relief is reduced by slumping to about 35-70 m.

Style D. The Ridge-parallel structural trends are dissected by cross faults of various trends. Thrusting and strike-slip faulting are suspected. The relief is about 35-55 m but shows up well because this ground, the 'Upper Plateau', is in relatively shallow water so that the shadows are correspondingly long.

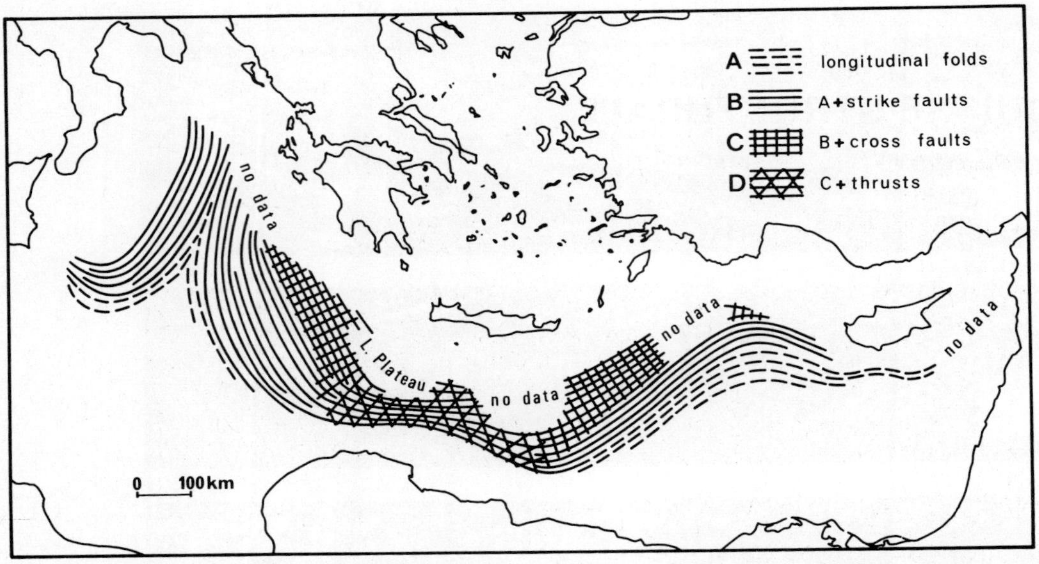

Figure 3 Geographical distribution of deformation styles A-D on the Outer Ridges. Deformation increases in intensity from south to north across the Outer Ridges, and is greatest on the 'Upper Plateau' between Cyrenaica and Crete. The structural trends are shown diagrammatically and the boundaries are often gradational.

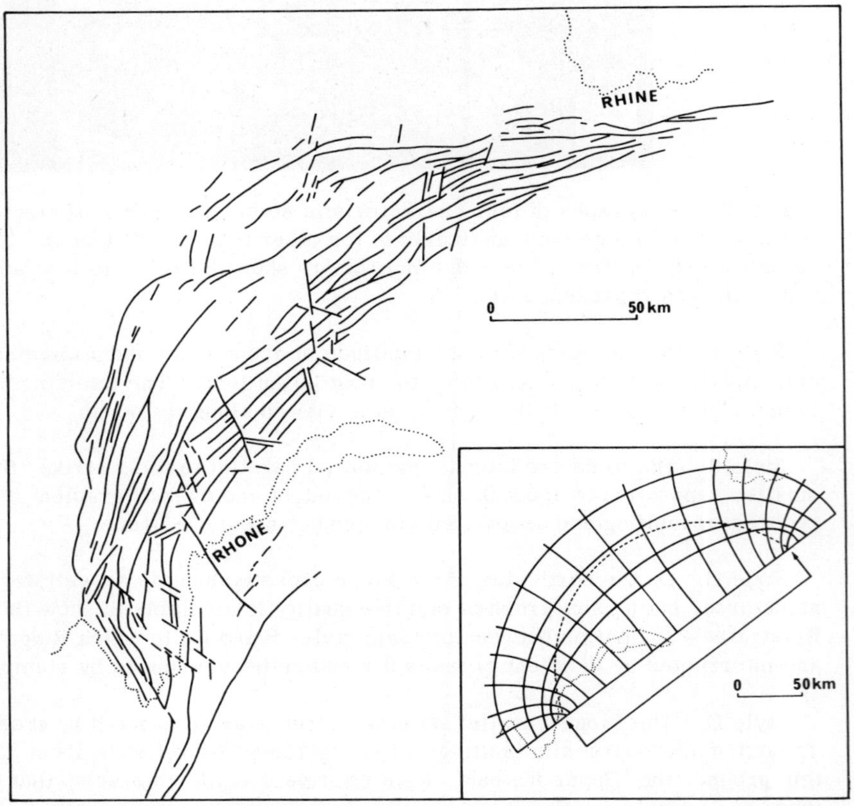

Figure 4 Simplified tectonic map of the Jura showing fold axes and strike-slip faults. Inset is an idealised theoretical model of the principal stress trajectories that would result from a "rigid punch" from the south-east. The pecked line is a circular arc passing approximately through the external termination of the larger strike-slip faults (after Laubscher, 1972).

INTERNATIONAL SYMPOSIUM ON THE STRUCTURAL HISTORY OF THE MEDITERRANEAN BASINS. SPLIT (YUGOSLAVIA) 25-29 OCTOBER 1976. B. BIJU-DUVAL AND L. MONTADERT, Eds. EDITIONS TECHNIP, PARIS 1977, pp. 241-256

ESQUISSE STRUCTURALE EN MÉDITERRANÉE ORIENTALE AU FRONT DE L'ARC ÉGÉEN

W.D. NESTEROFF [(1)], J. LORT [(2)], J. ANGELIER [(3)], M. BONNEAU [(3)] et A. POISSON [(4)]

L'un des buts de la Campagne Océanographique Medor 75 (N/O Le Suroît, octobre-novembre 1975) était de reconnaître les structures immergées de l'Arc Egéen. Nous avons recherché d'une part la prolongation en mer des structures et des grands accidents connus à terre et, d'autre part, les liaisons entre les flancs de l'arc et les structures océaniques profondes comme la Ride Méditerranéenne.

Nous avons effectué 20 profils de sismique-réflexion, le long de 840 miles. Notre équipement comprenait des canons à air Bolt ($4 dm^3$) et Sodera, une flûte AMG utilisant une seule voie et un enregistreur EPC 4100 (fig.1).

I. LES ENTOURS DE LA CRETE : CORRELATIONS AVEC DES STRUCTURES CONNUES A TERRE

Plusieurs profils de sismique-réflexion ont été réalisés au voisinage des côtes méridionale et orientale de la Crète, au travers des prolongements d'amples structures post-nappes étudiées à terre par deux d'entre nous (J.A., M.B.). Il s'agit principalement des profils 12 à 16, à un moindre degré des profils 11 Nord et 9 Nord (fig.1 à 7).

A terre, la structure post-nappe (mio-plio-quaternaire) de l'île de Crète est gouvernée par quatre systèmes principaux de failles, dont les rejets atteignent fréquemment plusieurs kilomètres.

Ces grandes failles, toutes normales, sont dirigées :

- Est-Ouest, notamment celles qui guident les littoraux crétois méridionaux ;
- Nord-Sud, avec une moindre continuité (bassin d'Héraklion par exemple) ;
- Nord-Est-Sud-Ouest, surtout dans la partie orientale de l'île (couloir de Sitia, fossé d'Ierapetra) ;
- Nord-Ouest-Sud-Est, surtout dans la partie occidentale de l'île (isthme de Réthymnon) ;
- ajoutons que les dispositifs E-W et NE-SW paraissent se raccorder, dans la moitié est de l'île, par des directions ENE-WSW (pied du Dictea).

La symétrie de ce dispositif doit être notée ; elle correspond à la courbure générale de l'arc égéen externe, la Crète étant à la jonction des branches ionienne et "libyenne". L'absence de rejets horizontaux latéraux notables pour la plupart des grandes failles explique que l'extension majeure demeure à peu près radiale (J.Angelier, 1976 a et b).

A. Le môle de Gavdos

Les îles de Gavdos et de Gavdopoula sont les sommets d'un môle NW-SE (fig.2) ; les grandes failles normales qui le limitent sont

(1) Laboratoire de Géologie Dynamique, Université Pierre et Marie Curie (Paris VI), 4 place Jussieu, 75230 Paris Cédex 05, France

(2) B.P. Petroleum Development Limited, Britanic House, Moor Lane, London EC 2Y 9BU, Great Britain

(3) Laboratoire de Géologie Structurale, Université Pierre et Marie Curie (Paris VI), 4 place Jussieu, 75230 Paris Cédex 05, France

(4) Laboratoire de Géologie Historique, Université de Paris-Sud, 91405 Orsay, France

submergées pour l'essentiel, mais quelques-unes se voient bien à terre, à Gavdos (extension NE-SW).

Les deux profils réalisés (15 et 16) mettent en évidence des escaliers de failles de part et d'autre du faîte de la structure, d'une part vers la fosse sud-Matapan au Sud-Ouest, d'autre part vers le fossé sud-crétois occidental, dirigé E-W, beaucoup moins profond, au Nord.

La bathymétrie reflète directement la tectonique de failles, la couverture sédimentaire paraissant affectée dans sa quasi-totalité. Les failles proches du faîte sont normales, comme à terre. Les corrélations entre les profils 15 et 16 montrent - ce que confirme la bathymétrie - que le môle de Gavdos tend à s'infléchir et à se rétrécir vers l'Ouest (fig.1 et 2).

B. Le môle de Chrysi

Son sommet est l'île de Chrysi ou Gaïdouronisi (fig.4 et 5). La structure est plus complexe que la précédente ; en effet, sur le littoral crétois d'Ierapétra s'entrecroisent de grandes failles normales dirigées NW-SE et WSW-ENE à W-E. Le môle est ainsi limité à l'Ouest par le grand fossé d'effondrement d'Ierapetra, qui coupe obliquement toute l'île et se prolonge, vers le Nord, par une fosse notable en mer de Crète (fosse de Kamiloni), que coupe le profil 9 Nord (fig.7). Au Nord, le môle est séparé par un petit fossé du littoral sud-crétois. Au Sud-Est, il jouxte la fosse de Pline (fig.5).

Les profils 11 Nord, 12, 13a et 13b ont été exécutés en fonction de cette disposition générale ; ils ont permis des corrélations bien meilleures que l'enchevêtrement des structures ne pouvait le laisser prévoir.

Le caractère normal des failles est flagrant çà et là sur les profils. Aucune faille inverse ne peut être soupçonnée. Toutes coupent la couverture sédimentaire récente, plio-quaternaire, comme d'ailleurs à terre (J.Angelier et M.Gigout, 1974). Des indices de mouvements synsédimentaires, voire de discordances, sont bien probables (profil 13).

Le trait essentiel commun aux profils 12 et 13 est le prolongement de la faille orientale du fossé d'Ierapetra, à regard NW, par un escarpement sous-marin important ; celui-ci, recoupé trois fois (profils 12, 13a, 13b), est dans l'alignement exact du grand escarpement d'Ierapetra-Kavoussi, bien connu à terre, où la faille est normale non décrochante, sur plus de dix kilomètres (fig.4 et 5). Au Sud, le môle de Chrysi est limité par un escalier de failles WSW-ENE bien corrélées dans l'ensemble, sinon dans le détail (profils 12, 13a, 13b, et prolongements sur le profil 11 Nord).

Le môle de Chrysi est donc nettement triangulaire ; le fossé modeste qui le limite au Nord (entre l'île et la côte crétoise) est le prolongement Est du fossé sud-crétois oriental. Celui-ci prend un grand développement vers l'Ouest : il existe des failles de même direction WSW-ENE que celles de la côte sud-crétoise, mais aux regards opposés (vers la Crète), à l'Ouest de Chrysi. Elles sont recoupées par les profils 12 et 13, et concourent à ménager un gradin triangulaire entre le môle et le fossé sud-crétois (fig.3 et 4).

Au total, le môle de Chrysi est, quant à sa structure sous-marine, un bon exemple de recoupements de systèmes de failles, d'ailleurs bien connus à terre (J.Angelier et M.Gigout, 1974). Il illustre bien à la fois la complexité de la structure faillée en extension des abords crétois, dans le détail, et leurs caractères généraux : passage des failles E-W (Crète) aux failles NE-SW (Dodécanèse), avec interférences et recoupements (J.Angelier, 1976 a et b).

C. Les fossés sud-crétois

Le profil 14 recoupe à la fois l'extrémité occidentale du fossé sud-crétois oriental et la partie orientale du fossé sud-crétois occidental, que prolonge à terre la Messara, profond fossé plio-quaternaire Est-Ouest entre les massifs de l'Ida au Nord et de l'Asteroussia au Sud (fig.3).

La partie nord du profil met en évidence cette suite sous-marine de la Messara, avec un remplissage plio-quaternaire très épais ; plus au Sud, au-delà d'un talus à regard méridional, prolongeant le versant méridional faillé de l'Asteroussia, le second fossé, au remplissage également très épais, est bordé au Sud par un môle rocheux au-delà duquel réapparaissent les niveaux plio-quaternaires, avec à la base peut-être du sel (fig.3). Ces fossés sud-crétois, avec leurs remplissages plio-quaternaires (et Miocène supérieur) très épais, assortis de discordances et d'indices de mouvements synsédimentaires, sont bien différents des grandes fosses (Pline et surtout Strabon) ; ils ont servi de piège à une sédimentation terrigène abondante, contemporaine de leurs effondrements, comme cela se voit d'ailleurs à terre.

D. Interprétation structurale d'ensemble

Peu de différences sont à noter entre la structure post-nappes de la Crète méridionale et celle des marges sud-crétoises : la bathymétrie, comme le relief de l'île, dépend directement de la tectonique de failles normales plio-quaternaire. Des interférences entre les différents systèmes de failles découpent des blocs triangulaires ou trapézoïdaux, les fossés ou les môles continus étant de ce fait assez rares, sauf au Sud de la Crète centrale (Asteroussia, Messara, fossés sud-crétois).

Le caractère récent de cette tectonique de failles (affectant fréquemment, à terre, des dépôts du Tyrrhénien ou du Würm) ressort bien sur les profils, de même que son influence sur la sédimentation. Somme toute, les profils sismiques mettent en évidence l'intensité de la néotectonique cassante au ventre sous-marin de l'arc égéen, dont les photographies par satellite donnent, pour les régions émergées, une image spectaculaire.

II. LE FLANC SUD DE L'ARC EGEEN

Dans la partie centrale de l'Arc le flanc interne, situé entre les îles de Crète-Kasos-Karpathos et les zones profondes de la Ride Méditerranéenne, présente une topographie compliquée. Très schématiquement, les auteurs y distinguent trois dépressions, les fossés sud-crétois, de Pline et de Strabon (Emery et *al.*, 1966) séparés par deux môles. La direction générale de ces structures est E-W dans la partie centrale de l'Arc, mais tourne avec celui-ci. Un système complexe de fractures délimite horsts et grabens. Cette organisation en compartiments fracturés dont certains sont effondrés se retrouve en mer Egée et dans l'Ouest de la Turquie. Nos profils 7-8, 9-10 et 11 (fig.7) qui traversent cette zone montrent que toutes les failles sont normales et délimitent, par des séries de marches d'escalier, des dépressions disloquées et assymétriques de part et d'autre de compartiments restés en position haute. Les petits bassins perchés sur les horsts, les pentes raides de leurs flancs, les strates horizontales qui viennent y affleurer et enfin les déformations qui affectent les remplissages sédimentaires des fosses suggèrent une lente subsidence synsédimentaire des zones déprimées.

Le premier des fossés, le Sud-Crétois, contient un millier de mètres de sédiments fortement affectés par la subsidence (fig.3 et 10). Il est séparé des fosses de Sud-Matapan et Pline par les môles de Gavdos et de Chrysi, ensemble décrits au paragraphe précédent (fig.2 à 6).

La continuité des différentes structures, aussi bien des horsts que des grabens, est difficile à mettre en évidence. Ainsi, le passage de la fosse de Sud-Matapan à celle de Pline est interrompu par des seuils, probablement individualisés par le jeu des blocs découpés par les réseaux de failles, qui transforment ce sillon en une suite de bassins ombiliqués. Le remplissage sédimentaire de la fosse de Pline, éloignée et parfois même isolée des apports terrigènes des îles de l'Arc, est beaucoup moins important que celui du fossé Sud-Crétois. Faible à l'Est (profil 8), il atteint 200 m au Sud de Kasos (profil 9) et 300 m ou plus au Sud de la Crète (profil 11). Les couches s'épaississent du Sud au Nord et, comme dans les autres dépressions, indiquent une subsidence synsédimentaire.

Un horst important culmine à 711 m (les Strabo Mountains d'Emery et *al.*, 1966) et sépare les fosses de Pline et de Strabon. Généralement libre de sédiments, il porte de petits bassins perchés contenant 100 à 200 m de dépôts à pendage nord. Son flanc nord présente une plateforme percée de diapirs que nous interprétons comme issus des formations évaporitiques messiniennes (Jongsma, 1975).

Enfin, encore plus au Sud, la fosse de Strabon sépare le horst précédent, limite de la marge immergée de l'Arc, de la Ride Méditerranéenne. Moins profonde (3000-3200 m) que Pline (plus de 4000 m), elle est pratiquement vide de sédiments (fig.7). Les deux systèmes de fosses, Pline et Strabon, sont découpées dans les séries évaporitiques et sont de formation plus récente.

Ainsi, nos profils présentent un certain nombre de caractères reconnus dans les Arcs Insulaires du Pacifique (Ludwig et *al.*, 1967), en particulier le large développement de failles normales. Toutefois, l'importance du remplissage sédimentaire de la Méditerranée orientale modifie les caractères morphologiques au front de l'Arc Egéen. Aussi est-il difficile de préciser si la subduction se produit dans la fosse de Pline ou plutôt de Strabon comme le suggérerait la répartition des séismes superficiels (Comminakis et Papazachos 1972). D'ailleurs, une migration de la zone de subduction a pu prendre place au cours de l'évolution de l'Arc.

En conclusion, le caractère le plus spectaculaire du front de l'Arc Egéen est la présence de fosses ouvertes postérieurement au dépôt des évaporites messiniennes et dont le remplissage est affecté de déformations synsédimentaires continues. Ces fosses témoignent de mouvements distensifs accompagnés d'une importante subsidence post-miocène, qui se sont exercés durant tout le Plio-Quaternaire et qui continuent peut-être encore de nos jours.

III. L'ARC A L'EST DE KARPATHOS

Un profil N-S a été levé à l'Est de l'île de Karpathos. La liaison avec les structures connues à terre (J.Aubouin et *al.*, 1976) n'est qu'hypothétique puisque nous n'avons pas pu établir la direction des grandes failles, nombreuses sur ce profil, par des coupes sériées. Cependant, plusieurs grandes structures ont pu être mises en évidence :

- au Nord de Karpathos, une petite fosse égéenne montre deux escarpements dissymétriques : tandis que le flanc nord n'est affecté que par de petites failles normales recoupant une structure monoclinale, le flanc sud est entièrement dû à de grandes failles. La sédimentation dans le bassin est gouvernée par cette dissymétrie : tandis qu'au Nord, les sédiments restent concordants sur un "socle" non défini, au Sud ils sont affectés par des failles synsédimentaires et de nombreuses petites discordances. Le grand môle faillé situé au Sud de la fosse correspond sans doute à un équivalent de l'île de Saria et des Monts Adra dans le Nord de l'île Karpathos. La structure et la morphologie de ces régions sont les résultats de grandes failles NE-SW qui coïncident avec les directions néotectoniques les plus caractéristiques de cette partie orientale de l'arc égéen (J.Angelier, 1976). Il est donc probable que les grandes failles NE-SW, récentes, soient responsables de l'individualisation de ce môle joignant le Nord de Karpathos à l'île de Rhodes ;
- le Sud du profil 7 montre un petit bassin "calé"le long d'une grande faille normale. Les sédiments y apparaissent perturbés, probablement par des évaporites. De telles évaporites sont connues à terre, justement dans la partie sud de Karpathos. Il s'agit de venues diapiriques traversant le flysch éocène et les calcaires Crétacé supérieur de la nappe d'Adra, et non d'évaporites messiniennes, inconnues dans l'île (J.Aubouin et *al.*, 1976).

IV. STRUCTURE DES MONTS ANAXIMANDRE

A. Enseignements des profils

Les profils exécutés à travers les monts Anaximandre (fig. 9) nous ont permis de mettre en évidence trois ensembles morphologiques nettement distincts : des horsts (à l'Est et au Sud), des fosses (à l'Est et au Nord) et un vaste plateau (au centre). Les raccords entre ces trois ensembles se font par l'intermédiaire de talus à pentes raides qui correspondent selon toute vraisemblance à des failles. Les directions d'allongements des horsts et des fossés sont donc étroitement liées au réseau de fractures.

Les horsts correspondent à des parties surélevées du substratum, empâtées de dépôts récents (les dragages effectués au sommet des pentes les plus fortes n'ont livré que des vases fines récentes, plus ou moins lithifiées). Ces horsts sont localisés à l'Est et au Sud en bordure du prolongement vers l'Est de la dépression de Strabon.

Les fosses sont localisées au Nord (fosse de Finike séparant les monts Anaximandre du promontoire lycien) et à l'Est (petites fosses entaillant le plateau). Elles présentent une morphologie assez caractéristique : des talus à forte pente et des fonds très plats correspondant à la surface d'importants remplissages de turbidites récentes. L'épaisseur de ces sédiments récents est estimée à 1000 m pour la fosse de Finike. Cette dernière comporte d'ailleurs au moins deux bassins séparés par un ensellement localisé vers 29°45 E bien visible sur les profils de Woodside (1975). Cet ensellement n'est recouvert que par une mince pellicule de dépôts. Dans ces fosses, les sédiments sont régulièrement stratifiés. Sur notre profil n°5, on remarque dans la fosse principale une inclinaison générale des couches vers l'Est, ainsi qu'une surface irrégulière du fond de la mer. Ces perturbations sont très certainement à mettre en relation avec une tectonique active. C'est ce que suggère Woodside pour expliquer le basculement général des couches le long de son profil n° 74/4.

Le plateau occupe la partie centrale au Nord et à l'Ouest des horsts et au Sud de la fosse de Finike. Sa surface, relativement plane, est inclinée vers le Nord et semble dépourvue d'une couverture importante de dépôts récents. On y observe tout près de la surface des structures diapiriques correspondant très vraisemblablement aux évaporites messiniennes.

B. Grands traits géologiques du promontoire lycien.

La géologie du promontoire lycien est connue dans ses grandes lignes. On y distingue un bâti autochtone (autochtone vrai ou relatif : Bey Dağları à l'Est et Susuz Dağ à l'Ouest, séparés par la vallée de Gokbük-Finike. Ce bâti est plissé en vastes anticlinaux d'axes NNE-SSE (Bey Dağları) à NE-SW (Susuz Dağ) séparés par des synclinaux pincés (vallée de Gokbük-Finike), et chevauché sur ses deux flancs par deux systèmes de nappes convergentes (nappes lyciennes au Nord-Ouest -nappes d'Antalya à l'Est). Les fronts de chevauchement correspondants ont des directions N-S à l'endroit où ils disparaissent en mer, aussi bien à l'Ouest qu'à l'Est.

Du point de vue stratigraphique, cet autochtone comporte une épaisse série mésozoïque et tertiaire, carbonatée jusqu'à l'Aquitanien marno-conglomératique au Burdigalien-Serravallien. Le Miocène supérieur n'est pas connu. Le Pliocène inférieur est marneux à microfaune marine dans la basse vallée de l'Esen Çay, conglomératique continental en amont de cette vallée. Il est par ailleurs conservé dans de petits vallons près de Kaş (marnes à Huîtres).

Du point de vue structural, l'autochtone est plissé comme nous l'avons vu plus haut, et affecté par un réseau de failles. A l'Ouest de la vallée de Gokbük-Finike, les directions des structures (plis et failles) sont typiquement celles de l'aile orientale de l'arc égéen, c'est-à-dire NE-SW. A l'Est de cette vallée, les directions sont nettement N-S ou NNE-SSW (Bey Dağları). Ce changement de direction est probablement lié au rebroussement vers le Nord de l'extrémité orientale de l'arc égéen sous l'effet du déplacement relatif vers l'Ouest du bloc de Konya (ou arc taurique de J.H.Brunn, 1975). Ce mouvement, responsable des directions N-S, est sensible dès le Tortonien (Poisson, 1974) et est responsable de l'écaillage vers l'Ouest de la bordure occidentale du golfe d'Antalya, du bombement ultime des Bey Dağları et du pincement E-W des synclinaux de la vallée de Gokbük-Finike. Ce mouvement est probablement encore actif.

Les nappes lyciennes comportent à leur front des unités essentiellement carbonatées et, en arrière, de grosses masses de péridotites. Les nappes d'Antalya sont plus diversifiées et plus compliquées structuralement du fait de l'écaillage signalé ci-dessus qui a provoqué une inversion structurale amenant le substratum en position haute le long de la chaîne côtière qui borde le golfe d'Antalya jusqu'au cap Kelidonya.

C. Corrélations possibles entre les Monts Anaximandre et le promontoire lycien.

Il est logique de chercher à retrouver, dans les monts Anaximandre, le prolongement du promontoire lycien. Dans cette hypothèse, le plateau correspondrait à un ancien bassin miocène de la marge continentale lycienne, comportant des évaporites, et qui serait le prolongement des bassins connus à terre jusqu'au Serravallien. Les horsts pourraient correspondre au substratum de ces bassins. Il est assez hasardeux de préciser la nature de ce substratum. Il est peu probable qu'il s'agisse des nappes lyciennes dont le front épouse la courbure de l'arc égéen et doit passer plus à l'Ouest (fosse de Rhodes) ; il est plus vraisemblable qu'il s'agisse en grande partie d'autochtone du type Susuz Dag (plateau et horsts sud). Le horst Est pourrait correspondre au prolongement de la chaîne côtière du cap Kelidonya, mais il est bien clair que ce sont des hypothèses, les fronts de chevauchement des nappes d'Antalya n'ayant pas été suivis en mer.

Ce qui nous semble net c'est que les monts Anaximandre se sont trouvés séparés du promontoire lycien au cours d'une phase d'effondrement post-messinienne. Le réseau de failles correspondant a entaillé les structures préexistantes : ainsi, le fossé de l'Esen Çay de direction N-S, recoupe le front des nappes lyciennes mais ne se prolonge pas très loin en mer. La faille de Demre traverse l'autochtone, recoupe le front des nappes d'Antalya dans le golfe de Finike et se perd vers le Sud-Est.

V. CONCLUSIONS

Deux observations générales peuvent être dégagées de nos profils. D'une part, les structures reconnues à terre, en particulier les grands réseaux de failles, peuvent souvent être suivies en mer sur d'assez grandes distances. Ainsi le découpage en horsts et grabens, caractéristique de certaines parties de l'arc, de la mer Egée, de la Turquie, se retrouve au Sud de l'arc où il conduit à la formation des fosses sud-crétoises, de Pline et de Strabon, séparées par des môles. D'autre part, ces structures témoignent d'importants mouvements tectoniques affectant toute la marge sud de l'arc depuis le Miocène. Des mouvements verticaux, d'ampleur kilométrique, ont découplé les horsts et les plateformes à évaporites messiniennes qui se trouvent de nos jours en position basse. Les fosses ont été ouvertes, elles aussi, après le dépôt des évaporites messiniennes. La plupart comportent d'épais remplissages d'apports terrigènes affectés de déformations synsédimentaires (fig. 10). Elles indiquent une tectonique distensive active ininterrompue du Miocène à nos jours, affectant les couches supérieures de la lithosphère, au front de l'arc, de la Crète à la Turquie. Cette tectonique serait l'expression superficielle de la subduction de la plaque africaine sous l'Arc Egéen, comme c'est le cas, dans des situations analogues, des flancs des fosses des arcs insulaires du Pacifique (Ludwig et *al.*, 1966).

Nous remercions très vivement MM.les Professeurs Aubouin et Brunn pour la lecture du manuscrit ainsi que le Centre National de la Recherche Scientifique et le Centre National pour l'Exploitation des Océans pour les soutiens financier et logistique qui ont rendu possible la Campagne Océanographique Médor 75.

BIBLIOGRAPHIE

ANGELIER (J.). - 1976a. Sur la néotectonique de l'arc égéen externe : les failles normales Est-Ouest et l'extension subméridienne crétoise. *C.R.Acad.Sc.Paris,* t.282, p.413-416.

ANGELIER (J.). - 1976b. La néotectonique cassante et sa place dans un arc insulaire : l'arc égéen méridional. *Bull.Soc.Géol.Fr.,* t.XVIII.

ANGELIER (J.) et GIGOUT (M.). - 1974. Sur les plateformes marines et la néotectonique quaternaire de la région d'Ierapetra (Crète, Grèce). *C.R.Acad.Sc.Paris,* t.282, p.2103-2106.

AUBOUIN (J.), BONNEAU (M.) et DAVIDSON (J.).- 1976. Contribution à la géologie de l'arc égéen : l'île de Karpathos (Grèce). *Bull.Soc. Géol.Fr.,* t.XVIII, n° 2, p.385-401.

BRUNN (J.H.). - 1976. L'arc concave zagrotaurique et les arcs convexes taurique et égéen : collision et arcs induits. *Bull.Soc. Géol.Fr.,* t.XVIII, p.553-567.

COMNINAKIS (P.E.). - 1972. Seismicity of the Eastern Mediterranean Ridge. *Bull.Geol.Soc. of Amer.,* v.83, p.1093-1102.

EMERY (K.O.), HEEZEN (B.C.) and ALLAN (T.D.). 1966. Bathymetry of the Eastern Mediterranean Sea. *Deep-Sea Res.,* v.13, p.173-192.

JONGSMA (D.). - 1975. A marine geophysical study of the Hellenic Arc. *Ph.D.Thesis, Univ. of Cambridge.*

LUDWIG (W.J.), HAYES (D.E.) and EWING (J.I.). 1967. The Manilla trench and West Luzon trough, 1, Bathymetry and sediment distribution. *Deep-Sea Res.,* v.14, p.533-544.

POISSON (A.). - 1974. Schéma structural des environs de Termessos. Eléments pour une chronologie des évènements tectoniques postérieurs à la mise en place des nappes d'Antalya (Turquie). *C.R.Acad.Sc.Paris,* t.279, p.1983-1986 et Rap.Com.int.Méd.,23,4a,207-209 (Congrès CIESM, Monaco 1974).

WOODSIDE (J.M.). - 1975. Evolution of the Eastern Mediterranean. *Ph.D.Thesis,Univ.of.Cambridge.*

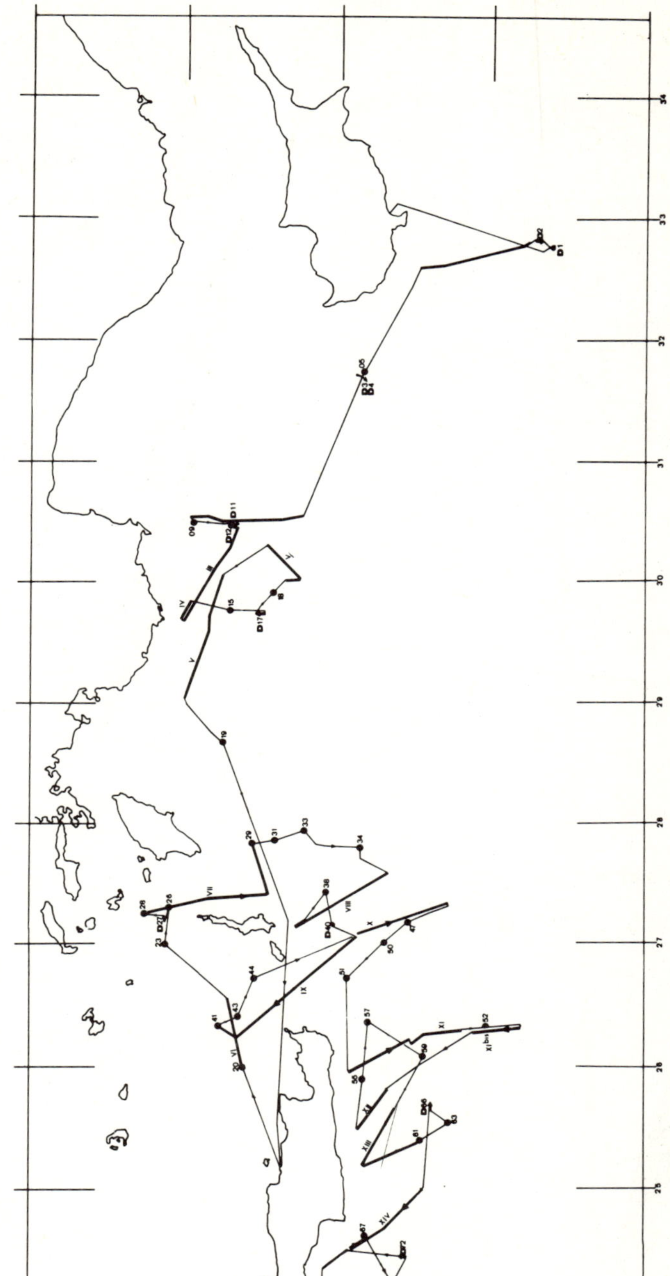

Fig. 1. - Position des profils sismiques.

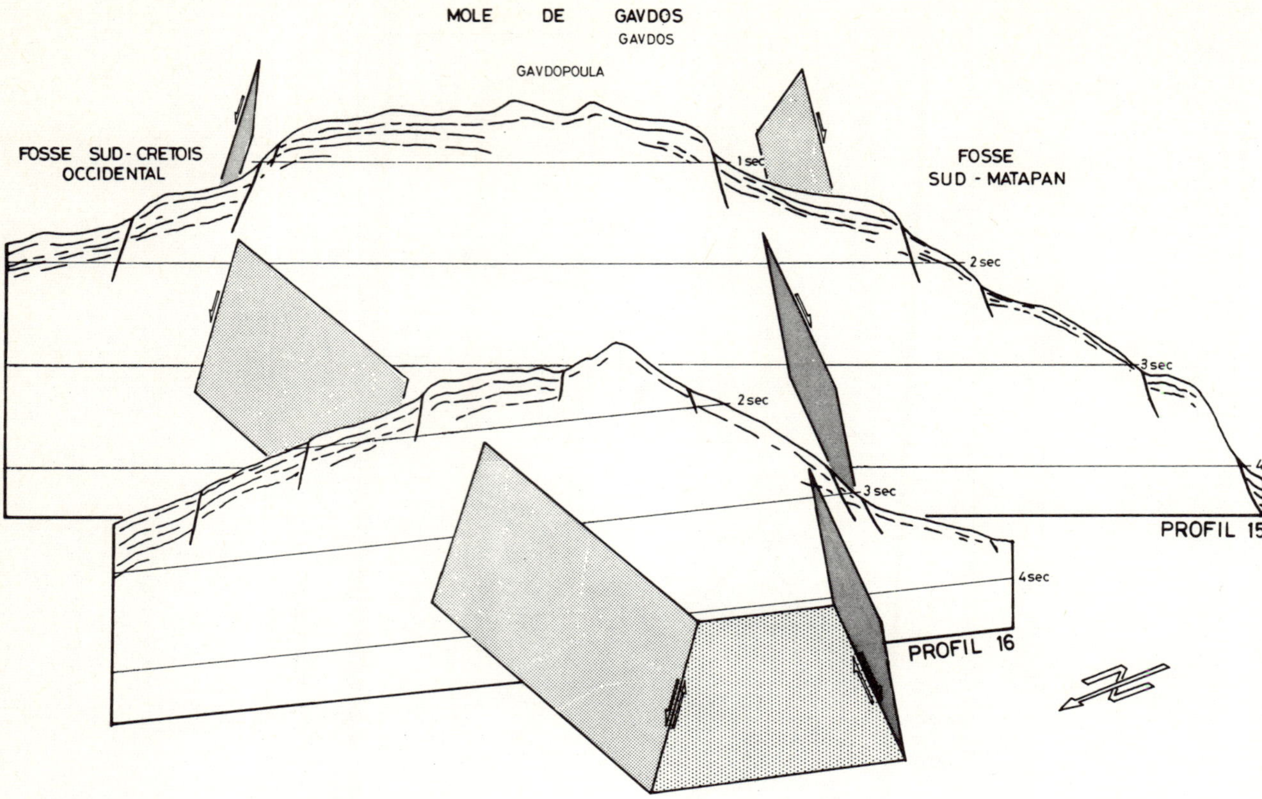

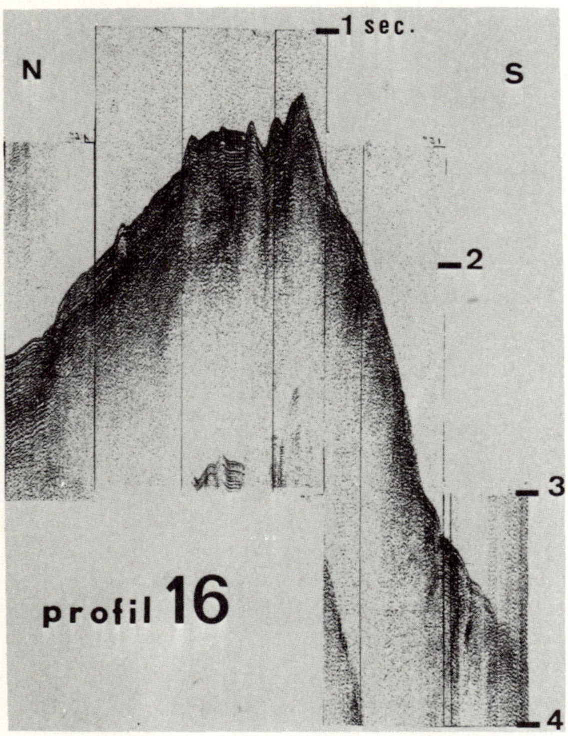

Fig. 2. - Le Môle de Gavdos

Bloc-diagramme montrant les profils 15 et 16 à travers le Môle.

- profil 14 : 34°26'N-25°00'E à 34°45'N-24°37'E
- profil 15 : 34°43'N-23°39'E à 35°05'N-23°58'E
- profil 16 : 34°50'N-23°37'E à 35°07'N-23°46'E

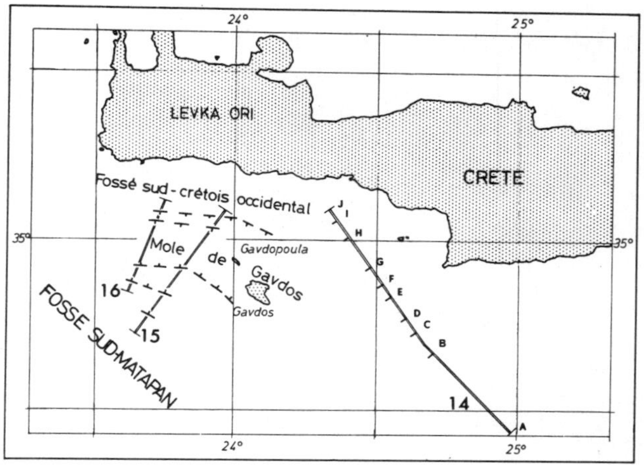

Fig. 3. - Localisation des profils 15 et 16 (Môle de Gavdos)

Failles en tireté gras, barbelé sur le compartiment supérieur.

Localisation et interprétation du profil 14 à travers les fosses sud-crétoises

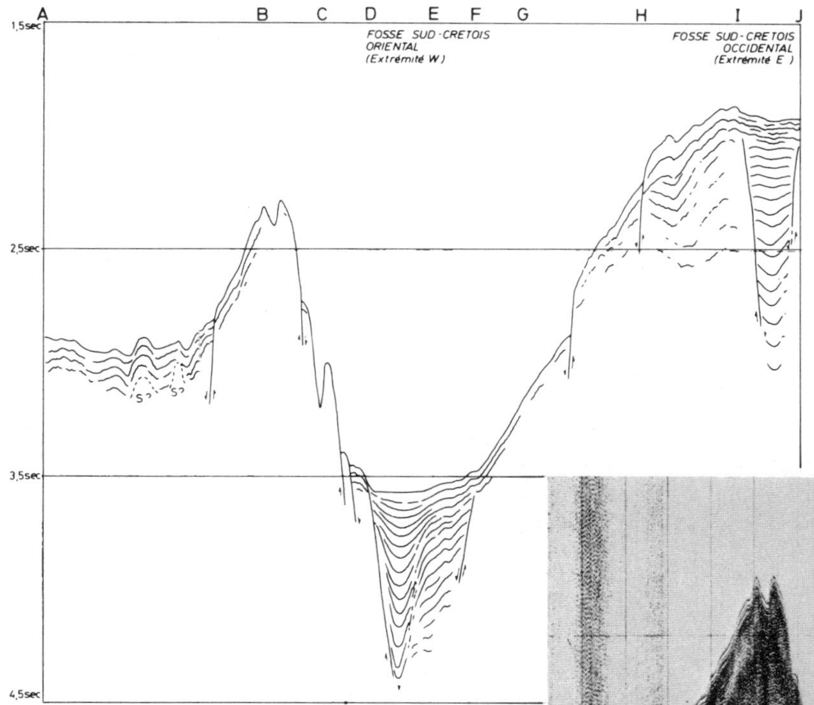

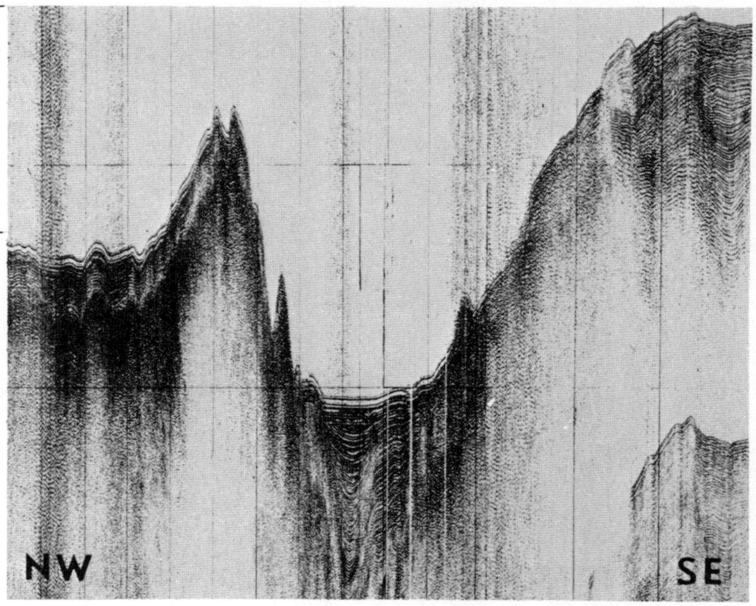

profil 14

S : sel messinien sous la couverture sédimentaire plio-quaternaire

Fig. 4. - Bloc-diagramme du Môle de Chrysi.

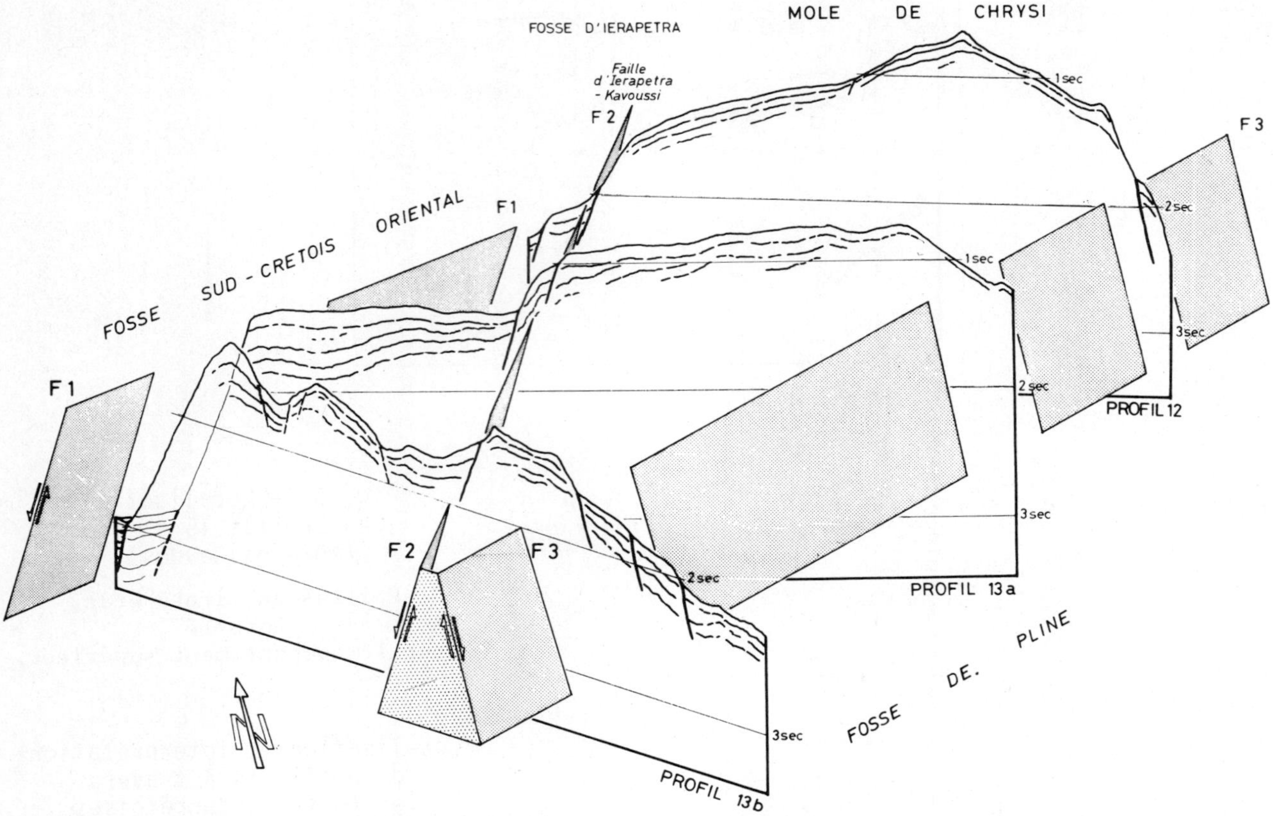

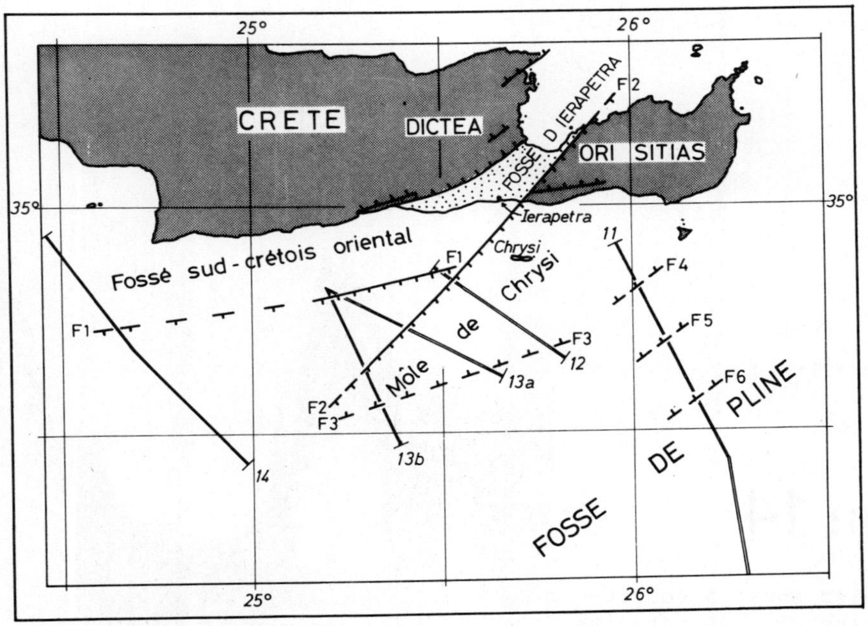

Fig. 5. - Le Môle de Chrysi.

Localisation des profils 11, 12, 13 et 14.
Failles en tireté gras, barbelé sur le compartiment supérieur.

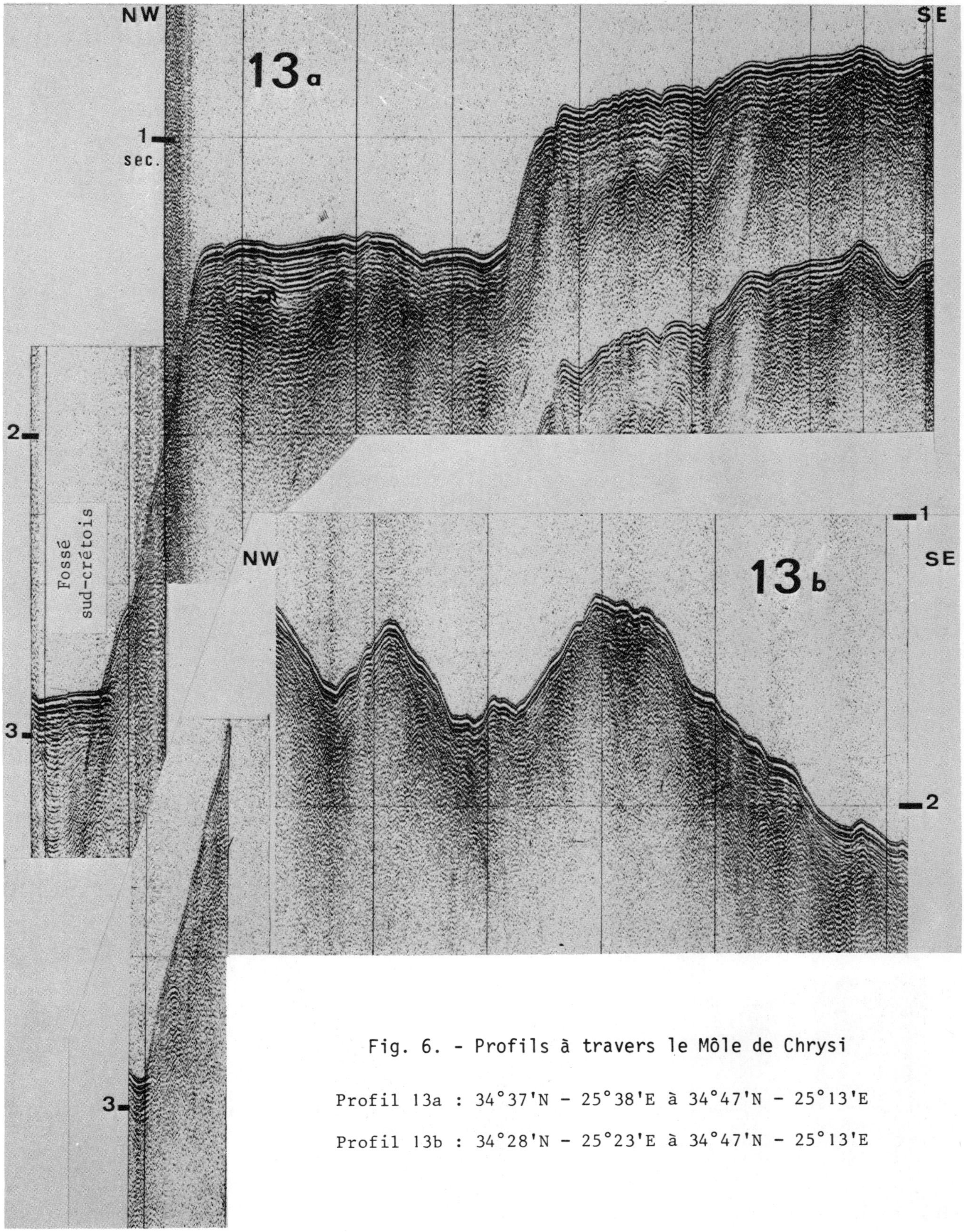

Fig. 6. - Profils à travers le Môle de Chrysi

Profil 13a : 34°37'N - 25°38'E à 34°47'N - 25°13'E

Profil 13b : 34°28'N - 25°23'E à 34°47'N - 25°13'E

Fig. 7. - Structure générale de l'Arc Egéen.

Les profils à travers l'arc montrent au Nord la structure en "bassin marginal" (profils 7 et 9) et au Sud le découpage du flanc interne en une série de môles et de fosses (sud-crétoise, Pline et Strabon, profils 9, 10 et 11).

- profil 7 : 35°27'N - 27°24'E à 36°14'N - 27°16'E
- profil 9 : 34°54'N - 27°01'E; c/c 35°41'5N - 26°15'E; 35°46'N - 26°20'E
- profil 10 : 34°17'N - 27°18'E à 34°52'N - 27°03'E
- profil 11 : 33°47'N - 26°20'E à 34°54'5N - 25°27'E

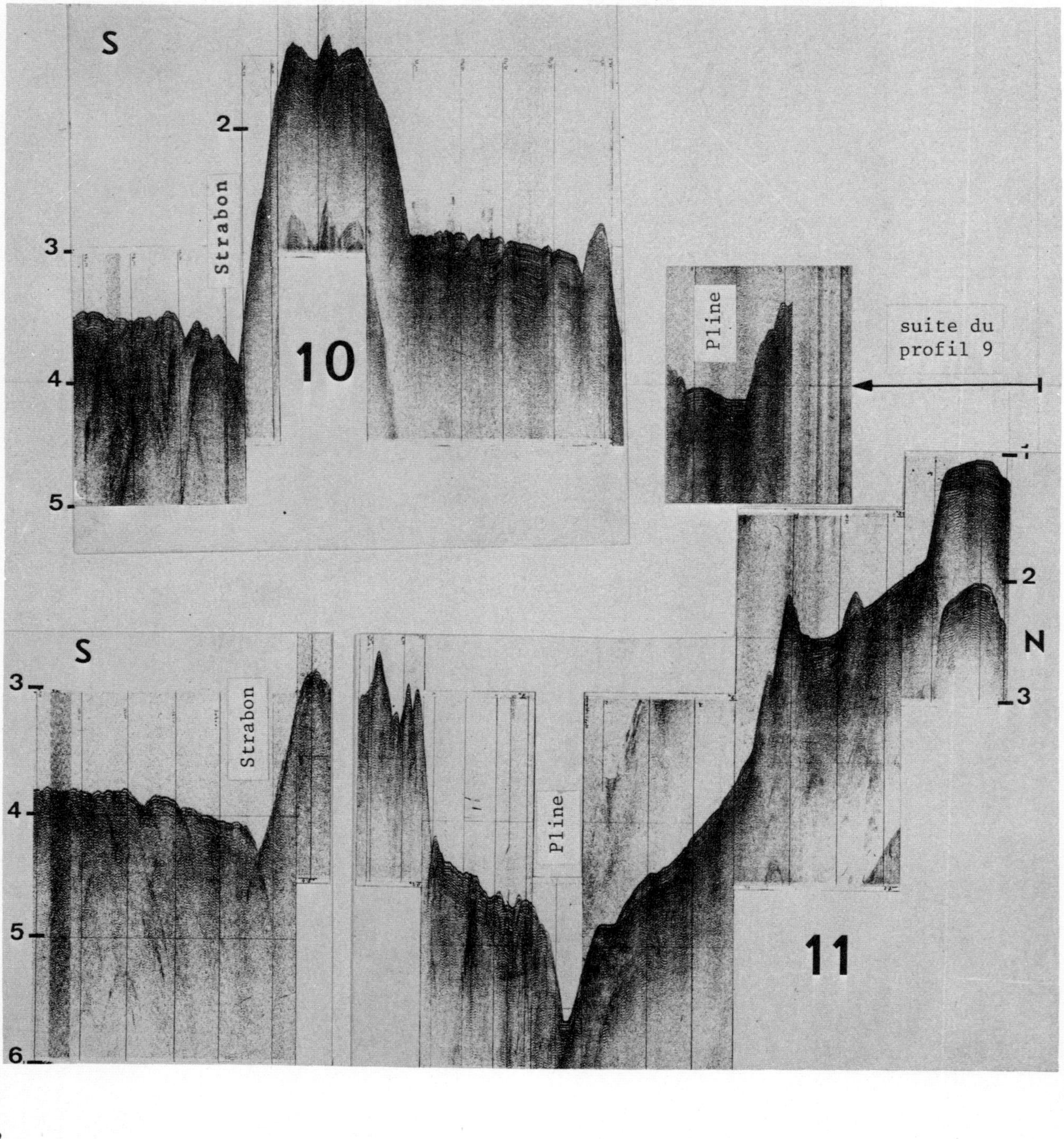

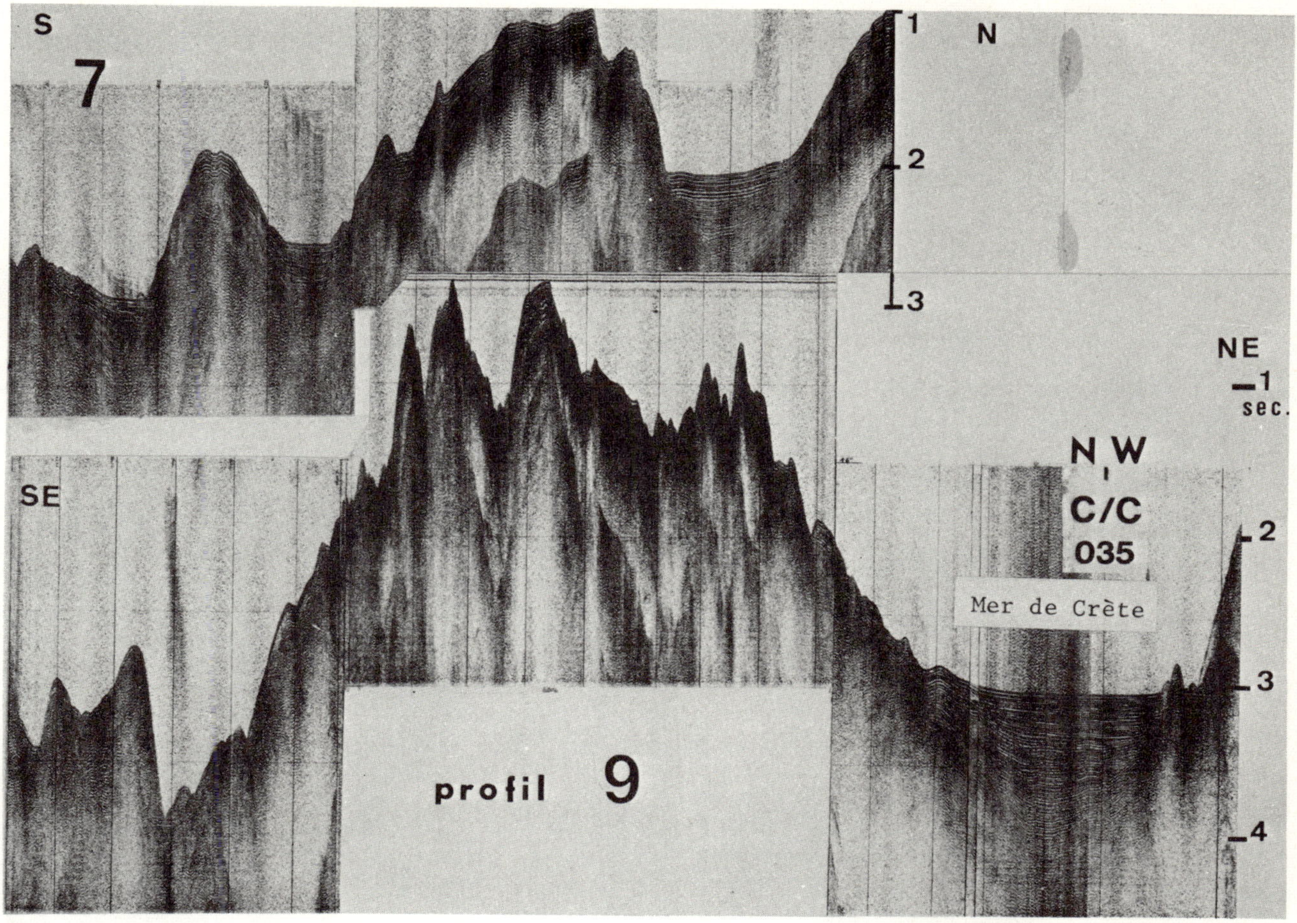
S
7
1
2
3
N
NE
1
sec.
N W
C/C
035
Mer de Crète
SE
2
3
4
profil 9

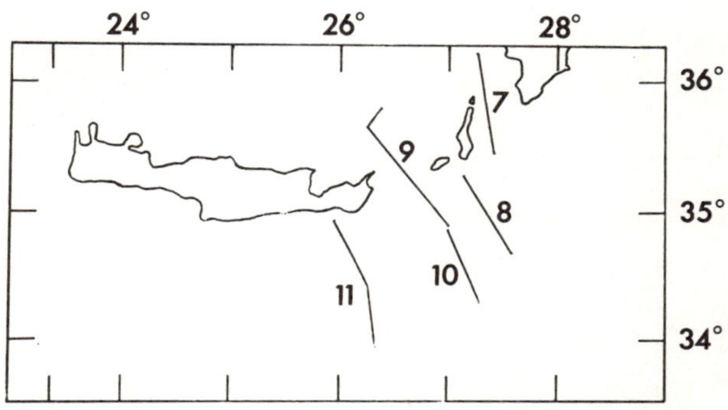
24°
26°
28°
36°
35°
34°
7
9
8
10
11

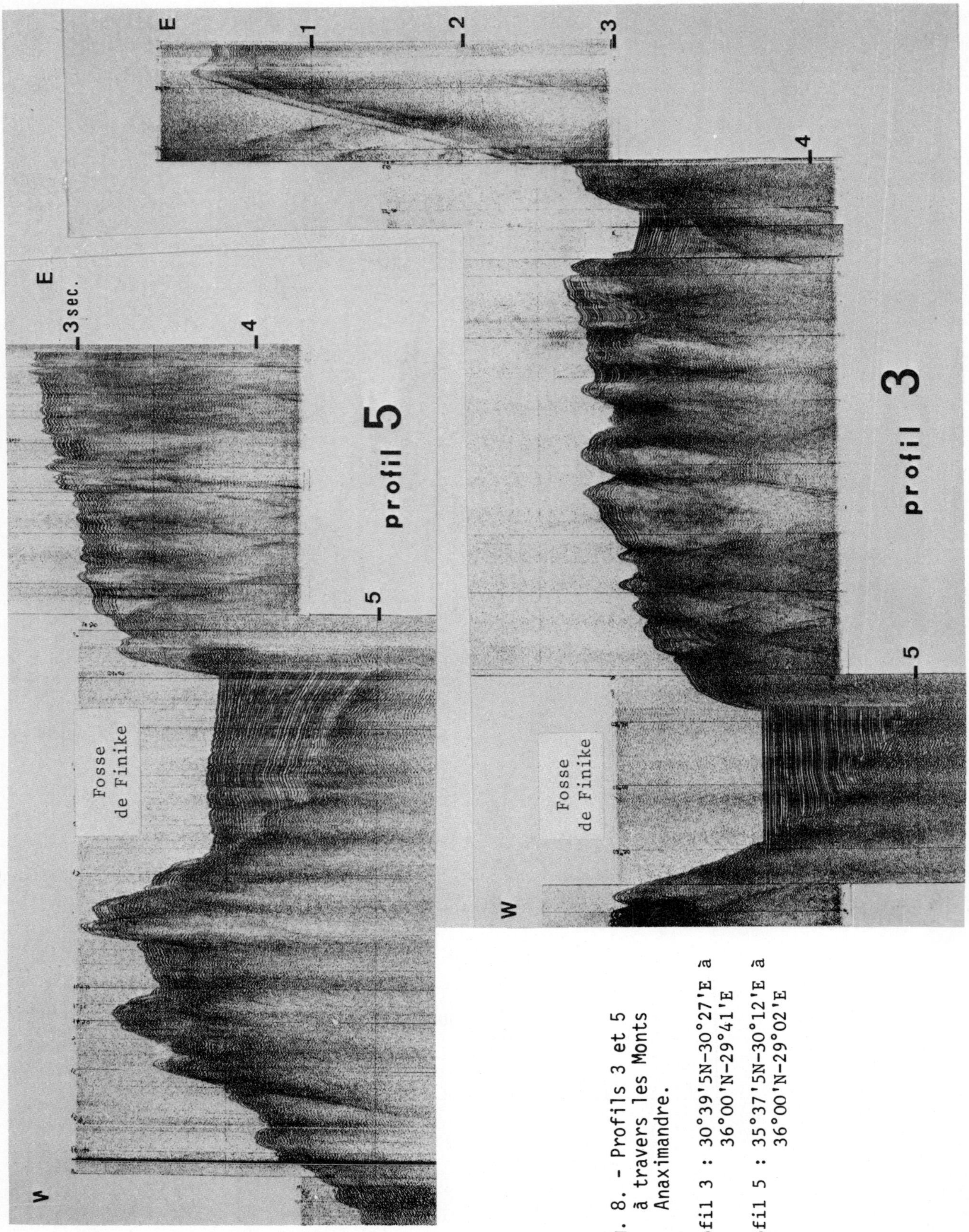

Fig. 8. - Profils 3 et 5 à travers les Monts Anaximandre.

profil 3 : 30°39'5N-30°27'E à 36°00'N-29°41'E

profil 5 : 35°37'5N-30°12'E à 36°00'N-29°02'E

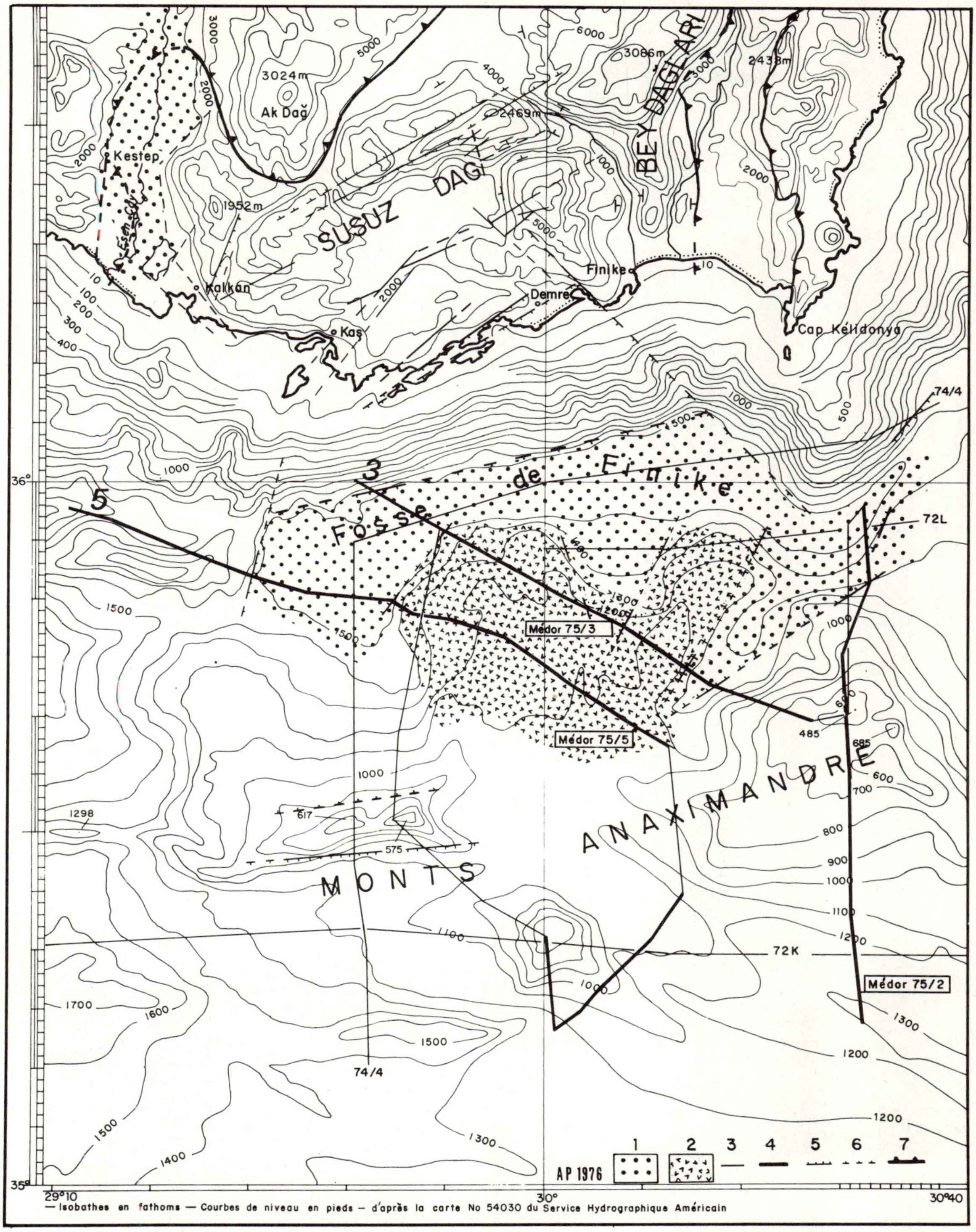

Fig. 9. - Schéma structural du promontoire lycien et des Monts Anaximandre

1 : bassins plio-quaternaires. - 2 : extension probable des évaporites messiniennes. - 3 : routes Medor 75 et Shackleton 72 et 74. - 4 : profils sismiques Medor 75. - 5 : failles et prolongements probables : (6). - 7 : front de charriage du promontoire lycien.

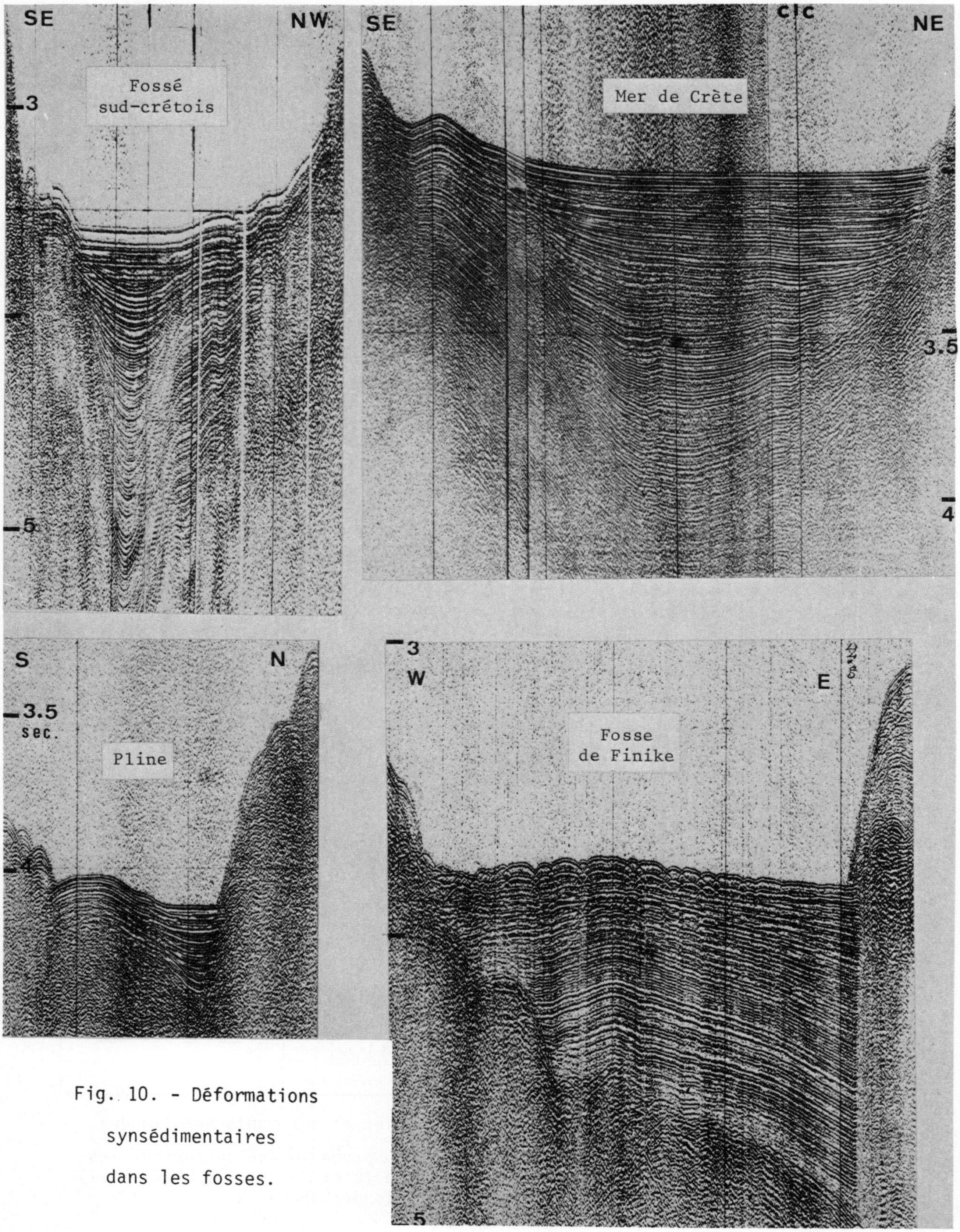

Fig. 10. - Déformations synsédimentaires dans les fosses.

INTERNATIONAL SYMPOSIUM ON THE STRUCTURAL HISTORY OF THE MEDITERRANEAN BASINS. SPLIT (YUGOSLAVIA) 25-29 OCTOBER 1976. B. BIJU-DUVAL AND L. MONTADERT, Eds. EDITIONS TECHNIP, PARIS 1977, pp. 257-262

third part

small ocean basins of the mediterranean

EDITORS' COMMENTS

For a long time, geologists and geophycists have been faced with the problem of the origin and structure of the small ocean basins in the Mediterranean area. In the first part of this volume, all reconstructions of the geological evolution of the Mediterranean-Alpine area have implied the existence of different subduction zones that were active from the Mesozoic to the present. Associated with these subduction zones, the formation of marginal basins in a back-arc position has been suggested. All geophysical techniques, land geology and studies of magmatism reinforce this interpretation of the Mediterranean basins with the exception of the eastern Mediterranean.

Heat flow measurements on the sea-bottom or in DSDP holes give strong arguments for the interpretation of the different small basins in the Mediterranean. For example, they already clearly show fundamental differences between a hot Cenozoic Western Mediterranean Basin and the cold eastern one and the Black Sea (Erikson et al. and Stegena and Horwath, in this volume; Panza and Calcagnile*(1). Studies of surface waves, seismicity, gravity and magnetics have been widely used to define the deep structure and the present tectonics. Seismic refraction would be of particular interest, but too few experiments have been performed up to now.

([1]) Names with the symbol* correspond to papers given during the Symposium, for which an abstract will be published in the "Rapports et Procès-Verbaux des Réunions" de la Commission Internationale pour l'Exploration Scientifique de la mer Méditerranée (CIESM) (in press).

The superficial offshore geology has made considerable progress in the last few years because of the intensive use of multichannel seismics (Mulder, 1973 ; Finetti and Morelli, 1973 ; Biju-Duval et al., 1974 ; Morelli, 1975 ; Finetti, 1976 ; Biju-Duval et al., 1977) and the drilling of 21 holes during both Legs 13 and 42 A by the Glomar Challenger. All these techniques have produced good agreement as to the structure of the small basins and their margins. Some of these basins can be interpreted as marginal basins of different ages, with a more or less normal oceanic crust (Western Mediterranean Basin, Tyrrhenian Sea, Black Sea) or with a more or less attenuated continental crust (Aegean, Pannonian). Some of them are still active today such as the Tyrrhenian or Aegean, while the eastern Mediterranean basins must be interpreted in a different way.

In the past, several publication based mainly on geophysical data, have discussed the structure of the eastern Mediterranean and especially the significance of the "Mediterranean Ridge" (Ryan et al., 1971; Sancho et al., 1973; Finetti and Morelli, 1973; Mulder, 1973; Biju-Duval, 1974; Biju-Duval et al., 1974, 1977; Lort, thesis, 1972; Jongsma, thesis, 1975; Woodside, thesis, 1976; Neev et al., 1976; Makris, 1975). Some of these previous data were discussed during the Monaco meeting in 1974 (see the discussion in the "Rapports et Procès-Verbaux des Réunions", CIESM, 1975, p. 213 and p. 225) and again at the Split symposium (Makris*, Matthews* and Finetti*). But few new data seem to be available at present to make a choice between the several hypotheses on the origin of the eastern Mediterranean. Broadly these basins are interpreted either as relicts of an oceanic proto-eastern Mediterranean formed during the Mesozoic with its stable old African margin southward or as recent basins formed by subsided continental crust. This problem has been discussed in Part One of this volume where the differences between Channel and Horwath's point of view appears (close to Argand's) and Biju-Duval et al. who assume an old oceanic crust presently being consummed. Generally, the Mediterranean Ridge is interpreted as a tectonic feature, south of the Hellenic Arc, with some differences between the authors, essentially because of the difficulty of interpreting the poor seismic results obtained in the area. Clearly, more progress in understanding the eastern Mediterranean needs a special effort in seismic refraction and drilling.

In the Western Basin, new and interesting data on the crustal and upper mantle structure between Corsica and the Northern Apennines have been provided by Morelli et al. in this volume; in its southern part, Ansorge et al.* (south of Spain) and Hatzfeld* (Gibraltar area) have presented some new information on the crustal structure. Galdeano* has given information about the new map of magnetic anomalies of the whole Western Basin (Galdeano, 1977). In a different way, from seismic reflection data Gennesseaux et al.* and Mauffret* have described models of the evolution of this basin in terms of a marginal basin initially uplifted during the rifted phase, then subsiding rapidly with the cooling of lithosphere and loading due to sediments. Very numerous detailed studies have now been made of this Western Basin, and the reader can obtain data from a special issue of the *Bull. Soc. Géol. France* (in press).

In the Tyrrhenian Sea, Panza and Calcagnile*, using Raleigh-wave phase velocities, show some correlations with heat flow data, thus improving our understanding of the upper mantle structure. Görler and Giese* have proposed a model of the evolution of the Tyrrhenian Sea and of the Calabrian Arc. Very interesting petrological studies presented by Barberi et al.* have demonstrated the oceanic nature of the volcanism in the abyssal plain (see also Barberi et al., 1977); while Wezel (this volume) analyzing sediments now tectonized in Sicily and Calabria, has shown that an early volcanism occurred in the Tyrrhenian area. On the Italian margins, several papers have been presented during the Symposium to illustrate the evolution of the Apennines on the northeastern margin of the Tyrrhenian sea (Ibbeken*, Grosscurth and Reuter*, on the relation between contemporaneous tectonics and sedimentation, Gunther and Wallbrecher*, on the superimposed fabric patterns in the metamorphic core of the Apuan Alps).

New data and attempts to make a synthesis now give us a better understanding of the Aegean Sea. Berckhemer (in this volume) presents a model of the evolution of marginal seas deduced from observations in the Aegean region. Different techniques have provided keys for interpretation : 1) deep seismic soundings characterize the crust and upper mantle in this area (Makris*); 2) Altherr et al. (abstract in this volume) give geochronological data on granitic rocks autcropping in the Aegean Sea, and this information is completed by measurements showing the age of metamorphism in this area which can be as old as late Jurassic or late Cretaceous (Seidel et al.*) or later (Altherr et al.*); 3) general geological papers have tried to present the structural evolution based mainly on land data (Jacobshagen*, Besenecker*), and they have been completed by more specific ones, i.e. Richter* concerning a transverse cross-section in the northern Peloponnesus, Mistardis* on the evolution of the Gulf of Corinth; 4) interpretation of recent seismicity by Papazachos and Comninakis has also characterized the lithospheric interactions in the Aegean area (in this volume).

It is interesting to compare these interpretations with those proposed for others areas close to the Mediterranean, i.e. the Pannonian Basin and the Black Sea Basin. Here both geophysical and geological data clearly reveal the evolution of these basins which are also included in the Alpine belt. A review of the evolution of the Pannonian basin interpreted as an interarc basin (by Horvath and Stegena, in this volume) has given supplementary data to the more general paper by Channel and Horvath (also this volume Part One). Airinei (in this part) also gives gravimetric interpretations which help to explain the plate tectonics of Romania. The Black Sea like the Pannonian basin, can be related to the evolution of a Tethysian arc. Here as in the other Mediterranean areas, land geology is necessary to interprete geophysical data. Eastern structures of the Black Sea partially outcrop between the Greater Caucasus and Lesser Caucasus, and Khain (this volume) makes a critical review of the data and hypotheses concerning the structural development of this region. A step is taken towards the sea with the combination of both land geology and seismic reflection profiling. Letouzey et al. (in this volume) try to explain the Black Sea Basin as a subsided marginal basin created in the Upper Cretaceous (age of the closure of the Tethys and of the ophiolite obduction). A third approach was made during the symposium with the results of DSDP leg 42 B presented by K. Hsu* (Ross, Neprochnov et al., 1977), which provided new data on the recent evolution of the Black Sea linked to the history of the para-Tethys since the Miocene.

BIBLIOGRAPHY

BARBERI (F.), BIZOUARD (H.), CAPALDI (G.), FERRARA (G.), GASPARINI (P.), INNOCENTI (F.), JORON (J.L.), LAMBRET (B.), TREUIL (M.), ALLEGRE (C.).- (1977). Age and Nature of Basalts from the Tyrrhenian Abyssal Plain. In : Initial Reports of the Deep Sea Drilling Project, vol. XLII, Washington, D.C., US Government Printing Office, in press.

BIJU-DUVAL (B.).- (1974). Commentaires de la carte géologique et structurale des bassins tertiaires du domaine Méditerranéen. Rev. Inst. Franç. du Pétrole, XXIX-5, 607-639.

BIJU-DUVAL (B.), LETOUZEY (J.), MONTADERT (L.).- (1977). Structure and Evolution of the Mediterranean Sea Basins. In : Initial Report of the Deep Sea Drilling Project, vol. XLII, Washington, D.C., US Government Printing Office, in press.

COMMISSION INTERNATIONALE POUR L'EXPLORATION SCIENTIFIQUE DE LA MER MEDITERRANEE.- (1975). Rapports et Procès-Verbaux des Réunions, Symposium Géodynamique, Monaco, 6-14 déc. 1974, 313 p.

FINETTI (I.), MORELLI (C.).- (1973). Geophysical exploration of the Mediterranean Sea. Boll. Geofis. Teorica. Appl., 15, n° 60, 263-341.

FINETTI (I.).- (1976). Mediterranean Ridge : a young submerged Chain associated with the Hellenic Arc. Boll. Geofis. Teorica. Appl., 18, n° 69, 31-65.

GALDEANO (A.), ROSSIGNOL (J.C.).- (1977). Assemblage à altitude constante des cartes d'anomalies magnétiques couvrant l'ensemble du Bassin occidental de la Méditerranée. Soc. Géol. France. Bull., 7, XIX, in press.

HSU (K.J.), MONTADERT (L.) et al.- (1977). Initial Reports of the Deep Sea Drilling Project, vol. XLII, Washington, D.C., US Government Printing Office, in press.

JONGSMA (D.).- (1975). A marine geophysical Study of the Hellenic Arc. Ph.D., Cambridge, 85 p.

LORT (J.M.).- (1972). The crustal structure of the eastern Mediterranean. Ph.D., Cambridge, 117 p.

MAKRIS (J.).- (1975). Crustal Structure of the Aegean Sea and the Hellenides obtained from Geophysical Surveys. Rapp. Proc. Verb. Réunions, CIESM, 23, 201-202.

MORELLI (C.).- (1975). Geophysics of the Mediterranean. Rapp. Comm. Int. Mer. Medit., Spec. Issue, 7, 29-111.

MULDER (C.J.).- (1973). Tectonic framework and distribution of Miocene chemical sediments with emphasis on the Eastern Mediterranean. In : Messinian events in the Mediterranean, Drooger (C.W.) Ed., Amsterdam, London, North Holland Publ. Co., 44-59.

NEEV (D.), ALMAGOR (G.), ARAD (A.), GINZBURG (A.), and HALL (J.K.).- (1976). The Geology of the Southeastern Mediterranean. Geol. Surv. Israel, 68, 1-51.

ROSS (D.A.).- (1974). The Black Sea. In : The Geology of Continental Margins. Burk (C.A.) and Drake (C.L.) Ed., New York, Heidelberg, Berlin, Springer Verlag, 669-682.

ROSS (D.A.), NEPROCHNOV et al.- (1977). Initial Reports of the Deep Sea Drilling Project, vol. XLII, Washington, D.C., US Government Printing Office, in press.

RYAN (W.B.F.), HSU (K.J.) et al.- (1973). Initial reports of the Deep Sea Drilling Project, vol. XIII, Washington, D.C., US Government Printing Office, 1447 p.

SANCHO (J.), LETOUZEY (J.), BIJU-DUVAL (B.), COURRIER (P.), MONTADERT (L.), WINNOCK (E.).- (1973). New data on the structure of the eastern Mediterranean Basin from seismic reflection. Earth Planetary Sci. Lett., 18, n° 2, 189-204.

SOCIETE GEOLOGIQUE DE FRANCE.- (1977). Géodynamique de la Méditerranée occidentale et de ses abords. Montpellier, 14-15 oct. 1976, Soc. Géol. France Bull., 7ème ser., XIX, in press.

WOODSIDE (J.M.).- (1976). Evolution of the Eastern Mediterranean. Ph. D., Cambridge, 185 p.

INTERNATIONAL SYMPOSIUM ON THE STRUCTURAL HISTORY OF THE MEDITERRANEAN BASINS. SPLIT (YUGOSLAVIA) 25-29 OCTOBER 1976. B. BIJU-DUVAL AND L. MONTADERT, Eds. EDITIONS TECHNIP, PARIS 1977, pp. 263-280

REVIEW OF HEATFLOW DATA FROM THE MEDITERRANEAN AND AEGEAN SEAS

A.J. ERICKSON [(1)], G. SIMMONS [(2)] and W.B.F. RYAN [(3)]

I. INTRODUCTION

The origin of small ocean basins, whether residual fragments of ancient oceans or newly formed by still poorly understood geological processes, is a major unsolved geophysical problem. Theories proposed for the origin of the Eastern and Western Mediterranean and Aegean Seas are varied, but all are ultimately dependent upon thermal energy to provide the uplift, warping, subsidence, or rifting called upon by various investigators. Knowledge of the temperature distribution beneath the Mediterranean and Aegean Seas provides an important constraint on models for the formation of these basins.

Several thousand measurements of the rate of heat flow through the seafloor have been made since Bullard and Day (1961) first reported on data from the Atlantic Ocean and Mediterranean Sea. In order for the measured heat flow through the near-surface sediments to be representative of the geophysically significant steady-state heat flow into the base of the sediments, certain conditions must pertain. Among the most important of these are:

- constant bottom water temperature;
- movement of heat entirely by molecular conduction;
- constant position of the seafloor with respect to a coordinate system within the earth, or, in other words, no erosion of, or deposition on, the seafloor;
- the absence of seafloor topography and variations in the thermal properties of the sediment beneath the seafloor;
- the absence of endothermic or exothermic processes within the crust.

In many areas of the deep ocean, the above conditions are met closely enough that the observed heat flow can be interpreted directly in terms of its geophysical significance. In shallower, enclosed, marginal basins such as the Mediterranean and Aegean Seas, sedimentation rates, seafloor sub-bottom topography, and bottom water temperature variations can cause the heat flow observed near the seafloor to be different from the geophysically significant heat flow at depth. Thus in order to draw meaningful conclusions regarding the temperature distribution in the mantle from the heatflow data, the measurements must be made in carefully chosen locations using techniques which minimize environmental effects, and the observed values must be corrected for whatever effects are believed to remain.

In this regard, the measurement of heat flow using temperatures determined in deep boreholes drilled during Leg 42A of the Deep Sea Drilling Project are extremely important because many of the environmental effects cited above are either less significant at depth (short-term bottom water temperature variations and local seafloor topography, for example) or can be estimated

(1) Geology Department, University of Georgia, Athens, Georgia 30602, U.S.A.

(2) Department of Earth and Planetary Physics, Massachusetts Institute of Technology, Cambridge, Massachusetts 02139, U.S.A.

(3) Lamont-Doherty Geological Observatory of Columbia University, Palisades, New York 10964, U.S.A.

from the drilling results themselves (erosion or sedimentation of the seafloor, movement of interstitial water).

II. HEATFLOW DATA

A. REVIEW OF HEATFLOW INVESTIGATIONS

Only a few isolated marine heatflow measurements, reported upon by Bullard and Day (1961), Lister (1963), Nason and Lee (1964), and Birch and Halunen (1966), had been reported from the Mediterranean Sea prior to 1969. Since that time additional marine heatflow measurements have been made in both the Eastern and Western Mediterranean by Ryan (1969), Erickson (1970), and Haenel (1972). The heatflow data obtained by Ryan (1969) and Erickson (1970), originally discussed in unpublished doctoral theses, are tabulated below in Table 1.

Additional marine heatflow measurements were made in the Adriatic Sea by Lavenia (1967) and in the Aegean Sea by Erickson (1970) and Jongsma (1974). Both the Adriatic and Aegean Seas, because of their shallow average depth, have extremely variable water temperature (Wüst, 1961; Pollack, 1951) and it would appear possible that conventional heatflow values could be seriously affected in either region. The fact that non-linear temperature-depth profiles were observed at all four heatflow stations made by Jongsma (1974) in the Southern Aegean further underscores the necessity for caution in attaching geophysical significance to the observed heat flow in these areas.

In addition to the above conventional marine heatflow data, five heatflow values, measured in boreholes drilled in the Eastern and Western Mediterranean and Aegean Seas during Leg 42A of the Deep Sea Drilling Project, have been reported upon by Erickson and Von Herzen, (in press, 1976). The measurements in the Mediterranean Sea are in generally good agreement with conventional heatflow data; however, the single borehole heatflow value determined in the Southern Aegean Sea is significantly higher than nearby marine heatflow values (see discussion below).

B. MEASUREMENT TECHNIQUES

1. <u>Conventional near-surface heatflow measurements</u>

Conventional heatflow measurements are made by inserting either a steel probe or hollow core barrel vertically into the sediment and measuring the temperature of the sediment at different depths below the seafloor (Langseth, 1965). The measurements reported here were made from temperatures determined at from two to eight depths over distances ranging from 1 to 12 m below the seafloor, depending upon both the type of equipment used and the ability of the temperature probe or corer to fully penetrate the sediment.

Thermal conductivity data are essential for calculation of heatflow values. Although the thermal conductivity of ocean sediments can be measured in several different ways, most of the heatflow values discussed in this paper were calculated using the results of shipboard conductivity measurements on sediment core determined by the transient needle-probe technique (Von Herzen and Maxwell, 1959). The thermal conductivity values measured aboard ship were, in most cases, corrected to values at the seafloor using temperature and pressure corrections reported by Ratcliffe (1960). The reader is referred to the references cited above for specific information on the method used by each investigator to compute and use conductivity data.

2. <u>Measurements in deep-sea boreholes</u>

Down-hole temperature measurements are made using an entirely self-contained instrument which fits in the bottom of a standard wire-line core barrel (Erickson <u>et al</u>., 1975). Temperature is measured every 8 seconds for about 1½ hours and the data are stored digitally on magnetic tape in the instrument. Data reduction is accomplished aboard ship using either a computer or tables to convert the digital data stored on the magnetic tape to downhole temperatures.

After drilling to the depth where a temperature measurement is desired, the downhole temperature instrument is lowered inside the core barrel to the bottom of the drillstring. Once the core barrel is latched into the drill bit, the drillstring is lowered slowly to the bottom of the previously drilled hole. The thermistor probe is then pushed

30 to 50 cm into the undrilled sediment and left with a minimum of disturbance for 15 minutes to permit it to approach thermal equilibrium with the sediment. At the end of the measurement interval the bit is raised a few meters above the bottom of the hole and the core barrel is retrieved using the wire line. Due to the low thermal diffusivity of sediment, the thermal effects of drilling do not penetrate more than 5 or 10 cm below the bottom of the hole in the length of time (usually 1 to 2 hours) from cessation of drilling until the temperature probe is emplaced. Thus the temperatures measured in the sediment are, in most cases, believed to be representative of the *in situ* sediment temperature.

Although the precision and accuracy of the downhole temperature system are 0.01 and 0.10°C, respectively, the actual error in estimating the *in situ* sediment temperature may be much larger, and, in some cases, difficult to estimate. The main source of uncertainty is whether or not the material in which the temperature probe is immersed is at *in situ* sediment temperature. If sediment sloughs off the walls of the hole and fills the bottom of the hole, it will have a lower temperature than the undrilled sediments beneath the hole. Depending upon the physical properties and quantity of the sloughed sediment, an erroneously low bottomhole temperature may be measured in the slumped sediment rather than ahead of the drill bit. Other problems arise if the undrilled sediment is either too hard for the probe to penetrate fully or is penetrated in such a way that cool water from the hole can flow down along the probe to cool the tip. If the bottomhole assembly containing the temperature probe is moving vertically due to ship motion, the temperature measurement will be disturbed by frictional heating and rapid cooling as the probe is reinserted into, or prematurely withdrawn from, the bottom of the hole during the temperature measurement. In such a case, the observed temperature may only provide a very rough estimate or set wide limits on the *in situ* sediment temperatures. Calm seas and easily penetrable sediment encountered during most measurements in the Mediterranean and Aegean Seas permitted measurements of downhole temperatures which are generally believed to be representative of *in situ* sediment temperatures. Thermal conductivity measurements were made aboard ship on sediment cores recovered during normal coring operations using the needle-probe technique (Von Herzen and Maxwell, 1959), and corrected to pressure and temperature conditions below the seafloor according to Erickson (1973).

C. PRESENTATION OF DATA

1. Western Mediterranean Basins

Twelve reliable marine heatflow values and one borehole heatflow value are available from the Alboran, Balearic, and Algero-Provencal Basins, as defined by Biju-Duval, *et al.* (1974) (Figure 1). Values vary from 3.15 to 0.79 HFU (1 HFU (heat flow unit) equals 1 μcal/cm^{2}s or 41.87 mW/m^2) and show no systematic pattern in their variation. Heatflow values significantly greater than the world average of about 1.45 HFU (Von Herzen and Lee, 1969) are found from the Alboran Sea to the northwestern Ligurian Sea. The mean and standard deviation of the 12 conventional heatflow values considered to be reliable is 1.83 $\pm$ 0.66 HFU, somewhat lower than the value of 2.45 $\pm$ 0.16 HFU determined from temperature measurements in a borehole drilled at the base of the East Menorca Rise (Erickson and Von Herzen, in press, 1976).

Heat flows of 1.35 and 3.15 HFU were measured in the western end of the Balearic Abyssal Plain, near the area affected by the turbidity currents associated with the Orleansville earthquake (Heezen and Ewing, 1955). Although submarine cables were not broken in the immediate vicinity of either station, the low heatflow value (1.35 HFU) may have been affected by the sudden deposition of a thick layer of turbidite material.

One heatflow measurement just west of the Balearic Abyssal Plain and two others on the Alboran Abyssal Plain yielded values of 0.79, 2.22, and 1.25 HFU, respectively. Their proximity to the steep, earthquake-prone North African Continental Slope and evidence for fault tectonics in the Alboran Basin suggests that some of the variability may be due to sedimentation, topographic, or refractive effects, in addition to (or instead of) local variations in the thermal properties of the upper mantle and crust.

Four reliable heatflow values are available from the Rhone Fan. Three are situated at the base of the fan, near the northern Balearic Abyssal Plain. The two measurements on the northeastern margin of the fan have values of 2.50 and 2.08 HFU, substantially higher than the heat flow of 1.2 HFU measured at a third station located either within or near a major channel at the southern edge of the fan. Nason and Lee (1964) noted that the value of 1.2 HFU was of questionable geophysical value because of the effects of instabilities in bottom water temperature and a probable high sedimentation rate. The observed value is lower than the regional average and may indeed reflect one or both of the perturbations suggested by Nason and Lee. A heat flow of 1.39 was measured at a station located well up on the fan in an area eroded by a system of large, disrupted, leveed channels that were possibly formed during the Pleistocene when Alpine glaciation and deglaciation introduced large volumes of relatively coarse sediment into the Rhone River (Menard et al., 1965). It is probable that the observed heat flow of 1.39 HFU also should be increased substantially in order to represent the steady state heat flux.

2. Tyrrhenian Sea

Twelve reliable conventional heatflow values are available from the Tyrrhenian Sea, 10 of which were obtained with the cooperation of Italian scientists aboard the M. N. Bannock (Figure 1). Although the Tyrrhenian Sea has a relatively small abyssal plain, it is surrounded on the northeast and south by a broad, deformed, gently sloping continental slope. Heatflow values from the Tyrrhenian Basin are strikingly high, with an average and standard deviation of 2.83 ± 1.07 HFU. The average of four measurements made in the basin southeast of the abyssal plain is 3.38 ± 0.16 HFU. Heatflow values on or near the abyssal plain range from 0.72 to 3.92 HFU with a mean of 1.83 ± 0.66 HFU. The lowest value (0.72 HFU) was measured in a narrow channel connecting the slightly deeper western lobe with the eastern lobe of the Tyrrhenian Abyssal Plain. It is possible that the unusually low heat flow is due to sediment slumping off of the adjacent topographic highs. In addition, heat would tend to flow out through the higher thermal conductivity areas on either side of the sediment filled channel, further reducing the heat flux measured in the channel. Heatflow values of 2.60 and 3.92 HFU were measured at stations in the southeastern corner and in the western lobe of the abyssal plain, respectively. The fact that significantly different heatflow values (4.14 and 1.38 HFU) were obtained at measurements only 10 km apart, less than 20 km northwest of the abyssal plain, indicates that if the source of the variation is geophysical rather than environmental, it is located within the crust or uppermost mantle.

A heatflow of 1.72 ± 0.25 HFU was measured in a borehole drilled through 370 m of sediment into the flank of a basaltic seamount in the central Tyrrhenian Abyssal Plain (Erickson and Von Herzen, in press, 1976). This value, although still greater than the average global flux (Von Herzen and Lee, 1969) is substantially lower than the average Tyrrhenian Sea heatflow (2.83 HFU) determined from conventional heatflow measurements. The location of the measurement on the flank of the seamount suggests the possibility that thermal refraction caused by the close proximity of the highly conductive basalt may have reduced the heat flow through the sediments at the flank of the seamount.

3. Straits of Sicily

A single, highly reliable heatflow measurement exists in the Linosa Basin south of Sicily. The measured value of 1.20 ± 0.13 HFU is lower than all but three of the 26 heatflow measurements made in the Western Mediterranean, but is higher than all but five (three of which were reported by Haenel (1972) and which may be systematically high) of the 38 heatflow values determined in the Eastern Mediterranean to be discussed in the next section. Abundant geophysical evidence reviewed by Morelli et al.(1975) exists to suggest that this region has been and still may be tectonically and magmatically active. The location of the measurement in a thickly sedimented depression may have preferentially reduced the heat flow through the sediment relative to the rate of heat flow through the surrounding area. It is interesting to note that this region appears to be thermally transitional between the high heat flow characteristic of the entire Western Mediterranean and the low heat flow found throughout the Eastern Mediterranean.

4. Eastern Mediterranean Sea

Thirtyeight measurements exist in the Eastern Mediterranean Sea (Figure 2), 31 of which are listed in Table 1. Four measurements in the Ionian Sea reported by Haenel (1972) and an additional heatflow value of 0.64 HFU determined by Birch and Halunen (1966) are included in Figure 2. The values reported by Haenel ranging from 1.01 to 1.37 HFU are significantly greater than all but one of the eight other heatflow values in the Ionian Sea which range from 0.32 to 1.76 HFU, with a mean of 0.82 HFU. Haenel notes in his paper that these data were obtained during the testing phase of a new heat probe, and that the observed temperature gradient has been corrected for bottom water temperature changes. The corrections applied range from +9.3% to -22.5% and the method and assumptions used in calculating the corrections are not given. It appears possible that these data are systematically high, and they are not considered further in the following discussion.

The most obvious characteristics of the observed heat flow in the Eastern Mediterranean are its generally low average value (0.74 ± 0.30 HFU) and its small variance across different physiographic provinces. There is no evidence for any regional heatflow anomaly associated with the Mediterranean Ridge, nor with the basin and trench provinces located north and south of it.

Where individual anomalously high or low values occur, the measurement is invariably located in an area where the anomalous heatflow value, whether high or low, can reasonably be attributed to the effect of nearby extreme topography and/or sedimentation processes. As an example, the station at which a heat flow of 1.76 HFU was measured was located at the bottom of a narrow, steep-walled valley in a highly deformed and fractured area of the Mediterranean Ridge. The effect of the local seafloor relief would be to increase the heat flux through the floor of the valley. Two of the three heatflow values (0.38 and 0.49 HFU) on the Nile cone are well below the Eastern Mediterranean average, probably as a result of the tectonic activity and/or locally high sedimentation rate caused by rapid deposition of the large amount of sediment carried into the Eastern Mediterranean by the Nile River (Ross and Uchupi, in press, 1976).

Further documentation of regionally low heat flow in the Mediterranean is provided by measurements made in boreholes in the central Ionian Sea and on the Florence Rise (Erickson and Von Herzen, in press, 1976). The measurement in the Ionian Sea is located in the central part of the Messina Abyssal Plain and should be representative of the regional heat flow. The well determined heat flow of 0.80 ± 0.10 HFU determined at this site is in excellent agreement with most other Ionian Sea heatflow determinations, (with the exception of the data noted above).

The heat flow of 0.94 ± 0.35 HFU measured on the Florence Rise west of Cyprus is less reliable in view of the observation of a decrease in temperature with depth below 74 m subbottom, presumably caused by downward flow of seawater into the drillhole due to penetration of a subsurface zone of high permeability (Erickson and Von Herzen, in press, 1976). However this value also is below the world average and is not significantly different from either the mean of the conventional marine heatflow measurements in the vicinity of Cyprus or from the average heat flow (0.67 ± 0.19 HFU) determined in boreholes on Cyprus by Morgan (in press, 1976).

5. <u>The Aegean Sea</u>

Fourteen conventional heatflow measurements and one borehole heatflow value exist from the Aegean Sea. Evidence for deep convective overturn of water in the Aegean has been presented by Miller (1972), suggesting that heatflow values obtained using conventional shipboard methods may be either increased or decreased relative to their steady state value. Indeed, nonlinear thermal gradients were observed at all four of the heatflow measurements made by Jongsma (1974) in the Southern Aegean Sea, from which Jongsma calculated heatflow values ranging from about 1.0 to 1.6 HFU. A single measurement (Erickson, 1970) near one of Jongsma's stations gave a well determined value of 1.05 HFU. Although the gradient was calculated using only the bottom water temperature and two sediment probes, the temperature appeared to increase linearly with depth from the seafloor to at least 5.2 m below the seafloor. This apparent linearity may simply be a consequence of the seasonal differences in the extent of warming which had taken place in the bottom water at the times (late August and November) that heatflow measurements were made by Erickson and Jongsma, respectively.

Nine additional heatflow values determined by Jongsma (1974) in the Northern and Central Aegean Sea ranged from 1.24 to 2.73 HFU, with a mean and standard deviation of 1.98 ± 0.24 HFU. Despite the fact that more than half of these measurements were made in water depths of less than 1 km, constant thermal gradients were determined over the two 1 m intervals using three outrigger-type temperature sensors. In view of the presence of non-linear temperature-depth relationships observed in the generally deeper Southern Aegean measurements, the reason why the measurements in the Northern and Central Aegean were not similarly affected is difficult to understand.

A reliable borehole heatflow measurement was made in the Southern Aegean and yielded a value of 2.12 ± 0.11 HFU (Erickson and Von Herzen, in press, 1976). This value is significantly greater than any of the conventional heatflow measurements in the Southern Aegean, but is nearly equal to the mean of the nine heatflow values determined in the Northern and Central Aegean.

III. INTERPRETATION

A. ENVIRONMENTAL EFFECTS

As noted earlier, local and regional geological, oceanographic, and tectonic factors are capable of causing the measured, near-surface heat flow to be either greater or smaller than the regional, geophysically relevant heat flow. Some of the local environmental effects which must be considered were discussed in the previous section. The main purpose of this section of the paper will be to document the existence of geophysically significant regional differences in the heat flow through the floors of the Western Mediterranean, Eastern Mediterranean, and Aegean Seas by demonstrating that the regional environmental effects are too small to significantly alter the measured heatflow values.

The general agreement between the conventional oceanographic and deep borehole heatflow measurements suggests that seasonal bottom water temperature variations have not seriously influenced the temperature distribution in the upper 10 m below the sea floor, except in the Southern Aegean Sea. Furthermore, persistence of nearly linear temperature-depth relationships to subbottom depths of 155 m, 304 m, and 226 m in the Western Mediterranean, Ionian, and Aegean Seas, respectively, suggests that bottom water temperatures have not undergone large amplitude, long-term changes during the last few thousand years.

Thermal refraction is caused by the preferential flow of heat through high thermal conductivity (rock or salt) materials and around low thermal conductivity (sediment) deposits. Thermal refraction can significantly affect the measured heat flow in the vicinity of shallowly buried, rough subbottom topography defining materials of strongly contrasting thermal conductivity (Selig and Wallick, 1966; Von Herzen and Uyeda, 1963; Lee and Henyey, 1974). The effect of thermal refraction is to alter the local distribution of heat flow, rather than the average regional heat flow. The presence of thermally conductive evaporite layers and associated salt domes and ridges (Biju-Duval et al., 1974; Mauffret et al., 1973) beneath much of the Western and Eastern Mediterranean Seas suggests that thermal refraction may be locally important in determining the measured heat flow, especially where there is substantial subbottom relief on the evaporitic layer and/or the structures are thinly buried. However, since the locations of the heatflow measurements were generally determined independently of the subbottom structure, the average of the heatflow values measured above, along side of, and far from highly conductive subbottom structures should yield a value close to the average regional heat flow.

The ability of long-term sedimentation to lower the measured heat flow is widely known (Von Herzen and Uyeda, 1963; Jaeger, 1965; Langseth, 1965). The generally thick and highly variable accumulations of sediment in the various Mediterranean basins suggest that sedimentation may have systematically decreased the measured heat flow by significant and variable amounts. In order to estimate the effects of sedimentation in the various regions, the sedimentary history (rate and duration of sedimentation) for each area must be known or estimated. These parameters are still poorly known, and thus the sedimentation corrections derived from these parameters are intended only to set limits on the heat flow through the various regions.

Sedimentation rates have been estimated at most of the Deep Sea Drilling Project boreholes drilled in the Mediterranean and Aegean Seas during Legs 13 and 42A. Although sedimentation rates as high as 31.3 cm/1000 yrs have been calculated for the Pleistocene, sedimentation rates during the Pliocene and Miocene are generally less than about 10 cm/1000 yrs (Cita, Ryan, and Kidd, in press, 1976). Averages of Pliocene and Pleistocene sedimentation rates in the Eastern and Western Mediterranean are approximately equal (6.69 and 5.58 cm/1000 yrs, respectively) and bracket the single sedimentation rate calculated for the Southern Aegean Sea (6 cm/1000 yrs). A major uncertainty arises as to whether or not it is reasonable to assume the same sedimentation rate for times earlier than Miocene, and just how far back in time one should extrapolate (i.e., how old are the basins? Furthermore, many of the drill sites were situated in areas chosen for their low sedimentation rates in order to maximize the possibility of drilling into older sediments, and thus probably are not representative for the basin as a whole.

A more reliable method of estimating the sedimentation correction is to use geophysical data on the average thickness and age of the sedimentary layers in the various basins to calculate average sedimentation rates and durations for each region. Unfortunately, this method also requires many questionable assumptions. Estimates of the ages of the various reflectors vary between investigators, and detection and/or identification of the base of the sedimentary sequence is often uncertain.

Generally good agreement exists among investigators on the thickness and age of the uppermost sediment layer, resting on strongly reflecting evaporitic and salt layers of Upper Miocene age throughout most of the Western and Eastern Mediterranean. Much less agreement is found on the subject of the maximum age and thickness of the older sediment (?) layers beneath the salt layer. The values used in Tables III and IV were chosen to represent a general average of the estimates presently in the literature.

Corrected heatflow averages were calculated for both the relatively well-known Plio-Pleistocene sedimentation (Table III), and for the less well-known pre-Pliocene sedimentary history (Table IV). In both cases, the heat flux through the Western Mediterranean and Tyrrhenian basins is at least three times as great as the heat flow through the floor of the Eastern Mediterranean Sea, principally due to the difference in the measured heat fluxes in the two regions rather than to the estimated effects of sedimentation. The small effect of sedimentation on heat flow in the Eastern Mediterranean is further suggested by the near equality of terrestrial heatflow values measured in boreholes on Cyprus (Morgan, in press, 1976) with marine heatflow measurements in the immediately surrounding regions of the Eastern Mediterranean, and the failure of the observed heatflow values in the Western Mediterranean basins and the Ionian Sea to be inversely related to sediment thickness beneath the measurement site as estimated from isopach maps presented by Auzende and Olivert (1974) and Finetti and Morelli (1972) suggests a similar result for this area as well.

The thinner sediment accumulation and higher measured heat flow in the Tyrrhenian Sea than in the other Western Mediterranean basins results in the calculation of approximately equal corrected heat flow through both regions. This is somewhat surprising in view of the evidence for active volcanism in and around the Tyrrhenian Sea as contrasted with the general absence of volcanism in the Mediterranean basins further west, and may suggest that the calculated sedimentation effect for the western Mediterranean Basins is too high.

The existence of heat flow out of the mantle greater than about 5 HFU is considered unlikely because of the unreasonably high temperature which would have to exist at the base of the 6 or 7 km thickness of sediment known to exist in the northwestern Mediterranean (Mauffret _et al._, 1973). If an average thermal conductivity of 5 mcal/cm s^{o}C (1 mcal/cm s^{o}C = 0.4187 W/mK), typical of consolidated terrigenous sedimentary rocks (Clark, Table 21-1, 1966) is assumed, temperatures at the base of 7 km of sediment should be well over 600°C. A positive indication that such is not the case is provided by the presence of well-defined, low amplitude magnetic anomalies throughout the Western Mediterranean basins (Bayer _et al._, 1973). The presence of magnetic anomalies, presumed to originate from beneath, rather than from within the sediments, requires that the basement temperatures are less than the Curie point of the common magnetic minerals ($\approx$550 C), and in fact must be less than the blocking temperature which, for basalt more than a few million years old, is in the range 300 to 500°C (Irving, 1970). Again assuming an average thermal conductivity of 5 mcal/cm s^{o}C, and that the temperature at the base of 7 km of sediment is 450°C hotter than the sea floor, a maximum heat flow of 3.21 HFU can be computed, somewhat higher than the value of 2.25 HFU arrived at through correcting the measured heatflow values for the effects of Plio-Pleistocene sedimentation (Table III).

The variability of the heatflow values in the Western Mediterranean Basin is at first surprising in view of the thick, generally horizontally-bedded sediments and the absence of any evidence of volcanic or igneous intrusive activity over much of the region.

Much of this variability is probably due to thermal refraction caused by salt domes and salt ridges (Selig and Wallick, 1966) made of dolomite, anhydrite, and rock-salt having thermal conductivities 4 to 5 times greater than those of near-surface unconsolidated sediments (Clark, Table XII-2, 1966). Diapiric structures are known to approach close to or even protrude above the sea floor at many places in the Balearic Basin (Stanley _et al._, 1974) and the northwestern Mediterranean (Mauffret _et al._, 1973). Heatflow values measured on the crests of diapiric structures on the continental slope west of Angola were systematically 2 or 3 times larger than those measured between the diapirs (Von Herzen _et al._, 1972). In addition to the effects of thermal refraction, vertical movement of the salt has, in some regions, produced sea floor relief of as much as several meters, which in itself may contribute to variability in the near-surface thermal gradient as well as modifying the sediment dispersal pattern (Stanley _et al._, 1974).

It is interesting to speculate why, in view of the widespread occurrence of evaporitic sediments beneath the Eastern as well as the Western Mediterranean (Biju-Duval _et al._, 1974), similar variation is not evident in the heatflow values measured in the Eastern Mediterranean. The much lower variability of the Eastern Mediterranean data is believed to be due to the generally more subdued subbottom topography of the evaporite layers. With the marked exception of the southeastern Mediterranean in the vicinity of the Nile Cone (Ross and Uchupi, in press, 1976), most of the evaporites in the Eastern Mediterranean are flat-lying, and thus are not effective in producing thermal refraction effects.

Lateral variations in the size, shape, and abundance of halokinetic structures in the Mediterranean have been noted by Stanley *et al*. (1974) and Mauffret *et al*. (1973). Mauffret *et al*. (1973), after considering the effects of variations in salt thickness, the thickness and the nature of the sedimentary overburden, and tectonic influence, conclude that there is apparently no simple explanation for the morphological variety of the salt structures in the Western Mediterranean, although they note that vertical faulting seems to play an important role. It is interesting to speculate that the much greater abundance of diapiric structures in the Western Mediterranean than in most of the Eastern Mediterranean, where the salt and evaporitic layers, though present, are generally flat-lying, is due to the higher temperature of the Western Mediterranean salt. Gussow (1968) states that "increased temperature greatly reduces the ultimate strength and eliminates work hardening. When salt is heated above 400°F (205°C), it becomes soft and plastic and flows indefinitely with a pressure gradient of about 33-100 kg/cm (460 to 1400 psi)."

Thermal gradients of 76 and 28°C/km were measured in deep boreholes in the western Provencal and Ionian basins respectively (Erickson and Von Herzen, in press, 1976). The Western Mediterranean salt layer, variously estimated from 0 to 4.0 km thick (Finetti and Morelli, 1972), is overlain by 1 to 1.5 km of Plio-Pleistocene sediments and 0.6 km of late Miocene evaporites (Mauffret *et al*., 1973). Even allowing for the high thermal conductivity of the evaporites and salt, temperatures within the salt layer could easily exceed 200°C.

Salt beneath the Eastern Mediterranean, with its lower thermal gradient and much thinner covering of Plio-Pleistocene sediments, will be at a lower temperature. Plio-Pleistocene sediments only 0.3 km thick are present throughout much of the Ionian sea over an evaporitic layer with an average thickness of about 2 km (Finetti and Morelli, 1972). Temperatures at the base of the thick, but highly conductive salt layer would be less than 50°C. Such low temperatures may well inhibit the development of diapiric structures except in areas where tectonic effects are important (Ross and Uchupi, in press, 1976).

IV. DISCUSSION

The main conclusion is that the observed regional difference between the observed heat flow through the floors of the Eastern Mediterranean and the Western Mediterranean Seas (Figure 3, Table II) is geophysically significant, and that it indicates the presence of hotter mantle material at shallower depths beneath the Western Mediterranean than beneath the Eastern Mediterranean.

The observation that heat flow is higher in the Tyrrhenian Sea than in the other Western Mediterranean Basins, in conjunction with the distribution of seismic and volcanic activity, indicates that the Tyrrhenian is currently underlain by hot, mobile mantle material and is tectonically active, both in its central and peripheral regions. These observations support the tectonic model proposed by many investigators in which the Tyrrhenian Sea is considered to be a back-arc basin produced as a consequence of subduction of lithosphere beneath the Calabrian Arc.

All of the remaining Western Mediterranean basins also are characterized by high heat flow, although somewhat lower than that measured through the floor of the Tyrrhenian Sea. Although sediment thicknesses in the Western Mediterranean basins are greater than in the Tyrrhenian Sea, the close approach to (or penetration of) the seafloor by salt and evaporitic structures having higher thermal conductivity than the overlying sediments can produce heat flow variability through thermal refraction. The observation of high heatflow values throughout the Western Mediterranean, even as far west as the Alboran basin, indicates that whatever process is producing the high heat flow is present beneath the entire Western Mediterranean. Production of the Western Mediterranean whether by rifting from a land-locked mid-ocean ridge segment (Bayer *et al*., 1973), induced flow behind an island arc (Andrews and Sleep, 1974), or in response to the translation and rotation of microcontinents (Alvarez *et al*., 1974), would be expected to be accompanied by high heat flow.

The well-documented uniformly low heat flow throughout the entire Eastern Mediterranean is not consistent (Morgan, in press, 1976) with the explanation that the low heat flow is due to depression of the isotherms as the floor of the Eastern Mediterranean is subducted beneath the Hellenic Arc (Erickson, 1970). Recent and continuing compression, thickening, and consequent uplift of the Mediterranean Ridge, as evidenced by the presence of elevated turbidite layers (Hsu and Ryan, 1973) could produce low heat flow, and in fact may be responsible for some of the exceptionally low heatflow values observed on the Mediterranean ridge. However this explanation also is incapable of explaining the regionally low heat flow, and in fact, heatflow values on the ridge are not systematically lower than heatflow values in the adjacent undeformed regions.

It appears more likely that the Eastern Mediterranean is a relict piece of the oceanic crust of the former Tethys seaway. The mean heat flow observed through the eastern Mediterranean (0.74 HFU) is somewhat lower than the commonly accepted equilibrium value of heat flow (1.1 to 1.2 HFU) (Sclater et al., 1976) through the oldest ocean floors. However, as noted, the assumption of reasonable sedimentation rates and durations permits calculation of a sedimentation effect which suggests that sedimentation may have reasonably reduced a steady state flux of 1.1 to 1.2 HFU to the observed mean value. The measurement of a similarly low heat flow in boreholes on Cyprus (Morgan, in press, 1976) supports suggestions that Cyprus was formerly a piece of Mesozoic ocean floor which was thrust up over itself during the convergence of the African and Eurasian lithospheric plates, but argues against the very low marine heat flow being entirely due to sedimentation effects.

A more speculative interpretation can be formulated based largely on controversial evidence presented by Jordan (1975) and Sipkin and Jordan (1975), and criticized by Okal and Anderson (1975), for the existence of thermal and compositional differences in the upper mantle beneath oceans and continents to depths exceeding 400 km. If the temperature and rate of radiogenic heat production in the asthenosphere beneath the continents is lower than that beneath oceanic lithosphere, then one of the possible effects of plate movement, and in particular of continent-continent collisions, might be the displacement of oceanic lithosphere onto or over continental asthenosphere (or visa versa). It is suggested that this has occurred, or is occurring beneath the Eastern Mediterranean, and that the present regionally low heat flow through that area is due to the northward movement of cool, poorly radiogenic "continental" asthenosphere from beneath Africa, resulting in the displacement of the hotter, oceanic asthenosphere which was originally beneath the Tethys northwards and downwards beneath the Aegean Sea, Turkey, and southeastern Europe, all of which are regions having had recent volcanic activity or characterized by above normal heat flow.

According to either of these models, the Eastern Mediterranean is underlain by what was formerly typical oceanic crust and lithosphere, which is in turn underlain by continental asthenosphere. However, the location of the remnants of this crust within the Alpine geosynclinal region has resulted in crustal thickening, deformation, accumulation of sediment, and possibly chemical alteration of the crust until it now has characteristics which are transitional between crust typical of oceanic regions and crust typical of continental regions.

Due to the presence of significant seasonal bottom water temperature variations in the Aegean, the geophysical interpretation of conventional marine heatflow values appears risky. It may be significant that the average of the 9 conventional measurements in the Northern and Central Aegean classified as reliable (Jongsma, 1974) is in excellent agreement with a single measurement in a deep borehole in the Southern Aegean, suggesting that the entire region may be characterized by heat flow of about 2.0 HFU, nearly three times higher than the heat flow south of Crete (Figure 4). As noted by Jongsma, correction for the effects of sedimentation and a post-glacial, long-term increase in bottom water temperature could increase this value by 10 to 30%. The determination of regionally high heat flow in the Aegean Sea as a whole supports its tectonic classification as a back-arc basin and would seem to preclude the possibility that the Northern Aegean region is undergoing compression (Papazachos, 1976), prior to the formation of a new subduction zone (Maley and Johnson, 1971).

V. SUMMARY

The high variability of the regional heat flow in the Eastern and Western Mediterranean and Aegean Seas indicates that few, if any, of these areas share a common style and/or time of origin, thus confirming the geophysical complexity of the Mediterranean area.

High heat flow, volcanism, the distribution of earthquakes, and geochemical data (Ninkovich and Hays, 1972) support the association--but not necessarily the origin--of the Tyrrhenian and Aegean Seas with the proposed subduction of lithosphere behind the Calabrian and Cretan arcs. Abundant geological and geophysical data exist to suggest the former existence of continental crust in both regions, thus the geodynamic models usually proposed for the formation of marginal seas in oceanic regions (Andrews and Sleep, 1974) probably cannot be applied to these seas without modification.

The contrast in heat flow through the Mediterranean Sea east and west of Sicily is especially striking (Figure 3). The fact that the entire Eastern and Western Mediterranean Seas are characterized by heat flow significantly lower and greater, respectively, than the world average suggests that these two regions have either originated in drastically different ways, and/or differ in age by lengths of time on the order of the thermal time constant of the lithosphere (100 m.y.).

It is tempting, as discussed earlier, to consider the low heat flow characteristic of the Eastern Mediterranean as indicating that this area is comprised of very old oceanic crust, thickened, altered, and deformed due to its position in the Alpine geosyncline.

Paleomagnetic and other lines of evidence suggest that the Western Mediterranean basins originated by a combination of subsidence, rifting, or rotation of both continents and pieces of continents (microcontinents) since Mid-Tertiary. These conclusions are in accordance with the determination of high heat flow in these basins, regardless of the details of the mechanism proposed. The source of the heat necessary to cause such reorganization is worthy of some speculation.

In general, regions of high heat flow are believed to be associated with zones of upwelling mantle material, usually manifested as "hot spots", mid-ocean ridges, or subduction zones. Despite evidence for the convergence of Africa and Europe, there is little geological or geophysical evidence for present or past subduction in the Algero-Provencal or Balearic basins, and even less for mantle plumes or midocean rises. Once again drawing upon the speculative (and controversial) results of Jordan (1975), the question is raised that *if* the oceanic upper mantle *is* hotter or more radioactive than the upper mantle beneath continents, then where is the thermally and compositionally distinctive oceanic upper mantle formerly beneath the Tethys, now that Africa and Europe are, in fact, touching? One possibility is that it has been forced northwards beneath what was once the edge of Europe, and has provided the thermal energy for the Alpine orogeny and for the formation of the various Mediterranean basins. The extremely high radioactivity (Imbo *et al.*, 1968) and geochemistry of southern Italian volcanoes is consistent with the assimilation of continental crust (Hurley *et al.*, 1966; Klerkx *et al.*, 1974) by anomalously hot, radioactive mantle in this region.

VI. ACKNOWLEDGEMENTS

The data listed in Table I were obtained during several different expeditions by various institutions. Most of the heatflow measurements in the Mediterranean were made either aboard the R/V Conrad of the Lamont-Doherty Geological Observatory of Columbia University, sponsored by Office of Naval Research grant N00014-67-A-0108-004 and National Science Foundation grant NSF-GA-27281, or aboard the R/V Chain of the Woods Hole Oceanographic Institution under Office of Naval Research grant Nonr 1841(74). Heatflow measurements in the Tyrrhenian Sea were made by George Harlow aboard the M/V Bannock with the cooperation of the Osservatorio Geofisico Sperimentale-Trieste. The borehole heatflow measurements were made aboard the D/V Glomar Challenger in cooperation with the Deep Sea Drilling Project and funded under National Science Foundation grant GA28504. We extend our sincere thanks and appreciation to the officers and crews of all these vessels, and to the multitude of other people who, through their assistance, made these measurements possible.

VII. REFERENCES CITED

Alvarez, W., T. Cocozza, and F. C. Wezel, 1974. Fragmentation of the Alpine orogenic belt by microplate dispersal: Nature, v. 248 (5446), p. 309-314.

Andrews, D., and N. Sleep, 1974. Numerical modeling of tectonic flow behind island arcs: Geophys. J.R. Astron. Soc., v. 38, p. 237-251.

Auzende, J. M. and J. L. Olivet, 1974. Structure of the Western Mediterranean basin: In *The Geology of Continental Margins*, ed. C. A. Burk and C. L. Drake, Springer-Verlag, p. 723-731.

Bayer, R., J. L. Le Mouel, and X. Le Pichon, 1973. Magnetic anomaly pattern in the Western Mediterranean: Earth Planet. Sci. Letters, v. 19, p. 168-176.

Biju-Duval, B., J. Letouzey, L. Montadert, P. Courrier, J. F. Mugniot, and J. Sancho, 1974. Geology of the Mediterranean Sea Basins: In The Geology of the Continental Margins, ed. C. A. Burk and C. L. Burk, Springer-Verlag, p. 695-721.

Birch, F. S. and A. J. Halunen, Jr., 1966. Heat-flow measurements in the Atlantic Ocean, Indian Ocean, Mediterranean Sea, and Red Sea: J. Geophys. Res., v. 71(2), p. 538-586.

Bullard, E. C. and A. Day, 1961. The flow of heat through the floor of the Atlantic Ocean: Geophys. J., v. 4, p. 282-292.

Cita, M. B., Ryan, W. B. F., and R. H. Kidd, 1976. Sedimentation rates in Neogene deep-sea sediments from the Mediterranean. Geodynamic implications of their changes: in Hsu, K. J., Montadert, L. C., et al., Initial Reports of The Deep Sea Drilling Project, v. 42A, U.S. Govt. Printing Office, Washington, D.C., in press.

Clark, S. P., 1966. Thermal conductivity: In Handbook of Physical Constants, Geol. Soc. America Memoir, 97, ed. S. P. Clark Geol. Soc. America, New York, p. 459-482.

Erickson, A. J., and R. P. Von Herzen, 1976. Downhole temperature measurements--Leg 42A: in Initial Reports of the Deep-Sea Drilling Project, v. 42A, U.S. Govt. Printing Office, Washington, D.C. in press.

Erickson, A. J., R. P. Von Herzen, J. G. Sclater, R. W. Girdler, B. V. Marshall and R. Hyndman, 1975. Geothermal measurements in deep-sea drill holes: J. Geophys. Res., v. 80(17), p. 2515-2528.

Erickson, A. J., 1973. Initial report on downhole temperature and shipboard thermal conductivity measurements, Leg 19: In Initial Reports of the Deep-Sea Drilling Project, v. 19, U.S. Govt. Printing Office, Washington D.C., p. 643-656.

Erickson, A. J., 1970. The measurement and interpretation of heat flow in the Mediterranean and Black Seas: unpub. Ph.D. thesis, Massachusetts Institute of Technology, Cambridge, Mass., 272 p.

Finetti, I., and C. Morelli, 1972. Wide scale digital seismic exploration of the Mediterranean Sea: Boll. Geof. Teorica ed Applicata, v. 14(56), p. 291-342.

Finetti, I., C. Morelli, and E. Zarudzki, 1970. Reflection seismic study of the Tyrrhenian Sea: Boll. Geof. Teorica ed Applicata, v. 12(48), p. 311-346.

Gussow, W. C., 1968. Salt diapirism: Importance of temperature, and energy source of emplacement: In Diapirism and Diapirs, American Assoc. Petrol. Geol. Memoir No. 8, ed. J. Braunstein and G. D. O'Brien, American Assoc. Petrol. Geol., Tulsa, Oklahoma.

Haenel, R., 1972. Heatflow measurements in the Ionian Sea with a new heatflow probe; Meteor Forsch.-Ergenbisse, Reihe C., v.11, p. 105-108.

Hamilton, E. L., 1976. Variations of density and porosity with depth in deep-sea sediments: J. Sed. Petrol., v. 46(2), p. 280-300.

Heezen, B. C., C. Gray, A. G. Segre, and E. F. K. Zarudski, 1971. Evidence of foundered continental crust beneath the Central Tyrrhenian Sea: Nature, v. 229 (5283), p. 327-329.

Heezen, B. C., and M. Ewing, 1955. Orleansville earthquake and turbidity currents: Bull. Am. Assoc. Petrol. Geol., 39(2), p. 2505-2514.

Hsu, K. J. and W. B. F. Ryan, 1973. Summary of the evidence for extensional and compressional tectonics in the Mediterranean: In Ryan, W. B. F., Hsu, K. J. et al., 1973, Initial Reports of the Deep Sea Drilling Project, v. 13, U.S. Govt. Printing Office, Washington D.C., 1011-1019.

Hurley, P. M., H. W. Fairbairn, and W. H. Pinson, Jr., 1966. Rb-Sr isotopic evidence in the origin of potash-rich lavas of western Italy: Earth Planet. Sci. Letters, v. 5, p. 301-306.

Imbo, G. P. Gasparini, G. Luongo, and A. Rapolla, 1968. Contributions to the volcanological researches by determination of the radioactivity of eruptive products: Bull. Volcanol., v. 32, p. 1-25.

Irving, E., 1970. The Mid-Atlantic ridge at 45°N. XIV. Oxidation and magnetic properties of basalt; review and discussion: Canad. J. Earth Sci., v. 7, p. 1528-1538.

Jaeger, J. C., 1965. Applied theory of heat conduction: In Terrestrial Heat Flow, American Geophys. Union, Washington D.C. p. 7-23.

Jongsma, D., 1974. Heat flow in the Aegean Sea: Geophys. J.R. Astron. Soc., v. 37, p. 337-346.

Jordan, T. H., 1975. The continental tectosphere: Rev. Geophys. and Space Physics, v. 13(13), p. 1-12.

Klerkx, J., S. Deutsch, J. Hertogen, J. De Winter, R. Gijbels, and H. Pichler, 1974. Comments on "Evolution of Eolian arc volcanism (Southern Tyrrhenian Sea)" by F. Barberi, G. Ferrara, F. Innocenti, J. Keller, and L. Villari: Earth Planet. Sci. Letters, v. 23, p. 297-303.

Langseth, M. G., 1965. Techniques of measuring heat flow through the ocean floor: In Terrestrial Heat Flow, American Geophys. Union, Washington D.C., p. 58-77.

Lavenia, A., 1967. Heatflow measurements through bottom sediments in the southern Adriatic Sea: Boll. Geof. Teorica ed Applicata, v. 9(36), p. 323-332.

Lee, T. C. and T. L. Henyey, 1974. Heatflow refraction across dissimilar media: Geophys. J.R. Astron. Soc., v. 39, p. 319-333.

Lister, C. L., 1963. Geothermal gradient measurement using a deep-sea corer: Geophys. J.R. Astron. Soc., v. 7(5), p. 571-583.

Maley, T. S., and G. L. Johnson, 1971. Morphology and structure of the Aegean Sea: Deep-Sea Res., v. 18, p. 109-122.

Mauffret, A., Fail, M. P., Montadert, L., Sancho, J., and E. Winnock, 1973. Northwestern Mediterranean sedimentary basins from seismic reflection profile: American Assoc. Petrol. Geol. Bull., v. 57(11), p. 2245-2262.

Menard, H. W., S. M. Smith, and R. M. Pratt, 1965. The Rhone deep sea fan: in The Colston Papers, Vol. XVII, ed. W. F. Whittard and R. Bradshaw, Butterworths Scientific Publications, London, p. 271-285.

Miller, A. R., 1972. Speculations concerning bottom circulation in the Mediterranean Sea: In The Mediterranean Sea: A Natural Sedimentation Laboratory, ed. D. J. Stanley, Dowden, Hutchinson and Ross, Inc., Stroudsburg, Pa., p. 37-42.

Morgan, P., 1976. Cyprus heat flow with comments on the thermal regime of the eastern Mediterranean: paper submitted for inclusion in the IUGG European Heat Flow Monograph, in press.

Morelli, C., G. Gantar, and M. Pisani, 1975. Bathymetry, gravity, and magnetism in the Strait of Sicily and in the Ionian Sea: Boll. Geof. Teorica ed Applicata, v. 17(65), p. 39-58.

Nason, R. D. and W. H. K. Lee, 1964. Heatflow measurements in the North Atlantic, Caribbean, and Mediterranean: J. Geophys. Res., v. 69(22), p. 4875-4883.

Ninkovich, D., and J. D. Hays, 1972. Mediterranean island arcs and the origin of high potash volcanoes: Earth Planet. Sci. Letters, v. 16, p. 331-345.

Okal, E. A., and D. L. Anderson, 1975. A study of lateral inhomogeneities in the mantle by multiple ScS travel-time residuals: Geophys. Res. Letters, v. 2(8), p. 313-316.

Papazachos, B. C., 1976. Seismotectonics of the Northern Aegean area: Tectonophysics, v. 33, p. 199-209.

Pollack, M. H., 1951. The sources of the deep water of the eastern Mediterranean Sea: J. Marine Res., v. 10(1), p. 128-151.

Ratcliffe, E. H., 1960. The thermal conductivities of ocean sediments: J. Geophys. Res., v. 65(5), p. 1535-1541.

Ross, D. A. and E. Uchupi, 1976. The structure and sedimentary history of the southeastern Mediterranean Sea - Nile Cone area: American Assoc. Petrol. Geologists Bull., in press.

Ryan, W. B. F., 1969. The floor of the Mediterranean Sea. Part I: The structure and evolution of the basins: PhD Thesis, Columbia University, N.Y.

Ryan, W. B. F., K. Hsu, et al., 1973. Mediterranean Ridge, Levantine Sea - Site 130, In Ryan, W. B. F., Hsu, K. J. et al., Initial Reports of the Deep Sea Drilling Project, v. 13, U.S. Govt. Printing Office, Washington D.C., p. 355-382.

Sancho, J., J. Letouzey, B. Biju-Duval, P. Courrier, L. Montadert and E. Winnock, 1973. New data on the structure of the eastern Mediterranean basin from seismic reflection: Earth Planet. Sci. Letters, v. 18, p. 189-204.

Sclater, J. G., J. D. Mudie, and C. G. A. Harrison, 1970a. Detailed geophysical studies on the Hawaiian arch near 24° 25'N, 157°40'W: A closely spaced suite of heatflow stations: J. Geophys. Res., v. 75(2), p. 333-348.

Sclater, J. G., E. J. W. Jones and S. P. Miller, 1970b. The relationship of heat flow, bottom topography, and basement relief in Peake and Freen deeps, northeast Atlantic: Tectonophysics, v. 10, p. 283-300.

Selig, F., and G. C. Wallick, 1966. Temperature distribution in salt domes and surrounding sediments: Geophysics, v. 31(2), p. 346-361.

Sipkin, S. A., and T. H. Jordan, 1975. Lateral homogeneity of the upper mantle determined from the travel times of ScS: J. Geophys. Res., v. 80(11), p. 1474-1484.

Sleep, N., 1975. Stress and flow beneath island arcs: Geophys. J. R. Astron. Soc., v. 45, p. 827-857.

Stanley, D. J., H. Gott, O. Leenhardt, and Y. Weiler, 1974. Subsidence of the Western Mediterranean Basin in Pliocene-Quaternary Time: further evidence: Geology, v. 2(7), p. 345-350.

Stanley, D. J., F. W. McCoy, and L. Diester-Haass, 1974. Balearic abyssal plain: An example of modern basin plain deformation by salt tectonism: Marine Geol. v.17. p. 183-200.

Stanley, D. J., and E. Mutti, 1968. Sedimentological evidence for an emerged land mass in the Ligurian Sea during the Paleogene: Nature, v. 218(5136), p. 32-36.

Storetvedt, K. M., 1973. Genesis of west Mediterranean basins: Earth Planet. Sci. Letters, v. 21, p. 22-28.

Vogt, P. R., and R. H. Higgs, 1969. An aeromagnetic survey of the eastern Mediterranean Sea and its interpretation: Earth and Planet. Sci. Letters, v. 5, p. 439-448.

Von Herzen, R. P., and A. E. Maxwell, 1959. The measurement of thermal conductivity of deep-sea sediments by a needle probe method: J. Geophys. Res., v. 64(10), p. 1557-1563.

Von Herzen, R. P., and S. Uyeda, 1963. Heat flow through the eastern Pacific Ocean floor: J. Geophys. Res., v. 68(14), p. 4219-4250.

Von Herzen, R. P., and W. H. K. Lee, 1969. Heat flow in oceanic regions: The Earth's Crust and Upper Mantle, Geophysical Monograph No. 13, ed. P. J. Hart, American Geophys. Union, Washington D.C., p. 88-95.

Von Herzen, R. P., H. Hoskins, and T. H. van Andel, 1972. Geophysical studies in the Angola diapir field: Bull. Geol. Soc. Am., v. 83, p. 1901-1910.

Wüst, G., 1961. On the vertical circulation of the Mediterranean Sea: J. Geophys. Res., v. 66(10), p. 3261-3271.

Table I. Heatflow data in the Mediterranean and Àegean Seas obtained during cruises of the R/V Chain (CH61 prefix), the R/V Conrad (C9 prefix), and the M/N Bannock (CT prefix).

Station Number	Position Lat.	Long.	Depth (corr. m)	Penetration meters	probes	Gradient ($^{o}C/m \times 10^{-2}$)	Conductivity ($mcal/cm^2\ s\ ^{o}C$)	Observed Heat Flow ($\mu cal/cm^2\ s$)
CH61-19	35°47'N	13°09'E	1525	11.5	5	5.45 ± 0.33	2.20 ± 0.16	1.20 ± 0.13
CH61-29	34°00'N	19°39'E	3955	12.3	5	2.73 ± 0.38	2.30 ± 0.11	0.63 ± 0.12
CH61-37	35°44'N	25°17'E	1862	5.3	2	5.21 ± 0.38	2.01 ± 0.04	1.05 ± 0.13
CH61-39	34°22'N	22°32'E	2679	3.8	2	2.74 ± 0.47	2.78 ± 0.07	0.76 ± 0.17
CH61-40	33°17'N	23°10'E	2470	6.0	3	5.70 ± 0.14	2.38 ± 0.29	1.36 ± 0.40
CH61-42	34°15'N	25°01'E	3500	5.8	3	2.50 ± 0.73	2.47 ± 0.17	0.62 ± 0.21
CH61-46	34°17'N	26°13'E	3140	4.6	2	2.30 ± 0.90	2.58 ± 0.18	0.59 ± 0.26
CH61-50	35°31'N	28°12'E	2450	5.4	2	2.88 ± 0.39	2.51 ± 0.14	0.73 ± 0.14
CH61-53	34°12'N	28°58'E	2603	7.8	5	3.02 ± 0.46	2.10 ± 0.25	0.63 ± 0.13
CH61-54	35°20'N	30°07'E	2017	2.5	2	3.81 ± 0.61	2.39 ± 0.12	0.91 ± 0.19
CH61-55	34°49'N	31°04'E	2466	7.0	4	2.67 ± 0.61	2.23 ± 0.14	0.59 ± 0.16
CH61-56	33°00'N	31°06'E	2167	9.0	5	2.31 ± 0.63	2.12 ± 0.16	0.49 ± 0.12
CH61-57	33°11'N	32°29'E	1926	7.5	4	1.60 ± 0.31	2.37 ± 0.31	0.38 ± 0.09
CH61-59	33°52'N	33°17'E	2301	7.4	4	3.75 ± 0.63	2.19 ± 0.19	0.82 ± 0.18
CH61-62	36°01'N	31°46'E	2500	4.4	3	2.84 ± 0.49	2.71 ± 0.18	0.77 ± 0.14
CH61-63	35°41'N	32°58'E	1240	7.3	4	2.33 ± 0.32	2.62 ± 0.11	0.61 ± 0.11
CH61-64	35°56'N	34°14'E	956	7.3	5	2.19 ± 0.72	2.52 ± 0.70	0.55 ± 0.21
CH61-65	35°32'N	35°26'E	1392	7.4	4	2.19 ± 0.40	2.23 ± 0.15	0.49 ± 0.11
CH61-66	34°19'N	34°42'E	1981	7.6	4	3.68 ± 0.81	2.32 ± 0.58	0.85 ± 0.23
CH61-67	34°11'N	33°37'E	1966	7.4	4	1.15 ± 0.19	2.23 ± 0.08	0.25 ± 0.05
CH61-68	34°53'N	34°53'E	1591	7.8	5	3.14 ± 0.26	2.13 ± 0.04	0.67 ± 0.09
C9-115	32°58'N	32°24'E	1398	10.5	4	4.02 ± 0.36	(2.60)	1.04 ± 0.15
C9-116	33°50'N	32°16'E	2641	11.5	2	2.87 ± 0.48	2.09	0.60 ± 0.13
C9-117	33°44'N	30°07'E	2775	9.1	2	2.88 ± 0.23	2.22	0.64 ± 0.08
C9-118	33°46'N	27°54'E	2622	8.5	2	1.78 ± 0.23	2.50	0.45 ± 0.08
C9-119	34°19'N	27°11'E	2627	9.5	1	1.10 ± 0.17	(2.25)	0.25 ± 0.05
C9-121	34°30'N	23°25'E	2690	3.5	2	5.38 ± 0.88	2.26	1.22 ± 0.26
C9-123	34°29'N	20°04'E	2860	7.0	2	2.92 ± 0.42	2.66	0.78 ± 0.15
C9-124	33°57'N	19°13'E	3889	10.5	3	3.36 ± 0.46	2.48	0.83 ± 0.16
C9-125	35°48'N	19°25'E	3304	5.5	2	3.31 ± 0.29	(2.65)	0.88 ± 0.12
C9-126	36°11'N	19°31'E	3290	6.5	2	6.61 ± 1.48	2.64	1.76 ± 0.48
C9-127	36°52'N	19°39'E	3373	9.0	3	2.83 ± 0.56	2.63	0.74 ± 0.21
C9-129	38°11'N	17°58'E	2348	12.0	3	1.30 ± 0.24	2.49	0.32 ± 0.07
C9-131	40°19'N	12°33'E	3579	10.0	3	3.22 ± 0.55	(2.24)	0.72 ± 0.19
C9-132	40°46'N	11°07'E	2338	13.5	4	6.23 ± 1.12	2.49	1.55 ± 0.35
C9-133	41°33'N	08°01'E	2699	10.1	4	6.35 ± 0.32	3.06	1.94 ± 0.19
C9-134	41°59'N	06°54'E	2703	11.5	4	7.48 ± 1.50	2.78	2.08 ± 0.52
C9-135	41°37'N	05°18'E	2395	6.3	2	5.61 ± 0.56	(2.49)	1.39 ± 0.28
C9-136	39°03'N	04°26'E	2332	11.2	4	6.39 ± 0.95	3.02	1.93 ± 0.38
C9-137	38°24'N	03°58'E	2659	11.9	4	7.20 ± 0.36	2.91	2.10 ± 0.21
C9-138	37°12'N	01°58'E	2783	10.9	3	11.20 ± 1.17	2.81	3.15 ± 0.48
C9-139	37°50'N	00°44'E	2701	12.0	3	4.82 ± 0.60	(2.81)	1.35 ± 0.24
C9-140	36°10'N	02°01'W	1286	10.2	3	3.08 ± 0.77	2.55	0.79 ± 0.24
C9-141	35°59'N	04°02'W	1375	11.2	3	9.51 ± 0.75	(2.43)	2.22 ± 0.40
CT-17	38°58'N	14°11'E	3504	6.8	5	15.10 ± 0.20	2.20 ± 0.17	3.32 ± 0.33
CT-29	39°20'N	13°44'E	3497	6.3	5	17.50 ± 0.11	2.06 ± 0.15	3.61 ± 0.41
CT-31	39°28'N	13°20'E	3545	10.3	7	15.20 ± 0.40	2.12 ± 0.19	3.23 ± 0.32
CT-32	39°35'N	12°53'E	3590	16.0	8	12.20 ± 0.40	2.13 ± 0.18	2.60 ± 0.26
CT-50	40°28'N	12°10'E	3416	8.8	3	18.60 ± 2.19	2.22 ± 0.16	4.14 ± 0.70
CT-53	41°30'N	11°59'E	3502	8.4	5	5.60 ± 0.24	2.47 ± 0.31	1.38 ± 0.14
CT-54	40°05'N	12°11'E	3617	10.6	5	18.20 ± 1.80	2.15 ± 0.15	3.92 ± 0.58
CT-55	40°21'N	11°16'E	2889	11.9	6	13.40 ± 0.50	2.30 ± 0.19	3.09 ± 0.31
CT-78	39°24'N	10°41'E	2785	12.0	7	12.60 ± 0.71	2.41 ± 0.18	3.04 ± 0.32
CT-79	39°11'N	13°00'E	3441	9.9	5	15.80 ± 0.30	2.12 ± 0.16	3.35 ± 0.34

Table II. Regional summary of all of the reliable observed heatflow values in the Mediterranean and Aegean Seas. Four conventional marine heatflow values in the Southern Aegean Sea (Jongsma, 1974) and four values in the Ionian Sea (Haenel, 1972) are not included in this table for reasons discussed in the text. The tabulation includes data other than those obtained by the authors, as referenced in Section II. The values given in the table represent the mean and standard deviation of the n values available from each region, except where n = 1 in which case the probable error is given in place of the standard deviation.

REGION	OBSERVED HEAT FLOW (μcal/cm^{2}s) CONVENTIONAL	BOREHOLE
WESTERN MEDITERRANEAN BASINS	1.83 ± 0.66, n = 12	2.46 ± 0.16, n = 1
TYRRHENIAN SEA	2.83 ± 1.07, n = 12	1.72 ± 0.25, n = 1
EASTERN MEDITERRANEAN BASINS	0.74 ± 0.30, n = 31	0.87 ± 0.07, n = 2
AEGEAN SEA	1.88 ± 0.59, n = 10	2.12 ± 0.25, n = 2

Table III. Parameters assumed for the calculation of the regional sedimentation correction applied to the observed heatflow values in the Mediterranean Sea, taking into account only sediment deposited on top of the well-defined evaporitic layer since the Upper Miocene (10 m.y.). The mean thermal diffusivity of the sediment would be expected to increase with the thickness of the sediment due to compaction. The original sediment thickness has been calculated from the observed thickness according to Hamilton, 1976. Observed thicknesses used were reported by 1. Finetti *et al.*(1970), 2. Mauffret *et al.*(1973), 3. Finetti and Morelli (1972), and 4. Sancho *et al.*(1973).

REGION	SEDIMENT OBSERVED (km)	THICKNESS EXPANDED (km)	SEDIMENTATION RATE (m/m.y.)	THERMAL DIFFUSIVITY (cm^2/s)	CORRECTED HEAT FLOW (μcal/cm^{2}s)
Tyrrhenian Sea	0.2 to 0.5[1]	0.5	50	0.004	2.97
Western Mediterranean Basins	1.0 to 1.5[2,3]	2.5	250	0.006	2.25
Eastern Mediterranean Basins	0.2 to 0.3[3,4]	0.4	40	0.004	0.77

Table IV. Parameters assumed for the calculation of the regional sedimentation corrections applied to the observed heatflow values in the Mediterranean Sea, taking into account the total average thickness of sediment deposited in the basin since its formation, as determined by the following investigators: 1. Finetti *et al.*(1970), 2. Mauffret *et al.*(1973), 3. Finetti and Morelli (1972), and 4. Sancho *et al.*(1973). The expanded thickness of the sediment was estimated from the observed thickness following Hamilton (1976), and thermal diffusivity was assumed to increase with sediment thickness due to compaction.

REGION	SEDIMENT OBSERVED (km)	THICKNESS EXPANDED (km)	BASIN AGE (m.y.)	SEDIMENTATION RATE (m/m.y.)	THERMAL DIFFUSIVITY (cm^2/s)	CORRECTED HEAT FLOW (μcal/cm^{2}s)
Tyrrhenian Sea	2.5[1]	6.0	20	300	.006	4.05
Western Mediterranean Basins	5 to 6[2]	15.0	20	750	.008	4.09
Eastern Mediterranean Basins	4 to 6[3,4]	12.5	62.5	200	.007	1.07

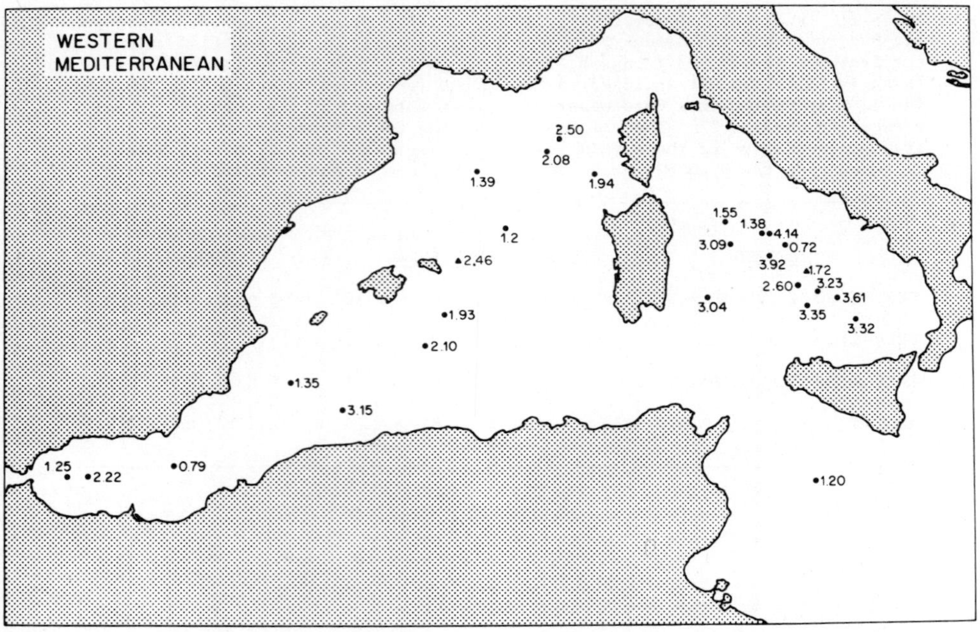

Figure 1. Compilation of reliable heatflow values in the Western Mediterranean Sea. Data, from various sources as referenced in the text, are given in $\mu cal/cm^2 s$. Values shown near circles were measured using conventional marine heatflow apparatus, those near the triangles were measured in boreholes drilled during Deep Sea Drilling Project Leg 42A. The reader is advised to consult the original references before attempting to deduce the geophysical significance of these data.

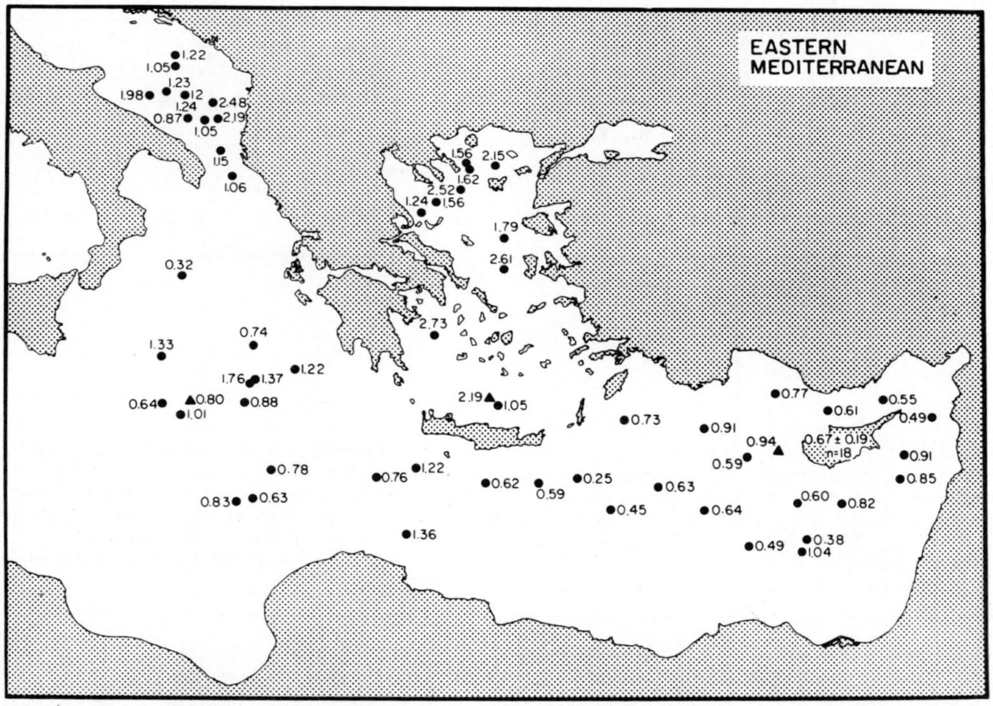

Figure 2. Compilation of reliable heatflow values in the Eastern Mediterranean and Aegean Seas. Data, from various sources as referenced in the text, are given in $\mu cal/cm^2 s$. Values shown near circles were measured using conventional marine heatflow apparatus, those near the traingles were measured in boreholes drilled during Deep Sea Drilling Project Leg 42A. The heatflow value shown for Cyprus is the mean of 18 reliable measurements made in boreholes by Morgan (in press, 1976). The reader is advised to consult the original references before attempting to deduce the geophysical significance of these data.

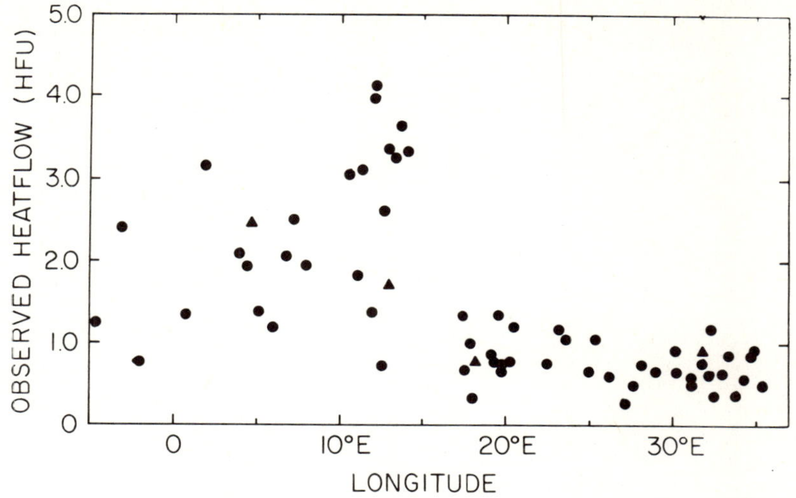

Figure 3. Plot of observed heat flow in the Mediterranean Sea plotted versus longitude, showing the significant decrease in heat flow east of Sicily (located near 15°E longitude). Values shown as circles were measured using conventional oceanographic techniques; those measured in boreholes drilled during Deep Sea Drilling Project Leg 42A are shown as triangles.

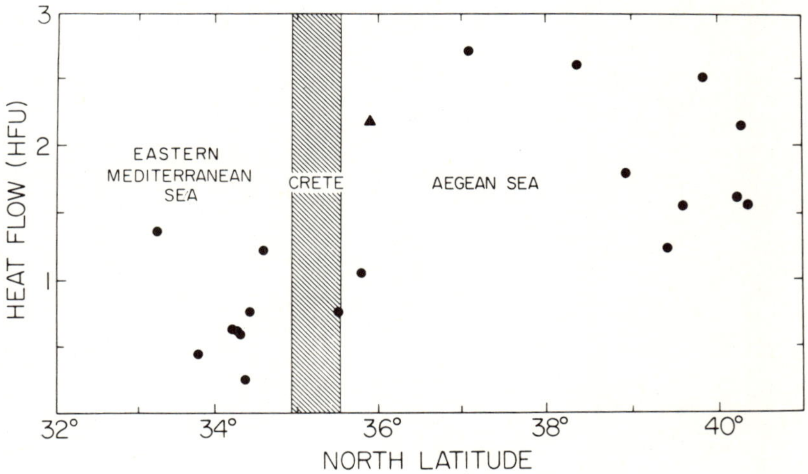

Figure 4. Plot of observed heat flow between 20° and 30°E longitude as a function of latitude, showing the abrupt increase in heat flow northward from the Eastern Mediterranean across the Cretan Arc into the Aegean Sea. Heatflow values designated by circles were obtained using conventional oceanographic techniques; a single value measured in a borehole drilled during Deep Sea Drilling Project Leg 42A is shown as a triangle.

INTERNATIONAL SYMPOSIUM ON THE STRUCTURAL HISTORY OF THE MEDITERRANEAN BASINS. SPLIT (YUGOSLAVIA) 25-29 OCTOBER 1976. B. BIJU-DUVAL AND L. MONTADERT. Eds. EDITIONS TECHNIP, PARIS 1977, pp. 281-286

SEISMIC INVESTIGATIONS OF CRUSTAL AND UPPER MANTLE STRUCTURE OF THE NORTHERN APENNINES AND CORSICA

C. MORELLI [(1)], P. GIESE [(2)], A. HIRN [(3)], B. COLOMBI [(4)], C. EVA [(5)], I. GUERRA [(6)], H. LETZ [(2)], R. NICOLIC [(7)], C. REICHERT [(2)], S. SCARASCIA [(4)] and P. WIGGER [(2)]

I. SUMMARY

In 1974, the crustal structure between Corsica and the Northern Apennines was investigated by seismic refraction measurements within a French, German, and Italian cooperation. Corsica shows a typically continental crust, 30 km thick. Elba and the NE Ligurian Sea are characterized by two discontinuities, both having the properties of a crust/mantle boundary. The deeper boundary (40 - 50 km) belongs to the Corsica-Sardinia block whereas the shallow one (15 - 25 km) must be associated with the Adriatic microplate.

II. INTRODUCTION

In autumn 1974, under participation of French, German, and Italian geophysical institutions, a large seismic refraction program was carried out in the area between the Northern Apennines, Elba, and Corsica aiming to investigate the structure of crust and upper mantle.

From the geological point of view, four regions, different in morphology, structure, and development can be distinguished.

1. The sea W of Corsica is characterized by a positive Bouguer anomaly indicating a high position of the upper mantle.

2. The Corsica-Sardinia block, being of Hercynian and Prehercynian age, shows a continental-like behaviour. The NE part is the only one that carries Alpine units.

3. The Ligurian Sea between Corsica and the Northern Apennines must contain Alpine and Apenninic parts as well. In this zone, the roots of the Apenninic nappe system have to be sought. Thus just here, a very complex crustal structure has to be expected.

4. The fourth region, the Northern Apennines, belongs to the Adriatic microplate. The mountain system of the Northern Apennines is mainly composed of sedimentary nappes. Only on Elba granitic rocks are exposed.

From the gravimetric point of view, the area under investigation is situated between the Bouguer gravity high of the Balearic Sea and the gravity minimum in front of the Apennines.

(1) Istituto di Miniera e di Geofisica Applicata Viale R, Gessi 4, I 34123 Trieste, Italy

(2) Institut für Geophysikalische Wissenschaften Rheinbabenallee 49, D 1000 Berlin 33, Federal Republic of Germany

(3) Institut de Physique du Globe, 4 place Jussieu, 75230 Paris Cédex 05, France

(4) Istituto di Geofisica della Litosfera, Via Mario Bianco 9, I 20131 Milano, Italy

(5) Istituto Geofisico e Geodetico, Via Balbi 30, I 16100 Genova, Italy

(6) Osservatorio Vesuviano, Ercolano (Napoli), Italy

(7) Osservatorio Geofisico Sperimentale, Viale R. Gessi 4, I 23134 Trieste, Italy

These seismic refraction measurements were planned as a combined land-sea program. All shots, fired on sea by the Italian research vessel "Marsili" were spaced 2.5 or 5 km. They were recorded on land by French, German, and Italian groups. Fig. 1 shows the position of the recording and shooting lines. The gaps on the shooting line are due to crossing telephone cables. Table 1 indicates where the various groups operated and who interpreted each line. In total, more than 100 shots were fired and about 40 recording stations took part in this large experiment.

Besides this project, extensive seismic reflection measurements were conducted in this area as early as in 1969 by the Osservatorio Geofisico Sperimentale (O.G.S.), Trieste [FINETTI and MORELLI 1974]. A reinterpretation of these data was done by one of the authors (NICOLIC).

III. RESULTS OF THE INTERPRETATION

A. THE MEDITERRANEAN SEA W OF CORSICA

The first seismic refraction measurements in this area, conducted by FAHLQUIST and HERSEY [1969], yielded a depth of 11 km for the crust/mantle boundary. A detailed velocity depth function could be realized from a wide-angle reflection study in 1972 by the Osservatorio Geofisico Sperimentale (O.G.S.) in the area about halfway between Menorca and Corsica. Total crustal thickness amounts to 16 km [FINETTI and MORELLI 1973]. An extensive seismic refraction program was carried out by a French-Spanish research group in autumn 1974, a couple of weeks before the project described here.

HIRN et al. [1976] found three P_n branches. The first P_n branch can have been generated by a wave penetrating a positive gradient layer beneath the Mohorovičić discontinuity which is situated between 13-16 km depth. An increase of velocity, starting at 15 km depth, to 8.0 km/s at 22 km accounts for the observation. The further branches give information on the detailed structure of the depth range down to 50 km. In Fig. 2 a simplified cross-section is presented with an homogeneous crust having a velocity of 6.0 km/s. Thus the crust/mantle boundary is situated at 18 km depth.

B. CORSICA AND THE CORSICA-ELBA CHANNEL

Corsica is regarded as a fragment split from the European continent during Tertiary. Its crustal structure is a typically continental one. In the middle of Corsica, crustal thickness is 30 km as derived from reflected waves [HIRN et al. 1976]. An intermediate crustal interface at about 20 km depth could be detected under the western part of Corsica. The transition from the oceanic crust in the Balearic Sea to the continental crust of Corsica takes place within 40 km offshore.

Between Corsica and Elba - today no more than 50 km apart - the transition zone between the Alpidic and Apenninic systems is hidden. As evidenced by reflection seismics, more than 5000 m postorogenic sediments have been deposited in this channel (reinterpretation by NICOLIC). A previous publication mentions more than 8000 m sediments [FINETTI et al. 1970].

The seismic refraction data obtained in 1974 show that an uplift of the crust/mantle boundary corresponds to the young depression mentioned above. This uplift of the mantle is present not only E of Corsica but also N of this island. This result is in agreement with considerations based on gravimetric data. CAROZZO (personal communications) calculated gravimetric models which require compensation of the low-density sediments by high-density material in the lower crust or an uplift of the crust/mantle boundary.

C. ELBA AND NORTHERN APENNINES

A line passing the westside of Elba and running northwards to the islands of Capraia and Gorgona bounds approximately the western margin of the Apennines. Here, tectonic structure and development are quite different from those of the previously mentioned area. The results derived here from the seismic measurements reveal a very complex crustal structure. The most detailed profile can be presented along the W-E line Corsica - Elba - W coast of Tuscany which is due to the fact that just here the shotpoints are nearest to the coast (Fig. 2).

Under Elba and the region of the NE Ligurian Sea, two boundaries could be detected showing the characteristics typical for a crust/mantle boundary.

Evidenced by strong reflections and first arrivals showing a velocity of 7.5 - 8.0 km/s, an interface in 20 - 25 km depth could be found. It can be stated that this interface has no connection to the Mohorovičić discontinuity previously described for the Corsica area.

In accordance with the complete change of the tectonic structure E of Elba, this shallow boundary vanishes under the Corsica-Elba channel which is revealed by the fact that it could not be detected under Corsica. Its eastward continuation under western Tuscany must, so far, remain unknown because the distance from shotpoints to observation points becomes too large.

Such a shallow boundary could also be discovered in the NE part of the Ligurian Sea. Here, however, the boundary velocity is reduced to 7.4 km/s (Fig. 3).

Returning to the Corsica-Elba profile, the eastward continuation of the crust/mantle boundary of Corsica has to be discussed: From the shots W of Corsica, first arrivals penetrating the upper mantle were recorded on Elba and in the western Tuscany up to distances of 120 km. The resulting travel-time curve gives velocity values that are, on an average, greater than 8 km/s. The constructed boundary, beginning under Corsica at 30 km depth, shows the uplift under the Corsica-Elba channel already mentioned previously and, in eastern continuation, plunges under Elba.

A late reflection recorded in the western Tuscany from the shots in the Corsica channel yields a reflector depth of about 60 km E of Elba and an average velocity of 6.3-6.4 km/s. This reflection horizon can be associated with the deep crust/mantle boundary. The resulting interval velocity between the shallow and the deep crust/mantle boundary amounts to 6.7 - 6.8 km/s. The minimal velocity within this depth range of 25 to 60 km is probably even smaller but not determinable. Values of 6 km/s or smaller in narrow zones cannot be excluded. Thus sialic material may exist partly in this depth range which is a possible source of the granitic rocks of Elba.

In the NE-part of the Ligurian Sea, the same structure is, in principle, existing but it has not been cleared up in detail up to now.

D. THE NORTHERN APENNINES

Due to the positions of the shotpoints on sea and the recording stations on land, the crustal and upper mantle structure under the Apennines themselves could be revealed only in the coastal range.

The top of the basement was clearly recognized by first arrivals being observed up to distances of 120 km. The culmination of the basement, situated about 20 km S of La Spezia, shows a sedimentary cover of 3,5 km thickness. The boundary between sediments and basement plunges down in southern but stronger still in northern direction (Fig. 3).

As established by refracted waves, showing a velocity of 7.4km/s, the crust/mantle boundary dips down towards the coast. Here, this boundary, at a depth of 35 km, could be detected by reflected waves. The northern downdip continues to the axis of the Apennines but it remains open whether the deepest point has already been met. From the gravimetric point of view, the greatest crustal thickness has to be expected under the outer margin of the Apennines.

The high crustal velocity of 6.35 km/s is of interest, indicating a higher content of basic material in the crust than normal. An intensive crustal low-velocity layer could not be recognized.

IV. CONCLUSIONS

The crustal structure derived and presented here is very complex and understandable only when considering simultaneously the tectonic structure and development of the southern Western Alps and the Northern Apennines. Two different continental units have to be distinguished. The Corsica-Sardinia crust, being a fragment of the European continent, is of normal continental composition and extends, under reduction of its thickness, some ten kilometers east- and northwards under the crust of the Adriatic microplate.

The Adriatic crust has a minimum thickness of 20 km under Elba and 15 - 20 km in the northern Ligurian Sea. Proceeding towards the crest of the Apennines, crustal thickness increases at least to 35 km.

Comparing the structure of the area under investigation with that of other Mediterranean orogenic zones [GIESE and PAVLENKOVA 1976], a distinct difference is recognizable. In the Ligurian region the crust of the hinterland, that is Corsica, plunges under the Apenninic and Adriatic crust. In the Alps and the Carpathians, however, the crust of the hinterland overlies orogenis series and the crust of the foreland. The different situation in the Ligurian Sea has to be seen in a larger frame, that is the transition from the Alps to the Apennines. Just in this area the Adriatic microplate behaves as hinterland in respect to the Western Alps whereas it is foreland with regard to the Northern Apennines.

REFERENCES

FAHLQUIST, D.A. and J.B. HERSEY: Seismic Refraction Measurements in the Western Mediterranean Sea. Bull. Inst. Océanographique, Monaco, 67, 1386, 1969.

FINETTI, J. and C. MORELLI: Geophysical Exploration of the Mediterranean Sea. Boll. Geofis. teor. ed appl. 15, 263-341, 1973.

FINETTI, I. and C. MORELLI: Esplorazione geofisica dell'aera mediterranea circostante il Blocco Sardo-Corso. Rend. del Sem. Fac. Scienze Univ. Cagliari, 1974

FINETTI, I, C. MORELLI, and E. ZARUDZKI: Reflection Seismic Study of the Tyrrhenian Sea. Boll. Geofis. teor. ed appl. 12, 48, 1970

GIESE, P. and N.I. PAVLENKOVA: Some Particularities of Crustal Structure in Young Orogenic Systems. Geol. Rundschau, 65, 3, 1109 - 1124, 1976

HIRN, A., L. STEINMETZ, and M. SAPIN: A long range seismic profile in the Western Mediterranean Basin structure of the upper mantle. In press 1976.

Table 1

Profile	Group	Interpreter
profile 1b	Milano, Napoli	Colombi, Guerra, Scarascia
profile 1	German, Trieste	Letz
profile 2	German, Milano, Napoli, Trieste	Colombi, Guerra, Scarascia
profile 3 (Corsica)	French	Hirn, Steinmetz, Sapin
profile 3 (Elba, Tuscany)	German, Milano, Napoli, Trieste	Reichert, Wigger
areally distributed moving and fixed recording stations NW of Genova	Genova	Eva
reflection measurements in the NE-Ligurian Sea	O.G.S., Trieste	reinterpretation: Nicolic

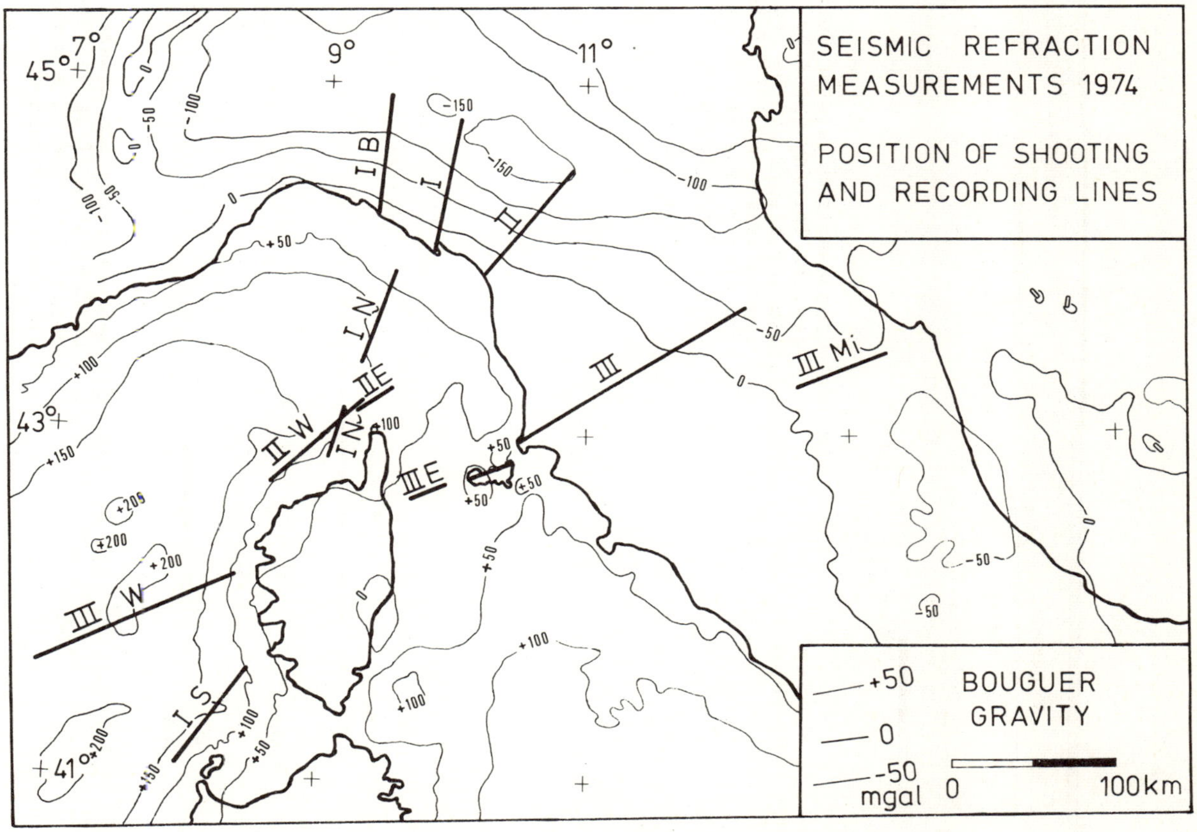

Fig. 1 Position map of shooting and recording lines. NW of Genova, the stations of the institute of Genova, some mobile, some permanent - not displayed in this map - recorded the shots of line 1.

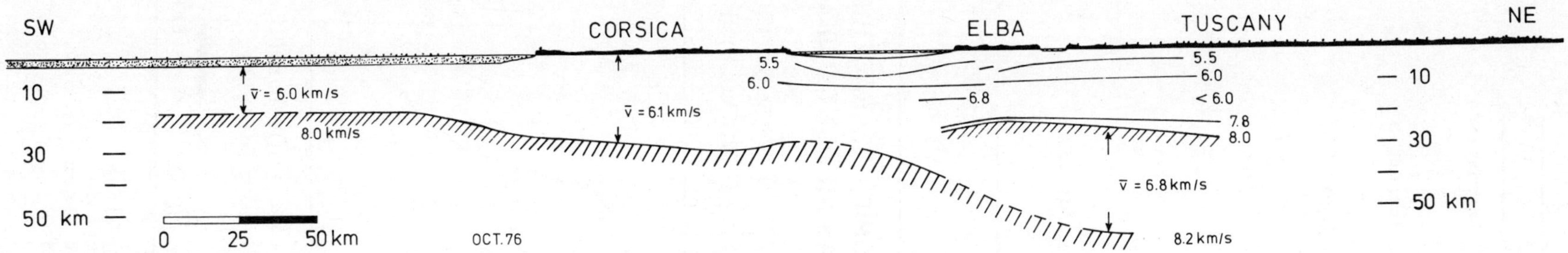

Fig. 2 Cross-section between Corsica - Elba - and western Tuscany (profile 3). The velocity isoline 6 km/s can be regarded in first approximation as top of the basement, whereas the line 8 km/s can be associated with the crust/mantle boundary

$\overline{v}$: average velocity

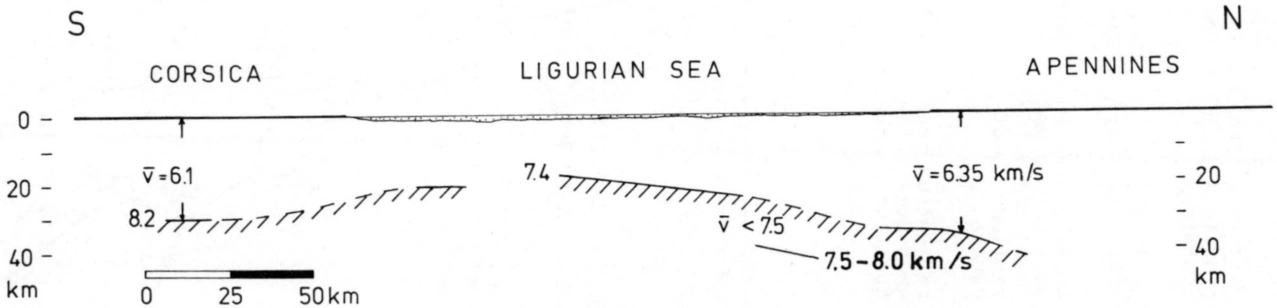

Fig. 3 N-S cross-section between Corsica and W of La Spezia (profile 1). Here, the upper discontinuity showing a boundary velocity of 7.4 km/s is regarded as base of the Apenninic crust.

INTERNATIONAL SYMPOSIUM ON THE STRUCTURAL HISTORY OF THE MEDITERRANEAN BASINS. SPLIT (YUGOSLAVIA) 25-29 OCTOBER 1976. B. BIJU-DUVAL AND L. MONTADERT, Eds. EDITIONS TECHNIP, PARIS 1977, pp. 287-302

WIDESPREAD MANIFESTATIONS OF OLIGOCENE-LOWER MIOCENE VOLCANISM AROUND WESTERN MEDITERRANEAN

F.C. WEZEL [(1)]

I. INTRODUCTION

In the Oligocene-lower Miocene of the Mediterranean region have been found, on land and in the sea, frequent traces of an impressive volcanic andesitic-dacitic-rhyolitic activity. These manifestations are fairly concomitant with large tectonic movements. All the data tend to show that the volcanism represents a symptom of a geologic "revolution" which produced marked changes in the paleogeography of the Mediterranean region.

One of the focal problems is an analysis of the time-space relationships between the volcanic activity and the tectonic phases. The knowledge of these relationships can permit the understanding of the significance of the volcanism in the geotectonic terms and furnish objective indications about the generation of andesitic magmas as well. The accurate dating of volcanic activity is a fundamental problem, solution of which may permit formulation of a more probable geotectonic hypothesis for the Mediterranean area.

Here I feel the need of stressing the importance of stratigraphic studies for better understanding of geological events. It was only through detailed stratigraphy that the advance and magnitude of tectogenesis can be estimated (Wezel,1975). Also the realisation of which of the two types of activity (volcanic or tectonic) starts first cannot be considered a banal problem. Actually, this question is a direct test of the veracity of the plate tectonics assumption that the andesitic magmas are formed by fusion of the mantle material along the Benioff planes. This concept implies a certain time delay, of on order of a few million years, between the beginning of the crustal subduction (i.e., the onset of the compressional tectonics) and the start of the andesitic activity. The previous studies too often accepted as sufficient a simple coincidence and a gross correlation existing between the periods of time in which the two activities have developed, without investigating in detail the actual synchronisms.

This paper presents the author's

(1) Istituto di Geologia dell'Università, 61029 Urbino and Istituto di Geologia e Paleontologia dell'Università, Via Zamboni 67, 40127 Bologna, Italy

thoughts on the geology of western Mediterranean, without entering into the problems of igneous petrogenesis, which are outside of his competence.

II. DESCRIPTION OF VOLCANIC MANIFESTATIONS AND THEIR AREAL EXTENT

In western Mediterranean the widespread manifestations of an Oligo-Miocene volcanic activity occur around the peripheral orogenic belts and on the concave sides of the orogenic arcs. Rocks of such volcanic nature are found in the Betic, Maghrebian and the Apennine mountain chains. They are also known in the Carpathians and in the Parathetys region.

The three main types of volcanic rocks recognized are: (1) the lava flows and pyroclastics, (2) the volcaniclastic rocks and, (3) the siliceous deposits. Rocks of the first type outcrop mainly in the western Sardinia. The lavas and pyroclastic products accumulate in tectonic troughs cut during the Tertiary into underlying Paleozoic rocks. Dupuy et al.,(1974) state that about two thirds of the volcanics are andesites, one third are basalts and a small remainder are dacites and rhyolites. These vulcanites are well known today representing, by far, the easiest volcanic formations to follow.

However, an as yet unknown volume of volcanic products has also been noted in volcaniclastic rocks, mainly volcanic sandstones, occurring within the peri-Mediterranean chains. These geosynclinal sandstones, graywackes, show up in the literature under different local names: "tuffaceous sandstones", "tuffites", etc. They outcrop in the Apennines (i.e., the Aveto and Petrignacola Sandstones), in the Calabro-Sicilian arc (Tusa Flysch) and in the Gibraltar arc (Beni-Ider and Algesiras Flysch). At time of this writing I am unaware of their presence in Tunisia and Algeria, within the zone of the Kabylian flysch. More widespread and younger are the tuff-ash layers and cinerite levels, generally epiclastic rather than pyroclastic, abundant in the Central (Umbrian-Marche) Apennines, i.e., the Bisciaro Formation. Volcanic sandstones are often green or greenish in colour as result of a chloritic alteration of the ferromagnesian minerals; they are a mixture, in all proportions, of the volcanic and non-volcanic constituents. The volcanic detritus appears to be freshly (penecontemporaneously) eroded from pyroclastic deposits rather than from lavas. Some sandstones like the Algesiras Flysch in Andalusia, do not contain volcanic fragments but only the mineral grains (volcanic quartz, zoned plagioclase, etc.) indicating their volcanic origin.

Most of strata I have examined in the peri-Mediterranean outcrops, are fine-grained volcanic turbidites indicating deposition in abyssal plains (e.g. Tusa Flysch in Sicily). Coarser and channelized sequences are also present suggestive of deep-sea fan environment (e.g. Aveto Sandstone in northern Apennines). The conglomeratic sequences outcropping in the Aveto Valley consist of 80% plutonic-metamorphic pebbles, 10% volcanic fragments and 10% sedimentary materials.

Also, some recently found pebbles represent fragments of andesitic lavas (work in progress). The clastics show their derivation from an emergent, granitic-metamorphic paleo-massif unknown today, in the northern Apennines. This source area I believe to be a fragment of an ancient Corso-Sardinian massif or, even of alpine origin. The volcanic sandstones are frequently associated with gray or bluish-gray marls in thick turbidite units (Tusa Flysch). Quite similar facies were found in the holes drilled at Sites 127 and 128 in the sea floor of the Hellenic Trench by the Glomar Challenger (see Bartolini et al.,1975). Successions are generally of "regressive" type with the upper strata deposited by currents of increasing energy. The stratigraphic evolution: marls ⟶ marls and siltite ⟶ arenaceous turbidites, reflects the gradual infiling of the trench caused perhaps by the steady approach of the migrating source.

The third type of volcanic manifestations is represented by various kinds of rocks containing marine siliceous organisms (i.e.: radiolarian tests, diatom frustules and sponge spicules), sometimes in association with volcanic materials (mainly vitric grains and shards). These rocks are: diatomite and benthonite horizons (e.g. within Numidian Flysch in Sicily), "tripoli" and tripolaceous sediments, "silexites" of the French workers (Didon et al.,1970), black chert levels, radiolarian cherts, etc. Diatomaceous deposits known as "moronitas" (Colom,1952) are widespread in southern Spain. Analysis of some of these deposits suggests that the silica is of volcanic origin and that the siliceous sediments were laid down during periods of volcanic activity. Altogether they are widely distributed along all of the peri-Mediterranean chains: e.g. Apennines, Maghrebides and Betides and also in the Carpathians (Patrut,1959; Kotlarczyk,1966).

III. STRUCTURAL POSITION AND REGIONAL SETTING

Each chain of the southern alpidic orogens consists of internal and external domains, subdivided in different paleogeographic zones. The orogens are formed by superposition of the internal domain nappes, involving the medial massif and bordering imbricate eugeosynclinal mass, upon the external domain units representing the former marginal parts of the African or Adriatic continental blocks (forelands). The volcanic rocks are found in the internal domain, represented by the medial Paleozoic massifs, of which the Corso-Sardinian and the Calabro-Peloritan sialic blocks are the two presently emergent fragments.

The volcaniclastic and siliceous rocks are, on the contrary, incorporated in the geosynclinal successions. The bulk of the volcanic sandstones outcrops in the more external part of the eugeosynclinal area, near the geosuture between the internal and external domains. The flysch units containing volcanic sandstones are tectonically separated by numerous thrust planes dipping toward the internal zones (i.e. toward the present western Mediterranean basins). In the successive thrust slices the flysch sequences are of a progressively younger age towards the foreland. A spectacular tectonic juxtaposition of thrust belts affecting late Oligocene volcanic flysch of Beni-Ider and early Miocene Numidian Flysch is present in northern Morocco between Tangiers and Ceuta (Wezel,1973b). This situation also applies to Sicily, where the late Oligocene-early Burdigalian Tusa Flysch is folded, overturned and subdivided by thrust planes in successive structural units. Starting from bottom, each of them comprises: 1) varicoloured clays ⟶ 2) chaotic strata ⟶ 3) volcaniclastic turbidites. This setting was interpreted (Wezel,1974a and 1975) as an evidence of the trench sequences progressively deformed and accreted against the peripheral margin (Peloritan) of the Tyrrhenian block in its advance towards the African foreland. Similarly to the Beni-

Ider Flysch and the Tusa Flysch also the Aveto Sandstones consist of a tectonic juxtaposition of the successive tectonic slices towards the Adriatic foreland. Briefly, this stratigraphy suggests that the tectonic imbrication of volcanic flysch is a time-transgressive (diachronous) process. Volcanic detritus is also present in the transgressive sedimentary cover overlying the massif and the eugeosynclinal imbricated tectonic units (e.g., Reitano-Capo d'Orlando Flysch in Sicily).

The siliceous deposits, on the other hand, are also present in the external domain units. Along all the northern and central Apennines the siliceous levels outcrop in the marginal portions of the Adriatic foreland (i.e., the "Tripoli of Contignaco" and the tripolaceous levels in Bisciaro). Similar episodes of silica sedimentation occur in the Numidian Flysch (Campisi,1962; Broquet,1973) which suggest underwater volcanic activity (Broquet,1973) in the lower Burdigalian. Thus, the siliceous deposits display vast lateral extension and constitute excellent guide-levels, particularly in the Burdigalian and Langhian strata. They appear during somewhat more recent times as compared with the age of the volcanic sandstones and can be related to the volumetrically large volcanic manifestations associated with eruptions which occurred in the waning stages of volcanic activity.

The sites of volcaniclastic sedimentation were the basins lying mainly within the internal domains. Volcanic detritus was deposited immediately before the contact and suturing between the internal and the external domains. Such contact occured in Morocco during Aquitanian, about 7 My before it reached Sicily during Langhian.

IV. COMPOSITION OF VOLCANICLASTIC ROCKS

Geosynclinal volcanic fragments are predominantly andesitic and dacitic. Volcanic detritus is accompanied by low quartz content (less than 25%), by common rock fragments (40 to 70%) by the plagioclase grains (20 to 50%) and by the micas in variable quantites. Crook (1960) calls such sandstones feldspatholitic. Microscopic studies also identified the abundant metamorphic-plutonic detritus, suggesting a derivation from erosion of pyroclastics overlying massif (e.g., the Peloritan massif). In fact, some of lithic polycrystalline grains show the same characteristics as the grains found in the sandstones transgressive over the medial (internal) massif. This suggests that the active Oligo-Miocene volcanoes were located on the internal blocks. The Beni-Ider and Algesiras Flysch in the Gibraltar arc contain a small quantities of volcanic fragments (less than 7%), a notable quantity of the metamorphic lithic fragments, micas (muscovite, biotite and green mica), a common effusive quartz and plagioclases (unpublished petrographic studies of G. Ottonello). One can postulate for them an origin from pyroclastics, gneiss and micaschists.

Petrochemical analyses of some selected volcanic sandstones were performed by A. Mazzucotelli and G. Ottonello. They showed (Fig.1) that the majority of volcanic debris derived from andesitic lavas belongs to the calc-alkalic (CA) series. Only the volcanic fragments of Tusa (Sicily) and Petrignacola (in northern Apennines) belong to the tholeiitic (TH) series. No volcanic fragments of the shoshonitic (SH) series were found as yet. The diagram of the Figure 1 reveals a trend of increase of the potash from the geosyncline (trench) towards the internal zones of the orogenic system. Actually, the volcanic rocks of Sardinia demonstrate a higher content of potash for the same content of silica. I am inclined to correlate the content of a "continental" element like potash with the thickness of the underlying crust. Condie (1973) already formulated an empirical relationship between K content of volcanic rocks and the thickness of the crust in the island arc areas and in the active continental margins. The right side of Fig.1 shows such tentative correlation of the various volcanic series with the crustal thickness, using Condie's data. If this interpretation is correct we can infer: (1) that the basement beneath the Oligo-Miocene andesitic volcanoes was ensialic and constituted by a thinned continental crust and (2) a probable crustal thickening from the trench towards the medial massif. The crust was thinner but probably not to such extent as to give rise to typical oceanic tholeiitic lavas which are found in the present Tyrrhenian abyssal plain (Del Monte,1972; Selli,1974; and DSDP Site 373A). Thinning of the crust should be the product of distension.

Thus, the composition and petrochemistry indicate that source terrains of the andesitic detritus were located on the concave side of an arc within the Tyrrhenian (or the Alboran) sialic blocks. At present the only Oligo-Miocene volcanoes are found in western Sardinia. Present distance of the Sardinian eruptive centers varies from 300 km (central Apennines) to about 500 km (Sicily). However these distances resulted from a later crustal lengthening caused by the continental distension of the Tyrrhenian block (Selli & Fabbri,1971). In my opinion this process began in the upper Miocene. An estimate of the lower Miocene crustal configuration allows for foreshortening of up to 40%. Thus, the original distance between the trench and the active volcanoes was not much different from that which divides today the Hellenic Trench from the frontal chain of active volcanoes of the Cyclades.

V. CORRELATION BETWEEN VOLCANIC AND TECTONIC ACTIVITY

The magmatic phases seem to correlate well with the tectonic phases in the western Mediterranean region. In Sicily the writer (1973a and 1974a) was able to reconstruct a timetable of the tectonic events. On basis of chaotic strata, olistostromes and other sedimentary mixtures it was possible to identify the main regional episodes of compressional deformations. These episodes, stratigraphically determined, are contemporaneous with phases of volcanic activity independently dated radiometrically by Coulon et al.(1974) in northwestern Sardinia. Collision and suturing started at the "<u>Orbulina</u> time" about 14 My coinciding with the end of the calc-alkalic volcanism in the western Sardinia.

In Sicily three main tectogenetic phases I) 24 to 20 My, II) about 17 My and III) about 14-15 My were paleontologically dated. So far in the northern Apennines, only two of these phases (i.e., I and III) have been identified. It is important to stress that the metamorphism of the Apuan Alps in the northern Apennines was dated with the K/Ar method as about 14.5 My (Giglia & Radicati Di Brozolo,1970). The event of collision between the opposite margins of the Apenninic geosyncline should have occurred at about the same time. Furthermore, the age of the climax of the marine transgression in Sardinia (see below) is the same as that of the orogenic compression in Sicily and in the northern Apennines and of the end of calc-alkalic volcanic activity in Sardinia.

The oldest radiometric ages of the

lava flows in southwestern Sardinia are 30 to 28 My (Savelli,1975; Bellon, pers. comm.). The oldest ages of volcanic sandstones in northern Morocco (Beni-Ider Flysch) were paleontologically determined as also about 30-28 My. At the present writing I am unaware of the existence of compressional deformations geographically associated with such early volcanic activity. According to Trümpy (1973), about 35 to 30 My begun the post-paroxysmal uplift of the "Meso-Alps", the molasse deposition and subsequent movements along the Insubric and other faults. Thus, the beginning of western Mediterranean volcanic activity was contemporaneous with tectonic relaxation and the related post-alpine orogenic distensional movements. It should be stressed that the early andesitic volcanites of the southern Sardinia (age: 30 to 28 My) outcrop along the shoulders of the Campidano and the Cixerri grabens. The large thickness (up to 500 m) of lavas and tuffs filling these grabens, found in deep wells, show that the volcanism occurred following the horst-and-graben distensional movements.

A part the early volcanic evidence in Sardinia and the volcaniclastics in Morocco, all other radiometric ages of vulcanites, as well as the paleontologic dates of the volcaniclastic sediments, fall within the limits of uppermost Oligocene and the lower Miocene (Fig.2). Similarly timed volcanic phases occured also within the European block (e.g., in the Massif Central and Rhinegraben) and also in the Pannonian region.

VI. VOLCANIC ACTIVITY AND THE MEDITERRANEAN TRANSGRESSION

During the Oligo-Miocene time a great marine transgression affected practically the whole Mediterranean area. Many authors, in particular Selli (1957) in his classical work on southern Italy, pointed out to this occurrence throughout the Mediterranean region. This widespread event occurred not only over the Balearic area and the Tyrrhenian block , but also over the marginal parts of the African and Adriatic blocks. It is characterized by: 1) its great lateral extent, 2) having produced conformity between strata and, 3) the rapid rate of land submergence. This transgression was an expression of crustal subsidence affecting major areas in Mediterranean. The beginning of this transgression may correspond to the age of distensional opening of the Balearic Basin (Fig. 3). In Sardinia the transgression began in the south starting about 24 My (data in Cherchi et al.,1974). In Sicily its start was somewhat more recent. In Sardinia the earlier andesitic lavas, sandwiched between two continental deposits (Cixerri Formation at the base and Ussana Formation at the top), are overlain by marine succession with a marly facies, occasionally heteropic with biohermal limestones (Cocozza & Jacobacci,1975). Such a non-marine environment (clastics and volcanics) underwent next the extensive marine incursion that inaugurated the Balearic rifting stage. The pronounced regional subsidence may have been due to crustal thinning by normal faulting and/or stretching. The magmatic activity was initiated possibly by lowering of the melting point of the upper-mantle or deep crustal materials when their pressure was reduced during the "arching stage". The character of the magmatism was probably controlled, among others, by the thickness and structural pattern of the overlying crust (i.e., its internal layering, block mosaic, depth and form of fractures, etc.). The area affected by the transgression is, naturally, more extensive than the zones of volcanic activity. Similarily to volcanites, also the evaporitic sequences represent another expression of the distensive tectonic movements which have caused regional subsidence (Wezel, 1974b).

The onset of magmatic activity precedes by a few My either the first signs of a regional transgression or the compressive phases in the peripheral chains. The successive distensional pulses, demonstrated by the volcanic phases in Sardinia, are accompanied by similar tectogenetic events in the Sicilian and Apennine chains. Therefore the opening of the Balearic Sea complements the folding in the peripheral orogens. In the same way the opening of the Tyrrhenian rift in the upper Miocene was compensated by crustal shortening in the above mentioned chains (Fig.4).

VII. SOME MORE DATA FROM OUT-OF--MEDITERRANEAN AREAS

Not only in the Mediterranean region but also in other parts of the world exists the geologic evidence of an increased volcanic and tectonic activity during the Oligocene-lower Miocene, occuring at times following the folding of the Alps. I shall quote but a few examples.

A major magmatic phase was recognized as happening during the early Miocene time contemporaneously in the Canary Islands, in the Cape Verde Islands and in the Gulf of Guinea, in different geological settings (Grunau et al.,1975). This event can be correlated with a block-faulting phase and occurrence of a major unconformity on the Moroccan shelf and continental slope.

The Red Sea rift originated in two phases, the initial phase occurred in the Oligocene-early Miocene and was characterized by an Oligocene block-faulting and volcanism and also by a following regional subsidence (Lowell et al.,1975). It is possible that is corresponds to a first displacement (60 km long) of the transcurrent Jordan fault (Freund et al.,1970) and a compressive phase in the Iranian mountain chains. The second phase of the Red Sea occured between 3 and 5 My (Girdler & Styles,1974), therefore in lower Pliocene. It was, probably, accompanied by a new displacement of 45 km along the same Jordan fault and by a new tectogenetic pulse within the mountain chain of Iran.

In the Caribbean, the distensional movements and the following subsidence in the upper Oligocene on the concave side of the arc (see Fox & Hezen,1975) is accompanied along the Antillean arc by an intense volcanic activity. This manifested itself with a deposition of diatomites (e.g., "Naparima marls" in Trinidad) and with tripolaceous sedimentation represented by the well-known "Oceanic Formation" of Barbados which is of upper Oligocene-lower Miocene age.

In the Philippine Sea the opening of Shikoku basin occured during the upper Oligocene-Aquitanian (26 to 22 My), nearly coincidental with a period of intensified volcanic activity in Japan (Watts & Weissel,1975). Mitchell and Reading (1971) discuss a volcanic arc in New Hebrides active during lower Miocene and producing calc-alcalic lavas. They also point out that in the late Oligocene or early Miocene and starting from Pliocene occurred important "episodes of lithospheric descent beneath many island arcs" related to major tectonic events.

VIII. RELATIONSHIP WITH THE PLATE TECTONICS

The above chronological data question some aspects of the theory of plate tectonics. In the western Mediterranean the calc-alcalic volcanism appears to precede by a few My the onset of compressive subduction of the margins of the geosyncline. It is also associated with block-faulting and

crustal collapse caused by regional distension in the internal zones of the orogens. I also see some difficulties in the concept of the magma originating in the Benioff zone:
1. Generation of andesitic magmas at too great depths (150–200 km). Many well known workers have, on another hand, maintained hypotheses of these magmas originating from partial fusion of deep crustal material (e.g., Turner & Verhoogen,1960; Dickinson,1962; Rittmann,1967; Pichler & Zeil,1972, to name but few). According to Rittmann (1967) the orogenic andesitic volcanism is mainly caused by the anatectic magmas of the lower Sial, while the rhyolitic and dacitic magmas derive from materials in the upper parts of the crust. Thus the fusion progresses with time always towards the higher levels of the continental crust.
2. Friction along the Benioff zone between plates does not generate, according to Rittmann (1967), sufficiently high heat to cause the partial melting of mantle material.
3. Andesitic volcanism is also present in the areas of subduction of the continental crust (e.g., in the eastern Mediterranean) or the lower part of it.

IX. AN ALTERNATIVE HYPOTHESIS: POST-COMPRESSIONAL DILATION FROM THE CORE OF OROGENIC BELTS AND GENERATION OF NEW MOUNTAIN CHAINS

The tectonic patterns of western Mediterranean show great complexity because the zones of compression are often transformed into zones of distension and vice versa. After the formation of orogenic belts they became broken up in the internal parts by "distensional zones" (DZ) which are the surface expressions of deeper-seated phenomena. I propose to call them "dilatation nuclei" (DIN). Since middle Oligocene the newly formed alpine chain became fragmented by these DZ. The distension was generally bilaterally absorbed by peripheral overthrusts which are more pronunced where the crustal thicknesses differ greatly. The hot "nuclei" rising towards surface could have initiated the partial fusion at the lower levels of an heterogeneous "orogenic" crust creating andesitic magmas. Thus the magmas could not be the products of Benioff zones but manifestations of the rise of the DIN through the crust previously "orogenized" (i.e., with sialic roots).

During the Oligocene-lower Miocene the emergence of the DIN in the western Mediterranean has, presumably, produced two major phases whose effects are observable at the surface:
1. Early taphrogenic block-faulting. This phase is initiated at middle Oligocene and is characterized by horst-and-graben tectonics superimposed upon the previous compressive alpine structures. The resulting grabens were filled with continental deposits ("red-beds", lacustrine shales, etc.) and interbedded with volcanic rocks. The volcanism began about 30 My. The character of the lavas reflected the type of crust through which they ascended: andesitic in the orogenic parts, basaltic in the rigid continental masses external to the orogens and in the rigid segments between the chains. This stage was associated with a broad regional uplift.
2. Later phase of regional collapse. It started around the Oligo-Miocene boundary causing widespread deposition of transgressive shallow marine beds. The latter overlay the upper evaporites which indicate the earliest marine invasion of the continental. The collapse was accompanied by successive distensional pulses which were absorbed by compressional phases in the peripheral chains ingenerating the subduction planes dipping towards the internal DZ.

One of the present examples of such DZ is the Tuscan region, particularly south of the River Arno. The zone became active since the upper Miocene following the compressional Tortonian phase. The region contains numerous recent grabens and Plio-Quaternary volcanoes located inside seismically active orogenic arcs absorbing the distensional movements. These arcs are made up of overthrusts with vergence towards Adriatic (i.e., the "Monte Falterona-Orvieto arc" and the more external "Sibillini Mts-Rieti arc"). Compressional deformations in the "Sibillini arc" were accompanied by transcurrent displacements along the "Ancona-Anzio line" beginning in the Messinian (Castellarin, Colacicchi and Praturlon, in progress). Westward, towards Tyrrhenian the distensions are compensated by a major crustal overthrust of the Apennine zone over the Corsican crust (Morelli et al., this volume). A possible existence of subduction zones dipping beneath the Apennines was suggested by Alvarez (1973) on the basis of variation in the potash content in the Tuscan lavas. The entire northern Apennines seems to be in isostatic unbalance caused by the overthrusts on the padano-adriatic forethrough (Elter et al.,1975).

The nature of these DIN can be inferred for example from geological and gravity data of the Gibraltar arc (Bonini et al.,1973). Actually, in the internal zones of the arc outcrop large ultrabasic masses (Beni Bouchera and Ronda massifs: see Fig.2) correlable to two zones of "gravity high". These bodies were interpreted as the high-temperature ultramafic intrusions vertically emplaced during Oligocene-Miocene orogeny from the upper mantle along the margins of a graben of continental-type crust which, presently, forms the western Alboran Sea. At the inside of the arc there are also andesitic volcanoes which Bonini et al. (1973) related to the thinning of the Alboran crust.

The DIN ought to be a product of the mantle activity. Starting from certain depth numerous hot "bubbles" rise towards the crust. Their general form could resemble an atomic toroidal mushroom or a hot fountain.

In the Mediterranean the "nuclei" appear to have relatively small dimensions and to move in time to different zones. Actually, the "Alboran nucleus" and "Balearic nucleus" became active in Oligocene-Miocene while the Tyrrhenian, Tuscan and Aegean "nuclei" started their activity in the upper Miocene.

My hypothesis of DIN adds some new considerations to the genesis and life of the alpine geosynclines. They are formed through the activity of the "nuclei" which, later, caused their closure and formation of orogens. So the DIN act first as constructive them as destructive factors in the orogeny. The sequence of events produced by them (Fig.4) suggests that the "nuclei" became particularly active during three geological periods: late Cretaceous, late Oligocene-early Miocene and since the start of upper Miocene. The present Mediterranean structures result from different kinematic events which can be separated only through datailed stratigraphic studies.

Regarding plate tectonics theory my hypothesis, presented here in a very preliminary form, may be either a collateral local effect (e.g., in the marginal basins) or an indication of different process.

Acknowledgements.

I am most grateful to my colleagues Profs. A. Castellarin, G.B. Vai and E.F.K. Zarudzki for their valuable critiques of the text and helpful discussions. Prof. Zarudzki greatly assisted with the English translation and the editorial work.

REFERENCES

ALVAREZ W. (1973) - The application of plate tectonics to the Mediterranean region. In: "Implications of continental drift to the earth sciences", Academic Press, vol.2: 893-908.

BARTOLINI C., MALESANI P.G., MANETTI P. & WEZEL F.C. (1975) - Sedimentology and petrology of Quaternary sediments from the Hellenic Trench, Mediterranean Ridge and the Nile Cone from DSDP, Leg 13, cores. Sedimentology, 22: 205-236.

BERNOULLI D. & JENKYNS H.C. (1974) - Alpine, Mediterranean, and central Atlantic Mesozoic facies in relation to the early evolution of the Tethys. SEPM, Sp. Publ. No. 19: 129-160.

BONI A., BRAGA G. & MARCHETTI G. (1968) - Nuovi dati sull'età della "Formazione di Bobbio" e della "Formazione di Val d'Aveto". Ist. Lombardo, Acc. Sc. Let., Rendiconti 102: 737-743.

BONINI E.W., LOOMIS T.P. & ROBERTSON J.D. (1973) - Gravity anomalies, ultramafic intrusions, and the tectonics of the region around the Strait of Gibraltar. J. Geophys. Res., 78(8): 1372-1382.

BROQUET P. (1973) - Présence de silexites dans le flysch numidien de Finale (Madonies, Sicile). C.R. Acad. Sc. Paris, 276: 465-467.

BROUSSE R., BELLON H. & GILLOT P.Y. (1975) - Recognition of Mio-pliocene distensional phases in the North Mediterranean region by analysis of Cantallian volcanism, France. "Meeting of European Geol. Soc."

CAMPISI B. (1962) - Una formazione diatomitica nell'altipiano di Gangi (Sicilia). Geol. Romana, 1: 283-288.

CATI F. (1974) - The Oligo-Miocene boundary in the Mt. Arligo section (Ozzano dell'Emilia, Bologna). Giorn. Geol., 39(2): 481-502.

CHERCHI A., MAXIA C. & ULZEGA A. (1974) - Evoluzione paleogeografica del Terziario della Sardegna. Rend. Sem. Fac. Sc., Univ. Cagliari, suppl. 43: 73-89.

COCOZZA T. & JACOBACCI A. (1975) - Geological outline of Sardinia. In: "Geology of Italy", Earth Sc. Soc. Libyan Arab Republic: 49-81.

COLOM G. (1952) - Aquitanian-Burdigalian diatom deposits of the North Betic Strait, Spain. J. Paleont., 26(6): 867-885.

CONDIE K.C. (1973) - Archean magmatism and crustal thickening. Geol. Soc. Am. Bull., 84: 2981-2992.

COULON C., DEMANT A. & BELLON H. (1974) - Premières datations par la méthode K/Ar de quelques laves cénozoiques et quaternaires de Sardaigne nord-occidentale. Tectonophysics, 22: 41-57.

CROOK K.A.W. (1960) - Classification of arenites. Am. J. Sci., 258: 419-428.

DECIMA A. & SPROVIERI R. (1973) - Presenza di sedimenti tripolacei basso-miocenici in Sicilia. Riv. Min. Siciliana, 142-144: 202-218.

DEL MONTE M. (1972) - Il vulcanesimo del Mar Tirreno: nota preliminare sui vulcani Marsili e Palinuro. Giorn. Geol., 38(1): 231-252.

DICKINSON W.R. (1962) - Petrogenetic significance of geosynclinal andesitic volcanism along the Pacific margin of North America. Geol. Soc. Am. Bull., 73: 1241-1256.

DIDON J., FERNEX F., LORENZ C., MAGNE J. & PEYRE Y. (1970) - Sur un niveau remarquable de silexite dans le Néogène inférieur d'Espagne méridionale et d'Italia du Nord. Bull. Soc. Géol. France, 11(6): 841-853.

DUPUY C., McNUTT R.H. & COULON C. (1974) - Détermination de $^{87}Sr/^{86}Sr$ dans les andésites cénozoiques et les lavas associées en Sardaigne Nord occidentale (Italie). Geochim. Cosmochim. Acta, 38: 1287-1296.

ELTER P., GIGLIA G., TONGIORGI M. & TREVISAN L. (1975) - Tensional and compressional areas in the recent (Tortonian to present) evolution of the Northern Apennines. Boll. Geofis. Teor. Appl., 17(65): 3-18.

FOX P.J. & HEEZEN B. (1975) - Geology of the Caribbean crust. In: "The ocean basins and margins", vol. 3: 421-466.

FREUND R., GARFUNKEL Z., ZAK I., GOLDBERG M., WEISSBROD T. & DERIN B. (1970) - The shear along the Dead Sea rift. Phil. Trans. R. Soc. London, A 267: 107-130.

GIGLIA G. & RADICATI DI BROZOLO F. (1970) - K/Ar age of metamorphism in the Apuane Alps (Northern Tuscany). Boll. Soc. Geol. It., 89: 485-497.

GIRDLER R.W. & STYLES P. (1974) - Two stage Red Sea floor spreading. Nature, 247: 7-11.

GRUNAU H.R., LEHNER P., CLEINTUAR M.R., ALLENBACH P. & BAKKER G. (1975) - New radiometric ages and seismic data from Fuerteventura (Canary Island), Maio (Capo Verde Islands) and Sao Tome (Gulf of Guinea). In: "Progress in Geodynamics": 90-118.

HSÜ K.J. et al. (1975) - Glomar Challenger returns to the Mediterranean Sea. Geotimes, August: 16-19.

KOTLARCZYK J. (1966) - Diatomite horizon of the Krosno Beds in the Skole Nappe, Polish Carpathians. Studia Geol. Polonica, 19: 1-129.

LOWELL J.D., GENIK G.J., NELSON T.H. & TUCKER P.M. (1975) - Petroleum and plate tectonics of the Southern Red Sea. In: "Petroleum and global tectonics" : 129-153.

MEZZETTI R. & OLIVIERI R. (1964) - Intercalazioni cineritiche di età oligocenica e miocenica nell'Appennino centro-settentrionale. Miner. Petrogr. Acta, 10: 129-147.

MITCHELL A.H. & READING H.G. (1971) - Evolution of island arcs. J. Geol., 79 (3): 253-284.

PATRUT I. (1959) - Géologie et tectonique de la région Valenii de Munte-Cosminele-Bustenari. Ann. Comité Géol. Bucarest, 26-28: 203-232.

PECORINI G. & POMESANO CHERCHI A. (1969) - Ricerche geologiche e biostratigrafiche sul Campidano meridionale (Sardegna). Mem. Soc. Geol. It., 8(4): 421-451.

PICHLER H. & ZEIL W. (1972) - The Cenozoic rhyolite-andesite association of the Chilean Andes. Bull. Volcan., 35(2): 424-452.

RITTMANN A. (1967) - Die Bimodalität des Vulkanismus und die Herkunft der Magmen. Geol. Rdsch., 57: 277-295.

RYAN W.B.F. et al.(1973) - Initial Reports of the Deep Sea Drilling Project. 13(1-2): 1-1447.

SAVELLI C. (1975) - Datazioni preliminari col metodo K/Ar di vulcaniti della Sardegna sud-occidentale. Rend. Soc. It. Min. Petr., 31(1): 191-198.

SELLI R. (1957) - Sulla trasgressione del Miocene nell'Italia meridionale. Giorn. Geol., 26: 1-54.

SELLI R. (1974) - Appunti sulla geologia del Mar Tirreno. Rend. Sem. Fac. Sc. Univ. Cagliari, suppl. 43: 327-349.

SELLI R. & FABBRI A. (1971) - Tyrrhenian: a Pliocene deep sea. Acc. Naz. Lincei, Rend. Cl. Sc. f.m.n., 50(5): 104-116.

TRÜMPY R. (1973) - Timing of orogenic events in the Central Alps. In: "Gravity and Tectonics": 229-251, J. Wiley.

TURNER F.J. & VERHOOGEN J. (1960) - Igneous and metamorphic petrology. Mc Graw-Hill, 694 p.

VASS D., SLAVIK J. & BAGDASARJAN G.P. (1975) - Radiometric time scale for Neogene of Paratethys (a state to August 1, 1974). VIth Congr. Reg. Comm. Med. Neogene Strat., Bratislava: 293-297.

WATTS A.B. & WEISSEL J.K. (1975) - Tectonic history of the Shikoku marginal basin. Earth Planet. Sci. Lett., 25: 239-250.

WEZEL F.C. (1973a) - Nuovi dati sulla età e posizione strutturale del Flysch di Tusa in Sicilia. Boll. Soc. Geol. It., 92: 193-211.

WEZEL F.C. (1973b) - Diacronismo degli eventi geologici oligo-miocenici nelle Maghrebidi. Riv. Min. Siciliana, 142-144: 219-232.

WEZEL F.C. (1974a) - Flysch successions and the tectonic evolution of Sicily during the Oligocene and early Miocene. In: "Geology of Italy", Earth Sc. Soc. Libyan Arab Republic: 1-23.

WEZEL F.C. (1974b) - Primo riesame delle carote raccolte nel Leg 13 del DSDP (Mediterraneo). Giorn. Geol., 39(2): 447-468.

WEZEL F.C. (1975) - Diachronism of depositional and diastrophic events. Nature, 253: 255-257.

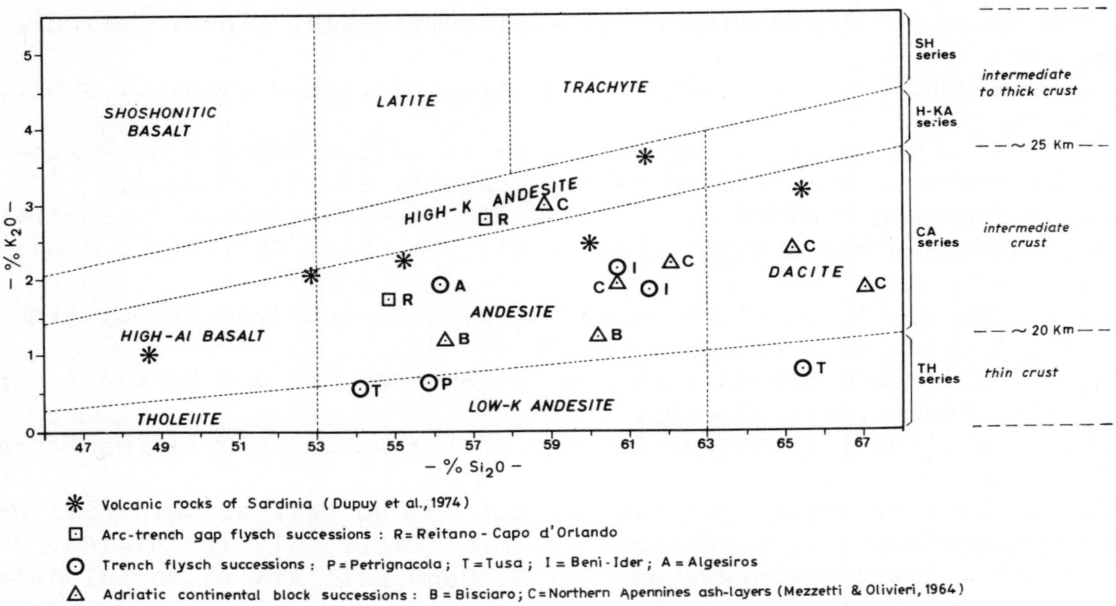

Fig.1 Variation diagram K_2O vs. SiO_2 from analyses of selected Oligo-Miocene volcanic sandstone in the peri-Mediterranean chains (analists: A. Mazzucotelli and G. Ottonello) compared with already published analyses of the Apenninic cineritic beds and the Sardinian lavas.

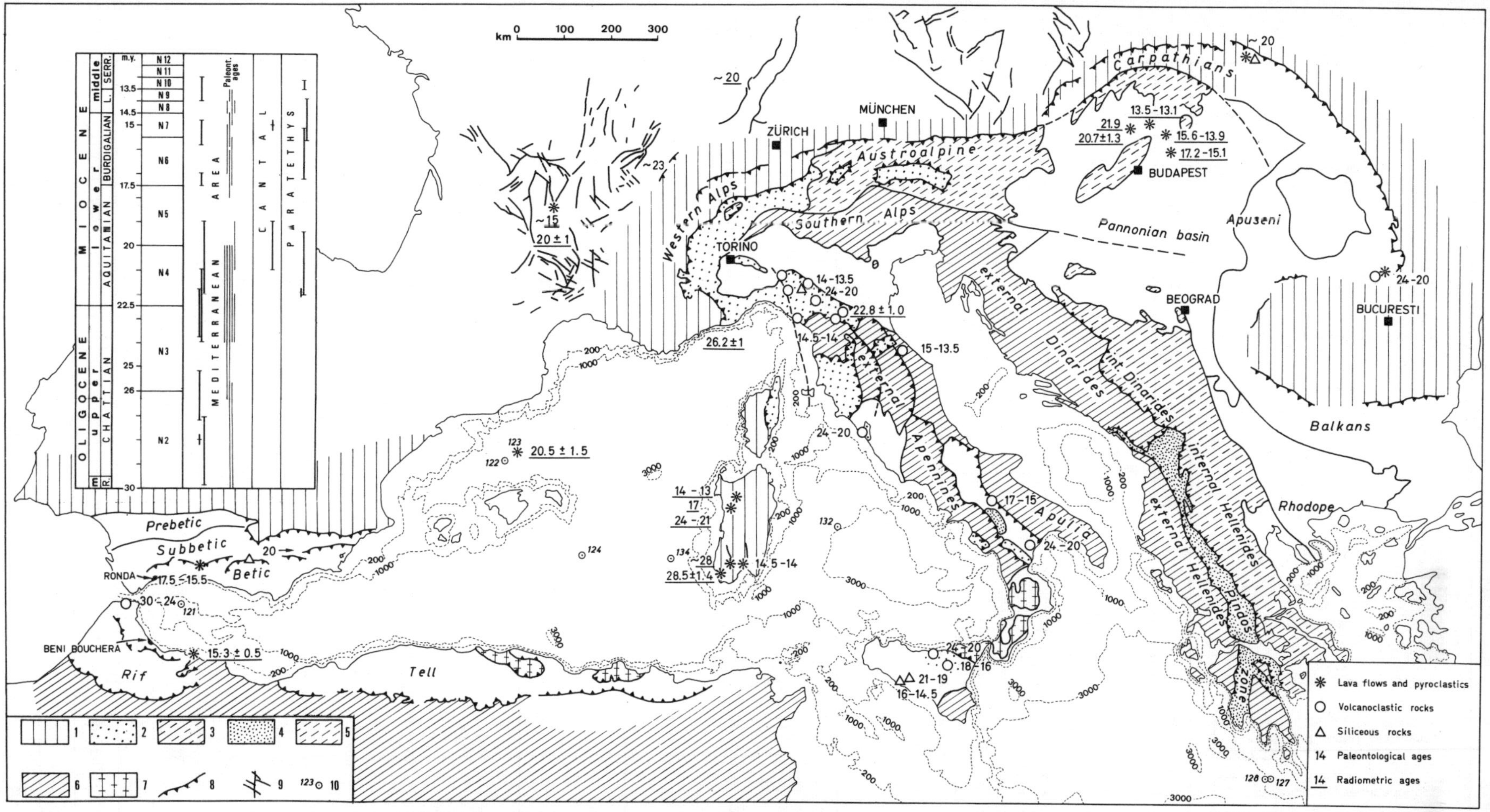

Fig.2 Map of the Alpine-Mediterranean region. Geology mainly after Bernoulli & Jenkins (1974). Radiometric ages (underlined) of the Oligo-Miocene volcanism based mainly on the data of: Bellon (1976, pers. comm.); Brousse et al.,(1975); Cati,(1974); Coulon et al.,(1974); Ryan et al., (1973); Savelli,(1975); Vass et al.,(1975). Paleontological ages were assigned their numerical values on the base of the Berggren's scale (see insert at top-left). Apart from the published and unpublished datings by the author and his collaborators, the data of the following workers were used: Boni et al.,(1968); Decima & Sprovieri,(1973); Didon et al.,(1970); Kotlarczyk,(1966); Patrut,(1959); Pecorini & Pomesano Cherchi,(1969).
Legend (insert bottom-left): 1) Neogene tectonic units of the northern margin of Tethys and the European foreland. 2) Pennine and Ligurian nappes. 3) Internal zones of the Dinarids and Hellenids. 4) Pindos and Lagonegro zones. 5) Tectonic units of the southern margin of the Tethys with a northern vergence. 6) Tectonics units of the southern margin of the Tethys with a "southern" vergence and the African foreland. 7) Calabrian and Kabylian basement nappes. 8) Overthrusts. 9) Deep fracture zones. 10) JOIDES drillsites.

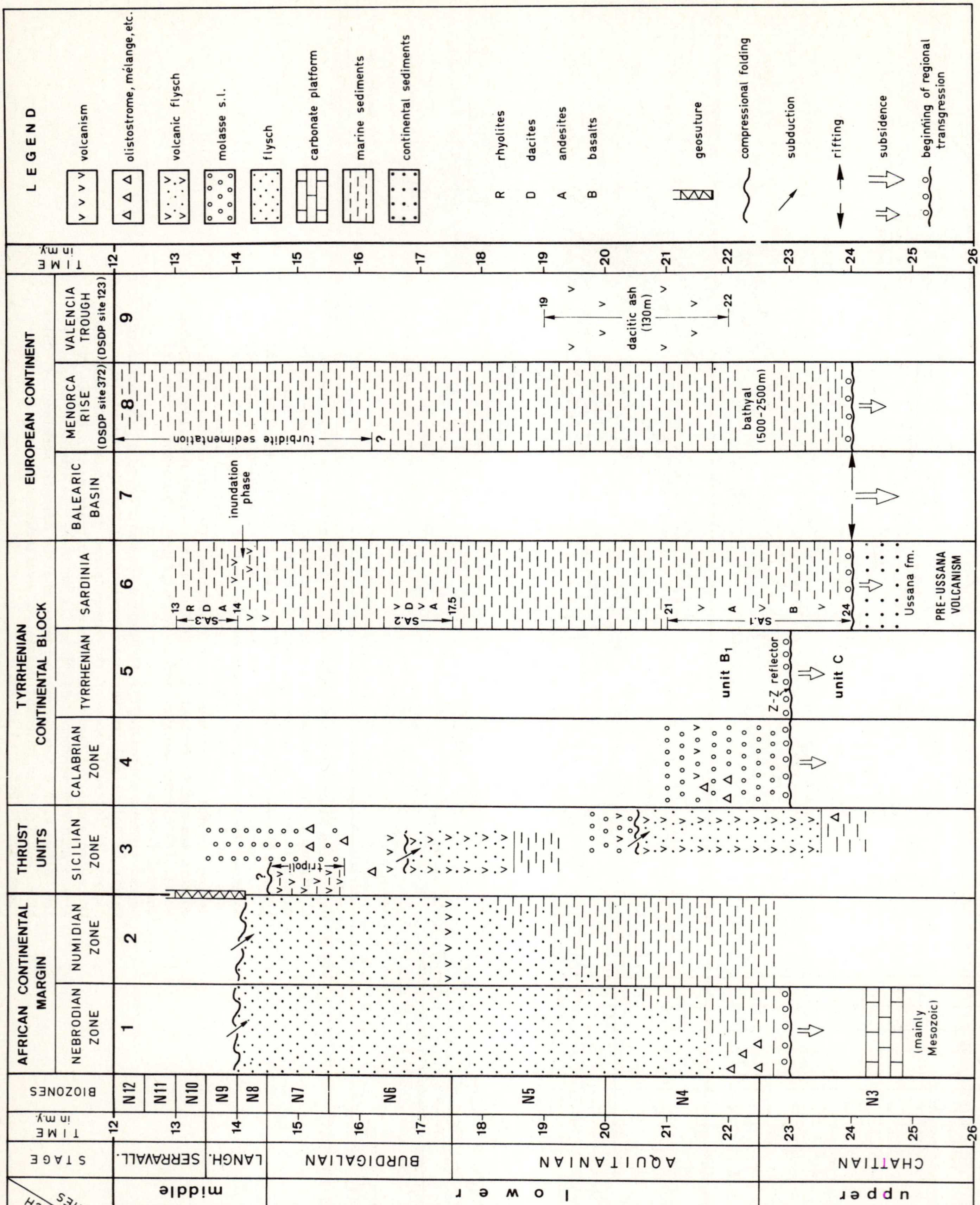
EPOCH
SERIES
STAGE
TIME in m.y.
BIOZONES
AFRICAN CONTINENTAL MARGIN
NEBRODIAN ZONE
NUMIDIAN ZONE
THRUST UNITS
SICILIAN ZONE
TYRRHENIAN CONTINENTAL BLOCK
CALABRIAN ZONE
TYRRHENIAN
SARDINIA
EUROPEAN CONTINENT
BALEARIC BASIN
MENORCA RISE (DSDP site 372)
VALENCIA TROUGH (DSDP site 123)
M I O C E N E
OLIGOCENE
middle
lower
upper
SERRAVALL.
LANGH.
BURDIGALIAN
AQUITANIAN
CHATTIAN
N12
N11
N10
N9
N8
N7
N6
N5
N4
N3
(mainly Mesozoic)
tripoli
unit B1
Z-Z reflector
unit C
SA.3
SA.2
SA.1
Ussana fm.
PRE-USSANA VOLCANISM
inundation phase
turbidite sedimentation
bathyal (500-2500m)
dacitic ash (130m)
LEGEND
volcanism
olistostrome, mélange, etc.
volcanic flysch
molasse s.l.
flysch
carbonate platform
marine sediments
continental sediments
R rhyolites
D dacites
A andesites
B basalts
geosuture
compressional folding
subduction
rifting
subsidence
beginning of regional transgression

Fig. 3

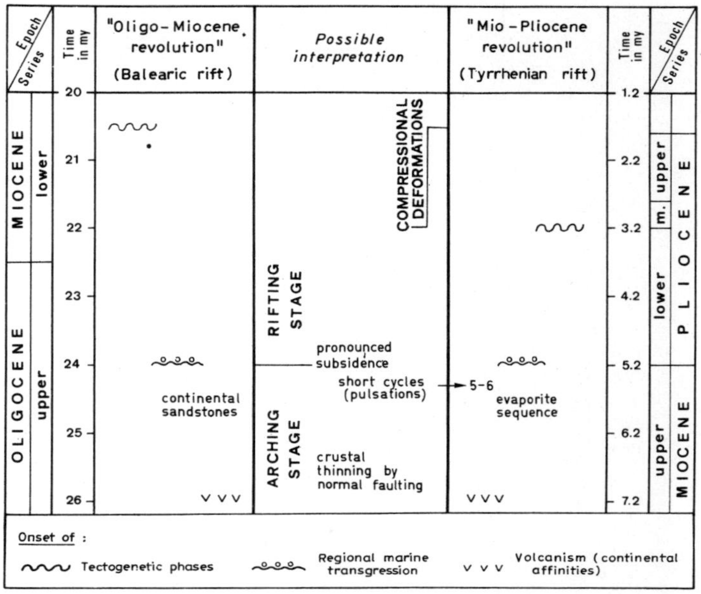

Fig.4 Stratigraphic diagram based mainly on data from Sicily and Sardinia in which are compared basic geologic events in Oligo-Miocene with similar events in Mio-Pliocene.

← Fig.3 Stratigraphic diagram showing the development in time of the geological events from Sicily to the Valencia Trough based on the following: 1-4) published data of the author; 5) Selli (1974); 6) Coulon et al.,(1974) and Pecorini & Pomesano Cherchi (1969); 7) author's interpretation; 8) Hsü et al.,(1975); 9) Ryan et al.,(1973).

INTERNATIONAL SYMPOSIUM ON THE STRUCTURAL HISTORY OF THE MEDITERRANEAN BASINS. SPLIT (YUGOSLAVIA) 25-29 OCTOBER 1976. B. BIJU-DUVAL AND L. MONTADERT, Eds. EDITIONS TECHNIP, PARIS 1977, pp. 303-314

SOME ASPECTS OF THE EVOLUTION OF MARGINAL SEAS DEDUCED FROM OBSERVATIONS IN THE AEGEAN REGION

H. BERCKHEMER (1)

Abstract. Based on a synopsis of available observations the Aegean region is characterized as a young, small, inactive marginal basin with high heat flow in a somewhat hampered stage of development. It offers in a unique way the possibility to study the interrelation between the evolution of an Alpine orogenic belt and the development of a marginal basin in it. Gravity instability caused by nonhydrostatic forces in a laterally inhomogeneous crust are regarded to be the initiator of disintegration in the marginal basin. Subduction and possible gravity instability in the mantle reinforce the process of marginal sea expansion. The intimate relation of young continental orogenies and marginal basins seems to be a general rule.

The Aegean region is a relatively small but dynamically very active part of the Alpine-Himalayan orogenic system. In recent years, stimulated by the international Geodynamics Project, comprehensive systematic research has been carried out in this area. It now can be regarded a comparatively well investigated region. Here an attempt is made to summarize and to evaluate some of the information available with the aim of a better understanding of the processes which have finally led to the present situation.

Physiographically the Aegean Sea is a marginal basin bordering or even penetrating the European Continent and being bounded toward the Eastern Mediterranean Sea by the Hellenic Island Arc. A first question arises: To what an extent is this structure related to Pacific, Indonesian, or Caribbean island arc systems? There exists a broad spectrum of different opinions. A systematic check may help us to find the answer.

The most comprehensively investigated part of the Hellenic Arc is the Peloponnesus region which, therefore, might serve for exemplification. From marine geophysics (Weigel 1974, Finetti 1976, Jongsma et al. 1977), neotectonics (Kelletat et al. 1977), precision hypocenter determinations and focal mechanisms (Leydecker et al. 1977, Ritsema 1974), deep seismic sounding (Makris 1977), and gravity (Makris et al. 1973) it may be justified to draw a generalized West-East cross section through the Ionian Sea, Hellenic Trench, Peloponnesus to the Aegean Sea as shown in Figure 1 (Berckhemer and Kovalczyk 1977). Underthrusting of the Ionian crust under the Peloponnesus is evidenced by gently inclined seismic boundaries, strong seismicity in

(1) Institut für Meteorologie und Geophysik der Universität, Feldbergstrasse 47, 6000 Frankfurt a.M.-1, Federal Republic of Germany

crustal and subcrustal depth in front of the island, almost horizontal compression perpendicular to the trench derived from fault plane solutions, air gun profiles in the trench region, and antithetic block rotations within the Peloponnesus. A Benioff zone branching off where the crust reaches its greatest thickness and dipping down under 4o° toward the central Aegean to a depth of 16o - 18o km is clearly visible from the focal distribution. The tension axis of some intermediate deep shocks point along the dip of the Benioff zone. Tensional tectonics in the Aegean crust is indicated by quarternary volcanism and normal faulting. Heavy lines indicate seismic reflectors or refractors, chiefly crust-mantle boundaries. A few reflecting elements remain unexplained by the geodynamic model shown in Figure 1.

Although one has to be careful in generalizing the situation at the Peloponnesus the following features, typical for marginal sea-island arc structures are found in the Aegean region:

- A nonvolcanic frontal arc (Hellenic Island Arc) stretching from Kephalinion via Peloponnesus, Crete to Rhodes.

- A deep sea trench with a pronounced negative free air gravity anomaly of - 13o to - 25o mgal (e. g. Fleischer 1964) and generally positive free air gravity of + 5o to + 1oo mgal in the Aegean Sea (Allan and Morelli 1971).

- High seismicity with foci in normal and intermediate depth (e.g. Papazachos and Comninakis 1971)

- A Benioff zone which is at least in pieces well recognizable (e.g. Papazachos 1973, Leydecker et al. 1977, Richter and Strobach 1977) dipping down under 35 - 45° toward the central Aegean.

- An inner island arc with quarternary volcanism (Aegina, Methana, Milos, Santorin, Nysiros, Antiparos, Kos) above hypocenters in 12o - 15o km depth. The K_2O/SiO_2 ratio being typical for andesitic volcanism related to seismicity in the corresponding focal depth (Ninkovich and Hays 1972).

- Distance trench-volcanic arc ca. 2oo km

- Tensional tectonics in the back arc region (see Fig. 3).

- High heat flow up to 2.7 HFU in the central Aegean Sea (Jongsma 1974) and low heat flow in front of the arc.

There are, on the other hand, differences in comparison with many other island arcs:

- Small dimensions. The radius of curvature of the frontal arc being only 3oo km. The structure, however, is young; its development began not before Upper Miocene.

- A thick, predominantly sialic crust of some 3o km below Crete and the Cyclades and 2o km below the Cretan Sea (Fig. 4) and a somewhat reduced submoho velocity v_p = 7.7 - 7.9 km/s (Makris 1977). There was, with the possible exception of the Saros trench, no true sea floor spreading.

- Maximum focal depth is 2oo km only and some scattered shocks of intermediate depth are found outside the Benioff zone indicated above.

- The nonvolcanic frontal arc is clearly dominating compared with the poorly developed volcanic arc.

- In the Aegean Sea shallow water depth and a field of sialic islands behind the volcanic arc.

- No typical oceanic crust (thickness approximately 2o km including some 1o km of Neogene) and a huge sedimentary chain (East Mediterranean Ridge) in front of the Hellenic Trench. In its eastern part the trench duplicates (Pliny - and Strabo Trench).

These differences from other marginal basins seem to be more of a gradual than of a principal character. According to Karig's classification (Karig 1971) the Aegean region can be considered a young, small, inactive marginal basin with high heat flow in a somewhat hampered state of development.

If this is true essentially the same physical and chemical processes should be responsible for the evolution of this island arc than for others and no special mechanism is required. On the other hand conclusions drawn from observations in the Aegean might be applicable to other marginal sea areas too.

As a unique case Alpine orogenic belts can be traced from the Dinarides through the island arc up to the Taurides (Fig. 2). The apparently intimate connection of an Alpidic orogeny and the successive island arc development in one and the same area is a fact which needs to be emphasized. The present frontal arc as well as the back arc islands were fully involved in the Alpine orogeny of the Dinarides - Hellenides - Taurides realm. This is briefly sketched below.

The geosynclinal stage is characterized by a broad stable shelf platform of the African Continent, some rather mobile depressions such as the Gavrovo-Triplitza and Pindos troughs and, separated by a crystalline anticline, the Vardar Basin representing the central Mesozoic Tethys. Together with the shelf region in the North the width of the whole marine area must have been some 15oo - 2ooo km. Corresponding to the process of rifting in the Atlantic Ocean lateral and convergent movements occured which led in course of the closure of the Tethys and the continent collision to the Alpine orogeny. Following Jacobshagen (1977) in the Hellenides five tectogenetic phases can be distinguished which migrated from the internal to the external zones and acted simultaniously or overlapping in time and space along the belt. Deformations and metamorphism in the circum Rhodope belt and the internal Vardar zone took place in Jurassic. During the Eohellenic Phase (Upper Jurassic/Lower Cretaceous) crustal shortening caused overthrusting of ophiolitic nappes from the external Vardar zone onto the Pelagonian platform. Flysch was deposited in the Pindos Trough. In Upper Cretaceous a similar phase of crustal shortening and subduction affected the southern and eastern Aegean and seems to be directly connected with the overthrusting of ophiolotic nappes in the Lycian Taurus. In Eocene a phase of folding traversed obliquely existing orogenic trends in the Aegean. In Oligocene subduction of the Pindos Trough under the Pelagonian Platform commenced. This culminated in Miocene (Burdigalian) in the high orogenic phase with rising of the Median Crystalline Massif and overthrusting of the Pindos nappes onto the external Hellenides perhaps by gravity sliding.
From the grade of metamorphism it can be deduced that the Median Crystalline Massif was rising and has subsequently been eroded 1o - 15 km deep (Dürr et al. 1977) but still the crustal thickness is more than 3o km in the central Aegean. The original proximity and continuous contact between the central Aegean and Crete is impressively demonstrated by piles of allochtoneous nappes transported far outward from internal zones and resting now on the Mesozoic carborate base of the Cretan crustal block. An elevated land mass in the Aegean is also indicated by an erosion surface observed all over the southern Aegean area (event A of Jongsma et al. 1977). Sediments overleying event A are considered younger than Middle Tortonian.

Based on these facts the existence of an orogenic mountain ridge crossing the present Aegean is postulated for that period. In Upper Miocene already distractive tectonics commenced with intensive normal faulting, vertical block movements, general subsidence and simultaneous extension in the Aegean uplift. The Aegean archipelagos proper developed in Upper Pliocene. The block tectonics was reactivated when some 3 m.y. ago the island arc volcanism began. A compilation of neotectonic structures in the Aegean region is shown in Figure 3. It is clearly seen that normal faulting and graben-like structures with the tendency of expanding the Aegean basin are abundant in particular in the southern part (Cretan Sea) and the northern part (Saros Trench) and penetrating deeply into western Anatolia. Miocene volcanism is widespread in northwestern Anatolia and volcanic intrusions within the Aegean Sea are marked by magnetic anomalies.

An already existing Orogenic ridge disintegrated and was destroyed. Why just here and not somewhere else in the Alpine Belt? This is our second question. The answer must be suspected from the fact that only this part of the Alpine chain is now bordered at one side by a pseudo-oceanic crust, perhaps the remainder of a Mesozoic Tethys Basin (Monod et al. 1974).

It is shown by a simple derivation in the appendix and has been proved and pointed out already by Artyushkov (1973) that a laterally inhomogeneous crust, even in isostatic equilibrium, exerts horizontal compressional forces acting from areas with high relief to those of low relief (Fig. 5). The surplus relief energy of a thick crust relative to a thin crust is in a 2-dimensional model:

$$U = 1/2\ g \cdot \rho_c \cdot \rho_m / (\rho_m - \rho_c) \cdot \int \zeta'^2 df$$

where ρ_c and ρ_m are the densities of crust and mantle, ζ' the surface elevation and df the surface element. About 2/3 of this energy is stored in the root of a mountain and 1/3 only in its surface relief. From the principle of virtual work the non-hydrostatic horizontal force K or the mean stress:

$$\bar{\sigma} = \frac{K}{h} = g \cdot \rho_c \cdot \zeta'$$

is found to be equal to the load of the surface relief. For example a mountain range of 3 km elevation relative to the adjacent sea floor produces $\bar{\sigma}$ = 840 bar which has to be taken up by counterforts provided by the adjacent crustal blocks or by the internal strength of the crust, otherwise it will flow apart with the tendency to equalize the crustal thickness. Low strength, however, is expected inside of a young mountain body because it is still under elevated temperature and the viscosity or any other kind of creep depends exponentially on temperature.

This gravity controlled disintegration is regarded the cause for the initiation of the island arc - marginal sea development. At first the eastern Mediterranean crust was passively overridden. Subsequently the suppressed lithosphere became actively pulling due to a density increase by high pressure phase transformation. The rate and direction of subduction is governed by the relative motion of Africa, Europe, and Arabia and by the expansion of the Aegean Sea. The expansion, in turn, is reinforced by upwelling mantle material displaced by the intruding downgoing lithosphere. This has been demonstrated in model experiments by the author and theoretically from viscous fluid dynamics by Andrews and Sleep (1964). In case of a pre-existing gravity instability in the mantle the disturbance in the lithosphere might cause an asthenospheric diapir to rise which could contribute as a third agent to the expansion.

In principle the process may continue until equilibrium is reached with the crust of the adjacent oceanic area. Because of the anomalously thick crust in the Eastern Mediterranean the expansion and crustal thinning in the Aegean is limited and no true seafloor spreading will occur. A crucial question is why once the crust bordering the Aegean region in the South was able to stand or even to provide the forces of the compressional orogenic phases but was later on subduced. A belt of Mesozoic carbonate series which bottomes the external Dinarides, Hellenides, Cretan Arc, Lycian Taurus (Bernoulli 1972) indicates the existence of a former relatively thick and stable shelf platform of Africa like the present pre-Apulian and Adriatic platform. In the rear of this platform a Mesozoic Tethian basin (The present Eastern Mediterranean) might have existed (Monod et al. 1974) which, with the outward migration of the Alpine orogeny approached the Aegean region and finally allowed for the island arc development.

The marginal sea-island arc development is considered an unsuccessful attempt of nature for orogenesis (in the sense of mountain building). The marginal basin has to wait for another continent collision in which it may play the role of an eugeosyncline basin (Dennies 1976).

By comparing the situation in the Aegean with other true marginal seas such as Andaman Sea, Bering Sea, eastern Caribbean, it becomes evident that in all these cases too the non-volcanic frontal island arc, even if only preserved in rudiments, merges continously into continental orogenic belts of Tertiary age. There are also in all these marginal seas areas with a thick epicontinental type of crust and basins with a thinner, more oceanic type of crust (Fig. 4). Also from this point of view the Aegean does not seem a completely unique structure. Therefore the evolutionary model outlined above might be applicable in modified form to other marginal sea areas too but this will be subject of future studies.

APPENDIX

1. Energy in a crust of laterally varying thickness.

The surplus in potential energy $u(\zeta')$ of an isostatically compensated crustal column of unit area with the surface elevation ζ' relative to a reference crust consists of two parts.

Energy in the relief

$$du_1 = g \cdot \rho_c \cdot \zeta \cdot d\zeta$$

Energy in the root

$$du_2 = \rho(\rho_m - \rho_c)\xi \cdot d\xi$$

where ξ' is the depth of the root (see Fig. 5)

Condition of isostatic equilibrium

$$\rho_c \cdot \zeta = (\rho_m - \rho_c) \cdot \xi \qquad 1$$

$$du = du_1 + du_2$$

$$= \zeta \cdot d\zeta \cdot g \cdot \rho_c \cdot \rho_m / (\rho_m - \rho_c) \qquad 2$$

$$u = \int_0^{\zeta'} du = 1/2\zeta'^2 \cdot g \cdot \rho_c \cdot \rho_m / (\rho_m - \rho_c) \qquad 3$$

Energy surplus of a crustal body of area F

$$\boxed{U = \frac{g \cdot \rho_c \cdot \rho_m}{2 \cdot (\rho_m - \rho_c)} \int_F \zeta'^2 \cdot df} \qquad 4$$

Example:

Assumed an orogenic ridge in the Aegean of $\zeta' = 3$ km devation relativ to the adjacent (eastern Mediterranean) sea floor (water ignored) and of an area F = 800 x 200 km^2.
$\rho_c = 2{,}8$ g/cm^3, $\rho_m = 3.3$ g/cm^3,
$g = 10^3$ cm/s^2.

$$U = 1.5 \cdot 10^{30} \text{ erg}$$

$$u \simeq 1 \cdot 10^{15} \text{ erg/cm}^3$$

If this energy is released at a constant rate during the process of disintegration from $5 \cdot 10^6$ yr ago to present

$$u' = 2 \cdot 10^8 \text{ erg/cm}^2 \text{ yr}$$

For comparison: mean seismic energy density annually released in the Aegean region (based on data from Galanopoulos 1971)

$$e \simeq 6 \cdot 10^6 \text{ erg/cm}^2 \text{ yr}$$

Mean heat flow density (2HFU) in the Aegean

$$w \simeq 2.5 \cdot 10^9 \text{ erg/cm}^2 \text{ yr}$$

It is seen that heat is still the strongest source of energy but the potential relief energy would be sufficient to account for the seismicity.

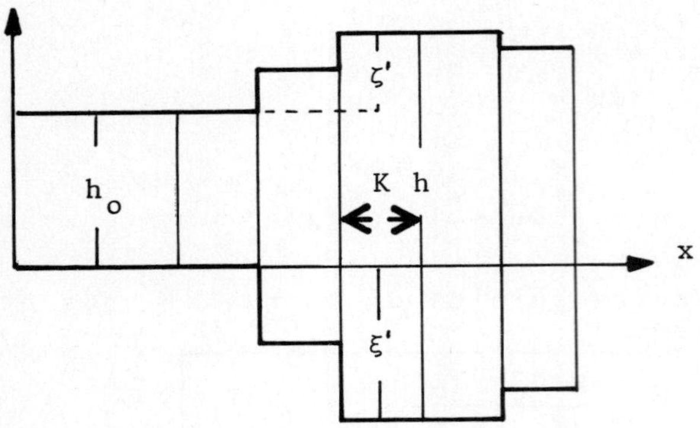

Figure 5

2. <u>Horizontal residual stress in a crust of laterally varying thickness (2-dimensional).</u>

Non-hydrostatic horizontal stress in a crust of zero-strength

$$\sigma = \sigma_{xx} - \sigma_{yy}$$

Residual horizontal force caused by the surface relief ζ'

$$K = \int_{-\xi'}^{h_o+\zeta'} \sigma \cdot dy = \bar{\sigma} \cdot h \qquad 5$$

$$h = h_o + \zeta' + \xi' \qquad 6$$

$\bar{\sigma}$ = mean residual stress

Work to be done for laterally narrowing a crustal column of thickness h and unit area by δx

$$\delta a = K \cdot \delta x \qquad 7$$

Conservation of volume and isostasy

$$dx \cdot h = (d\zeta + d\xi) \cdot 1 = d\zeta \cdot \rho_m / (\rho_m - \rho_c)$$
$$dx = d\zeta \cdot \rho_m / h \cdot (\rho_m - \rho_c) \qquad 8$$

Conservation of energy (plastic losses neglected):

$$\delta a = \delta u \qquad 9$$

From (8) and (2)

$$K \cdot d\zeta \cdot \rho_m / h \cdot (\rho_m - \rho_c) = \zeta' \cdot d\zeta \cdot g \cdot \rho_c \cdot \rho_m / (\rho_m - \rho_c)$$

$$K = g \cdot \rho_c \cdot \zeta' \cdot h$$

$$\boxed{\bar{\sigma} = \frac{K}{h} = g \cdot \rho_c \cdot \zeta'} \qquad 1o$$

valid only for moderate relief gradient

$\bar{\sigma}$ is equivalent to the vertical pressure caused by the weight of the relief.

Example:

As before $\zeta' = 3$ km, $\rho_c = 2.8$ g/cm^3

$\bar{\sigma}$= 84o bar

This value is much higher than the usual stress drop in crustal earthquakes and therefore probably higher than the strength of the crust.

LITERATURE

ALLAN, T. D., MORELLI, C.: A geophysical study of the Mediterranean Sea. Boll. Geofis. Teor. Appl. 14 (1972) 291 - 342

ANDREWS, D. J., SLEEP, N. H.: Numerical modelling tectonic flow behind island arcs. Geophys. J. R. astr. Soc. 38 (1974), 237 - 251.

ARTYUSHKOV, E. V.: Stresses in the lithosphere caused by crustal thickness inhomogeneities. Journ. Geophys. Res. 78, 32 (1973), 7675 - 7707

BERCKHEMER, H., KOVALCZYK, G.: Post-alpine geodynamics of the Peloponnesus. Alpes, Appennines, Hellenides. Edit: Closs, H., Roeder, D., Schmidt, K., Schweizerbart, Stuttgart (1977).(in press)

BOCCALETTI, M., MANETTI, P., PECERILLO, A.: The Balkanides as a instance of back - are thrust belt. Geol. Soc. Amerca. Bull. 85, 1077 - 1084

DENNIS, J. G.: Geosynklinale, Orogenese und Plattentektonik. Z. dt. geol. Ges. 127 (1976), 73 - 85.

blished but only suggested. The part of this model which concerns the northern Aegean area express the one of the two suggested below working hypotheses for the deep tectonics of this area. In the following we try to show how the very general seismic and other geophysical properties of the Aegean area can be interpreted.

The Mediterranean slab, which is part of the African lithsopheric plate, underthrusts the back - arc Aegean lithosphere at a mean angle of about 35°. A direct result of this interaction between the two lithospheres is the generation by compressional mechanism (thrust faulting) of shallow shocks in the convex side of the Hellenic arc. As this slab goes down frictional heat is produced along its top slip surface. A result of this heating is the generation of shear zones within the slab but nor far from its top surface. In this way, earthquakes of intermediate focal depths are produced by thrust faulting in the slab and form the rather narrow Benioff zone near the top of the slab. This description shows that subduction is the prevailing mode of interaction between the Mediterranean and the back-arc souther Aegean lithsopheres.

The heat generation along the top surface of the Mediterranean slab or hydrodynamic forces caused by the sinking of the slab, would trigger upward migration of hot material in the back - arc southern Aegean asthenosphere. This hot material (low Q) causes the inefficient transmission of the short period body waves in this region and the reduction of the wave velocity in this asthenosphere relative to the slab material. Due to the horizontal forces exerted at the bottom of the back-arc lithosphere by convective cells created in the mantle above the Mediterranean slab or on account of volume increase of this mantle caused by upward migration of thermal diapir, the back - arc southern Aegean lithosphere expands and breaks. Asthenospheric material intrudes the back - arc southern Aegean lithsophere, producing volcanic activity, high heat flow, magnetic anomalies and modification of the crustal structure in the southern Aegean. The fact that the volcanoes of the southern Aegean almost coincide with the isodepth of 150km (fig. 1) shows that the heat generation along the top surface of the Mediterranean slab produces melting at the depth of about 150 km and that material of these volcanoes come probably from that depth too. Results of the expansion are subsidence of crustal blocks and the occurrence of shallow shocks of normal faulting in the southern Aegean area. Due to this expansion, the southern Aegean lithsophere moves to the convex side of the Hellenic arc overthrusting the Mediterranean lithospheric slab. Therefore, extension is the prevailing mode of lithospheric interaction in the back-arc southern Aegean region.

The explanation of the extensional properties of the central and northern Aegean area can be probably based on a model suggested by Sleep and Toksöz (1971). Convective currents created above the Mediterranean slab may constitute of cells which drive the southern part of the back-arc Aegean lithosphere to the Hellenic trench and of cells which drive the central and northern part of this lithosphere to the north and expand it to this direction. This expansion can result to generation of shallow earthquakes by tensional mechanism, to magnetic anomalies, to high heat flow, etc in the central and northern Aegean area. This hypothesis explains also the thrust belt in the northernmost part of the Aegean area because the tensional stresses give gradually way to compressional stresses to the north. The occurrence, however, of intense extensional properties too far away (to north) from the Mediterranean slab and the generation of intermediate earthquakes (h>60km) in the northern Aegean cannot easily be explained by this hypothesis alone.

The occurrence of intermediate shocks in the northern Aegean leads to the hypothesis that a second lithospheric slab, which dips from the central to the northern Aegean area, may exists. This Aegean slab, which may include old oceanic crust trapped in this area, can be the cause of tectonic activity in the northern Aegean area similar to that caused by the Mediterranean slab in the southern Aegean area. However, the low deep seismicity (small number of intermediate shocks and low deep seismic energy) associated with the Aegean slab shows that the sinking of this slab is very slow now. It is probable that the downwards motion of the Aegean slab has been faster in the past and that the observed today high heat flow, magnetic anomalies and relatively high shallow seismic activity in the northern Aegean area are the results of deep geotectonic processes triggered by that motion. The occurrence of Neogene-Quaternary volcanics in the northern Aegean area shows that the deep tectonic activity was intense in the past.

According to this hypothesis, the northern Aegean trough is a still active remnant of a back-arc basin.

One objection about the existence of the Aegean slab could be the occurrence of shallow earthquakes produced by tensional mechanism on both sides of the region of convergence of this slab with the northern Aegean lithosphere. Such stress distribution, however, has been also observed in other areas of lithospheric convergence and has been attributed to accretion (Lomnitz 1974). In such areas, according to this opinion, parts of slabs sink into the mantle and the adjacent lithosphere moves by segmentation towards the area of sinking to fill the vacancy. In this way tensill stresses are produced in the surrounding area. Some other evidence for accretion in the northern Aegean also exist (Needham et al 1973). Such an assumption justifies the large scatter of the foci of the intermediate earthquakes in the northern Aegean because in areas of accretion such scattering is expected (Lomnitz 1974).

The interpretation that both the Cretan trough in the southern Aegean and the northern Aegean trough are the results of lithospheric expansion and that the former, which is closer to the Hellenic trench, is younger than the latter, is in accordance with the observation that the marginal basins increase in age away from the trenches and that within each basin the spreading zone remains at the axis of the basin (Karig 1971).

We can say, in conclusion, that the deep tectonic structure of the Aegean area is much complicated. Additional geophysical data, especially for the northern part of this area, are necessary to understand better its deep tectonic process. Only synthetic work based on both geophysical and geological reliable data can lead to a satisfactory theory on the deep tectonics of this area.

ACKNOWLEDGMENTS

The authors would like to express their sincere thanks to Drs V. Karnik and S. Gregersen for reading critically the manuscript of this paper.

REFERENCES

AGARWAL, N. K., JACOBY, W.R., and BERCKHEMER, H.,Teleseismic P - Wave Travel Time Residuals and Deep Structure of the Aegean Region, "*Tectonophysics*", 31, 33-57, 1976.

ALLAN, T.D., and MORELLI, C. , A Geophysical Study of the Mediterranean Sea, "*Boll.Geof. Teor. Appl.*", 13, 99-142, 1971.

BOCALLETTI, M., MANETTI, P., and PECCERILLO, A., The Balkanides as an Instance of Back-Arc Thrust Belt, "*Bull. Geol. Soc. Am.*", 85, 1077-1084, 1974.

COMNINAKIS, P. E. and PAPAZACHOS, B.C., Seismicity of the Eastern Mediterranean and Some Tectonic Features of the Mediterranean Ridge, "*Bull. Geol. Soc. Am.*", 82, 1093-1101, 1972.

COMNINAKIS, P. E., and PAPAZACHOS B. C., A Note on the Crustal Structure of the Eastern Mediterranean, "*Ann. di Geofis.*", 1976, (in press).

ERICKSON, A. J., The Measurement and Interpretation of Heat Flow in the Mediterranean and Black Sea, "*PhD Thesis, Mass.Inst. Tech.*", 1970.

FINETTI, I., Mediterranean Ridge: A young Submerged Chain Associated with the Hellenic Arc, "*Bollet. Geof. Teor. ed Applic.*", 19, 31-65, 1976.

GREGERSEN, S., P - Wave Travel Time Residuals Caused by a Dipping Plate in the Aegean Arc in Greece, "*Tectonophysics*", 1977,(in press).

JONGSMA, D., Heat Flow in the Aegean Area, "*Geophys. Jour. Roy. Astr. Soc.*", 37, 337-346, 1974.

KARIG, D. E., Ridges and Basins of the Tonga-Kermadec Island Arc System, "*Jour. Geophys. Res.*", 75, 239-254, 1970.

KARIG, D. E., Origin and Development of Marginal Basins in the Western Pacific, "*Jour. Geophys. Res.*", 76, 2542-2561, 1971.

KARNIK, V., A Note on the Morphology and Activity of Seismic Zones in the Aegean Region, "*Nature*", 239, 72-73, 1972.

KARNIK, V., Seismicity of the European Area, Part I, "*D. Reidel Publish. Company., Dordrecht-Holland*" pp. 364, 1969.

LOMNITZ, C., Global Tectonics and Earthquake Risk, "*Elsevier Scient. Publish. Comp.*", pp. 320, 1974.

MAASHA, N., and MOLNAR, P., Earthquake Fault Parameters and Tectonics in Africa, *"Jour. Geophys. Res."*, 77, 5731-5743, 1972.

MAKRIS, J., Some Geophysical Aspects of the Evolution of the Hellenides, *"Bull. Geol. Soc. Greece,"* 10, 206-213, 1973.

McKENZIE, D. P., The Relation between Fault Plane Solutions for Earthquakes and the Directions of the Principal Stresses, *"Bull. Seism. Soc. Am."*, 59, 591-601, 1969.

McKENZIE, D.P., Active Tectonics in the Mediterranean Region, *"Geophys. Jour. Roy.Astr. Soc."*, 30, 109-185, 1972.

MOLNAR, P., and OLIVER, J., Lateral Variations of Attenuation in the Upper Mantle and Discontinuities in the Lithosphere, *"Jour. Geophys. Res."*, 74, 2648-2682, 1969.

MORELLI, C., PISANI, M., and GANTAR, C., Geophysical Studies in the Aegean Sea and in the Mediterranean, *"Boll. Geof. Teor. ed Applic."*, 18, 127-167, 1975.

NEEDHAM, H. D., LE PICHON X., MELGUEN, M., PAUTOT, G., RENARD, V., AVEDIK, F., and CARRE, D., North Aegean Sea Trough: 1972 Jean Charcot Crusie, *"Bull. Geol. Soc. Greece "* 10, 152-153, 1973.

OXBURGH, E. R., and TURCOTTE, D. L., Origin of Paired Metamorphic Belts and Crustal Dilation in Arc Regions, *"Jour. Geophys. Res"*, 76, 1315-1327, 1971.

PAYO, G., Crustal Structures of the Mediterranean Sea, Part II, Phase Velocity and Travel Times, *"Bull. Seismol. Soc. Am."*,59, 23 - 42, 1969.

PAPAZACHOS, B. C., Phase Velocities of Rayleigh Waves in Southeastern Europe and Eastern Mediterranean Sea, *"Pure and Appl. Geophys."*, 75, 47-55, 1969.

PAPAZACHOS, B. C., Distribution of Seismic Foci in the Mediterranean and Surrounding Area and its Tectonic Implication, *"Geophys. Jour. Roy. Astr. Soc."*, 33, 419-428, 1973.

PAPAZACHOS, B. C., Seismotectonics of the Eastern Mediterranean Area, *"Engineer. Seismol. and Earthquake Engineer."*, Noordhoff Leiden, Ed. J. S. Solnes, 1-31, 1974.

PAPAZACHOS, B. C., Evidence of Crustal Shortening in the Northern Aegean Region, *"Boll. Geof. Teor. ed Applic."*, 13, 66-71, 1976a.

PAPAZACHOS, B. C., Seismotectonics of the Northern Aegean Area, *"Tectonophysics"*, 33, 199-209, 1976b.

PAPAZACHOS, B. C., A Lithospheric Model to Interprete Focal Properties of Intermediate and Shallow Shocks in Central Greece,*"Pure and Appl. Geophysics"*, 1976c (in press).

PAPAZACHOS, B.C., COMNINAKIS P.E., and DRAKOPOULOS, J. C., Preliminary Results of an Investigation of Crustal Structure in Southeastern Europe, *"Bull. Seismol. Soc. Am."*, 56, 1241-1268, 1966.

PAPAZACHOS, B. C., POLATOU, M., and MANDALOS, N., Dispersion of Surface Waves Recorded in Athens, *"Pure and Appl. Geophys."*, 67, 95-106, 1967.

PAPAZACHOS, B. C., and GIANNAKOPOULOS, P.A., Tectonic Stress Field in Very Active Seismic Regions, *"Bull. Greek Army Serv."*, 9, 27 - 37, 1968.

PAPAZACHOS, B. C. and DELIBASIS, N. D.,Tectonic Stress Field and Seismic Faulting in the Area of Greece, *"Tectonophysics"*, 7, 231 - 255, 1969.

PAPAZACHOS, B. C. and COMNINAKIS,P.E., Geophysical Features of the Greek Island Arc and of the Eastern Mediterranean Ridge,*"C. R. Acad. Sci. Paris"*, 16, 74-75, 1970.

PAPAZACHOS, B. C., and COMNINAKIS, P.E.,Geophysical and Tectonic Features of the Aegean Arc., *"Jour. Geophys. Res."*, 76,8517-8533, 1971.

RABINOWITZ, P. D., and RYAN, W. B.F.,Gravity Anomalies and Crustal Shortening in the Eastern Mediterranean, *"Tectonophysics"*, 5 - 6, 585 - 608, 1970.

RITSEMA, A. R., Seismotectonic Implications of a Review of European Earthquake Mechanism, *"Geologische Rundschau"*, 59, 37-55, 1969.

RITSEMA, A. R., The Earthquake Mechanism of the Balkan Region,*"Royal Netherl. Meteor. Inst."*, Scient. Rep. Nr. 74-4, 1-36,1974.

RITSEMA, A. R., General Trends of Fault Solutions in Europe, *"Proc. 14th Gen. Ass.*ESC, *Acad. Wiss. DDR, Nat. Com. Geod. Geophys. Berlin"*, 379-384, 1975.

SCLATER, J. G., Heat Flow and Elevation of the Marginal Basins of the Western Pacific *"Jour. Geophys. Res."*, 77, 5705-5719,1972.

SLEEP, N., and TOKSÖZ, M. N., Evolution of Marginal Basins, *"Nature"*, 233, 548-550, 1971.

SYKES, L. R., FLETCHER, J. P., and SBAR, M. L., Contemporary Stress, Interplate Earthquakes and Seismic Risk Associated with High Pressure Fluid Injection Wells in N. Y. State, *"Intern. Coll. Seism. Effe. of Reservoir Impounding"*, London, 27 to 29 March, 1973.

UTSU, T., Seismological Evidence for Anomalous Structures of Island Arcs with Special Reference to the Japanese Region, "

Rev. Geophys. and Space Phys.", 9, 839-890, 1971.

VOGT, P. R., and HIGGS, R. H., An Aeromagnetic Survey of the Eastern Mediterranean Sea and Its Interpretation, "*Earth Planet. Sci. Letters*", 5, 439-448, 1969.

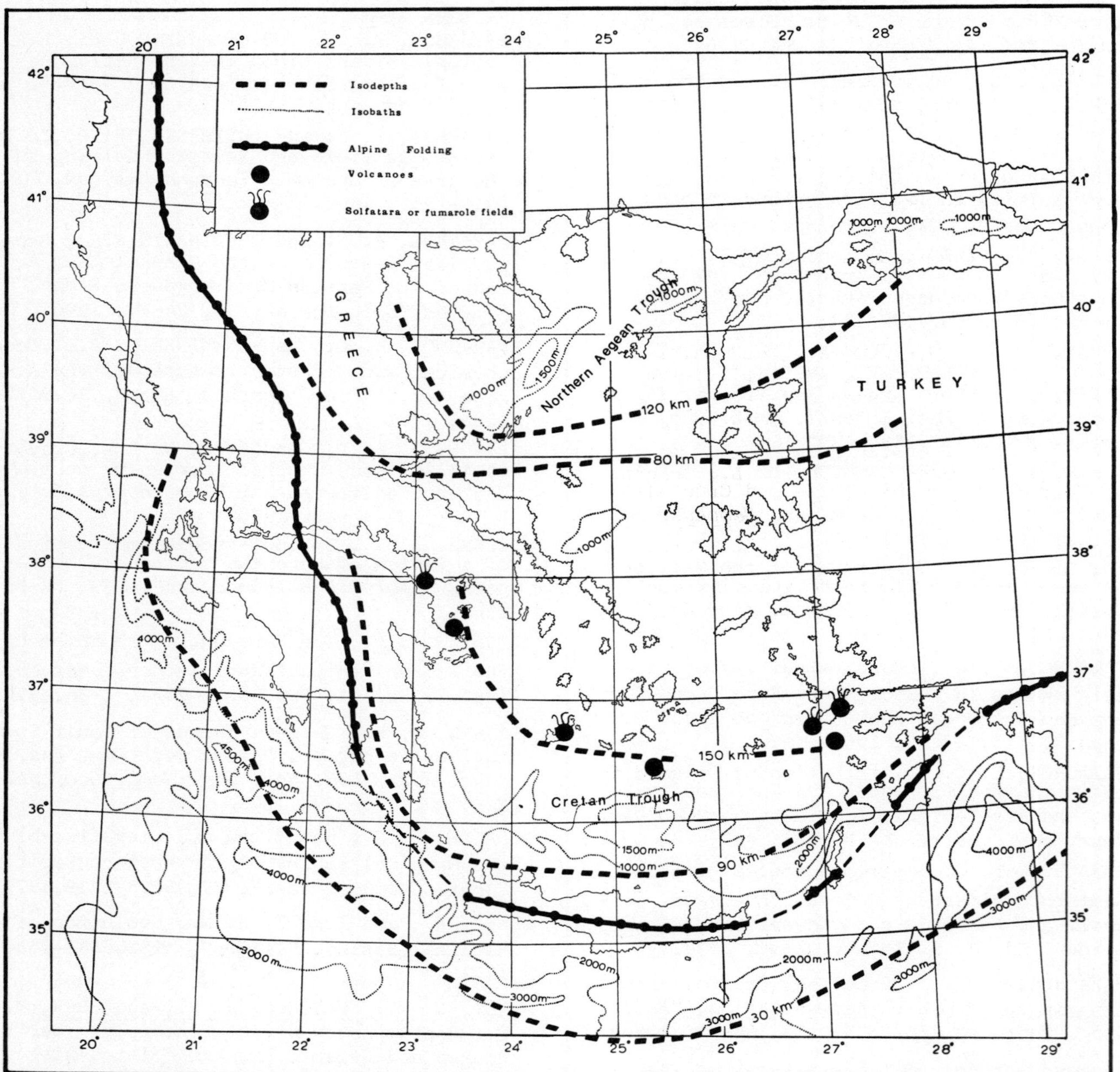

Fig. 1 Isodepths of earthquakes, recent volcanic activity and basic morphological features in the Aegean area.

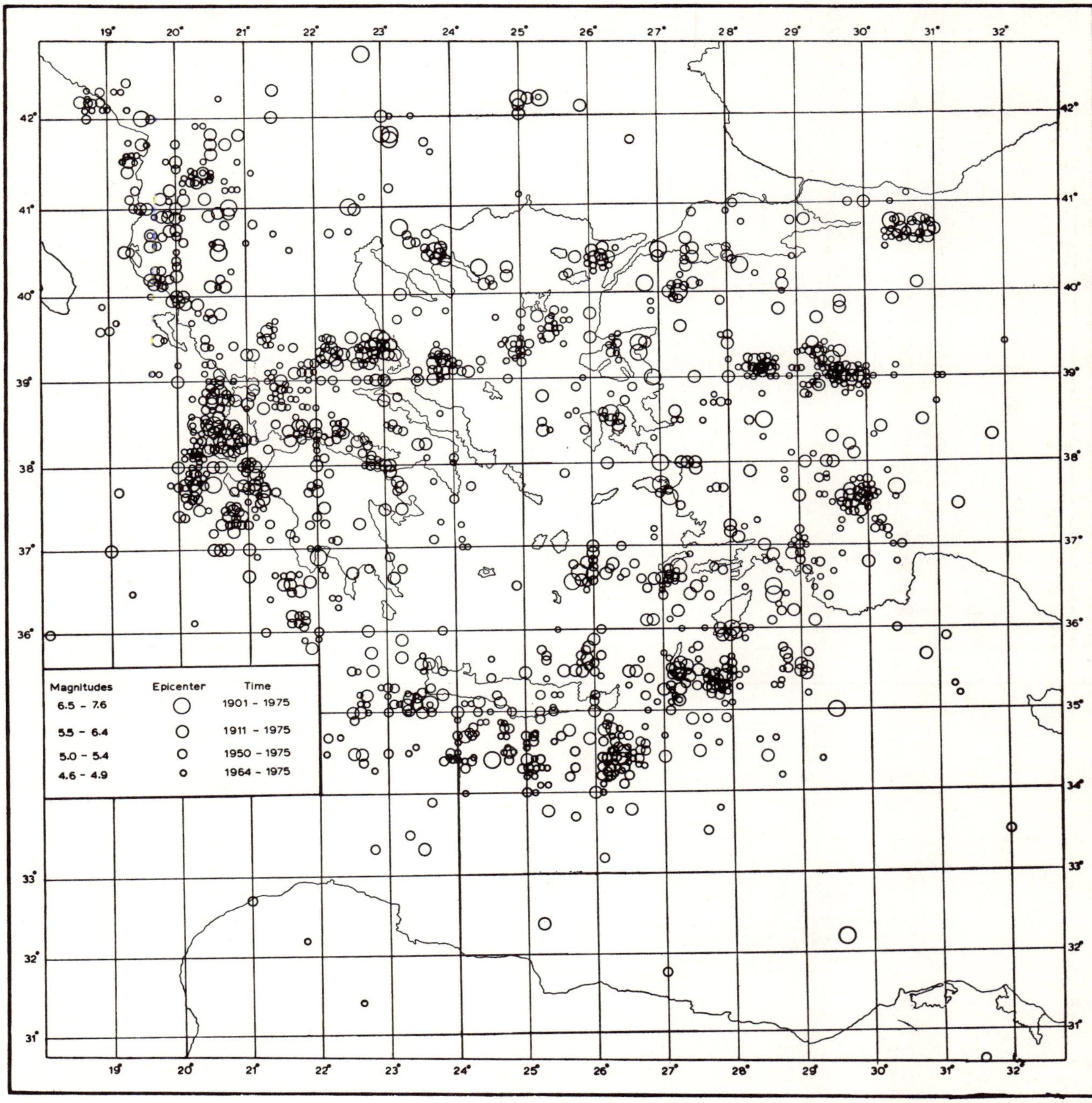

Fig. 2 Distribution of the epicenters of the shallow ($h<60$km) earthquakes in the Aegean and the surrounding area.

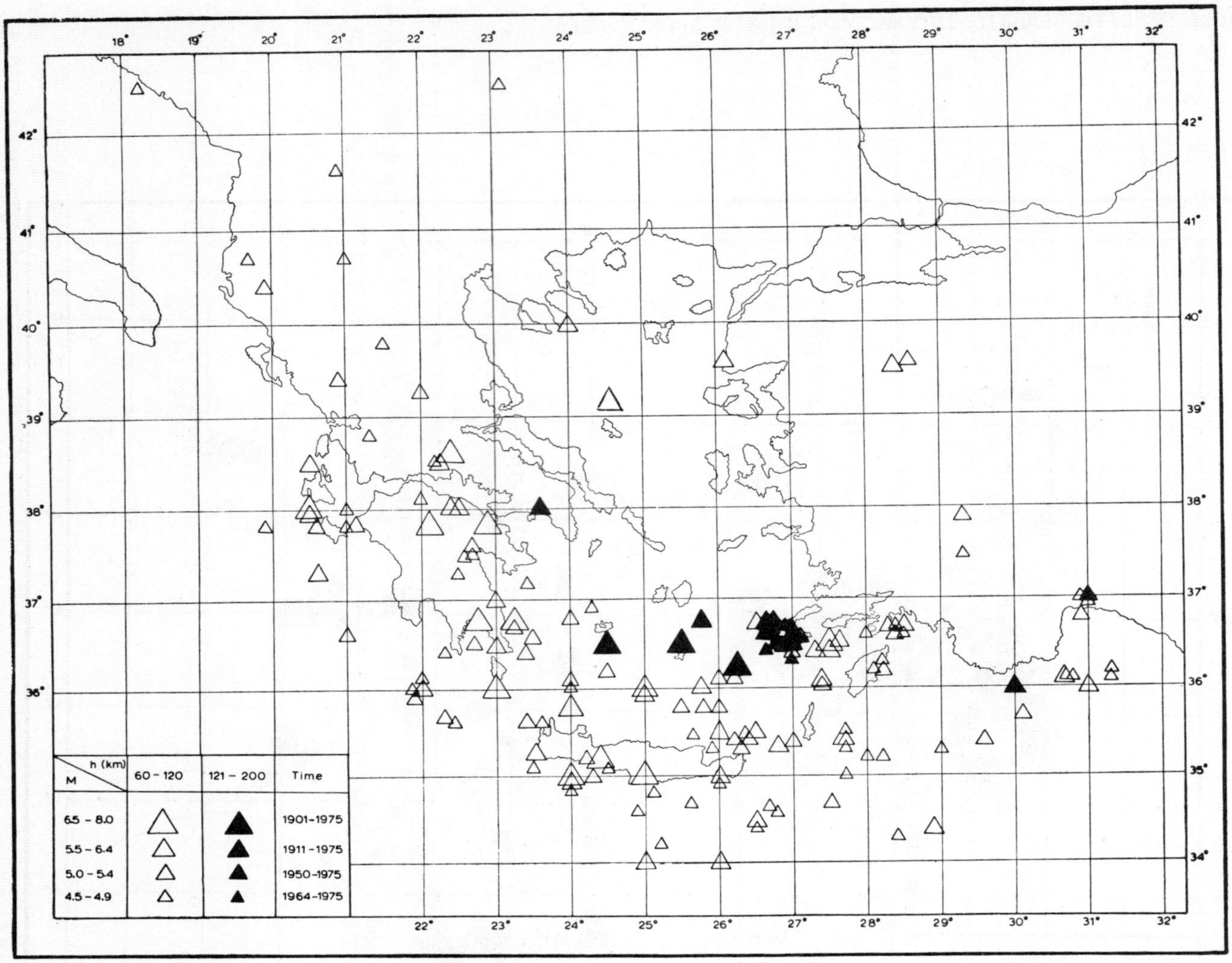

Fig. 3 Distribution of the epicenters of the earthquakes of intermediate focal depth in the Aegean and the surrounding area.

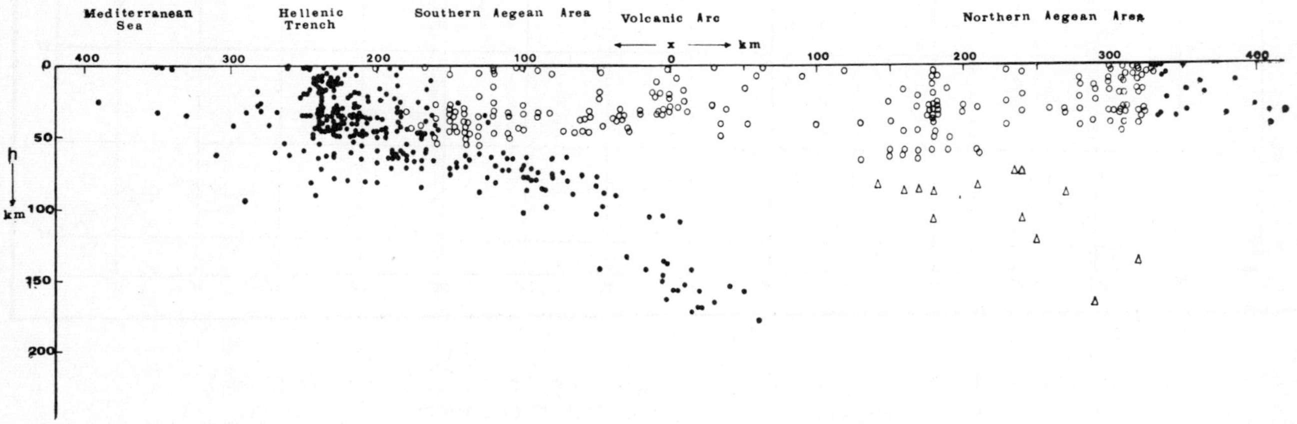

Fig. 4 Vertical distribution of the earthquake foci in the Aegean area. Black and white circles show areas of compressional and tensional focal mechanism , respectively, while triangles show areas of unknown focal mechanism.

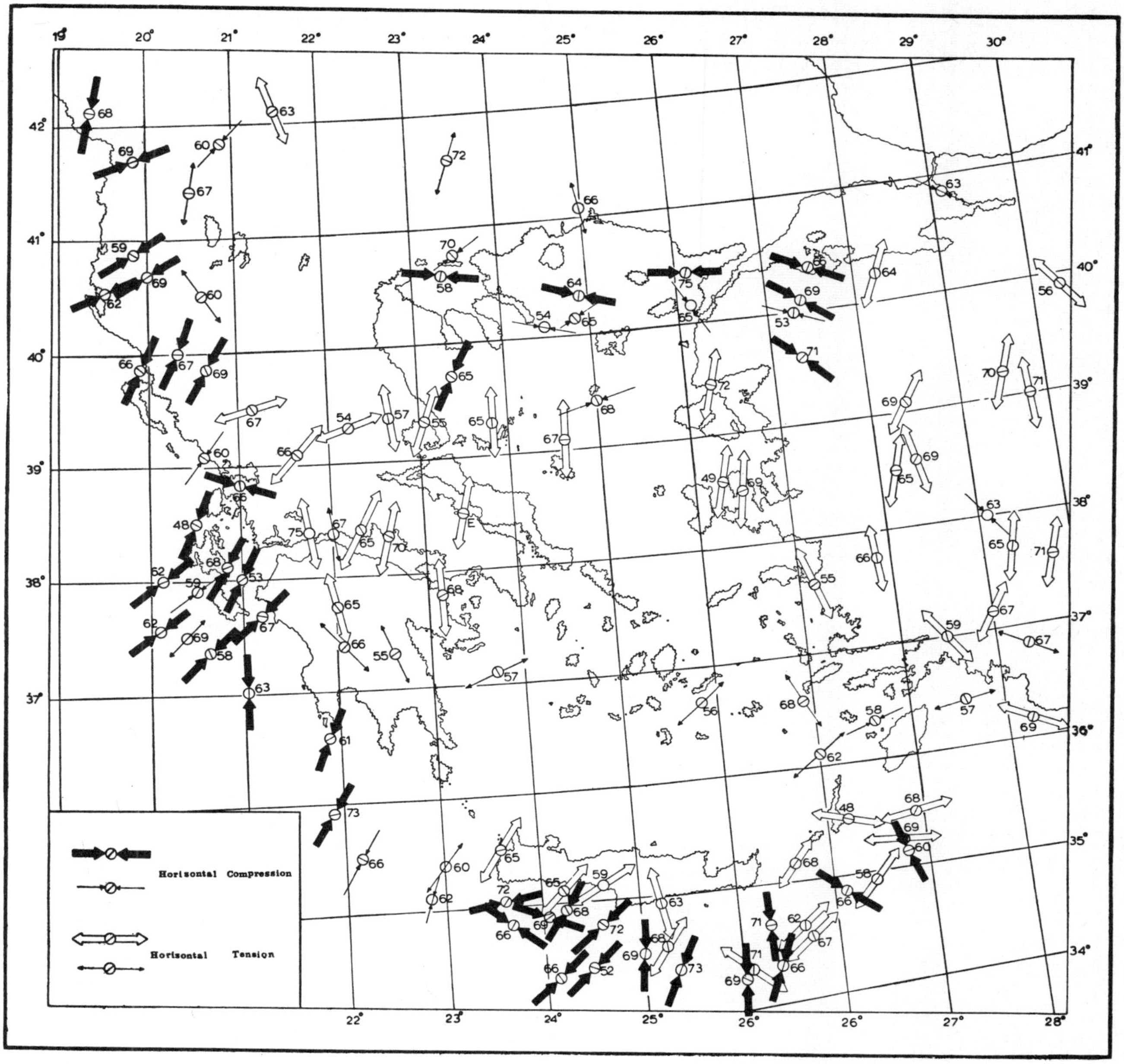

Fig. 5 Stress pattern in the Aegean area.

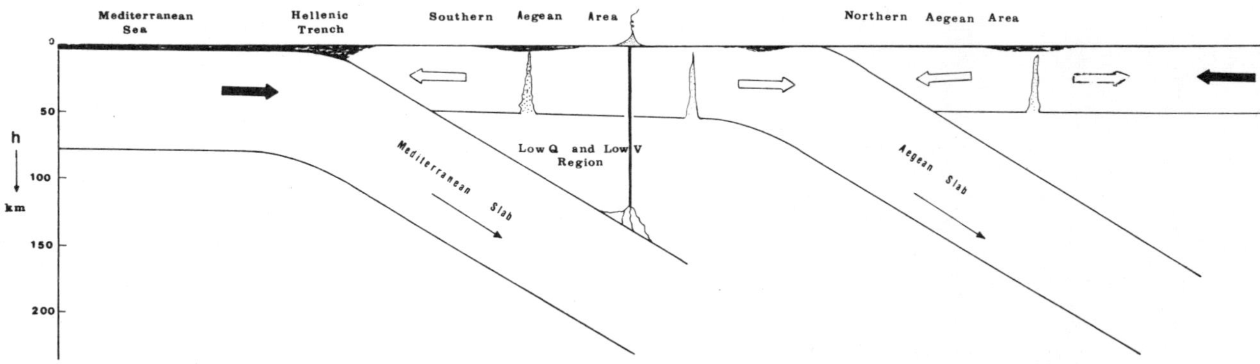

Fig. 6 Sketch of the suggested lithospheric interactions in the Aegean area on a vertical plane striking about NNE - SSW. Black arrows show compressive forces, white arrows show directions of expansion and thin arrows show motions of the slabs.

INTERNATIONAL SYMPOSIUM ON THE STRUCTURAL HISTORY OF THE MEDITERRANEAN BASINS. SPLIT (YUGOSLAVIA) 25-29 OCTOBER 1976. B. BIJU-DUVAL AND L. MONTADERT, Eds. EDITIONS TECHNIP, PARIS 1977, pp. 333-340

THE PANNONIAN BASIN : A MEDITERRANEAN INTERARC BASIN

F. HORVATH and L. STEGENA [1]

ABSTRACT

Geological and geophysical features of the Pannonian basin are reviewed. The Pannonian basin exhibits typical characteristics of an interarc /intermountain, retroarc/ basin. It is suggested that the formation of the basin has been controlled by upwelling partially molten mantle material. This mantle diapir was presumably generated by the Late Oligocene-Early Miocene subduction along the Carpathian arc.

1. INTRODUCTION

Interarc /retroarc, intermountain/ basins are mainly postorogenic depressions lying at the concave side of a mountain/ island arc. There are interarc basins with continental crust /ensialic interarc basins/ and with oceanic crust /ensimatic interarc basins/. These basins are mostly associated with the island/mountain arcs of the Circum-Pacific region and the Mediterranean. This article deals with one of the Mediterranean basins: the Pannonian basin.

2. PRE-BASINAL DEVELOPMENT

The Pannonian area may not be considered a Mesozoic or Cenozoic marginal basin of Western Pacific type as deep drillings usually find Precambrian and/or Paleozoic basement rocks.

The pre-Cenozoic geological history of the Pannonian area was complicated. Striking dissimilarities of the Mesozoic and earlier complexes suggest that the area inside the Outer Carpathians is the result of collision of two microplates having originally been situated on the southern and northern margin of the Mesozoic Tethys /Géczy, 1973; Stegena et al., 1975; Channel and Horváth, 1976/. The microplates have come into contact in the course of Tertiary presumably as late as the Oligocene. The contact zone exhibits the features of a suture zone.

3. SEDIMENTATION

Geological data indicate that the unified Pannonian region became emergent during the Late Oligocene-Early Miocene /Balogh and Körössy, 1974/. On the other hand this time interval was characterized by large compression in the surrounding mountains; the nappe structure of the Outer Carpathians has been formed /Mahel, 1974/.

(1) Eötvös University, 1083 Budapest, Kun B.t.2., Hungary

In the Pannonian region the period of uplift was followed by the beginning of basinal evolution. Major part of the region got flooded by the Paratethys during the Late Miocene. It was a very shallow sea, the average thickness of Miocene sediments is 230 m /Fig. 1/. In a NE-SW striking zone however the thickness of Miocene complex reaches 1000 m or more due to the large amount of volcanogenic material /Fig. 3/. The sea extended practically over the whole area during the Lower Pliocene. The average thickness of Pliocene sediments is 1200 m and in some region the subsidence amounted to 3-5000 m. The withdrawal of sea-water commenced in the Upper Pliocene and ended in the Quaternary. The mean velocity of subsidence, as calculated from the data of Figure 1 are shown on Figure 2. It is clearly seen that the velocity of subsidence was larger more than an order of magnitude in the Plio-Quaternary than before. The rate of the subsidence however varied in space and time giving rise to block-like /tectonique cassant/ structure of the basement, especially at the margin of the basin /Mehmed et al., 1976; Kullman, 1976/.

4. VOLCANISM

The Carpatho-Pannonian region is characterized by intensive calc-alkaline volcanic activity during the Mio-Pliocene. The areal distribution of the volcanism may be divided into two zones. A zone of mainly rhyolitic ignimbrites covered by young sediments streches through the Pannonian basin in NE-SW direction. The other zone follows the inner border of the Carpathian arc and it is made up of mainly andesitic rocks /Fig. 3/. The age of volcanic activity shows some variation. In the ignimbritic zone, northern Hungary and Slovakia the magmatism started in the Burdigalian--Helvetian and culminated in the Tortonian. In Rumania the volcanic activity started in the Tortonian and reached its maximum development during the Pliocene. Lexa and Konecny /1974/ realized that the outward migration of volcanic activity is accompanied by the same migration of tectonic mobility /block faulting, subsidence/.

There was a break in the volcanic activity of the Pannonian basin after the Sarmatian. The final, always basaltic volcanism took place in the Upper Pliocene-Pleistocene. This basaltic volcanism has always produced smallintrusions/Fig. 3/.

Several authors suggested that the calc-alkaline volcanism of the area was generated by subduction of oceanic lithosphere along the Carpathian arc /Szádeczky-Kardoss, 1971; Radulescu and Sandulescu, 1973; Stegena et al., 1975/. Lexa and Konecny /1974/ have argued on petrological and geochemical ground that the calc-alkaline volcanism is only indirectly related to the subducting slab; the primary source of volcanic material is the mantle diapir generated by the subduction. On the other hand, Peccerillo and Taylor /1976/ showed on the basis of the rare earth content of volcanic rocks, that the Calimani-Hargitha volcanism should be divided into two parts: the high-K andesite and rhyolite magmas probably represent primary melts of an undergoing oceanic slab, while the basalt, low-Si andesite, andesite and dacite magmas are considered to be produced by partial melting of the undergoing slab and the peridotitic mantle. They can accept however the possibility that high-K andesites and rhyolites were produced by partial melting of the lower part of the continental crust. We believe that the great amount of rhyolities in the central Hungarian volcanic zone was mainly produced in this way. It is supported by the fact that the lower crust of the Pannonian basin is thin.

The Plio-Pleistocene basaltic volcanism is apparently related to the fast subsidence of the region. The mantle origin of Hungarian basalts has also been suggested on the basis of microprobe analysis of lherzolite nodules in these rocks /Embey-Istin, 1976/.

5. CRUST AND UPPER MANTLE

Deep seismic soundings /Szénás, 1972/ have shown that the Outer Carpathians and foredeep is characterized by thick crust /40-65 km/ while the crust of the Inner--Carpathians and the Pannonian basin is thin /24-30 km/. In the latter areas thickness of the upper crust is of normal value /17-19 km/ and the lower crust is thinned out. The boundary of the thin and thick crustal regions is a deep fault system which is associated with the Pieniny Klippen belt /Fig. 4/. This belt is considered as the remnant of the subduction zone which was active during the Late Oligocene-Early Miocene /Lexa and Konecny, 1974; Channell and Horváth, 1976/.

The velocities of the upper crust, lower crust and uppermost mantle are normal

/5.6 - 6.5; 6.8 - 7.o; 8.o - 8.1 km/s resp./. Bisztricsány /1974/ demonstrated however that the low velocity layer /LVL/ of the upper mantle lies in an elevated position /about 75 km/, in agreement with the systematically positive seismic travel-time residuals /Stegena et al., 1975/.

Using forced reflection seismic technique Posgay /1975/ found that the velocity of P waves increases up to 9.1 km/s in the depth of about 5o km and then drops to 7.7 - 7.8 km/s at about 57 km. The deepest level from where reflections could be obtained /about loo km/ is presumably also characterized by 7.8 km/s velocity.

In concordance with these results, magnetotelluric soundings demonstrated the elevated position of the highly conducting layer /HCL/ beneath the Pannonian basin /Ádám, 1965/.

The Bouguer anomaly of the basin, is +1o — +15 mgal in average, which is strikingly low taking into consideration the thin crust, i.e. the elevated position of the upper mantle. Gravity model calculation shows that the upper mantle of the basin should have density lower than normal; the total mass defect amounts to $1o^{2o}$ g /Fig. 4/.

6. GEOTHERMICS

Both geotemperatures and heat flow values demonstrate that the Pannonian basin exhibits a marked geothermal high. Geotemperatures in the depth of 1 km are of 7o-8o°C in average, and near surface heat flows are in the range of 2.o - 2.6 HFU /Stegena, 1976; Boccaletti et al., 1976/.

Downward continnation of geothermal data /using some realistic heat conductivity /λ/ and heat production /A/ values, and the steady-state equation of heat conduction/ shows that the geothermal high is well developed in the upper mantle of the Pannonian basin too /Fig. 5 and 6/. The high values of temperature and heat flow in the depth of 6o km /1ooo - 16oo°C and 1.o - 1.8 HFU resp./ strongly suggest that the upper mantle is partially melted. All other features of the upper mantle /elevated LVL and HCL, low density/ support this conclusion.

7. TECTONOPHYSICAL CONSIDERATION

The Late Cenozoic geologic history and geophysical features of the Carpatho-Pannonian region may plausibly be explained in the following way.

During the Middle Oligocene-Early Miocene a southerly dipping subduction zone was active along the Carpathian arc. Presumably this subduction has induced the extra heat which has resulted in partial melting in the upper mantle above the subducing plate. The hot, upwelling mantle material expanded laterally at the base of the /continental/crust, thinned it out, and locally pierced where the large mass of andesites-rhyolites extruded. When the subduction terminated the thin but still continental crust began to subside and has filled by thick Pliocene-Quaternary sediments. The present aseismicity of the Carpathians and Pannonian basin, the lack of Recent volcanism show that the subducted lithosphere has been assimilated and the basin evolution has essentially terminated.

We think that the above model, which is similar to that of Karig /1971/ and Scholz et al. /1971/, may have some applicability for the evolution of other Mediterranean back-arc basins.

Acknowledgements

The authors are very grateful to Prof. G.Barta who kindly reviewed the manuscript.

REFERENCES

ÁDÁM, A., 1965: Einige Hypothesen über den Aufbau des oberen Erdmantels in Ungarn. Gerlands Beitr. Geophys., 74. /1/:2O.

BALOGH, K. and KÖRÖSSY, L., 1974: Hungarian Mid-Mountains and adjacent areas. In: M.Mahel /Editor/: Tectonics of the Carpathian-Balkan Regions, Geol. Institut, Bratislava, 391.

BISZTRICSÁNY, E., 1974: The depth of the LVL in Europe and some adjacent regions. Geofiz. Közl., Budapest, 22:61.

BOCCALETTI, M., HORVÁTH, F., LODDO, M., MONGELLI, F. and STEGENA, L., 1976: The Tyrrhenian and Pannonian basins: A comparison of two Mediterranean interarc basins. Tectonophysics, 35 /1-3/: 45.

BODRI, L., 1976: Deep temperature and heat flow in the Pannonian basin. Ph.D. thesis, Eötvös University, Budapest.

CHANNEL, J. and HORVÁTH, F., 1976: The African/Adriatic promontory as a palaeogeographical premise for Alpine orogeny and plate movements in the Carpatho-Balkan region. Tectonophysics, 35 /1-3/: 71.

EMBEY-ISTIN, A., 1976: Amphibolite/lherzolite composite xenolith from Szigliget, North of the Lake Balaton, Hungary. Earth Planetary Sci. Letters, 31: 297.

GÉCZY, B., 1973: The origin of the Jurassic faunal provinces and the Mediterranean plate tectonics. Ann. Univ. Sci. Budap. R. Eötvös Nom., Sec. Geol., 16:19.

KARIG, D.E., 1971: Origin and development of marginal basins in the Western Pacific. J. Geophys. Res. 76: 2542.

KULLMANN, E., 1976: The role of neotectonic movements in the development of ground water reservoirs in the North-Eastern part of the Vienna Basin. IASH-IAH Int. Hydr. Conf. Budapest, Abstracts: 53.

LEXA, J. and KONECNY, V., 1974: The Carpathian volcanic arc: A discussion. Acta Geol. Acad. Sci. Hung., 18 /3-4/: 279.

MAHEL, M., 1974: Introductory articles. In: Mahel /editor/: Tectonics of the Carpathian-Balkan regions, Geol. Inst. D. Stur, Bratislava.

MEHMED, A., BANDRABUR, T., CRATIUN, T., GHENEA, C., POLONIC, P. and VISARION, M., 1976: Contributions a la connaissance des structures a eaux thermales de la partie orientale de la Depression Pannonienne. IASH-IAH Int. Hydr. Conf. Budapest, Abstracts: 62.

PECCERILLO, A. and TAYLOR, S.R., 1976: Rare earth elements in the East Carpathian volcanic rocks. Earth Planet. Sci. Letters, 32: 121.

POSGAY, K., 1975: Mit Reflexionsmessungen bestimmte Horizonte und Geschwindigkeitsverteilung in der Erdkruste und im Erdmantel. Geofiz. Közl., Budapest, 23:13.

RADULESCU, D. and SANDULESCU, M., 1973: The plate-tectonics concept and the geological structure of the Carpathians. Tectonophysics, 16: 155.

SCHOLZ, C.H., BARAZANGHI, U. and SBAR, M.L., 1971: Late Cenozoic evolution of the Great Basin, Western United States, as an ensialic interarc basin. Geol. Soc. Am. Bull., 82: 2979.

STEGENA, L., GÉCZY, B. and HORVÁTH, F., 1975: Late Cenozoic evolution of the Pannonian basin. Tectonophysics, 26:71-90.

STEGENA, L., 1976: Geothermal temperature map of Central and Eastern Europe, p. 381-383. In: A. Ádám /editor/ Geoelectric and geothermal studies KAPG Geophysical Monograph, Akadémiai Kiadó, Budapest.

SZÁDECZKY-KARDOSS, E., 1971: The Carpatho-Dinarid area from the point of view of the new global tectonics. Geonomia és Bányászat, Budapest, 4: 82.

SZÉNÁS, Gy., 1972: /editor/: The crustal structure of central and southeastern Europe based on the results of explosion seismology. Geophys. Trans., Spec. Edit., Budapest.

VAN EYSINGA, F.W.B., 1975: Geological time table. Elsevier Publ. Co., Amsterdam.

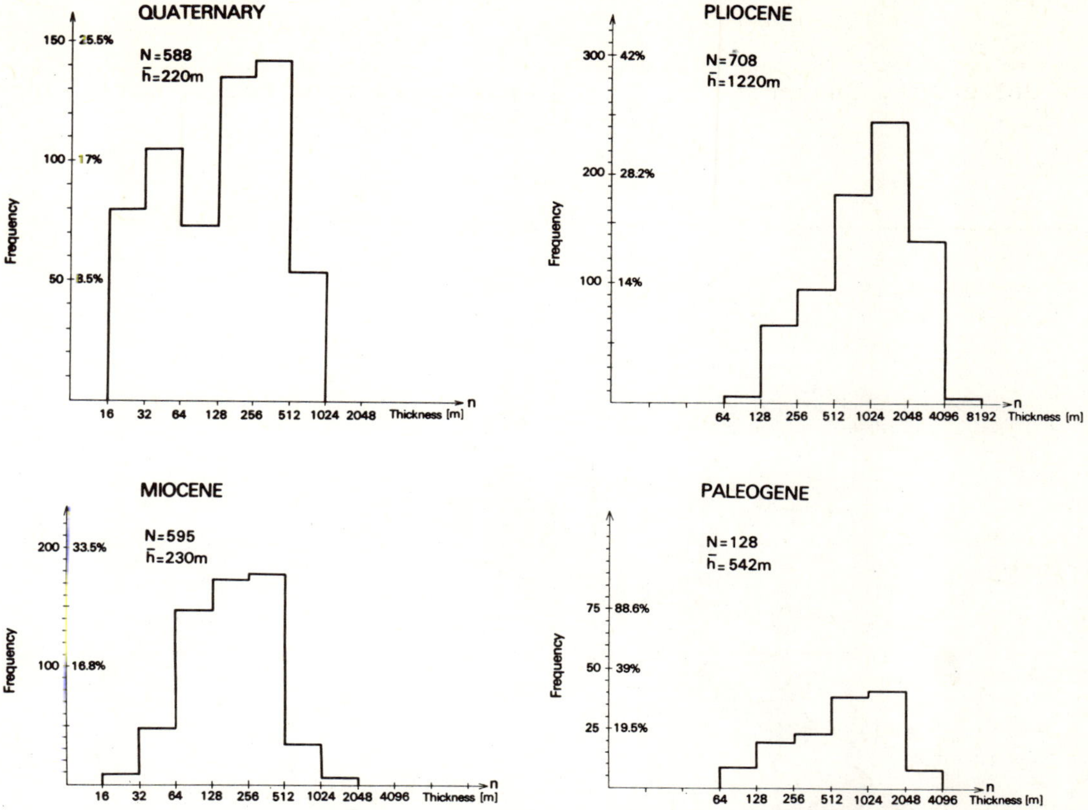

Figure 1. Frequency of thicknesses of Quaternary, Pliocene, Miocene and Paleogene sediments in the Pannonian basin, mean thicknesses / $\bar{h}$ / calculated from N data.

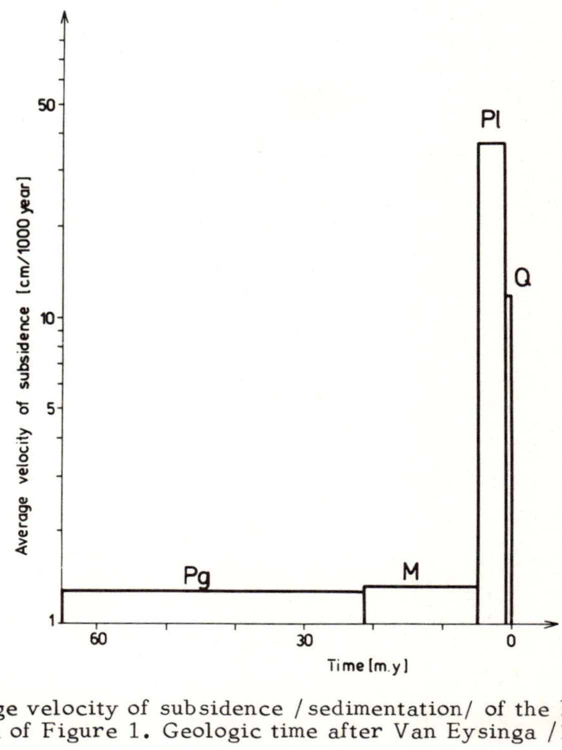

Figure 2. Average velocity of subsidence /sedimentation/ of the Pannonian basin, calculated from the data of Figure 1. Geologic time after Van Eysinga /1975/.

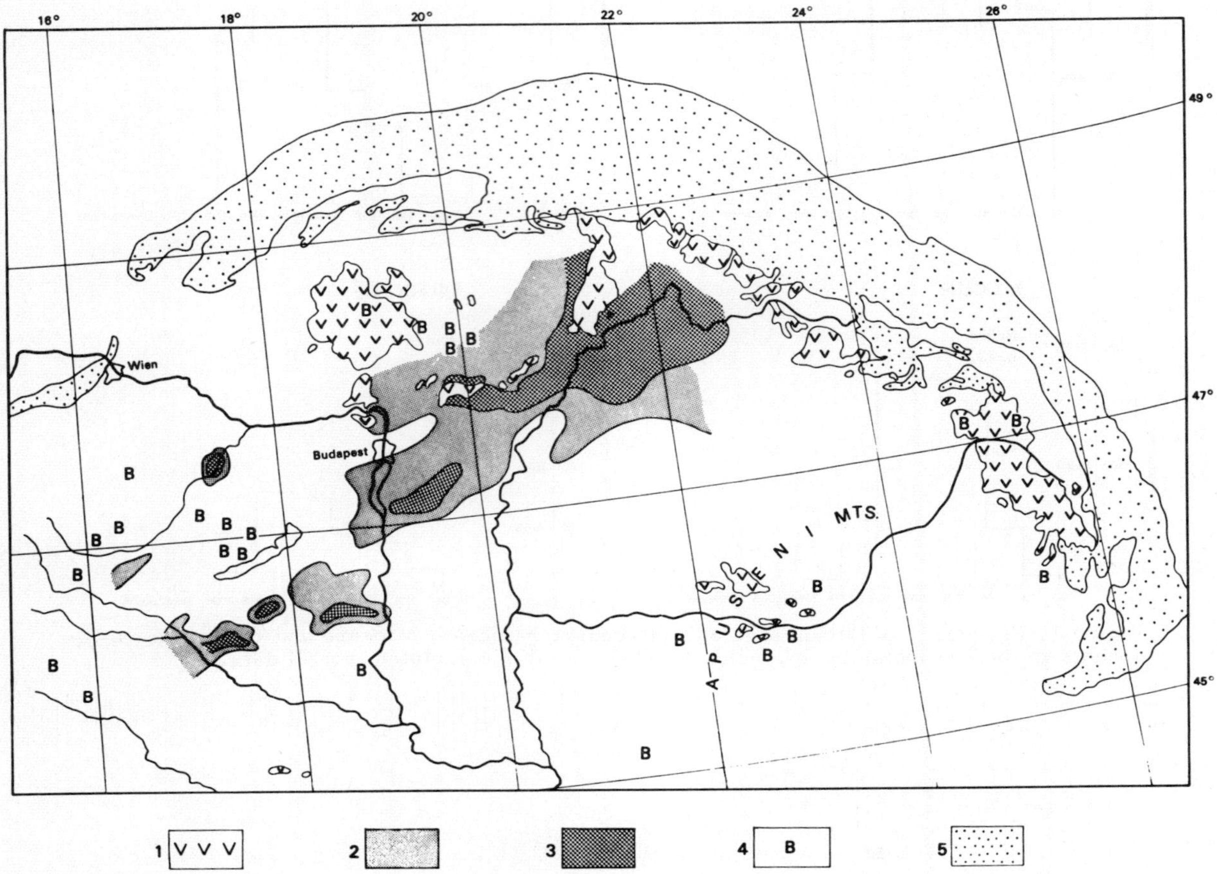

Figure 3. Neogene-Quaternary volcanism of the Carpatho-Pannonian region. Keys : 1 =outcrops of Mio-Pliocene, mainly andesitic volcanites; 2-3=Miocene ignimbritic volcanism /the thickness of which are o - 1ooo m and more than 1ooo m, resp./ covered by young sediments; 4=sites of Plio-Pleistocene basaltic outcrops; 5=Carpathian flysch complex.

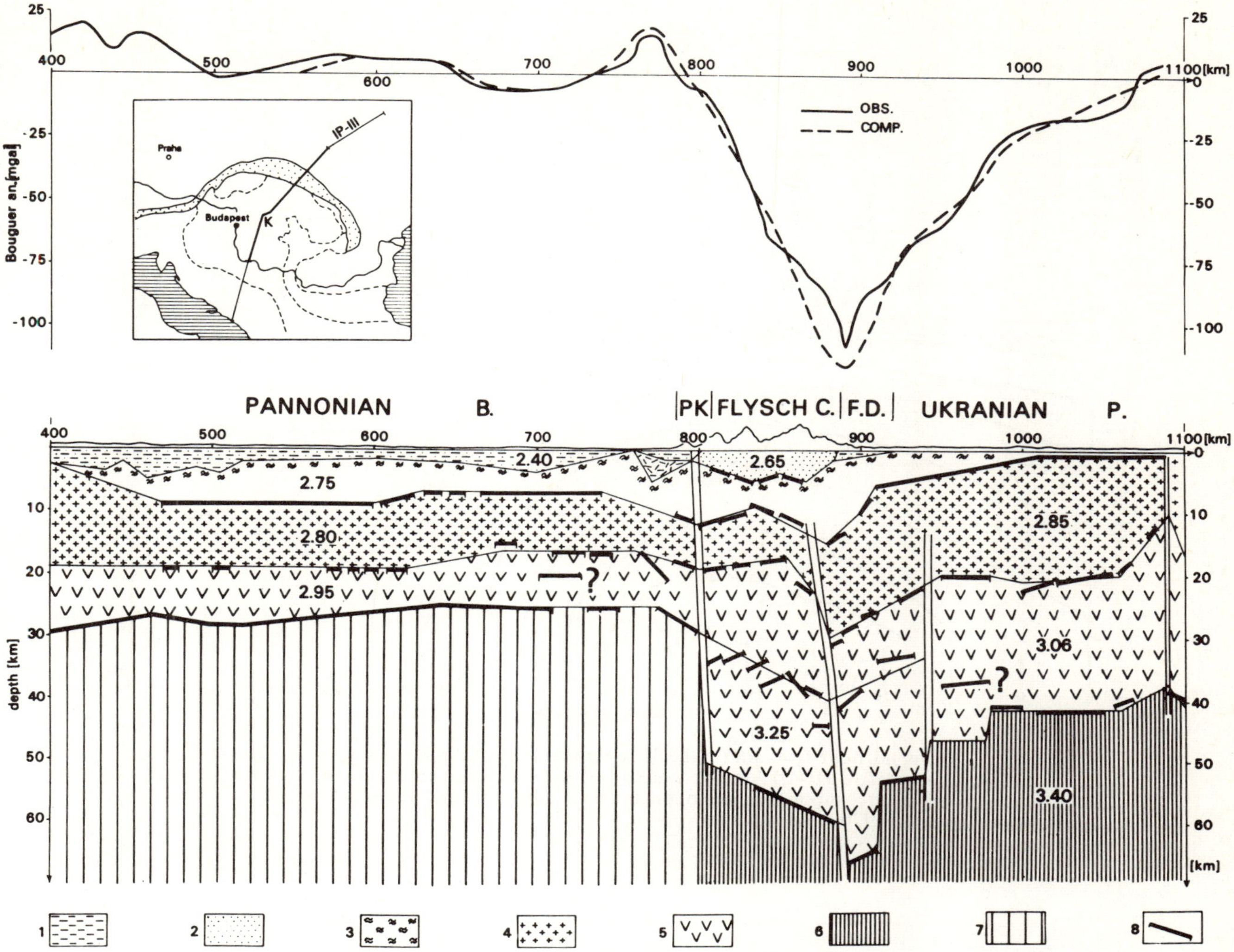

Figure 4. Observed and computed Bouguer anomalies along the IP-III seismic profile. Keys: 1 = sedimentary complex; 2 = flysch comples; 3 = basement; 4 = upper crust; 5 = = lower crust; 6 = normal upper mantle; 7 = low-density upper mantle; 8 = seismically derived boundaries.

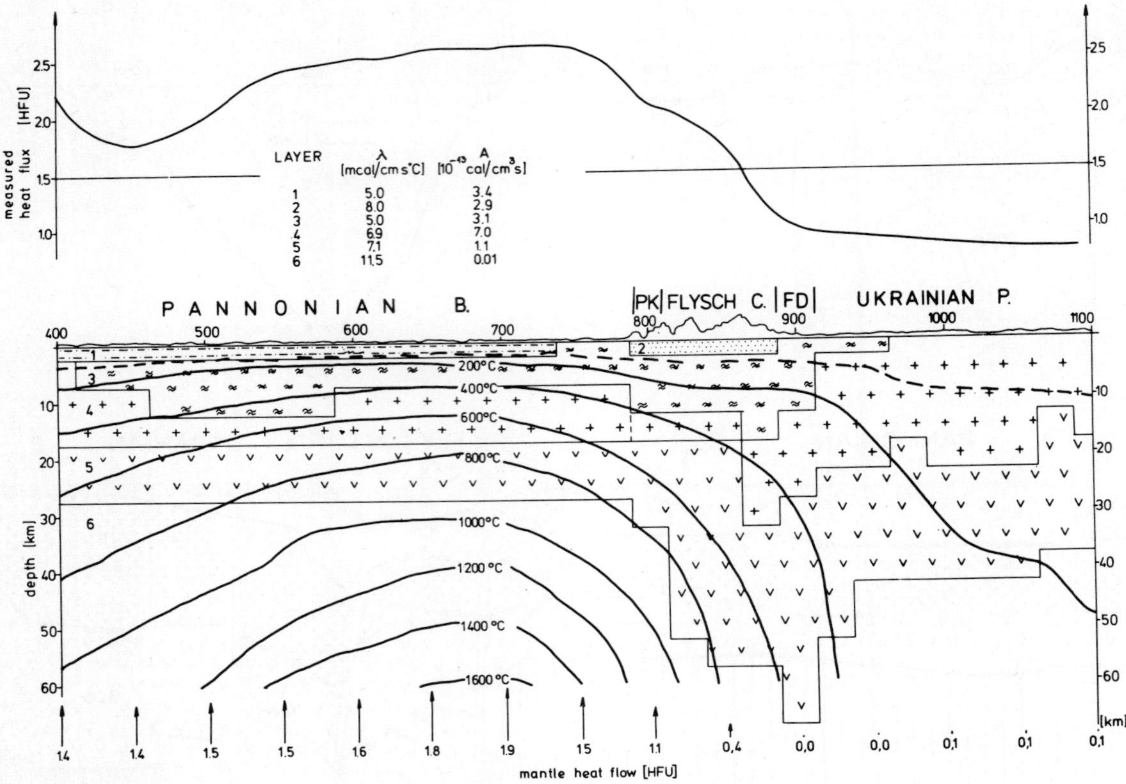

Figure 5. Measwured heat flow and calculated geoisoterms, and heat flow at 60 km depth along the seismic profile IP-III. In the course of calculation the given crustal model and parameters were used, supposing exponential decrease of heat production / A / in each layer /after Bodri, 1976/.

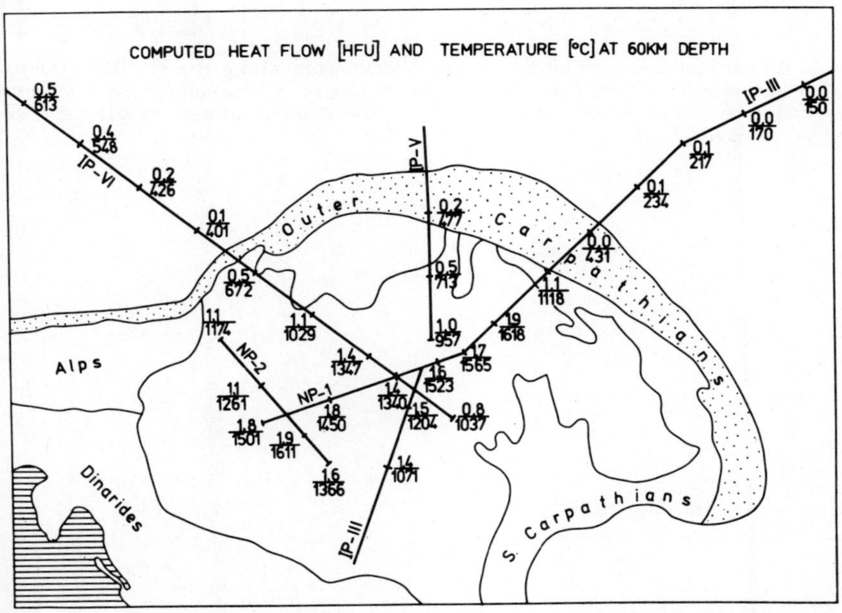

Figure 6. Location of deep seismic sounding profiles and calculated heat flow and geotemperature values at 60 km depth for the Pannonian basin and its environment /after Bodri, 1976/.

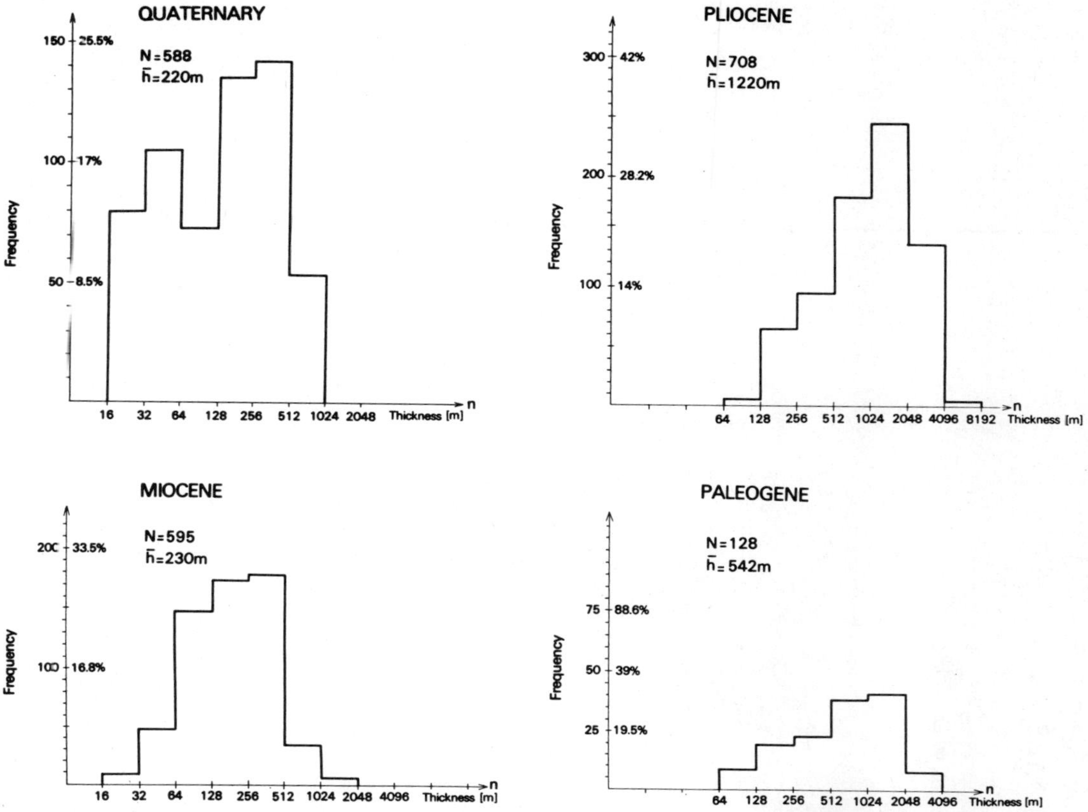

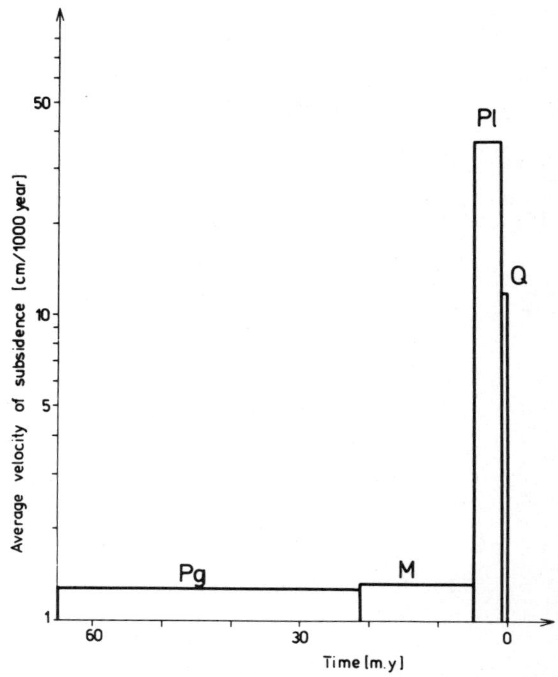

INTERNATIONAL SYMPOSIUM ON THE STRUCTURAL HISTORY OF THE MEDITERRANEAN BASINS. SPLIT (YUGOSLAVIA) 25-29 OCTOBER 1976. B. BIJU-DUVAL AND L. MONTADERT, Eds. EDITIONS TECHNIP, PARIS 1977, pp. 341-352

ANOMALIES GRAVIMÉTRIQUES RÉGIONALES POUVANT REFLÉTER DES SEGMENTS DE PLAQUES OU DE MICROPLAQUES DE LA LITHOSPHÈRE SUR LE TERRITOIRE DE LA ROUMANIE

S. AIRINEI [1]

Summary. The author took to submit a gravimetric contribution to the previous attempts of including the Romanian territory into plate-tectonics concept.The maximum regional anomalies reflect four Earth's crust segments: of the East-european plate and of the Inter-alpine, Moesian and Black Sea micro-plates; those of the minimum define zones having their contact limits between these segments.

Zusammenfassung. Die Arbeit stellt einen gravimetrischen Beitrag zu früheren Versuchen, den Boden Rumäniens in das Konzept der Plattentektonik einzurahmen, dar. Es wird dargelegt, dass die maximalen Regionalanomalien vier Plattensegmente der Lithosphäre reflektieren (die süd-westliche Extremität der osteuropäischen Platte; den östlichen Teil der interalpinen Mikroplatte; den nördlichen Teil der moesischen Mikroplatte und die WNW-liche Extremität der Schwarzen Meeres-Mikroplatte), während die der minimalen, die Zonen mit den Kontaktgrenzen zwischen diesen Segmenten, darlegen. Ferner spezifiziert der Verfasser die Bewegungsrichtungen dieser Plattensegmente, die konzentrische Weise der Bewegungen, die grössere Verrückungsgeschwindigkeit der Mikroplatte des Schwarzen Meeres, sowie die Konsequenzen die für den Subduktionsprozess in den östlichen Karpaten und dem Westgebirge aus diesen naheren Bestimmungen hervorgehen.

I. CONSIDERATIONS GENERALES D'ORDRE GEOLOGIQUE ET GEOPHYSIQUE

La couche sédimentaire, située dans l'échafaudage de la croûte terrestre du territoire de la Roumanie, contient des formations géologiques appartenant à des âges partant de l'Archaïque supérieur et probablement moyen aussi (situées dans les massifs cristallins centraux carpatiques et dans le "soubassement" métamorphique de l'avant-pays et de l'arrière-pays de la chaîne alpine) et allant jusqu'au Quaternaire (celles-ci couvrant 50,73 % de la superficie totale du pays dans les zones de plaines, de lacs et le long des cours d'eau). Les âges absolus les plus anciens, de deux milliards d'ans à peu près, ont été déterminés pour des roches métamorphiques du soubassement de la Dobroudja et de la Moldavie. La couche sédimentaire est organisée

(1) Université de Bucarest, Faculté de Géologie et Géographie, bd. N. Balcescu 1, Bucarest, Roumanie

dans un système structural complexe, dont 2/3 appartenant à l'orogenèse alpine et 1/3 seulement aux orogenèses plus anciennes, prèsentes sous la forme de cratons ensevelis pour la plupart sous des sédiments plus jeunes et soudés successivement aux boucliers de la Plateforme Moldave du NE et de la Plateforme Moesienne du S.

L'image de l'écorce terrestre de la Roumanie, conçue dans une première approximation sur des données géophysiques (gravimétriques, magnétiques et sondage sismique de profondeur) est relativement complexe, étant caractérisée par la présence des trois couches qui la composent (sédimentaire, granitique et basaltique), intersectées par de nombreuses fractures qui compartimentent la croûte en blocs distinctement caractérisés. D'après L. Constantinescu et al. (1975), les profondeurs des surfaces de séparation des couches de l'écorce terrestre oscillent entre des limites relativement larges: 1-17 km pour la surface du soubassement cristallin; 15-27 km pour la surface de la couche basaltique (la discontinuité Conrad ne pouvant être souivie continuellement en raison des particularités de transition qu'elle présente sur des espaces relativement grands); 25-54 km pour la discontinuité Moho (qui n'apparaît pas toujours comme une limite unique, mais en succession de 2-3 surfaces sismiques, relativement rapprochées entre elles, marquant l'existence de la zone de transition entre l'écorce et le manteau supérieur).

Le territoire de la Roumanie est intersecté par 15 champs de fractures crustales (I. Gavăţ et al., 1965) orientées principalement W-E et NW-SE mais aussi N-S ou SW-NE, chacune d'entre elles polarisant des successions d'événements géologiques datant des temps les plus anciens jusqu'aux plus récents. C'est sur elles que s'appuie tout l'échafaudage arborescent de failles, venant de plus en plus nombreuses des zones profondes de la lithosphère vers sa surface.

Des magmas subcrustaux ont envadé certains segments des champs de fractures crustalles et leurs ramifications, dont quelques-unes restées à des niveaux profonds dans l'écorce terrestre, tandis que d'autres ont monté à la surface. Les aires périmétrées à magmatites parvenues à la surface sont: 7370 km^2 d'âge quaternaire et néogène (3,1 % de la superficie du pays); 930 km^2 d'âge paléogène et crétacé supérieur (banatites - 0,4 %); 1482 km^2 d'âge mésozoïque et permien (0,6 %) et 2264 km^2 d'âge paléozoïque et précambrien (0,95 %) (St. Airinei, D.Sturdza, 1970). Les corps magmatiques, incorporés dans l'écorce terrestre à de différents niveaux de profondeur, augmentent le total des surfaces citées pour tous les âges de leur formation. Cette réalité est exprimée par la carte des anomalies magnétiques du territoire de la Roumanie, et la situation de ces distributions est synthétisée et interprétée dans l'image de la structure géologique profonde de ce territoire (I. Gavăţ et al., 1965). Cette dernière image est le messager d'un important nombre de messages géologiques concernant l'évolution de l'écorce terrestre du territoire de la Roumanie.

II. TENTATIVES D'ENCADRER LE TERRITOIRE DE LA ROUMANIE DANS LE CONCEPT DE LA TECTONIQUE DES PLAQUES

Le contexte général des transformations fondamentales dans la géologie planétaire et régionale a trouvé dans le vaste potentiel de faits et de création de la recherche géologique et géophysique roumaine un terrain tout préparé pour des tentatives d'encadrer le territoire de la Roumanie dans le concept de la tectonique des plaques. En ce sens on a élaboré trois catégories de modèles différenciés par le caractère et la valeur des éléments pris en considération, tous ces modèles étant appréciés par leurs auteurs comme des tentatives préliminaires: l'un appuyé uniquement sur des données géologiques (D. Rădulescu, M. Săndulescu, 1973), deux avec un supplément d'éléments géophysiques sporadiques (M. Breahu et al.,1973; N. Herz, H. Savu, 1974) et deux autres avec, en plus, des données séismologiques et séismotectoniques (C. Roman, 1970, 1971; L. Constantinescu et al., 1975).

Les particularités des deux premières catégories de modèles ont été l'objet d'une analyse comparative et critique relativement récente (St. Airinei, 1976, 1976). En dernière analyse, les auteurs des modèles cités ont été préoccupés par le déciffrement et la connaissance les plus complets et les plus unitaires possibles du "plan de situation actuel" concernant la formation, la distribution évolutive et le mouvement des plaques de la lithosphère, ayant comme moment géologique de début l'ouverture de la dorsale medio-atlantique, moment situé entre le Triassique supérieur et le Jurassique inférieur, c'est-à-dire il y a 200 millions d'ans.

Les idées et les tentatives roumaines d'encadrer le territoire de la Roumanie dans le concept de la tectonique des plaques, ont bénéficié des études régionales englobant le bassin méditerranéen comme une relique de l'Océan Tethys (P.D. McKenzie, 1970, 1972; M.J.Lort, 1971, F.J.Dewey et al., 1973 etc.). En ce qui concerne le territoire de la Roumanie, ces travaux admettent l'existence de plusieurs microplaques (dont certaines sont dénommés subplaques ou blocs), situées entre les plaques majeures Eurosiatique et Africaine, et qui auraient joué un rôle actif pendant l'évolution du "plan de situation actuel", telles que: la microplaque turque, en mouvement du SE au NW; la microplaque de la Mer Noire, avec un mouvement sensiblement semblable au précédent; la microplaque moesienne, avec un mouvement oscillant du sud vers le nord au vers l'ouest; la microplaque transylvaine, inactive ou avec un lent mouvement vers le SE etc. Vu que l'existence de certaines plaques n'a pas été prouvée par des arguments géologiques ou physiques, ces plaques étant admises comme une necessité par l'un ou l'autre des modèles élaborés, de nombreuses controverses continuent d'exister sur le rôle joué par ces plaques ou même sur la réalité de leur présence.

Un même caractère de point de départ dans cette matière peut être attribué aux deux derniers modèles, appuyés principalement sur des données séismologiques et séismotectoniques.

C. Roman (1973) - reprenant certaines conclusions de ses travaux précédents (1970 et 1971) et aussi l'idée lancée par McKenzie (1972) concernant l'existence d'une microplaque de la Mer Noire - considère que l'activité séismique de la région de Vrancea, située à la courbure des Carpates Orientales, est un résultat de la subduction de la microplaque de la Mer Noire, placé justement dans le substratum de cette zone. L. Constantinescu et al. (1973) établit que la distribution des foyers de tremblements normaux et intermédiaires de la région de Vrancea a lieu le long d'un dispositif ruptural du type Benioff, dont les caractéristiques sont définies par les solutions du mécanisme de production des tremblements dans leurs foyers (position des plans de faille, nature du cissaillement, type du tremblement, position du "vecteur nul", état de tension dans le foyers). L'image synthétique suggère qu'un ample processus de subduction a lieu dans le substratum de la région. En intégrant à ce tableau, relativement local, des informations obtenues par le "sondage sismique profond" sur le territoire de la Roumanie et sur ceux des pays voisins, L.Constantinescu et al. (1975) établit l'alignement d'une fracture crustale majeure et continue depuis la Mer Noire jusqu'au delà des Carpaten Septentionales, qui sépare deux domaines de l'écorce terrestre: l'un vers le SE où la croûte est plus épaisse (sur les aires de la Dépression Précarpatique, de la zone du flysch des Carpates Orientales et sur celle de la Plate-forme Moldave) et l'autre vers le SW, où l'écorce est plus mince (sur les aires de la Dépression Pannonienne, des Carpates Méridionales, de la Dépression Transylvaine et sur celle de la Plate-forme Moesienne). Au niveau de la discontinuité Moho, la fracture crustale majeure provoque un saut de 10 à 20 km dans le domaine continental et de 1 à 10 km sous la Mer Noire. La présence de nombreux foyers de tremblements normaux le long de cette fracture majeure suggère qu'elle se trouve encore à l'état actif.

D'après les derniers auteurs, on peut reconnaître sur le territoire de la Roumanie trois unités distinctes de la lithosphère, à savoir: la plaque Esteuropéenne, la sous-plaque interalpine avec ses secteurs transylvain et pannonien. Elles se rencontrent directement dans le substratum de la région de Vrancea. La limite de la plaque Esteuropéenne traverse le territoire de la Roumanie en direction NW-SE. Dans la région à trois confins de Vrancea, le contact mutuel des sous-plaques interalpine et moesienne présente des relations claires de subduction associée au systhème ruptural du type Benioff indiqué par la position des foyers des tremblements normaux et intermédiaires.

Le travail fondamental concernant la cinématique des plaques (X. Le Pichon, 1968) - qui relève que la limite entre les plaques européennes et africaines se trouve dans la région de la Méditerrannée et qu'elle est du type commpressif - a fourni la base d'une ample étude concernant la géodynamique de l'aire des Carpato-Balkano-Dinarides, en corrélation avec l'aire de la Méditerrannée centrale (R.Ciocârdel, M. Socolescu, 1972). Ce travail reprend, avec de nouveaux arguments, certaines idées antérieures des mêmes auteurs, en relation avec les blocs crustaux avec leurs mouvements dirigés par des énergies produits par les courants de convection subcrustaux par la dynamique du manteau supérieur pendant l'orogenèse alpine, par des phénomènes asténolytiques et de subduction, ainsi que par le rôle de l'équilibre isostatique des masses sous-poussée. Un rôle important est accordé aux mouvements de dérive continentale de la Plate-forme Podolienne et du Bouclier Africain, ainsi qu'aux fractures crustales dans le processus de l'évolution de l'écorce et des phénomènes magmatiques.

L'écorce existait déjà il y a au moins 4 milliards d'ans avant le Jurassique, ce qui oblige la pensée d'admettre que la lithosphère a évolué au pré-Jurassique dans une série de systèmes succesifs comportant des distributions et des mouvements de plaques crustales presque toujours renouvelées. Partout où il y a des volumes provenant de roches plus vieilles que les jurassiques, il est impérieusement nécessaire de résoudre un même problème: à quelle catégorie de "restes de plaques crustales" appartiennent ces volumes et à quels mouvements ont été soumises les plaques originaires avant et après leur formation? Aucune des personnes qui auront à opérer avec des problèmes de la tectoniques des plaques sur le territoire de la Roumanie ne saurait se soustraire de répondre de ces questions, de donner la solution de ces problèmes, et ne pourra expliquer complètement des ensembles d'événements plus anciens avant de savoir au mieux comment ont évolué les choses depuis le Jurassique jusqu'à présent,dans le cadre du "plan de situation actuel", du mouvement des plaques crustales. Dans la littérature roumaine de spécialité il existe une tentative d'incursion dans le pré-Jurassique, dans ces conditions mêmes, où l'on discute l'appartenance de quatre dislocations crustales du territoire de la Dobroudja à un probable système préjurassique de failles de transformation (St. Airinei, 1973).

La grande variété, comme âge et lithologie, des roches à partir du Précambrien et jusqu'à l'Actuel, ainsi que la complexité des structures dans lesquelles elles sont assises, constituent de bonnes preuves de la complexité et de la diversité, dans l'espace et dans le temps, des événements géologiques qui ont englobé le territoire de la Roumanie. Les cartes géologiques d'ensemble - telles, par exemple, la carte tectonique (I. Dumitrescu et al., 1962); la carte de la structure profonde (I. Gavăt et al., 1965); la carte des terrains cristallins (D. Giuşcă et al., 1969) etc. - contiennent des éléments de ce genre et chacune a suggéré ou appuyé une voie permettant de comprendre le film historique qu'elle déroule d'une manière cartographique et interprétative, conjugant ou non les caractéristiques des grandes unités géologiques connues avec les aspects géophysiques qui leur sont spécifiques.

III. ANOMALIES GRAVIMETRIQUES REGIONALES ET LEUR SIGNIFICATION GEOLOGIQUE PAR RAPPORT AU DECHIFFREMENT DU SYSTEME DE LA LITHOSPHERE EN ROUMANIE

Les cartes gravimétriques pour le territoire de la Roumanie et des régions limitrophes laissent voir que le soubassement dense s'enfonce progressivement et concentriquement de l'extérieur vers l'intérieur,offrant pour cette aire une image nouvelle et suggestive, dans laquelle se reflète la structure d'ensemble de l'écorce terrestre. La morphologie de la discontinuité Moho, déduite des données gravimétriques (M.Socolescu et al., 1964), exprime la même chose. La morphologie de la discontinuité Moho est complexe et rélève, pour l'épaisseur de l'écorce terrestre, des variations comprises entre 29 et 41 km. L'écorce terrestre présente des épaisseurs moindres dans la région qui se trouve devant les Carpates (30 à 34 km), dans la Dépression de Transylvanie, où elle se dresse sous la forme d'une coupole jusqu'à moins de 29 km par rapport à la surface du sol, et dans la Plaine Ouest du pays, où elle présente des épaisseurs variant entre 31 et 34 km. Les épaisseurs les plus fortes se trouvent le long des Carpates (atteignant des valeurs entre 39 et 41 km) et dans les Monts Apuseni (un peu plus de 39 km). Toute la surface de la discontinuité Moho est caractérisée par des déformations sinusoïdales des isobathes, parallelles aux méridiens géographiques et interprétées comme le résultat d'un système de structures "en harmonica" de l'écorce terrestre.

Sur la planche qui accompagne le texte on peut suivre les contours marqués des anomalies gravimétriques de haute régionalité du territoire de la Roumanie, séparées analytiquement par un réseau carré de 10 km, de la carte de l'anomalie Bouguer à l'échelle 1:500.000 (St. Airinei, Sc. Stoenescu 1964). On distingue quatre anomalies de maximum, séparées d'une vaste anomalie de minimum. Elles ne se trouvent que partiellement sur le territoire de la Roumanie. Trois des anomalies de maximum sont situées à l'extérieur de l'arc carpatique: l'une dans la partie sud de l'Olténie et de la Munténie, ouverte sur le Danube vers la R.P. de Bulgarie; une autre sur l'aire de la Dobroudja centrale et septentrionale, ouverte vers la Mer Noire et la partie sud de la République Moldave; la troisième dans la partie NE de la Moldavie, ouverte elle aussi vers le N et E, en même direction que la précédente. Enfin, la quatrième anomalie de maximum occupe entièrement l'espace placé à l'intérieur de l'arc carpatique et reste largement ouverte vers l'ouest; elle a une morphologie complexe suggérant un entassement de masses géologiques trois pour le moins, d'âges et d'origines diverses, se dressant en trois gradins successif de l'est vers l'ouest.

Les anomalies régionales de maximum correspondent à de grands volumes de la lithosphère, constitués de roches lourdes et englobant aussi dans certains cas, les effets subcrustaux produits par des masses denses montées du manteau supérieur et participant à l'amincissement même de l'écorce terrestre. Outre cela, les anomalies de maximum par devant les Carpates correspondent partiellement à des cratons précambriens (restes de vieilles chaînes érodées et successivement incorporées aux boucliers des plate-formes podolienne et moesienne, actuellement immobilisés sous l'échafaudage des morfostructures plus jeunes), figurés eux aussi sur la planche annexe. Aux termes de la "tectonique globale", les anomalies de maximum sont considérées comme illustrant les images d'un certain nombre de fragments de plaques ou de microplaques de la lithosphère qui ont été manoeuvrés par la géodynamique pré-et postjurassique déroulée dans cette partie de la planète.

Entre les anomalies gravimétriques régionales de maximum il y a une anomalie de minimum très étendue. Elle se compose de deux termes séparés à la hauteur du parallèle géographique Tg. Ocna: l'un situé devant les Carpates Méridionales et dans la région de courbure des Carpates Orientales, et l'autre longeant l'aire des dépôts de flysch et de molasse des Carpates Orientales. L'anomalie régionale de minimum présente deux apophyses vers l'extérieur qui séparent les trois anomalies de maximum par devant les Carpates: l'une vers le SE, en direction des localités Buzău-Mangalia, l'autre vers l'E, en direction Adjud-Cahul.

Les anomalies régionales de maximum correspondent à des épaisseurs moindres de l'écorce terrestre (26-35 km), tandis que l'anomalie de minimum couvre la partie la plus épaisse de l'écorce (35-55 km). Superposées aux anomalies régionales décrites ci-dessus, on connaît des centaines d'anomalies locales à degrés de régionalité successivement moindres, comme effets de nombreuses situations géologiques telles que: fractures de différentes ampleurs, rapports structuraux entre sédimentaire et soubassement cristallin, structures du sédimentaire ou du soubassement ainsi que des corps géologiques incorporés à celles-ci (volumes de volcanites, gisements de substances minérales utiles etc.)

La planche annexe présente également les champs de fractures crustales intersectant le territoire de la Roumanie et notés par des symboles allant de G_1 à G_{15}. Les champs de fractures crustales découpant l'écorce terrestre sur toute son épaisseur; ce sont des éléments structuraux très anciens représentant les alignements des zones de grande mobilité de la litosphère aussi bien pendant les événements géologiques qu'ils ont engendrés, que plus tard lorsqu'ils furent remobilisés par des événements géologiques plus récents. Certains champs de fractures crustales conservent jusqu'à nos jours une partie de la mobilité maxima qu'ils avaient atteinte dans leur passé géologique. Les signes (+) et (-) indiquent les positions hautes ou basses des compartiments de croûte séparés par les champs de fractures crustales.

L'une des grandes difficultés de la tectonique globale consiste à préciser la limite des plaques, surtout lorsque celles-ci traversent les continents. La difficulté devient encore plus grande en présence des microplaques qui, en principe et en sens géologique, se déplacent plus rapidement en comparaison du mouvement lent des plaques majeures. C'est justement sous ce rapport que les anomalies régionales de maximum fournissent, à notre avis, une importante contribution supplémentaire à côté de celle qui consiste dans le fait que leur forme et leur morphologie donnent la messure de l'extension et des caractères essentiels des segments de plaques et de microplaques de la lithosphère qui se trouvent dans le substratum profond du territoire de la Roumanie.

En effet, en suivant sur la planche annexe les contours des anomalies régionales de maximum, on peut distinguer: l'extrémité SW de la plaque esteuropéenne (A), le segment E de la microplaque interalpine (B), la partie N de la microplaque moesienne (C) et l'extrémité WNW de la microplaque de la Mer Noire

(D). Entre les anomalies de maximum se trouve l'anomalie de minimum avec ses deux apophyses étendues vers l'ESE. Eh bien, leurs axes de minimum peuvent constituer, du moins dans une première approximation, les limites ou les alignements de contact des segments de plaques et de microplaques de la lithosphère signalés. Sur les aires des segments de plaque et de microplaques sont également indiquées, par des flèches, les directions de mouvement et de déplacement de celles-ci, accordées avec les directions de mouvement et de déplacement des plaques et des microplaques évoluées depuis le Triasique supérieur jusque dans le post Miocène sur l'aire qui englobe partiellement l'Amérique du Nord, l'Europe (le bassin méditerranéen, y compris l'Arabie et l'Asie Mineure) et le Nord de l'Afrique (McKenzie, 1970 et 1973; Lort, 1971; Dewey et al., 1973).

Il est de toute évidence que le tableau de distribution des segments de plaque et de microplaques de la lithosphère sur le territoire de la Roumanie constitue la résultante actuelle du mouvement et du déplacement de ces segments dans le "plan de situation post-jurassique" de la tectonique des plaques à l'échelle planétaire. Il y a dans ce tableau une documentation géophysique qui confirme l'existence d'une microplaque de la Mer Noire pressentie par McKenzie (1972), ainsi que la critère de prolongement, dans le domaine euxinique, de la microplaque moesienne sur laquelle se trouvent installés les étages structuraux de la Plate-forme Moesienne dans le sens restreint (St. Airinei, 1969). Les distributions et les directions de mouvement et de déplacement des masses de la lithosphère telles qu'elles sont indiquées par les flèches, suggèrent certains rapports particulièrement intéressants au point de vue géodynamiques, tels par exemple:

- Les masses A, B et D ont un mouvement général sur des directions concentriques. Les vecteurs de déplacement peuvent être groupés deux par deux (A avec B et B avec D); vu que les intensités de ces vecteurs ne sont pas égales et que leurs directions respectives ne sont pas anti-parallèles, il ne saurait s'agir de couples de rotation proprement-dits pouvant conduire à des appréciations quantitatives capables d'expliquer l'ampleur des forces de compression, de tension et de cisaillement mises en jeu surtout dans le substratum de la région de Vrancea;

- Les données géologiques et géophysiques disponibles soutiennent que la plaque A suit un mouvement de subduction sous l'agglomérat de microplaques interalpin (B). Le processus de subduction semble avoir lieu aussi à présent, mais la zone qui connait l'activité la plus intense se trouve dans la région de courbure des Carpates Orientales, par suite de la vitesse supérieure de déplacement de la microplaque de la Mer Noire, poussée par la microplaque turque qui a le plus grande vitesse de déplacement de toute la partie E du bassin méditerranéen;

- A l'activité séismique et au caractère dynamique de l'équilibre isostatique associé aux ondulations négatives de la discontinuité Moho sous les Carpates Orientales et les Monts Apuseni (M.Socolescu et al., 1964), s'ajoute l'activité postvolcanique de ces deux régions (St. Airinei, A.Pricăjan, 1975), comme un argument de plus soutenant le caractère actif de la subduction le long des deux plans Benioff mis en évidence. La distribution des manifestations postvolcaniques (émanations mofettiques et solfatariennes, précipitations de roches carbonatées jeunes etc.) et la présence des anomalies géothermiques sur les deux aires à volcanites néogènes (St. Airinei et al., 1976) s'explique entre autres par la grande fréquence des ramifications, allant jusqu'à la surface, des champs de fractures crustales engendrés par les plans Benioff ainsi qu'en relation avec le processus de subduction qui se déroule dans le substratum des deux segments montagneux;

- La microplaque moesienne (C) a un déplacement presque parallèle et en même sens avec la microplaque de la Mer Noire (D), allant de l'ESE vers le WNW. La microplaque moesienne (C) semble glisser tangentiellement sous les Carpates Méridionales sur lesquelles elle exerce une compression maxima dans la zone de courbure des Portes de Fer. Conjointement avec la microplaque interalpine (B), elle semble réaliser un couple de rotation qui jouerait un rôle important pour accroître l'intensité des forces de cisaillement dans le substratum de la région de Vrancea;

- Le déplacement plus rapide de la microplaque de la Mer Noire (D), par rapport aux déplacements de la plaque esteuropéene (A) et de la microplaque moesienne (C) est attesté indirectement par la séismicité plus élevée manifestée le long des zones qui les séparent. En effet, sur les deux apophyses Est de l'anomalie de minimum, se prolongent, sous la forme d'inflexions avancées les isoséistes à 7 et 8 degrés de la zone épicentrale de Vrancea, point de séismicité maxima sur le territoire de la Roumanie (R. Ciocârdel et al., 1972; L. Constantinescu et al., 1975). De plus, le long des deux apophyses de minimum on connaît un certain nombre de foyers normaux, que nous pouvons expliquer, maintenant, comme un résultat du chargement produit par les frictions entre les plaques de la lithosphère; sur l'apophyse sud (entre D et C) se trouve la zone épicentrale Mangalia-Sabla qui se prolonge au NW jusque dans la zone des tremblements à foyers normaux Ploieşti-Cîmpina; sur l'apophyse nord (entre D et A) les zones bien connues de sensibilité séismique sont entretenues par les tremblements à foyers normaux du périmètre Bîrlad-Tecuci-Galaţi qui se prolonge vers l'Est en direction de la Crimée;

- Sur l'aire de la microplaque de la Mer Noire (D) ont été identifiés géophysiquement quatre champs de fracture crustales (G_5, G_4, G_2 et G_{15}). Partiellement, ils ont été identifiés aussi géologiquement, mais seulement sur les segments qui traversent le territoire de la Dobroudja. Le fait que leur distribution est presque parallèle et qu'ils séparent des blocs sialiques d'âges différents au niveau des formations préjurassiques, fut interprété dans le sens qu'ils font partie d'un probable système préjurassique de failles de transformation (St. Airinei, 1973). Leur présence semble confirmer aussi l'idée que la microplaque de la Mer Noire (D) est un agglomérat de restes de plaques et de microplaques qui ont évolué dans le domaine de l'Océan Tethys pendant les systèmes préjurassiques de la tectonique des plaques.

- Le mouvement plus rapide de la microplaque D, par rapport à la plaque A et à la microplaque C, présente un caractère visible de glissement affectant les zones intermédiaires (marquées par les apophyses de l'anomalie de minimum) par des sollicitations mécaniques complexes. De plus, simultanément avec les déformations qu'ils produisent, ces mouvements impriment une nouvelle activité à l'état de mobilité existant sur les limites de plaques sollicitées. Le glissement au WNW de la microplaque D se produit entre les champs de fractures crustales G_5, du sud, et S, du nord. L'alignement S a été mis en évidence par les géophysiciens soviétiques (I.S. Subbotin, 1960).

EN GUISE DE CONCLUSIONS

En synthétisant ce qui vient d'être exposé en relation avec la présence des anomalies gravimétriques régionales de maximum et de minimum sur le territoire de la Roumanie, nous pouvons ponctuer les contributions suivantes qui viennent s'ajouter aux tentatives antérieures d'encadrer cet espace planétaire dans le concept de la tectonique des plaques:

- Les anomalies régionales de maximum - confrontées avec les anomalies isostatiques, avec les éléments qui caractérisent la morfologie de la discontinuité Moho et avec les données fournies par le "sondage sismique profond" - reflètent quatre segments de plaques de la lithosphère, à savoir: l'extrémité SW de la plaque esteuropéenne (A), la segment Est de la microplaque interalpine (B), la partie N de la microplaque moesienne (C) et l'extrémité WNW de la microplaque de la Mer Noire (D);

- Les axes de l'anomalie régionale de minimum et de ses deux apophyses Est marquent les limites de contact entre les segments mentionnés de la lithosphère;

- Les directions de mouvement des segments de plaque sont établies en accord avec le tableau général des directions de déplacement des plaques et microplaques du bassin méditerranéen. La partie SW de la plaque est-européenne tend à se déplacer vers le WSW; les microplaques D et C se déplacent vers le WNW suivant des directions presque parallèles; la microplaque B se déplace ou reste stationnaire en oposant une résistance sur une direction ESE. La convergence de leurs mouvements vers la région de courbure des Carpates Orientales est frappante;

- En accord avec le tableau des vitesses de déplacement des plaques et microplaques dans la zone Est du bassin méditerranéen, il ressort que la microplaque de la Mer Noire (D), poussée par la microplaque turque, avance dans la direction WNW plus rapidement que la microplaque moesienne (C) et la plaque esteuropéene (A). Il paraît que nous assistons à un processus de glissement de la microplaque D entre la plaque A et la microplaque C le long des champs de fractures crustales G_5 du sud et S du nord. Le déplacement plus rapide de la microplaque de la Mer Noire (D) est attesté indirectement par les foyers des tremblements normaux situées sur ses flanc nord et sud, le long des apophyses de minimum qui sont également des aires de séismicité élevée;

- La présence des quatre champs de fractures crustalles (G_5, G_4, G_2 et G_{15}) sur l'aire de la microplaque D interprété comme appartenant à un probable système préjurassique de failles de transformation, semble confirmer aussi l'idée que la microplaque de la Mer Noire est un agglomérat de restes de plaques et microplaques qui ont évolué pendant les systèmes préjurassiques de la tectonique des plaques dans le domaine de l'Océan Tethys;

- La contribution des anomalies gravimétriques régionales du territoire de la Roumanie à la connaissance du processus proprement-dit de subduction dans les Carpates Orientales et les Monts Apuseni, mis en discussion dans des travaux antérieurs, résulte de ce qui vient d'être exposé ci-dessus. Elle s'ajoute aux arguments fournis par l'activité séismique, par le caractère dynamique de l'échilibre isostatique associé aux ondulations négatives de la discontinuité Moho, ainsi que par la présence des manifestations post-volcaniques et des anomalies géothermiques le long des aires à volcanites néogènes existant sur les deux secteurs montagneux des Carpates.

BIBLIOGRAPHIE

AIRINEI ST., STOENESCU SC. (1964), Relations concernant la rédaction de la carte gravimétrique de la R.S.R. (anomalie Bouguer) (éch. 1:1.500.000). D.S.Com.Geol. Rom., L/1 (1962-1963), 269-282, Bucureşti (en roumaine).

AIRINEI ST. (1969), Couples d'anomalies gravimétriques-magnétiques ouvertes sur le litoral roumain de la Mer Noire. Rapp. comm. int. Mer Médit., 19, 659-660, Monaco.

AIRINEI ST., STURDZA D. (1970), Les surfaces des formations géologiques du territoire de Roumanie planimétrées sur la carte géologique à l'echelle 1:200.000 . Bul.Soc.Stiinţ.Geol. Rom., XII, 225-233, Bucureşti (en roumaine).

AIRINEI ST. (1973), L'appartenance de quatre dislocations crustales dobrogéenes à un probable système préjurassique de failles de transformations. Rapp. comm. int. Mer Médit., 22, 2a, 158-159, Monaco.

AIRINEI ST., PRICAJAN A. (1975), Some geological connections between the mineral carbon and thermal waters and the post-volcanic manifestations correlated with the deep geological structure of the East Carpathians territory-Romania. I.G.G., Stud. tehn. econom., Seria E (Hidrogeologia) XII, 7-19, Bucureşti.

AIRINEI ST., PRICAJAN A., BANDRABUR T. (1976), Conceptual pattern concerning the study of geothermalism and the thermalization process of underground waters in Romania. Rev. roum. géol., géophys., géogr., Série de Géologie, 20, 2, Bucarest.

AIRINEI ST. (1976), Ideas concerning the integration of the Romanian territory into the plate-tectonics concept. St. cerc. géol. geofiz., geogr. Geofizica, 14 (in press) Bucureşti (in Romanian).

BLEAHU M., BOCCALETTI M., MANETTI P., PELTZ S. (1973), Neogen Carpathian arc: Island arc. Jour. Geophys. Rech., 78, 23, 5025-5032, Richmond, Virginia, USA.

CIOCARDEL R., SOCOLESCU M., POPESCU M., (1972), Eléments pour la sismicité de la Roumanie, Rev. roum. géol., géophys., géogr., Série de Géophysique, 16, 1, 21-34, Bucarest.

CIOCARDEL R., SOCOLESCU M.(1972), Aspects of the Carpathian-Balkannic-Dinaric and Central Mediterranean areas geodynamics. Rev. roum. géol., géophys., géogr., Série de Géophysique, 16, 2, 179-192, Bucarest.

CONSTANTINESCU L., CORNEA I., LAZARESCU V. (1973), An approach to the seismotectonics of the Romanian Eastern Carpathians. Rev. roum. géol., géophys. géogr., Série de Géophysique, 17, 2, 133-143, Bucarest.

CONSTANTINESCU L., CORNEA I., LAZARESCU V. (1975), Seismotectonics map of the Romanian Territory. I.G.G., Stud. tehn., econom., Seria D (Geofizica), X, 291-298, Bucureşti.

CONSTANTINESCU L., CONSTANTINESCU P., CORNEA I., LAZARESCU V. (1975) Recent seismic information on the lithosphere in Romania. Mem. of XIV-th General Assembly of the European Seismological Commision (Trieste, 16-22 September 1974), Ed. Nationalkommitee für Geodäsie und Geophysik bei der Akademie der D.D.R., p.411-415, Berlin.

DEWEY J.F., PITMAN III C.W., RYAN F.B., BONNIN J. (1973), Plate tectonics and the evaluation of the Alpine system. Geol. Soc. Am. Bull., 84, 3137-3180, Washington D.C., USA.

DUMITRESCU I., SANDULESCU M., LAZARESCU V., MIRAUTA O., PAULIUC S., GEORGESCU C. (1962), Mémoire à la carte tectonique de la Roumanie. An Com. Geol. Rom., XXXII, 5-96, Bucureşti.

GAVAT I., AIRINEI ST., BOTEZATU R., STOENESCU SC., SOCOLESCU M., VENCOV I. (1965), Contributions de la gravimétrie et de la magnétométrie à l'étude de la structure profonde du territoire de la R.P.Roumaine. Rev. roum. géol., géophys., géogr., Série de Géologie, 9, 1, 81-107, Bucarest.

GIUSCA D., SAVU H., BERCIA I., KRAUTNER H. (1969), Sequence of tectonomagmatic prealpine cycle on the territory of Romania. Acta Geologica Academiae Scientiarum Hungaricae, Intern. Geol. Congr. Bull., IX Congr. Carp.-- Balk. Geol. Assoc., I., 221-234, Budapest.

HERZ N., SAVU H. (1974), Plate Tectonics history of Romania. Geol. Soc. Am. Bull., 85, 1429-1440, Washington D.C., USA.

LE PICHON X. (1968), Sea-floor spreading and continental drift. Jour Geophys. Res., 73, 3661--3697, Richmond, Virginia, USA.

LORT J.M. (1971), The tectonics of the Eastern Mediterranean: a geophysical review. Rev. Geophys. Space Phys., 9, 189-216, Richmond, Virginia, USA.

McKENZIE D.P. (1970), Plate tectonics of the Mediterranean region. Nature, 226, 239-243, London Washington.

McKENZIE D.P. (1972), Active tectonics of the Mediterranean region, Geophys. J.R. Astr. Soc., 30, 109-116, London.

RADULESCU P.D., SANDULESCU M. (1973), The plate-tectonics concept and the geological structure of the Carpathians. Tectonophysics, 16, 155-161, Amsterdam.

ROMAN C.(1970), Seismicity in Romania - evidence for the sinking lithosphere. Nature, 228, 1176-- 1178. London - Washington.

ROMAN C. (1971), Plate tectonics in the Carpathians - A case in developement. Mem. XII-e Assemblée Générale de la Commission Séismologique Européenne, ed. J.M.Van Gils, 101, 37-40, Bruxelles.

ROMAN C. (1973), Rigid plates, buffers and subplates - Comment on "Active Tectonics of the Mediteeranean region" by D.P.McKenzie. Geophys. J.R. Astr. Soc., 33, 369-373, London.

SOCOLESCU M., POPOVICI D., VISARION M., ROSCA V. (1964), Structure of the crust in Romania as based on the gravimetric data. Rev. roum. géol., géophys., géogr., Série de Géophysique, 8, 3-11, Bucarest.

SOCOLESCU M., CIOCARDEL R., AIRINEI ST., POPESCU M. (1975). La physique et la structure de l'écorce terrestre de la Roumanie. Ed. tehnică, Bucureşti, (en roumaine).

SUBBOTIN I.S. (1960), The depth structure of the territory of the Ukrainian S.S.R. Inter. Geol. Congr., XXI Session, Rep. Sov. Geol., Part II Geophysics, 58-66,Moscow (in Russian).

STEFAN AIRINEI

ESQUISSE DES ANOMALIES GRAVIMÉTRIQUES RÉGIONALES SUR LE TERRITOIRE DE LA ROUMANIE ANALYSÉES DANS LE CONCEPT DE LA TECTONIQUE DES PLAQUES

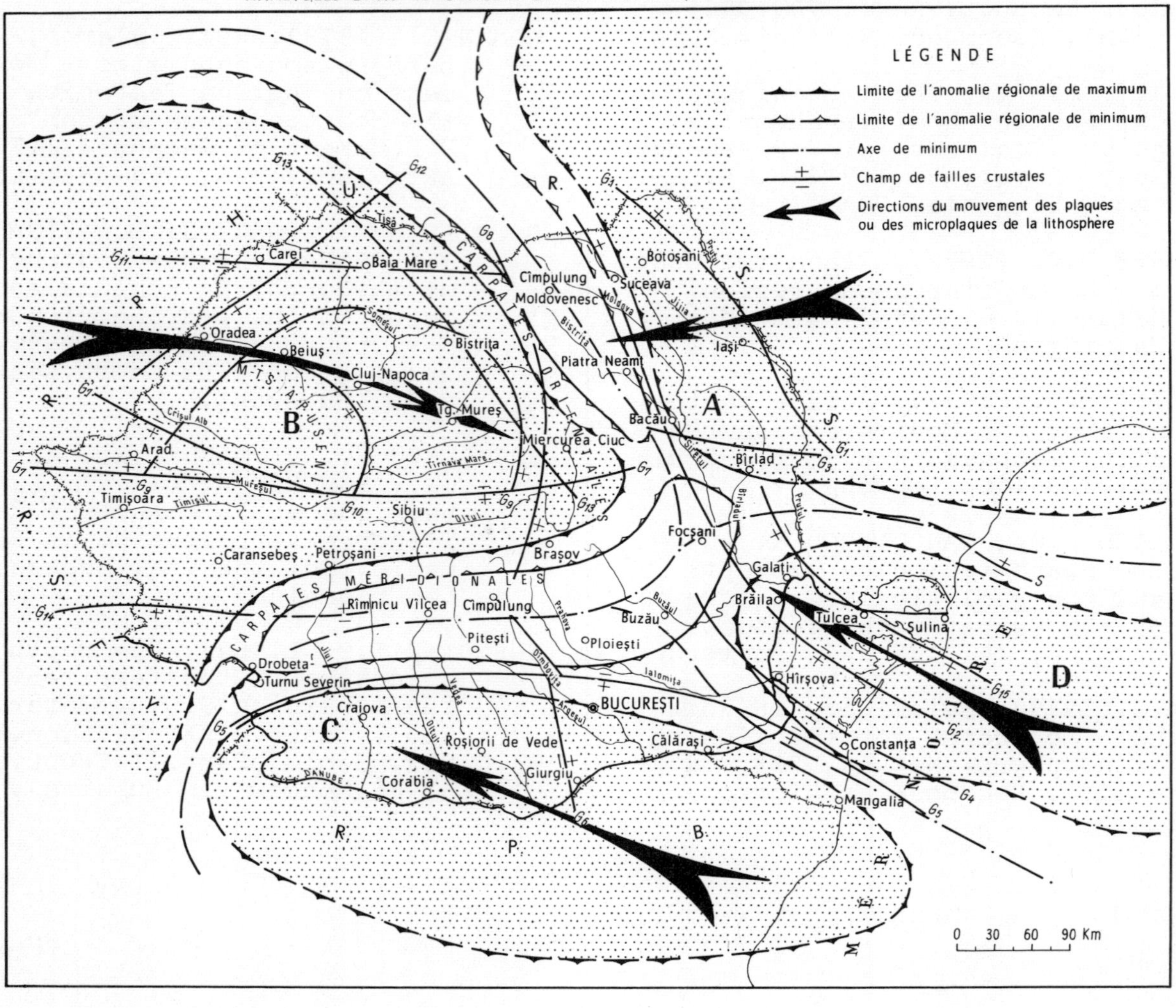

INTERNATIONAL SYMPOSIUM ON THE STRUCTURAL HISTORY OF THE MEDITERRANEAN BASINS. SPLIT (YUGOSLAVIA) 25-29 OCTOBER 1976. B. BIJU-DUVAL AND L. MONTADERT, Eds. EDITIONS TECHNIP, PARIS 1977, pp. 353-362

CRITICAL COMPARISON OF MOBILISTIC MODELS OF TECTONIC DEVELOPMENT OF THE CAUCASUS

V.E. KHAIN [(1)]

Up to the recent years and, partly, at the present time (KRASNOPEVTSEVA et al., 1976; SHEVCHENKO, REZANOV, 1974) the tectonic development of the Caucasus was considered from the fixist standpoint, the main attention being given to the reconstruction of the history of vertical movements, based on the study of distribution of facies and thickness of sediments and volcanics. A new trend in the investigation was begun by A.V.Peive (PEIVE, 1969) and developed rapidly first of all owing to A.L. Knipper (KNIPPER, 1975 and earlier publications) who worked in the central part of the Lesser Caucasus. Later on other works appeared dealing not only with the Lesser, but also with the Greater Caucasus and with the Caucasus as a whole (KHAIN, 1975; BELOV, 1976; ADAMIA et al., 1976; GAMKRELIDZE, 1976). The history of the Caucasus was considered by J. Dewey with co-authors (DEWEY et al., 1973) from the same standpoint against a background of the whole of the Mediterranean belt. In these works the ophiolite belts of the Greater and Lesser Caucasus are considered, in accordance with new concepts, relics of basins with the oceanic type of crust and the fold-thrust structure of the region, the result of the overall compression of the geosynclinal belt. It seems that such interpretation of the development of the Caucasus has at least three advantages as against the fixist models. Firstly, a fundamental difference between ophiolite belts and their separating sialic blocks is taken into consideration, as well as a similarity of ophiolite associations and the oceanic crust. Secondly, intense deformation of sedimentary rocks is explained more rationally, particularly in the Front Range of the Greater Caucasus, in its southern slope and in the central Lesser Caucasus where important nappes were discovered in recent years. Thirdly, a polarity in the variation of chemical composition of magmatic products noticed by Georgian geologists (ADAMIA et al., 1976) can be consistently explained.

At the same time it should be stated that theoretical points of departure and methodical bases being, on the whole, similar, definite paleotectonic reconstructions suggested in the abovementioned works noticeably differ, which is, certainly, due to insufficient data that cannot be interpreted unequivocally. The object of this report is, therefore, to compare the proposed models in order to select the most probable versions and to reveal questions requiring additional explanation.

(1) Department of Geology, Moscow State University, Moscow, USSR

As a preliminary, the data on which all the interpretations are based, should, apparently, be summed up. The Caucasus is the northern part of a narrow segment of the Alpine Mediterranean belt situated between the north-east Arabian prominence of the African craton and the south-east Rostov prominence of the East-European craton (Fig. 1). The boundaries of both cratons with the Mediterranean belt are obviously secondary, being sharply discordant to the interior structure of their Precambrian basement. An approximately N-S trend of younger elements of this structure, transversal to the Mediterranean trend has been long and persistently suggesting an idea of a superimposed, secondary nature of the Mediterranean belt. Comparison of ages of the youngest basement rocks of the cratons with those of the oldest obviously geosynclinal rocks of the mobile belt shows that the most probable time of origin of the latter in the region under review is about 850 m.y. ago. At the same time one must not disregard an essential difference between the Precambrian basement in the northern and southern framing of the Mediterranean belt, as is apparent from the comparison of the structures of the Rostov prominence of the Ukrainian shield and the Nubian-Arabian shield accessible to study. In the north the complete cratonization set in as early as at the end of Early Proterozoic, approximately at the 2000-1800 m.y. boundary, and in the south 1000 m.y. (!) later, 1000-800 m.y. ago. But there are good reasons to believe that the primary age of formation of most of the Nubian-Arabian basement is Early Precambrian, younger radiometric dates reflecting later reactivation. A similar phenomenon is observed in the western part of the East-European craton, in particular in the South-Scandinavian block of the Baltic shield. It is probable that relic Rb/Sr dates of about 850 m.y. in the crystalline nucleous of the Greater Caucasus are indicative of rocks here contemporary with the basement of the Arabian-Nubian shield. Granitoid pebbles 1100-850 m.y. old are found in the mountainous Crimea; apparently, they are derived from the western continuation of the Central Caucasus trending under the Black Sea (DOBROVOLSKAYA et al., 1970). Hence, the basement of the "Arabian" type may have originally extended north to the Main Range of the Greater Caucasus inclusive.

In the rest of the Caucasian-Arabian segment of the Mediterranean belt its oldest formations are apparently younger and contemporary with the initial period of the belt formation. These are sedimentary-volcanic rocks of the greenschist, more seldom amphibolite facies. They are most likely of Upper Riphean - Lower Vendian (850-620 m.y.) age and only in the Dzirula massif in Transcaucasia this complex comprises younger rocks up to the Lower Cambrian inclusive. Baikalian complex bilds up the basement of the Fore-Caucasus south of the Rostov prominence of the ancient craton, the North-Transcaucasian and the Central-Iranian sialic blocks; the northern prominence of the latter lies within the southern part of Transcaucasia, in the Araxes basin. This complex remobilized in Late Paleozoic may have been also included in the structure of the crystalline nucleous of the Greater Caucasus.

The Baikalian basement in Ciscaucasia is overlain by miogeosynclinal Mid-Paleozoic rocks forming a passive Paleozoic margin of the European continent and folded at the end of Paleozoic, in the Saalian epoch of the Hercynian tectogenesis. Among these formations individual grabens - taphrogeosynclines - are recognized, filled in with Triassic sediments and volcanics folded at the beginning of Jurassic. Above there is practically undeformed Jurassic and younger platform cover which permits to consider Ciscaucasia a part of a young platform that accreted the East-European craton at the end of Paleozoic - beginning of Mesozoic.

A more southern zone of the Paleozoic geosyncline already of the eugeosynclinal character trends through the Front Range of the Central Caucasus, its westerly and easterly continuations being traced by magnetic anomalies in the Southern Ciscaucasia. The structure of the Front Range that recently seemed rather simple, appeared very complicated, nappes bein developed here, composed of a typical ophiolite complex and of Silurian graptolite slates (BELOV, OMELCHENKO, 1976). Under the nappes there are exposed spilite-keratophyre volcanics and Devonian-Lower Carboniferous terrigene-carbonaceous rocks. The ophiolites are definitely Pre-Silurian (POTAPENKO, PRUTSKY, 1976) and, probably, Pre-Ordovician, pierced by granitoids about 500 m.y. of age. The nappes were transported, most likely, from south to north and their "roots" are, apparently, in the fault zone separating now the Front Range zone and the Main Range zone of the Greater Caucasus. The zone of the Front Range as a whole is thrust on the Pre-Baikalian marginal block of southern Ciscaucasia. The main deformations took place at the Early-Middle Carboniferous boundary. They were followed by the accumulation of Late Paleozoic molasse.

The Main Range zone is composed, as it was mentioned above, of crystalline schists, gneisses and migmatites, for the most part replaced by Hercynian granitoids. Age of the metamorphic complex is not yet clear: on the one hand it comprises Precambrian rocks older than 850 m.y., and on the other hand, Paleozoic rocks with crinoidal fragments and problematic remains of Devonian fauna. On the whole it is a remobilized block of the ancient continental crust.

To the south there is a zone of the southern slope of the Greater Caucasus, corresponding to the main Alpine geosynclinal trough of this folded system. The trough having originated at the end of Triassic - beginning of Jurassic is, in part, a descendant of the Paleozoic trough filled in with shallow and medium deep marine Silurian (?) - Permian. The maximal development of the south slope trough falls on Early-Middle Jurassic. Accumulation of thick shale and sandy-shale series was accompanied by outflows and intrusions of basic and even ultrabasic magmatites (spilites, diabases, picrites, etc.). This slate-diabase formation is covered by the Upper Jurassic - Cretaceous - Paleogene flysch formation related to the erosion of a non-volcanic island arc in the central Greater Caucasus. From Oligocene on, the zone was subject to folding reaching its climax in Late Miocene -Early Pliocene, a system of flysch nappes transported south for many dozens of kilometres being formed here. Northern Transcaucasia lying south was a relatively stable continental crust block - the Transcaucasian median mass (microcontinent) apparently from the beginning of Paleozoic (Middle Cambrian?). But in Late Paleozoic this block was essentially remobilized, which resulted in the formation of a volcanic-plutonic belt (acid volcanics, granitoids), the same phenomenon repeating on a larger scale in Jurassic and Cretaceous, accompanied by abundant andesite-basalt and, more seldom, acid and alkaline magma outpourings. At the middle of Cretaceous the western part of the block was rifted and the Adjaria-Trialeti taphrogeosyncline was formed (ADAMIA et al., 1974) the development of which ended in the latest Eocene with folding and uplift. Beginning with Oligocene and, particularly, with Late Miocene much of the North Transcaucasian block was subject to intense subsidence accompanied with the accumulation of thick molasse derived from newly formed mountainous structure of the Greater and Lesser Caucasus, while the southern part of the block itself was involved into the overall uplift of the Lesser Caucasus.

The most important and interesting zone of the Lesser Caucasus, the Sevan-Akera ophiolite belt, trends through its central part, being a direct continuation of the North-Anatolian ophiolite belt. At the present time it is known that the ophiolites of this belt occur everywhere in allochtonous position underlain by desintegration products of nappes - olistostromes - overlying, in their turn, an Albian-Cenomanian flysch series and relatively deep-sea Tithonian-Neocomian limestones lying on Mid-Jurassic volcanics similar to those in the neighbouring median mass zone to the north. Ophiolites were recently proved Pre-Upper Jurassic; indirect data permit to consider them Triassic [1] or, in part, still older.

---

1) The age of the North-Anatolian ophiolites was recently reconsidered as Triassic (BINGÖL, 1976; BRUNN et al., 1976; KAYA, 1976).

The latter fact is proved by Paleozoic radiometric datings of gabbroids and metamorphic basic volcanics in the melange.

The transport direction of the ophiolite nappes is still not known, its ascertainment being hindered by two circumstances. Firstly, in the north the ophiolite zone is overridden by the southern edge of the North Transcaucasian mass (or rather, the ophiolite zone is thrust under the massif, as was pointed out by K.N. Paffenholz long ago). The time of the displacement is Late Eocene, like that of the final deformations of the zone.

Secondly, south-west of the Sevan-Akera ophiolite belt there is the Kafan block with the sedimentary sequence similar with that of the Transcaucasian massif. The block is separated from the northern periphery of another median mass, the Central-Iranian one, by a large zone of compression and faults (the Guirratakh or Zangezur fault) where the sediments up to the Cretaceous inclusive are noticeably metamorphosed. "Roots" of ophiolite nappes are either in this zone (in this case they must have moved north and north-east) or beneath a younger overthrust of the south edge of the Transcaucasian massif.

Finally, it should be pointed out that the south-easterly continuation of the Sevan-Akera ophiolite belt is not yet known across the Araxes in Iran. In the Caspian regions of Iran outcrops of ophiolites were found, but their relations with the surrounding rocks are in favour of an older age, most probably Early Paleozoic. Therefore, they correlate rather with the ophiolites of the Front Range of the Greater Caucasus than with much younger ophiolites of the Sevan-Akera zone. It is not utterly impossible that this is, in fact, a south-easterly continuation of the Front Range belt shifted along a transcurrent (transform) fault. The Sevan-Akera belt is shifted, possibly, in a similar way. In this case its continuation may be the Sabzevar ophiolite zone between the East Alborz-Aladag and the Lut block of the Central-Iranian massif.

Within Southern Transcaucasia (the southern Lesser Caucasus) south of the Sevan-Akera ophiolite zone there is one more much narrower ophiolite zone, the Vedi one. It is separated from the former by an uplifted block of ancient, at least Pre-Devonian, continental crust. The Vedi zone dies out in the south-east, but continues westward into Northern Anatolia where it is apparently connected with the continuation of the Sevan-Akera zone. Ophiolites of the Vedi zone, like those of the Sevan-Akera one, are allochtonous (LOMIZE, 1971; KNIPPER, SOKOLOV, 1976) and it is not clear whether these nappes have common roots (most likely in the broader Sevan-Akera zone) or an independent basin underlain by oceanic crust corresponded to the Vedi zone. The last idea seems less probable, the more so as the time of overthrusting in both ophiolite zones of the Lesser Caucasus is the same - Coniacian.

South of the Vedi zone there is a vast Central-Iranian median mass (microcontinent) with a Baikalian basement and a persistent Vendian-Triassic cover. The mass was reactivated at the end of Triassic and later. A complex ophiolite belt of the Sanandadj-Sirdjan zone and of the Zagros Interior zone separates the mass from the Exterior Zagros folded system, that originated already at the periphery of the Arabian platform of the African craton (STÖCKLIN, 1974).

Such is the general geologic picture of the region under review, as it can be reconstructed on the basis of data available today.

Interpreting these data, it should be noted, first of all, that there is no positive information to state that as far back as at the earliest stage of the development, in the Late Riphean the discussed segment of the Mediterranean belt was underlain by oceanic crust; serpentinites that occur in the Dzirula massif may be younger. Nevertheless, from general considerations and some factual data (ZARIDZE, 1970) such an assumption is very probable.

Pre-Silurian, probably pre-Ordovician and even Vendian-Early Cambrian ophiolites are known in the Front Range of the Greater Caucasus, which corroborates the existence of a basin here with oceanic crust. But it remains unknown whether this basin was the axial zone of Paleotethys, separating Gondwana and Europe in Paleozoic (KHAIN, 1975), or it was a mere marginal sea according to Sh.Adamia and others (ADAMIA et al., 1976) who think that the axial part of Tethys in Paleozoic as well as in Mesozoic ran across the Sevan-Akera zone of the Lesser Caucasus (Fig. 2). Another hypothesis was proposed by A.A.Belov (BELOV, 1976); he places the axis of Paleotethys and, correspondingly, the border of Eoeurope and Gondwana along the Svanetia basin on the southern slope of the Greater Caucasus. But in this zone neither ophiolites are known, nor, practically, any other manifestations of initial magmatism; thus, by the character of its deposits this zone, rather than the Front Range one, can be considered a marginal sea.

The question whether the oceanic basin (more exactly, the basin with oceanic crust) of the Lesser Caucasus is a Mesozoic or Paleozoic, even Early Paleozoic, feature is, perhaps, the main one in the interpretation of the tectonic history of the Caucasus. Two arguments are in favour of the Paleozoic age of the basin: 1) Paleozoic K/Ar datings of the basic magmatites in the melange of the Sevan-Akera zone, 2) manifestations of Late Paleozoic acid magmatism in the Transcaucasian massif which may be due to the activity of a Benioff zone dipping under the massif from a probable Sevan-Akera oceanic basin. But K/Ar datings of the basic rocks are not quite reliable and the Late Paleozoic acid magmatism of the Transcaucasian massif may be due to a Benioff zone running along its northern edge and inclined in the opposite direction, to the south.

The ophiolite belt of the Lesser Caucasus forms a direct continuation of the North-Anatolian belt, the latter being, apparently, the continuation of the Vardar belt. Therefore, the time of its formation cannot be established in the framework of the Caucasus alone. From this point of view it is important that all evidence for Paleozoic ophiolites both in Anatolia and in the Dinarides have apparently fallen away at present. It is only O.Kaya (KAYA, 1976) who recently suggested the presence of pre-Silurian and Upper Paleozoic ophiolites in Northwestern Anatolia. Nevertheless, the fact that the Late Paleozoic acid magmatism is most extensive not only in the Caucasus, but also in Europe (MOSSAKOVSKY, 1972) needs explanation, and one of these may be an idea that the southern edge of Paleoeurope was a continental margin of the Andean type. In this case Mesotethys must have been formed already in Late Paleozoic, but not in Early Paleozoic when the area of a future Mesotethys may have belonged to the epi-Baikalian frame of Gondwana. However, at that time already the Central Iranian - Transcaucasian massif must have been separated from the main Gondwana supercontinent by the basin of the Sanandadj-Sirdjan zones, and in Northwestern Anatolia, according to O.Kaya (KAYA, 1976) there remnants of pre-Silurian ophiolites.

In any case the existence of areas of oceanic crust throughout the Paleozoic (and the terminal Precambrian) and Mesozoic in the Caucasian-Arabian segment of the Mediterranean belt is very doubtful. It is more probable that the evolution of the belt was the alternation of epochs of expansion of the oceanic crust areas and their reduction up to utter disappearance as a result of Hercanian and, earlier, Silurian-Caledonian collision of continental crust blocks. This conclusion must be correct of the present Black Sea area too. Sh.Adamia's (ADAMIA, 1974) criticism of the idea of J.F.Dewey and his coauthors that the Black Sea basin is an inherited Paleozoic structure seems quite true.

A new Alpine structural pattern of the Caucasian-Arabian segment appeared at the end of Triassic- beginning of Jurassic during the epoch of general breakdown of the lithosphere, that involved the whole territory of Ciscaucasia to Central Iran. At that time the main part of Tethys almost surely extended through the Lesser Caucasus. As to more northern regions, Ciscaucasia and the northern slope of the Greater Caucasus were, according to a probable assumption of Sh.Adamia and his colleagues, a continental margin of the Atlantic type, the southern slope of the Greater Caucasus being a marginal sea and Northern Transcaucasia, an island arc. A character of the crust in the marginal sea is not yet known. Apparently, it could not be normal continental. Taking into consideration sub-ophiolite magmatism and immense (10-15 km, probably more) thickness of relatively deep-sea sediments, the crust must have been either oceanic or continental thinned very much and pierced by abundant parallel basic-ultrabasic dykes, similar to the crust of the marginal parts of the Red Sea. Moreover, at the boundary of the shelf and the deep part of the marginal sea a non-volcanic island arc originated at the beginning of Malm, providing a source of flysch for the southern slope, while the volcanic arcs in Northern Transcaucasia form a superstructure on the ancient block of the continental crust (similar to the Japanese Islands) predominantly along its northern and southern peripheries at the boundary with the Lesser Caucasian (Sevan-Akera) oceanic basin a and a quasi-oceanic basin of the southern slope of the Greater Caucasus.

The last circumstances involve a new question: whether one or two Benioff zones existed in the frame of the Transcaucasian massif (Fig. 3). This question is analogous in many respects to a similar question concerning Later Paleozoic situation, and the best way to solve it is to analyse polarity in the change of the composition of magmatic products. Such analysis has not been done of the Late Paleozoic; petrochemical constructions made by Georgian geologists (ADAMIA et al., 1976) for the Jurassic and Cretaceous seem to point to a uniform increase of alkalinity and the K_2O/Na_2O ratio in comparable rocks north of the Sevan-Akera belt, which is in favour of a single Benioff zone along the northern frame of this belt. But a priori it is quite probable that double Benioff zones surrounded the median mass, which accounts still better for their intense heating and remobilization of the basement and to which recent development of Indonesia and the Philippines is similar.

The basic problem of the Mesozoic "flourishing" of oceanic environments in the discussed segment of the Mediterranean belt as a whole is that of relative importance of the Lasser Caucasian and South-Iranian deep-sea basins. This problem is not only of regional, but also of broader importance, as it concerns both the western and eastern segments of the belt. As to Anatolia, a group of French geologists (RICOU et al., 1973) advocate the idea of the allochtonous position of ophiolites in the southern ophiolite zone of Anatolia. It follows from this that the North-Anatolian basin alone (and its continuation in the Lesser Caucasus) is a true oceanic basin and there is no reason to distinguish an independent Iranian-Anatolian microcontinent. But it seems that though it is quite probable that South-Anatolian nappes are really of "ultra-Menderes" origin (BRUNN et al., 1976), it is hardly correct to negate a deep-sea basin with at least suboceanic crust in the present zone of the South-Anatolian and, particularly, South-Iranian nappes. This basin, apparently, widened and deepened eastwards. Paleontology indicates (SEYED-EMAMI, SARABY, 1976) that it was a more important biogeographical barrier than the North-Anatolian - Lesser Caucasian basin [1]), but it would be very desirable to obtain paleomagnetical data as well.

The last question to be dealt with here is that of the recent development of the region, At that stage no segments underlain by oceanic crust existed any more within the Caucasian isthmus; the "bridge" of the continental crust connecting the East-European and Arabian-African cratons was

---

1) On the other hand it appears that in Anatolia the main barrier was situated along the North-Anatolian ophiolite belt. Some changes may have occurred along the transform fault, separating Anatolia from Iran.

formed as early as by the end of Mesozoic. At the same time in the south (the Lesser Caucasus, Northern and Central Iran) in Eocene and further north (in the Greater Caucasus and even in Southern Ciscaucasia) in Neogene-Quaternary powerful volcanic manifestations took place, andesitic at the base and more acid, usually correlated with the activity of Benioff zones. Besides, the Lesser and Greater Caucasus have independent petrochemical polarity (ADAMIA et al., 1976). The last phenomenon may be indicative of the activity going on in at least two Benioff zones [1] after the collision of continental blocks of the cratons and microcratons, as it is recorded by seismic data in the present-day Hindu-Kush, Pamirs, and in the Vrancha zone of the Carpathians.

Here we could but briefly touch upon some main problems that arise as attempts are made at mobilistic interpretation of the tectonic history of the Caucasus and cause difference of opinion among scientists. But in spite of these difficulties we must not give up this methodic approach as a wrong one. It is obvious that throughout the Mediterranean belt (and not only in this belt) we have to choose between the environments of an open ocean and those of systems of marginal seas. In the last case a quantity of these basins and their separating microcontinents and island arcs must be determined and ancient Benioff zones, single or double, must be distinguishes. The cardinal way to solve these problems is to accumulate information on paleofacies, paleontology, paleomagmatism, as equivalents of recent environments of both types can be found, e.g. on the opposite sides of the Pacific Ocean.

---

1) In the southern slope of the Greater Caucasus and along the Zagros main thrust.

REFERENCES

ADAMIA (SH.A.). - 1975. Plate tectonics and the evolution of the Alpine system: discussion, Bull. Geol. Soc. Am., v. 86, no. 5, p. 719-720.

ADAMIA (SH.A.), GAMKRELIDZE (I.P.), ZAKARIADZE (G.S.) and LORDKIPANIDZE (M.G.). - 1974. Adjaria-Trialeti trough and the problem of formation of the Black Sea deep basin, Geotektonika, no. 1, p. 78-94. (In Russian).

ADAMIA (SH.A.), ZAKARIADZE (G.S.) and LORDKIPANIDZE (M.G.). - In press. Evolution of the ancient mobile continental margin, illustrated by the Alpine development of the Caucasus.

BELOV (A.A.). - 1976. Boundary between Gondwana and Eurasia in the Caucasian segment of the Mediterranean foldbelt. In: Tectonics and structural geology. Planetology, IGC, XXV Session, Reports of Soviet Geologists, Moscow, Nauka, p. 83-90. (In Russian).

BELOV (A.A.) and OMELCHENKO (V.L.).- 1976. Ophiolites in the structure of the Marukh nappe and some questions of stratigraphy and magmatism of the Paleozoic in the Front Range, N.Caucasus, Geotektonika, no. 2, p. 44-56. (In Russian).

BRUNN (J.H.), ARGYRIADIS (I.), RICOU (L.E.), POISSON (A.), MARCOUX (J.) et GRACIANSKY (P.-CH. de). - 1976. Elements majeurs de liaison entre les Taurides et les Hellenides, Bull. Soc. géol. Fr., (7), t. XVIII, no. 2, p. 481-498.

DEWEY (J.F.), PITTMANN III (W.C.), RAYN (W.B.F.) and BONNIN (J.). - 1973. Plate tectonics and the evolution of the Alpine system, Bull. Geol. Soc. Am., v. 84.

DOBROVOLSKAYA (T.I.), LEBEDINSKY (V.L.), MAKAROV (N.N.) and YURK (YU.YU.). - 1970. Geochronology of the Crimea according to radiological data, Proceedings of XV Session of the Commission on absolute age determination of rocks. Moscow, Nauka. (In Russian).

ENAY (R.). - 1976. Faunes anatoliennes (Ammonitina, Jurassiques) et domaines biogéographiques nord et sud Téthysiens, Bull. Soc. géol. Fr., t. XVIII, no. 2, p. 533-541. (7).

GAMKRELIDZE (I.P.). - In press. Tectonic development of the Anatolian-Caucasian-Iranian segment of the Mediterranean belt.

KAYA (O.). - 1976. Northwest Anatolian belt, Bull. Soc. géol. Fr., (7), t. XVIII, no. 2, p.477.

KHAIN (V.E.). - 1975. Structure and main stages in the tectono-magmatic development of the Caucasus: an attempt at geodynamic interpretation, Am. J. Sci., v. 275-A, p. 131-156.

KNIPPER (A.L.). - 1975. Oceanic crust in the structure of the Alpine folded area, Trudy GIN AN SSSR, issue 267, 207 pp. (In Russian).

KNIPPER (A.L.) and SOKOLOV (S.D.). - 1976. Vedi ophiolites (Armenia): autochtonous or allochtonous?, Geotektonika, no. 4, p. 54-66. (In Russian).

KRASNOPEVZEVA (G.V.), REZANOV (I.A.) and ŠEVČENKO (V.I.). - 1976. Tiefbau und Evolution der Erdkruste im Kaukasus, Geol. Rdsch, Bd. 65, N 1, S. 1-16.

LOMIZE (M.G.). - 1971. Vedi nappe, the Lesser Caucasus, Doklady Akademii Nauk SSSR, v. 198, no. 3, p. 672-675. (In Russian).

MOSSAKOVSKY (A.A.). - 1972. Paleozoic orogenic volcanism of Eurasia, Geotektonika, no. 1, p. 6-28. (In Russian).

PEIVE (A.V.). - 1969. Oceanic crust of the geological past, Geotektonika, no. 4, p. 5-23. (In Russian).

POTAPENKO (YU.YU.) and PRUTSKY (N.I.). - 1976. Ophiolite conglomerate in the Middle Paleogene of the Front Range, N.Caucasus, Doklady Akademii Nauk SSSR, v. 228, no. 5, p. 1179-1181. (In Russian).

RICOU (L.E.), ARGYRIADIS (I.) et LEFEVRE (R.). - 1973. Proposition d'une origine interne pour les nappes d'Antalya et le massif d'Alanya (Taurides occidentales, Turquie), Bull. Soc. géol. Fr., (7), XVI, p. 107-111.

ŠEVČENKO (V.I.) and REZANOV (I.A.). - 1974. Geotektonische Geschichte der Krim, des Kaucasus und des Kopetdag, Z. geol. Wiss., 2, N 11, S. 1287-1307.

SEYED-EMAMI (K.) and SARABY (F.). - 1976. An outline of Mesozoic in Iran, 25th Int. Geol. Congr., Abstracts, v. 1, p. 316-317.

STÖCKLIN (J.). - 1974. Possible ancient continental margins in Iran. In: Geology of continental margins, N.Y., Springer.

ZARIDZE (G.M.). - 1970. On the basalt substratum of the oldest geosynclinal belt of the Caucasus. In: Problems of structure of the earth crust and upper mantle. Moscow, Nauka. (In Russian).

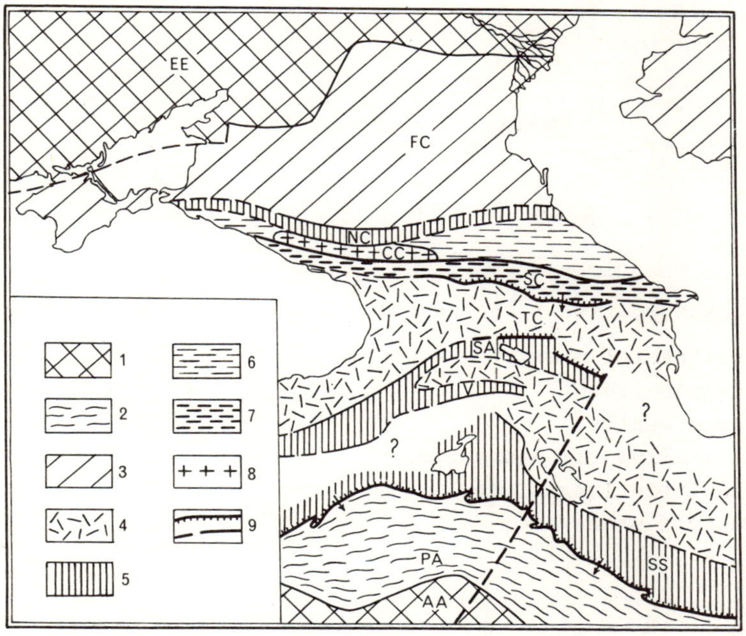

Fig. 1. Structural scheme of the Caucasus and adjacent areas.

1) Ancient cratons (EE, East-European; AA, African-Arabian).
2) Folded zone in the subsiding periphery of the African-Arabian craton (AA).
3) Young Ciscaucasian platform (FC).
4) Transcaucasian (TC) and Central-Iranian (CI) median masses (microcratons).
5) Ophiolite zones developed on oceanic crust (NC, North-Caucasian; SA, Sevan-Akera; V, Vedi; SS, Sanandadj-Sirdjan).
6-7) Marginal sea of the Greater Caucasus: 6, peripheral part; 7, the deepest part with oceanic or reduced continental crust (SC).
8) Crystalline nucleous of the Central Caucasus (CC).
9) Faults: a) fronts of nappes and overthrusts, b) other faults.

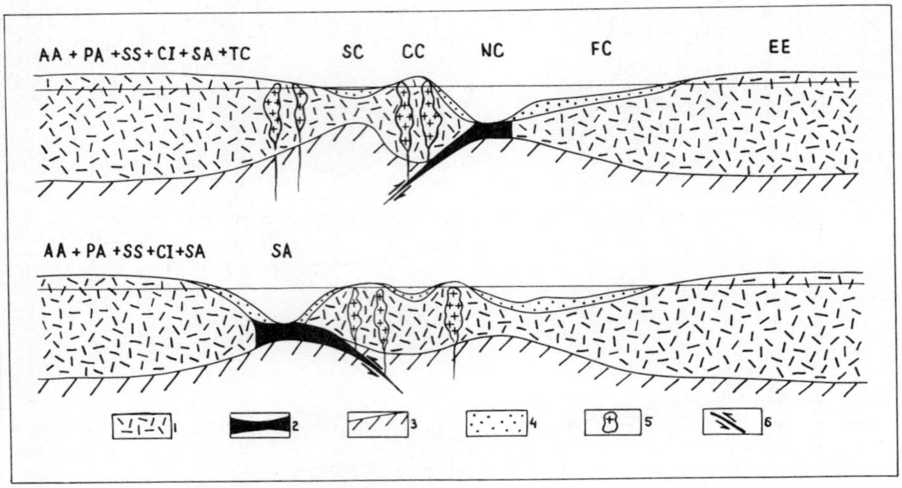

Fig. 2. Alternative paleotectonic profiles of the Late Paleozoic.

1) Continental crust.
2) Oceanic crust.
3) Upper mantle.
4) Epicontinental sediments.
5) Granitoid intrusions.
6) Benioff zones.

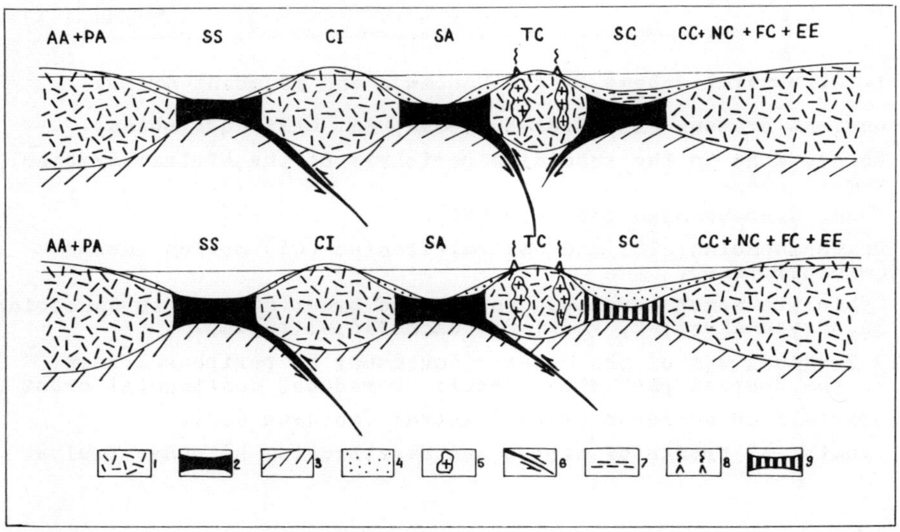

Fig. 3. Alternative paleotectonic profiles of the Middle Mesozoic.

1) Continental crust.
2) Oceanic crust.
3) Upper mantle.
4) Epicontinental sediments.
5) Granitoid intrusions.
6) Benioff zones.
7) Deep-sea sediments of marginal sea.
8) Volcanic arcs.
9) Reduced continental crust.

INTERNATIONAL SYMPOSIUM ON THE STRUCTURAL HISTORY OF THE MEDITERRANEAN BASINS. SPLIT (YUGOSLAVIA) 25-29 OCTOBER 1976. B. BIJU-DUVAL AND L. MONTADERT, Eds. EDITIONS TECHNIP, PARIS 1977, pp. 363-376

THE BLACK-SEA : A MARGINAL BASIN GEOPHYSICAL AND GEOLOGICAL DATA

J. LETOUZEY [1], B. BIJU-DUVAL [1], A. DORKEL [2], R. GONNARD [1], K. KRISTCHEV [3], L. MONTADERT [1] and O. SUNGURLU [4]

Résumé Les observations géologiques effectuées à terre montrent que le bassin océanique de Mer Noire pourrait avoir son origine dans la formation d'un "bassin marginal" (ou "Back Arc Basin"), créé entre le Lias et le Crétacé supérieur-Paléocène au nord de la marge pontique, en relation avec la disparition de l'Océan téthysien.

Les profils de sismique réflection à forte pénétration montrent :

1) La structure actuelle de la Mer Noire, divisée en deux sous-bassins séparés par une zone haute médiane;

2) La structure des marges, étroitement liée à l'évolution géologique des zones voisines;

3) La forte subsidence et le taux élevé de sédimentation, en particulier durant le Paléogène (probablement en partie liée au refroidissement de la croûte océanique) ainsi que l'importance de la sédimentation deltaïque durant le Plio-Quaternaire.

INTRODUCTION

Geophysical and geological data on the Black Sea and its margins show that the development of this basin is linked to the evolution of the Tethys as are the other Mediterranean Basins.

Geophysical reconnaissances surveys have determined the general features of the area (Garkalenko et al., 1971; Goncharov et al., 1972; Neprochnov et al., 1974-1975), and recently the Glomar Challenger has drilled several holes in the abyssal plain. Recent sedimentation and superficial structures observed by shallow-penetration seismic reflection have been described by Ross et al. (1974), but there is little information on the total sedimentary succession or the detailed structure of the margins.

In 1973 a multichannel seismic survey in the Black Sea (Flexichoc seismic source, 2 400 % stack) was carried out by French institutions.

In order to interpret the offshore data, IFP and CNEXO made some field studies in the Pontids and Thrace in collaboration with geologists from Turkish Petroleum, and in the Balkans with geologists of the Bulgarian Academy of Sciences. These surveys were used, together with sparker data from the Woods Hole Oceanographic Institution Atlantis cruise (Ross, 1973-1974), DSDP stratigraphic informations (leg 42 B), and results of oil drillings (Foose and Manheim, 1976) to produce a general scheme of evolution of the Black Sea (see also, Letouzey et al., 1977).

(1) Institut Français du Pétrole, 1-4 avenue de Bois-Préau, 92506 Rueil-Malmaison, France

(2) Société Nationale Elf Aquitaine (Production), 26 avenue des Lilas, 64001 Pau, France

(3) Institut Géologique de l'Académie Bulgare des Sciences, rue "36", Sofia-XIII, Bulgaria

(4) Turkiye Petrolleri A.O., P.O. Box 209, Bakanliklar, Ankara, Turkey

GENERAL FRAMEWORK

The Black Sea, like the Caspian Sea, represents a small marine depression, south of Europe, which has evolved more or less independently of the Mediterranean since the Neogene. Its size is about 450 thousand square kilometers. It is connected to the Mediterranean by the narrow Bosphorus and Dardanelles Straits via the Sea of Marmara. The whole northern part constitutes a broad shelf (up to 190 km west of the Crimea); in the northeast the Sea of Azov is only 10-14 m deep and communicates with the Black Sea Basin by the very shallow Kerg Straits. In opposition the southern and eastern shelves are very narrow (only 20 km) and the slope here is steep (up to 20° in some places) dissected by numerous canyons. The wide deep depression has a flat floor between 2 000 to 2 206 m (Ross, 1974). The Black Sea lies across several different structures between the old European craton and the Alpine folded belt on its northeast (Caucasus) and south (Pontides) margins (Fig. 1), appearing as a large depression filled by a thick undisturbed sedimentary series. During the Plio-Quaternary the sedimentation rate was especially high due to input by major rivers (Danube, Dniepr, etc.) (Fig. 2).

CRUSTAL STRUCTURE

Seismic refraction surveys (Neprochnov et al., 1975) have shown that the crustal structure under the abyssal plain is sub-oceanic; granitic crust is present only on the margins. Surface wave studies and gravimetric data (Balavadze et al., 1975) confirm the absence of continental crust in the abyssal plain. The Moho discontinuity rises up to 20 km below sea level towards the center of the basin. Nevertheless between Samsun (Turkey) and Kerg (Crimea) a basement high area appears to separate an eastern basin from a deeper and wider western one. Bouguer anomaly maps show two maxima under the two sub-basins.

Interpretation of magnetic data suggests that magnetic anomalies of the Pontids are related to the Cretaceous-Eocene volcanic rocks well known along the whole margin. The strong anomalies of the western basin do not correspond to high zones on seismic profiles; they could represent oceanic anomalies created by fields reversals during the formation of the basin; however characteristic linear trends are not observed :

- southeast of Crimea, positive anomalies could correspond to the basement highs observed on seismic lines;
- southwest of Caucasus a strong positive anomaly is located near a high zone defined by seismic refraction; it extends towards the Transcaucasian area where volcanic activity is known to have occurred during the Mesozoic.

Low heat flow values (Erikson, 1970; Lyubimova, 1973) in the abyssal plain are probably due to the thick sedimentary sequence and to the old age of formation of the basin, as confirmed by the absence of any recent volcanic activity. Some greater heat flow values south of Crimea and west of the Caucasus could be related to mud diapirism.

These data indicate that the Black Sea is a small oceanic basin (Fig. 6). It may be a relic of an old ocean or else a marginal sea developed behind an island arc, and analysis of geological and geophysical data enables a scheme to be suggested.

The structures of the different margins around the Black Sea are very varied. In the northwest the Scythian and Mysian platforms can be considered as stable since the Upper Jurassic. They are presently mainly marked by the Danube-Dniepr delta. The possible former relationships between Dobrudja, Crimea and Greater Caucasus are still subject to discussion. In the northeast the alpine Caucasus structures affect the margin of the basin. In the south a continuous fold belt runs from the Balkans to the Pontids and Lesser Caucasus. Our field observations in Turkish and Bulgarian areas together with important geological onland data now available allow us to study the southern part of the Black Sea and to discuss a model of formation of the basin.

NORTHWESTERN AREA

To the north of the Black Sea the Ukrainian Shield plunges southward beneath Mesozoic and Cenozoic formations forming the Scythian Platform and the floor of the deep Karkinit and Indol-Kuban Basins. These basins extend under the shelf, west and east of the Crimean Peninsula. More than 5 km of sediments fill the Karkinit Basin; they are largely undisturbed and thin towards a high area to the south which might well represent the westward extension of the Jurassic tectonized series known in the Crimean Mountains (Cimmerian orogenesis). The Dobrudja area which separates the Scythian and Moesian (or Mysian) platforms is also characterized by intense Cimmerian folding. But this whole area may be considered to have been a stable margin since the Cretaceous with a relatively thin Mesozoic-Cenozoic sedimentary cover.

On the margin between South Crimea and the Balkans, a part of the Mesozoic (Cretaceous) dips towards the basin, faulted and covered with thick Cenozoic deposits, deriving from the Danube and Dniepr. The subaerial Danube delta is relatively limited, but it is prolonged beneath the shelf and towards the abyssal plain (Fig. 2). During the Pliocene-Quaternary prograding sediments from the "pro-delta" extended widely into the basin, covering the entire western abyssal plain. In *DSDP* borehole 380 various sedimentary phenomena, and especially the "slump breccias", are probably related to these deltaic deposits. According to borehole dating, the thickness of the Plio-Quaternary section could be more than 2 500 m underneath the front of the Danube delta (Letouzey et al., 1977).

NORTHEASTERN AREA

The southern part of the Greater Caucasus folded belt lies between the Crimean Mountains and the Transcaucasian area along the northeast margin of the Black Sea.

Various onshore geological data have been published by Soviet geologists (Khain et al., 1963; Balavadze, 1968; Gamkrelidze, 1966; Khain, 1975; Lebedev et al., 1975). The tectonic development of the Greater Caucasus started before the Mesozoic. These data show that its more recent development is as follows :

1) During the Jurassic period the southern part was characterized by very thick clastic deposits and by important volcanism, while folding occurred in the Main Range as far as Crimea (Cimmerian folding).

2) Tectonic evolution continued during the Cretaceous and the Cenozoic, with flysch deposition, nappe systems and uplifts. Folds and overthrusts are mainly directed southwards in the direction of the Georgian basins (Transcaucasian area) and of the east Black Sea Basin. These structures could be related to a Cenozoic subduction zone, but they must be discussed with respect to the development of the southern area, Transcaucasia and Lesser Caucasus.

Shortening and deformation appear to be attenuated in the west where the Greater Caucasus plunges into the south of the Indol-Kuban Basin (Popovich, 1974); but deep structures are masked by extensive mud diapirism phenomena known in the Maikopian series (Oligocene).

The tectonic features offshore are difficult to make out on seismic records which are only single channel lines. South of Kerg (Crimea) some disturbance has been observed on seismic profiles. These are interpreted in terms of undercompaction phenomena and mud diapirism in comparison with the Maikopian argilokinesis. We have already stated that this could explain the abnormal heat flow values in this area.

Present seismicity (McKenzie, 1972; Riznichenko et al., 1975) attests that this northeast margin is still under compression.

EVOLUTION OF THE SOUTHERN MARGIN FROM THE BALKANS TO THE LESSER CAUCASUS

In the southwest and south margins of the Black Sea, the link between Balkan and Srednogorie zones and the Pontids is accepted by most authors (Boncev, 1951, 1975; Boccaletti et al., 1974; Brinkmann, 1974). These folded ranges were created during Upper Cretaceous and Eocene to the north of the zone where ophiolites and melanges were overthrust southwards ("ophiolitic suture zone"). Collapse of the margin and intense faulting (wrench North-Anatolian fault) then occurred in Miocene to recent times.

From land sections and geophysical profiling we will try to analyze the development of this margin :

From a general point of view the authors agree with the idea of a fundamental difference between the Mesozoic facies of the Pontids-Balkans and those of the Anatolids-Taurids (Hortinsk, 1971; Bergougnan, 1975; Fourquin, 1975). Close relationships with peri-European facies (Malm, Urgonian) have been emphasized, especially on the basis of the fauna. They enable us to distinguish a North Tethys province (south of Europe) from a South Tethys one (Enay, 1974).

An active margin may be deduced to have been in existence south of this European margin in which can be recognized superimposed tectonic trends, disconformities and volcanism.

In the Pontids detrital or volcano-detrital deposits of Middle Liassic age are unconformable on different series (Fig. 3) including basement, Trias and allochthonous units. North of Ankara these allochthonous units include Permian limestone thrust onto melanges or on sheets of "greenstones" (Fourquin, 1975). According to Kaya (1976) some of these greenstones can be interpreted as Paleozoic ophiolites.

In the Rhodopian massif some ultrabasic rocks are also recognized which may be Paleozoic oceanic fragments involved in the Hercynian orogeny or alternatively late Paleozoic-Triasic fragments which were tectonized in a late Triasic active margin. Such late Triasic tectonic activity is well known throughout Bulgaria, both in the Balkans and in the Moesian platform where Triassic deposits are also transgressive.

In the Pontids after a new tectonic event at the end of the Lias, the Dogger-Malm is transgressive on series as old as Paleozoic. Facies vary from limestones (Tithonian facies) to detrital or flysch deposits. Some olistostromes have been described (Sestini et al., 1968) near Sinop. Bellon and Letouzey (this volume) note the importance of the magmatic activity in this area from the Lias to the Upper Jurassic, especially developed in the Eastern part of the Pontids and in the Lesser Caucasus. Such volcanism is also known along the Stranja Ridge, but good chronological data are absent. In this area between Rhodope and the Balkans a great facies variation is known from flysch to Titonian limestone. As in the Pontids the thickness varies greatly. Some olistolites (Triassic to Lower Lias) are known in Dogger black shales in the Kotel area (Gocev, 1972).

In the northern part of the Pontids the sedimentation is continuous from the Malm to the Eocene (Fig. 3 B) with no important breaks, but in the southern part discontinuities are known between the Upper and Lower Cretaceous. This is also true in the Balkans where the Cenomanian is transgressive and also in the Moesian area. The exact importance of such a mid-Cretaceous tectonic phase is still uncertain (Mahel, 1975).

During the Upper Cretaceous the Pontids and the Srednogorie-Balkans are characterized by both flysch development and widespread volcanic activity, and there are great similarities from the Kraintides to the Adzhar-Trialet zone. The volcanism is mainly Campano-Maestrichian, but neither the age of the flysch nor the beginning of magmatism is synchronous throughout the whole area. The end of activity can be as late as the Eocene. This volcanism is described as "andesitic" but is probably more complex (see Bellon and Letouzey this volume). Volcanic rocks outcrop in different forms, *i.e.* flows, pillows, tuffs, agglomerates, conglomerates, tuffaceous sandstones, associated with flysch (Hajdvtov et al., 1974). Plutonism, granite to monzonite bodies are also described throughout the area. All these data, and especially this widespread upper Cretaceous magmatic activity along a linear zone south of the European craton and north of the Tethys suggest that this margin was a complex active margin during the whole Mesozoic Tethysian closure. In particular the volcanism could be connected with the formation of an "island arc-marginal basin" system on the sialic European margin. South of the Pontids the closure of the Tethys is marked by the obduction of ophiolites which were overthrust mainly southwards onto the continent of Anatolia during the late Cretaceous. This event marks the end of the Tethys Ocean and the beginning of continental collision (Biju-Duval et al., 1974, 1977, 1977). To the north of the ophiolitic suture zone, this tectonic event is not marked by strong deformation or folding and the volcanism continued after the collision.

During the Eocene, on the whole of the southern margin of the present Black Sea general open marine conditions existed towards the Black Sea. Clastic sedimentation including flysch can be seen easily in the Sinop area (fig. 3 B). On the present margin the basement seen on seismic section may correspond to Upper Cretaceous (Eocene ?) volcanoclastics, affected by folding and even overthrusts (Fig. 4). Such tectonics are well known inland where the overthrusts are mainly towards the north; the age of folding is generally admitted to be Upper Eocene, but some Oligocene-Early Miocene deposits are described near the margin. On the edges of the Stranja ridge and in Bulgaria (Srednogorie and Balkans) Upper Eocene folding is known but contemporaneity is not well established. Molassic undisturbed sediments of the Upper Eocene and Oligocene are known near Burgas and Varna. As confirmed by results of offshore drilling in the Kara Deniz Well and Igneada Well east of the Strandja Ridge (Foose and Mannheim, 1976), one of the main periods of subsidence of the Black Sea seems to correspond to the Upper Eocene-Oligocene period

(Figs. 4, 6). The deepest horizontal deposits seen in seismic sections under the abyssal plain probably correspond to such Eocene-Oligocene and Early Miocene deposits; but the deepest ones may include parts of Upper Cretaceous to Eocene series which are folded in the southern margin. The Cenozoic series generally overlap elevated areas on the margin of the western basin and the high zone separating the western and eastern basins.

Diapirism is observed under the abyssal plain west of the high zone and south of the Crimea (Fig. 4). It is considered to be due to mud-diapirism because of the low velocities of the flowing layer, the general slow increase in velocity in the total sedimentary column and because such phenomena are well known in the Maikopian series (Oligocene) in the Kerg Peninsula, south of Indol Kuban Basin (Letouzey et al., 1977).

The Strandja Ridge and Pontids have acted as a barrier between the Black Sea and other Mediterranean Basins since the Oligocene. The Kara Deniz and Igneada offshore wells have shown a large stretch of Miocene continental deposits, under the present margin. Marine Miocene deposits are unknown in the northern Pontids while transgression occurred in their southern part in Anatolia. In the subsided basins, under the present Abyssal Plain aqueous deposits of Pannonian (or Paratethys) type with temporary marine conditions are continuously deposited on the Oligo-Miocene. There may have been some relation to open marine conditions in the east in the direction of the Caspian (and Iran ?).

In seismic profiles in the Black Sea we have observed no thick Messinian salt layer showing halokinesis such as is present in abyssal plains in the Mediterranean. This is probably due to the very important influx of fresh water from the European craton.

The Plio-Pleistocene period probably corresponds to a new increase in the subsidence rate of the basins. Marine prograding series extend along the whole margin.

The present morphology of the Black Sea reflects fairly well the different mechanisms and types of recent sedimentation.

The narrow margins show steep prograding structures (Fig. 4) cut by sedimentary channels and ridges ([1]) that are affected by numerous sliding phenomena.

CONCLUSIONS

An active margin seems to have existed north of the present "tethysian ophiolitic suture zone". This margin would correspond to the southern edge of Europe and its activity would correspond to the northward subduction of the Tethys during the Mesozoic. It is characterized by intense magmatic activity since at least Liassic times, and by superimposed tectonics. Early Jurassic magmatism is well known from the Crimea-Caucasus to Lesser Caucasus-Pontids; it could be interpreted as the proof of a distensional area on the southern edge of the European plate, behind the subduction zone (Fig. 5).

The Upper Cretaceous is marked along the entire southern margin, by widespread volcanism, which could represent distensional magmatism corresponding to the opening of a marginal basin (or "Back Arc Basin") and island arc volcanism on the continental detached (Pontids) fragments moving to the south towards the Anatolian plate.

The oceanic Black Sea basins were thus definitely formed at that time, which corresponds to the final consumption of the Tethys marked by the obduction of ophiolites on the Anatolia plate.

During the Eocene-Oligocene, active subsidence of the basin may correlate to the cooling of the new oceanic crust which was accen-

([1]) On seismic sections, these channels and ridges can often be seen to migrate : the low or gently sloping edges of these channels move laterally at the same time as the deposit increases in thickness during sedimentation. On seismic lines this gives the illusion of an unconformity surface or erosion across the series. This is even further accentuated by the fact that the deposits inside the channel probably have a different lithology from that of the lateral deposits as shown by the seismic character.
This image which disturbs seismic correlations can often be seen either in the present or recent deposits on the edge of the abyssal plain (*e.g.* Fig. 4). This phenomenon is not caused by contour currents but by the discharge of the canyons into the abyssal plain. It might correspond to a period when rates of subsidence and sedimentation increased and is mainly observed in Plio-Quaternary series at the foot of the slopes of the Turkish margin.

tuated by the loading of thick sedimentary deposits.

Multichannel seismic profiles confirm that, after the continental collision stage at the end of the late Cretaceous, compression phases found onshore also affected the Black Sea margins, with overthrusting of the Pontids toward the north and of the Greater Caucasus toward the south. The seismic profiles show a burial of relief beneath the abyssal plain under a very thick sedimentary series. A high zone southeast of the Crimea is hidden under this overburden. No Messinian salt deposits are found in the Black Sea which contrasts with both the western and eastern Mediterranean. During this period the inflow of fresh water via the rivers must have brought about brackish water conditions in the Black Sea. According to observations on the margins, the rate of subsidence and sedimentation appear to have increased at the end of the Miocene and in the Pliocene-Quaternary. The tectonic phases affecting the margins do not seem to have appreciably disturbed the sediments on the abyssal plain. These sediments appear to have been disturbed only by mud diapirism phenomena.

In recent times, sediments from the prodelta prograde extensively toward the abyssal plain. The largest of these series is the Danube-Dnieper fan which covers the entire western basin. The thickness of the Plio-Quaternary (according to dating from the boreholes) could be more than 2 500 m underneath the front of the Danube delta.

Recent sedimentation rates, which are much higher in the Black Sea than in other Mediterranean basins, can easily be explained by the extensive catchment area of the Black Sea rivers, which is about five times greater than that of Mediterranean rivers (except for the Nile) as shown in Figure 2.

We would like thank *CNEXO*, *IFP*, *CFP*, *Elf/ERAP* and *SNPA* for permission to publish this paper.

We wish to thank J.P. Fail, R. Donatien and J. Cassand, geophysicists from *IFP* and the officers and crew of the "M.S. Florence" which made the seismic survey. We are grateful to J. Limond and B. Bailey for critical review and discussions.

BIBLIOGRAPHY

ADAMIYA (S.H.), GAMKRELIDZE (I.P.), ZAKARIDZE (G.S.), LORDKIPANIDZE (M.B.).- (1974). Adzhar-Trialet Trough and the Problem of the Black Sea Water Trough , Geotectonics, 1, 39.

ADAMIYA (S.H.).- (1975). Plate Tectonics and Evolution of the Alpine System : Discussion , Geol. Soc. of Am. Bull., 86, 719.

BALABANOV (V.F.).- (1975). Tectonics of the Pre-Black Sea Basin. Geotectonics, 9, 34.

BALAVADZE (B.K.).- (1968). Tectonics of the Black Sea and Azov area. Geotectonics, 4, 238.

BALAVADZE (B.K.), MINDELI (P.S.).- (1975). Methods of deep crustal Studies in the Black Sea Region. In : The Earth's Crust and the Development of the Black Sea Basin, Ed. Nauka, Moscou, p. 274.

BELLON (H.), LETOUZEY (J.).- (1977). Volcanism related to Plate-tectonics in the Western and Eastern Mediterranean (This volume).

BERGOUGNAN (H.).- (1975). Relations entre les édifices pontique et taurique dans le nord-est de l'Anatolie. Bull. Soc. Geol. Fr., (7), XVII, 1045.

BIJU-DUVAL (B.), LETOUZEY (J.), MONTADERT (L.), COURRIER (P.), MUGNIOT (J.F.) and SANCHO (J.).-(1974). Geology of the Mediterranean Sea Basins. In : The Geology of Continental Margins, Burk and Drake Ed., Springler-Verlag, p. 695.

BIJU-DUVAL (B.), GONNARD (R.).- (1975). Observations géologiques en Bulgarie sur la bordure de la Mer Noire et dans la chaîne du Balkan. Rapport IFP, n° 19893.

BIJU-DUVAL (B.), DERCOURT (J.), LE PICHON (X.).- (1976). La genèse de la Méditerranée. La Recherche, n° 71, 7, 811.

BIJU-DUVAL (B.), LETOUZEY (J.), MONTADERT (L.).- (1977). Structure and Evolution of the Mediterranean Basin. Rapport IFP, n° 24422. In : Initial reports of the Deep Sea Drilling Project, v. 42 A, Washington, D.C., U.S. Govt Printing Office (In press).

BOCCALETTI (M.), GOCEV (P.), MANETTI (P.).- (1974). Mesozoic isopic Zones in the Black Sea Region. Bull. Soc. Geol. It., 93, 547.

BONCEV (E.).- (1951). Tektonische probleme in Schwarzenmeergebiete. C.R. Acad. Bulg. Sci., 3, I, 25.

BONCEV (E.).- (1975). The Ponto-Caspian Plate and its tectonic Position. Geologica Balcanica, 5, I, 5.

BRINKMANN (R.).- (1974). Geologic Relations between Black Sea and Anatolia. In : The Black Sea; Geology, Chemistry, and Biology. Amer. Assoc. Petrol. Geol. Mem., 20, 63.

CANKOV (C.).- (1974). The Bulgarian Carpathian Balkan Area. In : Tectonics of the Carpathian Balkan Region, Geological Institute of Dionyz Stur, Bratislava, p. 303.

DOUST (H.) and ARIKAN (Y.).- (1974). The Geology of the Thrace Basin. In : Second Petroleum Congress of Turkey, H. Okay and E. Dileköz Ed., Ass. Turk. Petrol. Geol., Ankara, p. 119.

DRUMYA (A.V.).- (1962). On the Boundaries and Time of Formation of the western Part of the Black Sea Basin. Doklady Akad. Nauk SSSR, p. 145.

Deep Sea Drilling Project (Scientific Party).- (1975). Glomar Challenger drills the Black Sea (Leg 42 B). Geotimes, 20, n° 10, 18.

ENAY (R.).- (1976). Faunes anatoliennes (Ammonitina, Jurassique) et domaines biogéographiques nord et sud téthysiens. Bull. Soc. Geol. Fr., (7), XVIII, 533.

ERIKSON (A.J.).- (1970). The Measurement and Interpretation of Heath Flow in the Mediterranean and Black Sea. Ph. D. Thesis, Massachusetts Institute of Technology, 272 p.

FOOSE (R.M.) and MANHEIM (F.).- (1975). Geology of Bulgaria : a Review. Amer. Assoc. Petrol. Geol. Bull., 59, 303.

FOURQUIN (C.).- (1975). L'Anatolie du nord-ouest, marge méridionale du continent européen, histoire paléogéographique, tectonique et magmatique durant le Secondaire et le Tertiaire. Bull. Soc. Geol. Fr., (7), XVII, 1058.

GAMKRELIDZE (P.D.).-(1966).Main structural Features of the Caucasus. Geotektonika, 3, 3.

GARKALENKO (I.A.).- (1970). The deep-seated crustal Structure in the Western Part of the Black Sea and adjacent Areas : Seismic Reflection Measurement. Tectonophysics, 10, 539.

GARKALENKO (I.A.), MALOVITSKY (Ya.P.), NEPROCHNOV (U.P.).- (1971). Deep Structure of Western Black Sea (from DSS data). Acta Geol. Acad. Sc. Hungaricae, 15, 369.

GOCEV (P.G.).- (1972). On the Evolution of the Kimmerides on the Balkan Peninsula. Rev. Bulg. Geol. Soc., XXXIII, 1.

GONCHAROV (V.P.), NEPROCHNOV (U.P.), NEPROCHNOVA (A.F.).- (1972). Relief du fond et structure profonde de la dépression de la Mer Noire. Ed. Nauka, Moscou.

GOLOVINSKIJ (V.I.), GLUMOV (I.F.), GAGEL'GANTS (A.A.), DACHEV (K.I.), KAZAKOV (O.V.), YAKIMOV (Y.K.), LUTSUK (E.M.).- (1972). Quelques traits structuraux de la partie orientale de la dalle Moésienne et du plateau continental de la Mer Noire en République Populaire de Bulgarie (in Russian). Morskaya Geologiya i Geofizika, 3, 58.

GONNARD (R.), LETOUZEY (J.), DEVAUX (E.).- (1974). Observations géologiques sur les bordures turques de la Mer Noire, Anatolie Nord-Orientale et bassin de Thrace. Rapport IFP, n° 19172.

HORSTINK (J.).- (1971). The late Cretaceous and Tertiary geological Evolution of eastern Turkey. In : First Petrol Congress of Turkey, Kestin C. and Demirmem F. Ed., Ankara, Ass. Turkish Petrol. Geol., p. 25.

KAYA (O.).- (1976). Northwest Anatolian Belt. Bull. Soc. Geol. Fr., (7), XVIII, 477.

KHAIN (V.E.), MILANOVSKY (E.E.).- (1963). Structure tectonique du Caucase d'après les données modernes. Livre à la Mémoire du Pr P. Fallot. Mémoire hors-série, Soc. Géol. Fr., t. 2, 663.

KHAIN (V.E.).- (1975). Principal Stages of tectono-magmatic Development of the Caucasus. An experiment in geodynamical Interpretation. Geotectonic, 9, n° 1, 6.

KNIPPER (A.L.).- (1971). Constitution and Age of serpentinite Melange in the lesser Caucasus. Geotectonics, 5, 275.

LEBEDEV (T.S.), OROVETSKY (Y.P.).- (1975). The Mountainous Crimea. In : The Earth's Crust and the Development of the Black Sea Basin. Ed. Nauka, Moscou, p. 245.

LETOUZEY (J.), KHRISTCHEV (K.), GONNARD (R.), MONTADERT (L.), DORKEL (A.).- (1977). Black Sea : Geological Setting and recent Deposits Distribution from seismic Reflection data. In : Initial Report of DSDP, vol. XXXXII B (in Press).

LYUBIMOVA (Y.A.), SAVOSTIN (L.A.).-(1973).Teplovoi potok v tsentralnoi i vostochnoi chasti Chernogo Morya. Doklady Akademii Nauk SSSR, 212, n° 2, 349.

MAHEL (M.).- (1975). Carte tectonique du système Carpathes-Balkans. Inst. Geol. Dionys stur. UNESCO, Bratislava, 44 p.

MALOVITSKY (Y.P.), UGLOV (B.D.), OSIPOV (G.V.).- (1975). Geomagnetic Field Anomalies : The Black Sea Depression. In : The Earth's Crust and the Development of the Black Sea Basin. Ed. Nauka, Moscou, p. 184.

McKENZIE (D.P.).- (1972). Active tectonics of the Mediterranean Region. Geophys. Journ., 29, 300.

MURATOV (M.V.).- (1960). Tectonic Structures of the Alpine geosynclinal Area in Eastern Europe and Asia Minor and the History of their Development. 21th Intern. Geol. Congr., 18, 137.

MURATOV (M.V.).- (1972). History of the Development of the Deep-Water Trough of the Black Sea as compared with those of the Mediterranean. Geotectonics, 5, 269.

NEPROCHNOV (U.P.), NEPROCHNOVA (A.F.), MIRLIN (Ye.G.).- (1974). Deep Structure of Black Sea Basin. In : The Black Sea : Geology, Chemistry and Biology. Amer. Assoc. Petr. Geol. Mem., 20, 35.

NEPROCHNOV (U.P.), MALOVITSKY (Y.P.), BELOKUROV (V.S.), GARKALENKO (I.A.).- (1975). Crustal and upper mantle Structure of the Black Sea Region : Crustal cross Sections based on the Deep Seismic Sounding. In : The Earth's Crust and the Development of the Black Sea Basin. Ed. Nauka, Moscou, p. 284.

POPOVITCH (V.S.).- (1974). Structures of the Mesozoique Complex in the Kerch Peninsula. Geotectonics, 5, 312.

RIZNICHENKO (Y.P.), DRUMYA (A.V.), DZHIBLADE (E.A.).- (1975). Seismicity. In : The Earth's Crust and the Development of the Black Sea Basin. Ed. Nauka, Moscou.

ROSS (D.A.).- (1974). The Black Sea. In : The Geology of Continental Margins. Burk and Drake Ed., Springler-Verlag, p. 669.

ROSS (D.A.), DEGENS (E.T.).- (1974). Recent Sediments of Black Sea. In : The Black Sea; Geology, Chemistry, and Biology. Amer. Assoc. Petrol. Geol. Mem., 20, 183.

ROSS (D.A.), UCHUKI (E.), BOWIN (C.O.).- (1974). Shallow Structure of the Black Sea. In : The Black Sea : Geology, Chemistry, and Biology. Amer. Assoc. Petrol. Geol. Mem., 20, 11.

SESTINI (G.), CANUTI (P.).- (1968). Flysch Facies in the Pontic Mountains of Turkey. Bull. Soc. Geol. Ital., 87, 317.

SHIMKUS (K.M.), TRIMONIS (E.S.).- (1974). Modern Sedimentation in the Black Sea. Amer. Assoc. Petrol. Geol. Mem., 20, 279.

TEREKHOV (A.A.), MAMOSHINA (K.N.), MOSKALENKO (E.P.).- (1973). Continuation of Structures of the Northwest Caucasus in the Black Sea Basin. Geotectonics, 1, 54.

TSAGARELI (A.L.).- (1974). Geology of Western Caucasus. In : The Black Sea : Geology, Chemistry, and Biology. Amer. Assoc. Petrol. Geol. Mem., 20, 77.

ZENKOVITCH (V.P.).- (1969). Black Sea. In : The Encyclopedia of Oceanography, vol. 1, R.W. Fairbridge Ed., Reinhold Publ. Co., New-York, p. 145.

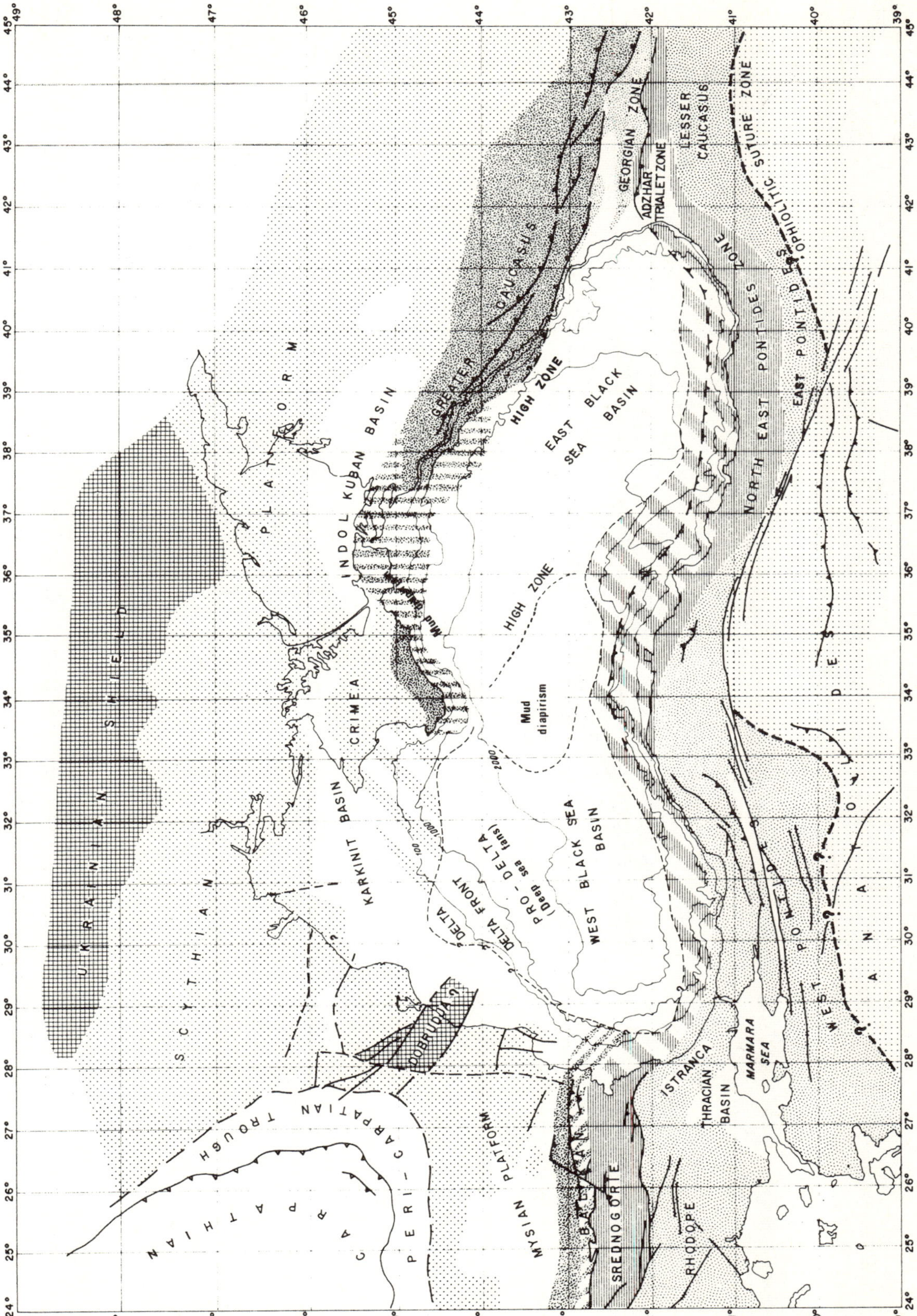

Figure 1 : Structural sketch map of the Black Sea region. Deep basins (> 5 km sediments) are unshaded.

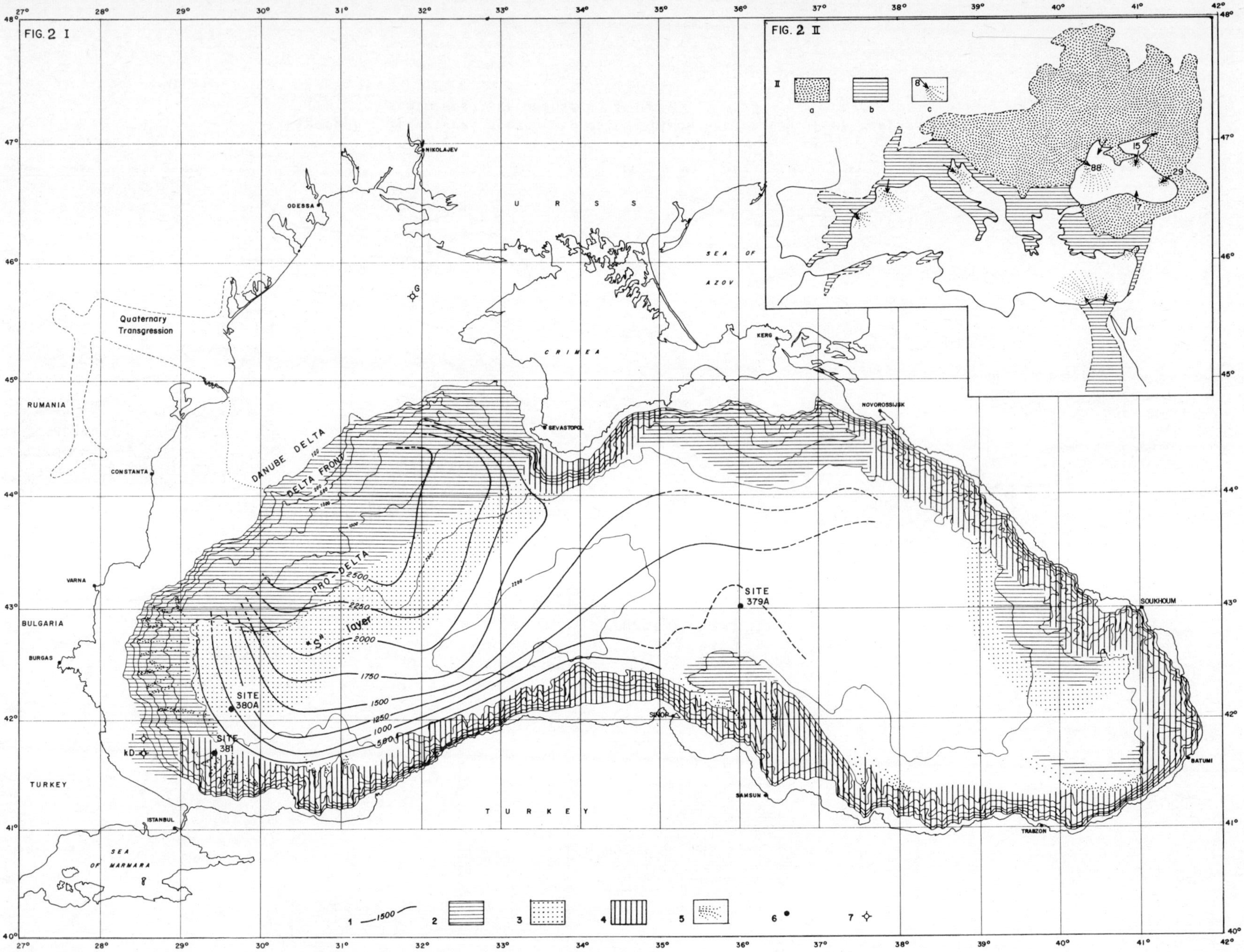

Figure 2 : I - Recent deposits distribution in the Black Sea : 1) - Thickness of Pliocene-Quaternary sediments (more recent than reflector k), contour interval 250 m ; 2) - Fans or gentle prograding formations ; 3) - Maximum extension of slump formations ; 4) - Steep marginal slope with channels, canyons and rises ; 5) - Main channels or canyons ; 6) - DSDP sites ; 7) - Oil boreholes.
II - (a) Black sea drainage basin ; (b) Mediterranean drainage basin ; (c) Solid decharge of rivers (106 t).

TECTONIC INTERPRETATIVE SECTION FROM STRANDJA TO MYSIAN PLATFORM

TECTONIC INTERPRETATIVE SECTION ACROSS THE PONTIDES

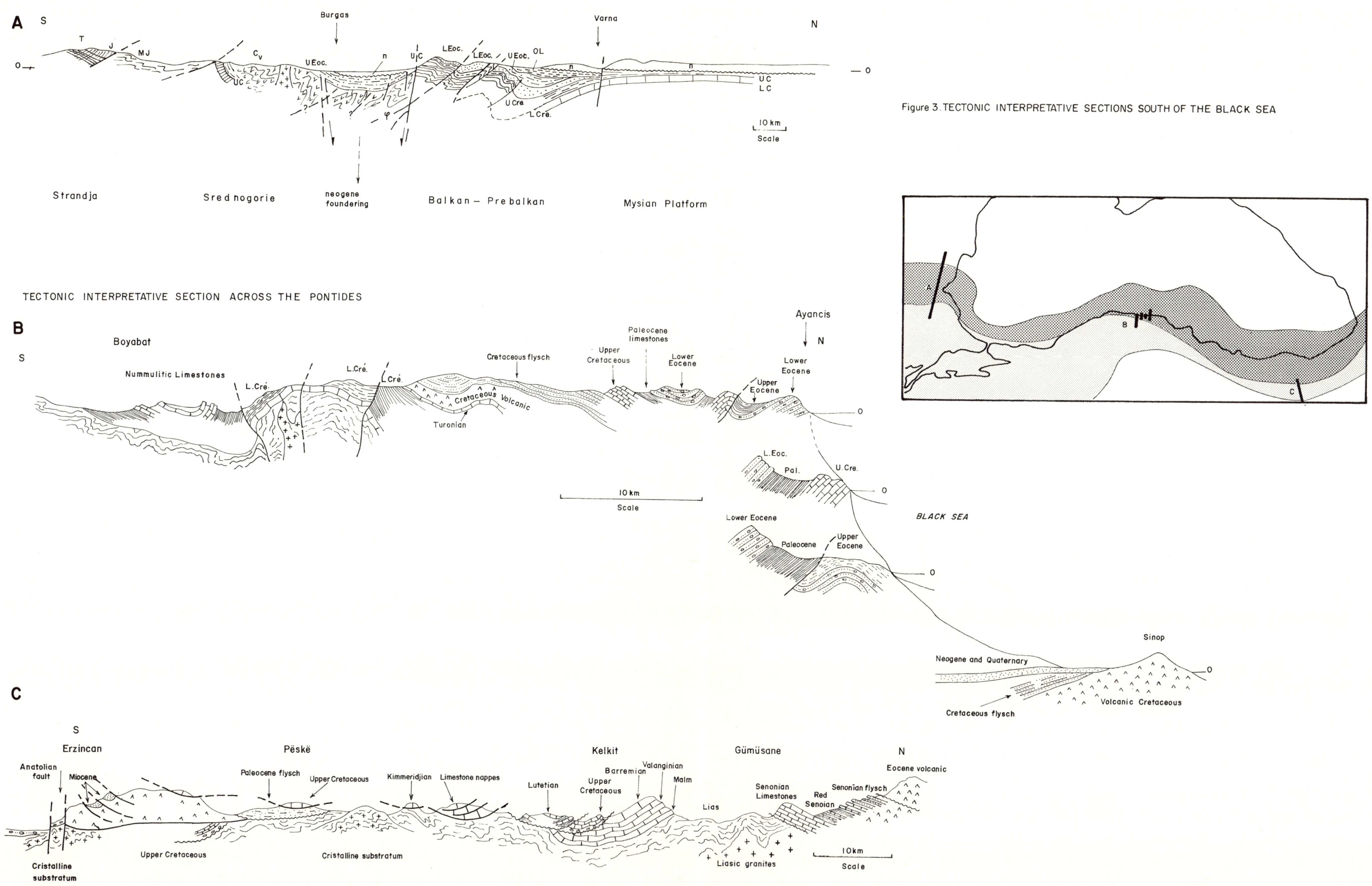

Figure 3. TECTONIC INTERPRETATIVE SECTIONS SOUTH OF THE BLACK SEA

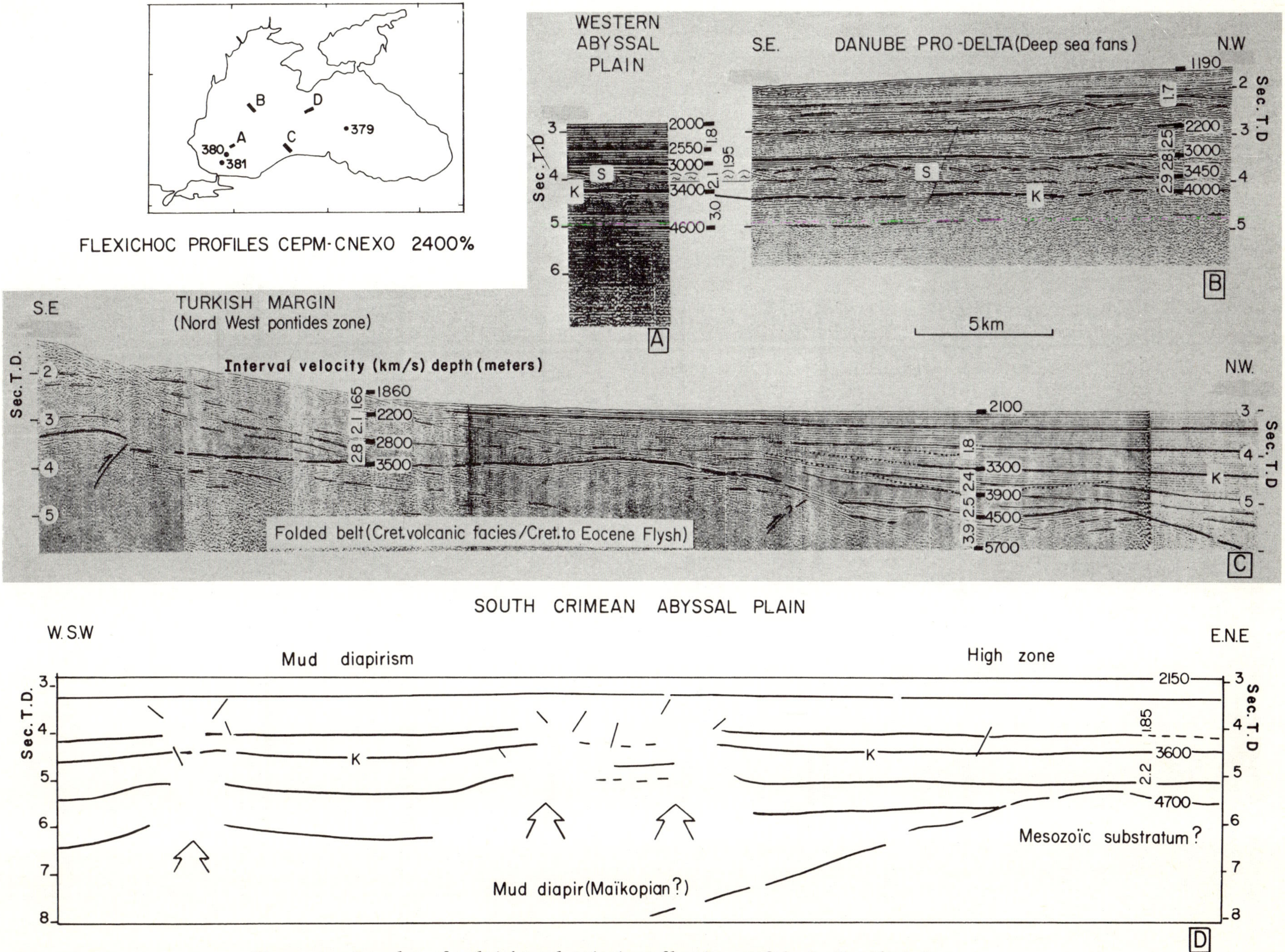

Figure 4 : Examples of multichannel seismic reflection profils in the Black Sea.

Figure 5 TENTATIVE GEOLOGICAL EVOLUTION OF THE BLACK SEA

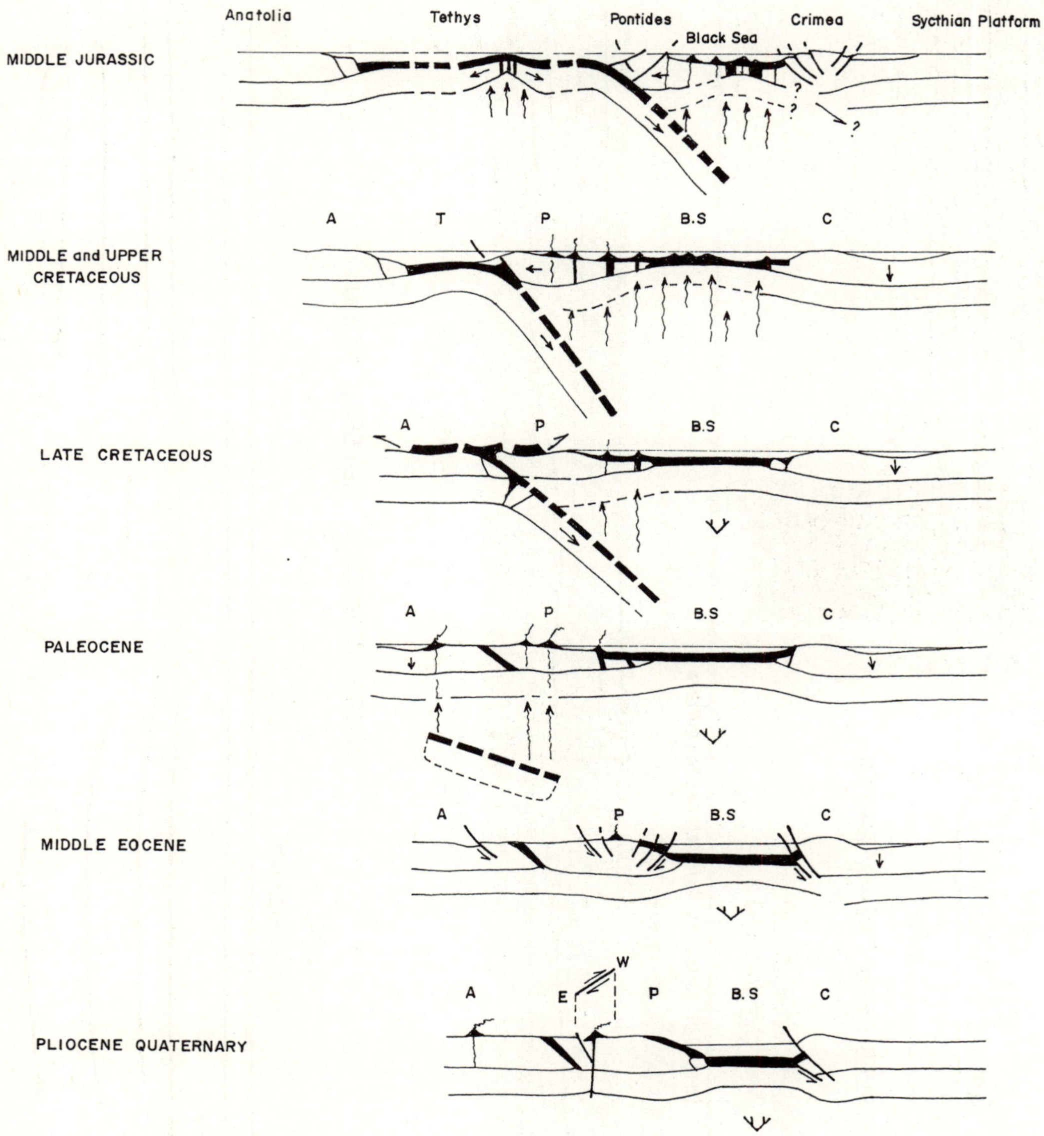

Figure 6 INTERPRETATIVE SECTION ACROSS THE BLACK SEA

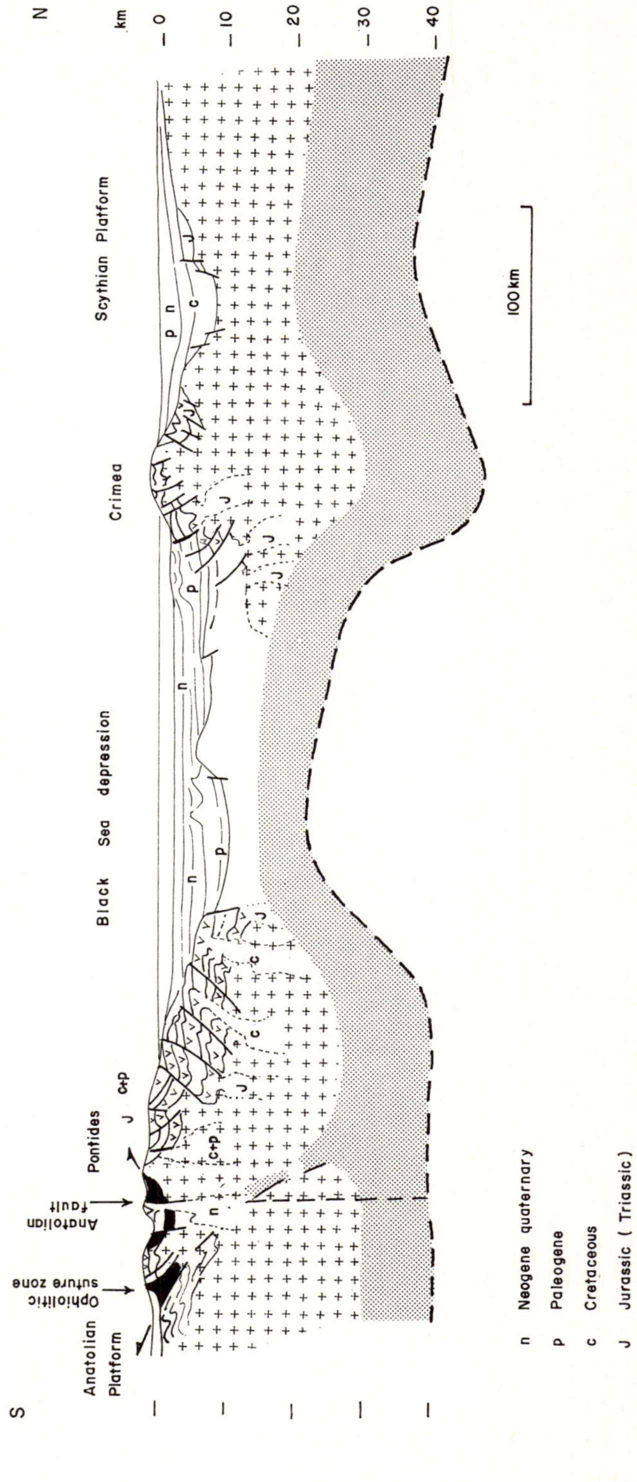

INTERNATIONAL SYMPOSIUM ON THE STRUCTURAL HISTORY OF THE MEDITERRANEAN BASINS. SPLIT (YUGOSLAVIA) 25-29 OCTOBER 1976. B. BIJU-DUVAL AND L. MONTADERT, Eds. EDITIONS TECHNIP, PARIS 1977, pp. 377-380

fourth part

the messinian event and recent sedimentation

EDITORS' COMMENTS

In addition to the active tectonic the recent history of the Mediterranean Basin is marked by a thick infilling of the basins, and especially by the well known Messinian Event which corresponded to widespread deposits of evaporites with a thick salt layer in the deepest parts of the basins. Its interpretation, discussed extensively after Leg 13 (Ryan, Hsü et al., 1973), was reinforced after Leg 42 A (Hsü, Montadert et al., 1977) and gave rise to numerous problems in sedimentology, stratigraphy and structural geology.

Due to the topic of the Symposium few detailed stratigraphic studies were presented during the Symposium. The paper by Zibrowius*[1] about deep fauna (Scleractiniaires) of Pliocene age again raised the question of paleobathymetry. A more general approach was the presentation by Bizon and Müller (in this volume) on biostratigraphic problems in the Mediterranean Neogene, with the presentation of correlations between nannoplankton and microforaminifer scales and an interpretation of paleotemperatures. Thaler* made interesting remarks on the geological significance of mammal fauna exchanges between Europe and Africa during the Miocene.

([1]) Names with the symbol* correspond to papers given during the Symposium, for which an abstract will be published in the "Rapports et Procès-Verbaux des Réunions" de la <u>Commission Internationale pour l'Exploration Scientifique de la mer Méditerranée</u> (<u>CIESM</u>) (in press).

As previously discussed (see the Special Issue, Messinian Events in the Mediterranean, Drooger Edit., 1973), agreement is not total about the conditions of deposition of the evaporites (Montenat et al., 1976 ; Fabricius et al., this volume), but some facts seem well established, i.e. existence of deep basins before the salinity crisis, deposition of the upper part of the evaporites in shallow water, extensive regression during deposition of the main salt layer marked by a strong erosion on all the margins, continued subsidence during the Pliocene-Quaternary (Hsü et al., abstract in this volume). Mülder and Parry (this volume), based on the results of a seismic survey, have presented data indicating, in their opinion, an early Messinian opening of the Strait of Gibraltar. This geophysical approach was completed by a presentation of seismic evidence of the Messinian Event by Montadert et al. in the whole Mediterranean (Montadert et al., 1977). Kracheninnikov and Tchoumakov discussed the different setting from shallow to deep conditions in the eastern Mediterranean. In this area Gvirtzman and Buchbinder (in this volume) have given new data about the Levantine margin which give support to the deep dessicated model, with some comparisons with the Mesopotamian Basin and the Red Sea where large evaporite deposits also exist.

Looking at the most recent history of the Mediterranean basins sedimentological studies of recent deposits clearly show that tectonic processes are still active in the Mediterranean (Stanley, Chamley et al., in this volume) in good agreement with the structural points of view discussed in Part Two of this book. Other numerous more regional papers presented during the Symposium provided new regional data about recent sedimentation and the geomorphology of different areas around the Mediterranean.

BIBLIOGRAPHY

DROOGER (C.W.) Ed.- (1973). Messinian Events in the Mediterranean. Amsterdam, London, North-Holland Publ., 272 p.

HSU (K.J.), MONTADERT (L.) et al. - (1977). Initial Reports of the Deep Sea Drilling Project, vol. XLII, Washington, D.C., US Government Printing Office, in press.

MONTADERT (L.), MAUFFRET (A.), LETOUZEY (J.).- (1977). Messinian Event : seismic contribution. In : Initial Reports of the Deep Sea Drilling Project, vol. XLII, Washington, D.C., op. cit.

MONTENAT (C.), BIZON (G.), BIZON (J.J.).- (1976). Continuité ou discontinuité de sédimentation marine mio-pliocène en Méditerranée occidentale. L'exemple du Bassin de Vera (Espagne méridionale). Rev. Inst. Franç. du Pétrole, XXXI-4, 613-663.

RYAN (W.B.F.), HSU (K.J.) et al.- (1973). Initial Reports of the Deep Sea Drilling Project, vol. XIII, Washington, D.C., US Government Printing Office, 1 447 p.

STANLEY (D.J.) Edit.- (1972). The Mediterranean Sea : a Natural Sedimentation Laboratory. Stroudsburg, Dowden, Hutchinson and Ross, 765 p.

INTERNATIONAL SYMPOSIUM ON THE STRUCTURAL HISTORY OF THE MEDITERRANEAN BASINS. SPLIT (YUGOSLAVIA) 25-29 OCTOBER 1976. B. BIJU-DUVAL AND L. MONTADERT, Eds. EDITIONS TECHNIP, PARIS 1977, pp. 381-390

REMARKS ON SOME BIOSTRATIGRAPHIC PROBLEMS IN THE MEDITERRANEAN NEOGENE

G. BIZON [(1)] and C. MULLER [(2)]

1. INTRODUCTION

Different zonations have been proposed for the Mediterranean Neogene; those based on planktonic foraminifera and calcareous nannofossils and/or combined reference scales of both microfossil groups appear to be the most suitable for timestratigraphic correlations. Nevertheless, we should keep in mind that ecological factors and dissolution may strongly affect the composition of the thanatocoenoses. This holds in particular for planktonic foraminifera; calcareous nannofossils are less sensitive in these respects and therefore their value for interregional correlations may be greater.

In this paper we will concentrate upon some major problems of Mediterranean Neogene bio- and chronostratigraphy; besides a tentative temperature curve, based on planktonic foraminifera and nannofossils will be presented. This is a preliminary report of our results and a first attempt of an interpretation. Our conclusions are based on the study of both microfossil groups from the same samples and on an evaluation of data from the literature. Furthermore a comparison is made with data from the Atlantic Ocean.

Figure 1 shows the location of the land sections studied and of the drilling sites of DSDP Leg 42 A and some of Leg 13. Our land sections from the westernmost part of the Mediterranean include sequences in Morroco, in the Vera Basin and at Alcoy, both in southeastern Spain, and outcrops on the Balears. The latter ones and the Alcoy section cover the upper part of the Burdigalian to Langhian or Serravallian (*Sphenolithus belemnos* Zone, NN3, to *Discoaster kugleri* Zone, NN7; N 5/N 6 to N 13 in terms of the planktonic foraminiferal standard zonation). The section of the Vera Basin, South East Spain, displays a probably continuous marine sequence ranging from the Tortonian into the Pliocene.

The Morrocan sections are situated in the eastern Rif basins (Melilla, Kert, Gareb, Boudinar, Guereif, Boured, Takar Souk, Taounate, Taza). In all the Morrocan sections the Upper Miocene (*Discoaster quinqueramus* Zone, NN 11, *Globorotalia acostaensis* Zone and *G. humerosa* Zone) is generally poor in nannofossils and planktonic foraminifera, while benthonics are abundant. Our Tortonian-Lower Pliocene data from Italy were collected in numerous sections, including stratotype localities (Rio Mazzapiedi, St. Agata Fossili, Rio Vindina, Ca Ciucco, Maccarone, Eraclea Minoa, Cap Rosello, Falconara and Capodarso.

Data from the eastern Mediterranean were derived from sections in Greece (island of Zakynthos), in the Turkish Antalya and Adana Basins and on Cyprus.

Another problem of our investigations is the Plio/Pleistocene boundary, which was studied in several sites of Leg 42 A and Leg 13 from the western and eastern Mediterranean, as

(1) Bureau d'Etudes Industrielles et de Coopération de l'Institut Français du Pétrole, 366 avenue Napoléon Bonaparte, 92500 Rueil-Malmaison, France

(2) Geologisch-Paläontologisches Institut, Johann Wolfgang Goethe Universität, Frankfurt a.M., Federal Republic of Germany, presently at Bureau d'Etudes Industrielles et de Coopération de l'Institut Français du Pétrole, Rueil-Malmaison, France

well as in land sections to test the value of the selected index fossils.

2. VALUE OF THE STANDARD ZONATIONS

The use of the foraminifera and nannoplankton standard zonations (Banner and Blow, 1965; Martini, 1971) in the Mediterranean, which are established in tropical regions, are more or less restricted to several intervals. This is due to the absence or scarcity of some index fossils (fig. 2 and 3). The often only sporadic occurrence of a species decreases considerably its value for a detailed zonation. Its presence or absence is strongly controlled by ecological factors and does not correspond necessarily with its total range.

Other difficulties are taxonomical problems and the uncertain classification of forms to one or the other species. Species of foraminifera as well as of nannofossils are often slightly different from the morphotypes as described from tropical areas, which is caused by different ecological conditions. The strong subdivision and description of numerous new species within the group of the foraminifera, which are probably often only varieties, makes the determination extremely complicated.

Furthermore the precise determination of zonal boundaries in the Mediterranean can be complicated by the high amount of reworking in several regions and stratigraphical intervals. This becomes very important if one uses the last occurrence of a species for the determination of zonal boundaries. Well-known examples for those difficulties are the type sections of the Langhian and Serravallian as well as Santa Maria di Catanzaro, the type section for the Plio/Pleistocene boundary.

However, figures 2 and 3 show that the limited value of the standard zonations is not restricted to the Mediterranean area but exists also in the Northeast Atlantic for the same stratigraphic intervals (the results from the Atlantic are mainly based on observations from Leg 48 in the Bay of Biscay and Rockall Bank, and on Leg 14 of the DSDP).

This fact indicates that worldwide events during the Neogene have to be kept in mind for an explanation.

3. INTERPRETATION OF THE PALEOTEMPERATURE CURVE

A Neogene paleotemperature curve for the Mediterranean area is constructed on the basis of calcareous nannofossils and planktonic foraminiferal data. Relatively cold periods are marked by a low species diversity, while higher temperatures were concluded from the occurrence of tropical and subtropical species and a high species diversity. In calcareous nannofossil assemblages cold and warm periods are inferred from amongst others, the number of discoasters : high numbers of these flora elements are supposed to be indicative of warm water, their repeated decrease or even disappearance in the Neogene successions would reflect lower temperatures.

In figure 4 the paleotemperature curve based on these fossil groups for the Mediterranean and a curve, based upon planktonic foraminifera only for the whole world, are plotted against the combined reference scale of both groups of organisms, as given by Ryan et al. (1974) and the Mediterranean plankton zonation of Bizon and Bizon (1972). Both paleoclimatic curves are in good agreement.

However, besides climatic changes, tectonic events, which caused a more and more isolated position of the Mediterranean during the Neogene must be taken into consideration.

Miocene.

Fossil assemblages of the lowermost Miocene (Aquitanian-lower part of the Burdigalian) are generally of lower species diversity, and the species are also often smaller and less evolved than in tropical areas, which may indicate lower water temperatures. This very distinct difference in size of the same species, present in the Oligocene and lowermost Miocene, was also observed in the Northeast Atlantic, where the lowermost Miocene is furthermore distinguished by a high level of the CCD and a higher production of siliceous microfossils. The absence or scarcity of some index fossils, known from the tropics like *Triquetrorhabdulus carinatus* and *Discoaster druggii* or *Globigerinatella insueta* may confirm this assumption. A lowering of the sealevel at this time, after a short transgressive phase during the Upper Oligocene, may be indicated by an unconformity in the lowermost Miocene. Such an unconformity was observed in many costal areas but does not exist in the deeper basins where a continuous sequence is represented from the Oligocene to Miocene.

However, the abundance of larger foraminifera and the vegetation indicate a warm climate at the beginning of the Miocene. Shallow water and/or the restricted connections with the open ocean, due to tectonic events, may be another explanation for the observed phenomena in the Mediterranean but not for the Northeast Atlantic. Temperatures are increasing from the Burdigalian (NN 3-N 5/6) to the Langhian (NN 5-upper part N 8 to N 10). This warming is connected with a widespread transgression in the Mediterranean as well as in Northern Europe (Martini and Müller, 1973). This Burdigalian transgression with slight differences in age (nannoplankton zone NN 3 ou NN 4) can be observed in many parts of the Mediterranean. The interval is marked by the occurrence of tropical species. The floral and faunal assemblages of the Atlantic, Mediterranean, Central Paratethys and Indo-Pacific are extremely uniform, indicating that during this time the Mediterranean was acting as a seaway connecting the Atlantic and Indo-Pacific.

A connection between the Mediterranean and the Central Paratethys through Northern Italy and Yugoslavia can be assumed for the Lower Serravallian (*Discoaster exilis* Zone, NN 6-*Globorotalia mayeri* Zone, N 11 to lower part of N 12) as suggested by the Mediterranean affinity of planktonic and benthonic foraminifera and molluscs of the Paratethys as well as by the occurrence of Atlantic-Mediterranean deep-water ostracods in Yugoslavia (Jiricek, 1974).

The elevation and emergence of the African and Arabian plate resulted probably in a separation of the Mediterranean from the Indo-Pacific at about 14 m.y. (Buchbinder and Gvirtzmann, 1976) i.e. within the Serravallian, approximately at the boundary of the *Discoaster exilis/Discoaster kugleri* Zone (NN 6/NN 7) and within the *Globorotalia mayeri* Zone (N 12) respectively. As a consequence of this separation the Mediterranean became an Atlantic marginal basin (Meulenkamp, 1975). The floral and faunal assemblages have to be compared with those of the Northeast Atlantic since this time.

However, a short influx from the Indo-Pacific into the eastern Mediterranean during the Middle Tortonian, within the nannoplankton zone NN 10 (*Discoaster calcaris* Zone, which corresponds to the foraminiferal zone N 16, *Globorotalia acostaensis* Zone) cannot be excluded with certainty.

Radiolarian, diatom and nannofossil assemblages from the Upper Badenian of the Paratethys, corresponding to the Upper Serravallian (*Discoaster kugleri* Zone, NN 7-upper part of the *Globorotalia mayeri* Zone, N 12), show still influences from the Indo-Pacific Ocean (Rogl et al., 1977), while at least the nannoplankton assemblages from the Mediterranean are comparable with those from the temperate zone of the Atlantic. The comparison of the foraminiferal assemblages of both regions is difficult because of the strong dissolution in the Atlantic at this time. Discoasters are indicative of warm water, strong fluctuations in their relative frequency suggest climatic fluctuations in Serravallian time.

The absence of tropical foraminifera and nannofossils in the Mediterranean since the upper part of the Serravallian, which is most conspicuous in the uppermost Serravallian and Lower Tortonian (*Catinaster coalitus-Discoaster hamatus* Zone, NN 8-NN 9, respectively upper part of the *Globorotalia mayeri-Globorotalia menardii* Zone, N 13-N 15) is consistent with observations in the Northeast Atlantic. Here, the Late Serravallian-Early Tortonian time-span is also characterized by a very high level of the CCD, which caused a very strong dissolution.

These observations indicate an important cooling at about 13 to 11 m.y (NN 7 to NN 9, respectively N 13 to N 15) with the maximum at about 12 to 11 m.y. This conclusion is supported by the results of palynological investigations by Benda (1973), who reported : "The most remarkable climatic changes to a drier and cooler climate must have taken place before the Tortonian or in the transitional interval between the Serravallian and Tortonian". A very distinct cooling at this time is also described from California from the Mohnian Formation at 12 m.y. (Casey, 1971; Casey et al., 1972).

A slight warming, probably connected with a short transgressive phase can be observed in the middle part of the Tortonian (*Discoaster calcaris* Zone, NN 10-*Globorotalia acostaensis* Zone, N 16). This interval is marked by the occurrence of species which seem to be less sensitive for lower temperatures like *Discoaster calcaris* and *Globorotalia acostaensis*. These species are also present throughout the Upper Miocene in the temperate zone of the Atlantic.

According to the foraminiferal and nannoplankton assemblages this slight warming is again followed by a cooler period during the

Upper Tortonian and Messinian. However, this cooling is less distinct than during the transitional interval between the Serravallian and Tortonian. This may also be confirmed by less dissolution in the Northeast Atlantic.

Nannofossil and foraminiferal assemblages of the Upper Miocene from the Mediterranean are comparable with those from the temperate zone of the Atlantic. *Discoaster quinqueramus* and ceratoliths are rare or missing. Therefore the suggested subdivision of the Upper Miocene-Lower Pliocene interval by using the ceratoliths (Rio et al., 1976) seems to be not very useful in the Mediterranean.

The repetition of marine incursions in the Mediterranean during the deposition of the upper evaporitic sequence of the Messinian may indicate climatic fluctuations. Fluctuations of the CCD and very distinct variations in the abundance of discoasters are typical for this interval in the Northeast Atlantic. Vertebrates, plants and clay minerals, however, indicate a warm humid climate with minor fluctuations. This contradiction probably can be explained by the impossibility for tropical marine species to enter into the Mediterranean, which in turn may be explained by their absence at corresponding latitudes of the Atlantic. However, regarding the faunal and floral assemblages of the Messinian the abnormal ecological conditions, which are an additional reason for the lower species diversity have to be kept in mind.

It can be assumed that the salinity crisis was not primarily caused by an abrupt climatic change at the beginning of the Messinian. So it seems very unlikely that the sea-level lowering has isolated the Mediterranean basin, until tectonic events had previously reduced the openings to relatively restricted size. If there was a complete separation of the Mediterranean from the Atlantic during the Messinian is still being discussed. The absence of the evolutionary lineages of keeled globorotalias close to the Mio/Pliocene boundary in the Mediterranean used by Cita (1976) to prove the complete separation and dessication of the Mediterranean is not obvious. These lineages were described from the tropical Atlantic but they were not observed in temperate areas, which have to be compared with the Mediterranean since the upper part of the Serravallian. It is known that those evolutionary processes are affected by ecological and climatic factors, which make it very unlikely to find them in the Mediterranean even in continuous marine sequences. On the other hand the evolutionary lineage from *Globorotalia miozea* to *Globorotalia cibaoensis* within the uppermost Messinian is known from the temperate zones of the Atlantic. *Globorotalia cibaoensis* is also known from the Messinian of the Vera Basin and in the type section of the Andalusian (Berggren and Haq, 1976).

Pliocene-Pleistocene.

The lower Pliocene with slight fluctuations indicated by the decrease in number of scyphospheres, ceratoliths and discoasters mainly within the nannoplankton zones NN 13/ NN 14 (*Ceratolithus rogosus* Zone/*Discoaster asymmetricus* Zone) is generally a period of slightly warmer water temperatures. Different water temperatures in the western and eastern Mediterranean may be indicated by larger and better evolved marine fossils in the eastern Mediterranean.

The beginning of the glaciation in the northern hemisphere is dated at about 3 m.y. and a most distinct faunal change in the southern oceans at about 2.5 m.y. (Hays and Opdyke, 1972; Theyer, 1973). This is in a very good agreement with the scarcity or absence of discoasters from the upper part of the *Discoaster surculus* Zone (NN 16) to the top of the *Discoaster brouweri* Zone (NN 18) in the western Mediterranean and Northeast Atlantic (Müller, 1977).

A short warming at the base of the Pleistocene may be inferred from the disappearance of the cool-water taxon *Coccolithus pelagicus* in the western Mediterranean for a short time (Müller, 1977).

Different water temperatures and the reduced influence from the Atlantic are the reason for different faunal and floral assemblages in the western and eastern Mediterranean since Upper Pliocene to Recent (Müller, 1977; Bizon and Müller, 1977).

4. STRATIGRAPHIC DISCUSSION

Many new stratigraphic data have been collected during the last years. However, some main problems like the Oligocene/Miocene, the Miocene/Pliocene, and the Pliocene/Pleistocene boundaries still have to be solved. These problems have been already discussed in the report of the R.C.M.N.S. working group of micropaleontology (Meulenkamp, 1975).

Oligocene/Miocene boundary.

Investigations of foraminifera and nannofossils assemblages of the same samples from Escornebéou, piston cores from the Basque shelf, Mallorca, Cyprus and Algeria give an Early Miocene age (Aquitanian, N 4) based on foraminifera, while the nannofossils still indicate a Late Oligocene age of the nannoplankton zone NP 25 (*Sphenolithus ciperoensis* Zone, fig. 5).

This observation confirms the assumption that *Globigerinoides primordius*, which is generally used to determine the Oligocene/Miocene boundary, occurs already in the Upper Oligocene based on nannofossils.

According to these results the previously assumed Early Miocene transgression in several parts of the Mediterranean has to be considered older, i.e. Late Oligocene.

Miocene/Pliocene boundary.

The Mediterranean area is not favourable to study the Mio/Pliocene boundary, due to the abrupt environmental change at this time. The *Sphaeroidinellopsis acme* Zone (MPL 1, Cita, 1975) at the base of the Pliocene seems to be only an ecozone below or within the range of *Globorotalia margaritae*. It was recognized at Site 376 above the evaporites, but here *Globorotalia margaritae* is not present. Onshore it is known from the Ionian Islands, South Italy, Sicily and Africa. It was not observed in the western part of the Mediterranean area (Sites 371 and 372, Morroco, Spain). However, the absence of this zone at the base of the Pliocene does not necessarily indicate an unconformity as supposed by several authors.

Pliocene/Pleistocene boundary.

The determination of the Pliocene/Pleistocene boundary still causes considerable problems. Contradicting interpretations of results obtained from the type section at Santa Maria di Catanzaro have shown that this sequence is not very useful for the definition of this boundary, which is, amongst others, due to the high amount of reworked species.

The Plio/Pleistocene boundary in deep-sea cores is determined by the evolutionary appearance of *Globorotalia truncatulinoides* from *Globorotalia tosaensis* and the extinction of *Discoaster brouweri*. Studies in different regions have shown that both biostratigraphic events are controlled by ecological factors and cannot be used in all areas.

Also the results from Legs 13 and 42 A show the limited value of these fossils for the precise determination of this boundary in the Mediterranean. Mainly in the western Mediterranean discoasters are rare or missing in the uppermost Pliocene (upper part of NN 16 to NN 18), due to low water temperatures (Müller, 1977). Also the first occurrence of *Globorotalia truncatulinoides* is very doubtful. Therefore the extinctions of *Cyclococcolithus macintyrei* and *Globigerinoides obliquus* were used to determine this boundary (Bizon and Müller, 1977). The extinction of both species takes place at the same time in the western as well as in the eastern Mediterranean (fig. 6).

Also investigations of land sections (e.g. the section below the Citadelle of Zakynthos, Bizon and Müller, in preparation) has proved the value of these both species. The first occurrence of *Hyalineabalthica* was used in this section for the determination of the Plio/Pleistocene boundary (Bizon, unpublished).

Same difficulties for the determination of the Plio/Pleistocene boundary are known from the Northeast Atlantic. Therefore also during Leg 48 the extinction of *Cyclococcolithus macintyrei* was used (Müller, in preparation).

Paleomagnetic measurements carried out on board of the Glomar Challenger (Site 400) by Dr Hailwood, Southampton, indicate that the extinction of this species is near the top of the Olduvai event, which is in good agreement with the extinction of *Discoaster brouweri* as calibrated in several areas. It confirms the value of *Cyclococcolithus macintyrei* for the determination of the Plio/Pleistocene boundary in regions where discoasters are rare or missing.

Acknowledgements The authors wish to thank Prof. Dr J.E. Meulenkamp, Utrecht, and Dr. J.V. Hinte and Dr. L. Smith, Bègles, for reviewing the written manuscript and their helpful advises for corrections.

REFERENCES

BANNER (F.T.) and BLOW (W.H.).- 1965. Progress in the Planktonic Foraminiferal Biostratigraphy of the Neogene. *Nature*, v. 208, n° 5016, pp. 1164-1166.

BENDA (L.).- 1973. Late Miocene sporomorph assemblages from the Mediterranean and their paleoclimatological implication, in : Messinian events in the Mediterranean, *Geodynamics sci.*, Rep. 7, pp. 256-259.

BERGGREN (W.A.) and HAQ (B.U.).- 1976. The Andalusian stage (Late Miocene) : biostratigraphy, biochronology and paleoecology. *Palaeogeography, Palaeoclimatology, Palaeoecology*, 20 (1976), pp. 67-129.

BIZON (G.) and BIZON (J.J.).- 1972. Atlas des principaux foraminifères planctoniques du bassin méditerranéen, Oligocène à Quaternaire, Paris, *Edition Technip*, pp. 316.

BIZON (G.) and MULLER (C.).- 1977. Remarks on the determination of the Plio-Pleistocene boundary in the Mediterranean, in : HSU (K.J.), MONTADERT (L.) et al., 1977, *Initial Reports of the Deep Sea Drilling Project*, vol. 42 A, Washington (U.S. Government Printing Office), in press.

BUCHBINDER (B.) and GVIRTZMAN (G.).- 1976. The breakup of the Tethys Ocean into the Mediterranean Sea, the Red Sea, and the Mesopotanian Basin during the Miocene : a sequence of fault movements and dessication events. *Abstr. 1st. Congr. Pacific Neog. Stratigr.*, pp. 32-35.

CASEY (R.E.).- 1971. Neogene radiolarian biostratigraphy and paleotemperatures, southern California, the Experimental Mohole, Antarctic Core E 14-8. *Palaeogeography, Palaeoclimatology, Palaeoecology*, v. 12, pp. 115-130.

CASEY (R.E.), PRICE (A.B.) and SWIFT (C.A.).- 1972. Radiolarian definition and paleoecology of the Late Miocene to Early Pliocene in southern California, in : Stinemeyer (E. H.) and Church (C.C.) (Eds.), *Soc. Economic Paleontologists and Mineralogists Pacific Coast Miocene Biostratigraphic Symposium Proc.*, pp. 226-238.

CITA (M.B.).- 1975. Planktonic foraminiferal biozonation of the Mediterranean Pliocene deep sea record. A revision. *Riv. Ital. Paleont.*, v. 81 (4), pp. 527-544.

CITA (M.B.).- 1976. Biodynamic effects of the Messinian salinity crisis on the evolution of planktonic foraminifera in the Mediterranean. *Palaeogeography, Palaeoclimatology, Palaeoecology*, 20 (1976), pp. 23-42.

HAYS (J.D.) and OPDYKE (N.D.).- 1967. Antarctic Radiolaria, magnetic reversals and climatic change. *Science*, v. 158, pp. 1001-1011.

JIRICEK (R.).- 1975. Biozonen der Zentralen Paratethys. *NAFTA*, Gbely (CSSR), pp. 20.

MARTINI (E.).- 1971. Standard Tertiary and Quaternary calcareous nannoplankton zonation. *Proc. IInd. Plankt. Conf. Roma*, 1970, v. 2, pp. 739-785.

MARTINI and MULLER (C.).- 1973. Nannoplankton-Gemeinschaften im Miozän und Pliozän des Nordseebeckens. *N. Jb. Geol. Paläont. Mh.*, 9, pp. 555-564.

MEULENKAMP (J.E.).- 1975. Report of the Working Group on Micropaleontology, in : *Report on activity of the R.C.M.N.S. Working Groups* (1971-1975), Bratislava, 1975, pp. 9-29.

MONTENAT (C.), BIZON (G.) and BIZON (J.J.).- 1976. Continuité ou discontinuité de sédimentation marine Mio-Pliocène en Méditerranée occidentale. L'exemple du Bassin de Vera (Espagne méridionale). *Rev. Inst. Franç. du Pétrole*, XXXI-4, pp. 613-663.

MULLER (C.).- 1977. Neogene calcareous nannofossils from the Mediterranean - Leg 42 A of the Deep Sea Drilling Project, in : HSU (K.J.), MONTADERT (L.) et al., 1977, *Initial Reports of the Deep Sea Drilling Project*, vol. 42 A, Washington (*U.S. Government Printing Office*), in press.

RIO (D.), MAZZEI (R.) and PALMIERI (G.).- 1976. The stratigraphic position of the Mediterranean Upper Miocene evaporites, based on nannofossils. Universita degli Studi di Parma, Istituto di Geologia, pp. 1-12.

ROGL (F.), STEININGER (F.F.) and MULLER (C.).- 1977. Middle Miocene Salinity Crisis and Paleogeography of the Paratethys (Middle and Eastern Europe), in : HSU (K.J.), MONTADERT (L.) et al., 1977, *Initial Reports of the Deep Sea Drilling Project*, vol. 42 A, Washington (*U.S. Government Printing Office*), in press.

RYAN (W.B.F.), CITA (M.B.), DREYFUS-RAWSON (M.), BURCKLE (L.H.) and SAITO (T.).- 1974. A paleomagnetic assignment of Neogene stage boundaries and the development of isochronous datum planes between the Mediterranean, the Pacific and Indian Oceans in order to investigate the reponse of the world ocean to the Mediterranean "Salinity Crisis" : a new approach. *Riv. Ital. Paleont. Stratigr.*, 80 (4), pp. 631-688.

THEYER (F.).- 1973. Globorotalia truncatulinoides datum plane; evidence for a Gauss (Pliocene) age in subantarctic cores. *Nature*, v. 241, pp. 142-145.

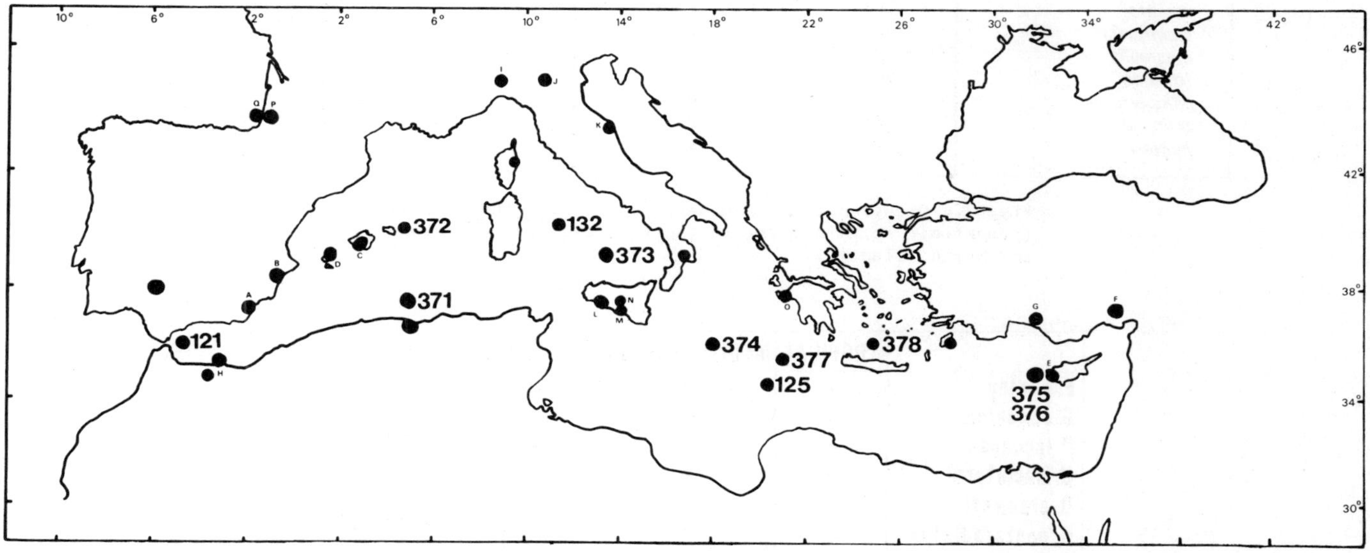

Fig. 1.- Location of the studied drilling sites of Legs 42 A and 13 and of the land sections.
A-Vera Basin, B-Alcoy section, C-Mallorca, D-Ibiza, E-Cyprus, F-Andana Basin, G-Antalya Basin, H-Eastern Rif basins Morroco, I-St. Agata Fossili and Rio Mazzapiedi, J-Rio Vindina, K-Ca Ciucco and Maccarone, L-Eraclea Minoa and Cap Rossello, M-Falconara, N-Capodarso.

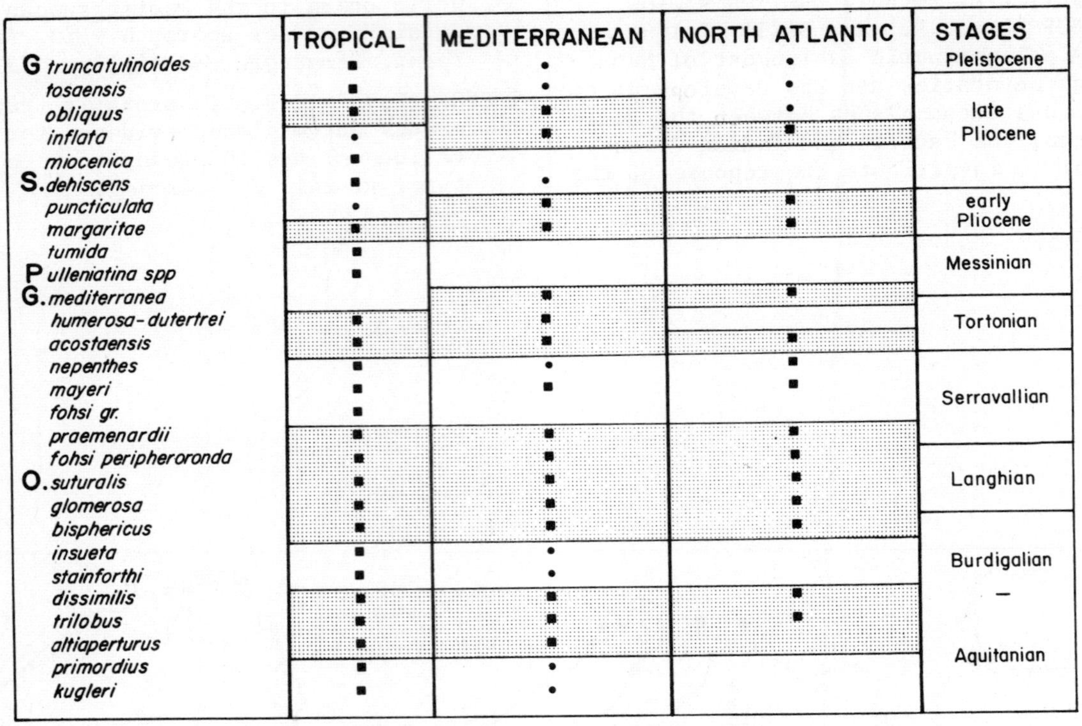

Fig. 2.- Occurrence and abundance of Neogene index fossils (planktonic foraminifers) in the tropics, Mediterranean and North Atlantic.

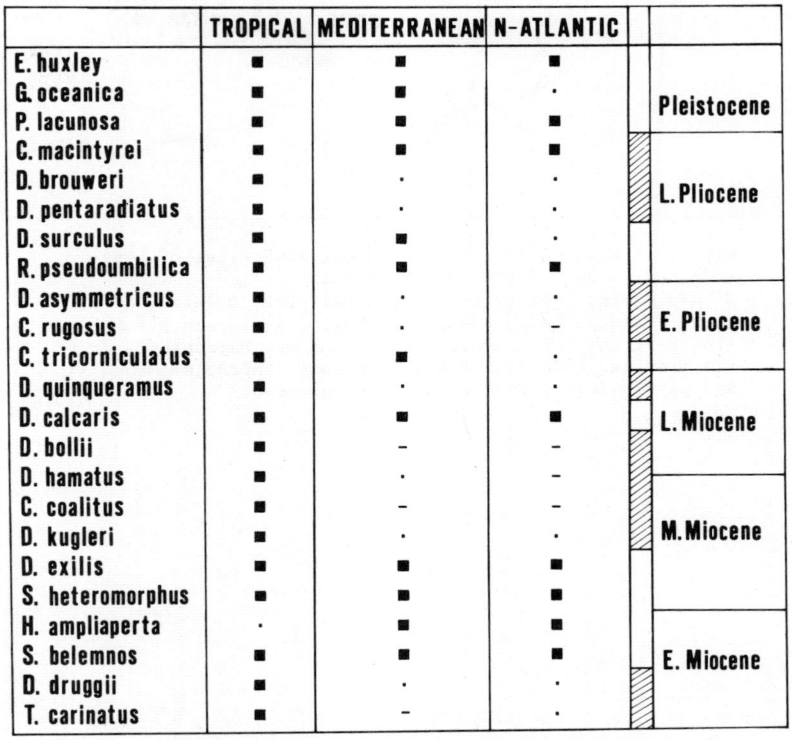

Fig. 3.- Occurrence and abundance of the Neogene index fossils (nannofossils) in the tropics, Mediterranean and North Atlantic.

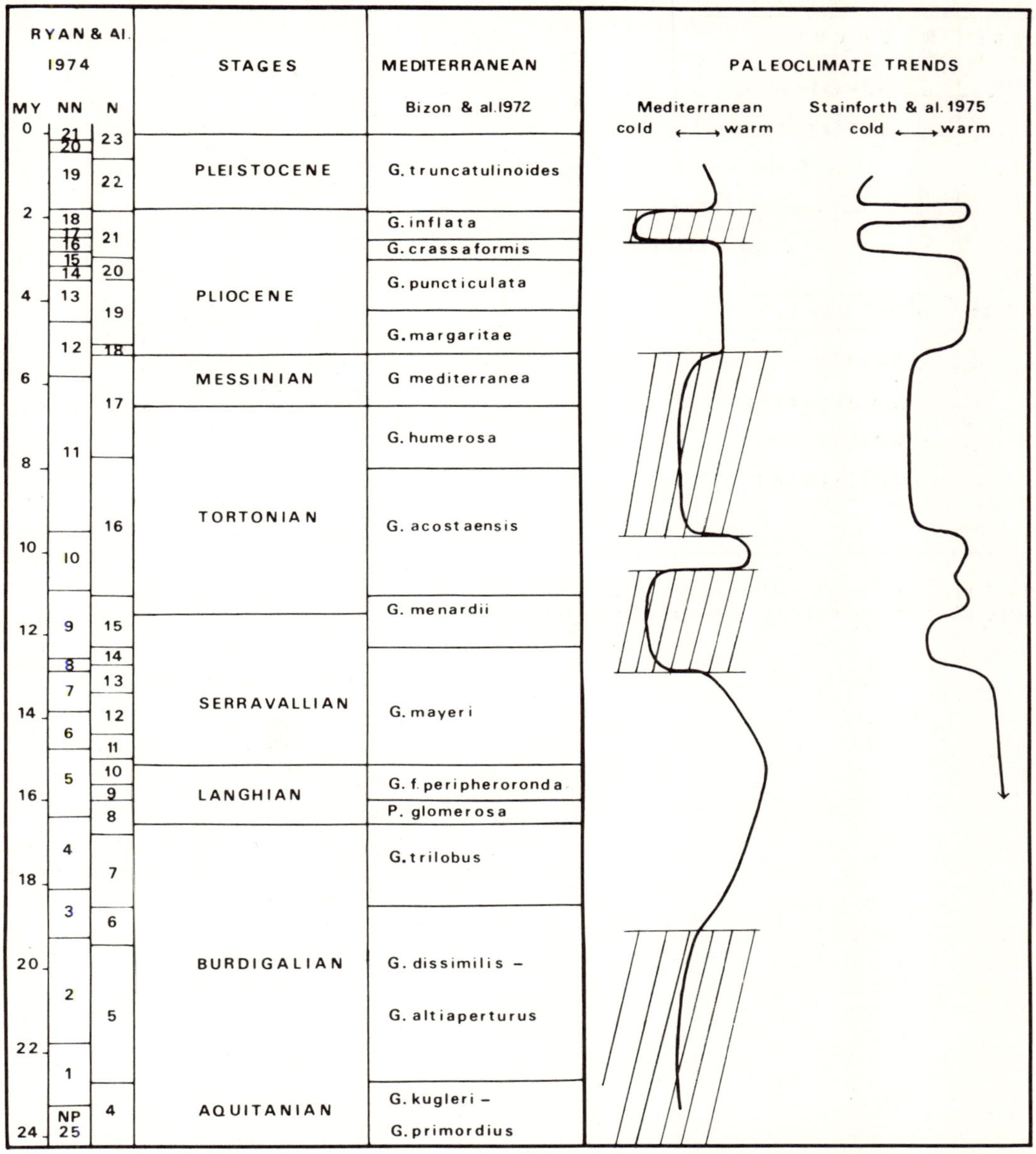

Fig. 4.- Paleotemperature curve for the Mediterranean Neogene and a general paleotemperature curve for the Neogene based only on planktonic foraminifers is plotted against the combined reference scale of both groups of organisms.

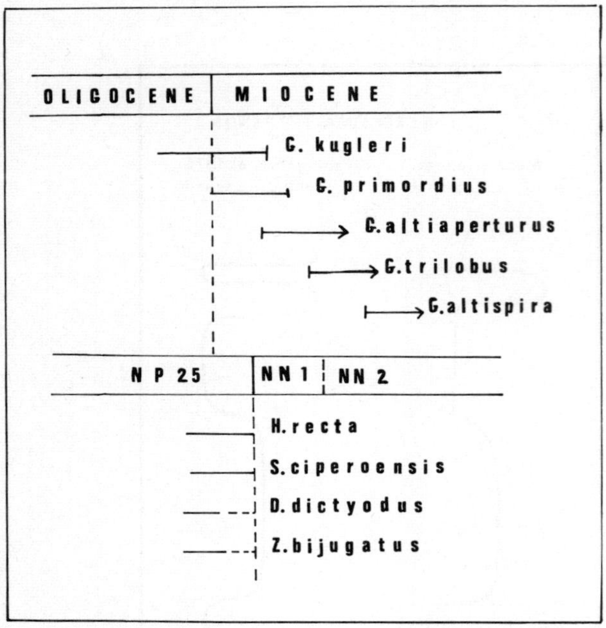

Fig. 5.- Stratigraphic distribution of planktonic foraminifers and nannofossils at the Oligocene/Miocene boundary.

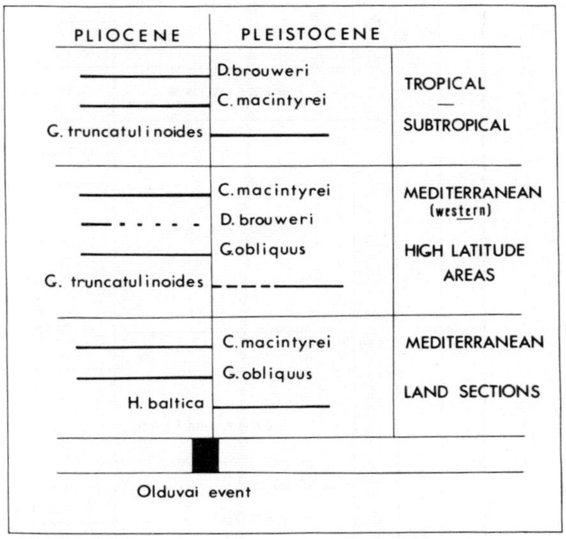

Fig. 6.- Extinction of *Cyclococcolithus macintyrei* and *Globigerinoides obliquus* at the Pliocene/Pleistocene boundary in different regions.

INTERNATIONAL SYMPOSIUM ON THE STRUCTURAL HISTORY OF THE MEDITERRANEAN BASINS. SPLIT (YUGOSLAVIA) 25-29 OCTOBER 1976. B. BIJU-DUVAL AND L. MONTADERT. Eds. EDITIONS TECHNIP, PARIS 1977, pp. 391-400

NEOGENE TO QUATERNARY DEVELOPMENT OF THE IONIAN BASIN (MEDITERRANEAN) : CONSIDERATIONS BASED ON A "DYNAMIC SHALLOW BASIN MODEL" OF THE MESSINIAN SALINITY EVENT

F.H. FABRICIUS and W. HIEKE [(1)]

Abstract

Investigations of Messinian to Quaternary sediments in the central Mediterranean region yielded evidence for a young age of the Ionian Basin. Subsidence (about 1m/1000 years) probably commenced in the Messinian and continued till Quaternary time.

During the Messinian, the Ionian area was a shallow basin, receiving oceanic water almost continuously. Deposition of evaporites compensated basinal subsidence. The outlined "dynamic model" is based mainly on the (i) existence of continuous marine sequences; (ii) on the cyclicity of the evaporitic facies; (iii) on the patterns of facies distribution. - From these evidences it is concluded that the "Messinian event" is a normal (non-catastrophic) development, accentuated only by the coincidence of critical factors.

The Miocene/Pliocene boundary often shows continuous facies developments, thus contradicting an all-Mediterranean and general transgression. The present configuration of some evaporites below the deep sea is indicative for post-Messinian movements, related to the foundering of the Ionian Basin.

Turbiditic material occurring in the Messina Abyssal Plain not before Quaternary time underlines the young age of the present day relief of the Ionian Basin.

1. INTRODUCTION

Most of our knowledge about marine Neogene sediments of the Mediterranean area originates from land sections, and only rather recently the JOIDES Deep Sea Drilling Project (DSDP) provided us with informations from 14 drill sites during Leg 13 in 1970 (1) and 9 sites during Leg 42A, 1975 (2). There can be no doubt about the scientific importance of these cores. Nevertheless, for a well balanced knowledge about the geological history of the different Mediterranean basins, far more bore holes are necessary, especially with continuous coring. Fortunately, some of the insular and peninsular outcrops, especially in Malta, Sicily, Crete, Cyprus and Italy respectively, yield informations from more basinal regions compared to the marginally situated land sections from the circum-Mediterranean continents. In other terms, the already existing DSDP cores, the few pre-Quaternary dredging samples, all these combined with the geophysical data, are not yet sufficient to cover the lack of informations especially as far as the deep parts of the Mediterranean are concerned. Therefore the understanding of the most complex geodynamic history of the entire Alpine-Mediterranean region is still fragmentary. Nevertheless, there seems to be clear already: There are no reasons whatsoever to impose the present morphology of the Mediterranean area upon the

(1) Lehrstuhl für Geologie, Abteilung Sedimentforschung/Meeresgeologie, Technische Universität München, Arcisstrasse 21, D-8000 München 2, Federal Republic of Germany

paleogeography of Pre-Quaternary time. The present morphology of the Mediterranean realm is more likely a transitional than a final and static one.

2. THE TORTONIAN BASEMENT

Only at two DSDP Sites of Leg 42A Tortonian sediments were recovered: At Site 372 (Balearic Sea, Menorca Rise) normal marine (hemipelagic?) nannofossil marls of a light bluish to greenish grey colour, without coarse turbidites, but showing decreasing burrowing towards the top of the Tortonian (2). — A very different lithofacies shows Site 375 (Florence Rise, Levantine Basin): Dark coloured dolomitic nannofossil marlstone with interbedded turbiditic arenites and siltites. Some thin interbedding of layers rich in organic matter are suggesting episodes of stagnation.

At Site 121 (Leg 13) a re-examination of the cores (3) revealed marine Messinian without evaporites or sedimentary discontinuities, but no Tortonian.

Even if paleobiobathymetry from these two sites indicated a "deep" depositional environment, it is no argument to postulate a situation comparable to the present one. Having also in mind the orogenic movements still active in the circum-Mediterranean area at Tortonian time, a complex Mediterranean paleogeography seems more likely to us and is proposed as working hypothesis.

Concerning the circum-Ionian area, Tortonian outcrops as well on the Sicilian as on the Greek side of the basin are interpreted as "not very deep" marine to neritic or even as near shore facies (4;5). This does not exclude the existence of troughs or local basins, having slopes steep enough to allow turbiditic transportation of sediment. However, the existence of turbiditic sediments does not necessarily imply an abyssal setting; nor does the existence of (local?) troughs indicate a deep basinal setting of the entire Tortonian "Mediterranean Sea", comparable to the present Mediterranean Sea.

For the time being the only statement we can be sure of is that the Mediterranean area was covered by a marginal sea which exchanged water with the open ocean(s).

3. THE MESSINIAN

The question, still much debated, is whether the uppermost Miocene evaporites were deposited within a deep basin or a shallow one. This can only be answered if informations from land sections as well as from sea can be interpreted synoptically. The aim of this paper is to add further arguments (6) in favour of Messinian paleography being dominated by a very shallow basinal setting.

Three observations are indicative for a shallow basinal situation of the Mediterranean area where evaporites were deposited:

- Continuous marine sedimentation in some areas;
- cyclicity of the Messinian evaporitic facies; and
- paleogeographic patterns of the lithofacies at the Miocene/Pliocene boundary.

3.1 Continuous Sedimentation

Of special importance to Mediterranean paleogeography is the discovery of a complete marine section comprising Messinian to Pliocene in the Vera Basin (Southern Spain) (7; 8) and of marine Messinian in DSDP Site 121 (3). This, together with other indications of a more or less normal marine sedimentation (sometimes interfingering with evaporites) at other places e.g. Sicily, Cephalonia, and Crete (some containing typical normal marine shallow water carbonates), underline the opinion that the Mediterranean area received oceanic water without important interruptions. Only the reflux-system was inhibited, probably by a reduction of the sill depth — perhaps due to eustatic sea level changes. Together with the unanimously accepted shallow-water to subaerial formation of some of the Messinian sulfates found in several DSDP Sites (124, 374 and others), the existence of a deep basinal setting should be excluded.

3.2 Sedimentary Cycles

In Messinian sediments as in other evaporitic sequences an increasing salinity is indicated by the following succession of facies:

- (a) On top of normal marine sediments (mainly argillaceous ones, but also limestones) generally follow
- (b) dolomitic marls, which are succeeded by
- (c) the group of sediments encompassing all the varieties of the sulfate facies, which in turn sometimes is followed by
- (d) the group of chloridic salts (including bitterns).

Reducing salinity is documented generally by the inverse sequence: d - c - b - a. Depending on the paleogeographic situation of an outcrop or drill site, such a "salinity cycle" can climax already at an earlier stage of evaporation, say at (c) or (b) instead of (d); or in areas with dominating oceanic influence, a salinity cycle is undetectable and sedimentation remains within the facies group (a), a case represented in the Vera Basin during the entire Messinian time (7).

Frequently, incomplete cycles can be observed where one or several members are omitted or very reduced. This can be due to local or regional variations in tectonical movements or other geological processes interferring with the process dictating the general cyclicity. It must be also mentioned that both sides of an evaporitic cycle, i.e. the "regressive" and the "transgressive" wing, are not always exactly identical, especially, if one compares thickness and facies variations. Facies variations may be also due to a decrease in salinity by fresh-water influx instead of sea-water. In that case a brackish to fresh-water facies represents marine dolomitic to normal marine sediments respectively.

We do notyet know the causes of these cycles. Eustatic sea level changes seem to be sufficient for their development (9). There must have been an additional factor responsible for the typical "Messinian" situation of the Mediterranean area, i.e. the lack of reflux of highly saline brines into the ocean(s). In this respect, the entire Messinian "salinity crisis" can be understood as a huge salinity cycle with an increasing beginning and a decreasing end. Not one or a series of "catastrophes" ("waterfalls" or a sudden collaps of a barrier down to abyssal depth) caused or ended respectively the ecological "crisis" of the Mediterranean Sea; it rather was a normal geological development which only was accentuated by the coincidence of some critical factors.

Besides this "mega salinity cycle" (=Messinian) and some ten to twenty cycles of second order ("séquences majeures"(10)), there are also cycles of a third order, only millimeters to centimeters thick. These "séquences mineures" (10) appear frequently at the beginning of sulfate deposition.

From the cyclicity of Messinian deposits we infer on the existence of events of a cyclic nature. However, this interpretation does not fit into a model based on a repeated filling of a deep basin "up to the brim" by means of a huge waterfall, which could have only produced polar, i.e. one-sided sequences of evaporitic sediments.

3.3 Facies Patterns

The facies patterns of a repeatedly desiccated deep basin should markedly differ from that of a shallow basin. Only the sediments of normal marine facies extend in both cases from a level at or near the world ocean sea level (WOSL) down to the center of the respective basin(s). In a desiccating deep basin the uppermost appearance of the sulfate facies must depend on the obvious lowering of the brine level depending on the state of concentration. At the end of this one-sided process, the halite or bitterns are precipitated in the deepest area of a basin. Hence, each of the three facies, normal marine, sulfate, and salt facies, should have a different upper limit on the flancs of the basin.

In a shallow basinal setting this pattern can be similar but with a much smaller vertical separation of facies. Nevertheless, in this case too, the sulfate sediments and the normal marine littoral facies should be separated. Also, Messinian sulfates and Pliocene littoral facies should not occur together.

However, in the Ionian area and elsewhere many outcrops show Messinian gipsiferous series concordantly topped by Pliocene arenaceous sediments containing abundant shallow-water pelecipods without signs of reworking. (The frequent intercalation of the so-called "Trubi" limestones does not discard this interpretation; to judge from the paleogeographic distribution of this facies, it is not typical for the center of a basin, (as indicated by the lack of it at Site 374) but for the margins (6).

From the superposition of sulfate-bearing Messinian series and Pliocene deposits of the upper shelf to littoral environment we conclude that during Messinian time the sea level has not been far from the (world-wide) sea level of the Early Pliocene. This does not exclude the possibility of intra-Messinian eustatic sea level fluctuations of the order of 100 to 150 m (maximum depositional depth of the mentioned Pliocene sediments). Adding perhaps another hundred meters for the depositional depth of halite, this indicates the morphology of a very shallow basin.

3.4 The Dynamic Model of the "Messinian Event"

For the Messinian time a deep basinal setting of the Mediterranean area can be excluded for at least three reasons:

- The continuous influx of oceanic water,
- the sulfate deposition at or near world ocean sea level, and
- the cyclic sedimentation all over the Mediterranean area.

On the contrary, only a very shallow basin can remain connected with the ocean(s) and yet dessicate to a great part down to its bottom. A combination of Late Miocen world-wide sea level changes and shallow sills are able to prevent the reflux of high salinity brines (9).

Evaporation under a "Mediterranean" climate (10; 11) caused deposition of evaporites in this area for about two million years. Although the exact location(s) of the inlet(s) from the open ocean(s) is (are) not known, a connection to the Atlantic is the most likely one; others may have existed in addition. From the inlet(s) to remote areas a network of waterways can be supposed. This can explain the (temporal) appearances of almost normal marine calcareous shallow water facies in the central Mediterranean area (Sicily (21); Ionian Islands (12)).

Under large parts of the Balearic as well as the Ionian-Levantine Basins seismic survey revealed very thick evaporitic sediments, sometimes exceeding two thousand meters. The top of these salt and sulfate deposits was cored at several DSDP sites of Leg 13 and 42 A, proving its Messinian age. (In central basinal setting the base of the evaporites never was reached. Therefore, the Early Messinian age of the basal strata is only assumed.)

The deposition of evaporites, probably mainly salt, more than one kilometer thick within a very shallow basin, claims for a considerable contemporaneous subsidence. During halite sedimentation, the depositional interface was perhaps covered by some hundred meters of water. But then, at the end of the Messinian, this water cover (although less concentrated) was considerably shallower: The growth of stromatolite-like algal mats on a sulfate-carbonate substratum (cf. Sites 124 (1) and 372 (2)) requires sedimentation within the photic zone.

In the center of the Ionian basinal area this subsidence was in the order of magnitude of 1 m per thousand years; in marginal areas perhaps only 10 to 1 cm per 1000 years. Extrapolated to the present time, this basinal subsidence would account almost precisely for the water depth of the Ionian Basin plus its Plio-Quaternary sediment cover.

4. THE MIOCENE/PLIOCENE BOUNDARY

As mentioned above, we understand the Miocene/Pliocene boundary as a comeback of normal marine conditions within the Mediterranean Sea. We regard the idea of a catastrophic collapse of the water way in the Gibraltar area down to more than 1500 m of water depth and its later rise to its present situation, as an overinterpretation of paleontological findings, the existence of which we do not deny. But this idea is far beyond any geological likelihood, and any implication of it has to be regarded with sceptical precaution.

As Sonnenfeld explained (9), a relatively small deepening of an already existing inlet (or the opening of another — not necessarily deep — one) would be sufficient to enable the revival of the reflux system. In case the formation (deepening) of the Strait of Gibraltar was a fast process (for example by fractural tectonics), the ecological change in the Mediterranean basins could have happened within a rather short time, too. But there is no need for such a fast turnover. The existence of transitional lithofacies in DSDP Sites 124, 372, 374 and 376 as well as in many land sections, e.g. in Spain (3,7,8), and in the circum-Ionian land sections (12), demonstrate gradual change: In deep sea cores frequently from the sulfate facies through dolomitic marls to normal marine calcareous sediments. In land sections the transitional dolomitic marls, although generally thin, are situated between sulfate series and the Trubi limestone.

Concerning the question of a general and "all-Mediterranean" transgression at the onset of the Pliocene, there certainly exist indications of a previous erosion and/or discordancies. Nevertheless, they are only of regional or local importance and should not be generallized. As mentioned already, there are numerous drill sites and land profiles without such features; and not all discordancies or erosional features at the Miocene/Pliocene boundary, published in literature, do sustain re-examination (12). Therefore, an all-Mediterranean Pliocene transgression has to be considered as an unfortunate generalization.

5. PLIO-QUATERNARY SEDIMENTATION AND ITS IMPLICATIONS ON THE FORMATION OF THE IONIAN BASIN

Drilling at DSDP Site 374 (Messina Abyssal Plain) penetrated 457 m of sediments. The stratigraphic boundaries are shown in Figure 1.

The thickness of the post-Messinian sediments is about 370 m. The thickness/time ratio for the Pliocene (about 80 m/ 3.8 million years) and the Quaternary (about 290 m/1.8 million years) indicate a considerable increase of the sedimentation rate during Quaternary. This increase is due to the turbiditic sedimentation characteristic for the Quaternary of the Messina Abyssal Plain, whereas the Pliocene is barren of turbiditic layers.

There are direct as well as indirect indications for turbidites: (a) Within the few cores recovered from the Quaternary interval of Site 374 a number of turbidite layers have been observed (2; 13). Moreover, in some piston cores from the Messina Abyssal Plain a 12 m thick turbidite layer with an age younger than 9.000 years B.P. has been found together with additional small turbidites (14). — (b) The rate of sedimentation of normal hemipelagic Late Quaternary sediments of the Mediterranean Ridge amounts to about 6 cm/1.000 years (15). In contrast to this, the rate of the Quaternary sedimentation at Site 374 is about 16 cm/1.000 years. — The question is, why the turbiditic sedimentation did not begin earlier than in Quaternary time. A causal connection between the development of the basin morphology and the onset of the turbiditic facies seems the most logical answer.

The distribution of thicknes of the Messinian evaporites (Figure 2) shows a pattern of maxima and minima irregular in respect to modern morphology (16). A marginal area (with actual water depths less than 2.000 and 3.000 m) is more or less barren of evaporites, with the only exception of the western Syrtis. Within the area of positive occurrence of evaporites, there is no direct relation between evaporite thickness and recent bathymetry.

To explain the differences in thickness we think of two alternatives:

- (a) Explaining these differences by a more static approach of the "Deep Basin Model" of the "Salinity Crisis", the preexisting irregular depths of the subbasins produced the irregular thickness of evaporites by filling up the basins to a certain more or less horizontal level. Thus, the maxima and minima of evaporite thickness indicate a distinct relief (within the deep basin) at the beginning of the Messinian desiccation period. — For example, the large area of 500 ms of evaporite thickness east of the Malta Escarpment must have been a flat sabkha-like area at the end of the Messinian, but now shows differences in elevation of about 900 m. At least for this area, we have to consider important Post-Messinian tectonic movements to obtain the actually tilted configuration of this particular evaporitic body.

- (b) The second alternative is based on the "Dynamic Shallow Basin Model" (see above). In this case sedimentation starts at the beginning of the Messinian in very shallow basins with a water level at or near WOSL. The rate of subsidence is supposed to have been regionally different, thus controlling the differences in the thickness of evaporites. — This alternative demands even more important tectonical movements to obtain the present relief but these movements started already at the beginning of Messinian time, i.e. some 6 to 7 million years ago.

The thickness distribution of the Post-Messinian sediments (Figure 3) (17) shows a distinct relation to the present morphology. The area of the thickness interval between 200 and 600 m coincides perfectly with the recent Messina Abyssal Plain; whereas most of the area north to it, is covered only by Post-Messinian sediments of less than 200 m thickness. (The exception at about 36° 30' N and 17° E, coincides with a small plain which is actually also somewhat deeper than its surroundings.) Therefrom we conclude that — at least since Quaternary time — the present Messina Abyssal Plain worked as a sediment trap, accumulating huge masses of turbiditic material. However, the formation of the basin has probably started already in Early Pliocene time, but obviously without supply of turbiditic material.

Summarizing: During Messinian time the area of the actual Messina Abyssal Plain as well as its surrounding regions, were covered by shallow water; the subsidence, however, had regional differences in velocity. At that time, the subsidence has been balanced by the accumulation of evaporites.

During Pliocene time the hemipelagic sedimentation was unable to balance this subsidence; hence, the Ionian Sea (in statu nascendi) became a starving basin, without the morphological feature of an important abyssal "plain". Simultaneously the basins' margins were uplifted (9): For instance, in Sicily marine sediments of the Siciliano on to 800 m above present sea level (18); in Calabria marine Pleistocene is situated at about 1.000 m above sea level (19).

Finally, all these movements produced a submarine relief which, since Quaternary time, enabled the frequent formation of turbidity currents strong enough to reach the central and deep area of the Ionian Basin. From this we conclude that the morphology of the Messina Abyssal Plain exists only since the beginning of Quaternary time.

6. NOTES AND REFERENCES

Titels of the communications given at the Symposium on the Structural History of the Mediterranean Basins at Split, October 1976:

- F.H. FABRICIUS: Geological History of the Eastern Mediterranean Since Late Miocene;
- W. HIEKE: The Possible Age of the Messina Abyssal Plain.

(1) RYAN,W.B.F., K.J.HSÜ et al.: Initial reports of the Deep Sea Drilling Project, vol. 13, Washington D.C., (U.S.Govt. Print. Office), 1972.

(2) HSÜ,K.J., L. MONTADERT et al.: Initial reports of the Deap Sea Drilling Project, vol 42A, Washington D.C., (U.S. Governmt. Print. Office), in print.

(3) MONTENAT, C., G. BIZON & J.J. BIZON: Remarques sur le Néogène du forage JOIDES 121 en mer d'Alboran (Méditerranée occidentale). - Bull.Soc. Geol.Franc., 7, XVII, 45-51, 1975.

(4) DAINA, A: Coralli hermatipici nel Tortoniano di Grotta (Agricento).- Atti Soc.Tosc.Sc.Nat. sér. A.,vol. LXXII, 238-243, 1965.

(5) CHEVALIER, J.P.: Recherches sur les madréporaires et les formation récifales miocenes de la Méditerranée occidentale. - Mém.Soc. Géol. France, (n.s.), vol. XL, 562 p., 1961.

(6) FABRICIUS,F.H., K.O. HEIMANN & K.BRAUNE: Comparison of Site 374 with circum-Ionian land sections.-Implications on the Messinian "Salinity Crisis" based on a "Dynamic Model".- In (2).

(7) BIZON,G., J.J. BIZON & C. MONTENAT: Le Miocene terminal dans le Levant espagnol (provence Alicante et de Murcia).- Rev.Inst.Franç.Pétrol, XXVII -6,831-862, 1972.

(8) MONTENAT,C., G. BIZON & J.J. BIZON: Continuité ou discontinuité de sédimentation marine Mio-Pliocene en Méditerranee occidentale.- Rev.Inst. Franc.Pétrol.,vol.31, 613-663, 1976.

(9) SONNENFELD, P.: The significance of upper Miocene (Messinian) evaporites in the Mediterranean Sea.- J.Geol. 83, 287-311, 1975.

(10) HEIMANN, K.O. & G. MASCLE: Les séquences de la série évaporitique messinienne. - C.R.Acad.Sc.Paris,279, ser.D, 1987-1990, 1974.

(11) HEIMANN,K.O. & W. JUNG: Palökologische und fazielle Untersuchungen an Gesteinen des Evaporitzyklus II bei Paghi/Nordkorfu (Griechenland).- Mitt. Bayer.Staatssamml.Paläont.hist.Geol., 16, 105-111, München, 1976.

(12) HEIMANN, K.O.: Die Fazies des Messins und untersten Pliozäns auf den Ionischen Inseln (Zakynthos, Kephallinia, Korfu/Griechenland) und auf Sizilien. - Unpubl. Dissertation Techn.Univ. München, 158 p., 1977.

(13) MÜLLER,J., W.HIEKE & F.FABRICIUS: Turbidites at Site 374: Their composition, provenance and paleobathymetric significance. - In (2).

(14) HIEKE,W. M. MELGUEN & F.FABRICIUS: Turbidites from the Messina Abyssal Plain and the Mediterranean Ridge,- Possible indications of young tectonics. - In preparation.

(15) HIEKE, W.: Problems of Eastern Mediterranean Late Quaternary stratigraphy -a critical evaluation of literature. - "METEOR" Forsch.-Ergebnisse, C 24, 68-88, 1976.

(16) FINETTI, I. & C. MORELLI: Geophysical exploration of the Mediterranean Sea.- Boll.Geofis.Teorica ed Appl., 15, 263-441, 1973.

(17) HINZ,K.: results of seismic refraction and seismic reflection measurments in the Ionian Sea.- Geol.Jahrb.,E 2, 33-65, 1974.

(18) RITTMANN, A.: Vulkanismus und Tektonik des Ätna. - Geol.Rundschau, 53, 788-800, 1964.

(19) GIESE,P., K.GÖRLER & H.IBBEKEN: Kalabrisches Massiv und Ionisches Meer. Modell einer rezenten Gebirgsbildung. - FU-Pressedienst Wissenschaft, H.3, Berlin 1971.

(20) MORELLI,C., G. GANTAR & M. PISANI: Bathymetry, gravity and magnetism in the Strait of Sicily and the Ionian Sea. - Boll.Geofis. Teorica ed Appl., 17, 39-58, 1975.

(21) SCHREIBER,B.Ch., G.M. FRIEDMAN, A. DECIMA & E. SCHREIBER: Depositional environments of the Upper Miocene (Messinian) evaporite deposits of the Sicilian Basin. - Sedimentology, 23, 729-760, 1976.

Acknowledgement

For many critical but fruitful discussions and the review of the manuscript we are much indebted to Drs. K. Braune, K. O. Heimann, J. Müller and W. Sigl.

The field work and the senior author's participation on DSDP Leg 42 A were financed by the Deutsche Forschungsgemeinschaft which we acknowledge gratefully.

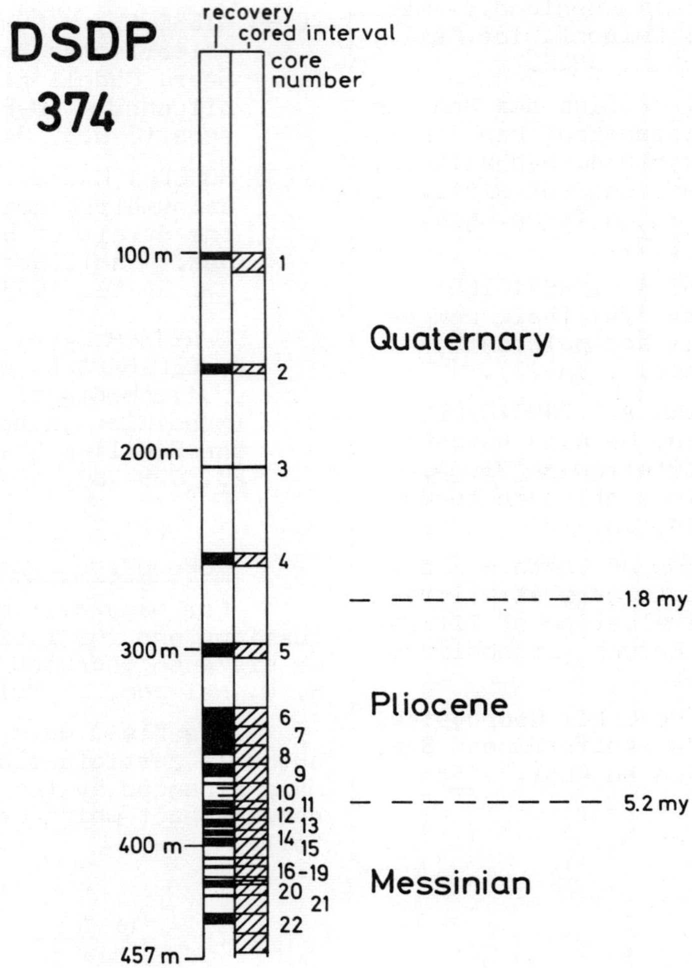

Figure 1. Schematic profile of DSDP Site 374 with stratigraphic boundaries as determined by the Shipboard Party (2).

Figure 2. Two-way time isochrone contours of the evaporitic interval (after Finetti & Morelli (16)). - Contour interval: 100ms; + maximum thickness; - minimum thickness; hatched area: thin or no seismic evidence of evaporites. Bathymetry after (20).

Figure 3. Thickness of unconsolidated sediments (Pliocene to Recent) after HINZ (1974, fig. 24) (17). Bathymetry after (20).

INTERNATIONAL SYMPOSIUM ON THE STRUCTURAL HISTORY OF THE MEDITERRANEAN BASINS. SPLIT (YUGOSLAVIA) 25-29 OCTOBER 1976. B. BIJU-DUVAL AND L. MONTADERT, Eds. EDITIONS TECHNIP, PARIS 1977, pp. 401-410

LATE TERTIARY EVOLUTION OF THE ALBORAN SEA AT THE EASTERN ENTRANCE OF THE STRAITS OF GIBRALTAR

C.J. MULDER and G.R. PARRY [(1)]

Abstract: Results of a 1975 seismic survey by Shell and the Moroccan government in the western Alboran basin correlated with drill site 121 of "Glomar Challenger" indicate an early Messinian opening of the Straits of Gibraltar. Numerous diapirs (clay or salt) have most probably a pre-Messinian origin.

Résumé: Les résultats d'une campagne sismique effectué en 1975 par le gouvernement Marocain et la Shell joints à ceux du forage 121 du Glomar Challenger indiquent que lé Détroit de Gibraltar s'est ouvert au début du Messinien. L'origine des nombreux diapirs (sel ou argile?) de la région est très probablement antérieur au Messinien.

A. INTRODUCTION

The Alboran basin on the concave innerside of the Gibraltar arc (Fig. 1) contains one of the most important accumulations of Neogene and Quaternary sediments in the Western Mediterranean. A reconnaisance ("Vaporchoc") seismic reflection survey carried out during 1975 on behalf of Shell and the Moroccan government agency BRPM shed some new light on the late Tertiary evolution of this area. The results and conclusions reached are illustrated with the help of a few selected seismic records depicted in figures 5-14.

B. GEOLOGICAL INTERPRETATION OF THE SEISMIC DATA

1. Age of seismic events

Stratigraphic correlations are based on the results of Glomar Challenger corehole DSDP-121 (Figs. 3 and 4) and on a comparison with stratigraphic and structural data of onshore Neogene basins, like the Murcia and Vera basins in Spain (Montenat, 1973; Montenat and Bizon, 1976) and the Chelif basin in North Africa (S.N. Repal, 1952). Palaeontological reinterpretation of DSDP-121 (Montenat, Bizon and Bizon, 1975) showed that the lower pelagic marl interval (± 720-867.2 m), which overlies crystalline basement, is of Messinian age and not of Tortonian age (Fig. 4). This necessitates a revision of our previous dating (Mulder, 1973), based on Ryan et al.(1973).

Four seismic events could be confidently correlated in large parts of the basin. The following approximate age was assigned to these events (cf. Figs. 5-15).

- horizon 1 : base Pliocene

(1) Shell Internationale Petroleum Maatschappij B.V., Gevestig'd te Den Haag H. reg, Postbus 162, The Hague, Holland

- horizon 2 : base Messinian
- horizon 3 : base Tortonian
- horizon 4 : Serravallian

The age of the interval between horizon 2 and 3 and that between horizon 3 and 4 must be considered as speculative since these intervals are stratigraphically below the levels reached in DSDP-121 (Fig. 5).

The top of the Messinian (or base of the Pliocene) as indicated by DSDP-121 is represented by a strong seismic event (horizon-1) which appears as an unconformity around the basin margins (Fig. 16). This reflector can be correlated with the M-reflector recognised elsewhere in the Mediterranean (Ryan et al. 1973).

A deeper low frequency reflector band of very distinct character (horizon-2, Fig. 17) can be correlated throughout the Alboran basin, although its continuity is interrupted by diapirs (Figs. 6-8) and by an intra-formational erosional feature (channel) which is discussed below. The outstanding acoustic contrast of the reflector band is possibly caused by evaporites (anhydrite or gypsum) which could represent a reduced equivalent of the Messinian evaporite interval of the Western Mediterranean (Ryan et al. 1973). Horizon 2 onlaps against the basement ridge on which DSDP-121 was drilled (Fig. 5). The interval between horizon 1 and 2 is partially present in DSDP-121, where it is represented by pelagic marls of Messinian age overlying crystalline basement (Montenat, Bizon and Bizon, 1975).

The next deeper seismic event traced in the larger part of the area (horizon 3, Fig. 18) represents locally a slight angular unconformity on the basin margins where it may overlap the underlying (older Miocene) sediments onto acoustic basement. It suggests a period of uplift and/or tilting prior to the later Miocene transgression. This tectonic event has been tentatively correlated with the angular unconformity at the base of the Upper Miocene at the margins of the Chelif basin (SN Repal, 1952) and with the compressional post-Serravallian phase in the Murcia and Vera basins (Montenat, 1973; Montenat and Bizon, 1976). The interval between horizons 2 and 3 consequently could essentially represent the Tortonian. The interval velocities and seismic character of this interval suggest a predominantly shaly or marly composition away from the basin margins. The sediments therefore could well correspond to the pelagic marls which in many places represent the lithology of the transgressive Tortonian in the western Mediterranean area.

Based on the above age interpretations it is thought that a still deeper seismic event (horizon 4, Fig. 19) is of pre-Tortonian intra-Serravallian age. The seismic records show, moreover, the presence of a thick sedimentary sequence below horizon 4 in large parts of the Alboran basin. These beds could include sediments of Lower Miocene or even Oligocene age.

2. Sub-marine erosional channel features

To the west horizons 2, 3 and 4 are terminated abruptly by an approximately east-west trending cliff, the southern edge of a sub-marine channel (Fig. 12-15 and Fig. 17-19). This earliest evidence of strong sub-marine currents is correlated with the opening of the Straits of Gibraltar in early Messinian time. The northern limit of the intra-Messinian erosional channel could not be accurately defined owing to a superposition by later (Pliocene) erosional events (see below) and because of an incomplete seismic coverage. The channel is filled by sediments pre-dating horizon-1 and consequently partly correlatable with Messinian pelagic marls encountered in the lower 140 m of DSDP-121 (Fig. 4). Sporadic low frequency reflectors within the channel suggest that the deeper parts may contain remnants of the evaporite sequence as described by Ricci Lucchi (1973) from the Messinian of the peri-Adriatic basin.

Structural time contours of the early Messinian channel bottom have been prepared in the area with sufficient seismic coverage (Fig. 20). Also an extrapolation has been made of the erosional pattern of this channel prior to its infill (Fig. 21). It appears that a conspicuous erosional depression with a depth of approximately 400 milliseconds two-way travel time occurred close to the eastern entrance of the Straits of Gibraltar. The existence of a strikingly similar sub-marine erosional pattern at the beginning of the Pliocene is demonstrated by the structural contours of horizon 1 (Fig. 16). The occurrence of an erosional channel, although less conspicuous than during the lower Messinian is clearly indicated on the seismic records (Figs. 12 and 13). Also the present-day bathymetry of the Straits of Gibraltar and western Alboran sea shows a very similar configuration of the sea floor with a marked depression at the eastern entrance of the Straits of Gibraltar (Fig. 2). Active sub-marine erosion is strongly indicated on the seismic records (Figs. 8-10 and Fig. 14).

These observations show that at the eastern entrance of the Straits of Gibraltar three main periods of strong sub-marine channel erosion can be recognised. The first approximately in early Messinian time, the second during the early Pliocene, whilst recent sub-marine erosion also appears quite active. In the intervening periods of the Messinian and Pliocene-Quaternary deposition within the channel appears to have dominated erosion. Messinian evaporite deposition may have just preceded the opening of the Straits of Gibraltar.

3. Diapiric features

Diapiric features are widespread in the central parts of the Alboran basin (Figs. 6-10). Study of seismic sections clearly shows that the diapiric movements were syn-sedimentary, commencing possibly during Middle or Lower Miocene time. In some places movements terminated in late Miocene or early Pliocene (Figs. 8-10), in other localities more recent movements cause upheaval of the sea floor (Fig. 7). In many places diapirs pierce the overlying sediments (Figs. 17-19) and in one locality in the more eastern part of the Alboran basin the mobile material reaches the present sea-floor.

The mobile material involved might be either clay or salt. The source is extremely deep and the base of the mobile layer is nowhere observed on the seismic records. It is speculated that its origin might be in early Miocene or older sediments. It certainly has no relation with the Messinian evaporites encountered elsewhere in the Mediterranean (cf. Auzende et al. 1973) and Pastouret et al. 1973).

C. CONCLUSION

The evolution of the Alboran sea area during late Tertiary and Quaternary times is envisaged to have shown approximately the following pattern.

In earlier Miocene (or Oligocene?) time extensional tectonics on the concave innerside of the Gibraltar orogenic arc created a system of inter-connected basins and troughs following ENE structural trends, still expressed by their present configuration (Fig. 1). The largest of these basins, which also underwent the greatest subsidence, is the Alboran basin closest to the inner margin of the Gibraltar arc. A connection with the Atlantic Ocean existed north of the Betics through the Guadalquivir basin and possibly through the Rif trough and Guerçif basin in the south.

Opening of the Straits of Gibraltar occurred during earlier Messinian time, possible facilitated by ENE-trending strike-slip faulting. Restricted environmental conditions with evaporite deposition may have existed during early Messinian, before the opening of the Straits of Gibraltar. This opening gave rise to strong sub-marine erosion and the formation of a sub-marine channel in which turbulent currents scoured a deep depression opposite the eastern entrance of the Straits of Gibraltar (Fig. 21).

Subsequent sedimentary infill of this channel during Messinian time was followed by renewed sub-marine erosion and sedimentary infill during Plio-Pleistocene time (Fig. 16). Recent sea-floor erosion still shows the same pattern, although a shift closer towards the entrance of the Straits of Gibraltar is noted (Fig. 2).

Layers of mobile material (clay or salt) deposited in lower Miocene or earlier time in the central parts of the

Alboran basin started their deformation probably during the Middle Miocene, giving rise to diapiric piercement features, which locally continued their growth until recent times.

ACKNOWLEDGEMENT

This paper is published by permission of the Royal Dutch/Shell Group, The Hague, Netherlands.

The authors wish to thank Messrs. A.J. Boomer and P. Lehner of Shell Internationale Petroleum Mij, B.V. for critically reading the manuscript. P. Lehner we are also indebted for his instigation to publish this paper. The authors were skilfully assisted with the preparation and drafting of the illustrations by Miss B.M.M. Ammerlaan, Miss W.A. Hilberding and Mr. N. de Water.

REFERENCES

- AUZENDE, J.M., RENAULT, J.P., PASTOURET, L., SZEP, B. and OLIVET, J.L. (1975) - Les bassins sédimentaires de la mer d'Alboran - Bull.Soc.Géol. Fr., (7), XVII, No. 1, p. 98-107.

- MONTENAT, C. (1973) - Le Miocène terminal des chaines bétiques (Espagne meridionale). Esquisse paléogéographique - In: "Messinian Events in the Mediterranean", Kon.Ned.Acad. van Wetenschappen, Amsterdam, p. 180-187.

- MONTENAT, C., BIZON, G. and BIZON, J.J. (1975) - Remarques sur le Néogène du forage Joides 121 en mer d'Alboran (Méditerranée occidentale) - Bull.Soc. Géol.Fr., (7), XVII, No. 1, p. 45-51.

- MONTENAT, C. and BIZON, G. (1976) - Apropos de l'évolution géodynamique mio-pliocène en Méditerranée occidentale. L'exemple du bassin de Vera (Cordillères bétiques, Espagne méridionale) - C.R. Somm.Soc.Geol.Fr., 1976, fasc. 1, p. 15-16.

- MULDER, C.J. (1973) - Tectonic framework and distribution of Miocene evaporites in the Mediterrranean - In: "Messinian events in the Mediterranean", Kon.Ned.Acad.van Wetensch., Amsterdam, p. 44-59.

- PASTOURET, L., OLIVET, J.L., AUZENDE, J.M. and REHAULT, J.P. (1975) - Remarques complémentaires sur le Néogène de la mer d'Alboran - Bull. Soc.Geol.Fr., (7), XVII, No. 6, 1168-1171.

- RICCI LUCCHI, F. (1973) - Resedimented evaporites: indicators of slope instability and deep-basin conditions in Peri-Adriatic Messinian (Apennines foredeep, Italy) - In: Messinian Events in the Mediterranean" Kon.Acad.van Wetensch., Amsterdam, p. 142-149.

- RYAN, W.B.F. et al. (1973) - Initial Reports on the Deep Sea Drilling Project, vol. 13, Washington (US Government Printing Office).

- SOCIÉTÉ NATIONALE DE RECHERCHE ET D'EXPLORATION DES PÉTROLES EN ALGÉRIE (S.N. REPAL) (1952) - Le bassin Néogène du Chelif - XIXe Congres Géologique Intern., Monogr. regionales, 1e Serie, No. 16.

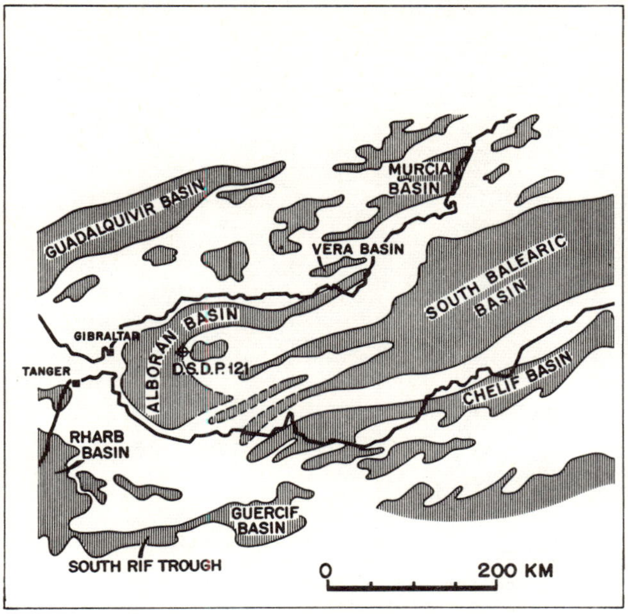

Fig. 1 - Neogene basin configuration in the southwestern Mediterranean area (schematic).

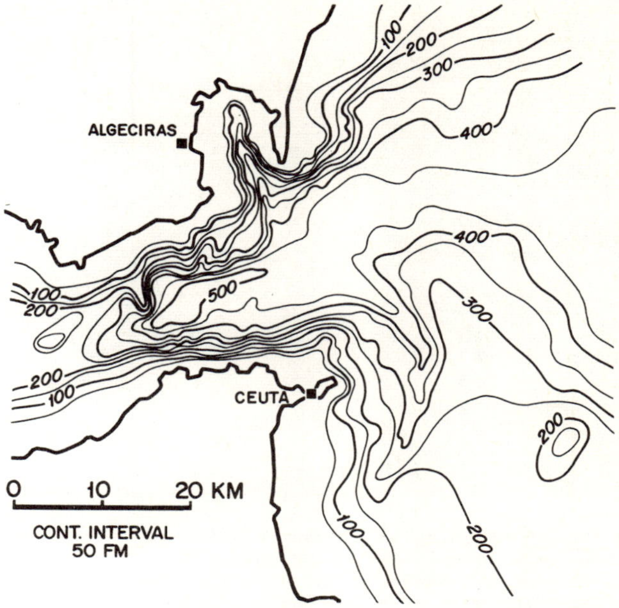

Fig. 2 - Bathymetry of the western Alboran Sea and Straits of Gibraltar Recent sub-marine erosional features can be inferred.

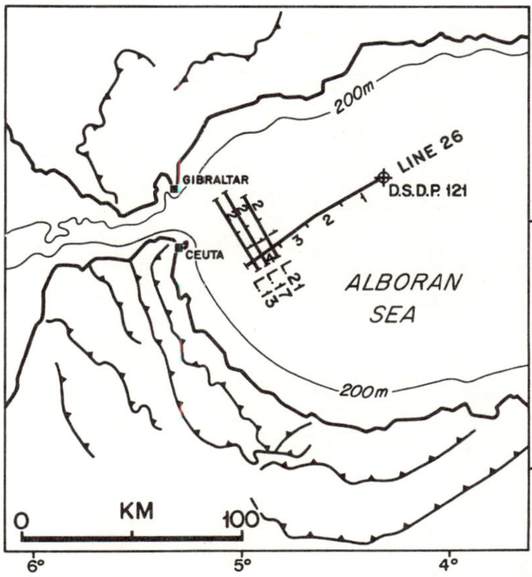

Fig. 3 - Location of Joides corehole DSDP-121 and seismic lines depicted in Figs. 5-14.

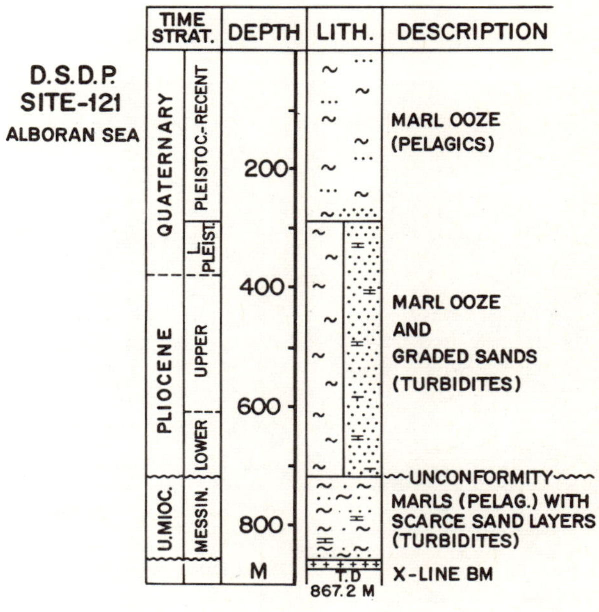

Fig. 4 - Results of corehole DSDP-121 (Ryan et al, 1972). Time stratigraphy mainly according to Montenat, Bizon and Bizon, 1975.

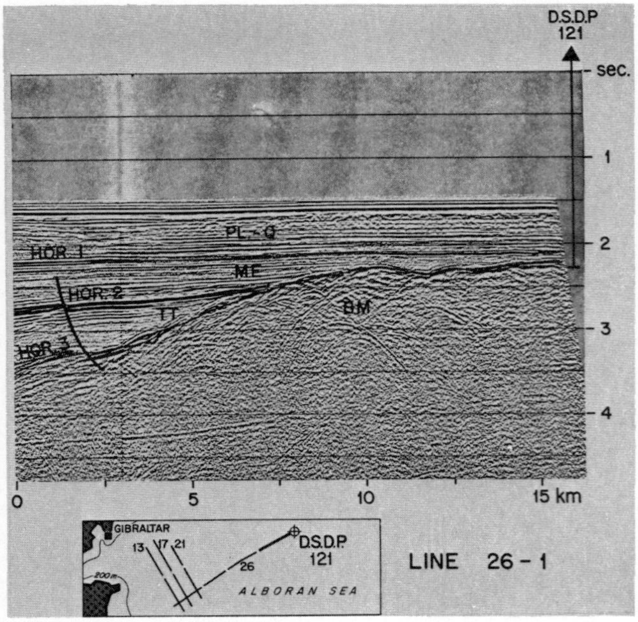

Fig. 5 - Geological interpretation of seismic line 26-1 showing onlap of Miocene sediments on the pre-Messinian basement high on which DSDP-121 was drilled. The age of the sediments is based on the corehole interpretation and regional geology. It must be considered as approximate only. Depth in seconds two-way travel time.
PL-Q: Pliocene-Quaternary; ME: Messinian; TT: Tortonian; BM: Basement.

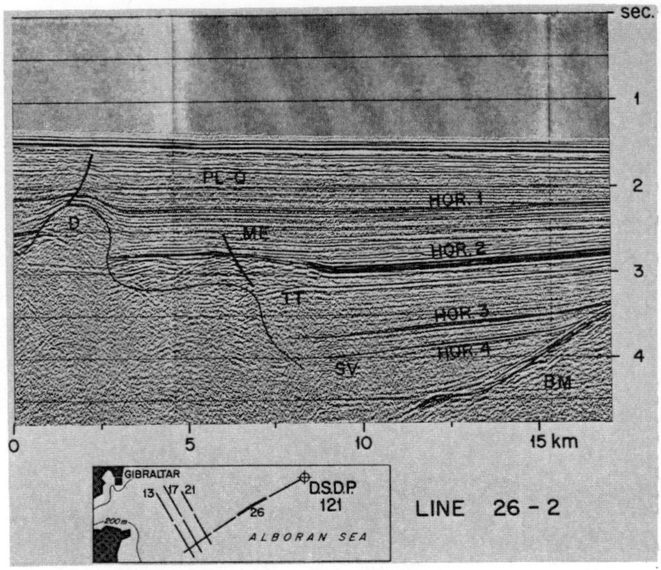

Fig. 6 - Geological interpretation of seismic line 26-2. Late Miocene as well as post-Miocene phases of diapirism are indicated. The strong low frequency reflector band interpreted as "lower Messinian" (Horizon 2) may represent a thin evaporite interval. PL-Q: Pliocene Quaternary; ME: Messinian; TT: Tortonian; SV: Serravallian; BM: basement; D: diapir (clay or salt). Cf. also comments below Fig. 5.

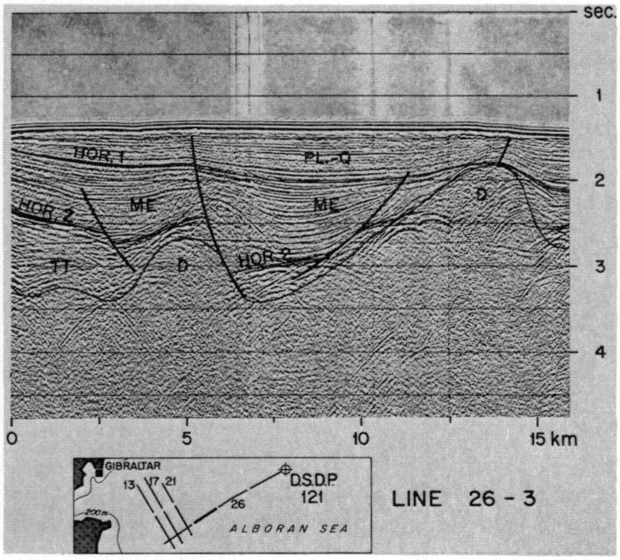

Fig. 7 - Geological interpretation of seismic line 26-3. PL-Q: Pliocene-Quaternary; ME: Messinian;
TT: Tortonian; D: diapir (clay or salt). Cf. also comments below Figs. 5 and 6.

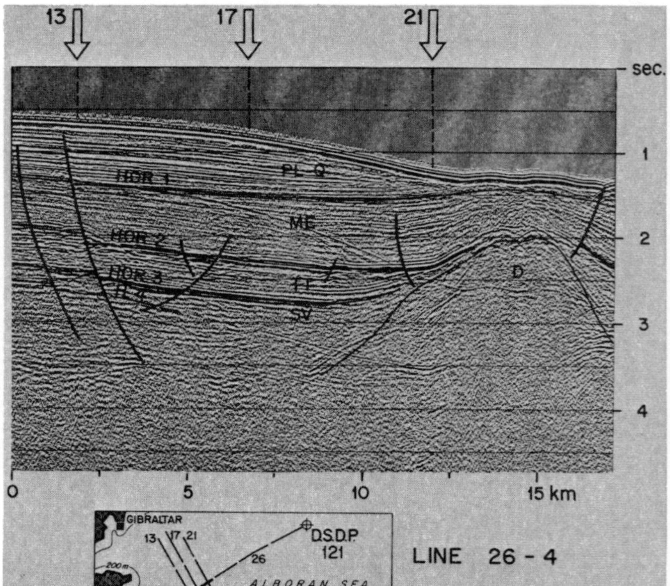

Fig. 8 - Geological interpretation of seismic line 26-4. PL-Q: Pliocene-Quaternary; ME: Messinian;
TT: Tortonian; SV: Serravallian;
D: diapir (clay or salt). Note recent sea-floor erosion. Cf. also comments below Figs. 5 and 6.

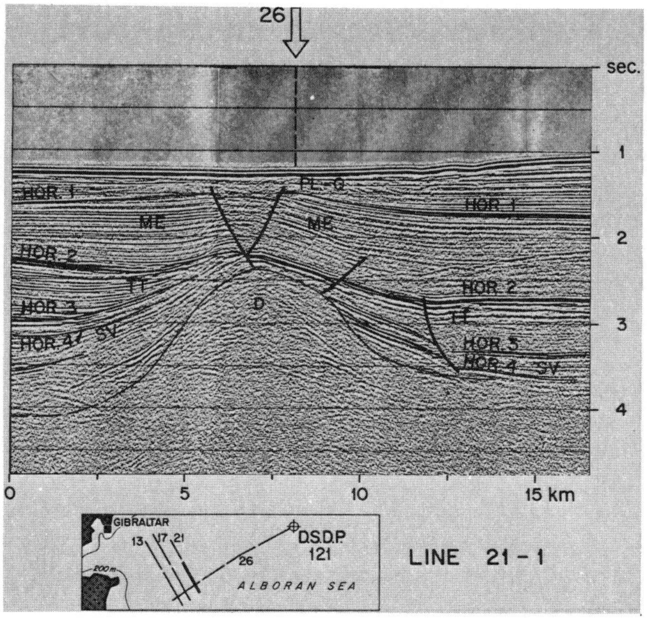

Fig. 9 - Geological interpretation of seismic line 21-2. PL-Q: Pliocene-Quaternary; ME: Messinian; TT: Tortonian; SV: Serravallian; D: diapir (clay or salt). Cf. also comments below Figs. 5 and 6.

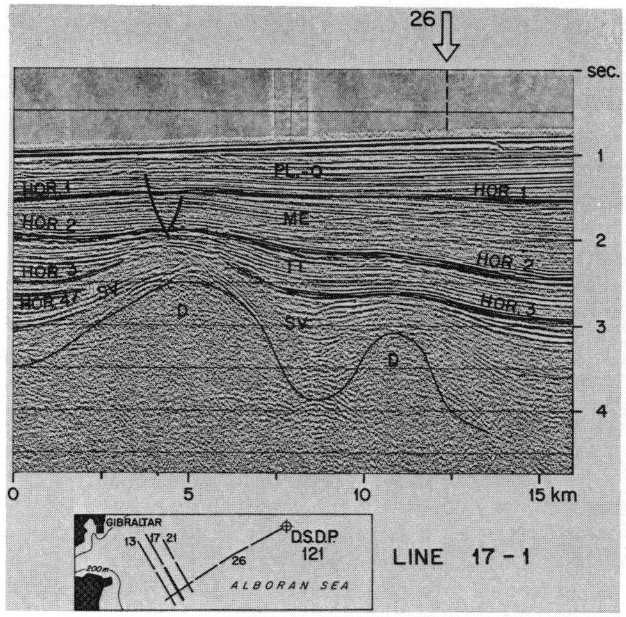

Fig. 10 - Geological interpretation of seismic line 17-1. PL-Q: Pliocene-Quaternary; ME: Messinian; TT: Tortonian; SV: Serravallian; D: diapir (clay or salt). Cf. also comments below Figs. 5 and 6. .

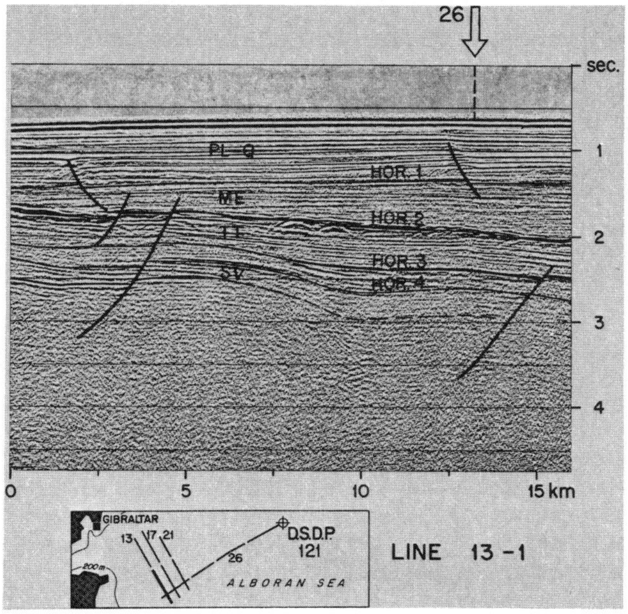

Fig. 11 - Geological interpretation of seismic line 13-1. PL-Q: Pliocene-Quaternary; ME: Messinian; TT: Tortonian; SV: Serravallian; Cf. also comments below Figs 5 and 6.

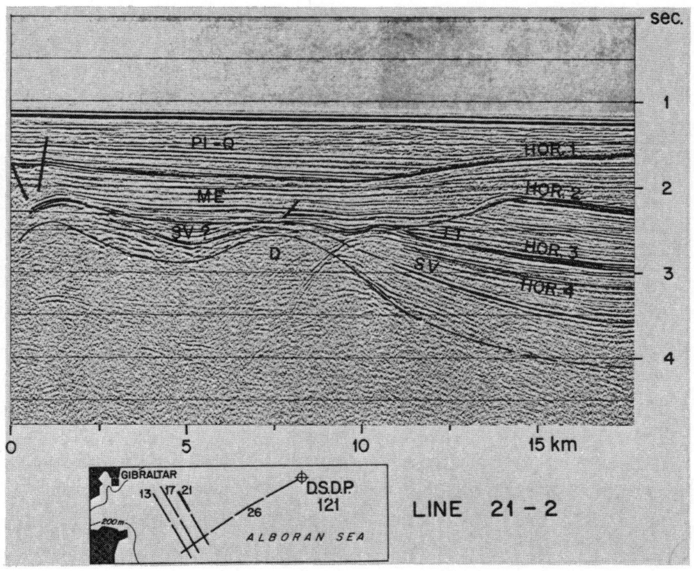

Fig. 12 - Geological interpretation of seismic line 21-2. A "lower Messinian" erosional channel feature as well as its less pronounced "Pliocene" successor are clearly seen. The "lower Messinian" channel base appears affected by diapiric clay or salt movements. PL-Q: Pliocene-Quaternary; ME: Messinian; TT: Tortonian; SV: Serravallian; D: diapir (clay or salt). Cf. also comments below Figs. 5 and 6.

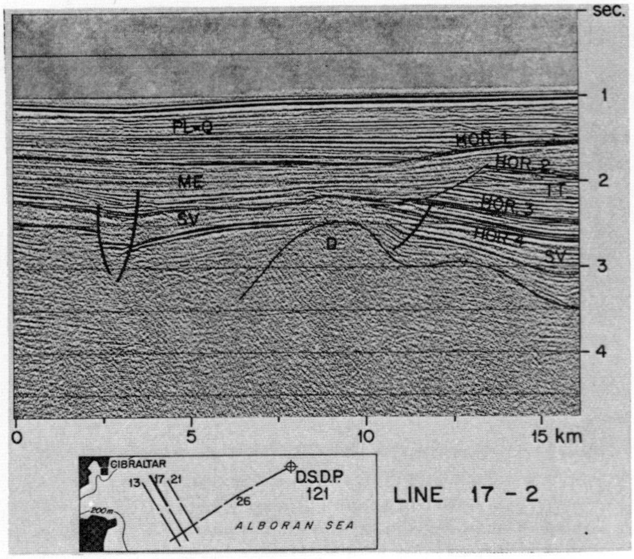

Fig. 13 - Geological interpretation of seismic line 17-2. Similar features are seen as on line 21-2 (Fig. 12). PL-Q: Pliocene-Quaternary; ME: Messinian; TT: Tortonian; SV: Serravallian; D: diapir (clay or salt). Cf. also comments below Figs. 5 and 6.

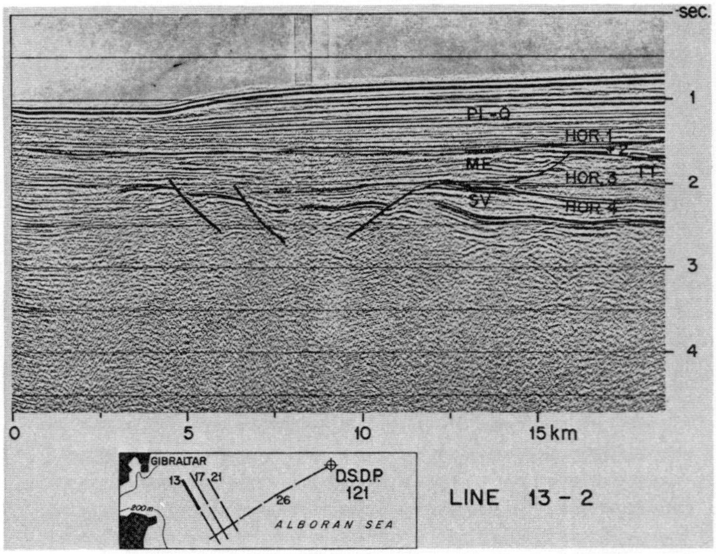

Fig. 14 - Geological interpretation of seismic line 13-2. "Lower Messinian" channel erosion is less deep than in the areas of line 21-2 and 17-2. The "upper Messinian" sediments are for a large part removed by erosion related to the "Pliocene" channel. Present-day sea-floor erosion can be seen on the left. PL-Q: Pliocene-Quaternary; ME: Messinian; TT: Tortonian; SV: Serravallian. Cf. also comments below Figs. 5 and 6.

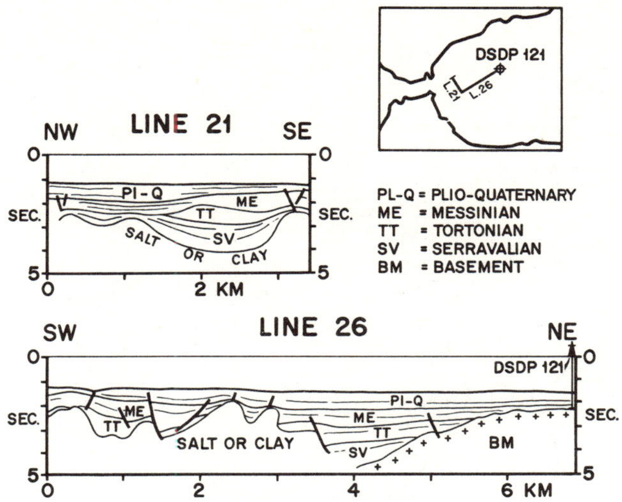

Fig. 15 - Regional structural profile based on seismic lines 21 and 26. Depth in two-way travel time. PL-Q: Pliocene-Quaternary; ME: Messinian; TT: Tortonian; SV: Serravallian; BM: basement.

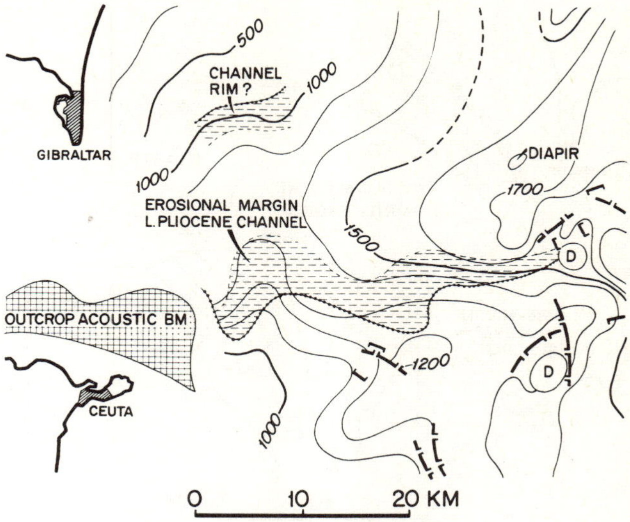

Fig. 16 - Structural contours of the "lower Pliocene" unconformity (hor. 1 in Figs. 5-14). Depths in metres b.s.l. The approximate position of the "lower Pliocene" erosional channel is also indicated. D: diapir.

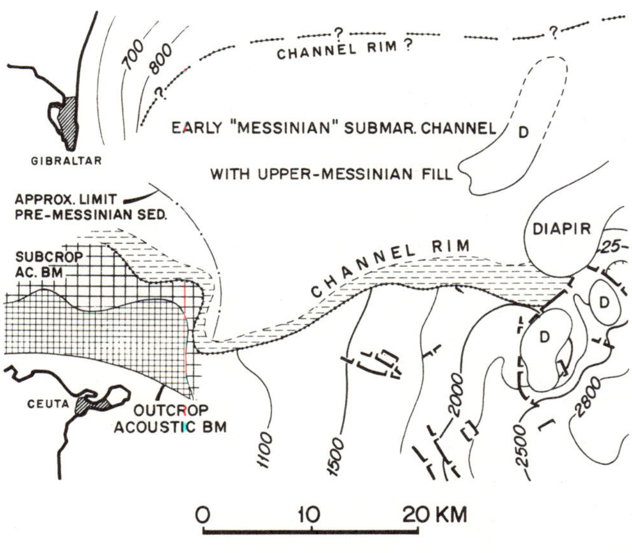

Fig. 17 - Structural contours of "base Messinian" (hor. 2 in Figs. 5-14) and location of "lower Messinian" erosional channel. In the east diapiric features pierce the channel bottom. Depths in metres b.s.l.

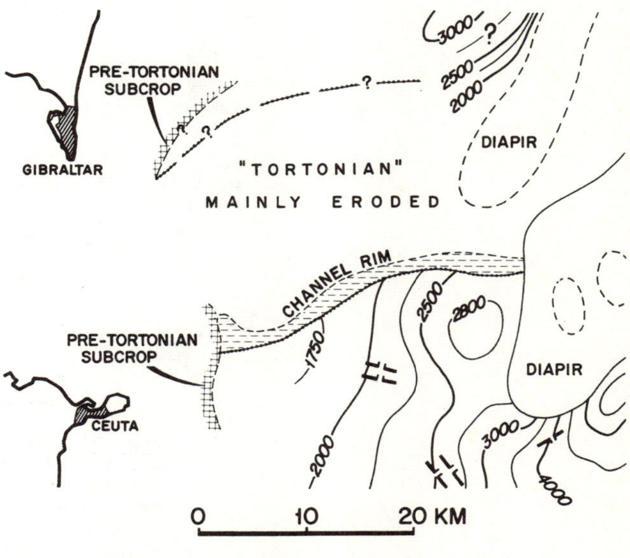

Fig. 18 - Structural contours of "base Tortonian" (hor. 3 in Figs. 5-14) and location of "lower Messinian" erosional channel. Depths in metres b.s.l.

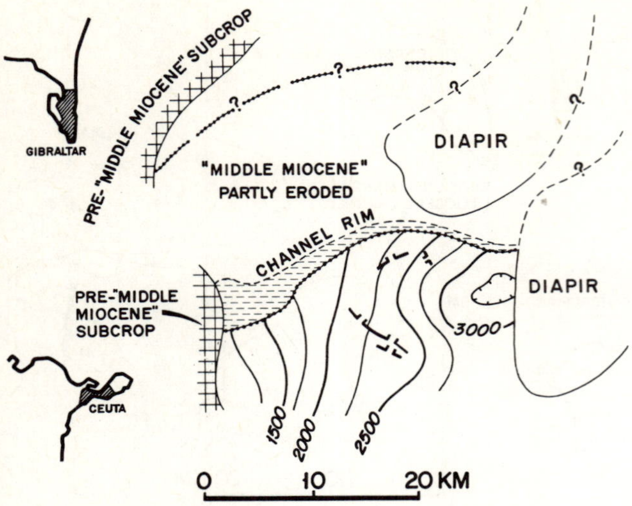

Fig. 19 - Structural contours of "intra-Middle Miocene" level (horizon 4 in Figs. 5-14) and location of "lower Messinian" erosional channel. Depths in metres b.s.l.

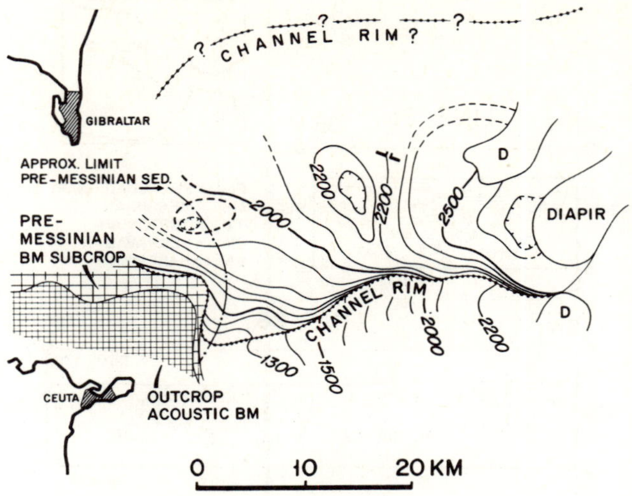

Fig. 20 - Structural contours of base "lower Messinian" channel. Depths in milliseconds two-way travel time b.s.l.

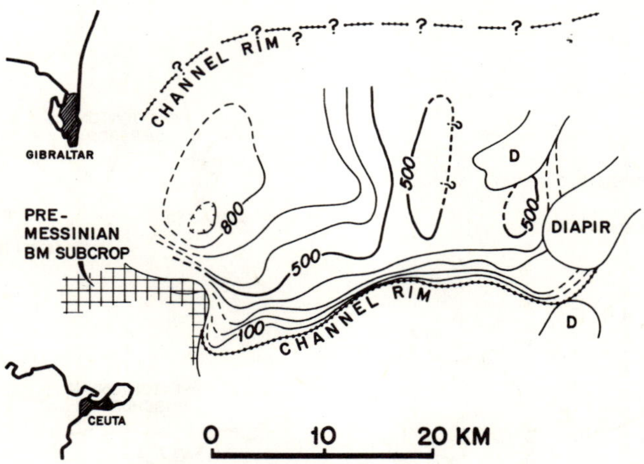

Fig. 21 - Extrapolated depth of erosion of "lower Messinian" channel. Contour interval 100 milliseconds two-way travel time. A marked erosional depression occurred close to the entrance of the Straits of Gibraltar, similar (but located farther east) to a present-day depression of the sea floor (Fig. 2).

INTERNATIONAL SYMPOSIUM ON THE STRUCTURAL HISTORY OF THE MEDITERRANEAN BASINS. SPLIT (YUGOSLAVIA) 25-29 OCTOBER 1976. B. BIJU-DUVAL AND L. MONTADERT, Eds. EDITIONS TECHNIP, PARIS 1977, pp. 411-420

THE DESICCATION EVENTS IN THE EASTERN MEDITERRANEAN DURING MESSINIAN TIMES AS COMPARED WITH OTHER MIOCENE DESICCATION EVENTS IN BASINS AROUND THE MEDITERRANEAN

G. GVIRTZMAN and B. BUCHBINDER [1]

ABSTRACT

Two phases of salinity crises occurred during the Miocene in the Tethys area in and around the Mediterranean. During the first phase, in the Middle Miocene (Langhian-Serravallian), 16-13 my BP, the desiccation occurred in the Mesopotamian Basin, in the Red Sea and in the Paratethys basins. The desiccation of the Mesopotamian Basin and the Red Sea was caused by the uplifting of the Arabian-African platform along the Levant fault system, which ran parallel to the present Levant and Sinai coasts. This movement caused the separation of the eastern arm of the Tethys (Syria-Mesopotamia-Indopacific Ocean) from the western arm (Mediterranean-Atlantic Ocean). In the Late Miocene (Messinian), 7-5 my BP, the Mediterranean basins were subjected to desiccation. A detailed study of the Israeli Coastal Plain and Continental Shelf, indicates that the desiccation events occurred in two stages. During the first stage, the Arabian and African platforms were elevated to about 2000 m above the Levant Basin and the resulting regression deposited anhydrite in sabkhas on an eroded relief of the sloping platforms. During the second stage, a cutoff of the Mediterranean from the global ocean system caused the formation of salt lakes in closed endorheic depressions which had regimes resembling that of the modern Dead Sea. The flow of waters and salts from the Paratethys lakes and from continental drainage systems, in equilibrium with the loss of water through intensive evaporation and continuous tectonic subsidence of the basin floor, brought about the accumulation of a huge volume of sodium and probably potassium salts. A notable similarity with respect to the volume of evaporites and the rates of accumulation exists between the various evaporation basins during the Miocene. The maximum time span of each desiccation event in the various basins was about 0.5-3.0 my. The average rate of accumulation varied between 0.1-0.75 km^3 per year. All the desiccation basins were formed due to the fragmentation of the Tethys and the formation of isolated shallow basins which were cut off from the ocean. The floors of the various basins subsided rapidly as a result of either subduction or spreading. All the basins are characterized by a peripheral belt of sabkha anhydrites and an inner core of chloride salts. It is probable that the Dead Sea endorheic salt lake model with arid climate and excess evaporation over precipitation and replenishment, can be applied to all the Miocene desiccation basins in and around the Mediterranean.

RESUME

La Téthys méditérranéenne et péri-méditérranéenne fut le théatre de deux phases de dessication (crises de salinité) durant le Miocène. Durant la première phase, au cours

(1) Oil Research Division, Geological Survey of Israel, 30 Malkhei Israel Street, Jerusalem, Israel

du Miocène moyen (Langhien-Serravallien) de -16 à -13 m.a., la dessication affecta les bassins de Mésopotamie, Mer Rouge et ceux de la Paratéthys. La dessication des bassins de Mésopotamie et de la Mer Rouge fut causée par le soulèvement de la plateforme arabo-africaine en bordure du système des failles du Levant, parallèle aux côtes du Levant et du Sinai. Ce mouvement est à l'origine de la séparation du bras oriental de la Téthys (Syrie-Mésopotamie-Océan Indopacifique) de son bras occidental (Méditerranée-Océan Atlantique). Au cours du Miocène supérieur (Messinien) de -7 à -5 m.a., la dessication affecta les bassins méditerranéens. Une étude détaillée des phénomènes de dessication en Méditerranée orientale, comprenant de nombreuses données de la plaine côtière et du précontinent d'Israël, montre que les évènements messiniens peuvent être subdivisés à leur tour en deux phases. Durant la phase la plus ancienne, un soulèvement d'environ 2000 m des plateformes arabique et africaine au-dessus du bassin interne du Levant, engendra une régression de la mer, laissant derrière elle un glacis erodé, dont le relief permit la formation de sabkhas à dépôts d'anhydrites. Pendant la phase plus récente la coupure de la Méditerranée du système océanique global entraina la formation de lacs salés dans des dépressions à régime aride et endorhéique, semblable à celui de la Mer Morte actuelle. L'influx des eaux et des sels en provenance des lacs de la Paratéthys et des systèmes de drainage continentaux s'est trouvé en équilibre constant avec les mécanismes d'évaporation et de subsidence tectonique, permettant ainsi le dépôt d'immenses volumes de sels de sodium et probablement de potasse. La similitude des volumes d'évaporites et de leur taux d'accumulation dans les divers bassins miocènes est frappante. La durée de chaque période de dessication dans les divers bassins varie de 0,3 à 0,5 m.a., le taux d'accumulation se situant entre 0,1 et 0,75 km^3/an. L'origine de tous les bassins de dessication est due au morcellement de la Téthys et à la formation de bassins indépendants peu profonds, coupés de l'océan, tandis que la rapide subsidence de leurs fonds était produite par des mouvements de subduction ou d'expansion. Une ceinture périphérique d'anhydrites de sabkhas autours d'un noyau central de chlorures caractérise chaque bassin. Le modèle de la Mer Morte, lac salé endorhéique de climat aride et à évaporation excessive par suite de l'insuffisance des précipitations et du remplissage, peut être vraissemblablement appliqué à tous les bassins de dessication du Miocène périméditerranéen.

I. INTRODUCTION

Analyses of recent data from deep drillings and seismic surveys in the Coastal Plain and the Continental Shelf of Israel (Gvirtzman and Buchbinder, 1977) resulted in a better understanding of the geological history of the Late Tertiary in the eastern Mediterranean (see Fig. 1). These data indicate that a tectonic uplifting of the Arabian Platform took place along the Levant fault system at the beginning of the Middle Miocene which influenced the barring of the Mesopotamian and the Red Sea basins, consequently triggering the Middle Miocene desiccation phase in these basins (Buchbinder and Gvirtzman, 1976). Although the younger Messinian desiccation phase was probably restricted to the Mediterranean basins only, we believe that the eastern Mediterranean Basin, which has been more intensively studied, can serve as an inductive model for a better understanding of the less known Middle Miocene desiccation basins around the Mediterranean. The following discussion focuses on the nature of the Messinian desiccation in the eastern Mediterranean Basin and attempts to compare it with the earlier Middle Miocene desiccation phase in the Mesopotamian and Red Sea basins and in the Paratethys.

II. THE MESSINIAN DESICCATION PHASE OF THE EASTERN MEDITERRANEAN

A. PALEOGEOGRAPHIC RECONSTRUCTION

During the Messinian, the eastern Mediterranean region was bordered on the south and east by the African and Arabian platforms and on the north by the orogenic Mediterranean Ridge (Fig. 2). The platforms were at least 2000 m higher than the floor of the inner basin due to tectonic uplifting along the Levant fault system (Gvirtzman and Buchbinder, 1977). In the areas now occupied by the Levant and Herodotus abyssal plains, two giant salt lakes developed in which about 2-3 km of sodium and probably potassium salts accumulated. The boundaries of these lakes (Fig. 2) are proposed in accordance with the distribution of seismic patterns indicating salt deformation or diapirs, as shown on the map of Biju-Duval et al. (1974)

and of Neev et al. (1976). According to the map of Biju-Duval et al. (1974), it can be assumed that the northern side of the Mediterranean Ridge was occupied by other salt lakes, namely, Iskenderun, Antalya and Cilicia. On the elevated areas around the salt lakes, anhydrite of sabkha origin was deposited and in local depressions salt and anhydrite were deposited under subaqueous conditions. In Israel the sabkha evaporites, evidenced by nodular anhydrites and algal mats, developed continuously over the slope of the platform (Gvirtzman and Buchbinder, 1977), and similarly, probably along the coasts of Lebanon and Syria. According to Said (1975), Messinian anhydrite developed below the sediments of the Nile Delta as well (see Gvirtzman and Buchbinder, 1977).

B. DRAINAGE SYSTEMS

The reconstructed pattern of the drainage system from the Arabian Platform into the Levant Basin (Fig. 2) shows a high degree of superposition between the various drainage systems in the Miocene and the younger drainage systems that have developed since then up to the present day. We therefore presume that the modern continental drainage system in the Levant area, at least from the water-divide seawards (pattern 1a, Fig. 2), is basically the headward continuation of the subsurface system in the Coastal Plain and the Continental Shelf area. It shows the convergence of high order distributaries (higher than fourth order) on the platform, into distributaries of third and fourth order on the slope, and finally, into main channels of first and second order in the Levant Basin. These main channels may possibly have entered the salt lakes through a series of alluvial aprons typical of desert landscapes. (For details, see Gvirtzman and Buchbinder, 1977.) Reconstruction of the Messinian Nile River (based on data of Said, 1975) shows incision of the platform edge by a canyon through the area which is today covered by the Pliocene-Pleistocene Nile Delta. The Messinian Nile probably discharged into the Herodotus salt lake. Unpublished subsurface data indicates that another drainage system in northwestern Sinai (west of the Bardawil lagoon), which is partly concealed by modern dune sands, probably flowed into the Nile channel and ultimately into the Herodotus salt lake.

C. DEEP DESICCATION VERSUS SHALLOW DESICCATION MODEL

There are two main contradicting opinions regarding the topography of the Mediterranean during desiccation times. One opinion, expressed by Hsu et al. (1973), assumes that during the Messinian the basin consisted of a deep dry hole of several thousand meters, more or less similar in depth to the present floor of the Mediterranean. The opposing opinion, as presented by Nesteroff (1973) and later by many other authors, regards the basin as a shallow depression of no more than a few hundred meters deep, whose present-day bottom topography resulted from a tectonic subsidence during Plio-Pleistocene times.

It seems likely that the nature of the marine intercalations within the Messinian evaporites in Israel holds a significant clue to the solution of this debate. There are two alternative explanations for these intercalations:

1) The intercalations are related to repeated filling of the Mediterranean Sea through the Strait of Gibraltar (Hsu, et al., 1973) during the desiccation event, whereby each marine intercalation represents a new fresh refill. Eight to ten repeated fillings were presumed by Hsu et al. (1973) in order to account for the total volume of the Messinian evaporites in the Mediterranean Basin. These authors found a reasonable correlation between the marine intercalations in the the Messinian sequence of Sicily and the theoretical 8-10 marine refilling cycles during the Messinian in the Mediterranean. The following is implied from this model: a. Every marine intercalation and every interbedded evaporitic layer is practically synchronous over the entire Mediterranean. b. Each marine intercalation and evaporite layer continuously paved the topography of the basin from its elevated margins down to its floor. c. A good correlation should exist between the number of marine intercalations and their individual characteristics in different and remote localities in the Mediterranean.

2) The Messinian regression and desiccation in the eastern Mediterranean is related to the tectonic uplifting of the surrounding platforms. This model calls for one continuous regression with the marine intercalations representing oscillations in sea level during the regression. The

higher marine intercalations at the edge of the platform would, accordingly, be the first to be formed, and the lower intercalations in the basin would have been formed last. The whole sequence of the marine intercalations probably represents deposition at an early stage of the Messinian event. The later stage was probably marked by the cutoff of the Mediterranean from the global ocean system and was characterized by evaporitic deposition only, in the salt lakes in the central basins. The following is implied from this model: a. During regression at the first stage, the salinity of the sea water remained more or less normal, having marine plankton and the potential to deposit bioclastic limestones along the shoreline. The evaporites accumulated only along coastal sabkhas and sea marginal flats. Any interruption in the regressive tendency due to either local small-scale eustatic or isostatic movements resulted in a short-term marine ingression with a related marine intercalation within the evaporite sequence. b. The marine intercalations are not synchronous over the entire basin. The number of intercalations at different localities is not constant. On the contrary, the proposed model requires diachronous relationships, i.e., the intercalations on the upper flanks of the basin are older than those on the lower topography.

The number and distribution of marine intercalations in the evaporitic sequence found in most boreholes in Israel supports the second interpretation. The number of intercalations is not constant on either the upper or the lower parts of the slope. There is also no correlation between their stratigraphic location and characteristics. It is therefore likely that the evaporites including the marine intercalations are diachronous, as proposed in the second model: the oldest are on the platform edge and the youngest are on the lower slope and in the Levant Basin.

D. WATER AND SALT BUDGETS

The assumption that the Gibraltar Strait is the exclusive or main gateway of the water and salt supply for the Messinian evaporites (Hsu et al., 1973) is over-simplified and is not unequivocal for the entire Mediterranean - particularly not for the eastern Mediterranean. There is no need to apply a mechanism of 8-10 repeated fillings and dryings. A minor discharge from rivers like the Nile, in addition to a continuous water supply from the Paratethys lakes (as proposed by Sonnenfeld, 1975), could have accounted for the Messinian evaporites. The salt lakes in the eastern Mediterranean could have existed continuously in equilibrium with an input from drainage systems and an output of evaporation. The eastern Mediterranean could have been a huge endorheic depression during Messinian times, in which the arid climatic conditions enhanced evaporation - a miniature example being the Jordan-Dead Sea catchment area in the Rift Valley (Neev and Emery, 1967), which has been in existence since the Early Pleistocene.

E. TWO STAGES IN THE DESICCATION OF THE EASTERN MEDITERRANEAN

We suggest that the Messinian history be divided into two stages. The first stage is characterized by a continuous, yet oscillatory, regression. The Arabian and African platforms and probably other surrounding platforms were elevated hundreds to thousands of meters above the global sea-level. Their pre-existing canyons were rejuvenated and paved by nodular anhydrite of coastal sabkha origin on sea marginal flats. In the central basins relatively shallow marine deposition prevailed with a possible connection to the global ocean system, allowing planktonic foraminifera to thrive. The second stage started with the cutoff of the Mediterranean from the global ocean system which resulted in the formation of the salt lakes in the depressions. Although the salt basins could have been lower than the ocean level, they were not necessarily very deep. The basins were very similar to the Dead Sea or to other depressions like Qattara, Dakhla and Farafra in Egypt. We believe that these two stages, namely, regression and desiccation, took place during the Messinian, 7-5 my BP (Ryan et al., 1974), although the duration of each stage is not yet clear.

III. THE MIDDLE MIOCENE DESICCATION PHASE AROUND THE MEDITERRANEAN

Prior to the Middle Miocene desiccation, the

Tethys Ocean extended, in the shape of a narrow elongated arm, from the Atlantic Ocean through the Mediterranean Sea and the Mesopotamian Basin into the Indopacific Ocean, with a bottleneck in the Levant area. As a result of the uplifting of the Arabian Platform along the Levant fault system, which roughly coincides with the present coastline of the Levant (Fig. 2), the Tethys was split into an eastern arm - from northwest Syria eastward into the Indopacific Ocean - and a western arm - from the Levant westwards through the Mediterranean into the Atlantic Ocean. Simultaneously, the Red Sea was also separated from the Mediterranean along the southwestern extent of the Levant fault system (Fig. 2). The cutting off of the Mesopotamian Basin from the Mediterranean and its restriction from the Indopacific at the Oman bottleneck resulted in its isolation and desiccation. The Red Sea was also isolated and desiccated since there was no connection with the Indian Ocean through the Bab-el-Mandeb Strait at that time. A more detailed paleogeographic reconstruction is now being prepared by the authors and will be published elsewhere.

According to the data from the Coastal Plain and Continental Shelf of Israel (Gvirtzman and Buchbinder, 1977) the absolute age of the tectonic movements which caused the above-mentioned events corresponds more or less to the first appearance of Orbulina or the "Orbulina-datum". The absolute age of this datum has been revised by different authors and the absolute ages of Units N8-N11 of Blow (1969), below and above the Orbulina-datum, have been changed accordingly. Berggren (1972, p. 210) attributed the age of 16 my BP to the Orbulina-datum. Later, Berggren and Van Couvering (1974; Fig. 11, pp. 93-96) revised the age of the Orbulina-datum to 14 my BP. Following their results, we attributed (in an earlier paper - Buchbinder and Gvirtzman, 1976) the movements of the Levant fault system to 14 my BP. However, at the end of Berggren and Van Couvering's paper (1974; Fig. 15, pp. 170-171) they suggested 15 my BP for this datum. Ikebe (1973) also proposed 15 $\pm$ 1 my BP for the same datum. Ryan et al. (1974) proposed the age of 16 my BP for the Orbulina-datum, as was originally proposed by Berggren (1972). We prefer the chronologic scale of Ryan et al. (1974) because of its comprehensive use of the interrelationships among paleomagnetic polarity changes, rates of oceanic sea-floor spreading, rate of sedimentation of biogenous ooze, biostratigraphic zonation of JOIDES cores and comparison with the Italian stratotypes (Gvirtzman and Buchbinder, 1977, and Fig. 1 herein). The age of 16 my BP for the Orbulina-datum is also recognized in the Paratethys basins. Vass (1975) and Vass et al. (1975) attributed an average age of 16 my BP to the base of the Badenian, or the base of the CPN 6-7 zones, in the Paratethys basins. They think that this datum corresponds to the N8/N9 boundary and therefore to the Orbulina-datum.

Since the uppermost sediments below the evaporites in the Mesopotamian Basin contain Orbulina and Borelis melo, the desiccation postdates the first appearance of these fossils, i.e., later than 16 my BP, according to Berggren's (1972, p. 210) time scale. Similarly, since the sediments which overlie the evaporites still contain Miogypsina, the desiccation ceased before the extinction of this fossil, i.e., approximately before 13 my BP, according to Berggren's time scale (1972, p. 210, the extinction of Miogypsina in the Far East). Detailed stratigraphic data and references will be given in a study now in preparation.

The desiccation of the Red Sea, according to the data from the Gulf of Suez, started after the first appearance of Orbulina (16 my BP). The main evaporite unit of the Gulf of Suez - the South Gharib Formation, was deposited before the termination of Unit N10 (Kerdany, 1968), approximately 15 my BP (time scale of Ryan et al., 1974). The age of the overlying Zeit Formation, which is also composed of evaporites, has not been definitely established, though it might also be related to the Messinian.

According to Rogl et al. (1977), the northward movement of plates which narrowed the Mesopotamian trough, almost sealed off the oceanic connection to the Paratethys and eventually resulted in the Late Badenian, 14.5-14.0 my BP, desiccation of the Paratethys basins (Transylvania, northeast Hungary and Slovakia within the Alpine orogenic belt; eastern and southern Romania, Bulgaria, Poland and the Ukraine in the Carpathian foredeep). In our opinion, these events also correspond to the Middle Miocene desiccation phase around the Mediterranean.

In conclusion, the desiccation events in the basins around the Mediterranean Sea are clustered within the time span of about 16-13 my BP, namely, Langhian to Serravallian, and are designated here as the Middle Miocene desiccation phase.

IV. TIME, VOLUME AND RATES OF MIOCENE EVAPORITE ACCUMULATION IN AND AROUND THE MEDITERRANEAN

The study of the desiccation processes in the basins around the Mediterranean and especially in the eastern Mediterranean Basin reveals notable similarities among the various desiccation basins. The fragmentation of the Tethys and its conversion into secondary shallow basins took place as a result of intensive tectonic activity during the Miocene. As shown in the eastern Mediterranean Basin, the process begins with a regression and shallowing of the basin, and ends with its cutoff from the global ocean system. This trend of events leads to a facies zonation which typifies most of the basins: a peripheral zone of nodular anhydrite of coastal sabkhas fringing the regressive coastline, and in the center of the basin, the accumulation of chloride salts which were precipitated from residual brines following the cutoff phase. There is also a striking similarity in the time spans of the desiccations in the various basins, in the volume of the accumulations in each basin, and in the rate of accumulation of the evaporites. It is therefore highly probable that the conditions prevailing in the eastern Mediterranean during the Messinian also existed in the surrounding desiccation basins during the entire Miocene. Some of these similarities, which have been extracted from a more comprehensive study now in preparation and which also includes the relevant references, are presented below.

The desiccation of the Mesopotamian Basin started with the deposition of sabkha anhydrite along the margins of the basin, after which chloride salt accumulated in the center of the basin. During a time span of no more than 3 million years, about 0.8 million km^3 of evaporites accumulated at an average rate of more than 0.27 km^3 per year. The space for this volume of evaporites was provided by the subduction of the Arabian plate under the Iranian plate along the Zagros-front.

The zonation of sabkha anhydrites with subordinate salt along the margins, and thousands of meters of salts in the center of the basin is also apparent in the Red Sea (according to data from the Gulf of Suez). During a time span of no more than 1 million years, about 0.11 million km^3 of evaporites accumulated at an average rate of more than 0.11 km^3 per year. The space for this huge accumulation of evaporites was provided by the rifting and the spreading of the Arabian plate away from the African plate.

In the Paratethys basins on both sides of the Carpathian chain, around 0.2 million km^3 of evaporites accumulated during a time span of no more than 0.5 million years. The details of the facies zonation are not clear to us. The average accumulation rate was more than 0.4 km^3 per year. The enormous space for this accumulation was provided by the subduction along the Carpathians.

Various Mediterranean basins such as the Balearic, the Ionian and others functioned as separate basins during the Messinian events. The anhydrite belt along all the margins of the Mediterranean is apparent, and the centers of the basins are marked by salt accumulation with diapir structures. During a time span of no more than 2 million years, 1.5 million km^3 of evaporites accumulated in the entire Mediterranean, at an average rate of 0.75 km^3 per year.

V. CONCLUSIONS

The following conclusions relate to the desiccation of the eastern Mediterranean during Messinian times:

1) The desiccation occurred in two stages: the first consisted of normal regression followed by the formation of sea marginal flats and sabkhas; and the second resulted from the cutoff of the Mediterranean basins from the global ocean system and the formation of endorheic salt lakes.

2) The endorheic basin, though below sea level, was not very deep. It received a continuous water and salt supply from the Paratethys lakes and from rivers.

3) The assumption that the Gibraltar Strait is the exclusive or main gateway for the water and

salt supply of the Messinian evaporites does not necessarily apply for the eastern Mediterranean Basin.

4) The climatic conditions were characterized by considerable excess of evaporation over precipitation and replenishment.

The following conclusions relate to the desiccation of all the Miocene basins in and around the Mediterranean:

1) The desiccation events were clustered in two Miocene phases: a. Langhian-Serravallian (16-13 my BP); b. Messinian (7-5 my BP).

2) The time span of each event, in almost every basin, was in the range of 0.5-3.0 my.

3) The average rate of accumulation of evaporites in almost every basin was in the range of 0.1-0.75 km^3/year. Although the figures of volume represent only rough restimates, it is believed that the order of magnitude of less than one cubic kilometer per year is a good estimate.

4) The major control of the salinity crises during the Miocene in the Tethys-Mediterranean belt is the tectonic shallowing and fragmentation of the Tethys into isolated basins, within or outside of the Alpine belt, and their disconnection from the global ocean system.

5) All the basins were subjected to continuous subsidence due to either subduction (Mesopotamia, Levant, Paratethys, etc.) of spreading (Red Sea), which kept pace with - or was faster than - the evaporite accumulation.

Salinity crises of such orders of magnitude as the Miocene crises are not so common in the geological history. The Miocene salinity crisis in the Tethys area resulted from the unique combination of specific tectonic and climatic conditions which prevailed during the breakup of the Tethys Ocean into separate isolated basins.

ACKNOWLEDGMENTS

This paper constitutes part of the results of Research Project No. 29371, carried out at the Geological Survey of Israel. Thanks are due to Mrs. Laura Buchbinder, Mrs. Beverly Katz and Dr. Francis Hirsch for their help in preparing this manuscript.

REFERENCES

BERGGREN, W.A., 1972, A Cenozoic time-scale - some implications for regional geology and paleobiogeography. Lethaia, v. 5, pp. 195-215.

BERGGREN, W.A.; VAN COUVERING, J.A., 1974, The Late Neogene: biostratigraphy, geochronology and paleoclimatology of the last 15 million years in the marine and continental sequences. Paleogeography, Paleoclimatology, Paleoecology, v. 16, pp. 1-216.

BIJU-DUVAL, B.; COURRIER, P.; APOSTOLESCU, V.; LETOUZEY, J ., 1974, Carte geologique et structurale de bassins tertiaires du domaine Mediterraneene, 1:2,500,000 en 2 coupures. Institut Francais du Petrole et Centre National pour l'Exploration des Oceans. Technip, Paris.

BLOW, W.H., 1969, Late Middle Eocene to Recent planktonic foraminiferal biostratigraphy. 1st International Congress on Planktonic Microfossils, Geneva, 1967, v. 1, P. Bronnimann and H.H. Renz, eds., Brill, Leiden, pp. 199-475.

BUCHBINDER, B.; GVIRTZMAN, G., 1976, The breakup of the Tethys Ocean into the Mediterranean Sea, the Red Sea and the Mesopotamian Basin during the Miocene: A sequence of fault movements and desiccation events. Abst., 1st Congress on Pacific Neogene Stratigraphy, Tokyo, pp. 32-35.

DERIN, B.; REISS, Z., 1973, Revision of marine Neogene stratigraphy in Israel. Israel J. Earth-Sci., v. 22, pp. 199-210.

GVIRTZMAN, G.; BUCHBINDER, B., 1977, The Late Tertiary of the Coastal Plain and Continental Shelf of Israel and its bearing on the history of the eastern Mediterranean. Initial Reports of Deep Sea Drilling Project, Gov't Printing Office, Washington, D.C., v. 42A (in press).

HSU, K.J.; CITA, M.B.; RYAN, W.B.F., 1973, The origin of the Mediterranean evaporites, pp. 1203-1232 in: Ryan, W.B.F., Hsu, K.J. et al. (eds.), Initial Reports of the Deep Sea Drilling Project, Gov't. Printing Office, Washington, D.C., v. 13, pt. 2, 1447 p.

IKEBE, N., 1973, Neogene biostratigraphy and

radiometric time-scale. Jour. Geosciences, Osaka City University, v. 16, pp. 51-67.

KERDANY, M.T., 1968, Note on the planktonic zonation of the Miocene in the Gulf of Suez region, U.A.R. Committee Mediterranean Neogene Stratigraphy Proc. IV Session, Bologna, 1967. Giornale di Geologia, Serie 2[a], v. 35, fasc. III, pp. 157-166.

NEEV, D.; ALMAGOR, G.; ARAD, A.; GINZBURG, A.; HALL, J.K., 1976, The geology of the southeastern Mediterranean. Geol. Surv. of Israel Bull. no. 68, 52 p.

NEEV, D.; EMERY, K.O., 1967, The Dead Sea, depositional processes and environments of evaporites. Geol. Surv. of Israel Bull. no. 41, pp. 1-147.

NESTEROFF, W.D., 1973, Mineralogy, petrography, distribution and origin of the Messinian Mediterranean evaporites, pp. 673-694 in: Ryan, W.B.F., Hsu, K.J. et al. (eds.), Initial Reports of the Deep Sea Drilling Project, Gov't Printing Office, Washington, D.C., v. 13, pt. 2, 1447 p.

ROGL, F.; STEININGER, F.F.; MULLER, C., 1977, Middle Miocene Salinity Crisis and Paleogeography of the Paratethys (Middle and Eastern Europe). Initial Reports of Deep Sea Drilling Project, Gov't Printing Office, Washington, D.C., v. 42A (in press).

RYAN, W.B.F.; CITA, M.B.; DREYFUS-RAWSON, M.; BURCKLE, L.H.; SAITO, T., 1974, A paleomagnetic assignment of Neogene stage boundaries and the development of isochronous datum planes between the Mediterranean, the Pacific and the Indian oceans in order to investigate the response of the world ocean to the Mediterranean "salinity crisis". Riv. Ital. Paleont., v. 80, no. 4, pp. 631-638.

SAID, R., 1975, The geological evolution of the River Nile, pp. 7-44 in: Wendorf, F. and Marks, A.E. (eds.), Problems in Prehistory: North Africa and the Levant, Southern Methodist Univ. Press, Dallas.

SONNENFELD, P., 1975, The significance of Upper Miocene (Messinian) evaporites in the Mediterranean Sea. Jour. Geol., v. 83, pp. 287-311.

VASS, D., 1975, Report on the working group on radiometric age and paleomagnetism, pp. 103-117 in: Senes, J. (ed.), Report on activity of the R.C.M.N.S. working group (1971-1975). IUGS Commission on stratigraphy, subcommission of Neogene stratigraphy, Bratislava, 154 p.

VASS, D.; SLAVIK, J.; BAGDASARJAN, G.P., 1975, Radiometric time-scale for the Neogene of Paratethys (a state to August 1, 1974), pp. 293-297 in: Senes, J. (ed.), Proceedings of the 6th Congress, Bratislava, September 4-7, 1975, v. 1; Commission on stratigraphy, subcommission on Neogene stratigraphy. Veda-Publishing House of the Slovak Academy of Sciences, Bratislava, 453 p.

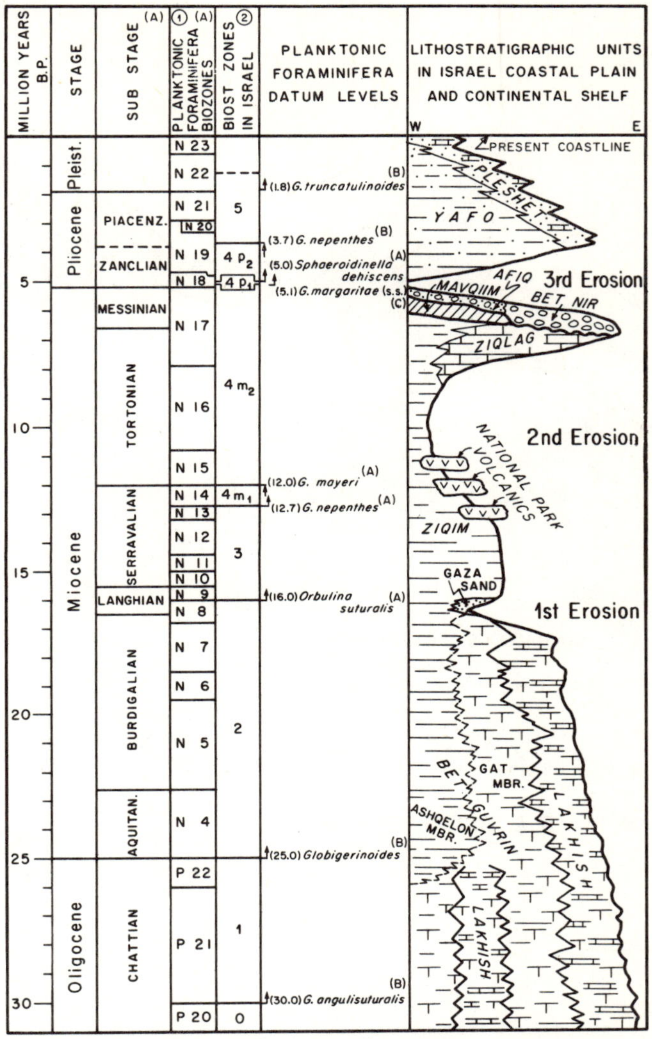

BIOSTRATIGRAPHIC UNITS AFTER: ① Blow 1969, ② Derin and Reiss 1973. POSITION OF BOUNDARIES AND DATUM LEVELS IN ABSOLUTE TIME SCALE AFTER: (A) Ryan *et al.* 1974, (B) Berggren and Van Couvering 1974, (C) Derin and Reiss 1973.

Figure 1 - Stratigraphy of the Oligo-Neogene in the Coastal Plain and Continental Shelf of Israel. Lithostratigraphic units are superimposed on a schematic E-W section perpendicular to the coastline. The stratigraphic sequence is divided by three erosional phases separating four sedimentary cycles. Each cycle is defined by an eastward transgression and a westward regression. The first erosion phase resulted from the tectonic uplifting of the Arabian and African platforms along the Levant fault system. The Messinian desiccation is associated with the third erosion phase.

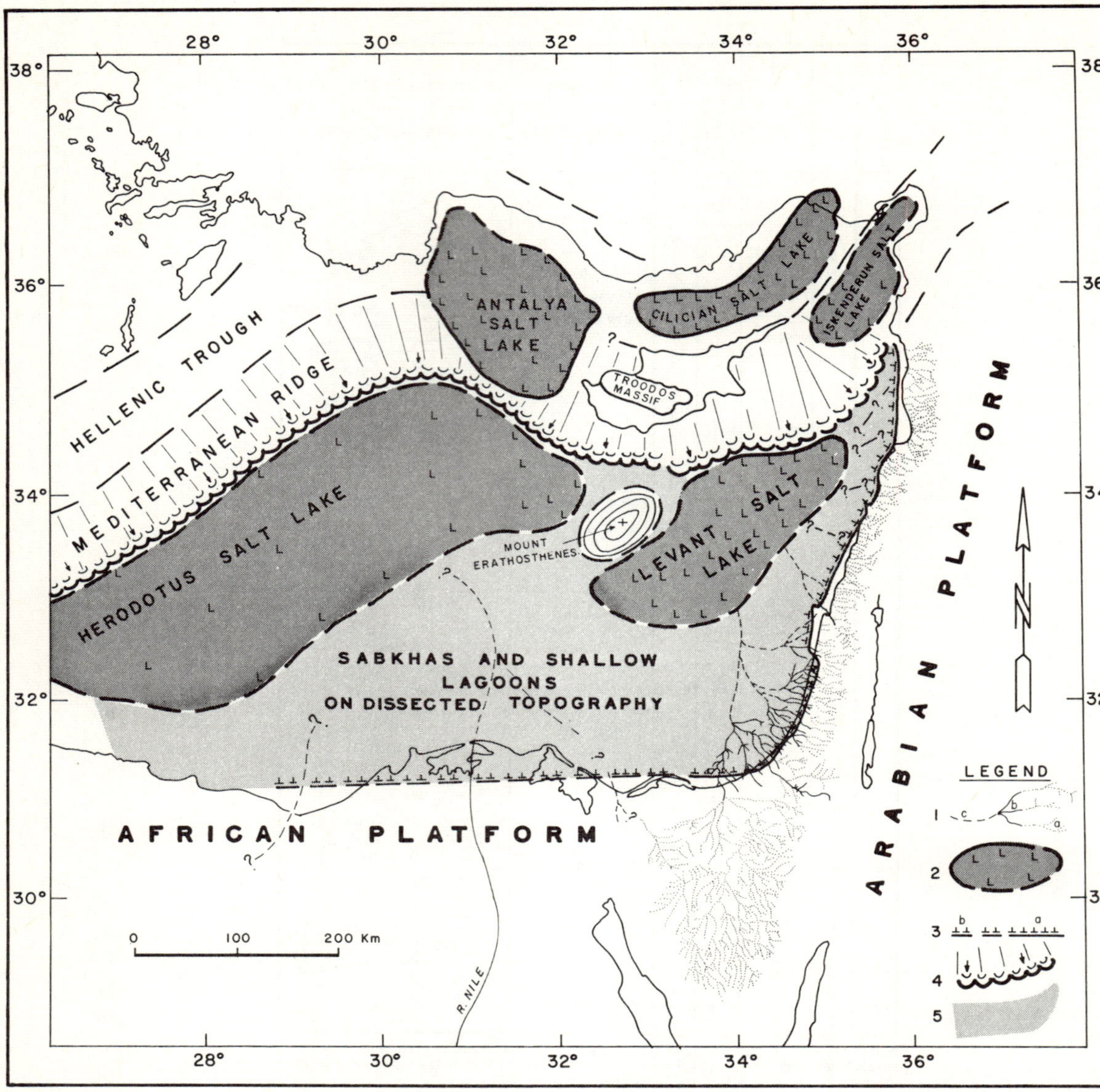

Figure 2 - Paleogeographic map of the eastern Mediterranean during the Messinian: 1 - Reconstructed drainage pattern: a-Recent drainage system on the Arabian Platform; b-Reconstructed drainage system from subsurface data of Israeli Coastal Plain and Continental Shelf; c-Postulated routes of the major channels in the basin area. 2 - Salt Lake; L-salt diapirs (after Biju-Duval et al., 1974). 3 - Edge of the Arabian and African platforms: a-inferred; b-hypothetical. 4 - The thrust front of the orogenic belt of the Mediterranean Ridge. 5 - Desiccated plains of the eastern Mediterranean Basin, covered by anhydrite.

INTERNATIONAL SYMPOSIUM ON THE STRUCTURAL HISTORY OF THE MEDITERRANEAN BASINS. SPLIT (YUGOSLAVIA) 25-29 OCTOBER 1976. B. BIJU-DUVAL AND L. MONTADERT, Eds. EDITIONS TECHNIP, PARIS 1977, pp. 421-422

HISTORY OF THE MEDITERRANEAN SALINITY CRISIS

K.J. HSÜ [(1)], L. MONTADERT [(2)], D. BERNOULLI [(3)], M.B. CITA [(4)], A.J. ERICKSON [(5)], R.E. GARRISON [(6)], R.B. KIDD [(7)], F. MELIERES [(8)], C. MÜLLER [(9)] and R. WRIGHT [(10)]

After Leg 13 of the glomar Challenger (Ryan (W.B.F.), Hsü (K.J.) et al., 1973) a model of a deep desiccated basin was proposed to explain the deposition of the Messinian evaporites in the Mediterranean. Nevertheless this explanation was still discussed by many geologists. Leg 42 A of the Glomar Challenger (Hsü (K.J.), Montadert (L.) et al., 1977) brought new results on the Messinian evaporites and pre-Messinian deposits, which combined with numerous new seismic data, allow to confirm and refine the hypothesis put forward after Leg 13.

An equatorial ocean was in existence during the Mesozoic between Africa and Eurasia; it was named Tethys. The Alpine Orogeny, culminating in the late Eocene and Oligocene, eliminated much of this ancient ocean. However, the relic Tethys, combined with the newly created post-orogenic basins, still managed to host a Mediterranean Seaway. The link was first broken in the Burdigalian with the collision of Africa and Eurasia along the Middle East front. This event started the gradual changes to cooler and drier climate in regions which were now situated on the west side of continent.

Middle Miocene movements cut off forever the opening on the east to the Indo-Pacific. Also the Perialpine Depression north of the Alps was raised and the communication between the Mediterranean and the Paratethys was severed. The considerable freshwater supply from Central Europe was now diverted to the Paratethys, and the reorganization of drainages placed considerable strain on the water budget of the Mediterranean. Finally, soon after the beginning of the Messinian stage, the last opening to the Atlantic, the Betic and Rif Straits were closed. Evaporative draw-down of the isolated Mediterranean sealevel, leading to desiccation, became unavoidable, when the

(1) Geological Institute, ETH, Sonneggstrasse 5, CH-8006 Zurich, Switzerland

(2) Institut Français du Pétrole, 1-4 avenue de Bois-Préau, 92506 Rueil-Malmaison, France

(3) Geological Institute at the University, Bernoullistrasse 32, 4056 Basel, Switzerland

(4) Istituto di Geologia, Universita, Milano, Italy

(5) Geology Department, University of Georgia, Athens, Georgia 30602, U.S.A.

(6) Earth Sciences Board, University of California, Santa Cruz, U.S.A.

(7) Institute of Oceanographic Sciences, Wormley, Great Britain

(8) Laboratoire de Géologie Dynamique, Université Pierre et Marie Curie (Paris VI), 4 place Jussieu, 75230 Paris Cédex 05, France

(9) Geologisch-Paläontologisches Institut, Johann Wolfgang Goethe Universität, Frankfurt a.M., Federal Republic of Germany, presently at Bureau d'Etudes Industrielles et de Coopération de l'Institut Français du Pétrole, Rueil-Malmaison, France

(10) Department of Geology, Beloit College, Beloit, U.S.A.

evaporative losses considerably exceeded the precipitation and the influx from rivers.

The Mediterranean Evaporite can be devided from seismic profiling, into two units. The deep sea drilling did not penetrate the Main Salt or the lower unit, nevertheless in two holes (372-374) it demonstrated that before the salinity crisis deep open marine conditions existed; this is confirmed by land sections in Italy. Thus the history of the onset of the salinity crisis and of the early salt deposition is not known by drilling. Nevertheless the land records in Italy indicate a very sudden change from deep, open marine conditions to environments of shallow-water carbonate evaporite deposition and of subaerial diagenesis. This first dessication was followed by the deposition of the Main Salt. A thick body of salt originated from the evaporition of seawater spilled over from the Atlantic into Mediterranean brine lakes of considerable depths.

Alternatively to this first desiccation, one can suggest also a continuous transition from deep open marine sediments to salt deposited in deep brine lakes.

An intra-Messinian desiccation led to widespread erosion demonstrated by seismic profiling and drilling, and recycling of primary halites. The Upper Evaporite deposition followed a marine flooding which may have again filled the Mediterranean to the brim. Evaporative draw-down of the Mediterranean sealevel and the gradual concentration of the brines led to formation of evaporites characterized by the "bull's eye" pattern of saline zonation : dolomite was the earliest mineral formed, followed by sulphate deposition. Halite was laid down on the bottom of playas, which are now abyssal plains, and the potash salts were restricted to the deepest and the very center of the Mediterranean basins. Repeated flooding and desiccation let to the accumulation of cyclically deposited sediments. Shallow-water diatoms and algal stromatolites proved that the basins were covered by very shallow waters during the flooding stage, and sedimentary structures indicate subaerial exposure during the desiccation stage, of those cycles.

The Mediterranean was underlain by a series of deserts and salt lakes towards the end of the Messinian, when the eastern Mediterranean was inundated by brackish water, probably from a Paratethys source. A number of lakes known collectively under the name Lago Mare came into existence. Marls, or dolomitic marls with an Ammonia-Cyprideis fauna were deposited in many of those lakes. The western Mediterranean may have remained desiccated before the Atlantic water began to spill over the western portal. The infill began with the deposition of marls with restricted marine, dwarf faunas.

The final and irrevocable flooding of the Mediterranean took place at the beginning of the Pliocene. Normal marine circulations prevailed ever since.

BIBLIOGRAPHY

RYAN (W.B.F.), HSU (K.J.) et al.- (1973). Initial reports of the Deep Sea Drilling Project, vol. XIII, Washington (US Government Printing Office, 1 447 p.

HSU (K.J.), MONTADERT (L.) et al.- (1977). Initial reports of the Deep Sea Drilling Project, vol. XLII, Washington (US Government Printing Office), to be published.

HSU (K.J.), MONTADERT (L.), BERNOULLI (D.), CITA (M.B.), ERIKSON (A.), GARRISON (R.E.), KIDD (R.B.), MELIERES (F.), MULLER (C.) and WRIGHT (R.).- (1977). History of the Mediterranean Salinity Crisis. To be published in Nature.

INTERNATIONAL SYMPOSIUM ON THE STRUCTURAL HISTORY OF THE MEDITERRANEAN BASINS. SPLIT (YUGOSLAVIA) 25-29 OCTOBER 1976. B. BIJU-DUVAL AND L. MONTADERT, Fds. EDITIONS TECHNIP, PARIS 1977, pp. 423-432

REPERCUSSIONS OF THE PLIO-PLEISTOCENE TECTONIC ACTIVITY ON THE DEEP-SEA CLAY SEDIMENTATION IN THE MEDITERRANEAN

H. CHAMLEY, Gh. GIROUD D'ARGOUD and Ch. ROBERT [1]

Abstract.

The deep-sea plio-pleistocene sediments of the Mediterranean show in their clay fraction some marked mineralogical changes that occured over a very short space of time and do not seem to be connected to variations in climate or current. We put forward a tectonical explanation : faster subsidence of marine areas at Early Pliocene ; raising of Sicily at Middle Pliocene ; uplift of Hellenides, Taurids and probably the other alpine chains at Early Pleistocene, perhaps at Lower Pliocene and Lower-Middle Pleistocene as well. The mineralogical interpretations are discussed with respect to the tectonic data originating from the surrounding land-masses.

Résumé.

La fraction argileuse des sédiments marins profonds de Méditerranée présente au Plio-Pléistocène des variations minéralogiques brusques et importantes, qui se superposent aux changements d'amplitude secondaire habituellement déterminés par les modifications du régime climatique ou courantologique. Ces variations sont interprétées par des changements structuraux du bassin méditerranéen : affaissement accéléré du domaine marin au Pliocène basal ; soulèvement de la Sicile au Pliocène moyen ; mouvements positifs des Hellénides, des Taurides et sans doute des autres chaînes alpines vers la base du Pléistocène, ainsi semble-t-il qu'au Pliocène inférieur et vers la base du Pléistocène moyen. Ces interprétations sont discutées dans le cadre des observations structurales effectuées à terre.

INTRODUCTION

The physiographical changes that occur during the ages in the sedimentary basins can leave some traces in the mineralogy of the deposits. The mineralogical composition of inherited materials can be affected by changes such as relief renewal, outcropping of new geological levels, or the opposite effects. We must be sure that the tectonic changes have not been thrown into confusion by diagenetic alteration, water circulation or climatic events.

We consider here the clay fraction (less than 2 μm) from Mediterranean basin sediments, deposited since the end of the Miocene. Such minerals, areas and ages are well-suited to the research of tectonic and mineralogical relationships for several reasons :

- The plio-quaternary sedimentation is hemipelagic and the detritic components of the deposits are essentially argillaceous. The

(1) Laboratoire de Géologie Marine, Centre d'Océanographie, Centre Universitaire de Luminy, 70 route Léon Lachamp, 13288 Marseille Cédex 2, France

dissemination properties of the clay particles encourage large-scale studies and the identification of moderate structural changes.

- The Mediterranean has not been affected by large communications of water from the world-ocean since the Pliocene. The deep-water circulation could only vary in small proportions, that is not sufficient to induce some important mineralogical changes. The clays from the Mediterranean areas, whose composition is well-known, are arranged according to the great geological zones, their disposition is reflected by recent sediments (Rateev et al. 1966, Chamley 1971).
- The Plio-Quaternary clays are devoid of burial diagenetic alteration, and the only modifications brought about by the chemistry of the deposit environment are easily recognizable and localized (sapropels, some volcanogenic sediments). Moreover, the climate has a global character established since the growth of the icecaps during the Upper Miocene ; the climatic changes induce only some progressive and moderate changes in the soil-shaped clays.
- At last, the studies carried out in the semi-closed Mediterranean basin on the relations between the circulation and mineralogy show the slight and progressive character of the current changes (see for inst. Chamley 1971, 1975a & b, Ciaranfi and Cita 1973, Cita et al. 1973, Ruddiman 1971, Ryan et al. 1974, Venkatarathnam and Ryan 1971).

In opposition to these moderate changes, the tectonic modifications described by many authors around the Mediterranean since the Upper Miocene sometimes correspond to important events liable to have an effect on the deep-sea sediments (see Ryan et al. 1971, Ryan, Hsü et al. 1973, Biju-Duval 1974, Biju-Duval et al. 1974).

We intend to restate the contribution of the major mineralogical changes established off-shore to the recent structural evolution of the Mediterranean area. Such an approach must be considered carefully because, in addition to its indirect applications, it refers to only a few sites. The results discussed here concern the X-ray diffraction and electro-microscopic study of about 800 samples, distributed in five sites (fig. 1) : Tyrrhenian sea (DSDP 132), South Sicily (Capo Rossello section), Messina abyssal plain (DSDP 374), East Mediterranean rise (Ionian Sea, DSDP 125-125A), Florence rise (West Cyprus, DSDP 376). The main data considered here in a chronological order are summarized in figure 2. The detailed results and discussions can be consulted in Chamley (1975a & b, 1976) and Mélières et al. (1977).

I. EARLY PLIOCENE

The basal hemipelagic Pliocene shows some important modifications in the clay mineralogy, compared to the evaporitic Late Miocene :

- The inherited fine particles increase considerably, indicating a spreading of the continental erosion.
- The nature of the minerals brought to the basin is greatly modified : where smectite, characteristic of the Messinian, seems to be chiefly of a pedogenic origin (Chamley et al. 1977a & b), it gives way to some varied associations, varying with the geographical location. These associations, whose components are inherited from the continent, always show a marked increase of attapulgite. The changes reach especially large proportions close to Africa, in the Messina abyssal plain (DSDP 374, cores 10 & 9) where the fibrous mineral increases from less than 5 % to about 60 % of the clay fraction (Mélières et al. 1977). West from Cyprus (DSDP 376) the base of the Plio-Quaternary is marked by numerous modifications but at each level, the biostratigraphy gives a Pliocene age (Hsü, Montadert et al. 1977), we note a noticeable attapulgite content (10-35 %).

In fact attapulgite appears in a special distribution in the rocks of the Mediterranean : missing or localized in the outcropping successions of Europe and Turkey, it exists in abundance in the mainly Paleogene sedimentary rocks which overlap the African and Arabian platforms (Millot 1964, Chamley 1971, Chamley and Millot 1975). Therefore the Mediterranean sediments reflect, during the basal Pliocene, the preferential erosion of these platforms, as well as the sharp decrease of the soil-formed products (smectite especially). That can be explained by an increase in the marine area subsidence at the beginning of the Pliocene, inducing the deceleration of the soil development and the acceleration of the erosion in the countries which are more advanced (southern platforms). Such an interpretation is in agreement with the speedy subsidence of the Mediterranean basin (see Burollet and Byramjee 1974a) with respect to the Lower Pliocene. It would be difficult to explain it without subsidence, when you have only incoming water in a very deep basin (see Drooger 1973), it is more likely to cause a decrease in erosion of the advanced countries.

A detailed mineralogical syccession is given by hole 132 from the Tyrrhenian sea, that probably includes the lowermost part of the Pliocene (extension of the *Sphaeroidinellopsis* zone, Cita 1973). Attapulgite, quite abundant when the hemipelagic deposits begin, is mainly replaced by iron oxides and kaolinite, and then increases slightly again within a more common clay suite (cores 21 & 20). Attapulgite is a mineral formed in the confined environments of the depressed zones of the basins, and its large-scale erosion in the lowermost part of the Pliocene suggests that the coastal plains contributed to the sedimentation. On the other hand iron oxides and kaolinite are well-drained environment minerals, which tend to indicate the erosion of the upper parts of the continental basins. A single climatic change at the beginning of the Pliocene would not really explain such a succession ; on the other hand it agrees with a fast subsidence of the marine area, inducing the erosion of the coastal, then the far off continental zones (Chamley 1975a). Note that site 374, in the Messina abyssal plain, gives a comparable succession of attapulgite then kaolinite enriched levels.

II. LOWER PLIOCENE

More or less represented in the five sites the Lower Pliocene remains characterized by noticeable amounts of attapulgite, with the exception of site 132 : at this site the removal of the African sources and the development of new soils and rocks around the Tyrrhenian area induces a progressive decrease in the abundance of attapulgite.

With the exception of the Capo Rossello section the attapulgite content in the whole Lower Pliocene is no longer as high as in the Early Pliocene : its abundance decreases very early, then becomes practically constant. This evolution suggests a decrease in the African countries contribution , and an increase in the northern contribution,perhaps related to the compression stage identified in the outer Aegean arc (Mercier 1976).

III. MIDDLE AND UPPER PLIOCENE

The Capo Rossello section in Sicily is the only one in which there is no decrease in the attapulgite content in the Lower Pliocene. Such a mineralogical change occurs later, towards the base of the Upper Pliocene (Chamley 1976). At the same time, the white hemipelagic marly limestones ("Trubi") turn to grey marls, which contain more fine detritic components ("Monte Narbone"). At once the attapulgite content decreases while the smectite and kaolinite content increases. Attapulgite is especially abundant in Africa and both the other minerals are characteristic of many sedimentary rocks of Sicily (Mascle 1973), also shown by the distribution of the three minerals in the recent sediments from the Sicily-Tunisia strait (Blanc-Vernet et al. 1975). So, the lower part of the Upper Pliocene shows an increase in erosion in Sicily, suggesting a strong uplift of the island at this time. Such an interpretation agrees with the structural observations of Mascle's thesis (1973) as well as the lithologic ones concerning shallower water sediments. Right through the Upper Pliocene the mineralogical change in content of the Monte Narbone marls increases, and the planctonic carbonates content decreases further (Cita and Decima 1975) : that reflects the continuation of the Sicily uplift from the North towards the South, leading to the appearance in a coastal position of Capo Rossello and the whole Agrigento country at the Lower Pleistocene.

In studies of other sites, the Middle and Upper Pliocene do not show any important mineralogical change,though some tectonic events have been described onshore in several countries, especially in the Western Mediterranean area (see for instance Burollet and Byramjee 1974b). The reason maybe the characteristic situation of the Sicilian section, relatively near to the coast of the time and a long way from the deep-sea wells. However some rapid mineralogical changes occur in the Upper and Late Pliocene of sites 132 and 125. These changes could express some physiographical modifications in Southern Italy, Peloponnese or on the East Mediterranean rise. But their size is close to those induced by climatic changes on the clay minerals, and it does not seem possible to make a decision once and for all.

IV. LOWER PLEISTOCENE

The Tyrrhenian sea site only indicates some minor mineralogical fluctuations, chiefly attributed to the continental climatic effects. The three East Mediterranean sites show, almost on the Pliocene-Pleistocene boundary, a major change : the attapulgite content decreases steeply to be replaced by some more common minerals like smectite, kaolinite, illite, chlorite, according to the location. Furthermore quartz and feldspars are more frequently associated with the clay minerals. The size and the rapidity of this change evokes a physiographical modification. That is all the more probable as the climatic changes recorded towards 1,8 MA BP have little intensity and few are consistent with the sediment observations (Ciaranfi and Cita 1973, Chamley 1975a).

As far as site 376 (West Cyprus) is concerned, the explanation of the change does not seem difficult. The attapulgite formed on the African and Arabian platforms (Israel, Lebanon, Syria) decreases when the smectite issued from Taurids increases (Turkey, Cyprus : various ultramafic and volcanic rocks; see Desprairies and Lapierre 1973, Venkatarathnam and Ryan 1971). That implies an incrase in erosion in the second area and can indicate an uplift stage. Chlorite, which increases in small amounts at the same time as smectite, reflects the contribution of the classical crystalline rocks from alpine origin.

The Ionian sea presents a more complex case because when the attapulgite content decreases in the Early Pleistocene, it is not replaced by the same minerals in the two sites (374 and 125), about 200 km away from each other. The minerals concerned are smectite and kaolinite in the Messina abyssal plain, and illite and chlorite on the East Mediterranean rise.

The relative abundance of smectite and kaolinite in Sicily can justify their sudden increase in the Messina abyssal plain (site 374) if you consider that an active uplift stage occured in the lowermost part of the Pleistocene. That agrees with the observations onshore (see Mascle 1973). But site 374 is also theoretically dependant on the Peloponnese, which supplies a great flow of illite and chlorite, often associated with quartz and feldspars (Chamley 1971). Now, both these clay minerals decrease in the Messina abyssal plain whereas they increase quickly at site 125, located in a comparable distance of the Greek coasts. The explanation of this difference can lie in the existence, as early as the Lower Pleistocene, of the East Mediterranean rise : located between the Greek areas, purveyors of illite and chlorite, and the Messina abyssal plain, this high submarine relief would form a barrier against the dispersal of the northern and eastern deposits towards the South. That would favour in an indirect way the hemipelagic and turbiditic Sicilian supplies.

On the East Mediterranean rise, the replacement of attapulgite by illite and chlorite might be connected to a single increase in erosion on the emerged Greek area. But that cannot explain why, the Mediterranean rise already in a bathymetric position near the one today, the Pliocene sediments cored North of the rise in the Matapan trench contain an important amount of attapulgite (30%; see Chamley and Millot 1975). On the other hand the large deposits of attapulgite coming from the stable African platform, present on this high area, is not easily explained for the whole Pliocene. Towards the Plio-Pleistocene boundary, the Eastern Mediterranean rise can be expected to go through a major uplift stage. This uplift, which could be accompanied by an elevation of the Greek area (Keraudren 1970 - 1972) and perhaps by a deepening of the Hellenic trench, combined with the stability of the African platform, could induce the recorded mineralogical change. This hypothesis, when compared with the other ones (Chamley 1975b), now seems to be the most nearest to explaining most of the facts. It agrees with the recent estimates of several authors as to when the growth of the East Mediterranean rise occured (Ryan et al. 1973, Rossignol-Strick 1974).

Nevertheless there remains a problem. The tectonic mechanisms considered here seem to have occured earlier than those identified in the Greek countries. The neotectonicians describe a marked compression stage during the Lower Pleistocene in the Ionian islands, Peloponnese, Creta,... This phenomenon on the whole seems to occur after the Early Pleistocene, at least in most southern countries (see for instance Mercier 1976, Dufaure et al. 1975, Mercier et al. 1975, Sorel 1975, Angelier 1976, Péchoux et al. 1976). Indeed it does not seem very conceivable for the main characteristics of the mineralogical changes recorded offshore not to be the result of changes recorded onshore. On the other hand it would be curious if the major Pleistocene compression stages had no effect on the composition of the marine sediments. Perhaps the solution lies in the beginning of the tectonic phenomenon onshore, preceding the main event and having an early influence on the nature of the fine materials eroded then deposited. Perhaps a better correlation of the chronological references used onshore and offshore could also help the understanding of the apparent shifting. The best way to approach this question is by a collaboration between structuralists and sedimentologists from the same sedimentary sections.

V. MIDDLE AND UPPER PLEISTOCENE

A marked mineralogical change concerning the more recent Eastern Mediterranean sediments occured over an indefinite period (discontinued cores) but localized around the lower part of the Middle Pleistocene. Less pronounced than the break occuring at the Basal Pleistocene, this change shows an increase in the Alpine contribution to the deposits : illite, chlorite and quartz of Greek (or Calabrian) origin on the Mediterranean rise (site 125, - 17 m) and to a minor degree in the Messina abyssal plain (site 374, - 100 m), chlorite of Turkish or Cyprian origin on the Florence rise (site 376, - 24 m). The attapulgite of African or Arabic origin, soon present in small amounts, decreases further, especially on the East Mediterranean rise (5 % or less). That could be a repercussion of the tectonic events identified during the Middle Pleistocene in several areas of the outer Aegean arc (Keraudren 1975, Mercier 1976, Péchoux et al. 1976). Until the Upper Pleistocene and the Holocene, no major event disturbs the mineralogical successions of deep-sea sediments, which are mainly dependant on climatic fluctuations (Cita et al. 1977).

CONCLUSION

The clay minerals deposited during the Plio-Pleistocene in the Mediterranean can contribute to the identification of the major structural events which have affected the basin on account of several propitious facts : petrographic already well-known changes on a large scale, the Mediterranean being relatively independant with respect to the world ocean, moderate currents, absent or localized diagenesis, a fairly variable climate, and active tectonics. The mineralogical studies on sediments from South Sicily and deep-sea drills (Glomar Challenger, Legs 13 & 42A, DSDP - JOIDES) favour the consideration of some global movements, onshore and offshore, the present day knowledge of which is as follows :

- Basal Pliocene : acceleration in subsidence of the deeply-submerged marine areas at the present time, with successive erosion of the lower then the upper zones of the continental basins, chiefly of North Africa.
- Lower Pliocene : possible uplift of the northern Alpine areas, the erosion of which balances the African contribution.
- Middle Pliocene : uplift stage of Sicily, from the North to the South, continuing by some progressive movements into the Upper Pliocene.
- Basal Pleistocene : dilution of the sedimentary components inherited from African and Arabic platforms, because of an increase in erosion in the Hellenides and Taurids, probably in Sicily and Southern Italy as well. The cause could be an uplift of the northern areas, associated with one of the submarine East Mediterranean rise, and perhaps with the deepening of the Hellenic trench.
- Middle Pleistocene : positive movements of the outer Aegean arc, more restrained than in the lower part of the Pleistocene.

The most of these events agree with the fields observations and measurements on the Mediterranean periphery. Some others, especially in the lower part of the Pleistocene, occur before the events described on the northern land-masses. This sort of mineralogical study, extended to various sections of the Mediterranean and to larger periods of its history, will help to give a better understanding of the limits of the method used here.

AKNOWLEDGMENTS

The sampling was obtained thanks to the National Science Foundation (USA) for the Glomar Challenger Legs 13 and 42A holes, also to the boarding team for Leg 42A. She Sicilian samples were given by M. B. Cita. Funds for this research were provided by C.N.E.X.O. (France) under grants n° 74/962 and 75/5155. The technical assistance was provided by M.

Acquaviva and C.-H. Froget. We gained a great deal from critical observations of G.A. Auffret, J.J. Blanc, B. Keraudren, J. Letouzey, G. Mascle, F. Mélières, L. Montadert, L. Pastouret and D. Sorel.

BIBLIOGRAPHY

ANGELIER (J.).- 1976. Esquisse néotectonique d'une transversale de l'arc Egéen ; de la mer Egée à la mer de Libye. *4è réun. ann. Sci. Terre*, Paris, p. 11.

BIJU-DUVAL (B.).- 1974. Commentaires de la carte géologique et structurale des bassins tertiaires du domaine méditerranéen. *Rev. Inst. franç. Pétrole*, vol. XXIX, n° 5, p. 607-639.

BIJU-DUVAL (B.), LETOUZEY (J.), MONTADERT (L.), COURRIER (P.), MUGNIOT (J.F.) and SANCHO (J.).- 1974. Geology of the Mediterranean Sea Basins. in " The Geology of continental Margins". *Drake and Burk Edit.*

BLANC-VERNET (L.), CHAMLEY (H.), FROGET (C.), LE BOULICAUT (D.), MONACO (A.) et ROBERT (C.).- 1975. Observations sur la sédimentation marine récente dans la région siculo-tunisienne. *Géol. Méditerr.*, t. 2, p. 31-48.

BUROLLET (P.F.) et BYRAMJEE (R.).- 1974a. Evolution géodynamique néogène de la Méditerranée occidentale. *CR Acad. Sci.*, Paris, sér. D, t. 278, p. 1321-1324.

BUROLLET (P.F.) et BYRAMJEE (R.).- 1974b. Réflexions sur la tectonique globale. Exemples africains et méditerranéens. *Comp. franç. Pétroles*, Paris, Notes et Mém., n° 11, p. 71-120.

CHAMLEY (H.).- 1971. Recherches sur la sédimentation argileuse en Méditerranée. *Sci. géol.*, Strasbourg, Mém. 35, 225 p.

CHAMLEY (H.).- 1975a. Sédimentation argileuse en mer Tyrrhénienne au Plio-Pléistocène d'après l'étude du forage JOIDES 132. *Bull. Groupe franç. Argiles*, t. XXVII, p. 97-137.

CHAMLEY (H.).- 1975b. Sédimentation argileuse en mer Ionienne au Plio-Pléistocène d'après l'étude des forages 125 DSDP. *Bull. Soc. géol. France*, sér. 7, vol. XVII, n° 6, p. 1131-1143.

CHAMLEY (H.).- 1976. Minéralogie des argiles, lithologie et tectonique dans le Pliocène du Capo Rossello (Sicile). *CR somm. Soc. géol. France*, fasc. 2, p. 39-41.

CHAMLEY (H.), DUNOYER DE SEGONZAC (G.) et MELIERES (F.).- 1977 (a). Clay minerals in the Messinian (Upper Miocene) from the Mediterranean Area. *Init. Rep. Deep Sea Drill. Proj.*, Washington (U.S. Government printing office), 42A (in press).

CHAMLEY (H.), GIROUD D'ARGOUD (G.) et ROBERT (C.).- 1977 (b). Genèse des smectites messiniennes de Sicile. Implications paléoclimatiques. *Géol. Méditerr.* (in press).

CHAMLEY (H.) et MILLOT (G.).- 1975. Observations sur la répartition et la genèse des attapulgites plio-quaternaires de Méditerranée. *CR Acad. Sci.*, Paris, sér. D, t. 281, p. 1215-1218.

CIARANFI (N.) and CITA (M.B.).- 1973. Paleontological evidence of changes in the Pliocene climates. *Init. Rep. Deep Sea Drill. Proj.*, Washington (U.S. Government printing office), Leg 13, vol. 2, p. 1387-1399.

CITA (M.B.).- 1973. Pliocene biostratigraphy and chronostratigraphy. *Init. Rep. Deep Sea Drill. Proj.*, Washington, (U.S. Government printing office), Leg 13, vol. 2, p. 1343-1379.

CITA (M.B.), CHIERICI (M.A.), CIAMPO (G.), MONCHARMONT (M.), D'ONOFRIO (S.) and RYAN (W.B.F.).- 1973. The Quaternary record in the Tyrrhenian and Ionian basins of the Mediterranean Sea. *Init. Rep. Deep Sea Drill. Proj.*, Washington (U.S. Government printing office), Leg 13, vol. 2, p. 1263-1339.

CITA (M.B.) and DECIMA (A.).- 1975. Rossellian : proposal of a superstage for the marine Pliocene. *VIth Congr. reg. Mediterr. Neog. Strat.*, Bratislava, p. 217-227.

CITA (M.B.), VERGNAUD-GRAZZINI (C.), ROBERT (C.), CHAMLEY (H.), CIARANFI (N.) and D'ONOFRIO (S.).- 1977. Paleoclimatic record of a long deep-sea core from the Eastern Mediterranean. *Quater. Res.* (in press).

DESPRAIRIES (A.) et LAPIERRE (H.). 1973. Les argiles liées au volcanisme du massif du Trodos (Chypre) et leur remaniement dans sa couverture. *Rev. Géogr. phys. Géol. dyn.*, vol. 15, fasc. 5, p. 499-510.

DROOGER (C.W.) (Editor).- 1973. Messinian events in the Mediterranean. *Kon. Nederl. Akad. Weten.*, North Holl. Publ. Comp., Amsterdam, 272 p.

DUFAURE (J.J.), KADJAR (M.H.), KERAUDREN (B.), MERCIER (J.-L.), SAUVAGE (J.) et SEBRIER (M.).- 1975. Les déformations plio-pléistocènes autour du golfe de Corinthe. *CR somm. Soc. géol. France*, sér. 7, vol. XVII.

HSÜ (K.J.), MONTADERT (L.) et al.- 1977. *Initial Reports of the Deep Sea Drilling Project*, Washington (U.S. Government printing office), Leg 42 A (in press).

KERAUDREN (B.).- 1970-1972. Les formations quaternaires marines de la Grèce. *Bull. Mus. Anthrop. préhist.*, Monaco, fasc 16, 17, 18.

KERAUDREN (B.)- 1975. Essai de stratigraphie et de paléogéographie du Plio-Pléistocène égéen. *Bull. Soc. géol. France*, sér. 7, vol. XVII, n° 6, p. 1110-1120.

MASCLE (G.).- 1973. Etude géologique des monts Sicani (Sicile). *Thèse, Sci. nat.*, Paris VI, 691 p. dact.

MELIERES (F.), CHAMLEY (H.), COUMES (F.) and ROUGE (M.).- 1977. X-Ray routine analyses. *Init. Rep. Deep Sea Drill. Proj.*, Washington (U.S. Government printing office), Leg 42A (in press).

MERCIER (J.-L.).- 1976. La néotectonique, ses méthodes et ses buts. Un exemple : l'arc Egéen (Méditerranée orientale). *Rev. Géogr. phys. Géol. dyn.* (in press).

MERCIER (J.-L.), CAREY (E.), PHILIP (H.) et SOREL (D.).- 1975. Les déformations plioquaternaires dans l'arc Egéen et leurs relations avec la séismité. In : *5è Coll. Géol. Rég. Egéennes*, Orsay, France, Abstracts.

MILLOT (G.).- 1964. Géologie des Argiles. *Masson Edit.*, Paris, 499 p.

PECHOUX (P.Y.), PHILIP (H.) et SEBRIER (M.). - 1976. Les mouvements plio-pléistocènes de la côte nord du golfe de Corinthe. *4è réun. ann. Sci. Terre*, Paris, p. 323.

RATEEV (M.A.), EMELYANOV (E.M.) and KHEIROV (M.B.).- 1966. Particularités de la formation des minéraux argileux dans les sédiments actuels de la Méditerranée. *Litol. polzn. Iskopaem. Akad. Nauk SSSR*, vol. 4, p. 6-23 (in Russian).

ROSSIGNOL-STRICK (M.).- 1974. Tectonique de plaques et sédimentation palynologique marine sur la ride méditerranéenne. *Sci. Géol.*, Strasbourg, t. 27, p. 135-145.

RUDDIMANN (W.F.).- 1971. Pleistocene sedimentation in the equatorial Atlantic : Stratigraphy and faunal paleoclimatology. *Bull. Geol. Soc. America*, vol. 82, p. 283-302.

RYAN (W.B.F.), CITA (M.B.), DREYFUS RAWSON (M.), BURCKLE (L.H.) and SAITO (T.).- 1974. A paleomagnetic assignment of neogene stage boundaries and the development of isochronous datum planes between the Mediterranean, the Pacific and Indian oceans in order to investigate the response of the World ocean to the Mediterranean "Salinity crisis". *Riv. Ital. Paleont.*, vol. 80, n° 4, p. 631-688.

RYAN (W.B.F.) HSÜ (K.J.) et al.- 1973. *Initial Reports of the Deep Sea Drilling Project*, Washington (U.S. Government printing office), Leg 13, vol. 1 & 2, 1447 p.

RYAN (W.B.F.), STANLEY (D.J.), HERSEY (J.B.), FAHLQUIST (D.A.) and ALLAN (T.D.).- 1971. The tectonics and geology of the Mediterranean Sea. in "The Sea", Part IV, *Maxwell A.E. Edit.*, 387-492.

RYAN (W.B.F.), VENKATARATHNAM (K.) and WEZEL (F.C.).- 1973. Mineralogical composition of the Nile cone, Mediterranean ridge and Strabo trench sandstones and clays. *Init. Rep. Deep Sea Drill. Proj.*, Washington (U.S. Government printing office, Leg 13, vol. 2, p. 731-746.

SOREL (D.).- 1975. Tectonique et néotectonique de la zone préapulienne. In : *5è Coll. Géol. Rég. égéennes*, Orsay, France, Abstracts.

VENKATARATHNAM (K.) and RYAN (W.B.F.).- 1971. Dispersal patterns of clay minerals in the sediments of the Eastern Mediterranean. *Mar. Geol.*, vol. 11, p. 261-282.

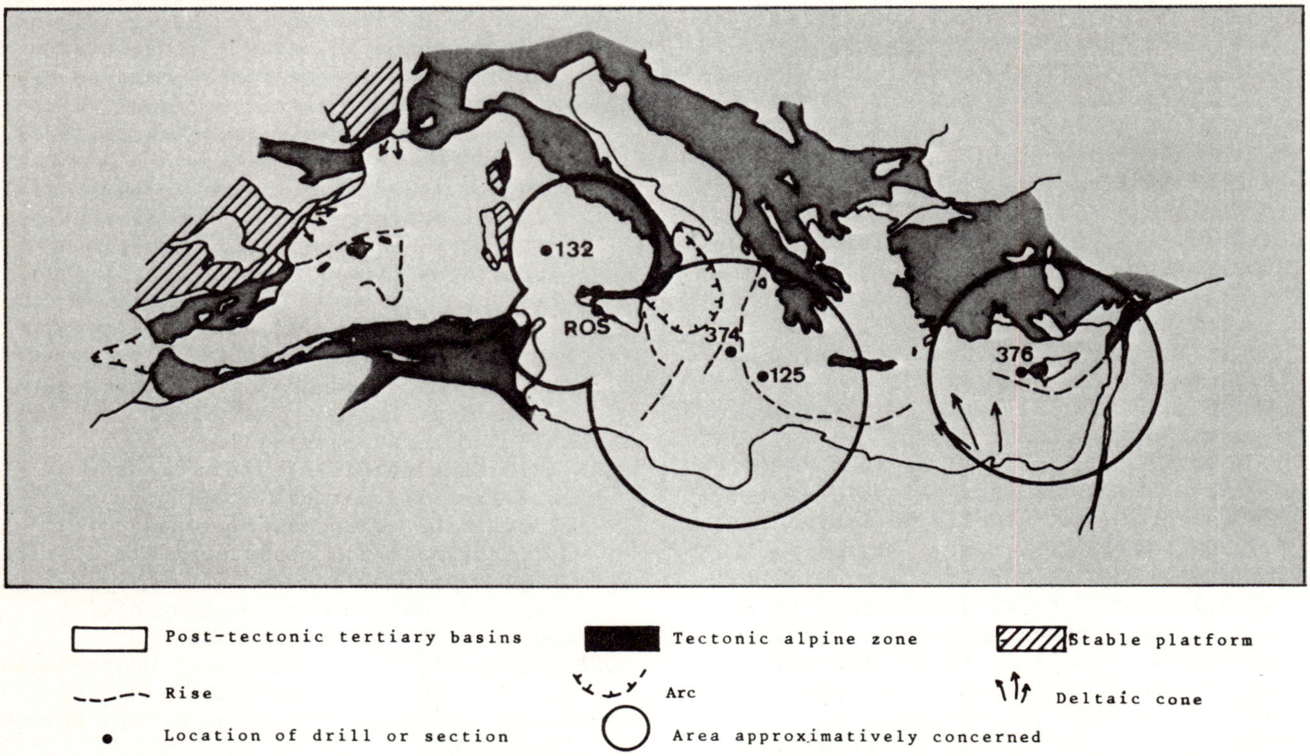

Figure 1. Location map.

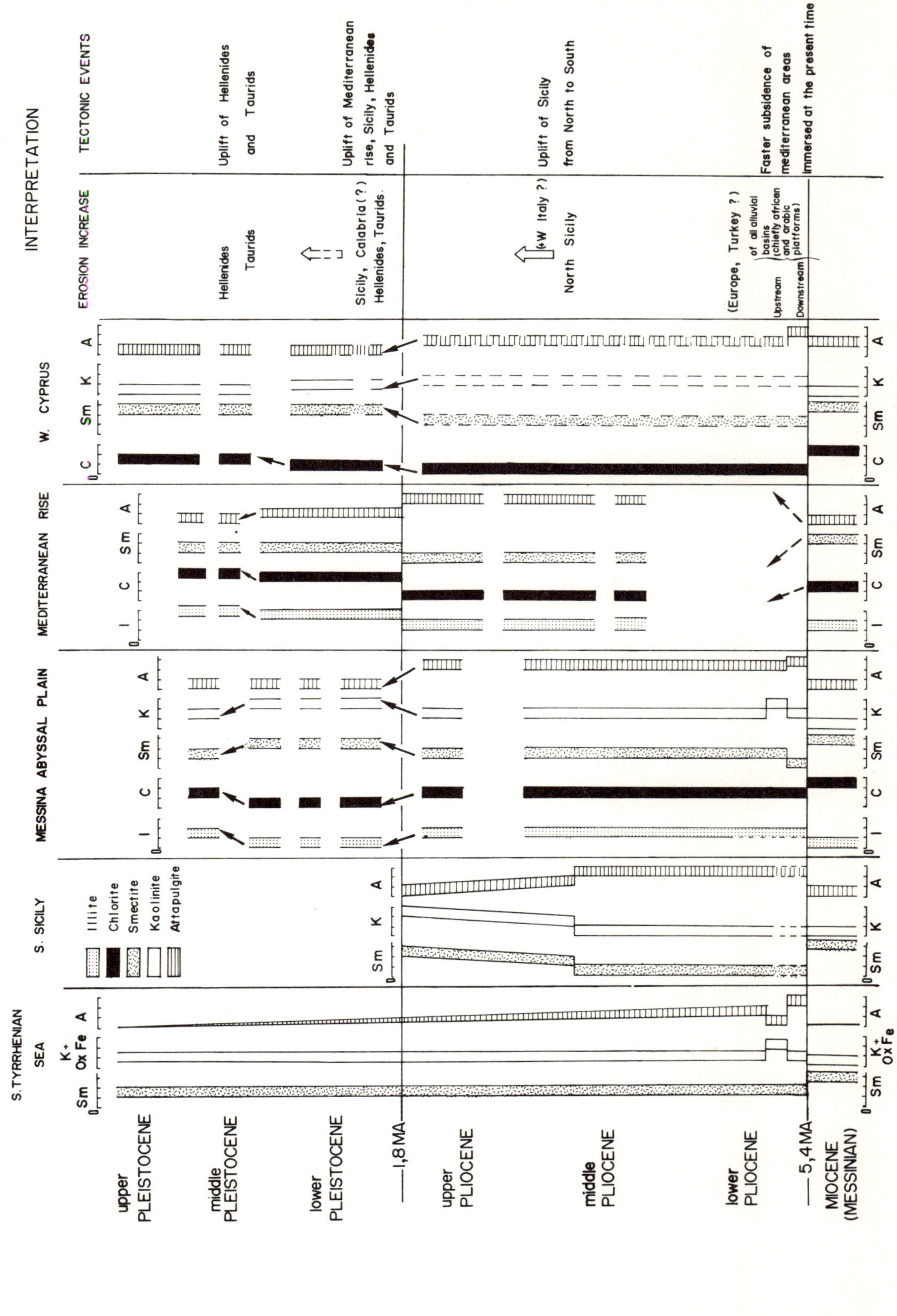

Figure 2. Tectonic interpretation of plio-pleistocene clay successions.

INTERNATIONAL SYMPOSIUM ON THE STRUCTURAL HISTORY OF THE MEDITERRANEAN BASINS. SPLIT (YUGOSLAVIA) 25-29 OCTOBER 1976. B. BIJU-DUVAL AND L. MONTADERT, Eds. EDITIONS TECHNIP, PARIS 1977, pp. 433-446

RECENT TECTONIC OVERPRINT ON COBBLESTONE DEPOSITION IN THE NORTHWESTERN HELLENIC ARC

D.J. STANLEY (1)

ABSTRACT

A 3.5 kHz subbottom survey in the northwestern Hellenic Arc west of the island of Zakinthos in the Ionian Sea reveals the interplay between geologically recent structural displacement and deposition. The high-resolution records indicate that vertical offset rates, as high as 1 mm per year, are considerably higher than rates of deposition. The preferential landward tilt of sediment between cobblestone knolls, asymmetry and pinchout of ponded sections, unconformities, and enhanced proportions of slumps are some of the features attesting to the tectonic overprint on both physiography and sedimentation.

I. INTRODUCTION

The Hellenic Arc in the eastern Mediterranean is a physiographically complex region directly influenced by geologically recent tectonic activity. An air-gun survey in the northwestern sector of the Hellenic Arc, west of the Peloponnesus and southwest of the island of Zakinthos in the eastern Ionian Sea, has revealed the presence of intensely deformed unconsolidated terrigenous sequences of Pliocene and Quaternary age (Got et al., 1977). This deformation is substantiated by other geological and subbottom investigations in the Ionian Sea (Hinz, 1974; Mercier et al., 1976; Sorel et al., 1976).

The major offshore structural trends are NE-SW to NNE-SSW and NW-SE. The attitude of the unconsolidated sequences recorded on seismic profiles corresponds to similarly trending structures on the island of Zakinthos. The complex physiography and associated asymmetric displacement of the unconsolidated sediment blanket, which locally may be as thick as 800 m, tend to support prevailing theories favoring compression and possible subduction in this sector of the Hellenic Arc (Papazachos and Comninakis, 1971; McKenzie, 1972).

This study, based on a 3.5 kHz subbottom survey, calls attention to the nature and rate of structural deformation displayed by the surficial sediment cover west and southwest of Zakinthos (Fig. 1). Particular attention is paid to sectors characterized by a hummocky, or cobblestone, topography. The very high-resolution subbottom data, some of which are presented here, clearly reveal the importance of recent tectonic activity and its influence on cobblestone depositional patterns.

(1) Division of Sedimentology, Smithsonian Institution, Washington, D.C. 20560, U.S.A.

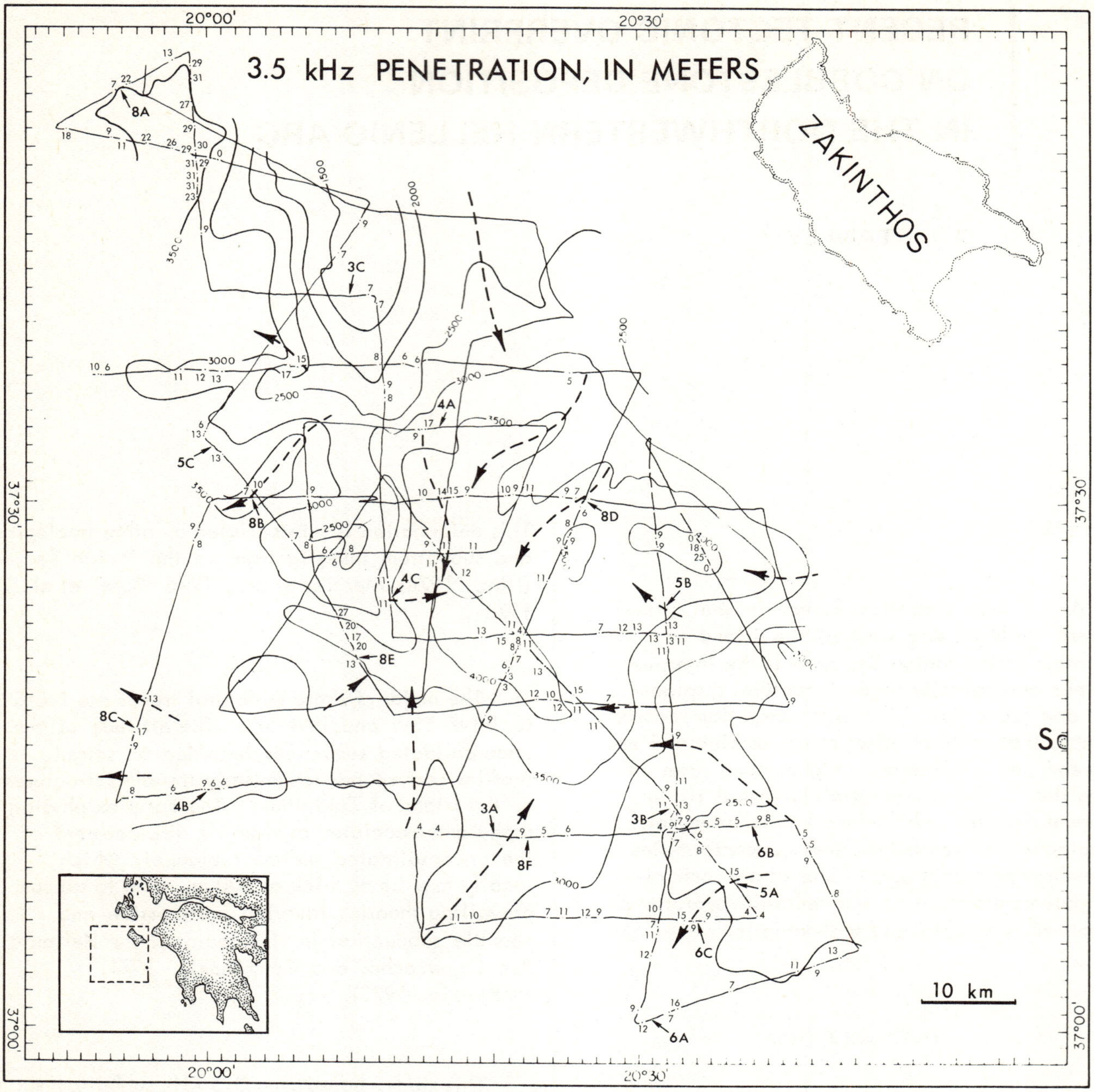

Figure 1. Chart of region west of the islands of Zakinthos and Strofadhes (S) showing track-lines of the R/V BANNOCK (September 1975) along which continuous 3.5 kHz subbottom profiles were collected. Small numbers along the lines represent penetration depth in meters; the larger number-letter designations show the position of 3.5 kHz profiles in Figures 3 to 6 and 8. Large arrows denote submarine valley trends.

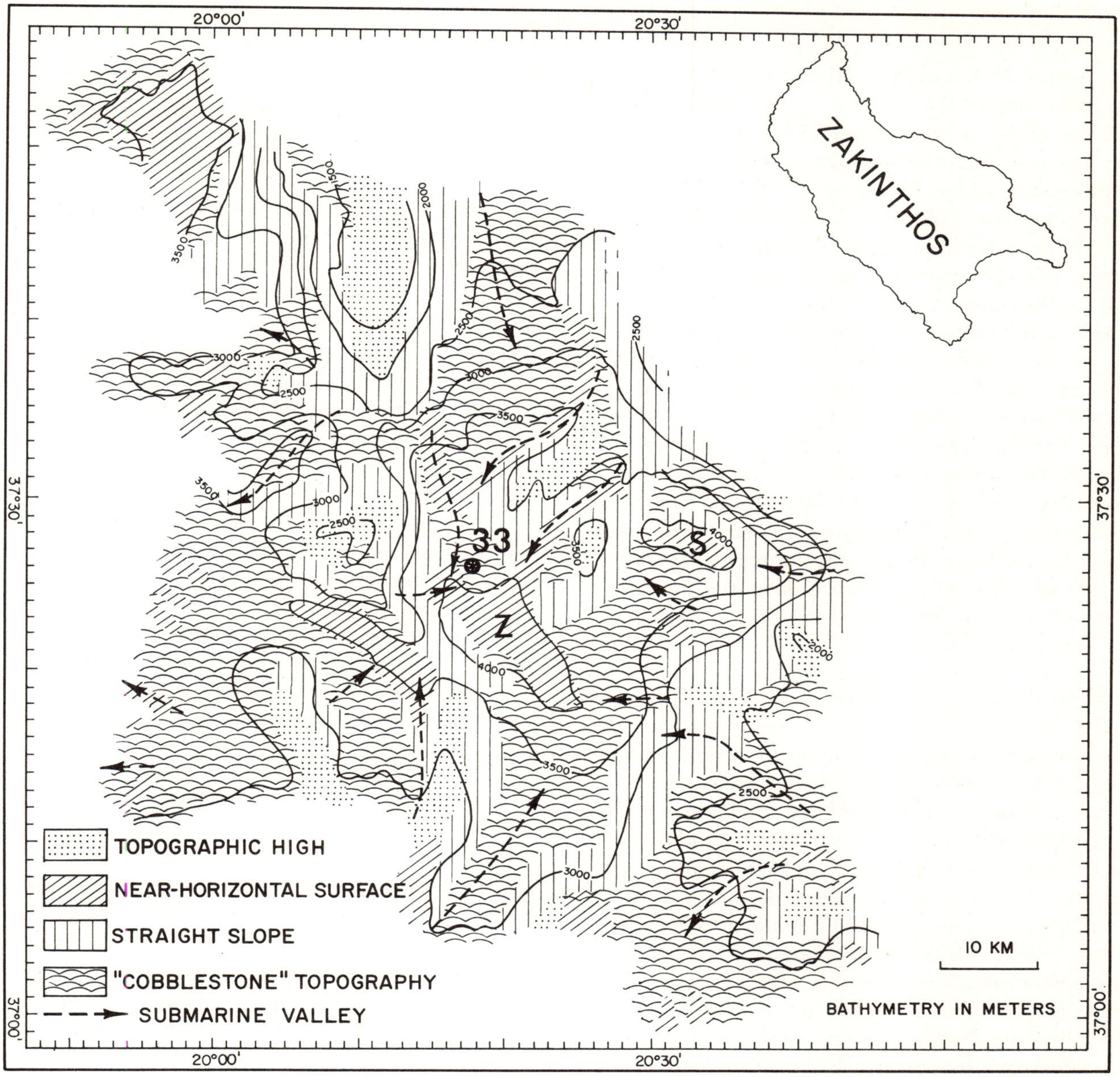

Figure 2. Chart showing major physiographic provinces, including cobblestone terrains, in the area west of Zakinthos. Z = Zakinthos Basin; S = Strofadhes Basin; 33 = core TR 172-33 shown in Figure 7. Large arrows show downslope direction. Depth in meters.

II. PHYSIOGRAPHIC COMPLEXITY OF THE STUDY AREA

A series of 3.5 kHz profiles (total of 1020 km) were collected on the C.N.R. R/V BANNOCK (MEDIBA cruise I, September 1974) in an area lying between the northern extension of the Mediterranean Ridge to the west and the islands of Strofadhes and Zakinthos to the east (Fig. 1). The 3500 m-isobath on Figure 2 highlights two major trends: NW-SE, parallel to the major structural strike on the island of Zakinthos, and NE-SW, related to a predominant compressive direction (Got et al., 1977). Within the rectangular area bound by the 3500 m-contour lies the NW-SE oriented Zakinthos Basin (18 km long, 8 km wide) and the smaller WNW-ESE trending Strofadhes Basin (10 km long, 3 km wide). These two depressions, both deeper than 4000 m and identified by the letters Z and S on Figure 2, are considered part of the Hellenic Trench physiographic-tectonic province which consists of numerous, deep isolated basins and troughs distributed along an arcuate belt (Stanley, 1974).

A detailed physiographic chart (Fig. 2) based on this 3.5 kHz survey shows the complex morphology characterized by ridges, topographic highs, steep slopes (some exceeding 20°) and a network of submarine valleys. The latter are for the most part short and discontinuous. More than half of the area surveyed is formed by a highly irregular, hummocky topography consisting of knolls and depressions; this type, termed cobblestone terrain (Hersey, 1965) has been traversed in many sectors of the eastern Mediterranean (Emery et al., 1966; Ryan et al., 1970), including the region to the south adjoining the study area (Hieke et al., 1973). The cobblestone topography is shown by multiple-crescent echos observed on echograms which are produced by the irregular relief that acts as numerous point scatters. Narrow beam echo-soundings on a similar morphology in other sectors of the eastern Mediterranean have shown that in many instances rounded cobblestone knolls are actually blocky in form and constitute a seafloor surface offset by a dense network of faults (Belderson et al., 1972). This interpretation is applicable to the study area as demonstrated by a seismic survey conducted on a subsequent cruise (Got and others, 1977). On air-gun profiles, the characteristic cobblestone seafloor generally appears as bedrock dislocated by faults and covered by sediment of variable thickness (usually less than 100 m).

III. NATURE OF COBBLESTONE TOPOGRAPHY

The 3.5 kHz records reveal details of the surficial cover not apparent on either conventional echograms or deeper-penetration continuous seismic profiles. The depth of penetration is highly variable throughout the study area and ranges to a maximum of about 30 m in some near-horizontal basin plains (Fig. 1). The cobblestone terrain, which frequently appears devoid of sediment on some air-gun records (Got et al., 1977, their figs. 3 and 5), reveals a thin (< 5 m) to moderately thick (>10 m) depositional cover on the much higher resolution 3.5 kHz profiles. In evaluating the 3.5 kHz records in sectors of cobblestone terrain, it is more accurate to measure the actual seafloor depth from the strongest, rather than from the first, return (Belderson et al., 1972). All of the traverses were made with the ship moving at approximately 8 knots; vertical exaggeration of the records is about x20 (Figs. 3-6, 8).

The profiles show gradations from low-relief fault-offset or gently folded sediment-covered terrains (Fig. 3A, B) to intensely faulted, high-relief knolls covered by little or almost no sediment (Fig. 3C). Relief between knolls is highly variable, but generally less than 100 m. Some sectors display low step-like blocks (Fig. 3A), while others are typified by broad, high knolls where the relief exceeds 60 m (Fig. 4A). The relatively narrow depressions between knolls generally range in width from about 500 to 1000 m. These may be V-shaped (Fig. 4) or near-horizontal (Fig. 5); more commonly, however, cobblestone depositional ponds display tilted surfaces with slopes of about 1° to 3° (Fig. 6).

IV. RECENT DISPLACEMENT OF THE DEPOSITIONAL COVER

Cores collected on a subsequent cruise in the study area (MEDIBA II; Got et al., 1977) include

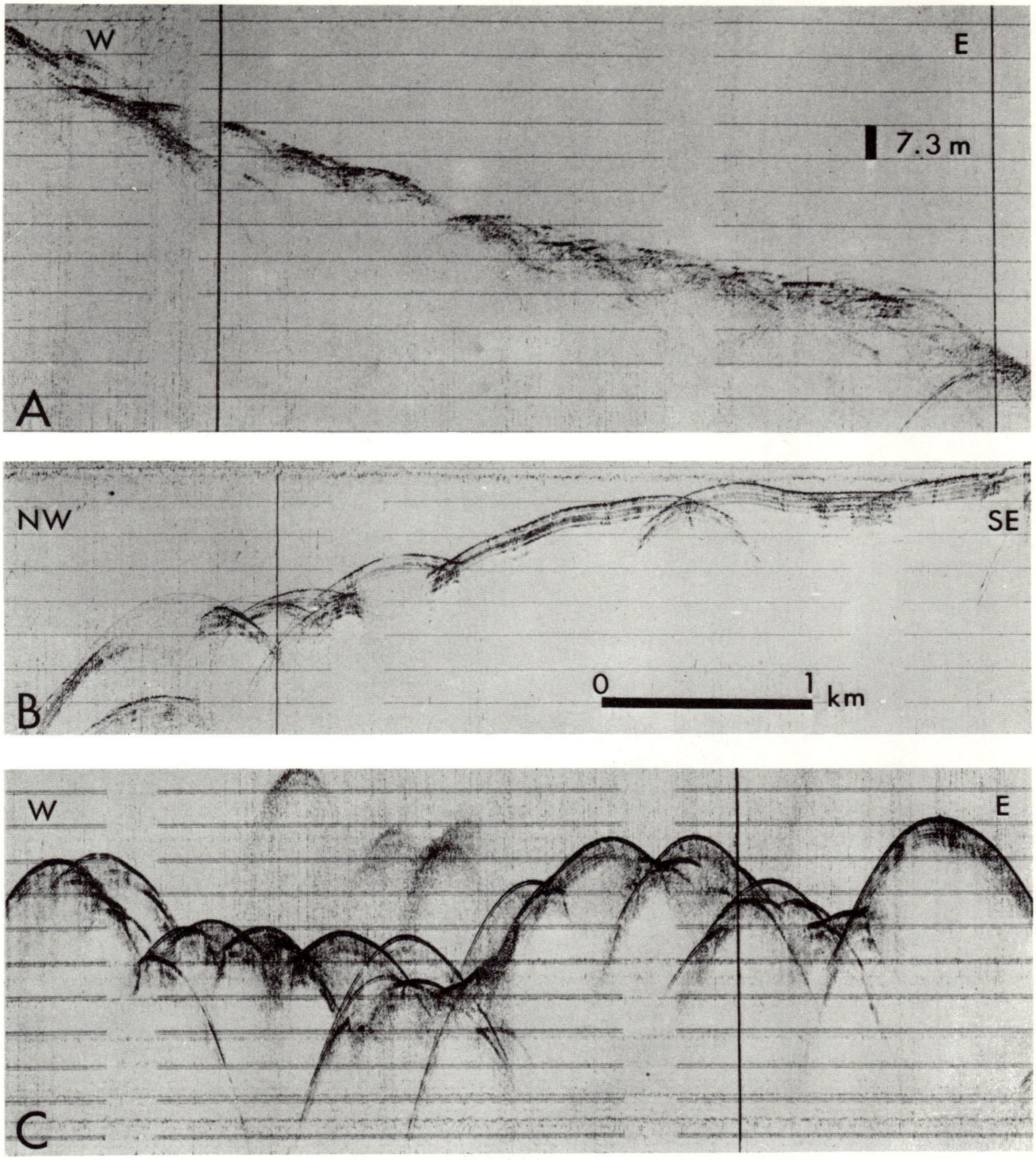

Figure 3. Selected 3.5 kHz profiles showing the variation of cobblestone terrain in the study area. The distance between horizontal lines on these records, and those in Figures 4 to 6 and 8, is 4 fathoms, or 7.3 meters.

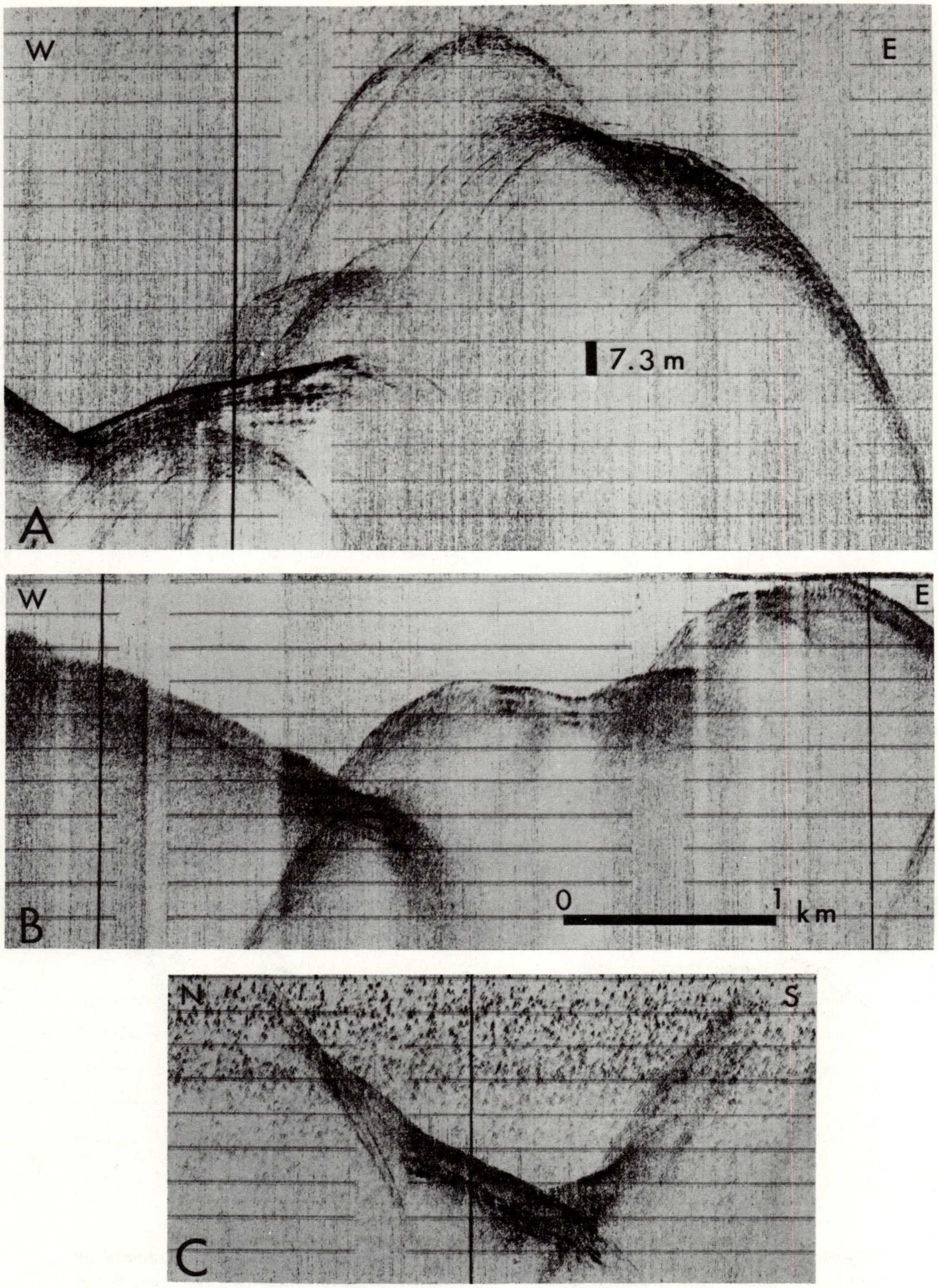

Figure 4. Selected 3.5 kHz profiles showing V-shaped depressions between cobblestone knolls. Some of these are valleys that probably served as by-ways for the downslope movement of sediment (see large arrows in Figure 2).

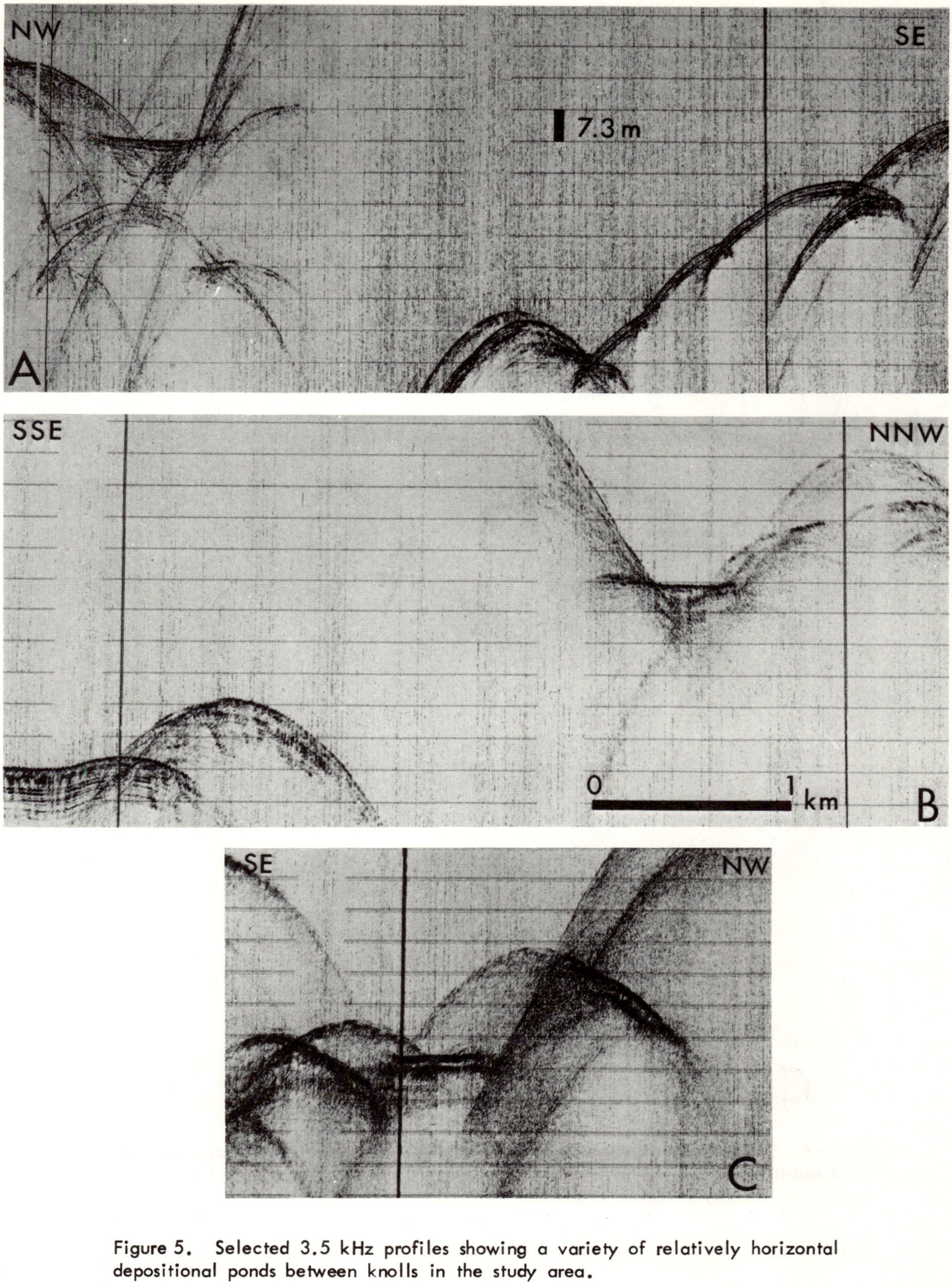

Figure 5. Selected 3.5 kHz profiles showing a variety of relatively horizontal depositional ponds between knolls in the study area.

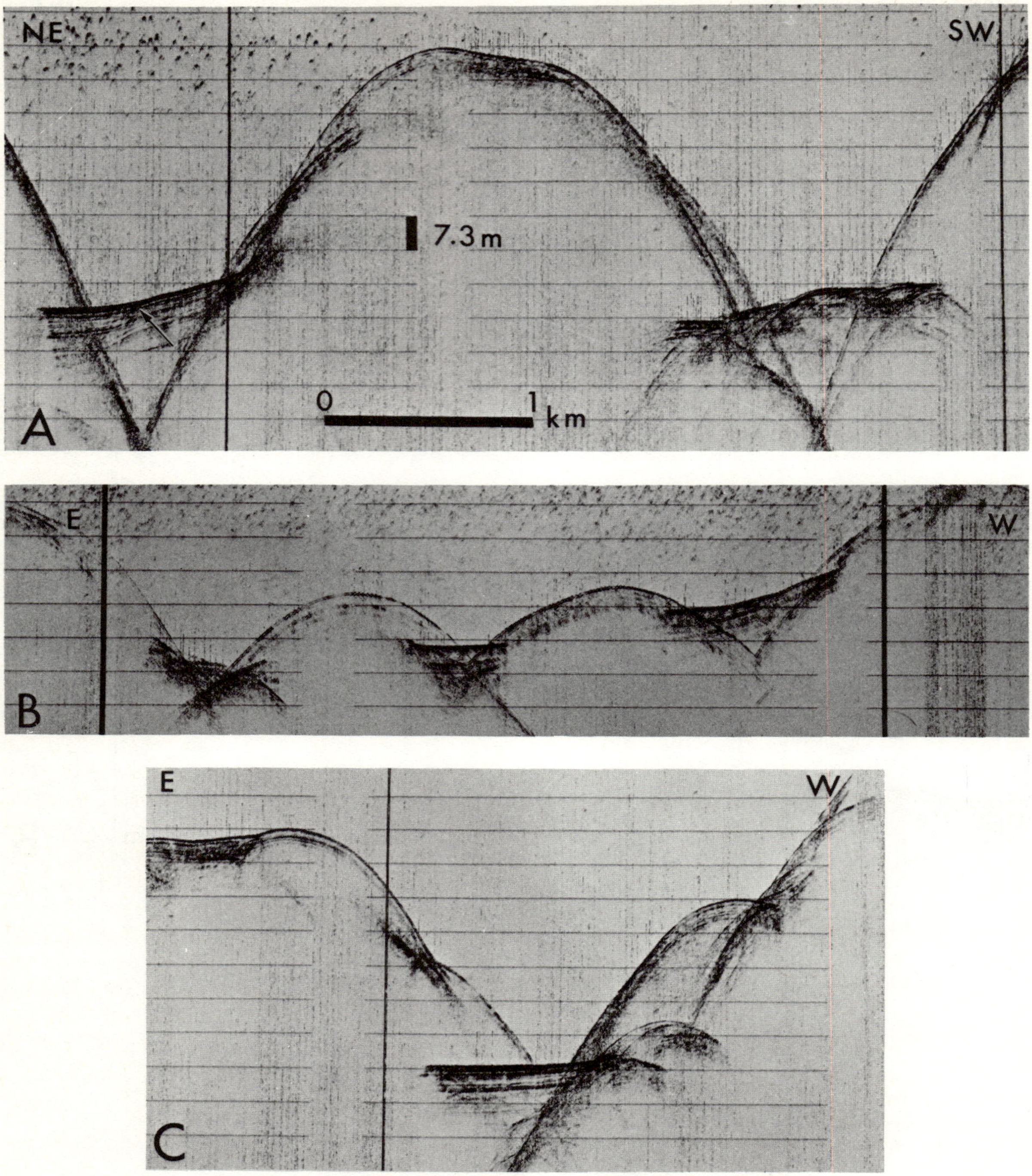

Figure 6. Selected 3.5 kHz profiles showing displacement of reflectors and preferential landward slope of cobblestone pond surfaces toward the east and northeast. Arrow in A shows unconformity between reflectors.

hemipelagic and gravity-induced deposits, including slumps (Fig. 7). The latter appear as disrupted or indistinct wavy reflectors on 3.5 kHz profiles (Fig. 8B, E, F). The proportion of slumps, debris flow deposits and turbidites is higher in cobblestone ponds and valleys than on the contiguous topographic highs. The seismic activity on the seafloor would enhance the probability of sediment failure on knolls by slumping and other gravity mechanisms. The study area west of the Peloponnesus is characterized by frequent earthquakes (Papazachos and Comninakis, 1971), many of which are intense enough to initiate failure and maintain slump and mass flow of sediment downslope toward depressions.

The intense seismicity not only has resulted in accelerated ponding rates, but also in the asymmetric accumulation of sediment between cobblestone knolls. The sediment sequences trapped between knolls slope preferentially toward the east or northeast (Figs. 6, 8A, B, D, F), that is, along the predominant compressional trends cited earlier. Reflectors within sections of ponded sediment are occasionally parallel banded, but more often display marked discontinuities (arrows, Figs. 6A, 8C). These stratigraphic breaks are associated with pinch-out of section (arrow, Fig. 8A), uplift of reflectors along pond margins (Figs. 6, 8A, C, D), and the inclination of layers in cores (Fig. 7i), all of which are the result of seafloor displacement during deposition.

The discontinuities record fluctuations in the relative amounts of sediment input and in the rate of movement at the site on which sediment accumulate. The rate of tilt is estimated by evaluating the amount of vertical uplift of the seaward margin, as indicated by subbottom reflectors, relative to depositional rates. Radiocarbon dates of core samples (R. Stuckenrath, pers. comm.) and lithostratigraphic correlation of core sections (R. J. Knight, in progress) show that Holocene sedimentation rates are approximately 10 cm/1000 years. Late Pleistocene accumulation rates in ponds were higher (to 20 cm/1000 years or higher). A conservative measurement of offset, based on a minimal average depositional rate of 10 cm/1000 years, indicates a Holocene tilting rate of about 1 mm per 4 to 6 years. It is estimated that vertical displacement may have been as high as 1 mm/year in the late Quaternary. This tilting phenomenon has resulted in the preferential accumulation of redeposited sediment toward the landward side of the depositional ponds.

V. DISCUSSION AND CONCLUSIONS

The structural overprint in the seismically active northwestern Hellenic Arc sector is recorded by the landward displacement of sediment within cobblestone depressions which has resulted in asymmetry of depositional ponds, discontinuous seismic reflector horizons, and gravitational slumps associated with a fault-displaced topography. One possible explanation for the preferential direction of tilt is a regional downbowing of the underlying lithosphere and rotation of thrust slices, perhaps in a manner similar to that postulated for various Pacific trenches (cf., Scholl et al., 1970; Kulm and Fowler, 1974; Prince and Kulm, 1975; Moore and Karig, 1976; Schweller and Kulm, 1977). The recent displacement of sediments on the horst-graben and thrust-fault physiography in the study area may be related in some way to the collision of lithospheric plates as suggested by Hsü and Ryan (1973). These authors propose that the Hellenic Trench near Deep Sea Drilling Project core sites 127 and 128 about 275 km southeast of Zakinthos presents a regional lithospheric curvature oriented in a manner compatible with a subduction hypothesis; the average convergence rate postulated is the order of 1.4 cm per year. However, the data detailing surficial sections presented here do not prove or disprove subduction, nor do they exclude the possible role of evaporite solution mechanics and halokinesis which may be effective in this sector. The latter hypothesis would be given some credibility if it can be demonstrated that Miocene evaporites of the type recorded on Zakinthos (Braune and Heimann, 1973) extend offshore west of the island.

It is apparent that deposition has played but a minor role in modifying the cobblestone topography, except in ponds, although the study area is close to land and has had access to a reasonable input of sediment. The highly irregular, tectonically-controlled physiography precludes long distance transport of mass-gravity deposits, and fosters the restricted extension of individual turbidite sheets and debris and grain flow layers. Furthermore, tilting of pond surfaces has resulted in the asymmetric

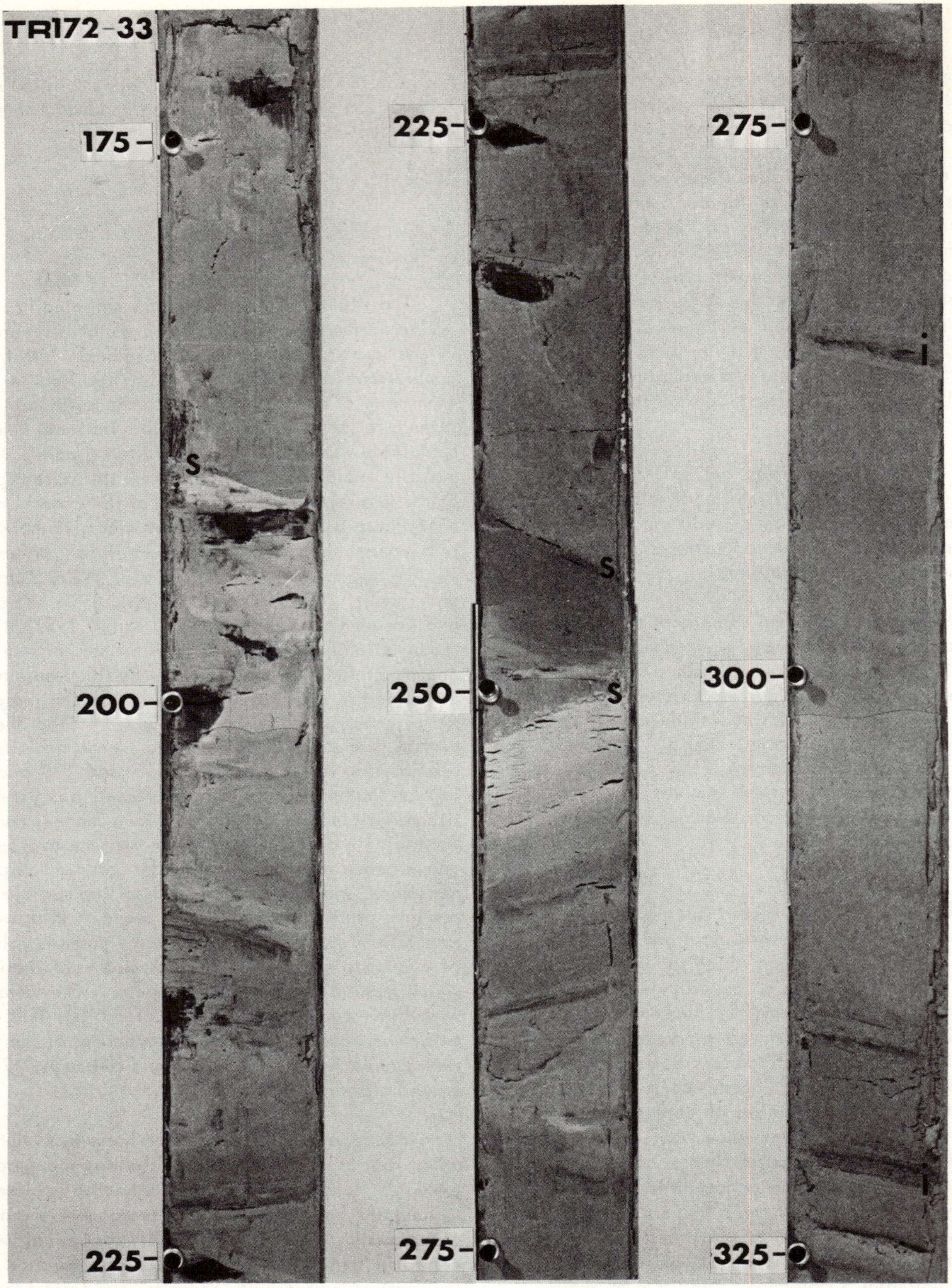

Figure 7. Sections of core TR 172-33 collected in the cobblestone terrain north of Zakinthos Basin (position shown in Figure 2). The slump unit between 175 and 275 (depth in centimeters from the top of core) shows disrupted layering and high-angle failure surfaces (S); note also the inclined stratification (i) of sandy layers between 275 and 325.

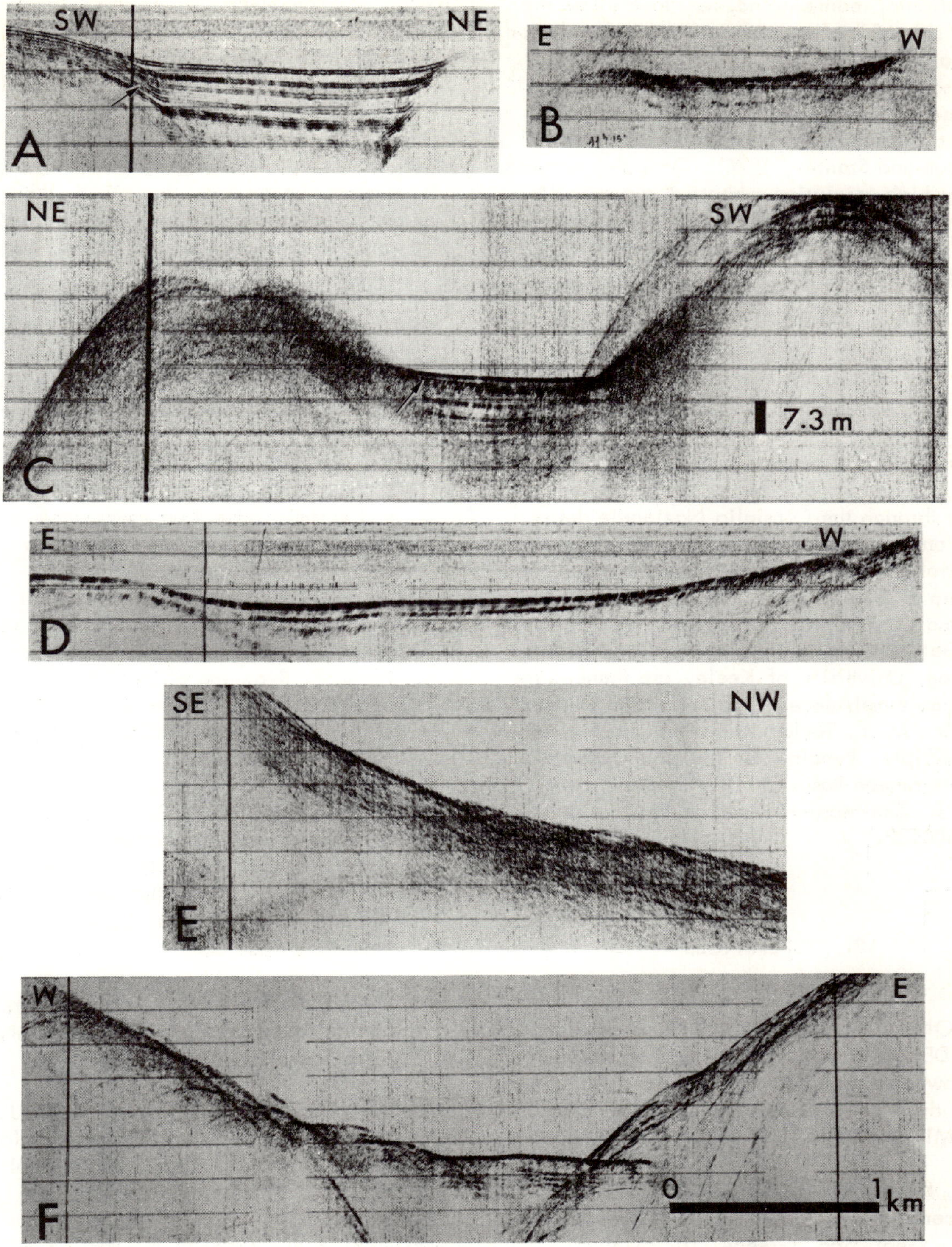

Figure 8. Selected 3.5 kHz profiles showing asymmetry of ponded deposits, pinch-out of section (arrow in A), uplift of reflectors at pond margins (A, C, D), unconformities between reflectors (arrow in C) and disrupted reflectors probably denoting slumps (B, E, F).

accumulation of sediment-induced flows along the landward side of knolls. Pond-to-pond correlation of surficial lithofacies can be achieved, at best, by using sapropels, tephra, oxidized horizons and carbonate-rich layers, many of which attain a broad distribution in the eastern Mediterranean (Maldonado and Stanley, 1976). On-going programs combining high-resolution subbottom surveys with coring are needed to further elucidate the interplay between geologically recent tectonics and sedimentation in the Hellenic Arc.

VI. ACKNOWLEDGMENTS

Appreciation is expressed to Professor A. Brambati, University of Trieste, for arranging ship-time through the Consiglio Nazionale delle Ricerche on the R/V BANNOCK in September 1974. Professor Brambati and his colleagues of the Istituto di Geologia e Paleontologia, Drs. M. Collins and G. Ferentinos, University of Wales, Dr. H. Got, University of Perpignan, and Professor G. Kelling, University of Keele, are thanked for their valuable assistance in the collection of data at sea. Dr. M. E. Taylor, U.S.G.S., reviewed the manuscript. Funding for this study, part of the Mediterranean Basin (MEDIBA) Project, was provided by Smithsonian Research Foundation grant 71500532.

VII. REFERENCES

BELDERSON, R. H., KENYON, N. H. and STRIDE, A. H., 1972. Comparison between narrow-beam and conventional echo-soundings from the Mediterranean Ridge. Mar. Geol., 12: M11-M15.

BRAUNE, K. and HEIMANN, K. O., 1973. Miocene on the Ionian islands (Greece). Bull. Geol. Soc. Greece, 10: 25-30.

EMERY, K. O., HEEZEN, B. C. and ALLAN, T. D., 1966. Bathymetry of the eastern Mediterranean Sea. Deep-Sea Res., 13: 173-192.

GOT, H., STANLEY, D. J. and SOREL, D., 1977. Northwestern Hellenic Arc: concurrent sedimentation and deformation in a compressive setting. Mar. Geol., 23 (in press).

HERSEY, J. B., 1965. Sedimentary basins of the Mediterranean Sea. In: W. F. Whittard and R. D. Bradshaw (Editors), Submarine Geology and Geophysics. Colston Papers, Butterworths, London, 17: 75-89.

HIEKE, W., SIGL, W. and FABRICIUS, F., 1973. Morphological and structural aspects of the Mediterranean Ridge SW off the Peloponnesus (Ionian Sea). Bull. Geol. Soc. Greece, 10: 109-126.

HINZ, K., 1974. Results of seismic refraction and seismic reflection measurements in the Ionian Sea. Geol. Jb., E2: 33-65.

HSU, K. J. and RYAN, W. B. F., 1973. Summary of the evidence for extensional and compressional tectonics in the Mediterranean. In: W. B. F. Ryan and others (Editors), Initial Reports of the Deep Sea Drilling Project. U. S. Govt. Print. Office, Washington, 13: 1011-1019.

KULM, L. D. and FOWLER, G. A., 1974. Oregon continental margin structure and stratigraphy: a test of the imbricate thrust model. In: C. A. Burk and C. L. Drake (Editors), The Geology of Continental Margins. Springer-Verlag, New York, 261-283.

McKENZIE, D. P., 1972. Active tectonics of the Mediterranean region. Geogr. Jour. Royal Astro-Society, 30: 109-185.

MALDONADO, A. and STANLEY, D. J., 1976. The Nile Cone: submarine fan development by cyclic sedimentation. Mar. Geol., 20: 27-40.

MERCIER, J. L., CAREY, E., PHILIP, H. and SOREL, D., 1976. La néotectonique plio-quaternaire de l'arc Egéen externe et de la mer Egée et ses relations avec la séismicité. Bull. Soc. Géol. France, 2 (in press).

MOORE, G. F. and KARIG, D. E., 1976. Development of sedimentary basins on the lower trench slope. Geology, 4: 693-697.

PAPAZACHOS, B. C. and COMNINAKIS, P. E., 1971. Geophysical and tectonic features of the Aegean Arc. Jour. Geophys. Res., 76: 8517-8533.

PRINCE, R. A. and KULM, L. D., 1975. Crustal rupture and the initiation of imbricate thrusting in the Peru-Chile Trench. Geol. Soc. Amer. Bull., 86: 1639-1653.

RYAN, W. B. F., STANLEY, D. J., HERSEY, J. B., FAHLQUIST, D. A. and ALLAN, T. D., 1970. The tectonics and geology of the Mediterranean Sea. In: A. E. Maxwell (Editor), The Sea, 4, Wiley-Interscience, New York, 387-492.

SCHOLL, D. W., CHRISTENSEN, M. N., VON HUENE, R. and MARLOW, M. S., 1970. Peru-Chile Trench sediments and sea-floor spreading. Geol. Soc. Amer. Bull., 81: 1339-1360.

SCHWELLER, W. J. and KULM, L. D., 1977. Depositional patterns and channelized sedimentation in active eastern Pacific trenches. In: D. J. Stanley and G. Kelling (Editors), Submarine Canyon, Fan and Trench Sedimentation. Dowden, Hutchinson and Ross, Stroudsburg (in press).

SOREL, D., LEMEILLE, F., LIMOND, J., NESTEROFF, W. D. and SEBRIER, M., 1976. Mise en évidence de structures compressives sous-marines plio-pléistocènes dans l'arc égéen externe au large de Levkas (Iles Ioniennes, Grèce). C.R. Acad. Sci. Paris (in press).

STANLEY, D. J., 1974. Modern flysch sedimentation in a Mediterranean island arc setting. In: R. H. Dott, Jr. and R. H. Shaver (Editors), Modern and Ancient Geosynclinal Sedimentation. S.E.P.M. Sp. Publ. 19: 240-259.

index

ACHEVÉ D'IMPRIMER
EN JUIN 1977
PAR L'IMPRIMERIE
HEMMERLÉ, PETIT et Cie, PARIS

N° d'impression : 3197
N° d'éditeur : 431

Dépôt légal : 2e trimestre 1977

IMPRIMÉ EN FRANCE